ADVANCED ASPECTS OF SPECTROSCOPY

Edited by **Muhammad Akhyar Farrukh**

INTECHOPEN.COM

Advanced Aspects of Spectroscopy
http://dx.doi.org/10.5772/2757
Edited by Muhammad Akhyar Farrukh

CBS, Edition 2016

Contributors

S. Lakshmi Reddy, Tamio Endo, G. Siva Reddy, Claudia Maria SimonescuWieslawa Urbaniak-Domagala, Jesús Rodarte Dávila, Jenaro C. Paz Gutierrez, Ricardo Perez Blanco, Taesam Kim, Chhiu-Tsu Lin, Lidia Martínez, Elisa Román, Roman Nevshupa, Vitaliy Tinkov, George Sarau, Arne Bochmann, Renata Lewandowska, Silke Christiansen, Daisuke Kosemura, Motohiro Tomita, Koji Usuda, Atsushi Ogura, Todd M. Alam, Janelle E. Jenkins, Kavian Cooke, Wenjun Liang, Jian Li, Hong He, Tatsuo Fujii, Ikko Matsusue, Jun Takada, Leonardo M. Moreira, Juliana P. Lyon, Ana Paula Romani, Divinomar Severino, Maira Regina Rodrigues, Hueder P. M. de Oliveira, Sam-Rok Keum, So-Young Ma, Se-Jung Roh, Luxi Li, Xianbo Shi, Cherice M. Evans, Gary L. Findley, E.S. Estracanholli, G. Nicolodelli, S. Pratavieira, C. Kurachi, V.S. Bagnato, Denys Kurbatov, Anatoliy Opanasyuk, Halyna Khlyap

Published by InTech

Janeza Trdine 9, 51000 Rijeka, Croatia

Publishing Process Manager Iva Simcic

Technical Editor InTech Prepress, Novi Sad

Cover InTech Design team

Additional hard copies can be obtained from orders@intechopen.com

Advanced Aspects of Spectroscopy, Edited by Muhammad Akhyar Farrukh
p. cm.
ISBN 978-953-51-0715-6

INTECH

open science | open minds

Contents

Preface

Spectroscopy is the interpretation of spectra of absorption and emission of electromagnetic radiation due to the interaction between matter and energy that energy depends on the specific wavelength of electromagnetic radiation.

Spectroscopy has proven invaluable research tool in a number of areas including chemistry, physics, biology, medicine, and ecology. Lasers have unlimited application potential and this book will help researchers and scientists to get potential new applications of spectroscopy. The research field is quite large, but the goal of this book to cover maximum advanced areas like atomic spectroscopy, laser spectroscopy, molecular spectroscopy, nanostructure spectroscopy, physical, and organic spectroscopy.

The book would have keen attention of PhD students, medical researchers, industrial researchers, molecular scientists, biologists, ecologists, physicists and material scientists who will find innovative original research and reports on experimentation, new spectroscopic techniques & instrumentation, and application of spectroscopy.

Common spectroscopic techniques like UV-Visible, Atomic Absorption, Flame Emission, Molecular Fluorescence, Phosphorescence, Infrared, Nuclear Magnetic Resonance, Mass, and Electron are being used to study the qualitative as well as quantitative analysis of samples. However, the spectroscopic field is growing day-by-day and scientists are exploring new areas in this field by introducing new techniques. The main purpose of this book is to highlight new spectroscopic techniques like Magnetic Induction Spectroscopy, Laser-Induced Breakdown Spectroscopy, X-ray Photoelectron Spectroscopy, Low Energy Electron Loss Spectroscopy, Micro- to Macro-Raman Spectroscopy, Liquid-Immersion Raman Spectroscopy, High-Resolution Magic Angle Spinning (HR-MAS) Nuclear Magnetic Resonance (NMR) Spectroscopy, Injection and Optical Spectroscopy, and Nano Spectroscopy etc. This book is divided into five sections including General Spectroscopy, Advanced Spectroscopy, Nano Spectroscopy, Organic Spectroscopy, and Physical Spectroscopy which cover topics from basic to advanced levels which will provide a good source of learning for teaching and research purposes.

Muhammad Akhyar Farrukh
Department of Chemistry
GC University Lahore
Pakistan

Section 1

General Spectroscopy

Chapter 1

Electronic (Absorption) Spectra of 3d Transition Metal Complexes

S. Lakshmi Reddy, Tamio Endo and G. Siva Reddy

Additional information is available at the end of the chapter

http://dx.doi.org/10.5772/50128

1. Introduction

1.1. Types of spectra

Spectra are broadly classified into two groups (i) emission spectra and (ii) absorption spectra

i. *Emission spectra* Emission spectra are of three kinds (a) continuous spectra,(b) band spectra and (c) line spectra.

Continuous spectra: Solids like iron or carbon emit continuous spectra when they are heated until they glow. Continuous spectrum is due to the thermal excitation of the molecules of the substance.

Band spectra: The band spectrum consists of a number of bands of different colours separated by dark regions. The bands are sharply defined at one edge called the head of the band and shade off gradually at the other edge. Band spectrum is emitted by substances in the molecular state when the thermal excitement of the substance is not quite sufficient to break the molecules into continuous atoms.

Line spectra: A line spectrum consists of bright lines in different regions of the visible spectrum against a dark background. All the lines do not have the same intensity. The number of lines, their nature and arrangement depends on the nature of the substance excited. Line spectra are emitted by vapours of elements. No two elements do ever produce similar line spectra.

ii. *Absorption spectra:* When a substance is placed between a light source and a spectrometer, the substance absorbs certain part of the spectrum. This spectrum is called the absorption spectrum of the substance.

Electronic absorption spectrum is of two types. d-d spectrum and charge transfer spectrum. d-d spectrum deals with the electronic transitions within the d-orbitals. In the charge – transfer spectrum, electronic transitions occur from metal to ligand or vice-versa.

2. Electronic spectra of transitions metal complexes

Electronic absorption spectroscopy requires consideration of the following principles:

a. *Franck-Condon Principle:* Electronic transitions occur in a very short time (about 10^{-15} sec.) and hence the atoms in a molecule do not have time to change position appreciably during electronic transition .So the molecule will find itself with the same molecular configuration and hence the vibrational kinetic energy in the exited state remains the same as it had in the ground state at the moment of absorption.
b. *Electronic transitions between vibrational states:* Frequently, transitions occur from the ground vibrational level of the ground electronic state to many different vibrational levels of particular excited electronic states. Such transitions may give rise to vibrational fine structure in the main peak of the electronic transition. Since all the molecules are present in the ground vibrational level, nearly all transitions that give rise to a peak in the absorption spectrum will arise from the ground electronic state. If the different excited vibrational levels are represented as υ_1, υ_2, etc., and the ground state as υ_0, the fine structure in the main peak of the spectrum is assigned to $\upsilon_0 \rightarrow \upsilon_0$, $\upsilon_0 \rightarrow \upsilon_1$, $\upsilon_0 \rightarrow \upsilon_2$ etc., vibrational states. The $\upsilon_0 \rightarrow \upsilon_0$ transition is the lowest energy (longest wave length) transition.
c. *Symmetry requirement:* This requirement is to be satisfied for the transitions discussed above.

Electronic transitions occur between split 'd' levels of the central atom giving rise to so called d-d or ligand field spectra. The spectral region where these occur spans the near infrared, visible and U.V. region.

Ultraviolet (UV)	Visible (Vis)	Near infrared (NIR)	
50,000 - 26300	26300 - 12800	12800 -5000	cm^{-1}
200 - 380	380 - 780	780 - 2000	nm

3. Russel-Saunders or L-S coupling scheme

An orbiting electronic charge produces magnetic field perpendicular to the plane of the orbit. Hence the orbital angular momentum and spin angular momentum have corresponding magnetic vectors. As a result, both of these momenta couple magnetically to give rise to total orbital angular momentum. There are two schemes of coupling: Russel-Saunders or L-S coupling and j-j coupling.

a. The individual spin angular momenta of the electrons, s_i, each of which has a value of ± ½, combine to give a resultant spin angular momentum (individual spin angular momentum is represented by a lower case symbol whereas the total resultant value is given by a upper case symbol).

$$\sum s_i = S$$

Two spins of each ± ½ could give a resultant value of S =1 or S= 0; similarly a resultant of three electrons is 1 ½ or ½ .The resultant is expressed in units of h/2 π . The spin multiplicity is given by (2S+1). Hence, If n is the number of unpaired electrons, spin multiplicity is given by n + 1.

b. The individual orbital angular momenta of electrons, l_i, each of which may be 0, 1 ,2, 3 , 4 ..... in units of h/2π for s, p, d, f, g, ...orbitals respectively, combine to give a resultant orbital angular momentum, L in units of h/2π . $\sum l_i = L$

The resultant L may be once again 0, 1, 2, 3, 4.... which are referred to as S, P, D, F G,... respectively in units of h/2π.The orbital multiplicity is given by (2L+1).

0	1	2	3	4	5
S	P	D	F	G	H

c. Now the resultant S and L couple to give a total angular momentum, J. Hence, it is not surprising that J is also quantized in units of h/2π.The possible values of J quantum number are given as

$$J = (L+S), (L+S-1), (L+S-2), (L+S-3), \ldots. |L-S|,$$

The symbol | | indicates that the absolute value (L – S) is employed, i.e., no regard is paid to ± sign. Thus for L = 2 and S = 1, the possible J states are 3, 2 and 1 in units of h/2π.

The individual spin angular momentum, s_i and the individual orbital angular momentum, l_i, couple to give total individual angular momentum, j_i. This scheme of coupling is known as spin-orbit coupling or j -j coupling.

4. Term symbols

4.1. Spectroscopic terms for free ion ground states

The rules governing the term symbol for the ground state according to L-S coupling scheme are given below:

a. The spin multiplicity is maximized i.e., the electrons occupy degenerate orbitals so as to retain parallel spins as long as possible (Hund's rule).
b. The orbital angular momentum is also maximized i.e., the orbitals are filled with highest positive m values first.
c. If the sub-shell is less than half-filled, J = L– S and if the sub-shell is more than half – filled, J = L +S.

The term symbol is given by $^{2S+1}L_J$. The left-hand superscript of the term is the spin multiplicity, given by 2S+1 and the right- hand subscript is given by J. It should be noted that S is used to represent two things- (a) total spin angular momentum and (b) and total angular momentum when L = 0. The above rules are illustrated with examples.

For d^4 configuration:

↑	↑	↑	↑	
+2	+1	0	-1	-2

m_l

Hence, L = 3 -1 = 2 i.e., D; S = 2; 2S+1 = 5; and J = L- S = 0; Term symbol = $^5D_0$

For d^9 configuration:

↓↑	↓↑	↓↑	↓↑	↑
+2	+1	0	-1	-2

m_l

Hence, L = +2+1+0-1 = 2 i.e., D ; S = 1 /2 ; 2S+1 = 2 ; and J = L+ S = 3/2 ; Term symbol = $^2D_{5/2}$

Spin multiplicity indicates the number of orientations in the external field. If the spin multiplicity is three, there will be three orientations in the magnetic field.- parallel, perpendicular and opposed. There are similar orientations in the angular momentum in an external field.

The spectroscopic term symbols for d^n configurations are given in the Table-1. The terms are read as follows: The left-hand superscript of the term symbol is read as singlet, doublet, triplet, quartet, quintet, sextet, septet, octet, etc., for spin multiplicity values of 1, 2, 3, 4, 5, 6, 7, 8, etc., respectively.$^1S_0$ (singlet S nought); $^2S_{1/2}$ (doublet S one–half); $^3P_2$ (triplet P two); $^5I_8$ (quintet I eight). It is seen from the Table-1 that d^n and d^{10-n} have same term symbols, if we ignore J values. Here n stands for the number of electrons in d^n configuration.

d^n	Term	d^n	Term
d^0	$^1S_0$	d^{10}	$^1S_0$
d^1	$^2D_{3/2}$	d^9	$^2D_{5/2}$
d^2	$^3F_2$	d^8	$^3F_4$
d^4	$^5D_0$	d^6	$^5D_4$
d^5	$^6S_{5/2}$		

Table 1. Term symbols

It is also found that empty sub -shell configurations such as p^0, d^0, f^0, etc., and full filled sub-shell configurations such as p^6, d^{10}, f^{14}, etc., have always the term symbol $^1S_0$ since the resultant spin and angular momenta are equal to zero. All the inert gases have term symbols for their ground state $^1S_0$ .Similarly all alkali metals reduce to one electron problems since closed shell core contributes nothing to L , S and J; their ground state term symbol is given by $^2S_{1/2}$. Hence d electrons are only of importance in deciding term symbols of transition metals.

5. Total degeneracy

We have seen that the degeneracy with regard to spin is its multiplicity which is given by (2S+1). The total spin multiplicity is denoted by M_s running from S to -S. Similarly orbital degeneracy, M_L, is given by (2L+1) running from L to -L. For example, L= 2 for D state and so the orbital degeneracy is (2x2+1) =5 fold. Similarly, for F state, the orbital degeneracy is seven fold. Since there are (2L+1) values of M_L, and (2S+1) values of M_s in each term, the total degeneracy of the term is given by: 2(L+1)(2S+1).

Each value of M_L occurs (2S+1) times and each value of M_s occurs (2L+1) times in the term. For ^3F state, the total degeneracy is 3x7 =21 fold and for the terms ^3P, ^1G, ^1D, ^1S, the total degeneracy is 9,9,5,1 fold respectively. Each fold of degeneracy represents one microstate.

6. Number of microstates

The electrons may be filled in orbitals by different arrangements since the orbitals have different m_l values and electrons may also occupy singly or get paired. Each different type of electronic arrangement gives rise to a microstate. Thus each electronic configuration will have a fixed number of microstates. The numbers of microstates for p^2 configuration are given in Table-2 (for both excited and ground states).

m_l -1		↑	↑		↓	↓		↓	↓			↑			↑↓
0	↑		↑	↓		↓	↓		↑	↑	↑	↓		↑↓	
+1	↑	↑		↓	↓		↑	↑		↓	↓		↑↓		
m_L	+1	0	-1	+1	0	-1	+1	0	-1	+1	+1	-1	+2	0	

Table 2. Number of microstates for p^2 configuration

Each vertical column is one micro state. Thus for p^2 configuration, there are 15 microstates. In the above diagram, the arrangement of singlet states of paired configurations given in A (see below) is not different from that given in B and hence only one arrangement for each m_l value.

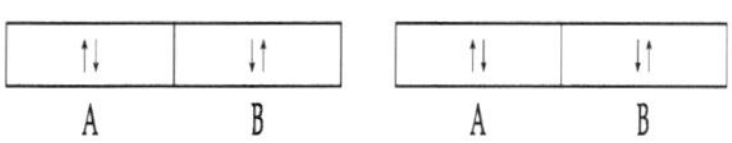

The number of microstates possible for any electronic configuration may be calculated from the formula,

Number of microstates = n! / r! (n - r)!

Where n is the twice the number of orbitals, r is the number of electrons and ! is the factorial.

For p^2 configuration, n= 3x2 =6; r = 2; n – r = 4

6! = 6 x 5 x 4 x 3 x 2 x 1 = 720; 2! = 2 x 1 =2; 4! = 4 x 3 x 2 x 1 = 24

Substituting in the formula, the number of microstates is 15.

Similarly for a d^2 configuration, the number of microstates is given by 10! / 2! (10 – 2)!

$$\frac{10\times9\times8\times7\times6\times5\times4\times3\times2\times1}{2\times1(8\times7\times6\times5\times4\times3\times2\times1)}=45$$

Thus a d^2 configuration will have 45 microstates. Microstates of different d^n configuration are given in Table-3.

d^n configuration	d^1,d^9	d^2,d^8	d^3,d^7	d^4,d^6	d^5	d^{10}
No.of microstates	10	45	120	210	252	1

Table 3. Microstates of different d^n configuration

7. Multiple term symbols of excited states

The terms arising from d^n configuration for 3d metal ions are given Table-4.

Configuration	Ion	Term symbol
d^1	Ti^{3+},V^{4+}	
d^9	Cu^{2+}	^2D
d^2	Ti^{2+},V^{3+},Cr^{4+}	
d^8	Ni^{2+}	^3F, ^3P, ^1G, ^1D, ^1S
d^3	Cr^{3+},V^{2+},Mn^{4+}	
d^7	Ni^{3+},Co^{2+}	^4F, ^4P, ^{2}(H, G, F, D, D, P)
d^4	Cr^{2+},Mn^{3+}	
d^6	Fe^{2+},Co^{3+}	^5D , ^{3}(H, G, F, F, , D, P, P), ^{1}(I, ,G, G, F, D, D, S,S)
d^5	Mn^{2+}, Fe^{3+}	
d^{10}	Zn^{2+}	^6S, ^{4}(G, F, D, P), ^{2}(I, H, G, G, F, F), ^{2}(D, D, D, P, S)
		^6S

Table 4. Terms arising from d^n configuration for 3d ions (n=1 to10)

8. Selection rules

8.1. La Porte selection rule

This rule says that transitions between the orbitals of the same sub shell are forbidden. In other words, the for total orbital angular momentum is $\Delta L = \pm 1$. This is La Porte allowed transitions. Thus transition such as $^1S \rightarrow {}^1P$ and $^2D \rightarrow {}^2P$ are allowed but transition such as $^3D \rightarrow {}^3S$ is forbidden since $\Delta L = -2$ .That is, transition should involve a change of one unit of angular momentum. Hence transitions from *gerade* to *ungerade* (*g* to *u*) or vice versa are allowed, i.e., $u \rightarrow g$ or $g \rightarrow u$ but not $u \rightarrow u$ or $g \rightarrow g$. In the case of p sub shell, both ground and excited states are odd and in the case of d sub shell both ground and excited states are even. As a rule transition should be from even to odd or vice versa.

The same rule is also stated in the form of a statement instead of an equation:

Electronic transitions within the same p or d sub-shell are forbidden, if the molecule has centre of symmetry.

8.2. Spin selection rule

The selection Rule for Spin Angular Momentum is

$$\Delta S = 0$$

Thus transitions such as $^2S \rightarrow {}^2P$ and $^3D \rightarrow {}^3P$ are allowed, but transition such as $^1S \rightarrow {}^3P$ is forbidden. The same rule is also stated in the form of a statement,

Electronic Transitions between the different states of spin multiplicity are forbidden.

The selection Rule for total angular momentum, J, is

$$\Delta J = 0 \text{ or } \pm 1$$

The transitions such as $^2P_{1/2} \rightarrow {}^2D_{3/2}$ and $^2P_{3/2} \rightarrow {}^2D_{3/2}$ are allowed, but transition such as $^2P_{1/2} \rightarrow {}^2D_{5/2}$ is forbidden since $\Delta J = 2$.

There is no selection rule governing the change in the value of n, the principal quantum number. Thus in hydrogen, transitions such as 1s $\rightarrow$ 2p, 1s $\rightarrow$ 3p, 1s $\rightarrow$ 4p are allowed.

Usually, electronic absorption is indicated by reverse arrow, $\leftarrow$, and emission is indicated by the forward arrow, $\rightarrow$, though this rule is not strictly obeyed.

8.3. Mechanism of breakdown of selection rules

8.3.1. Spin-orbit coupling

For electronic transition to take place, $\Delta S = 0$ and $\Delta L = \pm 1$ in the absence of spin-orbit coupling. However, spin and orbital motions are coupled. Even, if they are coupled very weakly, a little of each spin state mixes with the other in the ground and excited states by an amount dependent upon the energy difference in the orbital states and magnitude of spin –orbit coupling constant. Therefore electronic transitions occur between different states of spin multiplicity and also between states in which Δ L is not equal to ± 1. For example, if the ground state were 99% singlet and 1% triplet (due to spin– orbit coupling) and the excited state were 1% singlet and 99 % triplet, then the intensity would derive from the triplet –triplet and singlet-singlet interactions. Spin-orbit coupling provides small energy differences between degenerate state.

This coupling is of two types. The single electron spin orbit coupling parameter ζ, gives the strength of the interaction between the spin and orbital angular momenta of a single electron for a particular configuration. The other parameter, λ, is the property of the term. For high spin complexes,

$$\lambda = \pm \xi / 2S$$

Here positive sign holds for shells less than half field and negative sign holds for more than half filled shells. S is the same as the one given for the free ion. The λ values in crystals are close to their free ion values. Λ decreases in crystal with decreasing Racah parameters B and C. For high spin d^5 configuration, there is no spin orbit coupling because ^6S state is unaffected by the ligand fields. The λ and ζ values for 3d series are given in Table-5.

Ion	Ti(II)	V(II)	Cr(II)	Mn(II)	Fe(II)	Co(II)	Ni(II)
Ξ (cm^{-1})	121	167	230	347	410	533	649
λ(cm^{-1})	60	56	57	0	-102	-177	-325

Table 5. λ and ζ values for 3d series

8.3.2. La Porte selection rule

Physically 3d (even) and 4p (odd) wave functions may be mixed, if centre of inversion (i) is removed. There are two processes by which i is removed.

a. The central metal ion is placed in a distorted field (tetrahedral field, Tetragonal distortions, etc.,) The most important case of distorted or asymmetric field is the case of a tetrahedral complex. Tetrahedron has no inversion centre and so d-p mixing takes place. So electronic transitions in tetrahedral complexes are much more intense, often by a factor 100, than in a analogous octahedral complexes. *Trans* isomer of $[Co(en)_2Cl_2]^+$ in aqueous solution is three to four times less intense than the *cis* isomer because the former is centro-symmetric. Other types of distortion include Jahn –Teller distortions.
b. Odd vibrations of the surrounding ligands create the distorted field for a time that is long enough compared to the time necessary for the electronic transition to occur (Franck Condon Principle).Certain vibrations will remove the centre of symmetry. Mathematically this implies coupling of vibrational and electronic wave functions. Breaking down of La Porte rule by vibrionic coupling has been termed as "Intensity Stealing". If the forbidden excited term lies energetically nearby a fully allowed transition, it would produce a very intense band. Intensity Stealing by this mechanism decreases in magnitude with increasing energy separation between the excited term and the allowed level.

9. Splitting of energy states

The symbols **A**(or **a**) and **B** (or **b**) with any suffixes indicate wave functions which are singly degenerate. Similarly **E** (or **e**) indicates double degeneracy and **T** (or **t**) indicates triple degeneracy. Lower case symbols, a_{1g}, a_{2g}, e_g, etc., are used to indicate electron wave functions(orbitals) and upper case symbols are used to describe electronic energy levels. Thus $^2T_{2g}$ means an energy level which is triply degenerate with respect to orbital state and also doubly degenerate with respect to its spin state. Upper case symbols are also used without any spin multiplicity term and they then refer to symmetry (ex., A_{1g} symmetry). The subscripts **g** and **u** indicate *gerade* (even) and *ungerade* (odd).

d orbitals split into two sets - t_{2g} orbitals and e_g orbitals under the influence crystal field. These have T_{2g} and E_g symmetry respectively. Similarly **f** orbitals split into three sets - a_{2u} (f_{xyz}) , t_{2u} ($f_{x(y^2-z^2)}$, $f_{y(z^2-x^2)}$, $f_{z(x^2-y^2)}$) and t_{1u} (f_{x^3} , f_{y^3} , f_{z^3}). These have symmetries A_{2u}, T_{2u} and T_{1u} respectively.

Splitting of **D** state parallels the splitting of **d** orbitals and splitting of **F** state splits parallels splitting of **f** orbitals. For example, **F** state splits into either T_{1u}, T_{2u} and A_{2u} or T_{1g}, T_{2g} and A_{2g} sub-sets. Which of these is correct is determined by **g** or **u** nature of the configuration from which **F** state is derived. Since **f** orbitals are **u** in character ^{2}**F** state corresponding to f^1 configuration splits into $^2T_{1u}$, $^2T_{2u}$, and $^2A_{2u}$ components; similarly ^{3}**F** state derived from d^2 configuration splits into $^3T_{2g}$, $^3T_{1g}$ and $^3A_{2g}$ components because **d** orbitals are **g** in character.

9.1. Splitting of energy states corresponding to d^n terms

These are given in Table-6.

Energy	Sub- states
S	A_1
P	T_1
D	$E + T_2$
F	$A_2 + T_1 + T_2$
G	$A_1 + E + T_1 + T_2$
H	$E + T_1 + T_1 + T_2$
I	$A_1 + A_2 + E + T_1 + T_2 + T_2$

Table 6. Splitting of energy states corresponding to d^n terms

The d-d spectra is concerned with d^n configuration and hence the crystal field sub-states are given for all the d^n configuration in Table -7.

Configuration	*Free ion ground state*	*Crystal field substates*	*Important excited states*	*Crystal field state*
d^1, d^9	^2D	$^2T_{2g}$, ^2E_g		
d^2, d^8	^3F	$^3T_{1g}$, $^3T_{2g}$, $^3A_{2g}$	^3P	$^3T_{1g}$
d^3, d^7	^4F	$^4T_{1g}$, $^4T_{2g}$, $^4A_{2g}$	^4p	$^4T_{1g}$
d^4, d^6	^5D	$^5T_{2g}$, ^5E_g		
d^5	^6S	$^6A_{1g}$		

Table 7. Crystal field components of the ground and some excited states of d^n (n=1 to 9) configuration

10. Energy level diagram

Energy Level Diagrams are described by two independent schemes - Orgel Diagrams which are applicable to weak field complexes and Tanabe –Sugano (or simply T-S) Diagrams which are applicable to both weak field and strong field complexes.

11. Inter-electronic repulsion parameters

The inter-electronic repulsions within a configuration are linear combinations of Coulombic and exchange integrals above the ground term. They are expressed by either of the two ways: Condon - Shortley parameters, F_0, F_2 and F_4 and Racah parameters, A, B and C. The magnitude of these parameters varies with the nature of metal ion.

11.1. Racah parameters

The Racah parameters are A, B and C. The Racah parameter A corresponds to the partial shift of all terms of a given electronic configuration. Hence in the optical transition considerations, it is not taken into account. The parameter, B measures the inter electronic repulsion among the electrons in the d-orbitals. The decrease in the value of the interelectronic repulsion parameter, B leads to formation of partially covalent bonding. The ratio between the crystal B^1 parameter and the free ion B parameter is known as nephelauxetic rato and it is denoted by β. The value of β is a measure of covalency. The smaller the value, the greater is the covalency between the metal ion and the ligands. The B and C values are a measure of spatial arrangement of the orbitals of the ligand and the metal ion.

Racah redefined the empirical Condon –Shortley parameters so that the separation between states having the maximum multiplicity (for example, difference between is a function of ^3F and ^3P or ^4F and ^4P is a function of a single parameter, B. However, separations between terms of different multiplicity involve both B and C

12. Tanabe –Sugano diagrams

Exact solutions for the excited sate energy levels in terms of Dq, B and C are obtained from Tanabe-Sugano matrices. However, these are very large (10 x 10) matrices and hand calculations are not feasible. For this reason Tanabe-Sugano have drawn energy level diagrams known as T-S diagrams or energy level diagrams. The T-S diagrams are valid only if the value of B, C and Dq ae lower for a complex than for the free ion value.

Quantitative interpretation of electronic absorption spectra is possible by using Tanabe – Sugano diagrams or simply T-S diagrams. These diagrams are widely employed to correlate and interpret spectra for ions of all types, from d^2 to d^8. Orgel diagrams are useful only qualitatively for high spin complexes whereas T-S diagrams are useful both for high spin and low spin complexes. The x-axis in T-S diagrams represent the ground state term. Further, in T-S diagrams, the axes are divided by B, the interelectronic repulsion parameter or Racah Parameter. The x-axis represents the crystal field strength in terms of Dq/ B or Δ / B and the Y-axis represents the energy in terms of E/B.

The energies of the various electronic states are given in the T-S diagrams on the vertical axis and the ligand field strength increases from left to right on the horizontal axis. The symbols in the diagram omit the subscript, g, with the understanding that all states are *gerade* states. Also, in T.S. diagrams, the zero of energy for any particular d^n ion is taken to be the energy of the ground state. Regardless of the ligand field strength, then, the horizontal axis represents the energy of the ground state because the vertical axis is in units of E/B and x-axis is also in units of Δ /B. Thus, the unit of energy in T-S diagram is B, Racah Parameter.

The values of B are different for different ions of the same d^n (or different d^n configuration) which is shown on the top of each diagram. One T-S diagram is used for all members of an isoelectronic group. Also some assumption is made about the relative value of C/B.

13. Electron spin resonance

Electron Spin Resonance (ESR) is a branch of spectroscopy in which radiation of microwave frequency is absorbed by molecules possessing electrons with unpaired spins. It is known by different names such as Electron Paramagnetic Resonance (EPR), Electron Spin Resonance (ESR) and Electron Magnetic Resonance (EMR). This method is an essential tool for the analysis of the structure of molecular systems or ions containing unpaired electrons, which have spin-degenerate ground states in the absence of magnetic field. In the study of solid state materials, EPR method is employed to understand the symmetry of surroundings of the paramagnetic ion and the nature of its bonding to the nearest neighbouring ligands.

When a paramagnetic substance is placed in a steady magnetic field (H), the unpaired electron in the outer shell tends to align with the field. So the two fold spin degeneracy is removed. Thus the two energy levels, $E_{1/2}$ and $E_{-1/2}$ are separated by gβH, where g is spectroscopic splitting factor and is called gyro magnetic ratio and β is the Bohr magneton. Since there is a finite probability for a transition between these two energy levels, a change in the energy state can be stimulated by an external radio frequency. When microwave frequency (υ) is applied perpendicular to the direction of the field, resonance absorption will occur between the two split spin levels. The resonance condition is given by, hν = gβH, where h is Planck's constant.

The resonance condition can be satisfied by varying υ or H. However, EPR studies are carried out at a constant frequency (ν), by varying magnetic field (H). For a free electron, the g value is 2.0023. Since h and β are constants, one can calculate the g factor. This factor determines the divergence of the Zeeman levels of the unpaired electron in a magnetic field and is characteristic of the spin system.

In the crystal systems, the electron spins couple with the orbital motions and the g value is a measure of the spin and orbital contributions to the total magnetic moment of the unpaired electron and any deviation of magnetic moment from the free spin value is due to the spin-orbit interaction. It is known that the crystal field removes only the orbital degeneracy of the ground terms of the central metal ion either partially or completely. The strong electrical fields of the surrounding ligands results in "Stark Splitting" of the energy levels of the paramagnetic ion. The nature and amount of splitting strongly depends on the symmetry of the crystalline electric field. The Stark splitting of the free ion levels in the crystal field determines the magnetic behaviour of the paramagnetic ion in a crystal. Whenever there is a contribution from the unquenched orbital angular momentum, the measured g values are isotropic as a result of the asymmetric crystal field since the contribution from the orbital motion is anisotropic. To decide the ultimate ground state of a paramagnetic ion in the crystal, the two important theorems, Kramers and Jahn-Teller, are useful. Using group theory, one can know the nature of the splitting of the free ion levels in the crystal fields of various symmetries.

Jahn-Teller theorem states that any nonlinear molecule in an electronically degenerate ground state is unstable and tends to distort in order to remove this degeneracy. The direction of distortion which results in greatest stabilization can often be deduced from EPR and other spectroscopic data.

Kramers' theorem deals with restrictions to the amount of spin degeneracy which can be removed by a purely electrostatic field. If the system contains an odd number of electrons, such an electrostatic field cannot reduce the degeneracy of any level below two. Each pair forms what is known as a Kramers' doublet, which can be separated only by a magnetic field. For example, Fe(III) and Mn(II) belonging to d^5 configuration, exhibit three Kramers' doublets labeled as |±5/2>, |±3/2> and |±1/2>.

If the central metal ion also possesses a non-zero nuclear spin, I, then hyperfine splitting occurs as a result of the interaction between the nuclear magnetic moment and the electronic magnetic moment. The measurement of g value and hyperfine splitting factor provides information about the electronic states of the unpaired electrons and also about the nature of the bonding between the paramagnetic ion and its surrounded ligands. If the ligands also contain non-zero nuclear spin, then the electron spin interacts with the magnetic moment of the ligands. Then one could expect super hyperfine EPR spectrum.

The g value also depends on the orientation of the molecules having the unpaired electron with respect to the applied magnetic field. In the case of perfect cubic symmetry, the g value does not depend on the orientation of the crystal. But in the case of low symmetry crystal fields, g varies with orientation. Therefore we get three values g_{xx}, g_{yy}, and g_{zz} corresponding to a, b and c directions of the crystal. In the case of tetragonal site $g_{xx} = g_{yy}$ which is referred to as $g_{\perp}$ and corresponds to the external magnetic field perpendicular to the Z-axis. When it is parallel, the value is denoted as $g_{\parallel}$. Hence one can deduce the symmetry of a complex by EPR spectrum i.e., cubic, tetragonal, trigonal or orthorhombic. Anyhow, it is not possible to distinguish between orthorhombic and other lower symmetries by EPR.

13.1. EPR signals of first group transition metal ions

Transition metal ions of 3d group exhibit different patterns of EPR signals depending on their electron spin and the crystalline environment. For example, $3d^1$ ions, VO^{2+} and Ti^{3+} have s = 1/2 and hence are expected to exhibit a single line whose g value is slightly below 2.0. In the case of most abundant ^{51}V, s = 1/2 and I = 7/2, an eight line pattern with hyperfine structure of almost equal intensity can be expected as shown in Fig-1. In the case of most abundant Ti, (s = 1/2 and I = 0), no hyperfine structure exists. However, the presence of less abundant isotopes (^{47}Ti with I = 5/2 and ^{49}Ti with I = 7/2) give rise to weak hyperfine structure with six and eight components respectively. This weak structure is also shown in Fig-1.

Cr(III), a d^3 ion, with s = 3/2 exhibits three fine line structure. The most abundant ^{52}Cr has I = 0 and does not exhibit hyperfine structure. However, ^{53}Cr with I = 3/2 gives rise to hyperfine structure with four components. This structure will be weak because of the low abundance of ^{53}Cr. Thus each one of the three fine structure lines of ^{53}Cr is split into four weak hyperfine lines. Of these, two are overlapped by the intense central line due to the most abundant ^{52}Cr and the other two lines are seen in the form of weak satellites.

Mn(II) and Fe(III) with d^5 configuration have s = 5/2 and exhibit five lines which correspond to a $|\pm5/2\rangle \rightarrow |\pm3/2\rangle$, $|\pm3/2\rangle \rightarrow |\pm1/2\rangle$ and $|+1/2\rangle \rightarrow |-1/2\rangle$ transitions. In the case of ^{55}Mn, which has I = 5/2, each of the five transitions will give rise to a six line hyperfine structure. But in powders, usually one observes the six-hyperfine lines corresponding to $|+1/2\rangle \rightarrow |-1/2\rangle$ transition only. The remaining four transition sets will be broadened due to the high anisotropy. Fe^{3+} yields no hyperfine structure as seen in Fig -1.

Co^{2+}, a d^7 configuration, with s value of 3/2 exhibits three fine structure lines. In the case of ^{59}Co (I = 7/2), eight line hyperfine pattern can be observed as shown in the Fig-1.

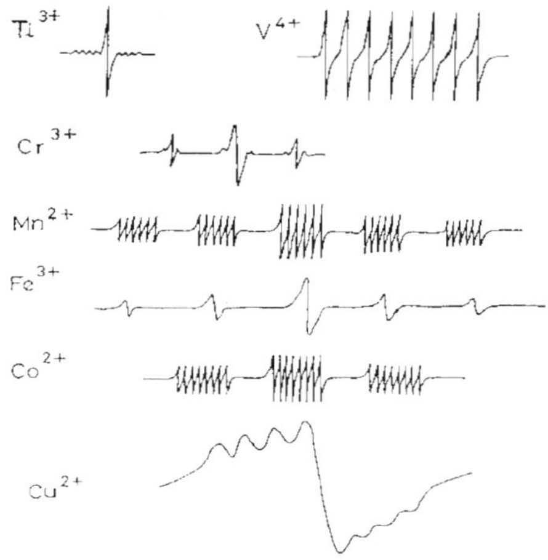

Figure 1. EPR signal of 3d ions

14. Survey of experimental results

14.1. Titanium

Titanium is the ninth most abundant element in the Earth's crust (0.6%). There are 13 known isotopes of titanium. Among them five are natural isotopes with atomic masses 46 to 50 and the others are artificial isotopes. The synthetic isotopes are all radioactive. Titanium alloys are used in spacecraft, jewelry, clocks, armored vehicles, and in the construction of buildings. The compounds of titanium are used in the preparation of paints, rubber, plastics, paper, smoke screens ($TiCl_4$ is used), sunscreens. The main sources of Ti are ilmenite and rutile.

Titanium exhibits +1 to +4 ionic states. Among them Ti^{4+} has d^0 configuration and hence has no unpaired electron in its outermost orbit. Thus Ti^{4+} exhibits diamagnetism. Hence no d-d transitions are possible. The ionic radius of Ti^{3+} is the same as that of Fe(II) (0.76 A.U). Ti(I) and Ti(III) have unpaired electrons in their outermost orbits and exhibit para magnetism

14.2. Electronic spectra of titanium compounds

The electronic configuration of Ti^3 is [Ar] $3d^1\ 4s^2$. It has five fold degeneracy and its ground state term symbol is ^2D. In an octahedral crystal field, the five fold degeneracy is split into $^2T_{2g}$ and ^2E_g states. Thus only one single electron transition, $^2T_{2g} \rightarrow {}^2E_g$, is expected in an octahedral crystal field. The separation between these energies is 10Dq, which is crystal field energy. Normally, the ground $^2T_{2g}$ state is split due to Jahn-Teller effect and hence lowering of symmetry is expected for Ti(III) ion. This state splits into $^2B_{2g}$ and ^2E_g states in tetragonal symmetry and the excited term ^2E_g also splits into $^2B_{1g}$ and $^2A_{1g}$ levels. Thus, *three bands* are expected for *tetragonal (C_{4v})* symmetry. Energy level diagram in tetragonal environment is shown in Fig -2.

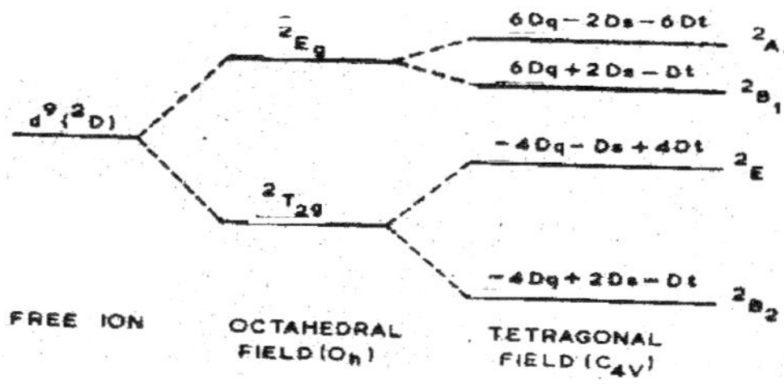

Figure 2. Energy level diagram of Ti^{3+} in octahedral and tetragonal fields

The transitions in the tetragonal field are described by the following equations.

$$^2B_{2g} \rightarrow {}^2E_g : \left[-4Dq - Ds + 4Dt - \left(-4Dq + 2Ds - Dt\right)\right] = -3Ds + 5Dt \quad (1)$$

$$B_{2g} \rightarrow {}^2B_{1g} : \left[6Dq + 2Ds - Dt - \left(-4Dq + 2Ds - Dt\right)\right] = 10Dq \quad (2)$$

$$^2B_{2g} \rightarrow {}^2A_{1g} : \left[6Dq - 2Ds - 6Dt - \left(-4Dq + 2Ds - Dt\right)\right] = 10Dq - 4Ds + 5Dt \quad (3)$$

In the above formulae, Dq is octahedral crystal field and Ds and Dt are tetragonal field parameters. The same sign of Dq and Dt indicates an axial elongation and opposite sign indicates an axial compression

14.2.1. EPR spectra of titanium compounds

When any Ti(III) compound in the form of powder is placed in a magnetic field, it gives a resonance signal. The single d-electron of Ti^{3+} has spin, s = 1/2. The abundance of isotopes is reported as $^{46}Ti \approx 87\%$, $^{48}Ti \approx 7.7\%$ and $^{50}Ti \approx 5.5\%$ and have nuclear spin I = 0, 5/2 and 7/2 respectively. Electron spin and nuclear spin interactions give rise to (2I+1) hyperfine lines (0,6 and 8) and appear as satellite. Since ^{46}Ti abundance is more, the EPR signal contains only one resonance line which is similar to the one shown in Fig-3. The g value for this resonance is slightly less than 2.0.

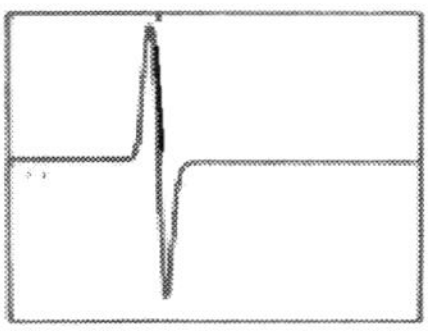

Figure 3. RT powered EPR spectrum of Ti(III).

14.2.2. Relation between EPR and optical absorption spectra

EPR studies for Ti^{3+} can be correlated with optical data to obtain the orbital reduction parameter.

$$\alpha = K\frac{\lambda_{(Covalency)}}{\lambda_{(Ionic)}} = \frac{(g_e - g_{11})\Delta E}{n\lambda_{(ionic)}} \tag{4}$$

where n is 8 for C_{4V} , ΔE is the energy of appropriate transition, λ is the spin-orbit coupling constant for Ti^{3+}, i.e., 154 cm^{-1} and k is the orbital reduction parameter.

14.2.3. Typical examples

EPR and optical absorption spectral data of selected samples are discussed as examples. The data chosen from the literature are typical for each sample and hence should be considered as representative only. For more complete information on specific example, the original references are to be consulted. X-band spectra and optical absorption spectra of the powdered samples are recorded at room temperature (RT).

14.2.4. Optical absorption studies

Ti(III) ion in solids is characterized by three broad bands around 7000, 12000 and 18000 cm^{-1}. These are due to the transitions from $^2B_{2g} \rightarrow ^2E_g$, $^2B_{2g} \rightarrow ^2B_{1g}$, and $^2B_{2g} \rightarrow ^2A_{1g}$ respectively. Three bands of titanite at 7140, 13700 and 16130 cm^{-1} and of anatase at 6945, 12050 and 18180 cm^{-1} are attributed to the above transitions. The optical absorption spectrum of lamprophyllite is also similar. The optical absorption spectrum of benitoite sample displays three bands at 8260, 10525 and 15880 cm^{-1}. From the observed band positions, the crystal field parameter in octahedral field, Dq and tetragonal field parameters, Ds and Dt, are given in Table-8.

Sample	Dq cm^{-1}	Ds cm^{-1}	Dt cm^{-1}
Titanite	1370	-1367	608
Anatase	1205	-1867	268
Lamprophyllite	877	-1426	1525
Benitoite	1050	-1945	485

Table 8. Crystal field parameters of Ti(III)

The magnitude of Dt indicates the strength of the tetragonal distortion. This is more in lamprophyllite when compared to the other samples.

i. X-band EPR spectra of the powdered sample of titanite shows a broad resonance line in the centre (335.9 mT). The measured g value is 1.957. Another resonance line is noticed at 341.4 mT with g =1.926. The central eight line transition is superimposed on the spectrum and the components are attributed to VO(II) impurity. The g value of Ti^{3+} is 1.957 and other g value is due to VO(II). The g value of 1.95 indicates that Ti^{3+} is in tetragonally distorted octahedral site.
ii. The EPR spectrum of anatase shows a large number of resonances centered around g value of 2 which is attributed to Ti^{3+}. The additional structures between g values of 2 and 4 are attributed to Fe(III) impurity in the compound. Both the ions are in tetragonally distorted environment.

iii. X band EPR of polycrystalline lamprophyllite sample indicates a broad resonance line with line width 56.6 mT and a g value of 2.0. This is due to the presence of Ti(III) in the compound. The broad line is due to the dipolar-dipolar interaction of Ti(III) ions. Even at liquid nitrogen temperature, only the line intensity increases indicating that Curie law is obeyed.

Using EPR and optical absorption spectral results of titanite, the covalency parameter is calculated using equation (4), $\alpha = \frac{(g_e - g_{11})\Delta E}{n\lambda_{(ionic)}}$. The α value obtained is 0.51, which indicates higher covalent character between ligand and metal ion.

15. Vanadium

Vanadium abundance in earth's crust is 120 parts per million by weight. Vanadium's ground state electron configuration is [Ar] $3d^3 4s^2$. Vanadium exhibits four common oxidation states +5, +4, +3, and +2 each of which can be distinguished by its color. Vanadium(V) compounds are yellow in color whereas +4 compounds are blue, +3 compounds are green and +2 compounds are violet in colour. Vanadium is used in making specialty steels like rust resistant and high speed tools. The element occurs naturally in about 65 different minerals and in fossil fuel deposits. Vanadium is used by some life forms as an active center of enzymes. Vanadium oxides exhibit intriguing electrochemical, photochemical, catalytical, spectroscopic and optical properties. Vanadium has 18 isotopes with mass numbers varying from 43 to 60. Of these, ^{51}V, natural isotope is stable:

15.1. Electronic spectra of vanadium compounds

Vanadium in its tetravalent state invariably exists as oxo-cation, VO^{2+} (vanadyl). The VO^{2+} ion has a single d electron which gives rise to the free ion term ^2D. In a crystal field of octahedral symmetry, this electron occupies the t_{2g} orbital and gives rise to ground state term $^2T_{2g}$. When the electron absorbs energy, it is excited to the e_g orbital and accordingly in octahedral geometry only one band corresponding to the transition, $^2T_{2g} \rightarrow {}^2E_g$, is expected. Because of the non-symmetrical alignment of the V=O bond along the axis, the site symmetry, in general, is lowered to tetragonal (C_{4V}) or rhombic (C_{2V}) symmetry. In C_{4V} site symmetry, $^2T_{2g}$ splits into $^2B_{2g}$ and ^2E_g, whereas ^2E_g splits into $^2B_{1g}$, $^2A_{1g}$. Hence three bands are expected in C_{4V} symmetry in the range of 11000 –14000, 14500 – 19000 and 20000 – 31250 cm^{-1}. The degeneracy of ^2E_g is also removed in C_{2V} symmetry resulting four bands. Energy level diagram of VO^{2+} in octahedral C_{4V} and C_{2V} symmetries are shown in Fig- 4. In the tetragonal C_{4V} symmetry transitions are described by the following equations.

$$^2B_{2g} \rightarrow {}^2E_g : \left[-4Dq - Ds + 4Dt - \left(-4Dq + 2Ds - Dt\right)\right] = -3Ds + 5Dt \tag{5}$$

$$B_{2g} \rightarrow {}^2B_{1g} : \left[6Dq + 2Ds - Dt - \left(-4Dq + 2Ds - Dt\right)\right] = 10Dq \tag{6}$$

$$^2B_{2g} \rightarrow {}^2A_{1g} : \left[6Dq - 2Ds - 6Dt - \left(-4Dq + 2Ds - Dt\right)\right] = 10Dq - 4Ds + 5Dt \tag{7}$$

In the above formulae, Dq is octahedral crystal field parameter and Ds, Dt are tetragonal field parameters. The same sign of Dq and Dt indicates an axial elongation and opposite sign indicates an axial compression.

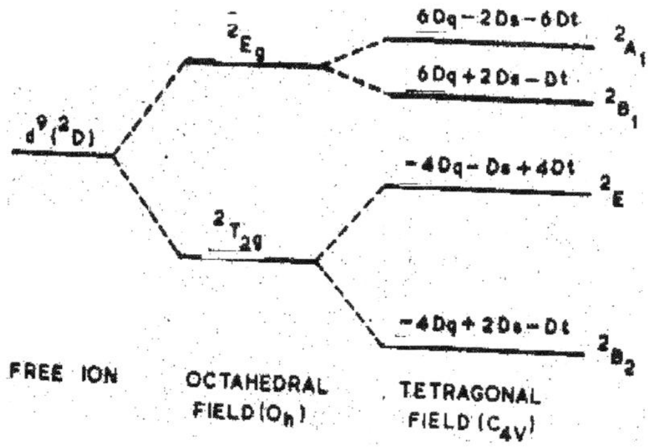

Figure 4. Energy level diagram indicating the assignment of the transitions in octahedral C_{4V} symmetry.

15.2. EPR spectra of vanadium compounds

The EPR signal is of three types. (i) is due to high concentration of vanadium. If the vanadium content in the compound is high, it gives a broad resonance line. Therefore the hyperfine line from ^{51}V cannot be resolved. The g value for this resonance is less than 2. (ii) VO^{2+} ion has s= ½ and I = 7/2. The EPR spectrum shows hyperfine pattern of eight equidistant lines. In C_{4v} symmetry two sets of eight lines are expected (sixteen-line pattern) whereas in C_{2v} symmetry three sets of eight lines are expected. Further in tetragonal distortion, $g_{11} < g_{\perp} < g_e$ which shows the presence of an unpaired electron in the d_{xy} orbital. This is characteristic feature of a tetragonally compressed complex.

Further lowering of symmetry gives rise to EPR spectrum which is similar to the one shown in g_{yy} and g_{zz} respectively. The hyperfine constants are designated as A_1, A_2 and A_3 respectively.

Using the EPR data, the value of dipolar term P and k term are calculated,

$$A_{11} = P\left[-\frac{4}{7} - k - \left(g_{11} - g_e\right) + \frac{3}{7}\left(g_{\perp} - g_e\right)\right] \tag{8}$$

$$A_{\perp} = P\left[\frac{2}{7} - k + \frac{11}{14}\left[g_{\perp} - g_e\right]\right] \tag{9}$$

$$g = \frac{1}{3}\left(g_{11} + 2g_{\perp}\right) \text{ and } A = \frac{1}{3}\left(A_{11} + 2A_{\perp}\right) \tag{10}$$

Using the EPR data, the admixture coefficients are calculated from the following formulae,

$$g_{11} = 2\left(3C_1^2 - C_2^2 - 2C_3^2\right) \tag{11}$$

$$g_{\perp} = 4C_1\left(C_2 - C_3\right) \text{ and } C_1^2 + C_2^2 + C_3^2 = 1 \tag{12}$$

$$A_{11} = P\left[g_{11} - \left(k + \frac{15}{7}\right)\left(1 - 2C_3^2\right) - \frac{3}{7}\left(1 + C_1C_2C_3\right)\right] \tag{13}$$

$$A_{\perp} = p\left[\frac{11}{14}g_{\perp} - 2C_1C_2\left(k + \frac{9}{7}\right)\right] \tag{14}$$

15.3. Relation between EPR and optical absorption spectra

The optical absorption results and EPR results are related as follows. EPR studies can be correlated with optical data to obtain the orbital coefficients β^{*2} and ε_{π}^{*2}.

$$g_{11} = g_e - \frac{8\lambda\beta^{*2}}{\Delta E_{xy}} \tag{15}$$

$$g_1 = g_e - \frac{2\lambda\varepsilon_{\pi}^{*2}}{\Delta E_{xz}} \tag{16}$$

Here g_{11} and $g_{\perp}$ are the spectroscopic splitting factors parallel and perpendicular to the magnetic field direction of g_e (i.e., 2.0023 for a free electron).

Δ E_1 is the energy of $^2B_{2g} \rightarrow {}^2B_{1g}$ and Δ E_2 is the energy of $^2B_{2g} \rightarrow {}^2E_g$.

λ is the spin-orbit coupling constant(160 cm^{-1}) for the free vanadium(VO^{2+}).

15.4. Typical examples

EPR and optical absorption spectral data of certain selected samples are discussed. The data chosen from the literature are typical for each sample. The data should be considered as representative only. For more complete information on specific example, original references are to be consulted. X-band spectra of the powdered samples and optical absorption spectra are recorded at room temperature (RT).

X-band EPR spectra of the vanadium(IV) complex with DMF recorded in solutions reveal a well-resolved axial anisotropy with 16-line hyperfine structure. This is characteristic of an interaction of vanadium nuclear spin (^{51}V, I = 7/2) with S. The observed EPR parameters are g_{11} =1.947, A_{11} = 161.3 x 10^{-4} cm^{-1} and $g_{\perp}$ =1.978, $A_{\perp}$ = 49.0 x 10^{-4} cm^{-1}. EPR parameters of several samples are available in literature and some of them are given in Table -9.

Mineral name	g_{11}	$g_{\perp}$	A_{11} mT	$A_{\perp}$ mT
Kainite	1.932	1.983	17.7	6.9
Apophyllite	1.933	1.982	18.02	6.02
Pascoite site I siteII	1.933 1.946	1.988 1.976	18.50 20.00	7.6 8.2
CAPH	1.933	1.993		

Table 9. Various EPR parameters of VO(II) in minerals

Using the EPR data, the admixture coefficients are calculated for apophyllite and pascoite minerals and are given in the Table -10.

Sample	C_1	C_2	C_3	K	P (x 10^{-4} cm^{-1})
Apophyllite	0.7083	0.7124	0.0028	0.86	122.7
Pascoite	0.7010	0.7116	0.0035	0.36	118.4
	0.7090	0.7285	0.03174	0.34	143

Table 10. Admixture coefficients of VO^{2+} ion

EPR spectrum of polycrystalline sample of wavellite with sixteen line pattern indicates the presence of VO^{2+} ion as an impurity. The EPR parameters calculated are g_{zz}= 1.933 and g_{yy}=g_{xx} = 1.970 and the corresponding A values are 19.0 and 6.2 mT.

15.5. Typical examples

a. (i) Divalent vanadium (V^{2+}) of d^3 configuration, containing halide and other ions in aqueous solutions, gives three transitions, i.e., $^4A_{2g}\rightarrow{}^4T_{2g}$, $^4A_{2g}\rightarrow{}^4T_{1g}$(F) and $^4A_{2g}\rightarrow{}^4T_{1g}$(P) in an octahedral geometry. In $\left[V\left(H_2O\right)_6^{2+}\right]$, the three bands are observed at 11400, 17100 and 24000 cm^{-1} along with some weak shoulders at about 20000 and 22000 cm^{-1}. The bands observed at 11400, 17100 and 24000 cm^{-1} are assigned to the transitions $^4A_{2g}\rightarrow{}^4T_{2g}$, $^4T_{1g}$(F) and $^4T_{1g}$(P) respectively. 10Dq is 11400 cm^{-1}. For divalent vanadium ion, Racah parameters are B = 860 and C = 4165 cm^{-1}. Calculated Racah parameters are expected to be less than the one in the free ion value. Accordingly, the weak shoulders observed at 20000 and 22000 cm^{-1} are assigned to $^4A_{2g}\rightarrow{}^2T_{2g}$, and $^4A_{2g}\rightarrow{}^2T_{1g}$, ^2E transitions.

(ii) The optical absorption spectrum of vanadium carboxylate tetrahydrate sample displays three bands at 11400, 17360 and 23920 cm^{-1}. These are assigned to the transitions, $^4A_{2g}\rightarrow{}^4T_{2g}$, $^4T_{1g}$(F) and $^4T_{1g}$(P) in an octahedral geometry.

b. Trivalent Vanadium *(V^{3+}) (d^2)* in aqueous solutions shows two stronger bands at about 15000 and 23000 cm^{-1} and some weaker bands at 11500, 18000 cm^{-1}. The stronger bands are assigned to the transitions, $^3T_{1g}(F) \rightarrow {}^3T_{2g}(F)$ and $^3T_{1g}(F) \rightarrow {}^3T_{1g}(P)$ in an octahedral environment. Since this ion contains two d electrons, it is not so easy to attribute to the other bands. Therefore T-S diagrams are used to identify the other bands. 10Dq is 16400 cm^{-1} and B = 623 (free ion B= 886 cm^{-1}and C =765 cm^{-1}). Third band could be expected at 32000 cm^{-1} due to $^3T_{1g}(F) \rightarrow {}^2A_{1g}$. This band corresponds to double electron transition and hence the intensity is expected to be lower than that of the first two bands. The weaker bands observed at 11500, 18000 cm^{-1} are attributed to the spin forbidden transitions, $^3T_{1g}(F) \rightarrow {}^1E_g$, $^1T_{2g}$ and $^1A_{1g}$.

c. (i) *Tetravalent vanadium* (V^{4+}) (d^1). The absorption spectrum of tetravalent vanadium compounds shows three transitions, $^2B_{2g} \rightarrow {}^2E_g$, $^2B_{2g} \rightarrow {}^2B_{1g}$ and $^2B_{2g} \rightarrow {}^2A_{1g}$. The $^2B_{2g} \rightarrow {}^2E_g$ is the most intense and $^2B_{2g} \rightarrow {}^2B_{1g}$ is the weakest. Accordingly, the bands observed in vanadium doped zinc hydrogen maleate tetrahydrate (ZHMT) at 13982, 16125 and 21047 cm^{-1} are assigned to the above three transitions respectively. The octahedral crystal field parameter, Dq (1613 cm^{-1}), and tetragonal field parameters, Ds (-2700 cm^{-1}) and Dt (1178 cm^{-1}), are evaluated.

(ii) The electronic absorption spectrum of the VO^{2+} in $CdSO_4.8H_2O$ recorded at room temperature shows bands at 12800, 13245, 14815, 18345 cm^{-1}. These bands are assigned to $^2B_{2g} \rightarrow {}^2E_g$, $^2B_{2g} \rightarrow {}^2B_{1g}$ and $^2B_{2g} \rightarrow {}^2A_{1g}$ transitions. The band observed at 12500 cm^{-1} is the split component of the band at 13245 cm^{-1}. The crystal field octahedral parameter, Dq (1465 cm^{-1}) and tetragonal field parameters, Ds (-2290 cm^{-1}) and Dt (1126 cm^{-1}) are evaluated.

Several examples are found in the literature. Some of them are given in the Table-11.

Sample	Transition from $^2B_2$ cm^{-1}			Dq cm^{-1}	Ds cm^{-1}	Dt cm^{-1}
	^2E,	$^2B_1$	$^2A_1$			
Cadmium ammonium phosphate hexahydradate(CAPH)	12270	16000	26625	1600	-3275	488
Aphophyllite	12500	15335	24385	1538	-2080	653
Pascoite site I	12255	14450	21415	1445	-2765	803
Site II	12255	16000	21415	1600	-2524	937

Table 11.

d. Pentavalent vanadium has no d electron and hence d-d transitions are not possible. Therefore, the observed bands in electronic absorption spectrum are ascribed to charge transfer bands. These appear around 37000, 45000 cm^{-1}. These are assigned to transitions from ligand orbitals to metal d-orbitals: $A_1 \rightarrow T_2$ ($t_1 \rightarrow 2e$) and $A_1 \rightarrow T_2$ ($3t_2 \rightarrow 2e$) in tetrahedral configuration for the ion VO_4^{3-}.

Vanadium doped silica gel also shows sharp band at 41520 cm^{-1} and shoulders at 45450 and 34480 cm^{-1}. These are also assigned to charge transfer transitions in tetrahedral environment of VO_4^{3-}. The minimum value of 10Dq for VO_4^{3-} is expected at about 16000 cm^{-1} in octahedral geometry. This is expected because the two bands at 34480 and 45450 cm^{-1} are from the ligand orbitals to two vacant d orbitals which are 10Dq apart. This would be about twice the energy separation (8000 cm^{-1}) observed for tetrahedral VO_4^{3-}.Hence the evidence does not satisfy the assignment of bands to d-d transitions. Therefore the bands are due to charge transfer transitions.

16. Chromium

Chromium is the 6th most abundant transition metal. Chromium is used in the manufacture of stainless steel and alloys. The ground state electronic configuration is [Ar] $3d^4 4s^2$. It exhibits +2 to +6 oxidation states. Most stable oxidation state are +2 (CrO), +3 (Cr_2O_3) and +6 ($K_2Cr_2O_7$).

16.1. Optical spectra

a. Divalent chromium(*d*^{2})

Cr^{2+} has a d^4 configuration and forms high spin complexes only for crystal fields less than 2000 cm^{-1}. The ground state term in an octahedral crystal field is ^5E_g belonging to the $t_{2g}^3 e_g^1$ configuration. The excited state $^5T_{2g}$ corresponds to promotion of one single electron to give $t_{2g}^2 e_g^2$ configuration. The d^4 electron is susceptible to Jahn-Teller distortion and hence Cr^{2+} compounds usually are of low symmetry. In lower symmetry, the excited quintet state of Cr^{2+} splits into three levels and the ground level quintet state splits into two levels. In the case of $Cr^{2+}(H_2O)_6$, the value of Dq is 1400 cm^{-1}. In spinels, $Cr^{2+}$is in the tetrahedral environment and Dq is about 667 cm^{-1}only.

b. Trivalent chromium(d^3):

In octahedral symmetry, the three unpaired electrons are in t_{2g}^3 orbitals which give rise to $^4A_{2g}$, ^2E_g, $^2T_{1g}$ and $^2T_{2g}$ states. Of these $^4A_{2g}$ is the ground state. If one electron is excited, the configuration is $t_{2g}^2 e_g^1$ which gives two quartet states $^4T_{1g}$ and $^4T_{2g}$ and a number of doublet states. When the next electron is also excited, the configuration is $t_{2g}^1 e_g^2$ which gives rise to one quartet state $^4T_{1g}$ and some doublet states.

$$^4F \rightarrow {}^4A_{2g}(F), {}^4T_{1g}(F), {}^4T_{2g}(F)$$

$$^4P \rightarrow {}^4T_{1g}(P)$$

$$^{2}G \rightarrow {}^{2}A_{1g}(G), {}^{2}T_{1g}(G), {}^{2}T_{2g}(G), {}^{2}E_{g}(G)$$

$$^{2}H \rightarrow {}^{2}E_{g}(H), 2{}^{2}T_{1g}(H), {}^{2}T_{2g}(H)$$

In both fields, $^4A_{2g}(F)$ represents the ground state. Hence, three spin allowed transitions are observed in high spin state $^4A_{2g}(F) \rightarrow {}^4T_{2g}(F)$ (υ_1), $^4A_{2g}(F) \rightarrow {}^4T_{1g}(F)$ (υ_2) and $^4A_{2g}(F) \rightarrow {}^4T_{1g}(P)$ (υ_3). These spin allowed bands split into two components when the symmetry of Cr^{3+} ion is lowered from octahedral to C_{4V} or C_{3V}. Generally, $^4A_{2g}(F) \rightarrow {}^4T_{1g}(P)$ occurs in the UV-Vis region.

The strong field electronic configurations for the ground state and their terms are given as follows:

$$(t_{2g})^{3}(e_{g})^{0} : {}^{4}A_{2g}(F), {}^{2}E_{g}(G), {}^{2}T_{1g}(G), {}^{2}T_{2g}(G)$$

$$(t_{2g})^{2}(e_{g})^{1} : {}^{4}T_{1g}(F), {}^{4}T_{2g}(F), {}^{2}T_{2g}(H)$$

$$(t_{2g})^{1}(e_{g})^{2} : {}^{4}T_{1g}(P)$$

Racah parameter, B, is calculated with spin allowed transitions using equation (17)

$$B = \left(2\nu_1^2 + \nu_2^2 - 3\nu_1\nu_2\right) \Big/ \left(15\nu_2 - 27\nu_1\right) \tag{17}$$

The octahedral crystal field parameter Dq is characteristic of the metal ion and the ligands. The Racah parameter, B depends on the size of the 3d orbital; B is inversely proportional to covalency in the crystal.

16.2. EPR spectra of chromium compounds

Cr^{3+} ion, splits into |±1/2> and |±3/2> Kramers' doublets in the absence of magnetic field, separated by 2D, D being the zero-field splitting parameter. This degeneracy can be lifted only by an external magnetic field. In such a case, three resonances are observed corresponding to the transitions, |-3/2> ↔ |-1/2>, |-1/2> ↔ |1/2> and |1/2> ↔ |3/2> at gβB – 2D, gβB and gβB + 2D respectively. In a powder spectrum, mainly the perpendicular component is visible. If all the three transitions are observed, the separation between the extreme sets of lines is 4D [gβB + 2D –(gβB - 2D) = 4D]. If D is equal to zero, a single resonance line appears with g ~ 1.98. If D is very large compared to microwave frequency, a single line is seen around g = 4.0.

16.3. Relation between EPR and optical absorption spectra

A comparison is made between the observed g_{eff} from EPR results and the calculated one from the optical spectrum. For Cr^{3+}, EPR and optical results are related by,

$$g_{11} = g_e - \frac{8\lambda}{\Delta E\left({}^4T_{1g}(F)\right)} \tag{18}$$

$$g_1 = g_e - \frac{8\lambda}{\Delta E\left({}^4T_{2g}(F)\right)} \tag{19}$$

Here g_{11} and $g_{\perp}$ are the spectroscopic splitting factors parallel and perpendicular to the magnetic field direction, g , the free electron value g_e, is 2.0023. These values give,

$$g_{eff} = \frac{1}{3}\left(g_{11} + g_1\right). \tag{20}$$

The value of D can also be estimated from the optical absorption spectrum. The ${}^4A_{2g}(F) \rightarrow {}^4T_{2g}(F)$ component in the optical spectrum is due to the lowering of symmetry which also includes the D term.

$$D = \left(\frac{2\lambda}{10Dq}\right)^2 \left(\Delta_z - \Delta_x\right). \tag{21}$$

The spin-orbit splitting parameter, λ [for free ion, Cr^{3+} is 92 cm^{-1}] is related to Racah parameter (B) by the equation,

$$\lambda = 0.11\left(B + 1.08\right)^2 + 0.0062 \tag{22}$$

16.4. Typical examples

The data chosen from the literature are typical for each sample. The data should be considered as representative only. For more complete information on specific examples, the original references are to be consulted. X-band spectra and optical absorption spectra of the powdered sample are recorded at room temperature (RT).

1. Trivalent chromium [d^3]: The optical absorption spectrum of fuchsite recorded in the mull form at room temperature shows bands at 14925, 15070, 15715, 16400, 17730 and 21740 cm^{-1}. The two broad bands at 16400 and 21740 cm^{-1} are due to spin-allowed transitions, ${}^4A_{2g}(F) \rightarrow {}^4T_{2g}(F)$ and ${}^4T_{1g}(F)$ respectively. The band at 17730 cm^{-1} is the split component of the ${}^4T_{2g}(F)$ band. This indicates that the site symmetry of Cr^{3+} is C_{4v} or C_{3v}. The bands at 16400 and 21700 cm^{-1} are responsible for the green color of the mineral. The additional weak features observed for the υ_1 band at 15715 and 15070 cm^{-1} are attributed to the spin-forbidden transitions, ${}^4A_{2g} \rightarrow {}^2T_{1g}(G)$ and ${}^4A_{2g} \rightarrow {}^2E_g(G)$. Using equation (17), Racah parameter, B, is calculated (507 cm^{-1}). Substituting Dq and B values and using T-S diagrams for d^3 configuration and solving the cubic field energy matrices ,another Racah parameter, C is evaluated (2155 cm^{-1}) which is less than the free ion value [C =3850 cm^{-1}].

Several examples are available in the literature. Some of them are given in the Table-12.

Compound	$^2E_g(G)$ cm^{-1}	$^2T_{1g}(G)$ cm^{-1}	$^4T_{2g}(F)$ (ν_1) cm^{-1}	$^2T_{2g}(G)$ cm^{-1}	$^4T_{1g}(F)$ (ν_2) cm^{-1}	$^2T_{1g}(H)$ cm^{-1}	$^4T_{1g}(P)$ (ν_3) cm^{-1}	Dq cm^{-1}	B cm^{-1}	C cm^{-1}	β	CFSE
Fuchsite quartz	15500	15995	19995	22720	27020	35700	43465	2000	677	3400	0.66	24000
Dickite	14690	15500	16260		23800		37000	1626	803		0.78	19512
Fuchite	15070	15715	16400 17730	14925	21740			1640	507	2155	0.49	19680
Chromate			17390		23810			1739				
Natural Ruby	14262 14296	16725 16919 17042	18170 17245	21012 21058 21389	24993			1830	732	2155	0.71	21960
Variscite			16660 18180	15380	21735	30295		1666	475	2200	0.46	19992
Synthetic Uvarovite			16670	18000	22730		28000					
$Sr_3Ga_2Ge_4O_{14}$ Garnet			16299		433.6			1629.9	712.3		0.69	19559
Ureyite			15600		22000				664		0.65	
Alexandrite	14000	-	16600	21000	25000							
Uvarovite			16600		23100							
Clinoclore								1834	668		0.728	63x350
Amesite								1782	737		0.899	58.0x
Muscovite								1610	737		0.89	55.6
Phlogopite								1690				58.0

1. The EPR spectrum of fuchsite recorded at room temperature (RT) clearly indicates a strong resonance line with a few weak resonances on either side of it. The g value for this centrally located strong line is 1.98. This is due to the main transition $|-1/2> \leftrightarrow |1/2>$ of Cr^{3+}. The calculated value of D is around 270 G. For weak lines, D is around 160 G. Since the lines are equally spaced on either side of the strong resonance, E is zero. The strong line at g (1.98) value is observed indicating a high concentration of chromium.
2. The EPR spectrum of chromate shows a broad EPR signal with g value of 1.903 which may be due to Cr^{3+} which is in high concentration in the mineral. The chromium ion is in octahedral coordination.
3. EPR spectrum of zoisite at LNT givesa g and D values of 1.99 and 42.5 mT respectively which are due to Cr^{3+} in octahedral environment.
4. EPR spectrum of chromium containing fuchsite quartz shows a g value of 1.996 which may due to Cr^{3+} which is in octahedral environment.
5. EPR spectrum of blue sapphire shows four Cr^{3+} sites with the same g value of 1.98 having different D values (130,105,65 and 34 mT) . Green sapphire also has the same g value but different D values (132,114, 94 and 35 mT). The results suggest that chromium content is slightly different in different sapphires.

Table 12. Assignment of bands for Cr(III) with $^4A_{2g}(F)$ ground state. All values are given in cm^{-1}

Several examples are given in the literature. Some of them are presented in the Table-13.

Compound	Observed			$^4T_{1g}(F)$	$^4T_{2g}(F)$	Calculated	
	$g_{\perp}$	g_{11}	g_{eff}	(ν_2) cm^{-1}	(ν_1) cm^{-1}	g_{eff}	λ (cm^{-1}
Varscite	1.958	1.9684	1.994	16660	21735	1.9615	75
Chromate			1.903				

Table 13. EPR parameters of Cr^{3+} compounds.

2. Tetravalent chromium (d^2):

Absorption spectra of Cr^{4+} in forsterite and garnet show the absorption band at 9460 cm^{-1} which is the typical of Cr^{4+} ions. It is attributed to the $^3A_{2g} \rightarrow {}^3T_{2g}$ transition. The absorption band at 19590 cm^{-1} is also attributed to $^3A_{2g} \rightarrow {}^3T_{1g}$ transition. The absorption band at 19590 cm^{-1}overlaps with the bands at 16130 and 23065 cm^{-1}.

17. Manganese

The atomic number of manganese is 25 and its outermost electronic configuration is [Ar] $3d^54s^2$. It exhibits several oxidation states, +2, +3, +4, +6 and +7, of which the most stable are +2 +4 and +7. The ionic radii of Mn^{2+} and Mn^{4+} are 0.80 and 0.54 A.U. respectively. Twenty three isotopes and isomers are known. A number of minerals of manganese exists in nature (~ 300 minerals) giving rise to an overall abundance of 0.106%. Twelve of the important among them are economically exploited and the most important of these are pyrolusite (MnO_2), manganite ($Mn_2O_3.H_2O$), hausmannite (Mn_3O_4) rhodochrosite ($MnCO_3$) and manganese(ocean) nodules. Much of the (85-90%) manganese is consumed in the manufacture of ferromanganese alloys. The other uses are: manganese coins, dry cell and alkaline batteries and glass. It is an essential trace element for all forms of life.

Octahedral complexes of Mn(III) are prone to Jahn-Teller distortion. It is of interest, therefore, to compare the structures of $Cr(acac)_3$ with $Mn(acac)_3$ since the former is a regular octahedron while the latter is prone to dynamic Jahn-Teller distortion.

17.1. EPR spectra of manganese compounds

1. Manganese(II): Manganese(II), being a d^5 ion, is very sensitive to distortions in the presence of magnetic field. Mn(II) has a total spin, S = 5/2. The six spin states labeled as ±5/2>, ±3/2> and ±1/2> are known as the three Kramers' doublets; in the absence of external magnetic field ,they are separated by 4D and 2D respectively, where D is the zero-field splitting parameter. These three doublets split into six energy levels by the application of an external magnetic field. Transitions between these six energy levels give rise to five resonance lines. Each of these resonance lines, in turn, splits into a sextet due to the interaction of the electron spin with the nuclear spin of ^{55}Mn, which is 5/2. Thus one expects a 30- line pattern. However, depending on the relative magnitudes of D and A (hyperfine coupling constant of manganese), these 30 lines appear as a separate bunch of 30 lines or 6 lines (if D = 0). The separation between the extreme set of resonance lines is approximately equal to 8D (first order). If D is very small compared to hyperfine coupling constant (A), the 30 lines are so closely packed that one could see only six lines corresponding -1/2 to +1/2 transition. If D = 0, the system is perfectly octahedral. Deviation from axial symmetry leads to a term known as E in the spin- Hamiltonian. The value of E can be easily calculated from single crystal measurements. A non-zero value of E results in making the spectrum unsymmetrical about the central sextet.

Further, the following parameters have been calculated from the powder spectrum using the Spin- Hamiltonian of the form:

$$H = \beta 1B_{3g}S + D\left(S_z^2 - \frac{1}{3}S(S+1)\right) + SA1 \tag{23}$$

Here the first term represents the electron-Zeeman interaction, the second term represents the zero field contribution and the third term represents the nuclear-Zeeman interaction. The extra set of resonances within the main sextet is due to the forbidden transitions. From the forbidden doublet lines, the Zero field splitting parameter, D is calculated using the formula,

$$\Delta H = \left(\frac{2D^2}{H_m}\right)\left[\frac{1+16(H_m - 8Am)^2}{9H_iH_m - 64Am}\right] \tag{24}$$

$$H_m = H_o - Am - \frac{\left[I(I+1) - m^2\right]A^2}{2H_o} \text{ or } H_m = H_0 - Am - (35 - 4m^2)\left(\frac{A^2}{8H_0}\right) \tag{25}$$

where H_m is the magnetic field corresponding to m ↔ m in HF line; H_o is the resonance magnetic field and m is the nuclear spin magnetic quantum number.

Percentage of covalency of Mn-ligand bond can be calculated in two ways using (i) Matumura's plot and (ii) electro negativities, X_p and X_q using the equation,

$$C = \frac{1}{n}\left[1 - 0.16(X_p - X_q) - 0.035(X_p - X_q)^2\right] \tag{26}$$

Here n is the number of ligands around Mn(II) ion; $X_p = X_{Mn} = 1.6$ for Mn(II) and $X_q = X_{ligand}$.

Also hyperfine constant is related to the covalency by,

$$A_{iso} = (2.04C - 104.5) \times 10^{-4} cm^{-1} \tag{27}$$

Further, the g value for the hyperfine splitting is indicative of the nature of bonding. If the g value shows a negative shift with respect to the free electron g value (2.0023), the bonding is ionic and conversely, if the shift is positive, then the bonding is said to be more covalent in nature.

17.2. Typical examples

1. Manganese(II): The EPR spectrum of clinohumite contains a strong sextet at the centre corresponding to the electron spin transition +1/2> to -1/2>. In general, the powder spectrum is characterized by a sextet, corresponding to this transition. The other four transitions corresponding to ±5/2> ↔ ±3/2> and ±3/2>↔ ±1/2> are not seen due to their

high anisotropy in D. However, in a few cases only, all the transitions are seen. Moreover, the low field transitions are more intense than the high field transitions. In addition, if $E \neq 0$, the EPR spectrum will not be symmetrical about the central sextet. In clinohumite, the spectrum indicates the presence of at least three types of Mn(II) impurities in the mineral.

The extra set of resonances within the main sextet is due to the forbidden transitions. From the powder spectrum of the mineral, the following parameters are calculated:

Site I: g = 2.000(1), A = 9.15(2) mT; and D = 43.8(1) mT.

Site II: g = 2.003(2), A = 9.23(2) mT; and D = 44.1(1) mT.

Site III: g = 2.007(1), A = 9.40(2) mT; and D = 44.1(1) mT.

This large value of D indicates a considerable amount of distortion around the central metal ion. Since EPR is highly sensitive to Mn(II) impurity, three such sites are noticed. These two sites have close spin- Hamiltonian parameters. A close look at the EPR spectrum indicates a non-zero value of E, which is very difficult to estimate from the powder spectrum.

2. Pelecypod shell EPR spectrum of powdered sample obtained at room temperature indicates the presence of Mn(II) and Fe(III) impurities. The spectrum contains a strong sextet at the centre of the spectrum corresponding to the electron spin transition +1/2> to -1/2>. Also, the powder spectrum indicates the presence of, at least, three types of Mn(II) impurities in the *pelecypod shell* which is noticed at the sixth hyperfine resonance line. The third Mn(II) site is of very low intensity. The extra set of resonances within the main sextet is due to the forbidden transitions. The variations of intensity are also due to the zero field splitting parameter. From the powder spectrum of the compound, the following parameters are calculated using the spin- Hamiltonian of the form:

$$H = \beta B g S + D\left[S_Z^2 - S^{(S+1)}\!\big/_3 + SAI\right] \quad (28)$$

where the symbols have their usual meaning.

Site I: g = 2.002(1), A = 9.33(2) mT; and D = 43.8(1) mT

Site II: g = 1.990(2), A = 9.41(2) mT; and D = 44.1(1) mT

Site III: g = 1.987(1), A = 9.49(2) mT; and D = 44.1(1) mT

This large value of D indicates a considerable amount of distortion around the central metal ion. A close look at the EPR spectrum indicates a non-zero value for E.

The hyperfine constant 'A' value provides a qualitative measure of the ionic nature of bonding of Mn(II) ion. The percentage of covalency of Mn-ligand bond is calculated using 'A' (9.33 mT) value obtained from the EPR spectrum and with Matumura's plot. It corresponds to an ionicity of 94%. Also, the approximate value of hyperfine constant (A) is calculated by using the equation (27).

The value obtained is 92x 10^{-4} cm^{-1}. This calculated value agrees well with the observed hyperfine constant (93.3 x 10^{-4} cm^{-1}) indicating ionic character of Mn-O bond in the shell under study.

Using the covalency, the number of ligands around Mn(II) ion is estimated using the equation (26)

$$C = \frac{1}{n}\left[1 - 0.16\left(X_p - X_q\right) - 0.035\left(X_p - X_q\right)^2\right]$$

Where X_P and X_q are the electronagativities of metal and ligand. Assuming $X_p = X_{Mn} = 1.4$ and $X_q = X_O = 3.5$, the number of ligands (n) obtained are 18. This suggests that Mn(II) may be surrounded by eighteen oxygens of six CO_3^{2-} ions.

i. Manganese (IV): This ion in biological samples gives rise to EPR signal around 3.30.
ii. (Mn^{7+} ion also gives EPR resonance signal at about 2.45 in ceramic materials and in biological samples.

17.3. Optical absorption studies

1. Manganese(II): The free ion levels of Mn^{2+} are ^6S, ^4G, ^4P. ^4D and ^4F in the order of increasing energy. The energy levels for Mn^{2+} ion in an octahedral environment are $^6A_{1g}(S)$, $^4T_{1g}(G)$, $^4T_{2g}(G)$, $^4E_g(G)$, $^4T_{1g}(G)$ $^4A_{1g}$, $^4T_{2g}(G)$, $^4E_g(D)$, $^4T_{1g}(P)$ respectively with increasing order of energy. The $^4Eg(G)$, $^4A_{1g}$ and $^4E_g(D)$ levels are less affected when compared to other levels by crystal field. Hence, sharp levels are expected relatively in the absorption spectrum which is the criterion for assignment of levels of Mn(II) ion. Since all the excited states of Mn(II) ion will be either be quartets or doublets, the optical absorption spectra of Mn(II) ions will have only spin forbidden transitions. Therefore, the intensity of transitions is weak.

Energy level diagram of Mn(II) is extremely complex. Exact solutions for the excited state energy levels in terms of Dq, B and C may be obtained from T-S matrices. These matrices are very large (up to 10 x10) and ordinary calculations are not feasible. For this reason, the T-S diagrams given in many places in the literature are not sufficiently complete to allow the assignment of all the observed bands. Therefore a set of computer programmes is written to solve the T-S secular equations for any selected values of B, C and Dq. With the computer program, it is only necessary to obtain values of B and C and the complete scheme for any Dq can be quickly calculated. Fortunately B and C can be obtained analytically, if a sufficiently complete spectrum is obtained using the transitions given below:

$$^4A_{1g}, {}^4E_g(G) \rightarrow {}^6A_{1g} = 10B + 5C = \nu_1$$

$$^4E_g(D) \rightarrow {}^6A_{1g} = 17B + 5C = \nu_2$$

If ν_1 and ν_2 are correctly observed and identified in the spectrum, B and C can be calculated. Identification is particularly easy in these cases because of the sharpness of the bands of these levels and are independent of Dq.

2. Manganese(III): This ion has four 3d electrons. The ground state electronic configuration is $t_{2g}^3 e_g^1$. It gives a single spin-allowed transition $^5E_g \rightarrow {^5T_{2g}}$ corresponding to one electron transition. This should appear around 20000 cm^{-1}. Mn^{3+} cation is subject to Jahn-Teller distortion. The distortion decreases the symmetry of the coordination site from octahedral to tetragonal (D_{4h}) or by further lowering the symmetry to rhombic (C_{2v}). Under the tetragonal distortion, the t_{2g} orbital splits into e_g and b_{2g} orbitals whereas the e_g orbital splits into a_{1g} and b_{1g} orbitals. Hence in a tetragonal site, three absorption bands are observed instead of one. Further distortion splits the e_g orbital into singly degenerate a_{1g} and b_{1g} orbitals. Thus four bands are observed for rhombic symmetry (C_{2v}).

The transitions in the tetragonal field are described by the following equations:

$$^2B_{1g} \rightarrow {^2A_{1g}} : \left[6Dq - 2Ds - 6Dt - \left(6Dq + 2Ds - Dt\right)\right] = 4Ds + 5Dt \tag{29}$$

$$^2B_{1g} \rightarrow {^2B_{2g}} : \left[-4Dq - 2Ds - Dt - \left(6Dq + 2Ds - Dt\right)\right] = 10Dq \tag{30}$$

$$^2B_{1g} \rightarrow {^2E_g} : \left[-4Dq - Ds + 4Dt - \left(6Dq + 2Ds - Dt\right)\right] = 10Dq + 3Ds \tag{31}$$

In the above equations, Dq is octahedral crystal field and Ds and Dt are tetragonal field parameters. The same sign of Dq and Dt indicates an axial elongation and opposite sign indicates an axial compression.

The optical absorption bands observed for Mn(III) in octahedral coordination with rhombic distortion (C_{2h}) in montmorillonite are given in Table -14.

Assignment		Localities		
D_{4h}	C_{2V}	(Mexico)	(Gumwood Mine)	(California)
$^5B_{1g} \rightarrow {^5A_{1g}}$	$^5B_{1g} \rightarrow {^5A_{1g}}$	10480	10276	10542
$^5B_{1g} \rightarrow {^5B_{2g}}$	$^5B_{1g} \rightarrow {^5A_{2g}}$	19041	18751	18560
	$^5B_{2g}$	20660	20496	20605
$^5B_{1g} \rightarrow {^5E_g}$	$^5B_{1g} \rightarrow {^5A_{3g}}$	21837	22127	22143

Table 14. Assignment of bands for Mn(III) in montmorillonite

18. Iron

The atomic number of iron is 26 and its electronic configuration is [Ar]$4s^2$ $3d^6$.. Iron has 14 isotopes. Among them, the mass of iron varies from 52 to 60 Pure iron is chemically reactive and corrodes rapidly, especially in moist air or at elevated temperatures. Iron is vital to plant and animal life. The ionic radius of $Fe^{2+}$is 0.76 A.U. and that of Fe^{3+} is 0.64 A.U. The most common oxidation states of iron are +2 and +3. Iron(III) complexes are generally in octahedral in shape, and a very few are in tetrahedral also.

18.1. EPR spectra of iron compounds

The EPR spectra of powdered Fe^{3+} compounds may be described by the spin- Hamiltonian,

$$H = gBS + D\left(S_z^2 - \frac{1}{3}S(S+1) + E\left(S_X^2 - S_y^2\right)\right) \tag{32}$$

The second and third terms in the equation (33) represent the effects of axial and rhombic components of the crystal field respectively. When D=E=0, it corresponds to a free ion in the magnetic field, H and if E= 0, it implies a field of axial symmetry. If λ (E/D) increases, it results in the variation of rhombic character. Maximum rhombic character is seen at a value of λ=1/3 and further increase in λ from 1/3 to 1 results in the decrease of rhombic character. When λ =1, the axial field situation is reached. When λ=1/3, the g value is around 4.27 and when λ is less than 1/3, g value is 4. Hence, the resonance is no longer isotropic and the powder spectrum in that region is a triplet corresponding to H along each of the three principle axes. For Fe^{3+}, in fields of high anisotropy, the maximum g value is 9. If g values are limited to 0.80 to 4.30, the Fe^{3+} ion is under the influence of a strong tetragonal distortion.

1. Iron (III): The iron (III) samples exhibit a series of g values ranging from 0 to 9. This is due to the fact that the three Kramers' doublets of |S=5/2> are split into |S±5/2>, |S±3/2> and |S±1/2> separated by 4D and 2D respectively where D is the zero field splitting parameter. Depending on the relative populations of these doublets, one observes g value ranging from 0 to 9.0. The line widths are larger in low magnetic field when compared to high magnetic field. If the lowest doublet, |S±1/2> is populated, it gives a g value of 2 to 6 whereas if the middle Kramers' doublet |S±3/2> is populated, a g value 4.30 is expected. If the third doublet |S±5/2> is populated, it gives a g value of 2/7 to 30/7. A few systems are known which exhibit resonances from all the three Kramers' doublets.

The iron(III) in the natural sample enters the lattice in various locations which may not correspond to the lowest energy configuration. After heating the sample, the impurity settles in the lowest energy configuration and the EPR spectrum is simplified. Thus, it is observed that heating the sample results in a simplification of the EPR spectrum and gives a g value of around 2.

18.2. Typical examples

1. The EPR spectrum of powdered red sandal wood obtained at room temperature contains a series of lines of various intensity and width. The g values obtained for these are 6.52, 2.63 and 1.92. These three peaks are attributed to Fe(III) impurity in the compound.
2. The EPR spectrum of prehnite at room temperature consists of two parts. The first part consists of the two strong lines (absorption and dispersion) and the second part

comprises a weak doublet within the strong doublet. The weak doublet also consists of two lines, absorption and dispersion line shapes. The g values of the strong doublet are 4.48 and 3.78 whereas the g values of the weak doublet are 4.22 and 3.96. The data reveal that there are two different centres of Fe(III) which are magnetically distinct.

3. The EPR spectrum of nano iron oxalate recorded at room temperature reveals three sets of four lines in low, medium and high fields corresponding to g_1, g_2 and g_3 respectively. From the positions of the peaks in the EPR spectrum, the following spectroscopic splitting factors are evaluated: g_1 = 2.130, g_2 = 2.026 and g_3 = 1.947. The hyperfine structure constants are A_1 = 78 mT, A_2 = 46 mT and A_3 =26 mT. The EPR spectrum is characteristic of Fe(III) ion or HCO_2^- or in rhombic symmetry. For the rhombic symmetry, g values follow in the sequence as $g_1 > g_2 > g_3$. Using the relation, spin-orbit coupling constant, λ is calculated. Resonant value of the magnetic field is given by the relation,

$$H_R(mT) = \frac{21419.49}{g\lambda(cm)} = \frac{0.07144775}{g}\nu(MHz) \tag{33}$$

λ calculated for each g tensor is 32.18.

For axial symmetry, λ is zero. If rhombic character in the crystal field is increased, it results in the increase of λ upto a maximum of $\frac{1}{3}$. In the present case, the observed λ is $\frac{1}{3}$ (32.18%). Thus the EPR studies indicate that the iron oxalate nano-crystal is in orthorhombic structure.

18.3. Optical absorption spectra of iron compounds

18.3.1. Trivalent iron

Trivalent iron has the electronic configuration of $3d^5$ which corresponds to a half-filled d - sub-shell and is particularly most stable. In crystalline fields, the usual high spin configuration is $t_{2g}^3 e_g^2$ with one unpaired electron in each of the orbitals and the low spin state has the t_{2g}^5 configuration with two pairs of paired electrons and one unpaired electron. The energy level in the crystal field is characterized by the following features. i) The ground state of d^5 ion, ^6S transforms into $^6A_{1g}$ - a singlet state. It is not split by the effect of crystal field and hence all the transitions are spin forbidden and are of less intensity. ii) In excited state, d^5 ion gives rise to quartets (^4G, ^4F, ^4D, ^4P) and doublets (^2I, ^2H, ^2G, ^2F, ^2D, ^2P, ^2S) . The transitions from the ground to doublet state are forbidden because the spin multiplicity changes by two and hence they are too weak. Thus sextet-quartet forbidden transitions observed are: $^6A_{1g} \rightarrow {}^4T_{1g}$ and $^6A_{1g} \rightarrow {}^4T_{2g}$. The transitions which are independent of Dq and which result in sharp bands are $^6A_{1g} \rightarrow {}^4E(^4D)$ $^6A_{1g} \rightarrow {}^4E_g + {}^4E_{1g}$ etc., iii) The unsplit ground state term behaves alike in both octahedral and tetrahedral symmetries and gives rise to same energy level for octahedral, tetrahedral and cubic coordination with usual difference,

$$\left[Dq_{Octa} : Dq_{Tetra} : Dq_{Cubic} = 1 : \frac{4}{9} : \frac{8}{9} \right]$$

18.3.2. Divalent iron

In divalent iron (d^6), the free ion ground term is ^5D and the excited terms are triplet states (^3H, ^3P, ^3F, ^3G, ^3D) and singlet states (^1I, ^1D). In an octahedral field, the ^5D term splits into an upper ^5E_g level and a lower $^5T_{2g}$ level of which the latter forms the ground state. The only allowed transition is $^5T_{2g} \rightarrow {}^5E_g$ which gives an intense broad absorption band. This band splits into two bands due to Jahn-Teller effect. The average of these two bands is to be taken as 10Dq band. The transitions arising from the excited triplet states are spin forbidden and hence are weaker than the 10Dq band.

18.3.3. Typical examples

1. For Fe^{3+}, there are three transitions: $^6A_{1g}(S) \rightarrow {}^4T_{1g}(G)$ (ν_1), $^6A_{1g}(S) \rightarrow {}^4T_{2g}(G)$($\nu_2$). ν_1 occurs between 10525 cm^{-1} and ν_2 occurs between 15380 to 18180 cm^{-1} usually as a shoulder. The bands corresponding to $^6A_{1g}(S) \rightarrow {}^4A_{1g}(G)$, $^4E_g(G)$ (ν_3) appear around 22000 cm^{-1} . The last transition is field independent. The ligand field spectrum of ferric iron appears as if the first (ν_1) and the third (ν_3) bands of octahedral symmetry are only present. The analysis of general features of the spectrum of Fe^{3+} containing plumbojarosite is discussed here. The first feature observed in the range 12000 to 15500 cm^{-1} is attributed to $^6A_{1g}(S) \rightarrow {}^4T_{1g}(G)$, the third band at 22730 cm^{-1} is sharp and is assigned to $^6A_{1g}(S) \rightarrow {}^4A_{1g}(G)$, $^4E_g(G)$ transitions respectively. A broad and diffused band at 19045 cm^{-1} is assigned to the $^6A_{1g}(S) \rightarrow {}^4T_{2g}(G)$ band. The other bands are also assigned to the transitions with the help of Tanabe-Sugano diagram. The assignments are given in the Table -15.
2. Optical absorption spectrum of prehnite recorded in the mull form at room temperature (RT) shows bands at 9660, 10715, 12100, 12610, 15270, 16445,17095, 23380 and 24390 cm^{-1} in the UV –Vis region. For easy analysis of the spectrum, the bands are divided into two sets as 12100, 15270, 23380 cm^{-1} and 12610, 16445, 17095, 24390 cm^{-1}. Accordingly the two bands observed at 12100 cm^{-1} in the first set and 12610 cm^{-1} in the second set are assigned to the same transition $^6A_{1g}(S) \rightarrow {}^4T_{1g}(G)$ whereas 15270 cm^{-1} in the first set and 16445,17095 cm^{-1} in the second set are assigned to $^4T_{2g}(G)$ transition. The third at 23380 cm^{-1} and 24390cm^{-1} is assigned to $^4A_{1g}(G)$, $^4E(G)$ (ν_3) transitions respectively. These two sets of bands are characteristic of Fe(III) ion occupying two different sites in octahedral symmetry. The broad and intense band observed at 10715 cm^{-1} with a split component at 9660 cm^{-1} is assigned to the transition $^5T_{2g} \rightarrow {}^5E_g$ for divalent iron in the sample. Using the Tree's polarization term, α = 90 cm^{-1}, the energy matrices of the d^5 configuration are solved for various B, C and Dq values. The evaluated parameters which give good fit are given in Table 15. A comparison is also made between the calculated and observed energies of the bands and these are presented in Table -15.

Prehnite						Plumbojarosite			Transition from $^6A_{1g}$
Site I			Site II						
Dq= 930, B= 600 and C =2475 cm^{-1} ,α= 90 cm^{-1}			Dq= 900, B= 600 and C =2500 cm^{-1} α= 90 cm^{-1}			Dq= 900, B= 700 and C =2800 cm^{-1} α= 90 cm^{-1}			
Wave length (nm)	Wave number (cm^{-1})		Wave length (nm)	Wave number (cm^{-1})		Wave length (nm)	Wave number (cm^{-1})		
	Observed	Calculated		Observed	Calculated		Observed	Calculated	
827	12100	793	793	12610	12528	800 650	12500	-- 15194	$^4T_{1g}(G)$
						525	15385	19379	
655	15270	608	608	16445	16331	440	19045	22766	$^4T_{2g}(G)$
430		585	585	17095		410	22730		
	23380	427	427	24390	23276	385		24815	$^4A_{1g}(G)$,
						330	24390	26474	$^4E(G)$
						265	25975	30656	
						240	30300	37710	$^4T_{2g}(D)$
							37735	41125	
							41665		$^4E_g(D)$

Table 15. Band headed data with assignments for Fe(III) in various compounds

19. Nickel

Nickel is the 7th most abundant transition metal in the earth's crust. The electronic configuration of nickel is [Ar]$4S^23d^8$. Nickel occurs in nature as oxide, silicate and sulphide. The typical examples are garnierite and pentlandite. Nickel exhibits +1 to +4 oxidation states. Among them divalent state is most stable. Nickel compounds are generally blue and green in color and are often hydrated. Further, most nickel halides are yellow in color. The primary use of nickel is in the preparation of stainless steel. Nickel is also used in the coloring of glass to which it gives a green hue.

19.1. Electronic spectra of nickel compounds

The electronic distribution of Ni(II) ion (d^8) is $t_{2g}^6 e_g^2$ which gives rise to ^3F, ^3P, ^1D and ^1S terms of which ^3F is the ground state. In a cubic crystal, these terms transform as follows:

$$^3F \rightarrow {}^3T_{1g}(F) + {}^3T_{2g}(F) + {}^3A_{2g}(F)$$

$$^3F \rightarrow {}^3T_{1g}(P)$$

$$^1D \rightarrow {}^3T_{2g}(D) + {}^1E_g(D)$$

$$^1G \rightarrow {}^1T_{1g}(G) + {}^1T_{2g}(G) + {}^1E_g(G) + {}^1A_{1g}(G)$$

$$^1S \rightarrow {}^1A_{1g}(S)$$

Of these crystal field terms, $^3A_{2g}(F)$ is the ground state. Hence three spin allowed transitions are possible and the others are spin forbidden The three spin allowed transitions are: $^3A_{2g}(F) \rightarrow {}^3T_{1g}(P)$, $^3A_{2g}(F) \rightarrow {}^3T_{1g}(F)$ and $^3A_{2g}(F) \rightarrow {}^3T_{2g}(F)$. These transitions are governed by linear equations as given below:

$$^3A_{2g}(F) \rightarrow {}^3T_{1g}(P) = 15Dq + 7.5B + 6B(1+\mu)^{1/2} = \nu_1 \tag{34}$$

$$^3A_{2g}(F) \rightarrow {}^3T_{1g}(F) = 15Dq + 7.5B - 6B(1+\mu)^{1/2} = \nu_2 \tag{35}$$

$$^3A_{2g}(F) \rightarrow {}^3T_{2g}(F) = 10Dq = \nu_3 \tag{36}$$

Here μ is of the order of 0.01. Dq and B are of similar magnitude. The spin allowed bands are calculated using the above equations whereas the spin forbidden bands are assigned using Tanabe-Sugano diagrams.

19.2. Typical examples

The data chosen from the literature are typical and representative for each sample. For more complete information on any specific case, original references are to be consulted. X-band spectra and optical absorption spectra of the powdered sample are recorded at room temperature (RT) only.

Divalent Nickel [d^8]: The optical absorption spectrum of falcondoite mineral recorded in the mull form at room temperature shows three intense bands at 9255, 15380 and 27390 cm^{-1} and a weak band at 24385 cm^{-1}. Using the equations 34 to 36, the calculated values of Dq and B are 925 and 1000 cm^{-1} respectively. Using these Dq and B values and T-S diagrams for d^8 configuration, the cubic field energy matrices and Racah parameter, C are evaluated (4.1B) .

Ni^{2+} also gives absorption bands in the NIR region. These bands suggest that Ni^{2+} is in tetrahedral site. In some of the samples, Ni^{2+}exbits both octahedral and tetrahedral coordination. Several examples are available in the literature. Some of them are given in the Table-16.

Compound	$^3T_{1g}(P)$ (ν_1)	$^3T_{1g}(F)$ (ν_2)	$^3T_{2g}(F)$ (ν_3)	$^1T_{1g}(G)$	$^1T_1g(D)$	$^1Eg(D)$	$^1T_{2g}(D)$	$^1T_{2g}(G)$	Dq	B	C
Falcondoite	27390	15380	9255				24385		925	1000	4100
Ullmannite	24993	14966	8618	25967	21546	12252	21546		860	840	3350
Takovite	26665	15380	8200 10000			14930	24095		910	940	4.25B
(Zn,Ni)KPO $_4$6H_2O	25967	15500	8770			14080	22216		900	890	3800
Ni(II) HZDT											
Garnierite	26300	15200	9100			13000					
Gaspeite	22730	13160 14705	7714 8685				20410	30300	810	800	3200
Annabergite		13885	8330								
Zartite	23805	14285	8195				21735		820	899	4.1B

Table 16. Assignment of bands for Ni(II) with $^3A_{2g}(F)$ as the ground state. All values are given in cm^{-1}.

19.3. EPR spectra

$Ni^{2+}(d^8)$ has no unpaired electron (square planer) in its orbit. Therefore it does not exhibit EPR signal at room temperature.

But in certain conditions, it shows EPR signal. The EPR data could be related with the optical data by the following equation $g = 2.0023 - \frac{8\lambda}{\Delta}$ where Δ is the energy of the transition of the perfect octahedral site. λ is 324 cm^{-1} for free Ni^{2+} ion.

20. Copper

Copper is one of the earliest known elements to man. The average percentage of copper in the earth's crust is 0.005%. Pure copper is soft and malleable. An important physical property of copper is its color. Most people refer copper colour as reddish-brown tint. Copper-63 and copper-65 are two naturally occurring isotopes of copper. Nine radioactive isotopes of copper are also known. Among them two radioactive isotopes, copper-64 and copper-67 are used in medicine. Copper easily reacts with oxygen and in moist air, it combines with water and carbon dioxide forming hydroxy copper carbonate ($Cu_2(OH)_2CO_3$).

Animals like crustaceans (shellfish like lobsters, shrimps, and crabs) do not have hemoglobin to carry oxygen through the blood but possess a compound called hemocyanin. This is similar to hemoglobin but contains copper instead of iron. Copper is an essential micronutrient for both plants and animals. A healthy human requires not more than about 2 mg of copper for every kg weight of the body. The main body parts where copper is found in animals are the tissues, liver, muscle and bone.

20.1. Copper compounds

Copper exists in two ionic states, Cu(I) and Cu(II). The ionic radius of Cu(II) is 0.73 A.U. The electronic configuration of Cu(I) is [Ar] $3d^{10}$ and hence has no unpaired electron in its outermost orbit. Hence it exhibits diamagnetism. The electronic configuration of Cu(II) is $[Ar]3d^9$ and has one unpaired electron which is responsible for its para magnetism. The main resources of copper are its minerals. Structural properties could be explored using electronic and EPR spectra which provides information on bonding between ligands and metal ion.

20.2. Electronic spectra of copper compounds

In optical spectroscopy, transitions proceed between the split orbital levels whereas in EPR spectroscopy they occur between spin sub- levels that arise due to the external magnetic field. Thus EPR spectroscopy is a natural sequel to optical spectroscopy.

20.3. Optical spectra

In octahedral crystal field, the ground state electronic distribution of Cu^{2+} is $t_{2g}^6e_g^3$ which yields ^2E_g term. The excited electronic state is $t_{2g}^5e_g^4$ which corresponds to $^2T_{2g}$ term. Thus only one single electron transition, i.e., $^2E_g \rightarrow {}^2T_{2g}$, is expected in an octahedral crystal field. The difference is 10Dq. Octahedral coordination is distorted either by elongation or compression of octahedron leading to tetragonal symmetry.

Normally, the ground ^2E_g state is split due to Jahn-Teller effect and hence lowering of symmetry is expected for Cu(II) ion. This state splits into $^2B_{1g}(d_{x^2-y^2})$ and $^2A_{1g}(d_{z^2})$ states in tetragonal symmetry and the excited term $^2T_{2g}$ also splits into $^2B_{2g}(d_{xy})$ and $^2E_g(d_{xz},d_{yz})$ levels. In rhombic field, ^2E_g ground state is split into $^2A_{1g}(d_{x^2-y^2})$ and $^2A_{2g}(d_{z^2})$ whereas $^2T_{2g}$ splits into $^2B_{1g}(d_{xy})$, $^2B_{2g}(d_{xz})$ and $^2B_{3g}(d_{yz})$ states. Thus, three bands are expected for tetragonal (C_{4v}) symmetry and four bands are expected for rhombic (D_{2h}) symmetry. Energy level diagram of d-orbitals in tetragonal elongated environment is shown in Fig. 5.

The transitions in the tetragonal field are described by the following equations:

$$^2B_{1g} \rightarrow {}^2A_{1g} : \left[6Dq - 2Ds - 6Dt - (6Dq + 2Ds - Dt)\right] = 4Ds + 5Dt \quad (37)$$

$$^2B_{1g} \rightarrow {}^2B_{2g} : \left[-4Dq + 2Ds - Dt - (6Dq + 2Ds - Dt)\right] = 10Dq \quad (38)$$

$$^2B_{1g} \rightarrow {}^2E_g : \left[-4Dq - Ds + 4Dt - (6Dq + 2Ds - Dt)\right] = 10Dq + 3Ds - 5Dt \quad (39)$$

In the above equations, Dq is octahedral, Ds and Dt are tetragonal crystalfield parameters. The same sign of Dq and Dt indicates an axial elongation [Fig. 5] and opposite sign indicates an axial compression .

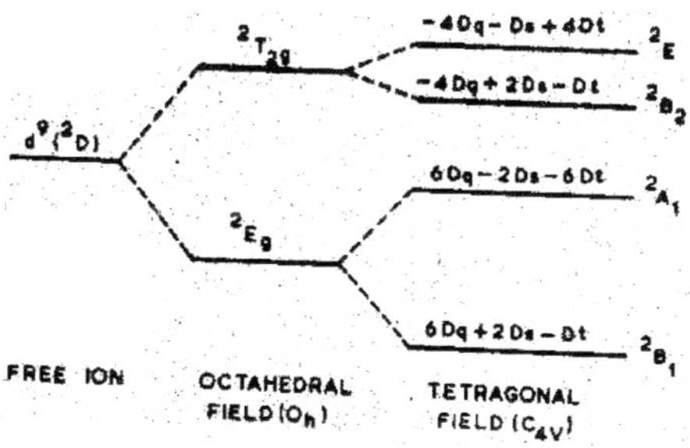

Figure 5. (a) Energy level diagram of Jahn-Teller distortion in d-orbital in octahedral and tetragonal elongation

The Jahn-Teller distortion is either tetragonal elongation along the Z axis or contraction in the equatorial xy plane which may ultimately result in a square planar environment in extreme cases as in D_{4h}.

The optical absorption bands observed for Cu(II) in octahedral coordination with rhombic (D_{2h}) symmetry are: $^2A_{1g}(d_{x^2-y^2}) \rightarrow {}^2A_{2g}(d_{z^2})$, $^2A_{1g}(d_{x^2-y^2}) \rightarrow {}^2B_{1g}(d_{xy})$, $^2A_{1g}(d_{x^2-y^2}) \rightarrow {}^2B_{2g}(d_{xz})$, $^2A_{1g}(d_{x^2-y^2}) \rightarrow {}^2B_{3g}(d_{yz})$ states respectively. This is shown in Fig.6. In rhombic (D_{2h}) field, i.e., C_{2V} symmetry, the strong band $^2A_{1g}(d_{x^2-y^2}) \rightarrow {}^2B_{1g}(d_{xy})$ gives 10Dq value which depends on the nature of the compound.

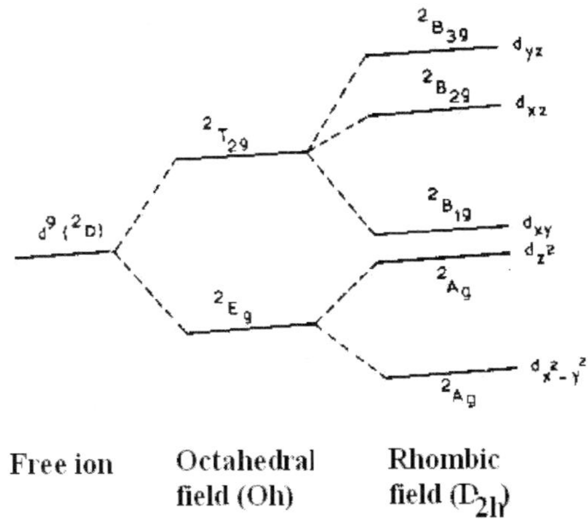

Figure 6. Energy level diagram of d-orbitals in rhombic distortion.

20.4. EPR spectra of copper compounds

When any Cu(II) compound in the form of powder is placed in a magnetic field, it gives a resonance signal. The signal is of three types. They are shown in Fig.-7:

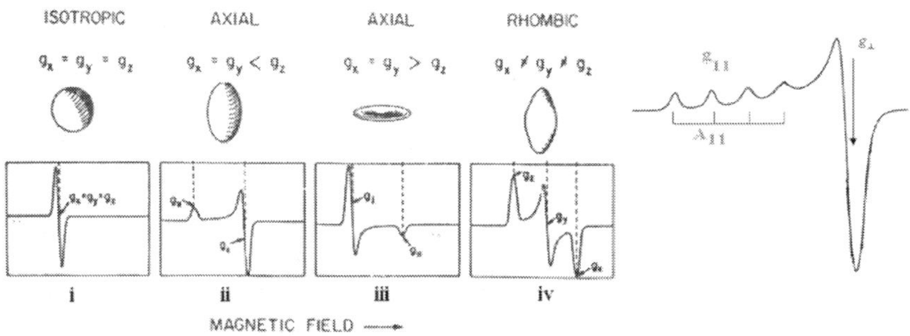

Figure 7. Different forms EPR spectra of Cu(II)

Fig.7(i) is due to high concentration of copper; if the copper content in the compound is high, it gives a broad resonance line. Therefore the hyperfine line from either ^{63}Cu or ^{65}Cu cannot be resolved. The g value for this resonance is around 2.2. (ii) Compression in the equatorial plane results in the elongation of Z axis .Elongation in the equatorial plane results in the compression of Z-axis. Thus there are two types of possibilities in the EPR spectrum. Hence an EPR spectrum similar to Fig. 7(ii) & (iii) is obtained. If $g_{11} > g_{\perp}$, the ground state is $^2B_{1g}$, [Fig. 7(a)] whereas if $g_{\perp} > g_{11}$ or $g_{11} = 2.00$, the ground state is $^2A_{1g}$ [fig.7(ii).]. The highest-energy of the half occupied orbital is $d_{x^2-y^2}$ as it has the largest repulsive interaction with the ligands in the equatorial plane. Here g_{11}(corresponding to the magnetic field oriented along the z axis of the complex) > $g_\perp$ > 2.00. This is a characteristic feature of $d_{x^2-y^2}$ ground state. Additionally, copper has a nuclear spin of (I)) 3/2 which couples with the electron spin to produce a four line hyperfine splitting of the EPR spectrum. This is shown in Fig-7(ii) and 7(v). Tetragonal cupric complexes generally have large A_{11} value than those of complexes with D_{4h} symmetry. If g_{11} > $g_\perp$, the ground state is $^2B_{1g}$ whereas if $g_\perp$ >g_{11} or g_{11} = 2.00, the ground state is $^2A_{1g}$. EPR results give rise to a new parameter, G which is defined as

$$G = \frac{(g_{11} - g_e)}{(g_\perp - g_e)} \tag{40}$$

If G value falls in between 3 and 5, the unit cell contains magnetically equivalent ions. If G value is less than 3, the exchange coupling among the magnetically non- equivalent Cu(II) ions in the unit cell is not very strong. If G is greater than 5, a strong exchange coupling takes place among the magnetically non -equivalent Cu(II) ions in the unit cell. Truly compressed structures are relatively rare when compared to elongated structures. In other words, $g_\perp$ > g_{11}, is an unusual observation and this implies two possibilities:

i. The concentration of copper in the complex is very high which results in the interaction between Cu(II) ↔ Cu(II) ions.
ii. The Cu(II) ion is a compressed octahedron. If the complex contains low copper content, it is assumed that Cu(II) ion is a compressed octahedron. Hence the ground state is $^2A_{1g}$ (d_{z^2}).
iii. Further lowering of symmetry gives rise to EPR spectrum which is similar to the one shown in Fig. 9(iv). This spectrum consists of three sets of resolved four lines in low, medium and high fields corresponding to g_1, g_2 and g_3 respectively. The hyperfine structure constants (A values) are designated as A_1, A_2 and A_3 respectively. Line width is estimated for simple cubic lattice using dipole-dipole equation;

$$H_p = 2.3 g_O \beta \rho \sqrt{s(s+1)} \tag{41}$$

where β is the Bohr magneton, s = spin, g_O = average value of g factor, ϱ = density (2.22 x 10^{21} spins/cc).

The calculated g values provide valuable information on the electronic ground state of the ion. If g_1> g_2 > g_3, the quantity R value is given by (g_2 –g_3) / (g_1-g_2) which is greater than unity and the ground state is $^2A_{2g}(d_{z^2})$; if it is less than unity, the ground state is $^2A_{1g}(d_{x^2-y^2})$. A large value of g_1 is indicative of more ionic bonding between metal and ligand. Further the structure of the compound is an elongated rhombus. From the spin –Hamiltonian parameters, the dipolar term (P) and the Fermi contact term (k) are calculated using the following expressions:

$$P = 2\gamma_{Cu} \beta_O \beta_N \left(r^{-3} \right) \tag{42}$$

$$k = \left(A_O / P \right) + \Delta g_e \tag{43}$$

Here γ_{Cu} is the magnetic moment of copper, β_o is the Bohr magneton, β_N is the nuclear magneton and r is the distance from the central nucleus to the electron, A_o is the average A value and $\Delta g_o = g_o - g_e$ where g_o is the average g value and g_e is the free electron g-value (2.0023). The Fermi contact term, k, is a measure of the polarization produced by the uneven distribution of d-electron density on the inner core s-electron and P is the dipolar term. By assuming either the value of P or k, the other is calculated. Using these values, the hyperfine constant is calculated. This is the average value of g_1, g_2 and g_3.

Using the data of EPR and dipolar term P, the covalency parameter (α^2) is calculated .

$$\alpha^2 = \frac{7}{6}\left[\left(\frac{A_3 - A_1}{P}\right) - (g_e - g_1) + \frac{11}{14}(g_e - g_3) - \frac{6}{14}(g_e - g_2)\right] \tag{44}$$

Thus the important bonding information is obtained. The bonding parameter, α^2, would be closer to unity for ionic bonding and it decreases with increasing covalency. Further the term, *k*, is calculated using the EPR data,

$$A_{11} = k\alpha^2 + P\left[-\frac{4}{7}\alpha^2 + \Delta g_{11} + \frac{3}{7}\Delta g_{\perp}\right] \tag{45}$$

$$A_{\perp} = k\alpha^2 + P\left[\frac{2}{7}\alpha^2 + \frac{11}{14}\Delta g_{\perp}\right] \tag{46}$$

20.5. Relation between EPR and optical absorption spectra

The optical absorption and EPR data are related as follows. In tetragonal symmetry, EPR studies are correlated with optical data to obtain the orbital reduction parameter in rhombic compression.

$$g_1 = g_e + \frac{8a^2k_1^2\lambda}{\Delta E_{xy}} \tag{47}$$

$$g_1 = g_e + \frac{2k_2^2\lambda\left(a + \sqrt{3}b\right)^2}{\Delta E_{xz}} \tag{48}$$

$$g_1 = g_e + \frac{2k_3^2\lambda\left(a + \sqrt{3}b\right)^2}{\Delta E_{yz}} \tag{49}$$

Similarly for rhombic elongation,

$$g_1 = g_e - \frac{8a^2k_1^2\lambda}{\Delta E_{xy}} \tag{50}$$

$$g_1 = g_e - \frac{2k_2^2\lambda\left(a+\sqrt{3}b\right)^2}{\Delta E_{xz}} \tag{51}$$

$$g_1 = g_e - \frac{2k_3^2\lambda\left(a+3b\right)^2}{\Delta E_{yz}} \tag{52}$$

where $a = \cos\theta$ and $b = \sin\theta$ which are coefficients for the mixing of the z^2 and x^2-y^2 orbitals. $a^2 + b^2 = 1$ and k_1, k_2, k_3 are the orbital reduction parameters. λ is the spin- orbit coupling constant for free Cu(II) ion = -830 cm^{-1}.

In equations (48) to (50), when a = 0, tetragonal compression is obtained [ground state is $^2A_{1g}(d_{z^2})$].

$$g_{11} = 2.0023 = g_e \tag{53}$$

$$g_\perp = g_e - \frac{6\lambda}{\Delta E_{\perp(xy,yz)}\left({}^2B_1 \to {}^2E\right) = \Delta_\perp} \tag{54}$$

Also in equations (51) to (53), when b is equal to zero, tetragonal elongation is obtained [ground state is $^2B_{1g}(d_{x^2-y^2})$].

$$g_{11} = g_e - \frac{8\lambda}{\Delta E_{11(xy)}\left({}^2B_1 \to {}^2B_2\right) = \Delta_{11}} \tag{55}$$

$$g_\perp = g_e - \frac{2\lambda}{\Delta E_{\perp(xy,yz)}\left({}^2B_1 \to {}^2E\right) = \Delta_\perp} \tag{56}$$

Further, if A_{11}, g_{11} and $g\perp$values are known, α^2 can be estimated using the equation [53]

$$\alpha^2 = -\left[\left(\frac{A_{11}}{0.036}\right) - \left(g_{11} - g_{e1}\right) + \frac{3}{7}\left(g_\perp - g_e\right) + 0.04\right] \tag{57}$$

20.6. Typical examples

EPR and optical absorption spectral data of selected samples are discussed. The data are chosen from the literature for each typical sample. However, it is to be noticed that the crystal field parameters, EPR parameters often depend on chemical composition, nature of ligands and temperature of the compound. The data should be considered as representative only. For more complete information on specific example, the original references are to be consulted. The X-band spectra and optical absorption spectra of powdered samples are mostly recorded at room temperature (RT).

1. The EPR spectrum of covellite is shown in Fig-9. It is similar to the Fig 8(i). It consists of a broad line with a small sextet. The g value for the broad line is 2.24 which is due to the presence of Cu(II) in the sample. The hyperfine line from either ^{63}Cu or ^{65}Cu could not be resolved since the copper content (Cu = 66 wt%) in the mineral is very high. Several copper compounds exhibit this type of EPR spectra.

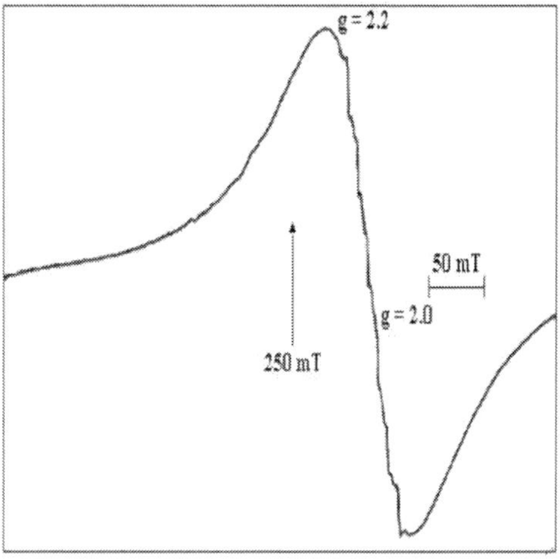

Figure 8. EPR spectrum of covellite at RT

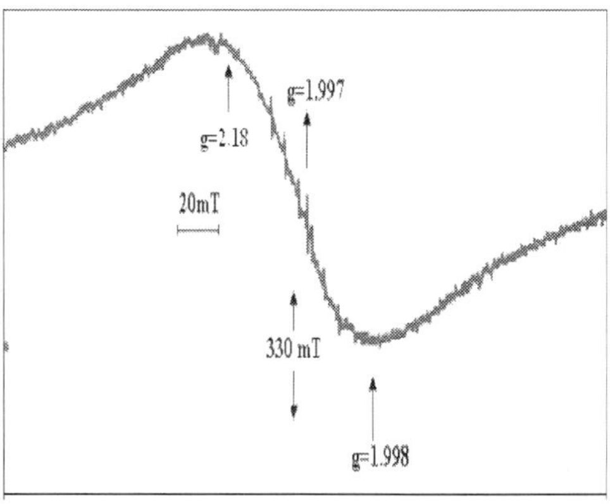

Figure 9. EPR spectrum of beaverite at RT

2. *Beaverite* [$Pb(Fe^{3+},Cu,Al)_3(SO_4)_2(OH)_6$]: X-band EPR spectrum of powdered sample recorded at RT is shown in Fig-9. This is similar to Fig-7(ii). The g values are: g_{11} = 2.42 and $g_\perp$= 2.097. In addition to the above, a g value of 2.017 is observed which is due to Fe(III) impurity. Fig.9 indicates expanded form of EPR spectrum of Cu(II) and is not resolved because of high copper percentage. Tetragonal cupric complexes with D_{4h} symmetry, possessing axial elongation have ground state $^2B_{1g}$ ($d_{x^2-y^2}$).The EPR results are in the order of $g_{11} > g_\perp > g_e$ and hence the ground state is $^2B_{1g}$. Though the optical absorption spectrum shows two sites for Cu(II) with same ground state, the same is not noticed in the EPR spectrum because the percentage of copper is high in the sample.

   A typical EPR spectrum of *enargite* is shown in Fig.10. The spectrum is symmetric with g_{11} = 2.289 and $g_\perp$= 2.048 which are due to Cu(II). Since $g_{11} > g_\perp > g_e$, the ground state for Cu(II) is $^2B_{1g}$ ($d_{x^2-y^2}$). Using EPR and optical absorption results, the orbital reduction parameters are evaluated, i.e., K_{11} = 1.03 cm^{-1} and $K_\perp$ = 1.93 cm^{-1}. Also G seems to be 5.0 which indicates that the unit cell of the compound contains magnetically equivalent ions.

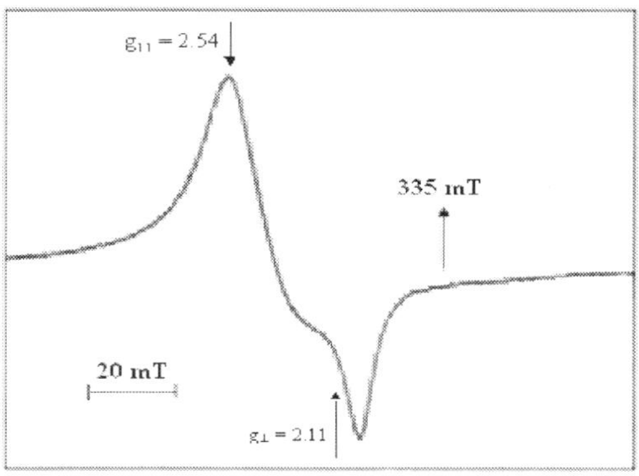

Figure 10. EPR spectrum of enargite at RT

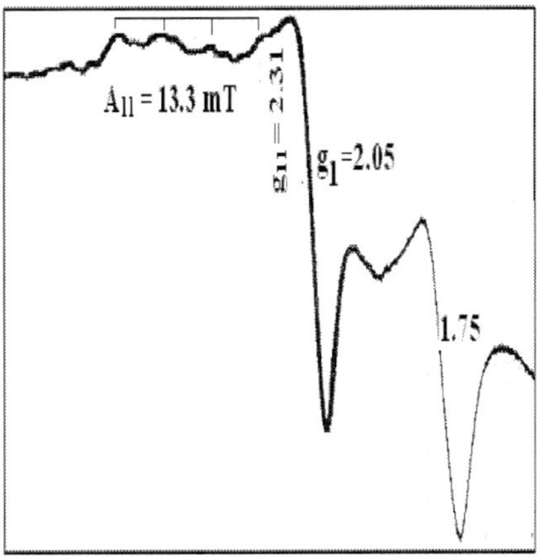

Figure 11. EPR spectrum of CuO-ZnO nano composite.

CuO-ZnO nano composite: EPR spectrum of CuO-ZnO nano composite recorded at room temperature is shown in Fig-11. The calculated g values are 1.76, 2.31 and 2.05. The g value of 1.76 is assigned to free radical of O^{2-}. Further g_{II} value of 2.31, $g_{\perp}$value of 2.05 are due to Cu(II) in tetragonal distortion. Also it has A_{11} =13.3 mT. These results show that the ground state of Cu(II) as $d_{x^2-y^2}$. Further, the covalency parameter, α^2 (0.74) suggests that the composite has some covalent character.

3. *Atacamite* [$Cu_2(OH)_3Cl$]: The EPR spectrum is shown in Fig.12. The g values corresponding to three sets of the resolved four lines in low, mid and high fields are g_1 = 2.191, g_2 =2.010 and g_3 = 1.92. The corresponding hyperfine structure constants are A_1 = 11.0 mT, A_2 = 3,0 mT and A_3 = 5.0 mT respectively. Since $g_1 > g_2 > g_3$, the quantity $R = (g_2-g_3)/(g_1-g_2) = 0.50$ which is less than unity. This indicates $^2A_{1g}(d_{x^2-y^2})$ is the ground state for Cu(II) which is in an elongated rhombic field. The optical absorption spectrum of the compound at RT shown in Fig-13 shows bands at 15380, 11083, 10296 and 8049 cm^{-1}. Using the EPR results, the energy states are ordered as $^2A_{1g}(d_{x^2-y^2}) < {}^2A_{2g}(d_{z^2}) < {}^2B_{1g}(d_{xy}) < {}^2B_{2g}(d_{xz}) < {}^2B_{2g}(d_{yz})$. Thus we have four bands with $^2A_{1g}(d_{x^2-y^2})$ as the ground state. Using the EPR results, the dipolar term (P) and the Fermi contact term (k) are calculated as 0.38 cm^{-1} and k = 0.3 respectively. The bonding parameter, α^2 is found to be 0.28 indicating reasonably high degree of covalent bonding between metal and ligands.

Synthetic copper doped *zinc potassium phosphate hexahydrate* (ZPPH), $ZnKPO_4 6H_2O$: It is similar to strubite, a bio-mineral. The g values are: g_1 = 2.372, g_2 =2.188 and g_3 = 2.032. The hyperfine structure constants are $A_1 = 78 \times 10^{-4}$ cm^{-1}, $A_2 = 48 \times 10^{-4}$ cm^{-1} and $A_3 = 63 \times 10^{-4}$ cm^{-1}respectively. It is seen that $g_1 > g_2 > g_3$ and the quantity $R = (g_2-g_3)/(g_1-g_2) = 0.85$. This confirms that the ground state for Cu(II) is $^2A_{1g}(d_{x^2-y^2})$ (elongated rhombic field). Using the EPR data and substituting free ion dipolar term [P= 0.036 cm^{-1}] for Cu(II) and g_e value in equation (57), the bonding parameter, α^2 = 0.55, is obtained. It indicates a predominant covalency in compound.

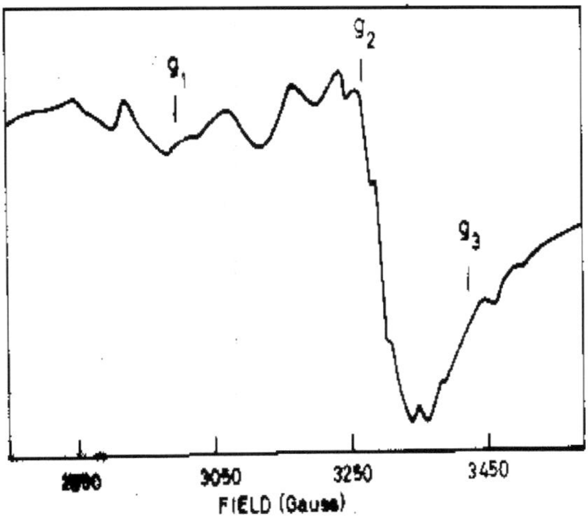

Figure 12. EPR spectrum of atacamite at RT

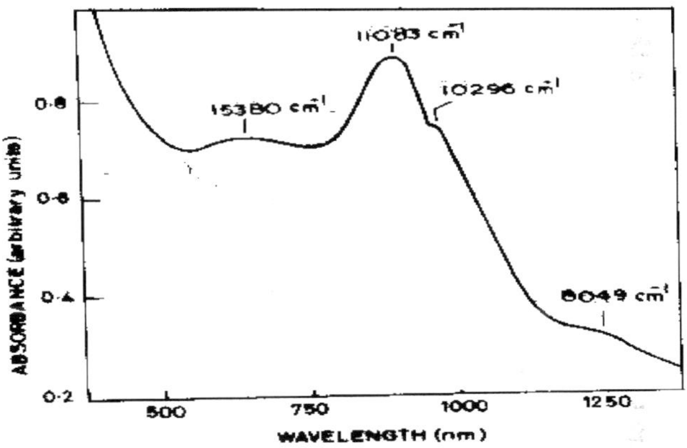

Figure 13. Optical absorption spectrum of atacamite

Author details

S.Lakshmi Reddy
Dept. of Physics, S.V.D.College, Kadapa, India

Tamio Endo
Dept. of Electrical and Electronics Engineering,
Graduate School of Engineering, Mie University, Mie, Japan

G. Siva Reddy
Dept. of Chemistry, Sri Venkateswara University, Tirupati, India

21. References

[1] B.N.Figgs,M.A.Hitchman, "Ligand Field Theory and Its Applications",Wiley-VCH, New York,(2000).

[2] A.Lund, M.Shiotani, S.Shimada,"Principles and Applications of ESR Spectroscopy", Springer New York (2011).

[3] C.J.Ballahausen, "Introduction to Ligand Field Theory", Mc Graw-Hill Book Co., New York (1962).

[4] P.B. Ayscough,"Electron Spin Resonance in Chemistry", Mathuen & Co., Ltd., London (1967).

[5] R.L.Carlin, "Transition Metal Chemistry", Marcel Dekker,New York (1969).

[6] Journal of "Coordination Chemistry Reviews".

[7] Journal of Spectrochimica Acta A Elsvier.

[8] J.S.Griffith, "Theory of Transition Metal Ions", Cambridge University Press,Oxford (1964).

[9] Journal of Solid State Communications.

Chapter 2

Application of FTIR Spectroscopy in Environmental Studies

Claudia Maria Simonescu

Additional information is available at the end of the chapter

http://dx.doi.org/10.5772/48331

1. Introduction

FTIR Spectroscopy is a technique based on the determination of the interaction between an IR radiation and a sample that can be solid, liquid or gaseous. It measures the frequencies at which the sample absorbs, and also the intensities of these absorptions. The frequencies are helpful for the identification of the sample's chemical make-up due to the fact that chemical functional groups are responsible for the absorption of radiation at different frequencies. The concentration of component can be determined based on the intensity of the absorption. The spectrum is a two-dimensional plot in which the axes are represented by intensity and frequency of sample absorption.

The infrared region of the electromagnetic spectrum extends from the visible to the microwave (Figure 1).

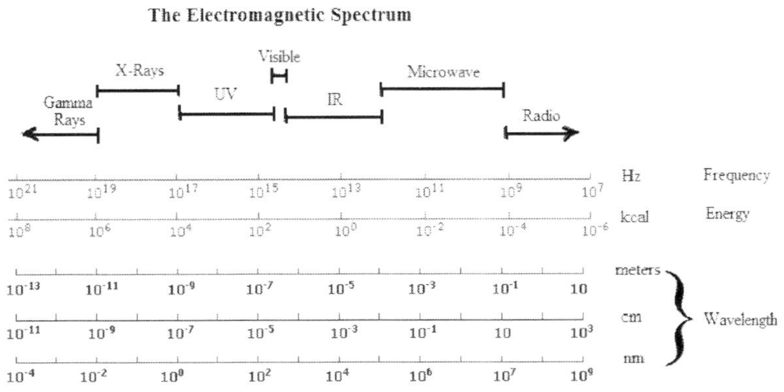

Figure 1. Schematic representation of the electromagnetic spectrum (adapted from http://www2.chemistry.msu.edu/faculty/reusch/VirtTxtJml/Spectrpy/UV-Vis/spectrum)?

Infrared radiation is divided into:

- near (NIR, ν = 10,000 – 4,000 cm^{-1});
- middle (MIR, ν = 4,000 – 200 cm^{-1}) and
- far (FIR, ν = 200 – 10 cm^{-1}).

Because all compounds show characteristic absorption/emission in the IR spectral region and based on this property they can be analyzed both quantitatively and qualitatively using FT-IR spectroscopy.

Today FT-IR instruments are digitalized and are faster and more sensitive than the older ones. FT-IR spectrometers can detect over a hundred volatile organic compounds (VOC) emitted from industrial and biogenic sources. Gas concentrations in stratosphere and troposphere were determined using FT-IR spectrometers (Puckrin *et al.*, 1996).

In case of environmental studies FTIR Spectroscopy is used to analyze relevant amount of compositional and structural information concerning environmental samples (Grube *et al.*, 2008). The analysis can be performed also to determine the nature of pollutants, but also to determine the bonding mechanism in case of pollutants removal by sorption processes. Techniques for measuring gas pollutants such as continuous air pollutants analyzer (SO_2, NO_2, O_3, NH_3), on-line gas chromatography (GC) used simple real-time instruments to quantify gas pollutants. They need to use several sensors in order to analyze multiple gas pollutants simultaneously.

FT-IR spectroscopy coupled with other spectroscopic techniques such as AAS (atomic absorption spectroscopy) have been used to assess the impact of industrial and natural activities on air quality (Kumar *et al.*, 2005; Childers *et al.*, 2001).

In addition to the traditional transmission FTIR (T-FTIR) methods (e.g. KBr-pellet or mull techniques), modern reflectance techniques are widely used today in environmental, agricultural, pharmaceuticals, and food studies. These modern techniques are attenuated total reflection FTIR (ATR-FTIR), and diffuse reflectance infrared Fourier transform spectroscopy (DRIFTS). The choice of the method to be used depends on many factors such as: the information needed (bulk versus surface analysis), the physical form of the sample, the time required for sample preparation (Majedová *et al.*, 2003).

In the following there will be presented some of the most important research studies related to the involvement of FTIR spectroscopy in environmental studies.

2. Traditional transmission FT-IR (T-FTIR) spectroscopy in environmental studies

Transmission spectroscopy is the oldest and most commonly used method for identifying either organic or inorganic chemicals providing specific information on molecular structure, chemical bonding and molecular environment. It can be applied to study solids, liquids or gaseous samples being a powerful tool for qualitative and quantitative studies.

FTIR instrument's principle of function is the following: IR radiation from the source that hits the beam splitter is partly directed towards the two mirrors arranged as shown in Figure 2. One of the two mirrors is stationary, and the other is moved at a constant velocity during data acquisition. As it can be seen in Figure 2 at first the IR beams are reflected by mirrors, after that are recombined at the beam splitter, and then passed through the sample and reach the detector. This records all wavelengths in the IR range. After the two beams reflected by the mirrors recombine, they will travel different distances, and the recombination will lead to constructive and destructive interference. The result will be an interferogram. After the recombined beam has passed through the sample the detector will record the Fourier transform of the IR spectrum of the sample. The data obtained are then processed by a computer that performs an additional Fourier transform to back-transform the interferogram into an IR spectrum (Smith *et al.*, 2011; Blum and John, 2012).

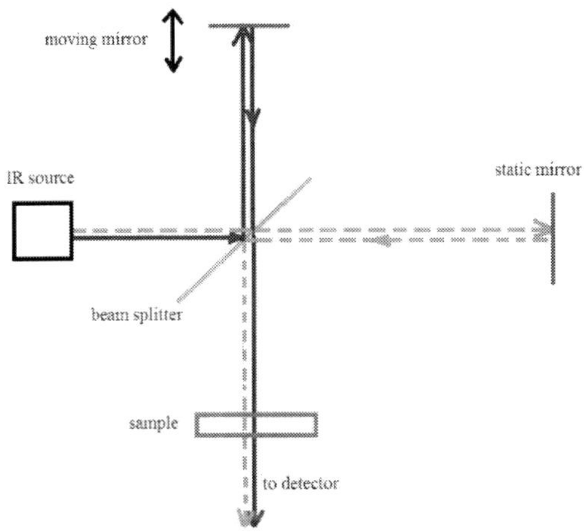

Figure 2. A schematic representation of an interferometer used in FTIR spectrometers (adapted from Blum and John, 2012 with permission (originally published in Drug Test. Analysis, DOI: 10.1002/dta.374))

The potential value of FTIR spectroscopy to a wide range of environmental applications has been demonstrated by numerous research studies. Some of them are presented below.

A review by McKelvy and coworkers containing 132 references at the chapter related to environmental applications of infrared spectroscopy (McKelvy *et al.*, 1998) covers the published literature about relevant applications of infrared spectroscopy for chemical analysis. The literature research was made for the period November 1995 to October 1997. The review contains aspects about infrared accessories and sampling techniques, infrared techniques, applications of infrared spectroscopy in environmental analysis, synthesis chemistry, food and agriculture, biochemistry and also the books and reviews appeared in that period for this subject (McKelvy *et al.*, 1998). An other review concerning the near-infrared and infrared spectroscopy was made by Workman Jr. This review covers the period 1993-1999 and presents the application of the near infrared spectral region to all types of analyses (Workman Jr, 1999).

The basic principle and methods of FTIR spectroscopy of the atmosphere are presented by Bacsik and coworkers in 2004 (Bacsik *et al.*, 2004). The same group of researchers published a review article related to the most significant and frequent applications of FTIR spectroscopy to the study of the atmosphere (Basick *et al.*, 2005). The authors summarized the basic literature in the field of special environmental applications of FTIR spectroscopy, such as power plants, petrochemical and natural gas plants, waste disposals, agricultural, and industrial sites, and the detection of gases produced in flames, in biomass burning, and in flares (Basick *et al.*, 2005).

Applications of FTIR spectroscopy to agricultural soils analysis were presented and discussed by Raphael in the book entitled "Fourier Transforms - New Analytical Approaches and FTIR Strategies" (Raphael, 2011). Chapter 19 of the same book presents the application of FTIR spectroscopy in waste management, and chapter 21 presents the study of trace atmospheric gases using Ground-Based Solar Fourier Transform Infrared Spectroscopy (Smidt *et al.*, 2011; Paton-Wals, 2011).

In case of air pollution the Fourier transform infrared (FTIR) instrument is used succesfully for measuring gas pollutants due to its many advantages such as: multiple gas pollutants will be monitored in real time, the IR spectra of sample can be analyzed and preserved for a long time, can be use to detect and measure directly both criteria and toxic pollutants in ambient air, measures also organic and inorganic compounds, can be also used to characterize and analyze microorganisms and monitor biotechnological processes, is generally installed at one location, but can be also portable and operated using battery for short-term survey, presents sensitivity from very low parts per million to high percent levels, can be applied to the analysis of solids, liquids and gases, no reagent is needed, and data acquisition is faster than with other physico-chemical techniques (Santos *et al.*, 2010).

The basic principle of FT-IR spectroscopy used in air pollutants detection and measuring is that every gas has its own „fingerprint" or absorption spectrum. The entire infrared spectrum will be monitored and FTIR sensor will read the different fingerprints of the gases present in the air sample. In case of determination of gas concentrations from stratosphere, the FT-IR spetrometers have to be designed with a fine resolution (0.01 cm^{-1}) due to the lower atmospheric pressure, and with a lower resolution between 0.05 cm^{-1} and 2 cm^{-1} for tropospheric gases determination. This is due to pressure broadening effects that result in broadened absorption lines. In troposphere water vapor concentrations are higher than those from stratosphere and they have a negative effect on the FT-IR spectrum measurements. The strong interference of water vapor in troposphere is overcome by detecting chemical substances in narrow bands of the IR spectrum where water absorption is very weak.

The total precipitable water vapour (PWV) from air which is responsible for the greenhouse effect being the most important trace gas can be measure using FT-IR spectroscopy. When it was compared with other instruments such as a Multifilter Rotating Shadow-band Radiometer (MFRSR), a Cimel sunphotometer, a Global Positioning System (GPS) receiver, and daily radiosondes (Vaisala RS92) it was estimated that FTIR spectrometer provides very precise trophospheric water vapour data, but when area-wide coverage and real-time data availability is very important, the GPS and the RS92 data are more appropriate. FTIR spectroscopy can be use also as a reference when assessing the accuracy of the other techniques, but those who use this technique have to be aware of the FTIR's significant clear sky bias (Schneider, 2010).

Animal farms are major sources of air pollution with ammonia and greenhouse gases. Air concentration of these pollutants may be higher or lower depending on the systems used. In addition, these systems have to correspond both in terms of animal welfare, and in terms of environmental protection. If it is considered animal welfare, the straw based systems are considered animal friendly systems, and when it is considered the environmental protection, the slurry based systems are preferred, due to lower ammonia (NH_3) and greenhouse gas (GHG) emissions. For slurry based systems air pollutants emissions were intensively researched, and the specific emission factors for several slurry-based housing systems for pigs are mentioned in the "Guidance document on control techniques for preventing and abating emissions of ammonia" developed by the UN/ECE "Expert Group on Ammonia Abatement" of the "Executive Body for the Convention on Long-Range Transboundary Air Pollution" (EB.AIR/1999/2). The straw based systems have not been extensively studied in terms of emissions of air pollutants. There are few research studies regarding these systems. Thus, high resolution FTIR spectrometry was used in order to determine the emissions of ammonia (NH_3), nitrous oxide (N_2O), methane (CH_4), and volatile organic compounds (VOC) at a commercial pig farm in Upper Austria using a straw flow system by Amon and coworkers (Amon *et al.*, 2007). The straw flow system is an animal friendly housing system for fattening pigs, being often equated with deep litter where there is no separation between the lying and the excretion areas. In deep litter systems most of the pigs welfare requirements are fulfilled. The main disadvatages of these systems are that there is a high straw consumption, the pigs are dirtier and the deep litter are characterized by high levels of NH_3 and greenhouse gases (GHG). Thus the level of NH_3 and greenhouse gases (GHG) has to be monitored in order to control and to avoid air pollution and to take appropriate measures for environment protection. For the pig farm monitored by Amon and coworkers it can be concluded that the straw flow system may combine recommendations of animal welfare and environmental protection (Amon *et al.*, 2007).

Environmental problems are also due to the incorect application of manure. The main air pollutants associated with manure application are ammonia, and nitrous oxides. In order to develop new environmentaly friendly methods for manure applications all aspects have to be investigated. For this purpose Galle and coworkers made some area-integrated measurements of ammonia emissions after spreading of pig slurry on a wheat field, based on gradient measurements using FTIR spectroscopy. They concluded that the gradient method is valuable for measurement of ammonia emissions from wide area, although the detection limits of the system limits its use to the relatively high emissions (Galle *et al.*, 2000).

In another study Jäger and coworkers reported that FTIR spectroscopy is capable of measuring low concentrations of CO_2, CH_4, N_2O and CO as well as isotope ratios (especially that of $^{13}CO_2$) in gas samples. The concentration levels of these gases are close to them in environmental air (Jäger *et al.* 2011). In the same paper the authors discussed also about the accuracy and stability of the FTIR instrument.

Volcanoes are considered important natural sources of air pollution. The most abundant gas typically released into the atmosphere by volcanoes is water vapor (H_2O), followed by carbon dioxide (CO_2) and sulfur dioxide (SO_2). Other gases such as hydrogen sulfide (H_2S), carbon monoxide (CO), hydrochloric acid (HCl), hydrofluoric acid (HF), hydrogen (H_2), helium (He), silicon tetrafluoride (SiF_4), carbon oxysulfide (COS) are released by volcanoes in small amounts. From the most dangerous to human, animals and agriculture are carbon dioxide, sulfur dioxide and hydrofluoric acid. Therefore it is important to monitor volcanic activities.

The first report about determination of HCl and SO_2 in volcanic gas dates from 1993 when Mori and coworkers used an FT-IR spectrometer during a stage of dome lava extrusion of the Unzen volcano (Mori *et al.*, 1993). Other gases including H_2O, CO_2, CO, COS, SO_2, HF were measured using a remote FT-IR spectral radiometer (Mori and Notsu, 1997; Francis *et al.*, 1996; Love *et al.*, 1998; Burton *et al.*, 2000; Mori and Notsu, 2008).

A telescope-attached FT-IR spectral radiometer was used to study the volcanic gases in seven active volcanoes from Japan. For one of the volcanoes monitored the authors have been used infrared radiation from hot lava domes, for three of them they used infrared radiation of the hot ground surface, and for the other three they used scattered solar light, as infrared sources. The observations over 15 years suggest that HCl/SO_2 and HF/HCl ratios are the most promising parameters reflecting volcanic activity among various parameters observable in remote FI-IR measurements (Notsu and Mori, 2010).

Oppenheimer and coworkers used thermal imaging and spectroscopic (FTIR) techniques to characterize phase-locked cycles of lava lake convection and gas plume composition of the Erebus volcano, Antarctica - a volcano continuously active for decades being now in steady-state. The authors identified a striking, cyclic correspondence between the surface motion of lava lake, and its heat and gas output. They concluded that this can be a reflection of unsteady, bi-directional magma flow in the conduit feeding the lake. It was also determined the ratio between gases emitted by volcanic lake, and the very tight correlation between CO_2 and CO was attributed to the redox equilibrium established in the lava lake. These results have a great contribution to the understanding of the laboratory models for magma convection degassing and volcanic gas geochemistry (Oppenheimer et al., 2009).

FTIR technique offers the potential for the non-destructive, simultaneous, real-time measurement of multiple gas phase compounds in complex mixtures such as cigarette smoke (Bacsik *et al.*, 2007a). Thus, in a study Bacsik and coworkers reported using of FTIR spectroscopy to study the mainstream cigarette smoke from cigarettes of different stated strengths (regular and various light cigarettes with different reported nicotine, tar and CO contents) (Bacsik, 2007b). The cigarette smoke is a very complex mixture that mainly consists of hydrocarbons and both carbon and nitrogen oxides. The results obtained by the authors reveal the fact that the strength of the cigarettes does not have a significant bearing on the quantity of the observed components obtained (Bacsik, 2007b).

An other anthropic source of air pollution is aircraft flight. The main pollutants released by aircrafts are unburnt hydrocarbons, carbon monoxide, and nitrogen oxides. The level of these pollutants is higher near the airport. For modern aircrafts the level of pollutants emissions is lower due to the using of more efficient turbine engine. Nevertheless the civil aviation authorities require the monitoring emissions from aircraft in airports and in the vicinity of airports. For this a non-intrusive Fourier Transform Infrared (FTIR) spectroscopy has been used to detect hydrocarbons in emissions from gas turbine engines (Arrigone and Hilton, 2005). The advantages of this mentioned techniques reported by Arrigone and Hilton are: it is non-intrusive—no sampling system is required and there is no physical interference with the exhaust plume while measurements are made; is useful for simultaneous monitoring of several species; the equipment is portable and can be simply set up and used outside the laboratory in engine test facilities, airfields (Arrigone and Hilton, 2005).

All these advantages encourage the use of FTIR spectroscopy as a valuable tool in monitoring emissions from aircraft in airports.

Quantitative information about air components and air pollutants is needed to study the impact of pollutants (gaseous, liquids or solids) on human health and atmospheric chemistry. To obtain these information an infrared spectral database was created. This database was completed with spectral information of gases emitted by biomass burning by Johnson and coworkers. The following classes of compounds: singly- and doubly-nitrogen-substituted aromatic, terpenes, hemi-terpenes, retenes and other pyrolysis biomarker compounds, carboxylic acids and dicarboxylic acids were identified in gases from biomass burning (Johnson *et al.*, 2010).

Throughout, latest years, the significance of bioaerosols has been discussed in environmental and occupational hygiene. Identification of microorganisms using cultivation and microscopic examination is time consuming and alone does not provide sufficient information with respect to the evaluation of health hazards in connection with bioaerosol exposure. FT-IR spectroscopy has widely been used for the characterization and identification of bacteria and yeasts, due to the fact that they are hydrophilic microorganisms and can easily be suspended in water for sample preparation (Essendoubi *et al.*, 2005; Duygu *et al.*, 2009). The identification of airborne fungi using FT-IR spectroscopy was described by Fischer and coworkers. They found that the method was suited to reproducibly differentiate *Aspergillus* and *Penicillium* species. The results obtained can serve as a basis for the development of a database for species identification and strain characterization of microfungi (Fischer *et al.*, 2006).

Studies on heavy metals and organic compounds removal from wastewaters using different natural and synthetic materials are many. The important role of FTIR spectroscopy in such studies is either related to the characterization of sorbents, chemical modified sorbents, or to establish the mechanism involved in sorption processes (Cheng *et al.*, 2012; Chen and Wang, 2012; Xu *et al.*, 2012; Ma *et al.*, 2012a; Wang *et al.*, 2011; Jordan *et al.*, 2011; Bardakçi and Bahçeli, 2010; Pokrovsky *et al.*, 2008; Parolo *et al.*, 2008).

Biosorption is considered as an alternative process for the removal of heavy metals, metalloid species, compounds and particles from aqueous solution by biological materials (Mungasavalli *et al.*, 2007). Biomaterials are adsorbent materials with high heavy metals adsorption capacity. They have many advantages such as reusability, low operating cost, improved selectivity for specific metals of interests, removal of heavy metals found in low concentrations in wastewaters, short operation time, and no production of secondary compounds which can be toxic (Mungasavalli *et al.*, 2007). FTIR spectroscopy can be used for characterization of biomaterials used in depolluting processes, but also to characterize materials obtained after chemical modification of them. Thus we used FTIR spectroscopy to characterize the material obtained after chemical modification of chitosan with glutardialdehyde in order to obtain a product with good sorption properties (Deleanu *et al.*, 2008), but also to characterize the materials obtained after alkaline treatment of bentonite to increase its capacity to retain ammonium ions from synthetic solutions (Simonescu *et al.*, 2005).

FT-IR spectroscopy has been used to identify the nature of possible sorbent (biosorbent) – pollutants (heavy metals, inorganic compounds, organic compounds) interactions.

For copper removal by fungal biomass to determine the characteristic functional groups that are responsible for biosorption of copper ions were made biomass's FTIR spectra before and after the bisorption process took place. The bonding mechanism between copper and biomass (fungal strain, cyanobacteria or other microorganism) (Yee *et al.*, 2004; Burnett *et al.*, 2006) can be determined by interpreting the infrared absorption spectrum.

We used in our studies fungal strains in order to remove heavy metals from synthetic waters which contain also copper in the form of copper sulfide nanoparticles, but also copper in dissolved state. In case of copper biosorption by *Aspergillus oryzae* ATCC 20423 the FTIR spectra registered are presented in Figure 3. The FTIR spectrum for *Aspergillus oryzae* ATCC 20423 before copper biosorption is presented in Figure 3a, the FTIR spectrum of *Aspergillus oryzae* ATCC 20423 after growth in the presence of copper solution with 25 mg copper/L is presented in Figure 3b, the FTIR spectrum of *Aspergillus oryzae* ATCC 20423 after growth in the presence of copper solution with 50 mg copper/L is presented in Figure 3c, the FTIR spectrum of *Aspergillus oryzae* ATCC 20423 after growth in the presence of copper solution with 75 mg copper/L is presented in Figure 3d, and the FTIR spectrum of *Aspergillus oryzae* ATCC 20423 after growth in the presence of copper solution with 100 mg copper/L is presented in Figure 3e.

From the Figure 3 it can be seen that all five FTIR spectra present distinct peaks in the following ranges: 3393 – 3418 cm^{-1}, 2926 – 2968 cm^{-1}, 1629 – 1638 cm^{-1}, 1404 – 1405 cm^{-1}, 1073 - 1077 cm^{-1}, and 529 – 533 cm^{-1}. The broad and strong band situated in the range 3393 – 3418 cm^{-1} can be attributed to overlapping of –OH and –NH stretching. The band from the range 2926 – 2968 cm^{-1} is attributed to the C-H stretching vibrations. The strong peak at 1629 – 1638 cm^{-1}

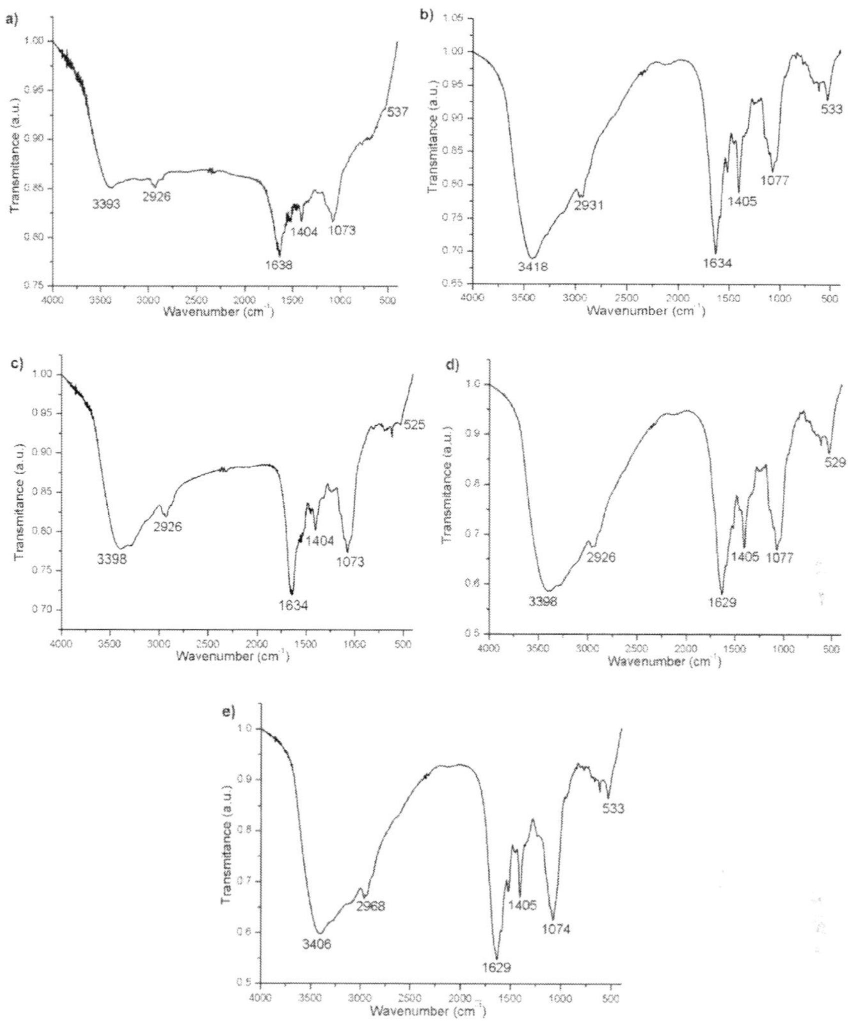

Figure 3. FT-IR spectra of *Aspergillus oryzae* ATCC 20423 unloaded (a) and loaded with Cu(II) ions (b-e)

can be due to a C=O stretching in carboxyl or amide groups. The peak at 1404 – 1405 cm^{-1} is attributed to N-H bending in amine group. The band observed at 1073 - 1077 cm^{-1} was assigned to the CO stretching of alcohols and carboxylic acids. Thus *Aspergillus oryzae* ATCC 20423 biomass contains hydroxyl, carboxyl and amine groups on surface.

From the Figures 3b-e it can be seen that the stretching vibration of OH group was shifted from 3393 cm^{-1} to 3418 cm^{-1} (3b), to 3398 cm^{-1} (3d), 3406 cm^{-1} (3e). These results revealed that chemical interactions between the copper ions and the hydroxyl groups occurred on the biomass surface. The carboxyl peak observed for unloaded biomass at 1638 cm^{-1} is shifted to 1634 cm^{-1} or 1629 cm^{-1}. This decrease in the wave number of the peak characteristic for C=O group from carboxylic acid revealed that interacts with carbonyl functional group are present between biomass and copper ions. These results indicated that the free carboxyl groups changed into carboxylate, which occurred during the reaction of the metal ions and carboxyl groups of the biosorbent.

No frequency changes were observed in the C-H and $-NH_2$ groups of biomass after copper biosorption. In addition, all FTIR spectrum of *Aspergillus oryzae* ATCC 20423 loaded with copper ions contain bands at 533, 529, 525 cm^{-1} which can be attributed to Cu-O stretching modes (Simonescu and Ferdes, *in press*).

The similar FT-IR results were reported for the biosorption Pb(II), Cd(II) and Cu(II) onto *Botrytis cinerea* fungal biomass (Akar *et al.*, 2005) and Pb(II) and Cd(II) from aqueous solution by macrofungus (*Lactarius scrobiculatus*) biomass (Anayurt *et al.*, 2009).

In our work we used also FT-IR spectroscopy in order to determine the characteristic functional groups which are responsible for biosorption of copper ions by *Polyporus squamosus*, *Aspergillus oryzae* NRRL 1989 (USA), *Aspergillus oryzae* 22343 (Simonescu *et al.*, 2012).

In case of biological degradation of pollutants a significant role can be attributed to biodegradation pathway due to the fact that different biodegradation pathways lead to different biodegradation products. Thus it is important to determine biodegradation pathways. For this purpose FTIR spectroscopy is a relevant tool for rapid determination of the resulting biotransformation product or mixtures. With this respect, Huang and coworkers investigated the ability of FT-IR to distinguish two different m-cresol metabolic pathways in *Pseudomonas putida* NCIMB 9869 after growth on 3,5-xylenol or m-cresol. From this study, it can be concluded that FT-IR spectral fingerprints were shown to differentiate metabolic pathways of m-cresol within the same bacterial strain and thus FTIR spectroscopy might provide a rapid, non-destructive, cost-effective approach for assessing of products resulted in biological degradation of pollutants (Huang *et al.*, 2006).

The main directions of use of FTIR spectroscopy in waste management are about getting information regarding the stage of organic matter for process and product control, and for monitoring of landfill remediation. For this purpose, Smidt and Meissl used FTIR spectroscopy to asses the stage of organic matter decomposition in waste materials (Smidt and Meissl, 2007). The results obtained confirm that FTIR spectroscopy represents an appropriate tool for process and quality control, for the assessment of abandoned landfills and for monitoring and checking of the successful landfill remediation (Smidt and Meissl, 2007).

The structural changing in biodegradation processes can be determined by FTIR analysis. Thus Tomšič and coworkers studied structural changes of cellulose fabric modified by imidazolidinone biodegradation after different period using electron microscopic and spectroscopic analyses (Tomšič *et al.*, 2007). Also FT-IR spectroscopy is a quick and useful method to monitor the composting process (Grube *et al.*, 2006). The aim of them study was to elucidate the typical IR absorption bands and correlation of band growth rates with the compost maturity or degradation degree. The results of this study revealed that IR spectroscopy is a simple, quick and informative method that can be used instead of several time consuming chemical methods for monitoring of routine composting processes.

Soil is a complex medium with important ecological functions. Its functions depend on its characteristics. FTIR spectroscopy can be used to describe soil characteristics in the form of complex multivariate data sets. Thus FTIR spectroscopy has been used by Elliott and coworkers to investigate soils at different stages of recovery from degradation following opencast mining and from undisturbed land (Elliott *et al.*, 2007). When a FT-IR spetrometer was used to determine gases from soils and rock formations no other gases than CO_2 have been detected except CO in the open-path compartment dedicated to atmosphere analysis (Pironon *et al.*, 2009).

The use of living organisms to manage or remediate polluted soils named bioremediation represents an emerging technology. This technology is defined as the elimination, attenuation or transformation of polluting or contaminating substances by the use of biological processes. The results *in situ* bioremediation depend by microbial strains from contaminated site. The biodegradation process can be monitored by FTIR spectroscopy. For this purpose Bhat and coworkers performed a study about remediation of hydrocarbon contaminated soil through microbial degradation. The bacterial strains involved in bioremediation process were collected to be isolated from contaminated soil. FTIR spectra of untreated and treated soil samples revealed that the isolated bacterial strains have a substantial potential to remediate the hydrocarbon contaminated soils (Bhat *et al.*, 2011).

Biomineralization has an important role for pollutants removal from environment. It has been known the mecanism involved in such processes to establish the nature of intemediates and final compound formed. FTIR spectroscopy is well-suited for such investigations, because it provides simultaneously molecular-scale information on both organic and inorganic constituents of a sample. Consequently FTIR spectroscopy was used in several complementary sample introduction modes as transmission (T-FTIR), attenuated total reflectance (ATR-FTIR), diffuse reflectance (DRIFTS) to analyze the processes of cell adhesion, biofilm growth, and biological Mn-oxidation by *Pseudomonas putida* strain GB-1 by Parikh and Chorover (Parikh and Chorover, 2005).

Fourier Transform Infrared (FT-IR) and Attenuated Total Reflectance (ATR) spectroscopy in the mid infrared (MIR) wavelength range (2500 – 16,000 nm) have been also developed for contaminant detection in water (Gowen *et al.*, 2011a). The authors tested the near infrared spectroscopy (NIRS) for the detection and quantification of pesticides including Alachlor and Atrazine in aqueous solution. Calibration models were built to predict pesticide concentration using PLS regression (PLSR). The proposed method shows potential for direct measurement of low concentrations of pesticides in aqueous solution. The research was performed in the laboratory conditions, and it is well known that the NIR spectrum of aqueous samples is susceptible to changes in the environment (e.g. temperature, humidity) and sample (e.g. pH, turbidity). Thus further experiments are necessary to test the effect of such perturbations on predictive ability (Gowen et al., 2011b).

By joining FTIR spectroscopy with two dimensional correlation analysis (2DCORR) there will be obtained a device with improved performance in the study of complex environmental systems (Noda and Ozaki, 2005). The two dimensional correlation analysis (2DCORR) is a method to visualize the dynamic relationship between the variables in multivariate data set with application of the complex cross-correlation function. With the help of this analysis there will be identified the spectral features which change in phase (i.e. linearly correlated among them) and out of phase (partially or not at all correlated among them) (Mecozzi *et al.*, 2009). This technique can be applied to study the evolution of environmental complex systems. Mecozzi and coworkers applied FTIR spectroscopy joined with two dimensional correlation analysis (2DCORR) to identify the aggregation pathways of extractable humic substance from marine sediments, and to compare the molecular modifications determined by the actions of different pollutants on the marine algae *Dunaliella tertiolecta* that is a biomarker of environmental quality (Mecozzi *et al.*, 2009). From this study it can be concluded that FTIR spectroscopy joined with 2DCORR analysis can be an important tool for evaluating toxic effects on the marine life.

3. Attenuated Total Reflection – Fourier Transform Infrared (ATR-FTIR) spectroscopy in environmental studies

Attenuated Total Reflection – Fourier Transform Infrared (ATR-FTIR) Spectroscopy was introduced in 1960s (Harrick, 1967), and now is widely used in many areas.

The principle of this is FTIR technique is that light introduced into a suitable prism at an angle exceeding the critical angle for internal reflection develops an evanescent wave (a special type of electromagnetic radiation) at the reflecting surface. Interaction of this evanescent wave with the sample determines ATR spectrum recording. The main charactesistic of this techniques is the fact that particle samples are deposited on the surface of a horizontal ATR crystal for spectroscopic analysis (Figure 4). Zinc selenide (ZnSe) or Ge crystals are the most commonly used in ATR-FTIR spectroscopy.

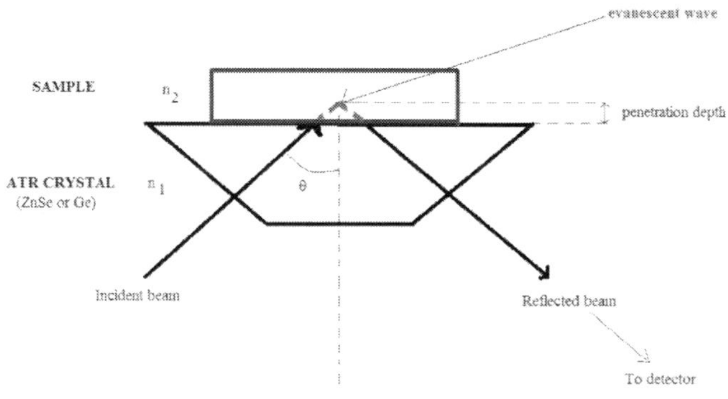

Figure 4. The principle of ATR-FTIR where n_1 and n_2 are the refractive indices of the crystal and the sample, respectively.

The main advantages of ATR-FTIR spectroscopy are: can be applied to a large variety of materials such as: powders, liquids, gels, pastes, pellets, slurries, fibers, soft solid materials, surface layers, polymer films, samples after evaporation of a solvent being a versatile and non-destructive technique; is useful for surface characterization, opaque samples; faster sampling being a non-destructive technique; is considered an extremely robust and reliable techique for quantitative studies involving liquids; excellent sample-to-sample reproducibility.

All these advantages make that ATR-FTIR spectroscopy to be used for: analysis of processes at surfaces (Freger and Ben-David, 2005), surface modification (Lehocký et al. 2003; Janorkar et al., 2004), surface degradation (Bokria et al., 2002), study of enzymatic degradation of a substrate film attached to a solid surface (Snabe et al. 2002), study of sunscreens on human skin (Rintoul et al., 1998), research of cereal, food and wood systems (D'Amico et al. 2012), detection of microbial metabolic products on carbonate mineral surfaces (Bullen et al., 2008), self-assembled thin films (Gershevitz et al. 2004), grafted polymer layers (Granville et al. 2004), adsorption processes (Sethuraman et Belfort 2005; Al-Hosney et Grassian 2005) of biological (Jiang et al. 2005; Mangoni et al. 2004) and synthetic (Freger et al. 2002) materials.

The followings are some examples of *in situ* ATR-FTIR spectroscopy's application in environmental studies.

In recent years adsorptive removal of heavy metals from aqueous effluents have received much attention because numerous materials such as: clays, zeolites, activated carbon can be used as adsorbents. The adsorption of inorganic ions on metal oxides and hydroxides was resolved using *in situ* ATR-FTIR spectroscopy. In a review Lefèvre describes and discusses *in situ* ATR-FTIR used in order to obtain information on the sorption mechanism of sulfate, carbonate, phosphate, perchlorate on hematite, goetite, alumina, silica, TiO_2 (Lefèvre, 2004). This is due to the fact that FTIR technique allows to analyze the sorption/desorption phenomena *in situ* being helpful in determining of the speciation of sorbed inorganic anions or ternary inorganic complexes formed. In addition this technique offers the possibility to distinguish outer-sphere and inner-sphere complexes. In this regard Yoon and coworkers used *in situ* ATR-FTIR spectroscopy and quantum chemical methods to determine the types and structures of the adsorption complexes formed by oxalate at boehmite (γ-AlOOH)/water and corundum (α-Al_2O_3)/water interfaces (Yoon *et al.*, 2004). They found that the adsorption mechanism of a aqueous HO_x^- species involves loss of protons from this species during the ligand-exchange reaction. The results obtained are useful in establishing the transport model of toxic species in natural waters, and remediation of liquid wastes.

Contamination of soils and groundwater by radioactive wastes containing uranium and other actinides is a significant problem. The fate and transport of these kind of pollutants in aquifers, design of cost-effective remediation techniques for radioactive-contaminated soils, and developing of materials proper for encapsulation and disposal of nuclear waste require knowledge of mechanism of radioactive pollutants – sorbent interactions. For radioactive waste depositories one of the most important factors which has to be considered is the long-term safety of them. For this, natural or anthropogenic barriers for sorption of radionuclides around the depositories are placed. Sorption data at the laboratory scale are useful to predict the behaviour of real systems. For this purpose Lefèvre and coworkers used ATR-IR spectroscopy to study the sorption of uranyl ions onto titanium oxide (mixture of rutile and anatase) and hematite. They found that the uranyl sorption on titanium oxide in the pH range 4-7 occurs by formation of one surface complex with uranium atoms bounded by two different chemical environments (Lefèvre *et al.*, 2008), and in case of sorption on hematite they concluded that the same surface species is responsible for the uranyl sorption in the pH range 5-8 (Lefèvre *et al.*, 2006). Due to the fact that experiments were reversible the authors concluded that reaction of hematite deposit with uranyl ions is the same with the reaction of it in dispersed suspensions (Lefèvre *et al.*, 2006). The sorption of U(V) on different forms of titanium dioxide was also studied using ATR-IR spectroscopy by Comarmond and coworkers. They showed the effect of different sources of sorbent and its surface properties on radionuclide sorption (Comarmond *et al.*, 2011). On the same subject Müller and coworkers used the high sensitivity of the *in situ* ATR-FTIR spectroscopy to establish the mechanism of sorption processes of U(VI) onto TiO_2 even at concentrations down to the low micromolar range. The Mid-IR spectra of U(VI) aqueous solutions and of U(VI) sorption onto different TiO_2 samples is presented in Figure 5.

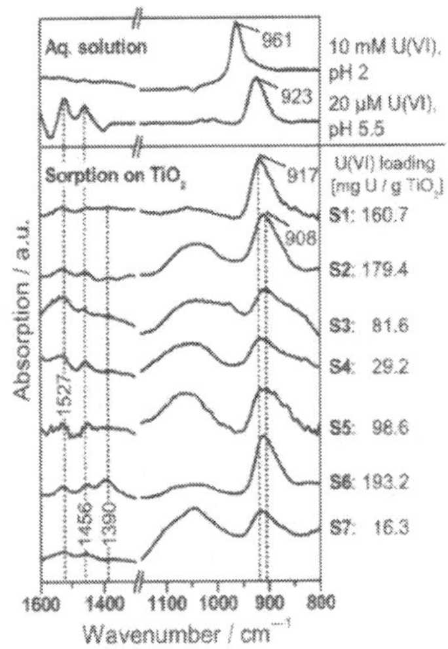

Figure 5. Mid-IR spectra of U(VI) aqueous solutions and of U(VI) sorption onto different TiO_2 samples (the values on the IR spectra are in cm^{-1}) (S1-S7 are different titania samples with different content of anatase and rutil, different particle size, and different origins) (from Müller *et al.*, 2012 used with permission (originally published in Geochimica et Cosmochimica Acta, http://dx.doi.org/10.1016/j.gca.2011.10.004))

By comparing the spectrum of the aqueous species spectra with the spectra of samples obtained after U(VI) sorption on TiO_2 it can be seen that the frequencies of the $\nu_3(UO_2)$ modes presented at 961 cm^{-1} for the aqueous species are significantly shifted (with 53-44 cm^{-1}) which suggests that uranyl surface complexes are formed at all titania samples.

This study is one complex due to the fact that authors performed researches to establish the influence of: stages of *in situ* sorption experiments (conditioning, sorption, and flushing), the contact time of U(VI) with the mineral, the initial U(VI) concentration, pH values, the origin and manufacturing procedure of TiO_2 samples and the absence of atmospheric-derived carbonate on the species formed in sorption processes of U(VI) on TiO_2. The results obtained by authors are relevant to the most environmental scenarios (Müller *et al.*, 2012).

Sorption of Np(V) onto TiO_2, SiO_2 and ZnO was investigated using ATR-FTIR spectroscopy. The results showed obtaining structurally similar bidentate surface complexes for all sorbents used (Müller *et al.*, 2009).

ATR-FTIR spectra confirmed formation of actinyl-carbonato complexes from interaction of actinide with hematite at a specific pH value. This can control the actinide transport in numerous subsurface receptors due to the abundance of carbonate in aquifers (Bargar *et al.*, 1999).

The influence of dissolved CO_2 on UO_2^{2+} sorption process was determined by Foerstendorf and Heim using ATR-FT-IR spectroscopy. They obtained a similar surface complex of the uranyl ion at the ferrihydrite-phase irrespective of the presence of atmospheric CO_2. Sorption of actinide ion on mineral phase determines a change of the carbonate ion from a monodentate to a bidentate ligand (Foerstendorf and Heim, 2008).

ATR-FTIR and FT-IR spectroscopy together with other techniques were used to determine the fate and transport of radionuclides in natural environments. The main mechanisms that are responsible for these are: sorption on organic (living matter and humic materials), sorption on inorganic materials (soil media and minerals), precipitation of them under oxic conditions, reduction in presence of microorganisms, and structural incorporation in different mineral host phases (Duff et *al.*, 2002).

Citric acid being a naturally-occuring acid commonly found in soils, and also a strong complexant of UO_2 is often found as a component of radioactive waste. Advantages such as: its biodegradability and complexing efficiency make from it a good candidate for remediation of uranium contaminated soils (Kantar and Honeyman, 2006). Factors with influence on the uranyl adsorption process to oxide minerals in presence of citric acid were determined by Logue and coworkers. Redden and coworkers have proposed formation of a ternary uranyl-citrate complexes on goethite (Redden *et al.*, 2001). Establishing the interactions between UO_2, citrate and mineral surfaces on a molecular level represents a key factor for modeling adsorption phenomena affecting transport in soils. For this purpose Pasilis and Pemberton used ATR-FTIR to elucidate the mechanism of UO_2 adsorption on aluminium oxide in the presence of citrate. They found that there is an enhanced citrate adsorption to Al_2O_3 in the presence of uranyl. This result suggests that uranyl may be the central link between two citrate ligands, and the uranyl is associated with the surface through a bridging citrate ligand. One other observation is that uranyl citrate complexes interact with citrate adsorbed to Al_2O_3 through outer shere interactions (Pasilis and Pemberton, 2008).

In recent years it ATR-FTIR spectroscopy has been used to investigate the atmospheric heteregenous reactions. Thus Al-Hosney and Grassian (2005) used this technique to investigate water adsorption on the surface of $CaCO_3$. They further used T-FTIR in order to investigate the role of surface adsorbed water in adsorption reactions of SO_2 and HNO_3 (Zhao and Chen, 2010). In other study Schuttlefield and coworkers (2007a) used ATR-FTIR spectroscopy to provide detailed information about water uptake and phase transitions for atmospherically relevant particles. To determine the factors involved in water uptake on the large fraction of dust present in the Earth's atmosphere, Schuttlefield and coworkers (2007b) used a variety of techniques, including ATR-FTIR. They concluded that water uptake on the clay minerals depends on the type and the source of the clay. These results are important because mineral dust aerosol provides a reactive surface in trophosphere being involved in reactions for atmosphere. The role of halogens in the aging process of organic aerosols was determined by Ofner and coworkers (2012) using long-path FTIR spectroscopy (LP-FTIR), attenuated-total reflection FTIR (ATR-FTIR), UV/VIS spectroscopy, and ultrahigh resolution mass spectroscopy (ICR-FT/MS). They concluded that the aerosol-halogen interaction might strongly contribute to the influence of organic aerosols on the climate system (Ofner *et al.*, 2012).

Khalizov and coworkers (2010) investigated the heterogeneous reaction of nitrogen dioxide (NO_2) on fresh and coated soot surfaces to assess its role in night-time formation of nitrous acid (HONO) in the atmosphere using ATR-FTIR (Khalizov *et al.*, 2010).

Segal-Rosenheimer and Dubowski (2007) combined two setups of FTIR for the parallel analysis of both condensed and gas phases of products resulted at the oxidation of cypermethrin (a synthetic pyrethroid being one of the most important insecticides in wide-scale use both indoors and outdoors) by gaseous ozone (Segal-Rosenheimer and Dubowski, 2007).

ATR-FTIR and T-FTIR methods provide detailed information on the composition of PM (particulate matter) samples. Both techniques can be used for qualitative and quantitative studies of particulate samples. Thus Veres (2005) used both methods to analyse particulate matter collected on Teflon Filters in Columbus – Ohio. He mentioned that ATR spectroscopy has limited applications in quantitative studies since it has a penetration depth of only a few microns, and this method can be replaced by transmission spectroscopy which penetrates into the bulk of substance (Veres, 2005).

Several groups of researchers used ATR-FTIR to particulate matter analysis. Thus Shaka and Saliba (2004) used ATR-FTIR spectroscopy in order to determine the concentration and the chemical composition of particulate matter at a coastal site in Beirut, Lebanon. Kouyoumdjian and Saliba (2006) determined the levels of the coarse (PM10-2.5) and fine (PM2.5) particles in the city of Beirut using ATR-FTIR spectroscopy. They also showed that nitrate, sulfate, carbonate and chloride were the main anionic constituents of the coarse particles, whereas sulfate was mostly predominant in the fine particles in the form of $(NH_4)_2SO_4$. Ghauch and coworkers (2006) used the same technique for the determination of small amounts of pollutants like the organic fraction of aerosols in the French cities of Grenoble and Clermont-Ferrand.

The applications of ATR-FTIR cover a wide range of subjects such as estimating of soil composition and fate of some soil components.

Monitoring of nitrate in soil is very important for managing fertilizer application and controlling nitrate leaching. This monitoring help to adjust nitrate level in soils in order to mantain the soil fertility, or to detect soil pollution. Due to the technological limitations, *in situ* or near real-time monitoring of soil nitrate is currently not feasible. In this purpose can be used the following methods: nitrate selective electrodes (Sibley, 2010), ion sensivitive field effect transistor (Birrell and Hummel, 2001), mid-infrared spectroscopy, and more particularly attenuated total reflectance (ATR) with Fourier transform infrared (FTIR) spectroscopy. Thus Raphael Linker submitted a report to the Grand Water Research Institute about simultaneous determination of $^{15}NO_3$-N and $^{14}NO_3$-N in aqueous solutions, soil extracts and soil pastes. The results obtained show that a combination of ATR-FTIR analysis with appropriate chemometrics can be successfully used to monitor $^{15}NO_3$-N and $^{14}NO_3$-N concentrations in soil during an incubation experiment (Linker, 2010). From the studies performed about measurement of nitrate concentration in soil pastes it can be concluded that ATR-FTIR appears to be a promising tool for direct and close to real-time determination of nitrate concentration in soils, with minimal treatment of the soil samples (Linker *et al.*, 2004; Linker *et al.* 2005; Linker *et al.*, 2006; Linker *et al.*, 2010). The same technique was used by Du and coworkers in order to evaluate net nitrification rate in *Terra Rosa* soil (Du *et al.*, 2009). ATR-FTIR spectroscopy was the technique preferred to mass spectrometry due to reduced cost, it is not time consuming, and doesn't require long and laborious preparation procedures. The results obtained have made major contributions for the estimation of the contribution of applied nitrogen and mineralized nitrogen to net nitrification rates ((Du *et al.*, 2009).

Soil paste was used by Choe and coworkers in order to improve the contact between sample and ATR crystal in case of using of the ATR-FTIR spectroscopy to determine the level of nitrate in soils. By comparing the nitrate peak intensity of soil pastes and their supernatant, it was shown that the nitrate dissolved in soil solution of the paste mainly responded to the FTIR signal. The results obtained are useful for the monitoring of nutrients in soils (Choe *et al.* 2010).

4. Diffuse Reflectance Infrared Fourier Transform (DRIFTS) spectroscopy in environmental studies

DRIFTS spectroscopy is considered a technique more sensitive to surface species than transmission measurements and is an excellent *in situ* technique. The principle is simple one: when incident light strikes a surface, the light that penetrates is reflected in all directions. This reflection is called diffuse reflectance. If the light that leaves the surface will pass through a thin layer of the reflecting materials, its wavelength content will have been modified by the optical properties of the matrix. The wavelength and intensity distribution of the reflected ligth will contain structural information on the substrate (Analytical Spectroscopy available at: http://www.analyticalspectroscopy.net/ap3-11.htm) (Figure 6).

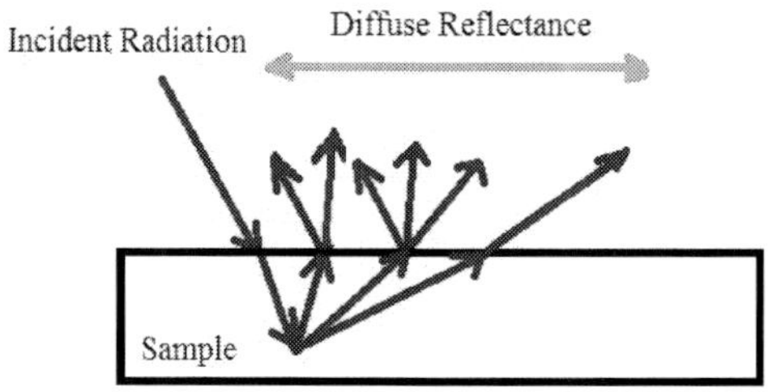

Figure 6. The principle of Diffuse Reflectance Infrared Fourier Transform Spectroscopy (adapted from Analytical Spectroscopy available at: http://www.analyticalspectroscopy.net/ap3-11.htm)

The main advantages of DRIFTS spectroscopy are: fast measurement of powdered samples, minimal or no sample preparation, ability to detect minor components, ability to analyze solid, liquid or gaseous samples, is one of the most suitable method for the examination of rough and opaque samples, high sensitivity, high versatility, capability of performing of the measurements under real life conditions.

In the environmental studies diffuse reflectance Fourier transform infrared (DRIFTS) spectroscopy is considered an alternative methodology for the quantitative analysis of nitrate in environmental samples (Verma and Deb, 2007a). It is considered a new, rapid and precise analytical method for the determination of the submicrogram levels of nitrate (NO_3^-) in environmental samples like soil, dry deposit samples, and coarse and fine aerosol particles. The DRIFTS method is a feasible nondestructive and time saving method for quantitative analyses of nitrate in soil, dry deposit and aerosol samples.

It is well known that soil can act as sinks as well as sources of carbon. A major fraction of carbon in soils is contained in the soil organic matter (SOM). It contributes to plant growth through its effect on the physical, chemical, and biological properties of the soil. Characterization of soil organic matter (SOM) is important for determining the overall quality of soils. For this DRIFTS spectroscopy can be used. This method only takes a few minutes, and is much faster than fractionating of soil samples using chemical and physical methods and determining the carbon contents of the fractions (Zimmermann *et al.*, 2007). In another study, Rumpel and coworkers tested diffuse reflectance infrared Fourier transform (DRIFT) spectroscopy in combination with multivariate data analysis [partial least squares (PLS)] as a rapid and inexpensive means of quantifying the lignite contribution to the total organic carbon (TOC) content of soil samples (Rumpel *et al.*, 2001). DRIFTS spectroscopy is also considered to be one of the most sensitive infrared technique to analyze humic substances (Ding *et al.*, 2000). Studies by Ding and coworkers demonstrate that both DRIFT and ^{13}C NMR are suitable for examining the effect of tillage on the distribution of light fraction in soil profile (Ding *et al*, 2002). More recently Ding and coworkers examined the effect of cover crops on the chemical and structural composition of SOM using chemical and DRIFT spectroscopic analysis. From this study it was concluded that both organic carbon (OC) and light fraction (LF) contents were higher in soils under cover crop treatments with and without fertilizer N than soils with no cover crop. Thus cover crops had a profound influence on the SOM and LF characteristics (Ding *et al.*, 2006). In other study Janik and coworkers (1995) showed that the use of diffuse reflectance infrared Fourier-transformed Spectroscopy (DRIFT) in combination with partial least squares algorithm (PLS) is a fast and low-cost method to predict carbon content and other soil properties such as clay content and pH. Zimmermann and coworkers evaluated the possibility of using of DRIFT-spectroscopy to estimate the soil organic matter content in soil samples from sites across Switzerland (Zimmermann *et al.*, 2004). It was concluded that DRIFT spectroscopy is a tool to predict changes in soil organic matter contents in agricultural soils resulting from changes in soil management. In other study Nault and coworkers used DRIFT spectroscopy to compare changes in organic chemistry of 10 species of foliar litter undergoing *in situ* decomposition for 1 to 12 years at four forested sites representing a range of climates in Canada (Nault *et al.*, 2009). This study demonstrated that DRIFT spectroscopy is a fast and simple analysis method for analyzing large numbers of samples to give good estimates of litter chemistry. Thus DRIFTS spectroscopy is considered a more faster technique to analyse the composition and the dynamics of organic matter in solis compared with FTIR spectroscopy (Tremblay and Gagné, 2002; Spaccini *et al.*, 2001).

Earth's atmosphere contains aerosols of various types and concentrations devided in: anthropogenic products, natural organic and inorganic products. The negative effects of these components refers to interaction with Earth's radiation budget and climate. In direct way aerosols scatter sunlight directly back into space, and indirect aerosols in the lower atmosphere can modify the size of cloud particles, and consequently changing the way in which clouds reflect and absorb sunlight. Aerosols act also as sites for chemical reactions to take place. As an exemple of these kind of reactions can be mentioned destruction of stratospheric ozone. The inorganic component of aerosols consist of inorganic salts (e.g. sulfate, nitrate, and ammonium). The most used method for analyzing these salts is ion chromatography (IC) (Chen *et al.*, 2003). The main disadvantages of this method are: time required for sample preparation and analysis that is up to 1 week, and the fact that this method is a destructive method of analysis. IR spectroscopy offers a simple and rapid alternative to IC for aerosols analysing, but it is imprecise and therefore only semi-quantitative. Advances in optics and detectors have allowed the development of more precise IR spectroscopy methods such as FTIR and DRIFT spectroscopy. FTIR spectroscopy was employed to determine on-site chemical composition of aerosol samples and to investigate the relationship between particle compositions and diameters (Tsai and Kuo, 2006). DRIFTS spectroscopy was used for quantitative analysis of atmospheric aerosols (Tsai and Kuo, 2006). The components of aerosols determined quantitative in area investigated were SO_4^{2-}, NO_3^- and NH_4^+. Compared with IC method, the DRIFT spectroscopy is a non-destructive, and quantitative method for aerosols analyzing.

Nitrogen dioxide, one of the key participants in atmospheric chemistry has been determined using DRIFT spectroscopy. Compared with other methods for nitrogen dioxide determination such as chemiluminiscence and fluorescence method that are multi-reagent procedure with the increased possibility of the experimental errors, the DRIFTS spectroscopy involves using NaOH–sodium arsenite solution as an absorbing reagent.

Another advantage of DRIFTS spectroscopy is that it can determine ambient nitrogen dioxide, in terms of nitrite, at submicrogram level (Verma *et al.*, 2008).

The feasibility of employing diffuse reflectance Fourier transform infrared (DRIFT) spectroscopy as a sensitive tool in the submicrogram level determination of sulphate (SO_4^{2-}) was checked by Verma and Deb in a study performed in 2007. The level of sulphate in environmental samples analysed like coarse and fine aerosol particles, dry deposits and soil was in range of ppb. The DRS-FTIR absorption spectrum of these real samples are presented in Figure 7.

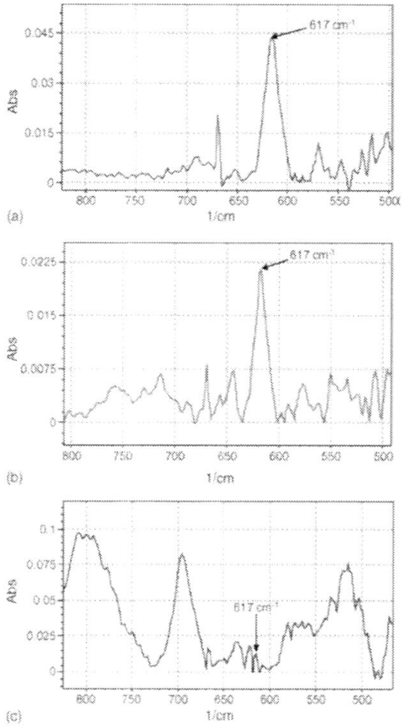

Figure 7. DRS-FTIR absorption spectrum of: (a) aerosol samples; (b) dry deposition sample; (c) soil sample (from Derma and Deb, 2007b used with permission (originally published in Talanta, doi:10.1016/j.talanta.2006.07.056))

For all real samples analyzed two-point baseline corrections were performed to obtain the quantitative absorption peak for sulphate at around 617 cm^{-1} (Verma and Deb, 2007b). The DRIFT method involved in this study did not require pretreatment of samples being reagent less, nondestructive, very fast, repeatable, and accurate and has high sample throughput value (Verma and Deb, 2007b). On the same topic Ma and coworkers have published paper entitled, "A case study of Asian dust storm particles: Chemical composition, reactivity to

SO_2 and hygroscopic properties". This paper presents a study about characterization of Asian dust storm particles using multiple analysis methods such as SEM-EDS, XPS, FT-IR, BET, TPD/mass and Knudsen cell/mass. The atmospheric dust particles are responsible by absorption and scattering of solar radiation and indirect acting as cloud condensation nucleus. The composition, source and size distribution of dust storm are important in predicting them impacts on climate and atmospheric environment. The dust particles can react with gaseous components or pollutants from the atmosphere such as sulfur dioxide. Thus numerous studies were performed to determine the role of dust in SO_2 chemistry (Prince *et al.*, 2007; Ullerstam *et al.*, 2002, 2003; Zhang *et al.*, 2006; Ma *et al.*, 2012b). The morphology, elemental fraction, source distribution, true uptake coefficient for SO_2 and higroscopic behavior were studied. The major components of Asian dust storm particles were aluminosilicate, SiO_2 and $CaCO_3$ mixed with some organic and nitrate compounds. The particles analyzed by Ma and coworkers are coming from anthropogenic sources and local sources after long transportation. Between SO_2 uptake coefficient and mass was established a linear dependence. Consequently DRIFTS and FTIR spectroscopy combined with other analitical methods will provide important information about the effects of dust storm particle on the atmosphere (Ma *et al.*, 2012b).

One of the most important application of DRIFTS spectroscopy is to investigate sorption-uptake processes on different materials in order to reduce the impact of pollutants. Thus Valyon and coworkers studied N_2 and O_2 sorption on synthetic and natural mordenites, and on molecular sieves 4A, 5A and 13X using DRIFT spectroscopy (Valyon *et al.*, 2003). Kazansky and coworkers used DRIFTS spectroscopy to study sorption of N_2, both pure and in mixtures with oxygen, O_2, by zeolites NaLSX and NaZSM-5 (Kazansky et al., 2004). Llewellyn and Theocharis studied carbon dioxide adsorption on silicate using DRIFTS spectroscopy (Llewellyn and Theocharis, 1991). Heterogeneous oxidation of gas-phase SO_2 on different iron oxides was investigated *in situ* using a White cell coupled with Fourier transform infrared spectroscopy (FTIR) and diffuse reflectance infrared Fourier transform spectroscopy (DRIFTS) by Fu and coworkers (Fu *et al.*, 2007). From this study it can be concluded that adsorbed SO_2 could be oxidized on the surface of most iron oxides to form a surface sulfate species at ambient temperature, and the surface hydroxyl species on the iron oxides was the key reactant for the heterogeneous oxidation (Fu *et al.*, 2007). Heterogeneous reaction of NO_2 with carbonaceous materials (commercial carbon black, spark generator soot, Diesel soot from passenger car and high-purity graphite) at elevated temperature (400°C) was studied using DRIFT spectroscopy. Different infrared signals appear when NO_2 is adsorbed either on aliphatic or graphitic domains of soot (Muckenhuber and Grothe, 2007).

Gas sensors are playing an important role in the detection of toxic pollutants such as CO, H_2S, NO_x, SO_2, and inflammable gases such as hydrocarbons, H_2, CH_4. Diffuse Reflectance Infrared (DRIFT) spectroscopy has been used to characterize them. Thus, the studies performed by Harbeck in him Dissertation have shown that thick film sensors can easily be characterised in different working conditions (at elevated temperatures, in the presence of humidity) using Diffuse Reflectance Infrared (DRIFT) spectroscopy. He characterized undoped and Pd-doped SnO_2 sensor surfaces at different temperatures using two different methods in parallel: DRIFT spectroscopy and electrical measurements. Simultaneous recording of the DRIFT spectra and the sensor resistance helped him to clarify the role of the individual surface species in the sensing mechanism. The results of his work show that several reactions take place in the presence of CO depending both on temperature and humidity. It was found that all surface species are involved in the reactions and it is supposed that parallel and consecutive CO reactions take place on the surface (Harbeck, 2005).

DRIFT spectroscopy is also suitable for application to studies of surface phenomena and large specific surface materials such as the sensing layers. In this purpose Bârsan and Weimar investigated the effect of water vapour in CO sensing by using Pd doped SnO_2 sensors obtained using thick film technology as an example of the basic understanding of sensing mechanisms applied to sensors. The results obtained show that all parts of the sensor (sensing layer, electrodes, substrate) have influence to the gas detection and their role has to be taken into consideration when one attempts to understand how a sensor works (Bârsan and Weimar, 2003).

All the examples mentioned above show the importance of DRIFT spectroscopy in analyzing of environmental samples either liquid, solid or gaseous.

5. Open Path FT-IR spectroscopy in environmental studies

The open-path FT-IR Spectroscopy is conventionally used for monitoring gaseous air pollutants, but can also be used for monitoring both the gaseous or particulate air pollutants. The principle of function is the same with classical FTIR Spectroscopy, except the cell into the sample will be injected which it is extended to open atmosphere (Minnich and Scotto, 1999). In this technique the infrared light sources can be either natural solar light, or light coming from a heated filament situated behind the target gas. The infrared signal passes through a sample and chemical vapors present in sample will absorb the infrared energy at different wavelengths. All compounds in the vapor will give unique fingerprints of absorbance features which will be compared to a library of spectra on the computer. This comparison will be useful to identify and quatify in real time.

The advantages of open-path FT-IR Spectroscopy include: no sample collection, handling ar preparation is necessary; good sensivity for certain species; real time data collection and reporting; ability to simultaneously and continuously analyze many compounds; remote, long-path measurements; *in situ* application; stored data can be used and re-analyzed for a divers range of volatile or non-volatile compounds; cost effectiveness (Marshall *et al.*, 1994).

The main disadvantage of OP-FTIR is considered to be the fact that it can be applied only to the cases with high concentrations of gases such as stack measurement, landfill measurement, and fence-line monitoring (Hong *et al.*, 2004). Thus Perry *et al.* (1995) and Tso and Chang applied OP-FTIR to determine the VOC and ammonia concentrations in industrial areas, the concentration of pollutants being in this area in the level of 0.1 ppm

(Perry *et al.*, 1995; Tso and Chang, 1996). Childers *et al.* applied OP-FTIR spectroscopy for the measurement of ammonia, methane, carbon dioxide, and nitrous oxide in a concentrated swine production facility. The pollutants concentration was in the reanges 0.1 – 100 ppm. The results have led authors to conclude that the confinement barns was the significant source of ammonia emission, and the waste treatment lagoon was the major source of methane (Childers *et al.*, 2001). A similar research was performed by Hedge et al. in oder to monitor methane and carbon dioxide emitted form a landfill in northern of Taiwan (Hedge *et al.*, 2003), and Thorn et al. used OP-FTIR to measure phosphine concentrations in the air surrounding the large fumigated structures of a tobacco warehouse (Thorn *et al.*, 2001). OP-FITR was used by Harris and coworkers to monitor ammonia and methane emissions from animal housing and waste lagoons due to the ability to detect multiple compounds simultaneously (Harris *et al.*, 2007).

Levine and Russwurm described in an article the use of the open-path FT-IR Spectroscopy in remote sensing of aiborne gas and vapor contaminants (Levine and Russwurm, 1994). Applying open-path Fourier transform spectroscopy for measuring aerosols was described by Wu and coworkers (Wu *et al*, 2007).

Air monitoring during site remediation using open-path FTIR Spectroscopy was reported by Minnich and Scotto (Minnich and Scotto, 1999), and monitoring trace gases from aircraft emissions using the same technique was reported by Haschberger (Haschberger, 1994).

The use of OP-FTIR spectroscopy for identification of fugitive organic compound (VOC) emission sources and to estimate emission rates at an Air Force base in United States was described by Hall (Hall, 2004). Galle *et al.* have demonstrated advantages of FTIR over traditional point-measurement methods by providing detection over large sampling areas (Galle *et al.*, 2001).

OP-FTIR was successfully applied by Walter et al., and Kagann *et al.* for the measurements of air quality criteria pollutants such as ozone, carbon dioxide, sulfur dioxide, and nitrogen dioxide in ambient air (Walter *et al.*, 1999; Kagan *et al*, 1999). Grutter and coworkers used OP-FTIR spectroscopy to measure trace gases over Mexico City. This was the first report on the concentration profiles of acetylene, ethylene, ethane, propane, and methane in this region. Specific correlation between the profiles and wind direction were made in order to determine the main sources that contribute to these profiles (Grutter *et al.*, 2003).

A comparison between different analysis techniques applied to ozone and carbon monoxide detection was made by Briz and coworkers. They compared classical least-squares (CLS) procedures with line-by-line method (SFIT) to analyze OP-FTIR spectra and concluded that discrepancies observed in CLS-based methods were induced by the experimental background reference spectrum, and SFIT results agreement well with the standard extractive methods (Briz *et al.*, 2007). The same author together with other coworkers proposed a new method for calculating emission rates from livestock buildings applying Open-Path FTIR spectroscopy (Briz *et al.*, 2009). The method was applied in a cow shed in the surrondings of La Laguna, Tenerife Island (Spain), and results obtained revealed that the livestock building behaves such as an accumulation chamber, and methane emission factor was lower than the proposed by Emission Inventory (Briz *et al.*, 2009).

As was described by Lin and coworkers an open-path Fourier transform infrared spectroscopy system can be used for monitoring of VOCs in industrial medium. They used this system to monitor VOCs emissions from a paint manufacturing plant, and they determined seven VOCs in ambient environment. The same system was also used to determine the VOCs in a petrochemical complex. The results obtained were correlated with meteorological data and were effective in the depiction of spatial variations in indentifying sources of VOC emissions. They also mentioned another important advantage of OP-FTIR spectroscopy such as the ability to obtain more comprehensive data than by using the traditional multiple, single-point monitoring methods. It can be concluded that OP-FTIR can be useful in both industrial hygiene and environmental air pollutat regulatory enforcement (Lin *et al.*, 2008).

Ammonia, CO, methane, ethane, ethylene, acetylene, propylene, cyclohexane, and O-xylene were identified as major emissions in a coke processing area from Taiwan using OP-FTIR system by Lin and coworkers (Lin *et al.*, 2007). Main gaseous byproducts (CO, CO_2, CH_4 and NH_3) of thermal degradation (pyrolysis) of biomass in forest fires were determined accurately using OP-FTIR. The results obtained in this study can help to improve the modelling of the pyrolysis processes in physical-based models for predicting forest fire behaviour (de Castro *et al.*, 2007). An other reasearch in this field was performed by Burling and coworkers who measured trace gas emissions from biomass burning of fuel types from the southeastern and southwestern United States (Burling *et al.*, 2010) with the help of OP-FTIR. The authors detected and quantified 19 gas-phase species in these fires: CO_2, CO, CH_4, C_2H_2, C_2H_4, C_3H_6, HCHO, HCOOH, CH_3OH, CH_3COOH, furan, H_2O, NO, NO_2, HONO, NH_3, HCN, HCl, and SO_2. The emission factors depend on the fuel composition and fuel types.

All the advantages of OP-FTIR spectroscopy and all the studies mentioned above demonstrate the utility of OP-FTIR in measuring and monitoring of atmospheric gases. This technique has increasingly been accepted by different environmental agencies as a tool in the measurement and the monitoring of the atmospheric gases (Russwurm and Childers, 1996; Russwurm, 1999).

6. Conclusion

All these presented above show the importance of FTIR spectroscopy in environmental studies. The major advantages of this technique are: real time data collection and reporting, excellent sample-to-sample reproductibility, enhanced frequency accuracy, high signal-to-noise ratios, superior sensitivity, analytical performance. In addition, the measurement is very rapid so that a large number of samples can be analyzed. Consequently FTIR spectroscopy coupled with other techniques is widely used to determine the nature of pollutants (gaseous, liquid or solid), to monitor environment, to asses the impact of pollution on health and environment, to determine the level of decontamination processes.

The modern techniques such as attenuated total reflection FTIR (ATR-FTIR), and diffuse reflectance infrared Fourier transform spectroscopy (DRIFTS), but also traditional transmission FTIR can be used for such studies according to the information needed, the physical form of the sample, and the time required for the sample preparation.

Author details

Claudia Maria Simonescu
Department of Analytical Chemistry and Environmental Engineering, Faculty of Applied Chemistry and Materials Science, „Politehnica" University of Bucharest, Romania

Acknowledgement

The author wants to thank all the authors who gave her the permission to cite them works, and to the publishers for reusing of some figures from the papers published by them.

7. References

Akar T, Tunali S, Kiran I (2005) *Botrytis cinerea* as a new fungal biosorbent for removal of Pb(II) from aqueous solutions. *Biochemical Engineering Journal* 25(30): 227-235.

Al-Hosney HA, Grassian VH (2005) Water, sulfur dioxide and nitric acid adsorption on calcium carbonate: A transmission and ATR-FTIR study. *Phys. Chem. Chem. Phys.* 7: 1266-1276.

Amon B, Kryvoruchko V, Fröhlich M, Amon T, Pöllinger A, Mösenbacher I, Hausleitner A (2007) Ammonia and greenhouse gas emissions from a straw flow system for fattening pigs: Housing and manure storage. *Livestock Science* 112: 199–207.

Anayurt RA, Sari A, Tuzen M (2009) Equilibrium, thermodynamic and kinetic studies on biosorption of Pb(II) and Cd(II) from aqueous solution by macrofungus (*Lactarius scrobiculatus*) biomass. *Chemical Engineering Journal* 151(1-3): 255-261.

Arrigone GM, Hilton M (2005) Theory and practice in using Fourier transform infrared spectroscopy to detect hydrocarbons in emissions from gas turbine engines. *Fuel* 84: 1052–1058.

Bacsik Z, Mink J (2007a) Photolysis-assisted, long-path FT-IR detection of air pollutants in the presence of water and carbon dioxide. *Talanta* 71: 149–154.

Bacsik Z, McGregor J, Mink J (2007b) FTIR analysis of gaseous compounds in the mainstream smoke of regular and light cigarettes. *Food and Chemical Toxicology* 45: 266–271.

Bacsik Z, Mink J, Keresztury G (2005) FTIR spectroscopy of the atmosphere part 2. applications. *Appl. Spectrosc. Rev.* 40: 327–390.

Bacsik Z, Mink J, Keresztur G (2004) FTIR spectroscopy of the atmosphere I. principles and methods. *Appl. Spectrosc. Rev.* 39: 295–363.

Bardakçi B, Bahçeli S (2010) FTIR study on modification of transition metal on zeolites for adsorption. *Indian Journal of Pure & Applied Physics* 48: 615-620.

Bargar JR, Reitmeyer R, Davis JA (1999) Spectroscopic Confirmation of Uranium (VI)-Carbonato Adsorption Complexes on hematite. *Environ. Sci. Technol.* 33: 2481–2484.

Bârsan N, Weimar U (2003) Understanding the fundamental principles of metal oxide based gas sensors; the example of CO sensing with SnO_2 sensors in the presence of humidity. *J. Phys.: Condens. Matter* 15: R813–R839.

Bhat MM, Shankar S, Shikha, Yunus M, Shukla RN (2011) Remediation of Hydrocarbon Contaminated Soil through Microbial Degradation- FTIR based prediction. *Advances in Applied Science Research* 2(2): 321-326.

Birrell S.J, and Hummel JW (2001) Real-time multi-ISFET/FIA soil analysis system with automatic sample extraction. *Comp. and Elect. in Agric.* 32(1): 45-67.

Blum M-M, John H (2012) Historical perspective and modern applications of Attenuated Total Reflectance – Fourier Transform Infrared Spectroscopy (ATR-FTIR). *Drug Test. Analysis* 4: 298-302.

Bokria JG, Schlick S (2002) Spatial Effects in the Photodegradation of Poly(acrylonitrile-butadiene-styrene): A Study by ATR-FTIR. *Polymer* 43: 3239-3246.

Briz S, Barrancos J, Nolasco D, Melián G, Padrón E, Pérez N (2009) New Method for Estimating Greenhouse Gas Emissions from Livestock Buildings Using Open-Path FTIR Spectroscopy, *Remote Sensing of Clouds and the Atmosphere XIV*, edited by Richard H. Picard, Klaus Schäfer, Adolfo Comeron, Evgueni Kassianov, Christopher J. Mertens, Proc. of SPIE 7475: 747510-1-747510-9.

Briz S, de Castro AJ, Diez S, Lopez F, Schaefer K (2007) Remote Sensing by Open-Path FTIR Spectroscopy: Comparison of Different Analysis Techniques Applied to Ozone and Carbon Monoxide Detection. *Journal of Quantitative Spectroscopy and Radiative Transfer* 103(2): 314-330.

Bullen HA, Oehrle SA, Bennett AF, Taylor NM, Barton HA (2008) Use of Attenuated Total Reflectance Fourier Transform Infrared Spectroscopy to Identify Microbial Metabolic Products on Carbonate Mineral Surfaces. *Applied and Environmental Microbiology* 74(14): 4553-4559.

Burling IR, Yokelson RJ, Griffith DW, Johnson TJ, Veres P, Roberts J, Warneke C, Urbanski SP, Reardon J, Weise DR, Hao W & De Gouw JA (2010) Laboratory measurements of trace gas emissions from biomass burning of fuel types from the southeastern and southwestern United States. *Atmospheric Chemistry and Physics* 10(22): 11115-11130.

Burnett P-GG, Daughney JC, Peak D (2006) Cd adsorption onto *Anoxybacillus flavithermus*: Surface complexation modeling and spectroscopic investigations. *Geochimica et Cosmochimica Acta* 70: 5253–5269.

Burton MR, Oppenheimer C, Horrocks LA, Francis PW (2000) Remote sensing of CO_2 and H_2O emission rates from Masaya volcano, Nicaragua. *Geology* 28: 915–918.

Chen Y, Wang J (2012) Removal of radionuclide Sr^{2+} ions from aqueous solution using synthesized magnetic chitosan beads. *Nuclear Engineering and Design* 242: 445-451.

Cheng C, Wang JN, Xu L, Li AM (2012) Preparation of new hyper cross-linked chelating resin for adsorption of Cu^{2+} and Ni^{2+} from water. *Chinese Chemical Letters* 23: 245–248.

Childers JW, Thompson Jr EL, Harris DB, Kirchgessner DA, Clayton M, Natschke DF, Phillips WJ (2001) Multi-pollutant concentration measurements around a concentrated swine production facility using open-path FTIR spectrometry. *Atmospheric Environment* 35(11): 1923-1936.

Choe E, van der Meer F, Rossiter D, van der Salm C, Kim K-W (2010) An Alternate Method for Fourier Transform Infrared (FTIR) Spectroscopic Determination of Soil Nitrate Using Derivative Analysis and Sample Treatments. *Water Air Soil Pollut.* 206:129–137.

Comarmond MJ, Payne TE, Harrison JJ, Thiruvoth S, Wong HK, Augtherson RD, Lumpkin GR, Müller K, Foerstendorf H (2011) Uranium Sorption on Various Forms of Titanium Dioxide – Influence of Surface Area, Surface Charge, and Impurities. *Environ. Sci. Technol.* 45(13): 5536-5542.

D'Amico S, Hrabalova M, Müller U, Berghofer E (2012) Influence of ageing on mechanical properties of wood to wood bonding with wheat flour glue. *European Journal of Wood and Wood Products*, DOI: 10.1007/s00107-012-0595-x

Deleanu C, Simonescu CM, Căpăţînă C (2008) Comparative study on the adsorption of Cu(II) ions onto chistosan and chemical modified chitosan. *Proceedings of 12th Conference on Environment and Mineral Processing* – Part III – 5.-7.6. Ostrava, Czech Republic, 201-207.

Ding G, Liu X, Herbert S, Novak J, Amarasiriwardena D, Xing B (2006) Effect of cover crop management on soil organic matter. *Geoderma* 130: 229–239, doi:10.1016/j.geoderma.2005.01.019

Ding G, Novak JM, Amarasiriwardena D, Hunt PG, and Xing B (2002) Soil Organic Matter Characteristics as Affected by Tillage Management. *Soil Sci. Soc. Am. J.* 66:421–429.

Ding, G, Amarasiriwardena D, Herbert S, Novak J, and Xing B (2000) Effect of cover crop systems on the characteristics of soil humic substances. p. 53–61. *In* E.A. Ghabbour and G. Davis (ed.) *Humic substances: Versatile components of plants, soil and water.* Press, Orlando, FL. The Royal Society of Chemistry, Cambridge.

Du CW, Linker R, Shaviv A & Zhou JM (2009) *In Situ* Evaluation of Net Nitrification Rate in *Terra Rossa* Soil Using a Fourier Transform Infrared Attenuated Total Reflection N-15 Tracing Technique. *Applied Spectroscopy* 63: 1168-1173.

Duff MC, Coughlin JU, Hunter DB (2002) Uranium co-precipitation with iron oxide minerals. *Geochimica et Cosmochimica Acta.* 66(20): 3533–3547.

Duygu D (Yalcin), Baykal T, Açikgöz İ, Yildiz K (2009) Fourier Transform Infrared (FT-IR) Spectroscopy for Biological Studies. G.U. *Journal of Science* 22(3): 117-121.

Elliott GN, Worgan H, Broadhurst D, Draper J, Scullion J (2007) Soil differentiation using fingerprint Fourier transform infrared spectroscopy, chemometrics and genetic algorithm-based feature selection. Soil Biology & Biochemistry 39: 2888–2896.

Essendoubi M, Toubas D, Bouzaggou M, Pinon J.-M, Manfait M, Sockalingum GD (2005) Rapid identification of *Candida* species by FT-IR microspectroscopy. Biochim Biophys Acta. 1724(3):239-47, http://dx.doi.org/10.1016/j.bbagen.2005.04.019

Fischer G, Braun S, Thissen R, Dott W. (2006) FT-IR spectroscopy as a tool for rapid identification and intra-species characterization of airborne filamentous fungi. Journal of Microbiological Methods 64: 63– 77, doi:10.1016/j.mimet.2005.04.005

Foerstendorf H, Heim K (2008) Sorption of uranium(VI) on ferrihydrite – Infleunce of atmospheric CO_2 on surface complex formation invetsigated by ATR-FT-IR spectroscopy. NRC7- Seventh International Conference on Nuclear and Radiochemistry, Budapest, Hungary

Freger V and Ben-David A (2005) Use of Attenuated Total Reflection Infrared Spectroscopy for Analysis of Partitioning of Solutes between Thin Films and Solution. Analytical Chemistry B. 77(18): 6019–6025.

Freger V, Gilron J, Belfer S (2002) TFC polyamide membranes modified by grafting of hydrophilic polymers: an FT-IR/AFM/TEM study. J. Membr. Sci. 209: 283-292.

Fu H, Wang X, Wu H, Yin Y, Chen J (2007) Heterogeneous Uptake and Oxidation of SO_2 on Iron Oxides. J. Phys. Chem. C 111: 6077-6085.

Galle B, Klemedtsson L, Bergqvist B, Ferm M, Törnqvist K, Griffith DWT, Jensen N-O, Hansen F (2000) Measurements of ammonia emissions from spreading of manure using gradient FTIR techniques. Atmospheric Environment 34: 4907-4915.

Gershevitz O, Sukenik CN (2004) In Situ FTIR-ATR Analysis and Titration of Carboxylic Acid-Terminated SAMs. J. Am. Chem. Soc. 126: 482-483.

Ghauch A, Deveau P-A, Jacob V, Baussand P (2006) Use of FTIR spectroscopy coupled with ATR for the determination of atmospheric compounds. Talanta. 68(15): 1294–1302.

Gowen A, Tsenkova R, Bruen M and O'Donnell C (2011a) Vibrational Spectroscopy for Analysis of Water for Human Use and in Aquatic Ecosystems. *Critical Reviews in Environmental Science and Technology*.

Gowen A, Tsuchisaka Y, O'Donnell C, Tsenkova R (2011b) Investigation of the Potential of Near Infrared Spectroscopy for the Detection and Quantification of Pesticides in Aqueous Solution. *American Journal of Analytical Chemistry* 2: 53-62. doi:10.4236/ajac.2011.228124

Granville AM, Boyes SG, Akgun B, Foster MD, Brittain WJ (2004). Synthesis and Characterization of Stimuli-Responsive Semi-Fluorinated Polymer Brushes by Atom Transfer Radical Polymerization. *Macromolecules* 37: 2790-2796.

Grube M, Muter O, Strikauska S, Gavare M, Limane B (2008) Application of FT-IR spectroscopy for control of the medium composition during the biodegradation of nitro aromatic compounds. *J. Ind. Microbiol. Biotechnol.* 35: 1545-1549.

Grube M, Lin JG, Lee PH, Kokorevicha S (2006) Evaluation of sewage sludge-based compost by FT-IR spectroscopy. *Geoderma* 130: 324– 333.

Grutter M, Flores E, Basaldud R, Ruiz-Suárez LG (2003) Open-path FTIR spectroscopic studies of the trace gases over Mexixo City. Atmos. Oceanic Opt. 16(3): 232-236.

Hall FE Jr (2004) Case Study: Environmental Monitoring Using Remote Optical Sensing (OP-FTIR) Technology at the Oklahoma City Air Logistics Center Industrial Wastewater Treatment Facility. *Fed. Facilities Environ. J.* 15: 21-37.

Harbeck S (2005) Characterisation and Functionality of SnO_2 Gas Sensors Using Vibrational Spectroscopy. Dissertation. available at: http://tobias-lib.uni-tuebingen.de/volltexte/2005/1693/pdf/Serpil_Harbeck_thesis_final_druck.pdf Accessed 2012 March 11.

Harrick NJ (1967) *Internal Reflection Spectroscopy*, Wiley-Interscience, New York.

Harris DB, Shores RC, Thoma ED (2007) Using Tnable Diode Lasers to Measure Emissions form Animal Housing and Waste Lagoons, U.S. EPA, NRMRL, Research Traingle Park,

NC. *16th Annual International Emission Inventory Conference – Emission Inventories: Integration, Analysis, and Commnications*. 14-17 May 2007, NC. 16 pp.

Haschberger P (1994) Remote measurement of trace gases from aircraft emissions using infrared spectroscopy, Mitt.- Dtsch. Forschungssanst. Luft-Raumfahrt (94-06, *Impact of Emisssions from Aircraft and Spacecraft Unpon the Atmosphere*): 100-105.

Hegde U, Chang TC, Yang SS (2003) Methane and Carbon Dioxide Emissions from Shan-chu-ku Landfill Site in Northern Taiwan. *Chemosphere* 52: 1275-1285.

Hong DW, Heo GS, Han JS, Cho SY (2004) Application of the open path FTIR with COL1SB to measurements of ozone and VOCs in the urban area. *Atmospheric Environment* 38: 5567-5576.

Huang WE, Hopper D, Goodacre R, Beckmann M, Singer A, Draper J (2006) Rapid characterization of microbial biodegradation pathways by FT-IR spectroscopy. *Journal of Microbiological Methods* 67: 273–280.

Janorkar AV, Metters AT, Hirt DE (2004) Modification of Poly(lactic acid) Films: Enhanced Wettability from Surface-Confined Photografting and Increased Degradation Rate Due to an Artifact of the Photografting Process, *Macromolecules* 37: 9151-9159.

Jäger F, Gluschke O, Doll R (2011) FTIR- and NDIR-Spectroscopy measurements on environmental air. How accurate are $^{13}CO_2$ isotope ratio and trace gas measurements with an outdoor instrument? *Geophysical Research Abstracts* Vol. 13, EGU2011-8715-1.

Jiang CH, Gamarnik A, Tripp CP (2005) Identification of lipid aggregate structures on TiO_2 surface using headgroup IR bands. *J. Phys. Chem.* 109: 4539-4544.

Johnson TJ, Profeta LTM, Sams RL, Griffith DWT, Yokelson RL (2010) An infrared spectral database for detection of gases emitted by biomass burning. *Vibrational Spectroscopy* 53: 97–102.

Jordan N, Foerstendorf H, Weiβ, Heim K, Schild D, Brendler V (2011) Sorption of selenium(VI) onto anatase: Macroscopic and microscopic characterization. *Geochimica et Cosmochimica Acta* 75(6): 1519-1530.

Kagann RH, Wang CD, Chang KL, Lu CH (1999) Open-Path FTIR Measurement of Criteria Pollutants and Other Ambient Species in an Industrial City. *Proc. SPIE* 3534: 140-149.

Kantar C. and Honeyman B. D. (2006) Citric acid enhanced remediation of soils contaminated with uranium by soil flushing and soil washing. *J. Environ. Eng.* 132: 247–255.

Kazansky VB, Sokolova NA, Bülov M (2004) DRIFT spectroscopy study of nitrogen sorption and nitrogen–oxygen transport co-diffusion and counter-diffusion in NaLSX and NaZSM-5 zeolites. *Microporous and Mesoporous Materials* 67: 283–289.

Khalizov AF, Cruz-Quinones M, and Zhang R (2010) Heterogeneous Reaction of NO_2 on Fresh and Coated Soot Surfaces. *J. Phys. Chem.* A 114: 7516–7524.

Kouyoumdjian H and Saliba NA (2006) Mass concentration and ion composition of coarse and fine particles in an urban area in Beirut: effect of calcium carbonate on the absorption of nitric and sulfuric acids and the depletion of chloride. *Atmos. Chem. Phys.* 6: 1865–1877.

Kumar R, Singh G, Pal AK (2005) Assessment of impact of coal and minerals related Industrial activities in Korba industrial belt of Chhattisgarh through spectroscopic techniques. *Mineral Processing Technology* 605-612.

Lefèvre G, Kneppers J, Fédoroff M (2008) Sorption of uranyl ions on titanium oxide studied by ATR-IR spectroscopy. *J. Colloid. Interface Sci.* 327(1): 15-20.

Lefèvre G, Noinville S, Fédoroff M (2006) Study of uranyl sorption onto hematite by in situ attenuated total reflection-infrared spectroscopy. *J. Colloid. Interface Sci.* 296(2): 608-613.

Lefèvre G (2004) In situ Fourier-transform infrared soectroscopy studies of inorganic ions adsorption on metal oxides and hydroxides. *Advances in Colloid and Interface Science* 107: 109-123.

Lehocký M, Drnovská H, Lapíková B, Barros-Timmons AM, Trindade T, Zembala M, Lapík L (2003) Plasma Surface Modification of Polyethylene. *Colloids Surf. A* 222: 125-131.

Levine SP, Russwurm GM (1994) Fourier-transform infrared optical remote sensing for monitoring airbone gas and vapor contaminants in the field, *TrAC. Trends Anal. Chem.* (Pers. Ed), 13(7): 258-262.

Lin C, Liou N, Sun E (2008) Applications of Open-Path Fourier Transform Infrared for Identification of Volatile Organic Compound Pollution Sources and Characterization of Source Emission Behaviors. *Journal of the Air & Waste management Association* 58(6): 821-828.

Lin C, Liou N, Chang PE, Yang JC, Sun E (2007) Fugitive coke oven gas emission profile by continuous line averaged open-path Fourier transform infrared monitoring. *J Air Waste Manag Assoc.* 57(4): 472-479.

Linker R (2010) Development of chemometric tools for FTIR determination of N-species in environmental systems, Final Report Submitted to the Grand Water Research Institute available at: http://gwri-ic.technion.ac.il/pdf/gwri_abstracts/2110/1.pdf. Accessed 2012 March10.

Linker R, Weiner M, Shmulevich I, Shaviv A (2006). Nitrate Determination in Soil Pastes using Attenuated Total Reflectance Mid-infrared Spectroscopy: Improved Accuracy via Soil Identification. *Biosystems Engineering* 94 (1): 111–118, doi:10.1016/j.biosystemseng.2006.01.014

Linker R, Shmulevich I, Kenny A, Shaviv A (2005), Soil identification and chemometrics for direct determination of nitrate in soils using FTIR-ATR mid-infrared spectroscopy. *Chemosphere* 61: 652–658.

Linker R, Kenny A, Shaviv A, Singher L, Shmulevich I (2004). FTIR/ATR nitrate determination of soil pastes using PCR, PLS and cross-correlation. Applied Spectroscopy, 58: 516–520.

Llewellyn PL, and Theocharis CR (1991) A diffuse reflectance fourier transform infra-red study of carbon dioxide adsorption on silicalite-I. *J. Chem. Technol. Biotechnol.* 52: 473–480. doi: 10.1002/jctb.280520405

Love SP, Goff F, Counce D, Siebe C, Delgado H (1998) Passive infrared spectroscopy of the eruption plume at Popocatépetl volcano, Mexico. *Nature* 396: 563–567; doi:10.1038/25109.

Ma X, Li L, Yang L, Su C, Wang K, Jiang K (2012a) Preparation of hybrid $CaCO_3$–pepsin hemisphere with ordered hierarchical structure and the application for removal of heavy metal ions *Journal of Crystal Growth* 338(1): 272-279

Ma Q, Liu Y, Liu C, Ma J, He H (2012b) A case study of Asian dust storm particles: Chemical composition, reactivity to SO_2 and hygroscopic properties. *Journal of Environmental Sciences* 24(1): 62–71.

Majedová J (2003) FTIR techniques in clay mineral studies – Review. *Vibrational Spectroscopy* 31: 1–10.

Mangoni ML, Papo N, Mignogna G, Andreu D, Shai Y, Barra D, Simmaco M (2003) Ranacyclins, a new family of short cyclic antimicrobial peptides: biological function, mode of action and parameters involved in target specificity. *Biochemistry* 42: 14023-14035.

Marshall TL, Chaffin CT, Hammaker RM, Fateley WG (1994) An intorduction to open-path FT-IR, Atmospheric monitoring. *Environ. Sci. Technol.* 28(5): 224A-232A.

McKelvy ML, Britt TR, Davis BL, Gillie JK, Graves FB, Lentz LA (1998) Infrared Spectroscopy. *Anal. Chem.* 70: 119R-177R.

Mecozzi M, Moscato F, Pietroletti M, Quarto F, Oteri F, Cicero AM (2009) Applications of FTIR spectroscopy in environmental studies supported by two dimensional correlation analysis. *Global NEST Journal*, 11(4): 593-600.

Minnich TR, Scotto RL (1999) Use of Open-Path FTIR Spectroscopy to Adrees Air Monitoring Needs During Site Remediations. Invited Article Published in „Remediati", Summer 1999, 1-16.

Mori T., Notsu K (2008) Temporal variation in chemical composition of the volcanic plume from Aso volcano, Japan, measured by remote FT-IR spectroscopy. *Geochem.* J. 42: 133–140.

Mori T, Notsu K (1997) Remote CO, COS, CO_2, SO_2, HCl detection and temperature estimation of volcanic gas. *Geophys. Res. Lett.* 24: 2047–2050.

Mori T, Notsu K, Tohjima Y, Wakita H (1993) Remote detection of HCl and SO_2 in volcanic gas from Unzen volcano, Japan. *Geophys. Res. Lett.* 20: 1355–1358.

Muckenhuber H, and Grothe H (2007) A DRIFTS study of the heterogeneous reaction of NO_2 with carbonaceous materials at elevated temperature. *Carbon*. 45 (2): 321–329.

Müller K, Foerstendorf H, Meusel T, Brendler V, Lefèvre G, Comarmond MJ, Payne TE (2012), Sorption of U(VI) at the TiO2–water interface: An in situ vibrational spectroscopic study. *Geochimica et Cosmochimica Acta.* 76: 191–205. http://dx.doi.org/10.1016/j.gca.2011.10.004

Müller K, Foerstendorf H, Brendler V, Bernhard G (2009) Sorption of Np(V) onto TiO_2, SiO_2, and ZnO: An in Situ ATR FT-IR Spectroscopic Study. *Environ. Sci. Technol.* 43(20): 7665–7670.

Mungasavalli DP, Viraraghavan T, Chunglin Y (2007) Biosorption of chromium from aqueous solutions by pretreated Aspergillus niger: batch and column studies. *Colloids. Surf. A Physicochem. Eng. Aspects* 301: 214-223.

Nault JR, Preston CM, Trofymow JAT, Fyles J, Kozak L, Siltanen M, Titus B (2009) Applicability of Diffuse Reflectance Fourier Transform Infrared Spectroscopy to the

Chemical Analysis of Decomposing Foliar Litter in Canadian Forests. *Soil Science.* 174(3): 130-142, DOI: 10.1097/SS.0b013e318198699a.

Noda I. and Ozaki I. (2005) Two-dimensional correlation spectroscopy. *Application in vibrational and optical spectroscopy*, John Wiley & Sons, UK

Notsu K, Mori T (2010) Chemical monitoring of volcanic gas using remote FT-IR spectroscopy at several active volcanoes in Japan. *Applied Geochemistry* 25: 505–512.

Ofner JN, Balzer N, Buxmann J, Grothe H, Schmitt-Kopplin P, Platt U and Zetzsch C (2012) Halogenation processes of secondary organic aerosol and implications on halogen release mechanisms. *Atmos. Chem. Phys. Discuss.*12: 2975–3017.

Oppenheimer C, Lomakina AS, Kyle PR, Kingsbury NG, Boichu M (2009) Pulsatory magma supply to a phonolite lava lake. *Earth and Planetary Science Letters* 284: 392–398, doi:10.1016/j.epsl.2009.04.04

Parikh SJ, Chorover J (2005) FTIR Spectroscopic Study of Biogenic Mn-Oxide Formation by *Pseudomonas putida* GB-1. *Geomicrobiology Journal*, 22:207–218.

Parolo ME, Savini MC, Vallés JM, Baschini MT, Avena MJ (2008) Tetracycline adsorption on montmorillonite: pH and ionic strength effects. *Applied Clay Science* 40: 179–186.

Pasilis SP, Pemberton JE (2008) Spectroscopic investigation of uranyl(VI) and citrate coadsorption to Al_2O_3. *Geochimica et Cosmochimica Acta* 72: 277–287.

Paton-Walsh C (2011). Remote Sensing of Atmospheric Trace Gases by Ground-Based Solar Fourier Transform Infrared Spectroscopy, *Fourier Transforms - New Analytical Approaches and FTIR Strategies*, Prof. Goran Nikolic (Ed.), ISBN: 978-953-307-232-6, InTech, Available from: http://www.intechopen.com/books/fourier-transforms-new-analytical-approaches-and-ftir strategies/remote-sensing-of-atmospheric-trace-gases-by-ground-based-solar-fourier-transform-infrared-spectroscopy Accessed 2012 March 12.

Perry SH, McKane PL, Pescatore DE, DuBois AE, Kricks RJ (1995) Maximizing the use of open-path FTIR for 24-h monitoring around the process area of an industrial chemical facility. *AWMA Conference on Optical Remote Sensing for Environmental and Process Monitoring SPIE* 2883: 333–344.

Pironon J, de Donato Ph, Barrès O, Garnier Ch (2009) On-line greenhouse gas detection from soils and rock formations. *Energy Procedia* 1: 2375–2382.

Pokrovsky OS, Martinez RE, Golubev SV, Kompantseva EI, Shirokova LS (2008) Adsorption of metals and protons on Gloeocapsa sp. cyanobacteria: A surface speciation approach. *Applied Geochemistry* 23: 2574–2588.

Prince A P, Kleiber P, Grassian V H, Young M A (2007) Heterogeneous interactions of calcite aerosol with sulfur dioxide and sulfur dioxide-nitric acid mixtures. *Physical Chemistry & Chemical Physics* 9(26): 3432–3439.

Puckrin E, Evans WFJ, Adamson TAB (1996) Measurement of tropospheric ozone by thermal emission spectroscopy. *Atmospheric Environment* 30(4): 563-568.

Raphael L (2011). Application of FTIR Spectroscopy to Agricultural Soils Analysis, Fourier Transforms - *New Analytical Approaches and FTIR Strategies*, Prof. Goran Nikolic (Ed.), ISBN: 978-953-307-232-6, InTech, Available from:

http://www.intechopen.com/books/fourier-transforms-new-analytical-approaches-and-ftir-strategies/application-of-ftir-spectroscopy-to-agricultural-soils-analysis. accessed 2012 March 12.

Redden G, Bargar J, and Bencheikh-Latmani R (2001) Citrate enhanced uranyl adsorption on goethite: an EXAFS analysis. *J. Colloid Interface Sci.* 244: 211–219.

Rintoul L, Panayiotou H, Kokot S, George G, Cash G, Frost R, Bui T, Fredericks P (1998) Fourier transform infrared spectrometry: a versatile technique for real world samples. *Analyst* 123: 571–577.

Rumpel C, Janik LJ, Skjemstad JO, Kögel-Knabner I (2001) Quantification of carbon derived from lignite in soils using mid-infrared spectroscopy and partial least squares. *Organic Geochemistry* 32(6): 831–839.

Russwurm GM (1999) Compendium of Methods for the Determination of Toxic Organic Compounds in Ambient Air – second edition. Long-Path Open-Path Fourier Transform Infrared Monitoring Of Atmospheric Gases – Method TO-16: 16-1-16-41. available at http://www.epa.gov/ttnamti1/files/ambient/airtox/tocomp99.pdf accessed 2012 March 12.

Russwurm GM, Childers JW (1996) FT-IR Open-Path Monitoring Guidance Document. U. S. Environmental Protection Agency, Research Triangle Park, NC, EPA/600/R-96/040, April 1996.

Santos C, Fraga ME, Kozakiewicz Z, Lima N (2010) Fourier transform infrared as a powerful technique for the identification and characterization of filamentous fungi and yeast. *Research in Microbiology* 161: 168-175.

Schneider M, Romero PM, Hase F, Blumenstock T, Cuevas E, Ramos R (2010) Continuous quality assessment of atmospheric water vapour measurement techniques: FTIR, Cimel, MFRSR, GPS, and Vaisala RS92. *Atmos. Meas. Tech.* 3: 323-338, doi:10.5194/amt-3-323-2010.

Schuttlefield J, Al-Hosney H, Zachariah A, and Grassian VH (2007a) Attenuated Total Reflection Fourier Transform Infrared Spectroscopy to Investigate Water Uptake and Phase Transitions in Atmospherically Relevant Particles. *Appl. Spectrosc.* 61: 283-292.

Schuttlefield JD, Cox D, and Grassian VH (2007b) An investigation of water uptake on clays minerals using ATR-FTIR spectroscopy coupled with quartz crystal microbalance measurements. *J. Geophys. Res.* 112. D21303, doi:10.1029/2007JD008973.

Scott RPW *Analytical Spectroscopy* available at: http://www.analyticalspectroscopy.net/ap3-11.htm Accessed 2012 March 11.

Segal-Rosenheimer M and Dubowski Y (2007) Heterogeneous Ozonolysis of Cypermethrin Using Real-Time Monitoring FTIR Techniques. *J. Phys. Chem.* C 111: 11682-11691.

Sethuraman A, Belfort G (2005) Protein structural perturbation and aggregation on homogeneus surfaces. *Biophys. J.* 88: 1322-1333.

Shaka H and Saliba N (2004) Concentration measurements and chemical composition of PM10-2.5 and PM2.5 at a coastal site in Beirut, Lebanon. *Atmospheric Environment* 38: 523 – 531.

Sibley KJ,. Brewster GR, Astatkie T, Adsett JF, and Struik PC (2010). In-Field Measurement of Soil Nitrate Using an Ion-Selective Electrode, *Advances in Measurement Systems,*

Milind Kr Sharma (Ed.), ISBN: 978-953-307-061-2, InTech, Available from: http://www.intechopen.com/books/advances-in-measurement-systems/in-field-measurement-of-soil-nitrate-using-an-ion-selective-electrode Accessed 2012 April 30.

Simonescu CM, Dima R, Ferdeş M, Meghea A (2012) Equilibrium and Kinetic Studies on the Biosorption of Cu(II) onto *Aspergillus niger* Biomass. Rev. Chim. (Bucharest). ISSN 0034-7752 63(2): 224-228.

Simonescu CM, Deleanu C, Bobirică L, Melinescu A, Giurginca M (2005) Bentonite and Na-bentonite used in ammonium removal from wastewaters. *Proceedings of 14th Romanian International Conference of Chemistry and Chemical Engineering* (RICCCE XIV). ISBN: 973-718-284-7, ISBN: 973-718-288-X, Ed. Printech, Bucharest 22 – 24 September 2005, 6: S06-262.

Simonescu CM, Ferdeş M Fungal biomass for Cu(II) uptake from aqueous systems. Polish Journal of of Environmental Studies. *in press*

Smidt E, Böhm K and Schwanninger M (2011) The Application of FT-IR Spectroscopy in Waste Management, Fourier Transforms - *New Analytical Approaches and FTIR Strategies*, Prof. Goran Nikolic (Ed.), ISBN: 978-953-307-232-6, InTech, Available from: http://www.intechopen.com/books/fourier-transforms-new-analytical-approaches-and-ftir-strategies/the-application-of-ft-ir-spectroscopy-in-waste-management Accessed 2012 March 12.

Smidt E, Meissl K (2007) The applicability of Fourier transform infrared (FT-IR) spectroscopy in waste management. *Waste Management* 27: 268–276.

Smith BC (2011) *Fundamentals of Fourier Transform Infrared Spectroscopy*, 2nd Edn, CRC Press, Boca Raton, FL, USA.

Snabe T, Petersen BS (2002) Application of infrared spectroscopy (attenuated total reflection) for monitoring enzymatic activity on substrate films. *Journal of Biotechnology* 95:145-155.

Spaccini R, Piccolo A, Haberhauer G, Stemmer M, and Gerzabek MH (2001) Decomposition of maize straw in three European soils as revealed by DRIFT spectra of soil particle fractions. *Geoderma* 99:245–260.

Thorn TG, Marshall TL, Chaffin CT (2001) Open-Path FTIR Air Monitoring of Phosphine around Large Fumigated Structures; *Field Anal. Chem. Technol.* 5: 116-120.

Tomšič B, Simončič B, Orel B, Vilčnik A, Spreizer H (2007) Biodegradability of cellulose fabric modified by imidazolidinone. *Carbohydrate Polymers* 69: 478–488.

Tremblay L, and Gagné J-P (2002) Fast quantification of humic substances and organic matter by direct analysis of sediments using DRIFT spectroscopy. *Anal. Chem.* 74:2985–2993.

Tsai YI, Kuo S-C (2006) Development of diffuse reflectance infrared Fourier transform spectroscopy for the rapid characterization of aerosols. *Atmospheric Environment* 40: 1781–1793.

Tso TL, Chang SY (1996) Unambiguous identification of fugitive pollutants and the determining of annual emission flux as a diurnal monitoring mode using open-path Fourier transform infrared spectroscopy. *Analytical Sciences* 12: 311–319.

Ullerstam M, Johnson M S, Vogt R, Ljungström E (2003) DRIFTS and Knudsen cell study of the heterogeneous reactivity of SO_2 and NO_2 on mineral dust. *Atmospheric Chemistry and Physics* 3(6): 2043–2051.

Ullerstam M, Vogt R, Langer S, Ljungström E (2002) The kinetics and mechanism of SO_2 oxidation by O_3 on mineral dust. *Physical Chemistry & Chemical Physics* 4(19): 4694–4699.

Valyon J, Lónyi F, Onyestyák G, Papp J (2003) DRIFT and FR spectroscopic investigation of N_2 and O_2 adsorption on zeolites. *Microporous and Mesoporous Materials* 61(1–3): 147–158, ZEOLITE '02 (Proceedings of the 6th Internationanl Conference on the Occurrence, Properties and Utilization of Natural Zeolites), http://dx.doi.org/10.1016/S1387-1811(03)00362-7.

Veres P (2005) FTIR Analysis of Particulate Matter Collected on Teflon Filters in Columbus, OH - A Senior Honors Thesis - The Ohio State University June 2005.

Verma SK, Deb MK, Verma D (2008) Determination of nitrogen dioxide in ambient air employing diffuse reflectance Fourier transform infrared spectroscopy. *Atmospheric Research* 90: 33–40.

Verma SK, Deb MK (2007a) Nondestructive and rapid determination of nitrate in soil, dry deposits and aerosol samples using KBr-matrix with diffuse reflectance Fourier transform infrared spectroscopy (DRIFTS). *Analytica Chimica Acta* 582: 382–389.

Verma SK, Deb MK (2007b) Direct and rapid determination of sulphate in environmental samples with diffuse reflectance Fourier transform infrared spectroscopy using KBr substrate. *Talanta* 71: 1546–1552. doi:10.1016/j.talanta.2006.07.056

Walter WT, Perry SH, Han JS, Park CJ (1999) Open-Path FTIR Ozone Measurements in Korea; *Proc. SPIE* 3534: 133-139.

Wang S-L, Lee J-F (2011) Reaction mechanism of hexavalent chromium with cellulose. *Chemical Engineering Journal* 174: 289– 295.

Workman Jr. JJ (1999) Review of Process and Non-invasive Near-Infrared and Infrared Spectroscopy: 1993–1999. *Applied Spectroscopy Reviews* 34(1&2): 1–89.

Wu C-F, Chen Y-L, Chen C-C, Yang T-T, Chang PE (2007) Applying open-path Fourier transform infrared spectroscopy for measuring aerosols, *Journal of Environmental Science and Health*, Part A. 42(8): 1131-1140.

Xu L, Wang JN, Meng Y, Li AM (2012) Fast removal of heavy metal ions and phytic acids from water using new modified chelating fiber. *Chinese Chemical Letters* 23: 105–108.

Yee N, Benning LG, Phoenix VR, Ferris FG (2004) Characterization of Metal-Cyanobacteria Sorption Reactions: A Combined Macroscopic and Infrared Spectroscopic Investigation. *Environ. Sci. Technol*. 38: 775-782.

Yoon TH, Johnson SB, Musgrave CB, Jr Brown GE (2004) Adsorption of organic matter at mineral/water interfaces: I. ATR-FTIR spectroscopic and quantum chemical study of oxalate adsorbed at boehmite/water and corundum/water interfaces. *Geochimica et Cosmochimica Acta* 68(22): 4505-4518.

Zhang X Y, Zhuang G S, Chen J M, Wang Y, Wang X, An Z Set al. (2006) Heterogeneous reactions of sulfur dioxide on typical mineral particles. *Journal of Physical Chemistry* B,110(25): 12588–12596.

Zhao Y and Chen Z (2010) Application of Fourier Transform Infrared Spectroscopy in the Study of Atmospheric Heterogeneous Processes. *Applied Spectroscopy Reviews* 45: 63-91.

Zimmermann M, Leifeld J, Fuhrer J (2007) Quantifying soil organic carbon fractions by infrared-spectroscopy. *Soil Biology & Biochemistry* 39: 224–231.

Zimmermann M, Leifeld J, Schmidt MWI, Fuhrer J (2004) Characterization of Soil Properties by DRIFT-Spectroscopy. Eurosoil Congress, Freiburg, D.Poster and Proceedings. available at http://www.bodenkunde2.uni-freiburg.de/eurosoil/abstracts/id101_Zimmermann_full.pdf Accessed 2012 March11.

Chapter 3

The Use of the Spectrometric Technique FTIR-ATR to Examine the Polymers Surface

Wieslawa Urbaniak-Domagala

Additional information is available at the end of the chapter

http://dx.doi.org/10.5772/48143

1. Introduction

The development of material engineering is accompanied by a growing demand for routine, nondestructive techniques for material and product testing. These techniques are to be used for the assessment of chemical and physical structure of new materials as well as for a systematic control of their manufacturing processes. Nowadays nanotechnologies fulfill a particular role in creating new materials of nanometric dimensions. The products of nanotechnology are made in various forms, mostly such as coatings and fibers. Coatings are of great practical importance while deposited on conventional substrates, such as metals, ceramics and polymers to impart new functions, e.g. anticorrosive, reflexive, sensory properties, etc., to them. Coatings are mostly made of polymers and hence their functional properties and durability mainly depend on polymer chemical and supermolecular structure. The current control tests of the chemical properties and supermolecular characteristic of materials are carried out with the use of IR absorption spectroscopy. Currently, these are dedicated to test the surface of materials. This paper concerns the spectroscopic technique FTIR used to test the surface of polymeric materials and coatings formed on polymeric substrates. The general characteristics, advantages and drawbacks of this technique in testing polymer surfaces have been presented.

2. Technique of infrared absorption spectroscopy

Material testing by the technique of IR spectroscopy consists in making a spectrum of radiation energy absorbed by material molecules and interpreting the spectrum obtained. IR radiation within the wavelength range from 2.5 mm to 15 mm (the wave number from 4000 cm^{-1} to 666 cm^{-1}) is selectively absorbed by material molecules and converted into their oscillatory energy. The oscillations of molecules are of various characters, connected with their chemical structure, and depend on the type of bonds (frequency increases with increasing bond energy), relative atomic weights (frequency decreases with increasing atomic weight), spatial position of atoms in a molecule, intra- and intermolecular interaction forces. During absorption, various vibration modes are generated that can be ranged with respect to energy in the following order: stretching vibration > bending vibration > oscillatory/torsional vibration. Vibration modes are active in IR only when the frequency of radiation coincides with the own frequency of molecule oscillation (resonance) and the dipole moments of molecules change in the same direction as the electric vector of IR radiation wave.

In the absorption spectroscopy techniques, IR radiation, after passing the material, where molecules selectively absorb radiation quanta, the absorption spectrum is recorded in the form of changes in the IR spectra radiation intensity as a function of radiation energy. The intensity of a beam after passing through sample (I), transmittance (T) or absorbance (A) is assumed as a measure of absorption. If the intensity of the primary incident beam on a sample is equal to I_0, the relation between intensity, transmittance and absorbance is as follows: $T = I/I_0$, $A = \log (I_0/I) = -\log T$. Energy is expressed in eV, but mostly practical parameters such as IR radiation wavelength (λ, nm), wave number ($\nu=1/\lambda$, cm^{-1}) and radiation frequency (υ, Hz) are used to express energy.

Originally, tests and recording the IR radiation absorption spectra of samples were performed by means of two-beam diffraction spectrometers. Modern technical solutions of IR spectrometers consist in replacing the reticular monochromators with interferometers, which considerably increases the sensitivity of spectrometers (a high value of the signal to noise ratio is obtained), making it possible to shorten the spectrum recording and to obtain its good definition. Moreover, there occur the transformation and ordering of the interferogram obtained to the frequency domain by the use of Fourier Transform (FTIR). The high resolving power of spectrometer makes it possible to record complicated spectra of materials, spectra mixing, the distinction of band derived from crystalline and non-crystalline areas and performing static and dynamic tests.

The IR spectrometry technique can be used in two variants: transmission and reflection. The transmission version is used to test the effects of IR radiation absorption in the volume of sample. It is possible to test samples in any form: solid, liquid and gaseous with the use of an appropriate procedure. Gases and liquids are placed in special cuvettes with windows, made of transparent materials for IR radiation (e.g. ionic crystals: KBr, NaCl). The spectra of solids can be measured using previously prepared specimens on quartz plates, in a suspension in liquid paraffin or in the form of tablets made of KBr. If the object tested is sufficiently thin and transparent, its spectra are measured directly on a sample. The transmission technique cannot be used for materials that strongly absorb IR radiation and to test local areas of sample such as surface. In the sixties of the last century, the reflection variant was developed, so-called Attenuated Total Reflection (ATR), which makes it possible to test specific version of samples. The ATR-IR uses the phenomenon of a complete reflection during the transition of IR radiation from an optically denser medium (prism) to thinner medium (sample). A sample is placed on the IR-transparent prism surface with a refractive index being always higher than that of the sample (Figure 1). The radiation beam is directed by one of the prism wall to the prism-sample interface at angle θ higher than the limiting. Under these conditions, a complete reflection occurs at the internal prism side and the beam reflected comes out through the second prism wall, where the beam intensity and absorption spectrum are recorded.

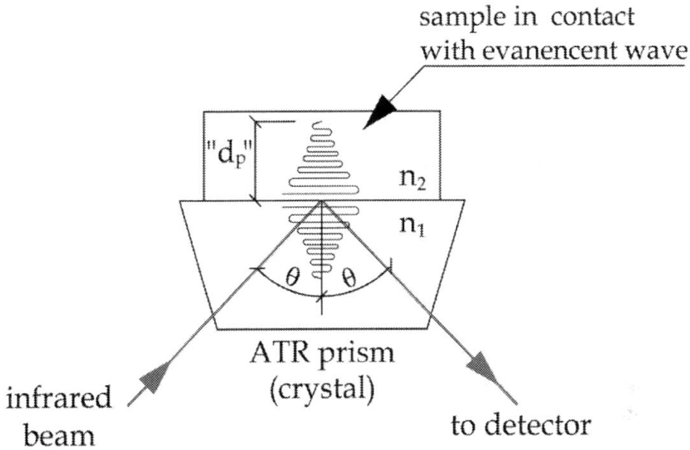

Figure 1. The schematic representation of infrared beam reflected on the crystal - sample interface in FTIR-ATR spectrometer. (on the base http://www.sprpages.nl/SprTheory/SprTheory.htm)

During the total internal reflection in the optically thinner medium (sample) is formed an electromagnetic wave, so-called evanescent wave that fulfills the condition of the continuity of electromagnetic field vectors at the interface of media with various wave refractive indices, n_1 and n_2 (Fornel, 2000). The IR evanescent wave has two wave vector components: parallel component to the interface of the contact between materials, under influence of which the wave propagates along surface resulting in the formation of so-called Goos-Hänchen's displacement (Goos&Hänchen, 1947), and perpendicular component, under the influence of which the wave propagates in the optically thinner medium in the direction perpendicular to the surface and exponentially disappears. The evanescent wave penetration depth, „dp", in sample depends on the IR radiation wavelength (λ), incident angle, (θ), prism refractive index, (n_1), and sample refractive index in relation to the prism ($n_{2,1}$) and is expressed by the following equation (Dechant, 1972):

$$d_p = \frac{\lambda / n_1}{2\pi\sqrt{(\sin^2\theta - n_{21}^2)}} \tag{1}$$

Along the path of IR evanescent wave the sample selectively absorbs energy to decrease the intensity of radiation. The weakened wave returns to the prism and then to an IR detector.

There the system generates an FTIR-ATR absorption spectrum characteristic of the given sample. The FTIR-ATR absorption spectrum slightly differs from that obtained by the transmission method. The differences concern the intensity and frequency of absorption peaks characteristic of chemical groups in view of the phenomenon of reflection, e.g. Goos-Hänchen's displacement. Thus it is necessary to take corrective action that can be realized automatically. The penetration depth of IR beam can be controlled within some range by selecting an appropriate prism (selection of the refractive index) and the incident angle of beam. The commonly used prisms are made of diamond, germanium, silicon and ZnSe, whose refractive indices are equal to 2.4, 4.0, 3.4 and 2.4, respectively, and the beam penetration depths: 2.03 μm, 0.67 μm, 0.84 μm and 2.03 μm, respectively, at ν = 1000 cm^{-1} (Material Thermo Scientific Smart ITR). During testing sub-micrometric coating, the beam penetrates a higher depth than the coating depth and also passes to the substrate, on which the coating is deposited. The absorption spectrum then constitutes a superposition of the spectrum of coating material and substrate. In such cases, qualitative analysis is carried out, which takes into account the absorption spectrum of substrate.

The basic requirement for ATR technique is to place a sample in direct contact with the prism as only such conditions allow the IR evanescent wave to penetrate the sample surface layer. Moreover, there should be a considerable difference between the refractive indices of prism and sample to get the phenomenon of internal reflection occurred.

The drawback of ATR technique is a relatively low sensitivity and susceptibility to the effect of environmental conditions, which makes it necessary to calibrate the IR spectrum. Modern spectrometers have an option of automatic computer-aided spectrum correction. ATR technique has numerous advantages. FTIR-ATR shows the features of a routine method for testing the chemical and physical surface structure of materials such as polymers, films and membranes provided that these well adhere to the crystal. Tests with a modern instrumentation are characterized by a high reproducibility (better than 0.1%) (Urbanczyk, 1988). FTIR-ATR makes it possible to record spectra within a wider frequency range of IR radiation than transmission spectroscopy owing to the lack of limitations caused by the absorption of cuvette windows. An important advantage of this technique is the possibility of recording spectra *in situ* and *in vivo*, e.g. in testing biological objects and using it as a diagnostic tool in medicine.

In this work, the FTIR-ATR technique was used to analyze the surfaces of modified polymers and to test the polymeric layers deposited on substrates.

3. Examples of testing polymers by FTIR-ATR

Tests were carried out by means of a single-beam FTIR-Nicolet 6700 spectrometer from Thermo Scientific, equipped with a diamond crystal (refractive index n = 2.4). IR spectra were recorded as changes in absorption as a function of wave number ranging from 600 cm^{-1} to 4000 cm^{-1}. A DTGS KBR detector was used. The following measurement technical conditions were used: measurement recording accuracy - 4 cm^{-1}, mirror travel rate - 0.31 cm^{-1}/s, aperture - 50, minimal scans number – 32.

3.1. Assessment of the modification effects on the surface of polypropylene (PP) films and nonwovens

Polypropylene products are commonly used in commodity production due to their special chemical properties (resistance to organic and inorganic solvent, hydrophobic properties) and physical characteristics (lightness, mechanical strength, electro- and thermal insulating capabilities). In the methods of making products such as composites with the use of PP films or fibers as reinforcing components, a serious drawback of these materials is their low free surface energy, which results in weak molecular interactions between the composite components. The free energy of PP material surface can be increased by creating new functional chemical groups in the material surface layer. This task has been fulfilled by exploring different approaches such as chemical, electrochemical, physical and plasma methods . The effectiveness of the methods used was assessed by means of the FTIR-ATR technique.

The moleculare structure of polypropylene is the same in the use of films and nonwovens products:

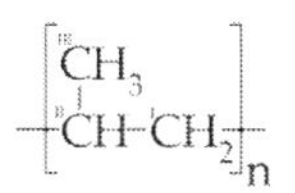

In one unit of PP molecule chain are tree atoms of carbon, in the form of different groups: -CH_2- ; >CH- ; and –CH_3. Each of them is correlated in IR spectra with the suitable absorption peak by definite wavenumber values (Figure 2). The proper characteristic, concerning this correlation is presented in Table 1.

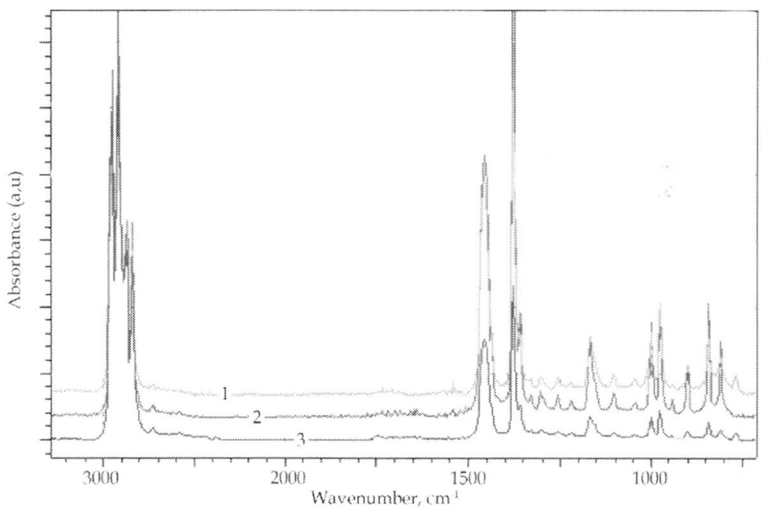

Figure 2. FTIR-ATR spectra of commercial PP films : non-oriented, non-crystalline PP Cast (Sample 1); bidirectionally oriented, crystalline PP AG (Sample 2) and PP nonwoven (Sample 3).

IR spectra of PP for film and nonwoven can differ between themselves only in defiles (the differences in shape and intensity of peaks), because in ATR technique, the contact of the samples with measure crystal, can be different for different structure of samples (film – continuous structure, nonwoven – porous structure). The explanation of this problems shown as an example at Figure 3.

Wave number, cm^{-1}	Absorbing group and type of vibration
2916	νa (CH_2)
2959	νa (CH_3)
2881	νs (CH_3)
2841	νs (CH_2)
1460	δa (CH_3)
1376	δs (CH_3)
1357,	γw (CH_2- CH)
1328	γw (CH_2 - CH)
1302, 1224, 941	Carbon lattice pulsation
1170, 1153	γw (CH_3), δ (CH2), δ (CH)
975, 899,	γr (CH_3), νr (CH2), νr (CH)
841, 810	γr (CH2), νr (CH), νr (CH_3)
765	γw (CH_2)

*) νs - stretching vibration symmetrical and νa -asymmetrical, δs - deformation vibration symmetrical and δa – asymmetrical, γw - wagging vibration, γr – rocking vibration

Table 1. IR absorption bands of Polypropylene (Urbanczyk, 1988; Rau, 1963)

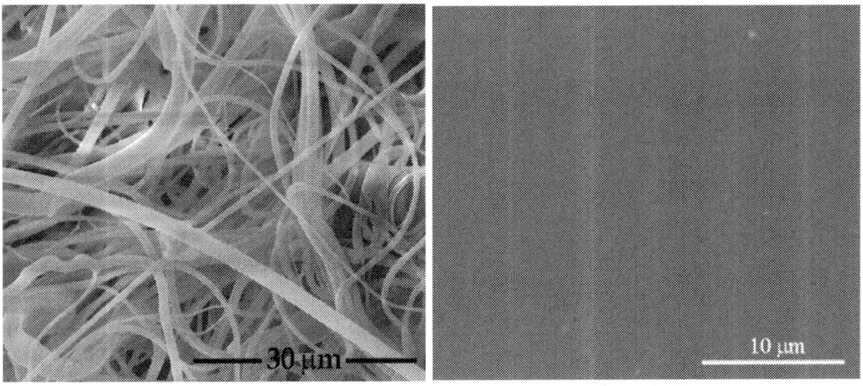

Sample A Sample B

Figure 3. SEM images of type surface structure : PP nonwoven fabrics (Sample A), PP Cast film (Sample B)

3.1.1. *Effects of PP film modification*

Commercial PP, non-oriented, non-crystalline (PP Cast) and bidirectionally oriented, crystalline (PP-AG) films were modified in media of strong oxidants, such as: 3M nitric acid, 30% hydrogen peroxide, and a saturated solution of potassium dichromate in 70% sulfuric acid ($K_2Cr_2O_7$+H_2SO_4). The electrochemical oxidation was carried out with the use of anolyte ($AgNO_3$ solution in nitric acid) and catholyte (nitric acid solution). The physical modification of PP was performed by means of a Xenotest apparatus, irradiating PP film with UV radiation according to EN ISO 105-B02:2006 (Urbaniak-Domagala, 2011). Plasma modification processes were carried out with the use of RF glow discharge of special gases under decreased pressure (Urbaniak-Domagala, 2011). Figures 4, 5 show the FTIR-ATR spectrograms of the PP film surfaces after oxidation compared with unmodified PP films. In the FTIR-ATR spectrograms of the PP film surface layer, one can observe absorption bands that are consistent with those of isotactic PP obtained by the authors mentioned in Table 1.

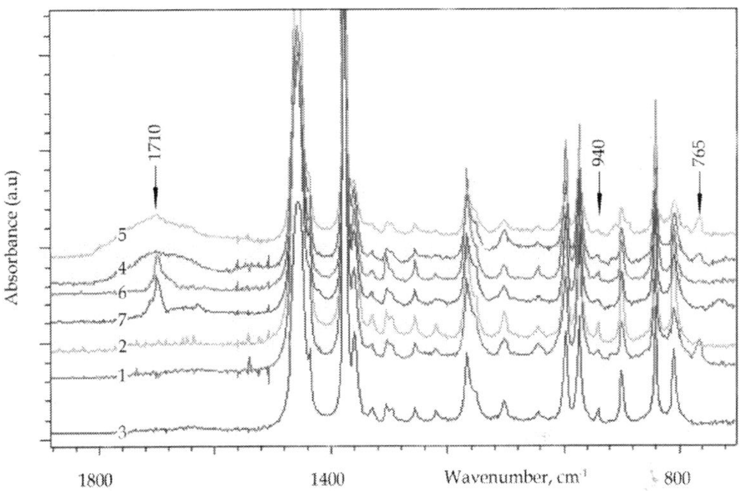

Figure 4. FTIR-ATR spectra of commercial PP films before and after oxidation. 1- PP Cast untreated, 2 - PP AG untreated. Samples 3÷7 PP Cast modified: by using electrochemical method, current intensity: 100 mA/cm^2, 30 min. (Sample 3), UV treatment (Xenotest) 170h (Sample 4), $K_2Cr_2O_7$+H_2SO_4 solution at 70°C, 3 min. (Sample 5), 3M nitric acid at 20°C, 24 h (Sample 6), 30% hydrogen peroxide at 20°C, 1 h (Sample 7).

Moreover, the spectrograms of PP surface layer oxidized by chemical methods show a new absorption band within the wave number range of (1730 – 1680) cm^{-1} that corresponds to a carbonyl group formed in a oxidizing medium as a results of the nucleophilic substitution of PP, mainly at the tertiary carbon atom: - CH_2 – C < R H – CH_2 – (the substitution susceptibility of the tertiary, secondary and primary carbon is 7000: 1100: 1, respectively) (Wiberg & Eisenthal, 1964). The absorption maximum of carbonyl group is slightly shifted depending on the type of oxidizing medium.

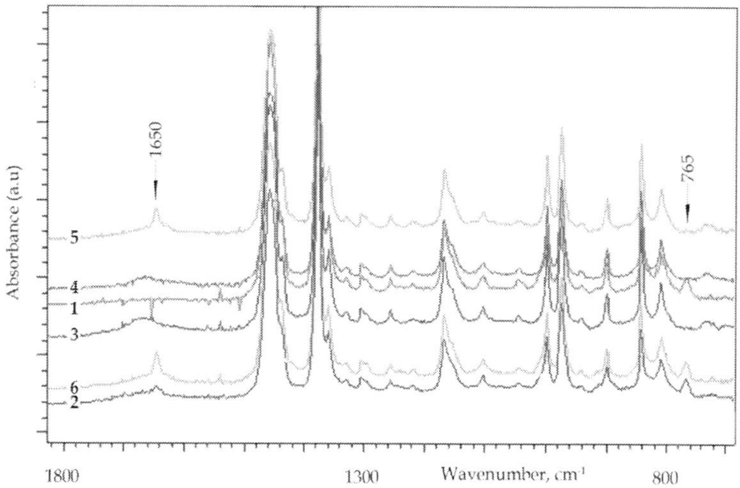

Figure 5. FTIR-ATR spectra of PP Cast films before and after plasma oxidation. Sample 1 - PP untreated, Sample 2. PP treated Ar plasma p=0.05Tr, power 300W, t=10 min., Sample 3 - PP treated Acetic Acid Vapour plasma: p=0.05Tr 300W, t=10 min., Sample4 - PP treated Water Vapour plasma: p=0.05Tr; 300W, 10min, Sample 5 - PP treated Air plasma: p=0.05Tr; 300W. t= 5min., Sample 6 - PP treated Air plasma: p=0.05Tr; 300W, t= 10min.

The absorption band of carbonyl group in the PP spectrum is broad, which can indicate the presence of carbonyl group in various products of oxidation, such as aldehydes and ketones (Carlsson & Wiles, 1969): 1700 cm^{-1}absorption (>C = CH-CO-OH), 1710 cm^{-1}absorption (-CO-OH), 1715 cm^{-1}absorption (>C = O), 1718 cm^{-1} absorption (-CCH_3 –CH_2 –CO- CH_2- CH_3), 1726 cm^{-1}absorption (-CCH_3 –CH_2 –CO- CH_3).

The spectrogram of oriented and crystalline PP AG shows no absorption band of carbonyl group despite the fact that the determination of the contact angle of PP surface wetted with polar liquids indicated an increase in free energy (Urbaniak-Domagala, 2011). One may assume that the active center concentration is too low for the FTIR-ATR method. The spectrogram of PP oxidized by chemical methods also indicates changes in two bands at 940 cm^{-1} and 765 cm^{-1} (Figure 4).

The first one indicates the skeleton vibration of mer links with a relative phase shift of 2/3, being mainly characteristic of the crystalline phase (Rau, 1963). In the case of PP AG film, this band is intensive, while in PP Cast, it decreases and after oxidation is absent, which can indicate that the PP surface layer becomes amorphous due to the oxidation process. The absorption band at 765cm^{-1} is characteristic of non-crystalline PP, caused by the deformation vibration of methylene group (–CH_2 –) (Kazicina.&Kupletska, 1976). This band is absent in the spectrogram of PP after oxidation, which can be due to the decrease in the number of methylene groups caused by the degradation of the polymer in its surface layer. This band is also absent in crystalline PP due to spherical limitations caused by a long-range order.

FTIR-ATR absorption spectra (Figure 5) present the chemical effects of plasma on PP film. The gases used in this process included: argon and air and vapors of acetic acid and water under optimal conditions of plasma treatment (time and power applied to the system).

The spectra of the plasma-treated PP film show a new absorption band within the range of wave numbers of (1640 ÷ 1660) cm^{-1} , which can indicate the formation of carbonyl group, >C=O, (valence vibration) as well as –C=C- groups (valence vibration) (Kazicina&Kupletska, 1976). The prolongation of plasma treatment and increase in power leads to the increase in the IR radiation intensity of the band of new functional groups. The position of IR absorption maximum slightly shifts depending on the plasma composition. These new active centers can be regarded as a result of PP surface oxidation with plasma particles. In the case of Ar plasma, the effect of surface functionalization can result from the so-called post-treatment process (Guruvenket et al, 2004). The results obtained indicate a particular activity of air plasma as oxidizing medium for polypropylene.

3.1.2. *Effects of PP nonwovens modification*

FTIR-ATR was also used to assess the effects of plasma-treated PP nonwovens. PP melt-blown nonwovens (surface weight: 80 g/m^2, average thickness: 1.5mm) made of PP fibers with an average thickness of 2.12 μm were modified by means of synthetic air plasma to form chemically active centers on the PP fiber surface.

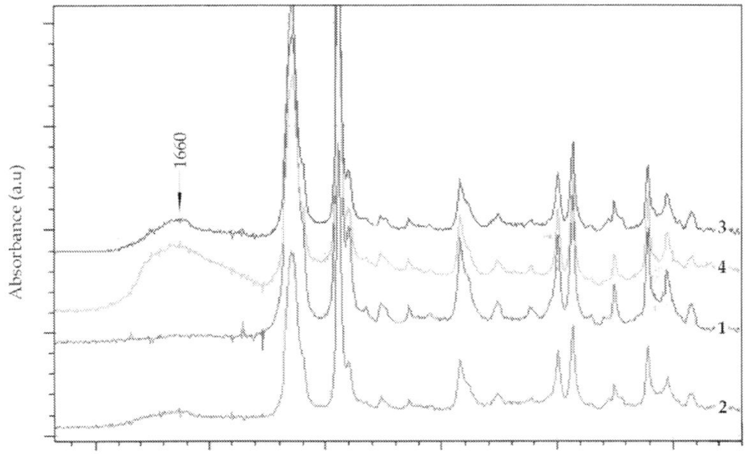

Figure 6. ATR IR spectra of PP nonwoves untreated (Sample 1) and air plasma treated, pressure 0.1Tr : Sample 2 - power 50W, time 5min. Sample 3- power 100W, time 5min., Sample 4 - power 100W, time 10min

The FTIR-ATR spectrogram of the air plasma-treated PP nonwoven shows two broad bands at 1660 cm^{-1} and 3320 cm^{-1} that indicate the formation of carbonyl group >C=O and hydroxyl group – OH (Kazicina.&Kupletska, 1976). One may assume that the air plasma oxidizes the fiber surface with the aid of reactive oxygen, peroxide and nitrogen groups that together with electrons react with the PP fiber surface causing not only the etching of surface layer but also its functionalization.

3.2. Polymeric coatings deposited on PP nonwovens

Nonwovens constitute a specific substrate for depositing thin polymeric layers. The nonwoven surface is developed to an extent dependent on the diameter of elementary fibers, density of their distribution and the formation technique used. The melt-blown PP nonwovens (see section 3.1.2.) were coated with thin layers of plasma polymers in a methane plasma and in hexamethyldisiloxane (HMDSO: O-(Si-$(CH_3)_2$) vapors [Urbaniak-Domagala et al, 2010). As a result of this process, the nonwoven surface was covered with a plasma polymer layer with a thickness of about 100 nm. SEM photographs (Figure 7) indicate that the coating obtained shows a character of a continuous film fitted to the uneven nonwoven surface covering only the elementary fibers in the near-surface nonwoven layer.

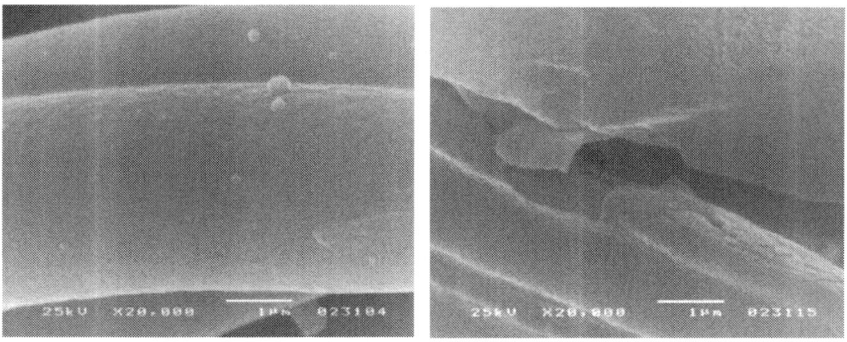

Sample A Sample B

Figure 7. SEM images of PP fibers at nonwoven fabrics, treated methane plasma (Sample A), treated HMDSO plasma (Sample B). Plasma process time 10 min., pressure 0.05Tr, power 100W.

The FTIR-ATR spectrogram of the methane plasma-modified nonwoven surface (Figure 8) indicates that the layer chemical structure has a character of a hydrocarbon polymer as the PP substrate. The broad band with a maximal absorption at 1650 cm^{-1} can be assumed as a post-treatment effect (Guruvenket et al, 2004).

The IR-ATR spectrogram of the HMDSO plasma-modified nonwoven indicates that the layer deposited has a chemical structure of a SiOC:H polymer (Creatore et al., 2002) and contains intensive absorption bands at 800 cm^{-1}, 841 cm^{-1}, 1040 $cm^{-1}$and 1256cm^{-1} (Table 2) being characteristic of chemical groups containing silicon (Borvon et al., 2002).

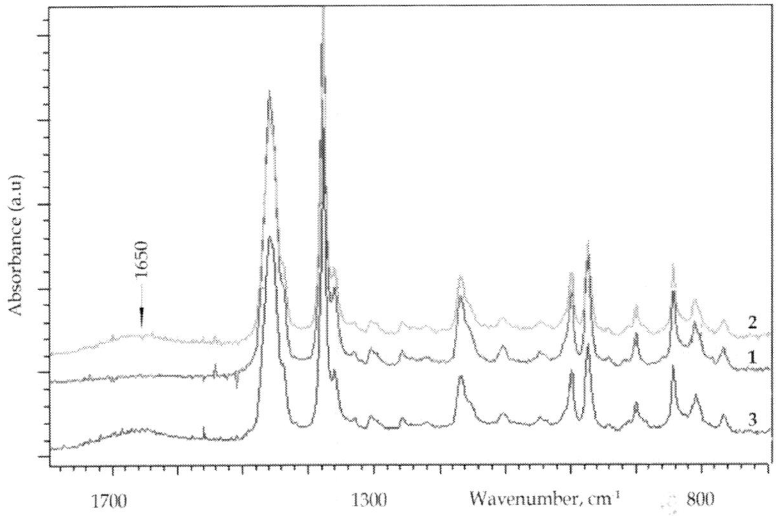

Figure 8. FTIR-ATR spectra of PP nonwoves. Sample 1- untreated, Sample 2 - methane plasma treated: power 25W, pressure 0.05Tr, Sample 3 - methane plasma treated: power 100W, pressure 0.05Tr.

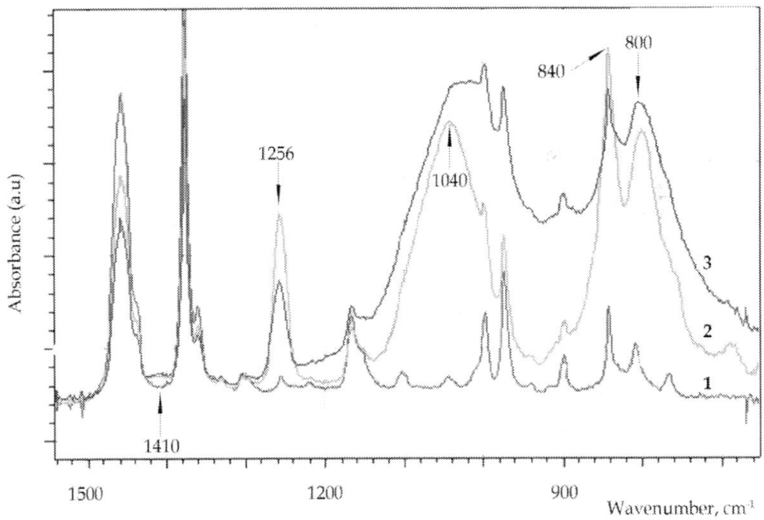

Figure 9. FTIR-ATR spectra of PP nonwoves. Sample 1 - untreated, Sample 2 - HMDSO plasma treated: power 25W, pressure 0.1Tr, Sample 3 - HMDSO plasma treated: power 100W, pressure 0.1Tr.

The assessment of the surface of samples was carried out in the diamond-sample system, in which the IR radiation beam penetrates the layer 2 μm in depth within the spectrum range discussed. The penetration depth of the IR radiation beam considerably exceeds the thickness of the p-HMDSO layer (~ 0.1 μm), hence characteristic bands of PP substrate also occur in the absorption spectrum.

Wave number, cm^{-1}	Absorbing group and type of vibration
1410	νs (CH_3), νa in $Si(CH_3)x$
1256	ν CHx, δ (CH_3)in $Si(CH_3)x$
1040	νa (Si-O-Si)
840	ν $Si(CH_3)x$, γr (CH_3)in $Si(CH_3)_3$
800	ν (Si-O-Si), γr (CH_3)in $Si(CH_3)_2$

*) νs - stretching vibration symmetrical and νa -asymmetrical, δ - deformation vibration, γr – rocking vibration

Table 2. IR absorption bands of p-HMDSO plasma layers (Aumaille et al., 2002; Agres et al., 1996)

3.3. Testing polymeric coatings containing polypyrroles

Polypyrrole is a polymer widely used in commodity production owing to its high thermal stability, resistance to atmospheric conditions and biocompatibility. Its important advantages include electric properties. Using appropriate synthesis conditions, one can obtain electro-conductive, semi-conductive or electro-insulating polypyrroles. In view of processing difficulties, polypyrroles are produced directly on material surfaces in the form of coatings by *"in situ"* chemical, electrochemical or plasma methods. Moreover, polypyrroles are used to make composites as reinforcing and functional materials. In this work, the FTIR-ATR technique was used to monitor the results of polypyrrole synthesis by chemical and plasma methods and the preparation of pyrrole - containing composites.

3.3.1. Formation of latex-pyrrole composites

Polypyrrole (PPy) was used to make an electro-conductive composite as a backing of textile floor coverings (TFC). The TFC piles are fixed in a standard procedure with the use of dressing containing a synthetic rubber and vinyl-acrylic thickeners. The standard latex coating shows electro-insulating properties and impedes the leakage of static charges generated on the TFC pile during exploitation. A functional dressing was prepared to facilitate the leakage of static charges from TFC. PPy microspheres in the form of an aqueous dispersion, prepared by polymerization in an aqueous solution of ferric chloride, were added to an aqueous dispersion of butadiene-styrene-carboxyl copolymer (LBSK 4148) (Urbaniak-Domagala, 2005).

The dispersion components were intermixed by means of an ultrasonic stirrer and the resultant dressing was applied on the bottom of a raw TFC followed by the cross-linking process. The volume resistance tests of the latex-PPy coat confirmed its antistatic properties already with a 3% (by wt.) content of PPy in relation to the dry copolymer mass in the dressing.

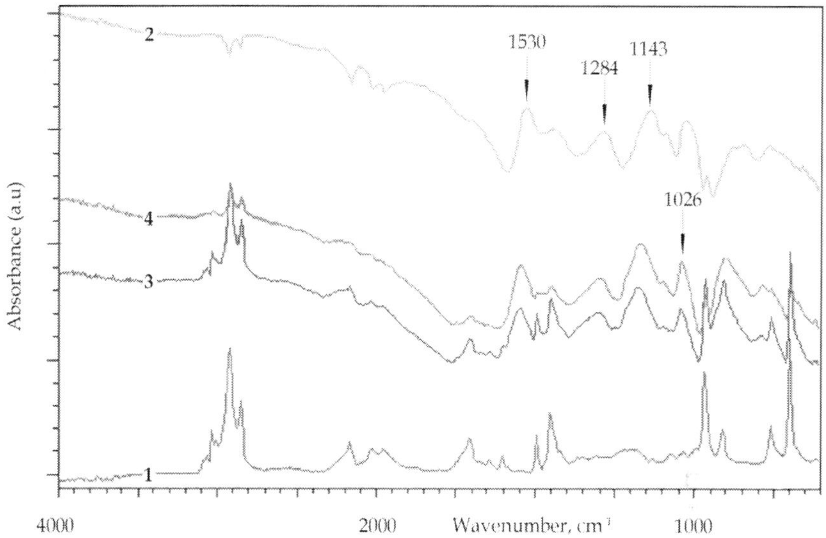

Figure 10. FTIR-ATR spectra pure latex LBSK 4148 (Sample 1), chemically synthesized polypyrrole (Sample 2), latex/PPy composite containing 2 wt% PPy (Sample 3), 3 wt% PPy (Sample 4)

The percolation of the coating electric conduction has a continuous character and the percolation threshold is relatively low. The coating formation on the TFC bottom was controlled by means of FTIR-ATR spectrometry

Changes in the IR radiation absorbance of the coatings were observed within the range from 600 cm^{-1} to 1700 cm^{-1}. Figure 10 shows the absorption spectra of pure LBSK 4148 latex (Sample 1), pure PPy (Sample 2) and two samples of LBSK-PPy composite containing 2% by wt. of PPy and 3% by wt. of PPy, respectively. The locations of absorption bands of LBSK, PPy and LBSK-PPy composite containing 3% by wt. of PPy are listed in Table 3.

The spectrum of LBSK indicates the presence of three types of butadiene isomeric units (1,4-cis, 1,4-trans and 1,2-vinyl), styrene PS and carboxyl (Molenda et al.,1998; Munteanu & Vasile, 2005) have carried out fundamental research of FT-IR spectra of butadiene-styrene copolymers with various structural arrangements (block and linear copolymers, block copolymers of the star type and statistic copolymers). The type of spectrum found for LBSK 4148 latex indicates the architecture of statistic copolymer. The spectrogram of the microspheres of chemically synthesized PPy (powder) indicates PPy rings in the polymer structure and groups connected with the ring being consistent with the results of authors (Eisazadeh, 2007; Cruz 1999; Ji-Ye Jin et al.,1991).

LBSK 4148		Chemically synthesized polypyrrole (PPy)		LBSK 4148/PPy
λ, cm^{-1}	Absorbing group and type of vibration	λ, cm^{-1}	Absorbing group and type of vibration	λ, cm^{-1}
-		659	δ(C-N) out of plane ring	659
699, 758	δ (CH) out of plane in the aromatic ring, PS units	-		699, 758
-		770	δ (C-H) out of plane pyrrole ring, γ(NH_2), ν(C-N-C), δ(C-N-C)	790
911	δ (CH) out of plane near the double bond of the vinyl-PB units	-		911
966	δ (CH) out of plane near the double bond in trans-PB units	-		966
-		1026	δ (N-H), δ(C-H), pyrrole rings pulsation ν(C-N) secondary amines	1040
-		1143	δ (C-H)	1170
-		1284	ν(C-N) secondary amines, ν (C-N) in pyrrole ring	1300
1451	δ (CH) in cis-PB, trans-PB, vinyl-PB units	-		1451
1492	ν (C=C) in aromatic ring PS units	-		-
-		1530	ν(C=C) in pyrrole ring ν(C=N) in pyrrole ring , pyrrole ring pulsation	1548
1600	ν (C=C) in aromatic ring PS units	-		-
1638	ν (C=C) PS units ν (C=C) in vinyl-PB	-		-
1700	ν (C=O)	-		1700

Table 3. The FTIR absorption bands for latex LBSK 4148, chemically synthesized polypyrrole and LBSK 4148/PPy composite containing 3 wag.% PPy (Kazicina.&Kupletska, 1976, Molenda et al.,1998; Munteanu & Vasile, 2005; Bieliński et al., 2009; Eisazadeh, 2007; Cruz 1999; Ji-Ye Jin et al.,1991)

The spectrograms of LBSK-PPy composite samples indicate the superposition of characteristic bands of the composite components: PPy and latex. As the PPy content in the composite increases, one can observe an increase in the intensity of characteristic peaks of PPy, but the quantitative analysis of the composite is difficult to perform due to great differences in the absorbance of the composite components (latex is white, PPy is black). The band maxima shown by the PPy powder are delocalized in the spectrum of latex-PPy composite. The band indicating the pyrrole ring vibration at 1530cm^{-1} is shifted towards a higher frequency to 1548 cm^{-1}. The bands of groups linked up to the pyrrole ring are also shifted: for CH deformation vibration (out of plane quinol PPy) from 1026 cm^{-1} to 1040 cm^{-1}, for C-N deformation vibration from 1143 cm^{-1} to 1170 cm^{-1}, and for the valence vibration of CN in pyrrole ring from 1284 cm^{-1} to 1300 cm^{-1}. The shifts of bands can be due to the scattering of IR radiation in the structure of PPy powder, but they can also indicate the occurrence of PPy - latex intermolecular interactions, with which the oscillatory excitation of chemical groups in PPy requires a higher energy.

3.3.2. *Synthesis of polypyrroles*

Below are presented examples of using the FTIR-ATR technique to assess the progress in the synthesis of PPy. The polymerization of pyrrole was carried out by chemical and plasma methods. Thin layers of PPy were formed on the surface of a PP film by the *in situ* technique.

3.3.2.1. *Chemical polymerization method*

Polymer layers were formed by the polymerization of pyrrole according to the redox mechanism. Two media of pyrrole oxidation were used: an aqueous solution of ferric chloride and aqueous solution of ammonium sulfate with p-toluenesulfonic acid as dopant. Based on the FTIR-ATR spectrum of the polymer, its synthesis progress and chemical structure were characterized. Figure 11 shows examples of the spectra of PPy synthesized in both media for 2h and 5 h.

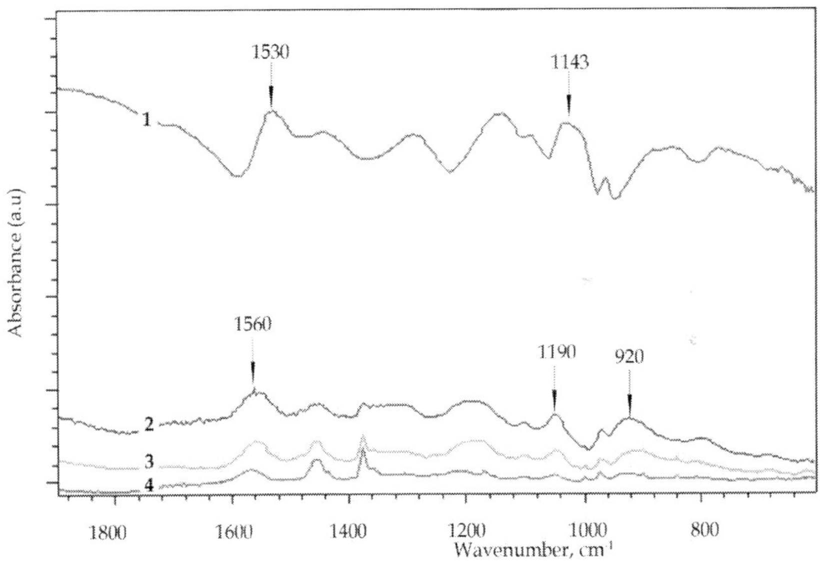

Figure 11. FTIR-ATR spectra chemically synthesized polypyrrole. Sample 1 - PPy powder (molar ratio of FeCl3:Py=2,3:1) - polymerization time 5h; PPy layers on the polypropylene foil: Sample 2 – molar ratio of FeCl3:Py=2,3:1, polymerization time 5h, Sample 3 – molar ratio of FeCl3:Py=2,3:1, polymerization time 2h; Sample 4 – molar ratio of $(NH_4)_2S_2O_8$: $CH_3C_6H_4SO_3H$: Py=0,2:0,25:1, polymerization time 2h.

The spectra of PPy are recorded on the PP substrate. Owing to the low thickness of layers (0.1 - 1μm), the spectrum additionally contains bands derived from the substrate. For comparison, the spectral characteristics of PPy synthesized in the form of powder were also presented. The absorption spectra of all the polymer samples within the wave number range of 600 cm^{-1} – 1800 cm^{-1} confirm the presence of pyrrole group (Table 3). The intensity of absorption bands increases with increasing polymerization time, which is due to the increased layer thickness. In the process of chemical synthesis, the aromatic character of pyrrole ring is maintained, which results in the formation of conjugated double bonds in the linear macromolecule chain. In the presence of admixtures intercalated to the system, the polymer is electro-conductive (incorporated dopants: Cl^- and $CH_3C_6H_4SO_3^-$). The oxidized form of conductive PPy obtained shows a considerable absorption of IR radiation (black color of the polymer). The spectrogram of PPy synthesized with the use of two different oxidants shows no differences between the polymer chemical structures. Differences concern the progress rate of the synthesis: the higher intensity of pyrrole group bands in the polymer synthesized in the aqueous solution of ferric chloride indicates a higher polymerization rate, which is confirmed by the higher rate of layer building up.

3.3.2.2. Plasma polymerization method

Polymer synthesis performed in glow discharge of monomer vapors is a dry, ecological, energy- and material-saving process. The polymerization process is initiated by means of electrons and radicals formed in the gas discharge. The polymerization of pyrrole was carried out in a flow reactor, in glow discharge of the induction type by means of RF field 13.56 MHz (Urbaniak-Domagala, 2008). PP film substrate was centrally and axially placed on a glass carrier in the reactor. The film surface was preliminary purified by means of argon plasma followed by the deposition of the plasma polymer. The FTIR-ATR technique was used to examine the effect of process parameters, such as deposition time, pressure in the reactor and power input to the reactor, on the chemical structure of plasma polymer. In order to impart semi-conductive properties, the plasma PPy was doped after the deposition process by two methods: *in situ* in the reactor in glow discharge of the vapors of organic iodine compounds, and *ex situ* after removal from the reactor in crystalline iodine vapors.

Figure 12 shows the FTIR spectra of the plasma polymer within the range of (600-1850) cm^{-1}. The spectrogram shows the superposition of the absorption bands of plasma polymer (thickness 0.3 μm) and PP substrate. The broad band at (1500 – 1800) cm^{-1} indicates different structure of plasma PPy compared to that of PPy synthesized by the chemical method. This band points to a possible occurrence of primary and secondary amines , secondary amides (Kazicina.&Kupletska, 1976), and carbonyl groups (Ji-Ye Jin et al.,1991) in the polymer. The broad absorption band of the polymer indicates a complex absorption caused by the products of broken pyrrole rings that initiate the branching and cross-linking of the polymer followed by various substitutions. Thus the plasma spectrograms can show secondary and tertiary amines that complicate the absorption in this range. Moreover, one cannot exclude the occurrence of the band at 1710cm^{-1} that, according to authors (Ji-Ye Jin et al.,1991) indicates the presence of carbonyl groups. This band is often observed in neutral or weakly doped forms of PPy, mainly due to their susceptibility to oxidation in air. The spectrum of

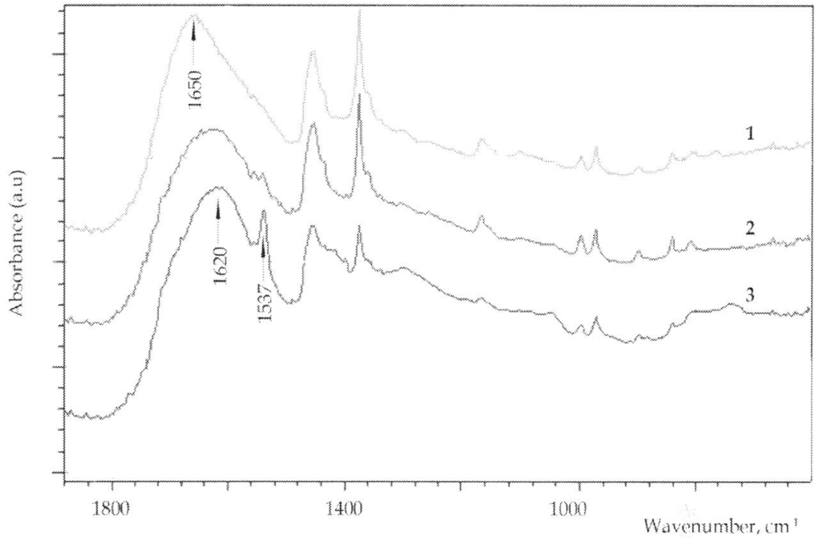

Figure 12. FTIR-ATR spectra plasma synthesized polypyrrole, pyrrole plasma: 50W, 10min, p=0,05Tr. Sample 1-polymer without dopand, Sample 2 - polymer dopanded at plasma CH_2J_2 : 25W, 30min., p= 0,05Tr, Sample 3- polymer dopanded at J_2 vapours, 30min

the polymer doped with iodine vapors contains an additional band at 1537cm^{-1}, induced by the vibration of pyrrole ring, especially intensive in the polymer doped with crystalline iodine vapors. (Groenewoud et al. 2002) observed an increase in the intensity of peak 1520 cm^{-1} under the influence of iodine vapors, which is connected with the formation of a new CH_2=J group in the reaction of iodine with radicals present in the surface layer of the plasma polymer.

The spectrum of the plasma polymer synthesized in the presence of nitrogen as a carrier of pyrrole proves how significant is the influence exerted by the process gas on the chemical polymer structure. This is particularly evident in the polymer synthesized for a longer time (the spectrum of polymer after a 1 h process – Figure 13). The FTIR-ATR spectrum of the plasma PPy at (500-1000) cm^{-1} contains numerous bands with a high absorption intensity that indicate the presence of primary amines (Kazicina.&Kupletska, 1976), and products of substituting chemical groups that were additionally formed in the polymer under the influence of the nitrogen plasma.

The absorption spectra obtained by the FTIR-ATR technique identify the chemical structure of PPy coatings and the structural changes that appear during changing the process parameters, such as pressure, power, type of doping agents, method of incorporating doping agents and the presence of process gas.

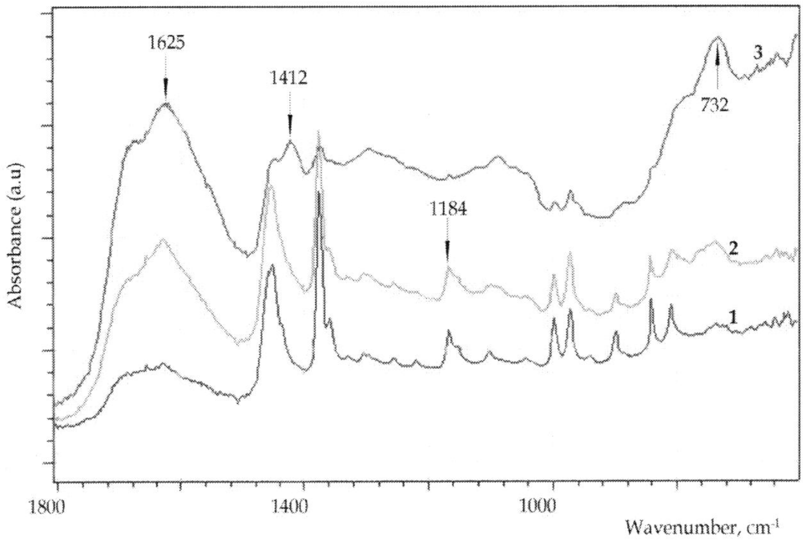

Figure 13. FTIR-ATR spectra plasma synthesized polypyrrole without dopands. Sample 1- plasma pyrrole 0,1Tr 10W, 15min, without processing gas, Sample 2- processing gas N_2 and pyrrole p= 0,15Tr. 10W, 15min. Sample 3 - processing gas N_2 and pyrrole p=0,15Tr 10W, 60min

4. Conclusions

The FTIR-ATR technique makes it possible to examine polymers in a simple, fast procedure avoiding sample destruction. It creates many opportunities for application to the chemical identification of the surface layer of polymers and thin polymeric layers. The analyses of test results of thin layers deposited on polymeric substrates can have rather qualitative character due to the penetration reach of the IR radiation beam being greater than the layer thickness. The examples of PP modification processes and deposition of coatings on polymeric substrates presented confirm that the FTIR-ATR method can be helpful in the examinations of the following:

- the chemical structure of the polymer surface layer and changes in the layer structure caused by the effects of chemical agents and electrochemical factors, UV radiation and low temperature plasma on polymers;
- the chemical structure of polymeric coatings deposited on substrates by chemical and plasma methods;
- the dependence of the chemical structure of polymeric coatings on the parameters of technological process.

Author details

Wieslawa Urbaniak-Domagala
Technical University of Lodz, Department of Material and Commodity Sciences and Textile Metrology, Poland

Acknowledgement

The author would like to thank Professor Barbara Lipp-Symonowicz for good advices and for helpful discussions.

5. References

F. de Fornel, (2000). *Evanescent waves—From Newtonian optics to atomic optics.* Berlin: Springer- Verlag, ISBN: 9783540658450

Goos F.; Hänchen H. (1947). Ein Neuer und fundamentaler Versuch zur Totalreflexion, Ann. Phys. (436) 7-8, 333-346

Dechant J. (1972) *Ultrarotspektroskopische Untersuchungen an Polymeren,* Berlin Available from [http://www.sprpages.nl/SprTheory/SprTheory.htm]

Urbańczyk G.W. (1988) *Mikrostruktura Włókna - Badanie Struktury Krystalicznej i Budowy Morfologicznej,* WNT, ISBN 83-204-1014-2, Warszawa

Urbaniak-Domagala W. (2011) Pretreatment of polypropylene films for the creation of thin polymer layers, part 1: The use of chemical, electrochemical, and UV methods" *Journal of Applied Polymer Science* Vol. 122, No. 3, 2071–2080, 5 November 2011

Urbaniak-Domagala W. (2011) Pretreatment of polypropylene films for following technological processes, part 2: The use of low temperature plasma method. *Journal of Applied Polymer Science,* Vol.122, No. 4, 2529–2541, 15 November 2011

Urbaniak-Domagala W., Wrzosek H., Szymanowski H.,Majchrzycka K., Brochocka A., (2010) Plasma Modification of Filter Nonwovens Used for the Protection of Respiratory Tracts, *FIBRES & TEXTILES in Eastern Europe* , Vol. 18, No. 6 (83) pp. 94-99

Rau J.H. (1963). *Melliand Textilberichte,* 44, pp.1102, 1197, p. 1320

Wiberg K.B.; Eisenthal R. (1964) On the mechanism of the oxidation of hydrocarbons with chromic acid and chronyl chloride. *Tetrahedron* 20, 1151-1161

Carlsson D.J.; Wiles, D.M. (1969) The Photodegradation of Polypropylene Films. II. Photolysis of Ketonic Oxidation Products *Macromolecules* 2, 587.

Kazicina L., Kupletska N.:(1976) *Metody spektroskopowe wyznaczania struktury związków organicznych* 2nd edition; PWN: Warsaw

Guruvenket S.; Rao G.M.; Komath M.; Raichur A.M. (2004) Plasma surface modification of Polystyrene and Polyethylene. *Appl. Surf. Sc.,* 236, 278-284

Creatore M., Palumbo F., d`Agostino R., (2002) Deposition of SiOx Films from Hexamethyldisiloxane/Oxygen Radiofrequency Glow Discharges: Process Optimization by Plasma Diagnostics *Plasmas and Polymers,* Vol.7, No. 3, 291-310

Borvon G., Goullet A., Granier, A., Turban, G., (2002) Analysis of Low-*k* Organosilicon and Low-Density Silica Films Deposited in HMDSO Plasmas *Plasmas and Polymers*, Vol.7, No. 4, 341-352

Aumaille K., Vallee C., Granier, A., Goullet, A., Gaboriau, F., Turban, G., Turban (2000) A comparative study of oxygen/organosilicon plasmas and thin SiOxCyHz films deposited in a helicon reactor. *Thin Solid Films* Vol. 359,188-196

Agres L., Segui Y., Delsol R., Raynaud, P. (1996) Oxygen Barrier Efficiency of Hexamethyldisiloxane/Oxygen Plasma-Deposited Coating *Journal of Applied .Polymer Science*. Vol. 61, 2015-2022

Urbaniak-Domagala W. (2005) Modyfikacja właściwości elektrycznych klejonki lateksowej stosowanej do włókienniczych pokryć podłogowych" Konference *ENP`2005, Elektrotechnologie w nowoczesnym przemyśle,* Bialystok 2005

Molenda J.,Grądkowski M., Makowska M., Kajdas C. (1998) Tribochemical characteristic of some vinyl-type compounds in aspect of antiwear interactions. *Tribologia,* No. 3, 318-329

Munteanu S.B., Vasile C. (2005). Spectral and thermal characterization of styrene-butadiene copolymers with different architectures. *Journal of Optoelectronics and Advanced materials,* Vol.7, No.6, 3135-3148

Bieliński D., Głąb P., Ślusarski L. (2009) FT-IR internal reflection study of migration and surface segregation of carboxylic acid in butadiene-styrene rubber. *Polimery,* 2009, Vol. 54, No. 11-12, 706-711

Eisazadeh H. (2007). Studying the Characteristics of Polypyrrole and its Composites. *World Journal of Chemistry* 2 (2): 67-74

Cruz G.J., Morales J.,Olayo R. (1999) Films obtained by plasma polymerization of pyrrole. *Thin Solid Films,* 342, 119-126

Ji-Ye Jin, Kumi.T., Ando, Teramae N., Haraguchi H. (1991) FT-IR Spectroscopy of Electrochemically Synthesized Polypyrrole. *Analytical Sciences* Vol. 7, 1593-1594

Urbaniak-Domagala W. (2007) Morphology of polypyrrole films formed in low-temperature plasma. *Proceedings Conference IMTEX 2007,* October 8-9, Lodz, 96-99, ISBN 978-83-911012-6-1

Groenewoud L.M.N.,Engbers G.H.M., White R., Feijen J., (2002) On the iodine process of plasma polymerized thiophene layers" *Synth. Met.* Vol. 125 , 429-440

Section 2

Advanced Spectroscopy

Chapter 4

Use of Magnetic Induction Spectroscopy in the Characterization of the Impedance of the Material with Biological Characteristics

Jesús Rodarte Dávila, Jenaro C. Paz Gutierrez and Ricardo Perez Blanco

Additional information is available at the end of the chapter

http://dx.doi.org/10.5772/48763

1. Introduction

The basic electrolytic experiment consists of a homogeneous[1] electrolytic solution with two identical electrodes (Fig 1). We know that a homogeneous solution hasn't boundaries or membranes, except the electrodes and the solution receipt.

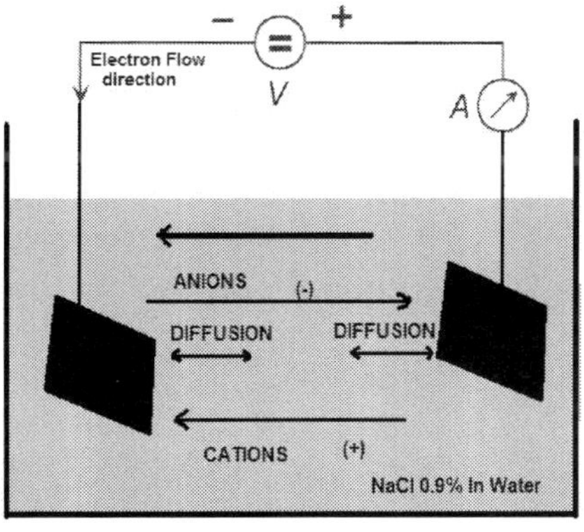

Figure 1. The basic electrolytic experiment

Electrolyte solution selected as the most important of the human body: NaCl aqueous solution with a concentration of 0.9% by weight [i].

[1] If NaCl is dissolved in water then NaCl is the solute (and the electrolyte), and water is the solvent, together form the solution

In a homogeneous conductive material the impedance (Z) is proportional to its length and inversely proportional to its cross sectional area (A) (Fig. 2).

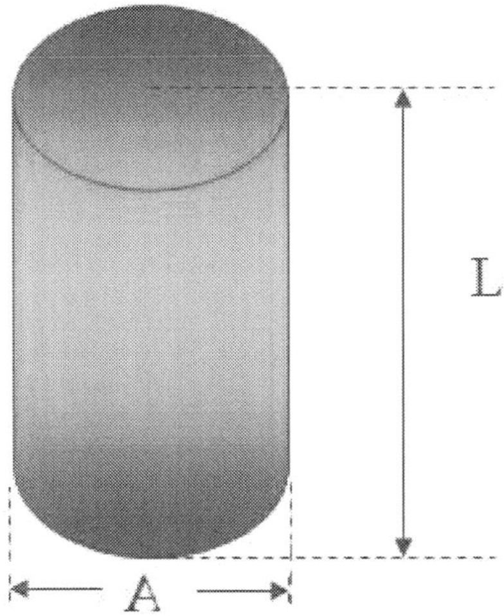

Figure 2. The impedance (Z) of homogeneous conductive material is directly proportional to its length and resistivity and inversely with its area

$$Z=\varrho L/A=\varrho L^2/V \tag{1}$$

According with the Figure 2 Z=impedance, L=length, A=area, V=volume ϱ=resistivity=1/σ (conductivity). An empirical relationship can be established between the ratio (L^2 / V) and the impedance of the saline solution which contains electrolytes that conduct electricity through the sample. Therefore impedance (Z) = ρ L / A = ρ L^2 / V.

Hoffer et al. [2] and Nyboer [3] were the first to introduce the technique of four surface electrodes Bio-impedance analysis.

A disadvantage presented by this technique is the use of a high current (800 mA) and a high voltage to decrease the volatility of injected current associated with skin impedance (10 000 Ω/cm^2)[4]

Harris et al.[5], (1987) uses a four terminals device to measure impedance for the purpose of eliminating the effect of electrodes in an aqueous medium.

Asami et al.[6] , (1999) used a pair of coils submerged for monitoring the current induced in the coil pair, which he called electrode-less method, however still requires physical connections between the coils and electronic instruments.

To measure the complex spectrum of the permittivity of a biological culture solution Ong et al.[7], uses a remote sensor resonant circuit, to obtain the impedance of the environment by observing the resonant frequency and the frequency of zero reactance.

In another case for monitoring the fermentation process Hofmann et al.[8], Use a sensor based on a transceiver, as a way of overcoming the effect of having two metal electrodes to measure the impedance of the culture broth in a fermentation process, as in such processes the behavior of living cells is as small capacitors, then measure the impedance represented by these small capacitors correlates with the number and size of living cells in the system (Hofmann et al., 2005).

2. Magnetic induction spectroscopy

"The Magnetic Induction Spectroscopy (MIS) aim is the non-contact measurement of passive electrical properties (PEP) σ, ε and μ of biological tissues via magnetic fields at various frequencies.[9]"

The basic requirements of this method are:

a. Creating a varying in time magnetic field, from an exciting coil to induce the field to the object under study.
b. Obtaining information generated from the disturbance or "reaction" of coils-environment system through the Receiver / Sensor coil.

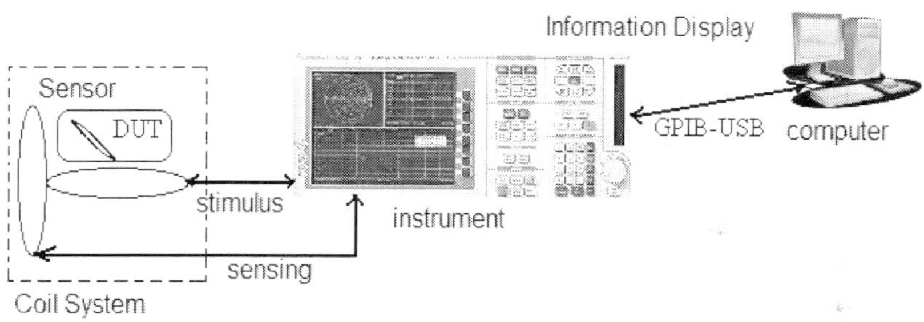

Figure 3. Measurement System: composed of a coil arrangement, network analyzer, and a pc

3. Measurement system

Following the protocol of the method used in Figure 3 we present our system Equipment-Interface into three sections:

a. Computer: Using the platform that provides National Instrument [10] - LabView V8.6, displays and processes the information obtained from the coil system from the Instrument.

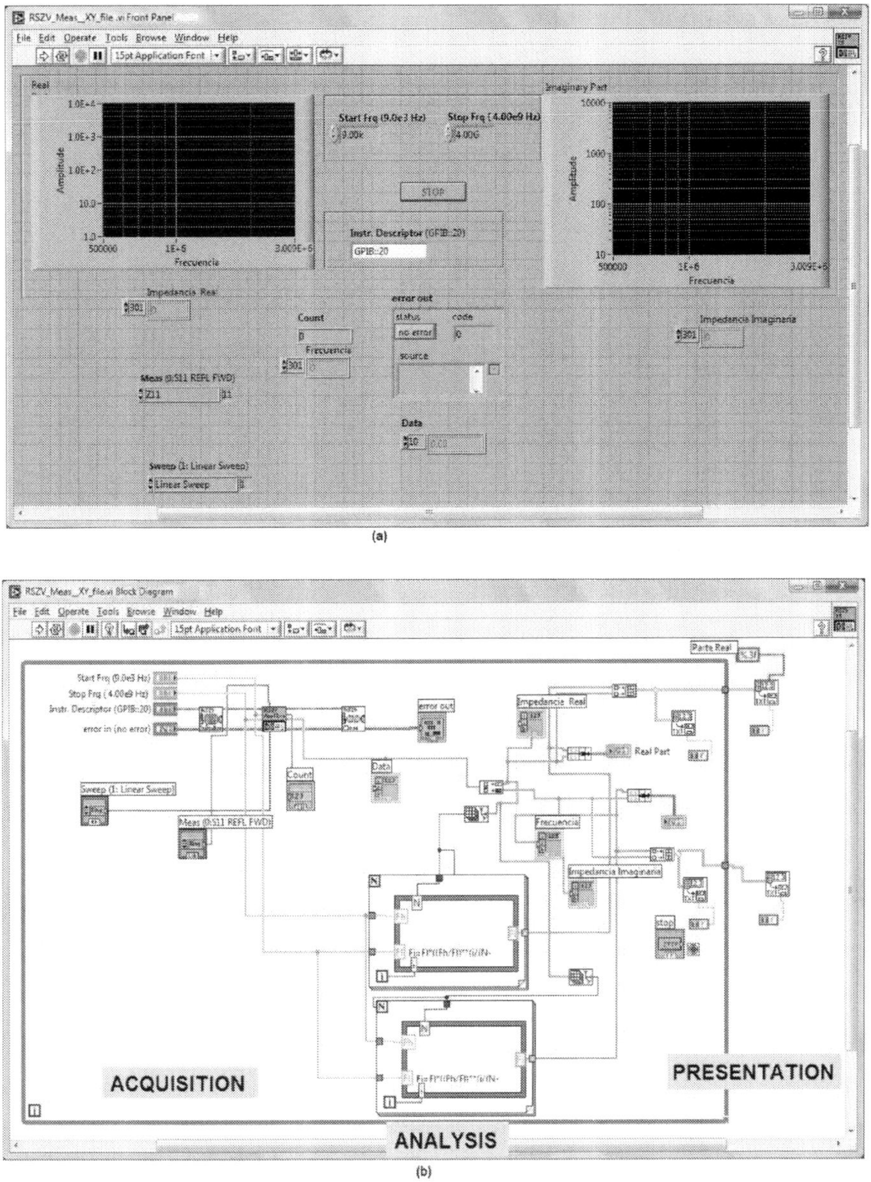

Figure 4. Virtual Instruments LabVIEW (National Instrument VI), a) Front panel (user), b) Block diagram panel (interconnections), uses them to automate the acquisition and management of information.

b. Instrument: Commercial Equipment, RS ZV Vector Network Analyzer [11] "Rohde & Schwartz" which performs the frequency sweep from 100 kHz to 4 MHz range, applying it to the exciting coil, the same equipment then captures the data or information from the receiver coil and sends to the computer via General-Purpose Interface Bus Universal Serial Bus or GPIB-USB.

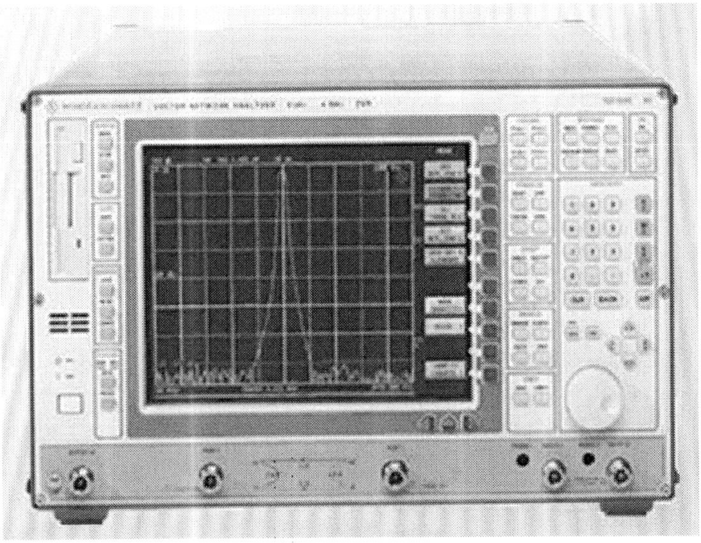

Figure 5. Network Analyzer "Rohde & Schwartz", as a frequency sweep source

c. Coil System: Fig. 6 show this system, consisting of three coils, an exciter, a receiver completely perpendicular to the exciting, adjusting it to the mutual inductance between two coils is minimal, and a third coil to function as "mirror-sensor" of the magnetic field generated by the exciting coil.

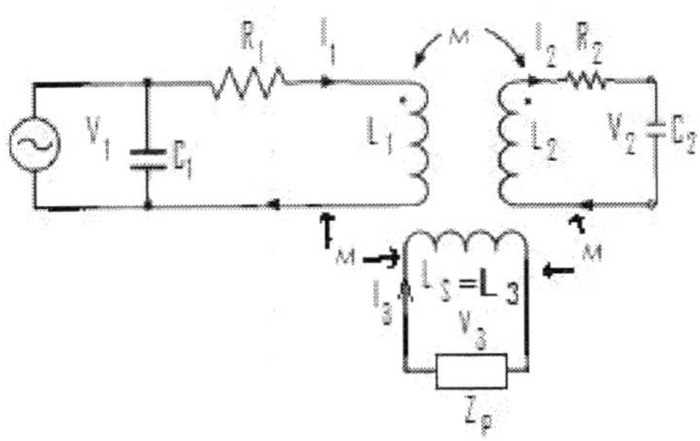

Figure 6. Representative schematic of the three coil arrangement

3.1. Drive coil L1

For generating a spectral magnetic field a flat coil was used as a transmitting antenna built on a phenol board as printed circuit with the following characteristics:

- Spiral coil [12] internal diameter of 2 cm and an outer diameter of 9.5 cm with 20 turns, with cooper tracks width of 500 μm , and an equal distance between them, see Figure 7
- 24 μH Inductance measured experimentally in a frequency range of 100 KHz to 5 MHz
- Shielding is used to minimize capacitive coupling at the top and bottom of the exciting coil, forming a sandwich, this shielding was "grounded".

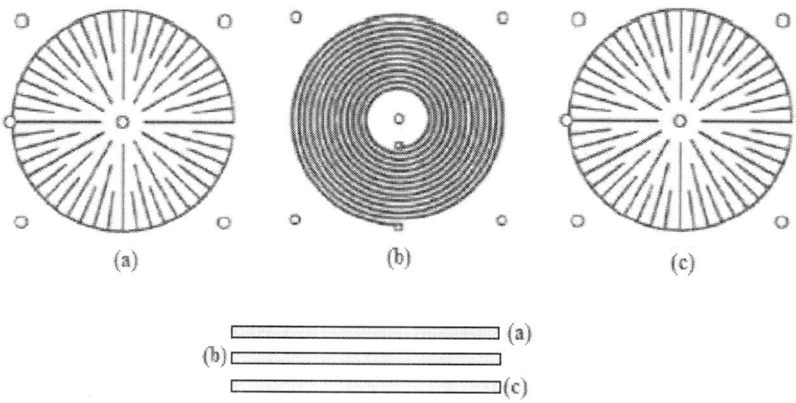

Figure 7. Coil and Shielding employee

The Network Analyzer employee served as a source of power in a frequency range of 100 KHz to 5 MHz applied. The combination coil and capacitance of the cables, that although were coaxial cables it presented a resonance frequency of 4.612 MHz This initial structure was proceeded with a series of measurements with actual physical capacitors with a dual role, first see the system answer at different values, and the second as reference calibration [13].

3.2. Receiver coil L2

For reception of the spectrum magnetic field, also use a flat coil as the receiving antenna, built on a phenol board as printed circuit with the following characteristics almost as similar to the transmitting antenna:

- Spiral coil internal diameter of 2 cm and an outer diameter of 9.5 cm with 20 turns, with cooper tracks width of 500 μm, and an equal distance between them, see Figure 8
- 24μH Inductance measured experimentally in a frequency range of 100 KHz to 5 MHz
- Shielding is used to minimize capacitive coupling in the bottom of the receiving coil, this shield was "grounded".

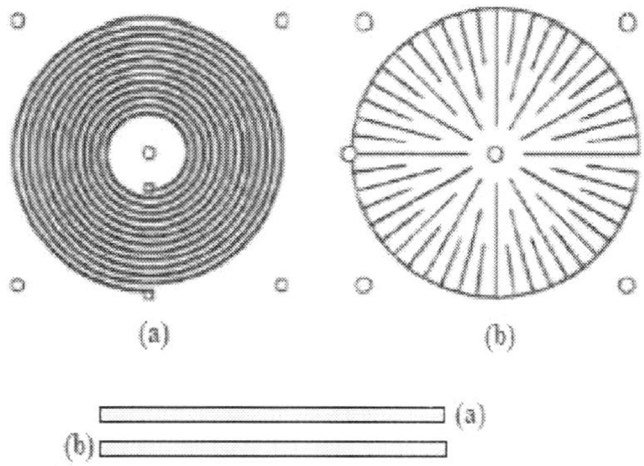

Figure 8. Coil and shield used

- In addition to the above conditions and in order to minimize inductive coupling both coils are placed in perpendicular way see Figure 9, so that the only magnetic field received were the projected by the coil sensor.

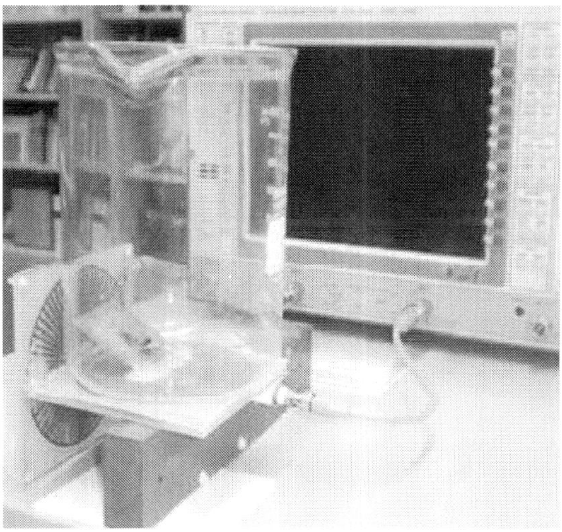

Figure 9. Mechanical available of the coils to minimize Inductive coupling between transmitter coil and receiver coils

Network Analyzer employee close the system, as seen in Figure 9, the application is through the coil L1, placed horizontally, which serves as basis for the deposition of both the saline and the samples biological tissue. Receiving and monitoring of the signal through the coil L2, vertically positioned and perpendicular to L1

3.3. Mirror-sensor coil

As a passive coil mirror-sensor used a flat square coil of about 7.92 μH, see Figure 10, with a measured value of 9.716 μH, at a frequency applied of 1 MHz, Coil with a capacitor of 330 nF added to make a 1 MHz resonant circuit, finally resulting a resonant frequency of 1.14 MHz, like the previous coils was constructed on phenol board as a printed circuit having the following characteristics:

- The inductance L, was determined by experimental measurements, confirming the calculations used in the approximation developed by Ong et al[14].

$$L = 1.39 * 10^{-6}(OD + ID) * N_L^{\frac{5}{3}} * log(4 * \frac{OD+IDD}{OD-ID}) \quad (2)$$

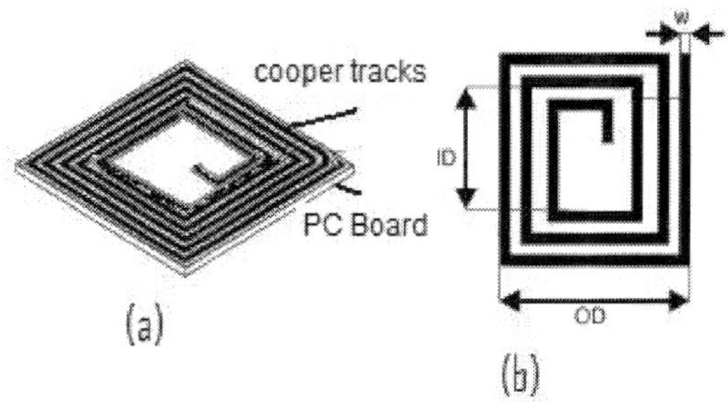

Figure 10. Layout of the Mirror-Sensor coil with interdigital capacitor (not shown)

- Square spiral coil internal diameter of 3.6 cm and an outer diameter of 4.7 cm with 10 turns, according to the expression [1] L = 7.92 μH
- To minimize resistive-capacitive effect between turns, when immersed in saline was used an insulating paint.

4. Characterization of the system

The technique of network development has in its whole with the element to explore interesting properties, which generates all possible parameters of the immittance associated with the two-port network, Figure 11.

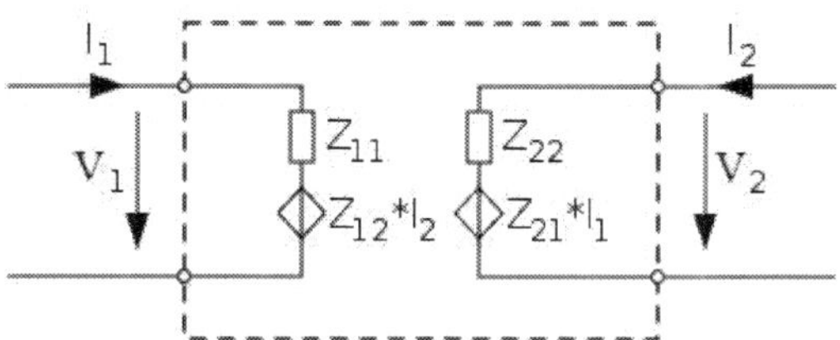

Figure 11. Electronic models of the transmitter coil and receiver coil

The formulation of network equations obtained by choosing voltages across the capacitors which are not physically in our arrangement, but it is necessary to include consideration coaxial cables within the coil system, which by their physical characteristics have a value of capacitance and current flows through the antenna-coils (inductors) as variables used in the description of the network, produces many equations as reactive elements present in a given network.

VI relations of our model, without loss, respecting the conventions of voltage and current shown in Figure 11, are[15]:

$$V_1 = Z_{11} * I_1 + Z_{12} * I_2 \tag{3}$$

$$V_2 = Z_{21} * I_1 + Z_{22} * I_2 \tag{4}$$

Since the primary reactive elements of this network are the inductors $Z_{11} = j\omega L_1$ y $Z_{12} = j\omega M$; $Z_{22} = j\omega L_2$ y $Z_{21} = j\omega M$ and since there is an inductive link between the coils, equations [2] and [3] are reconsidered:

$$V_1 = j\omega L_1 * I_1 + j\omega M * I_2 \tag{5}$$

$$V_2 = j\omega M * I_1 + j\omega L_2 * I_2 \tag{6}$$

in equations [4] and [5], M is the mutual inductance (dimension H) between the coils L_1 and L_2, defined by $M = k * \sqrt{L_1 * L_2}$ where k (dimensionless number, ≤ 1) is the coupling factor between the inductors (inductive link ratio), controllable with the relative position between them.

The model presented in Figure 6 allows the development of matrix impedance parameters of a bi-port, as expressed in equations [3-6], and involves the mutual inductance of the coil system, then presents the mathematical analysis of the inductive link.

The design and structure of the coil system aims to make the inductive link between the coils L1 and L2 is minimized by establishing a physical way the perpendicularity between L1 and L2, seeking with this strategy to establish a relationship of mutual inductance increased by engaging in two-port network a third coil that provides a "mirror" effect, the Figure 12 shows a representation of the this inductive link.

The mutual inductance M_{12} of the inductor L_2 in relation to L_1, which (according to the reciprocity theorem) it was found experimentally equal to the mutual inductance M_{21} of the inductor L_1 in relation to L_2, theoretically defined as the coefficient of mutual magnetic flux $\Phi_{12} = \Phi_{21}$ which spans both coils due to the current flowing in each coil respectively.

$$M_{12} = \frac{\Phi_{12}}{I_1} \tag{7}$$

$$M_{21} = \frac{\Phi_{21}}{I_2} \tag{8}$$

"The Biot-Savart law, states that if a small length of conductor δl carrying a current i, then the magnetic field strength at a distance r and angle θ is

$$\delta H = \frac{i\delta l sin\theta}{4\pi r^2} \tag{9}$$

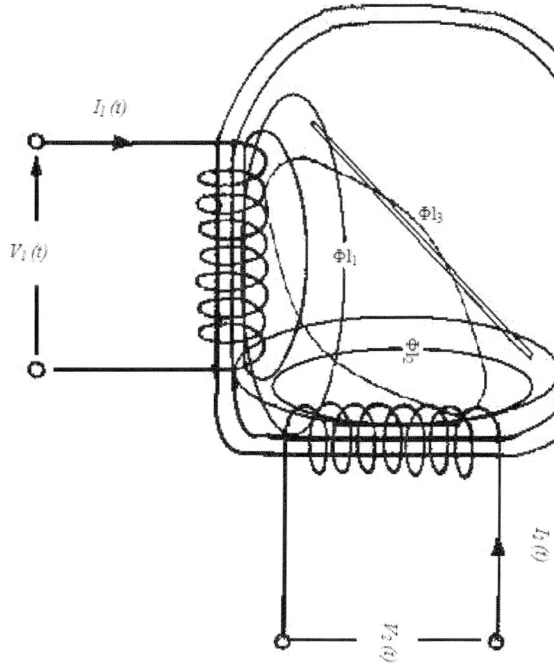

Figure 12. Approximate representation of variable in time magnetic induction in the coils system

(Sine θ merely states that if power is not in an optimal direction, then the field at that point decreases.) This is confirmed by Ampere's rule, $\int Hdl = i$; The line integral of the magnetic field in a closed loop is equal to the electric current flow through the closed loop" [16]. In particular, the magnetic field strength H in a circular path of radius *r* around an electric current *i* in the center is

$$H \cdot 2\pi r = i, o\ B = \frac{\mu i}{2\pi r} \tag{10}$$

Because L_1 and L_2 have very similar characteristics, and considering the expression [9] may approximate the magnetic field concentric loops through both coils (see Figure 8a), B_1 is the magnetic field of the first ring, the flow magnetic Φ_2 through second ring we can determine from B_1.

$$\Phi_2 = B_1 A_2 = \left(\frac{\mu_0 I_1}{2R_1}\right) \pi R_2^2 = \frac{\mu_0 \pi I_1 R_2^2}{2R_1} \tag{11}$$

where μ_0 is the magnetic permeability of free space, R_1 and R_2 are the radii of the rings which form the coils.

The mutual inductance is then:

$$M = \frac{\Phi_2}{I_1} = \frac{\mu_0 \pi R_2^2}{2R_1} \tag{12}$$

This expression shows that M depends only on geometric factors, R_1 and R_2, and is independent of the current in the coil. As regards the expression $M = k * \sqrt{L_1 * L_2}$ to obtain an approximation of mutual inductance between coils, since low values of coupling coefficients "k" with air-core coils are obtained usually in the order of 0.001 to 0.15 [17] experimentally is considered as the value of k_1 = 0.00167 (caused by a misalignment both angular and lateral between coils L_1 and L_2), and a value of k_2 = k_3 = 0.465 (caused by an angular misalignment between coils L_1, L_2 over L_3), so we get the following values:

$$M_{31} = M_{13} = 7.1\mu H$$

$$M_{23} = M_{32} = 7.1\mu H$$

$$M_{21} = M_{12} = 0.04\ \mu H$$

By obtaining these values, we proceeded to simulate the circuit in PSpice [18], see Figure 13, to verify that they were not far from the actual physical model, which was obtained following graphs.

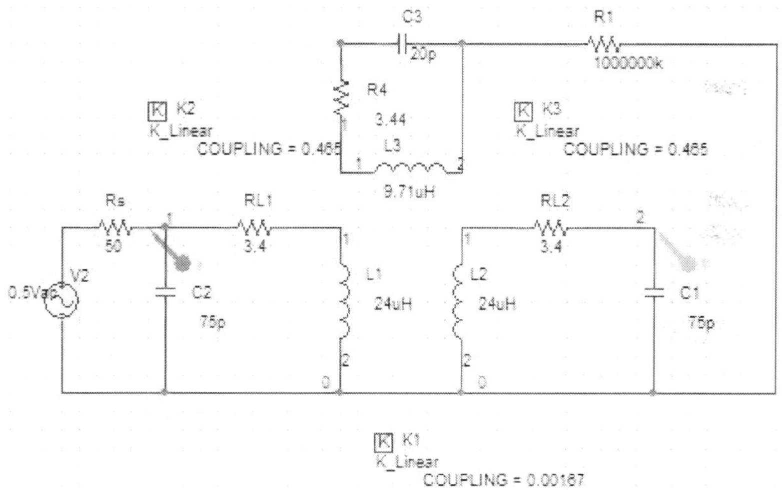

Figure 13. System coils with their respective coupling factors for achieving the simulation circuit had to be "grounded" L3 but with the highest permissible value, simulating a physical disconnection between the two coils.

In Figure 14 we can observe the system performance in terms of voltage induction refers to both L_2 and L_3.

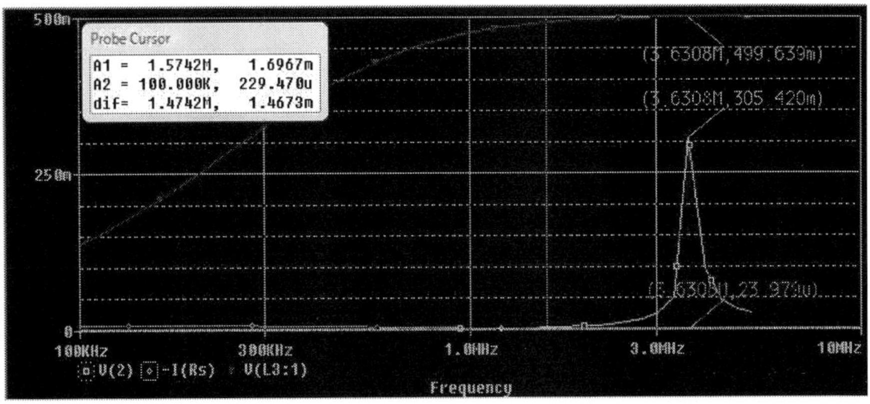

Figure 14. Response of the coil system to a frequency sweep from 100 kHz to 5 MHz, with estimated coupling factors.

The obtained simulation results permit the development equations which describe the model in Figure 6; these can be expressed in matrix form:

$$\begin{bmatrix} V_1 \\ V_2 \\ V_3 \end{bmatrix} = j\omega \begin{bmatrix} L_1 M_{12} M_{13} \\ M_{21} L_2 M_{23} \\ M_{31} M_{32} L_3 \end{bmatrix} \begin{bmatrix} i_1 \\ i_2 \\ i_3 \end{bmatrix} \tag{13}$$

Mij are mutual inductances, and the proportionality factors of the currents are the impedances so that the matrix can be rewritten in terms of the Z parameters.

$$\begin{bmatrix} V_1 \\ V_2 \\ V_3 \end{bmatrix} = j\omega \begin{bmatrix} Z_{11} Z_{12} Z_{13} \\ Z_{21} Z_{22} Z_{23} \\ Z_{31} Z_{32} Z_{33} \end{bmatrix} \begin{bmatrix} i_1 \\ i_2 \\ i_3 \end{bmatrix} \tag{14}$$

As experimentally determined that $Z_{12} = Z_{21}$ be as small as possible to perceive impedance changes as the sensor coil L_3, so we have:

$$Z_{12} = \frac{V_1}{i_2} | i_1 = 0; Z_{21} = \frac{V_2}{i_1} | i_2 = 0$$

$$\frac{V_3}{i_3} = Z_3 \tag{15}$$

If we consider the equations [14] and [15] in combination with the matrix [12], expressing V2 in terms of Z3, we have:

$$V_2 = j\omega M_{21} i_1 + j\omega M_{23} \frac{V_3}{Z_3} \tag{16}$$

5. Making a reference

Considering the previous development, and articles concerning the use of systems like ours (Ong. 2000, Hofmann 2005) is therefore important to have baseline measurements of physical elements such as capacitors, to be served at one time as models of reference and calibration, was determined to carry out the completion of the procedure detailed below.

Figure [16] the electrical circuit presented is the model of a capacitive type humidity sensor [19], the behavior of this sensor is comparable to the behavior of L_3 so we can take the development of the equation for Z_P that would be the Z_3 response, which could take the complex impedance as:

$$Z_P = R_S + \frac{R_P}{1+j\omega R_P C} \tag{17}$$

Misevich et al refers to the resistance changes predominantly in the range of 10-100 MΩ in a circuit as shown in Figure [15], fall about 0.5 Ω in a very humid environmental, is not our case so is taken as a criterion for considering a single value of R_P of 100Ω, likewise considering the impedance presented by R_P, varying ω in the range of $12e^{+6}$ to $25e^{+6}$ rad / s the impedance C in parallel with R_P take R_P value in the range of C values from 10pF-100pF, and goes to take the R_S value in the range of C values from 1nF-100nF, see table 1, that added to the XL values presented by ω is negligible this value so we can reduce our equation to:

$$Z_P = \frac{R_P}{1+j\omega R_P C} \tag{18}$$

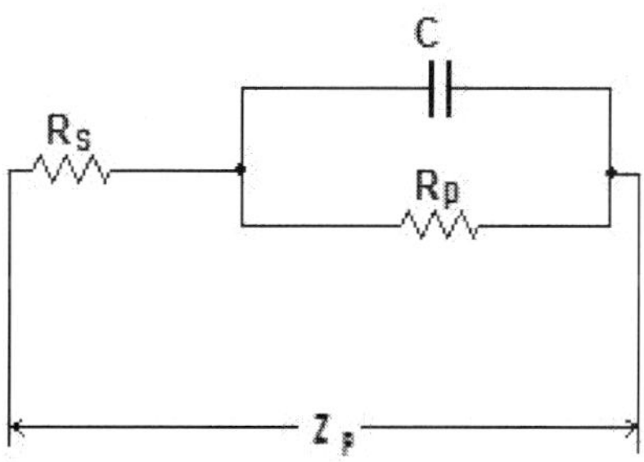

Figure 15. Misevich, K.W. shows the development of a capacitive humidity sensor element impedance equivalent to measuring the coil system, where R_S = 3.44Ω series resistance of the sensor, R_P = 100Ω, C multivalued see table 2

Val. Cap./Freq.	2.5MHz	3MHz	3.5MHz	4MHz
10pF	98.4Ω	98.1Ω	97.8Ω	97.5Ω
100pF	86.4Ω	84.1Ω	82Ω	80.0Ω
1nF	38.9Ω	34.6Ω	31.25Ω	28.5Ω
10nF	5.98Ω	5.04Ω	4.35Ω	3.82Ω
100nF	0.63Ω	0.53Ω	0.45Ω	0.4Ω

Table 1. $Z_P = Z_3$ values obtained by calculation.

$$Z_3 := \frac{R_3}{1+(j \cdot R_3 \cdot C_3)} = 97.459 - 17.73i \qquad (19)$$

Expression [19] used for calculating the impedance of Table 1 using MathCad [20]; Therefore the values obtained with small capacitors values by calculating means as measured by the analyzer is in a range of 5% in 3.09 MHz, the resonance frequency, and not higher capacitor values that are very de-correlated, this is due to the length of the coaxial cables, overriding the capacitance introduced by these.

Val. Cap./Freq.	2.5MHz	3MHz	3.5MHz	4MHz
10pF	96.18Ω	102.7Ω	107.8Ω	112Ω
100pF	100.9Ω	107Ω	110.5Ω	112.6Ω
1nF	60.2Ω	41.4Ω	29.5Ω	24.5Ω
10nF	22.6Ω	29.6Ω	36.2Ω	42.7Ω
100nF	29.6Ω	35.7Ω	41.7Ω	47.63Ω

Table 2. Z_3 values obtained by direct measurement by network analyzer.

If we consider the matrix equation [12] and equation [18], V_3 can be expressed in terms of $Z_P = Z_3$, we have then:

$$V_3 = \frac{j\omega M_{23} i_2}{1 - \frac{j\omega L_S}{Z_3}} \qquad (20)$$

Considering the matrix equation [12] and equation [19], we have V_1 expressed in terms of Z_3:

$$V_1 = j\omega M_{12} i_2 + j\omega M_{13}\left(\frac{j\omega M_{23} i_2}{Z_3 - j\omega L_S}\right) \qquad (21)$$

If we consider the equation [14 and 20] we express Z_{12} as:

$$Z_{12} = j\omega M_{12} - \omega^2 \left(\frac{M_{23} M_{31}}{\sqrt{\left(\frac{1}{R_P}\right)^2 + (\omega C)^2 - j\omega L_S}} \right) \qquad (22)$$

Figure 16 shows the values obtained by performing calculations using equation [22] implemented in Mathcad, it is appreciated also that around the two resonance frequencies in the system the magnitude of Z_{12} is the same.

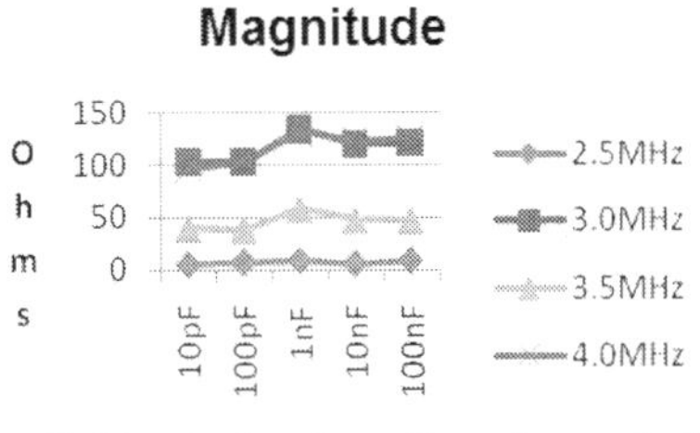

Figure 16. Comparison between Z_{12} Magnitude values of impedance calculation in Mathcad.

6. Validation and calibration

Figure 17a shows the graphical model used for our experiment, and Figure 17b refers as a practical way to the interface between the network analyzer and sensor description given above in a comprehensive manner.

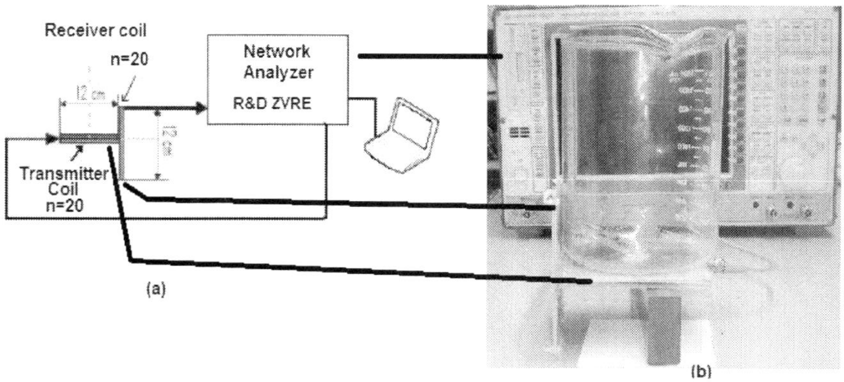

Figure 17. Magnetic Impedance Spectroscopy Method: a) block diagram, and b) practical implementation

Figure 18 a shows a block diagram of a three-port circuit, the voltages and currents are indicated, the upper terminal of the instance is positive with respect to the terminals of the bottom and the currents flowing inwards as indicated by KCL for each port, Figure 18b shows the block diagram of a real form.

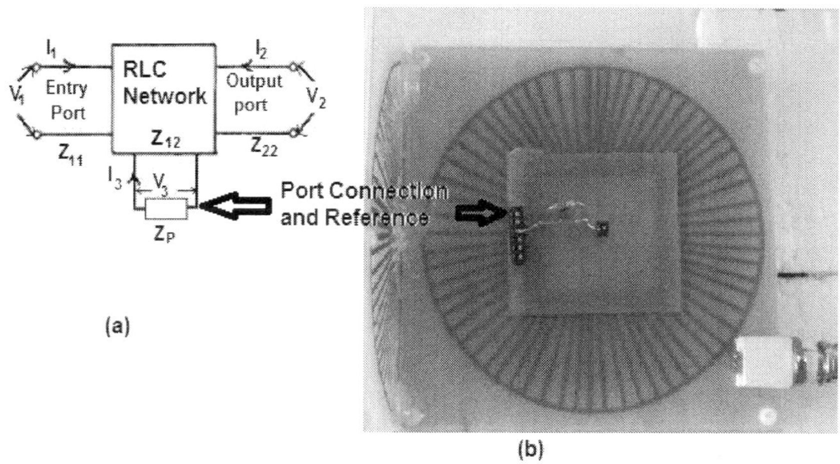

Figure 18. Three-port circuit in preparation for constant section cell measuring based on measurement of physical components.

The values obtained for both measurement and calculation as shown in Table 2 allows us to make the cell preparation of constant section for measuring conductivity. For this preparation had to consider the mechanical error sources such as physical dimensions of the saline cell, the effects of temperature on the cell, and error handling and positioning of it. It also ponder the discrepancy between the measurements of a real physical element such as a capacitor in parallel with a resistor, and the resistivity and permittivity expected of an electrolytic cell, it was therefore necessary to have a truer reference of known salt solution to evaluate deviation of our measurements and bring the measuring physical model to a virtual model and simulate their behavior in this way to "induce" their performance, come to perform these measurements with electrodes.

In Figure 19 shows a sample obtained from a series of meaşurements made under the scheme of Figure 18, changing the values of the capacitances in parallel with resistors first small values, reaching values used to 1 kΩ .

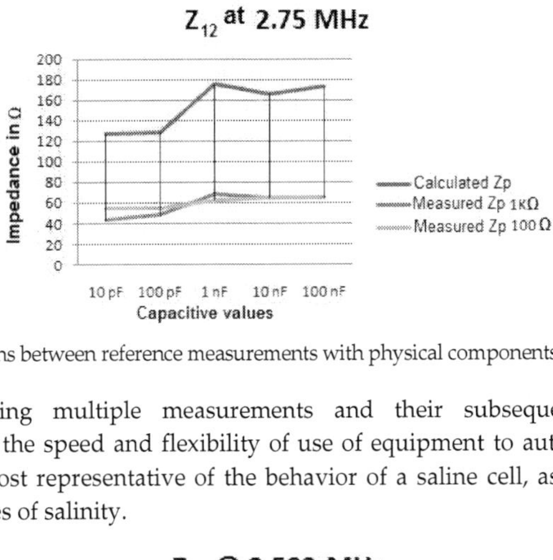

Figure 19. Comparisons between reference measurements with physical components and estimated values

Although performing multiple measurements and their subsequent acquisition of information due to the speed and flexibility of use of equipment to automate this process, include only the most representative of the behavior of a saline cell, as seen in Figure 20, with various degrees of salinity.

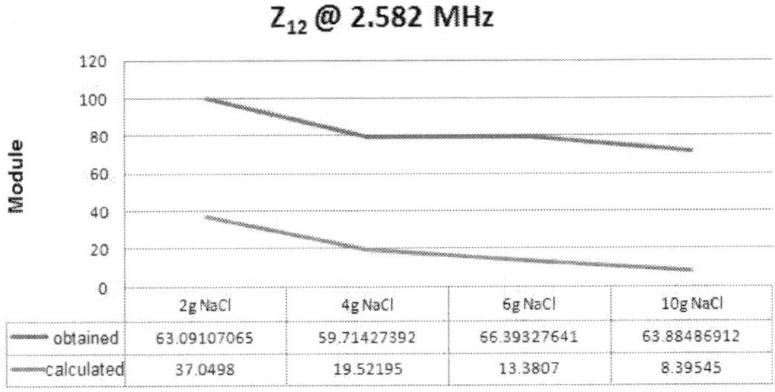

Figure 20. Measurement of Z_{12} of four saline cells

Values obtained with the saline cell, Figure 20, as well as the physical components are between 4% and 8% above the Z12 average obtained from the saline cell, provides that comparatively physical model is closer to cell model representing a saline cell of Figure 21, which is formed by a resistance of 100 Ω in parallel with a capacitor of 1 nF, with a response very "right" to the resonant frequency of the coil system which is approximately 3 MHz

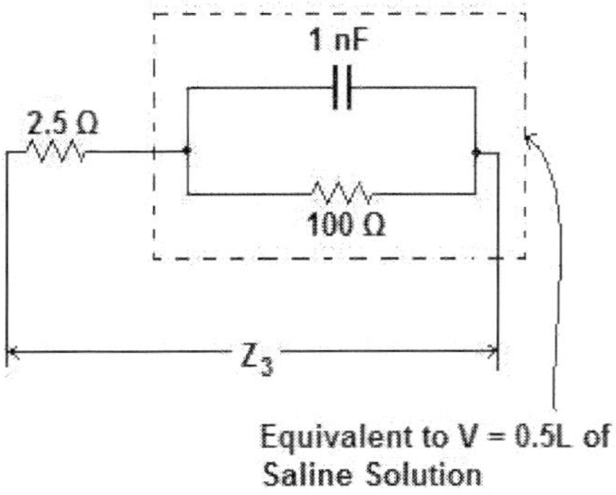

Figure 21. Physical model equivalents to a 500 mL saline cell at a frequency of ~ 3 MHz

The conductivity of a saline solution with 2 g of salt dissolved in 1L of deionized water having a conductivity measurement 5.8μS/cm, presents a σ = 3.9 mS / cm at 25 ° C and σ = 3.4 mS / cm at 20 ° C, this represents a 0.2% concentration and 0.034 M.

If we use this solution as the first reference electrolytic cell and according to equation [1], with an equal volume to 500 mL and a length of 85 mm, we calculate a 37.05Ω impedance (Z) , with these values and data from our physical model equivalent gives a σ = 2.29 mS / cm which leads us to obtain a correction term 3.9/2.29 = 1.7.

7. Measuring saline cell permittivity with mis

The resistivity of an aqueous sodium chloride dissolution (NaCl) [21] is obtained from considering the current density (J), determined this by the ions types (Na^+ and Cl^- are the most abundant ions) of the solution, this is directly proportional to the factor "α" of dissociation of molecules, approximately equal to 1 if the electrolyte is strong, with a concentration "c "gram-equivalent, and a mobility" μ "of the ions, also the electric field" E "as shown in the following relationship: J = F α c μ E.

A solution consisting of 9 grams of sodium chloride dissolved in one liter of water in medicine[2] is called Normal Saline, since the concentration of 9 grams per liter divided by 58 grams per mole (approximate molecular weight of sodium chloride) provides 0.154 moles per liter, that is, contains 154 mEq / L of Na^+ and Cl^-. The fact contain more solute per liter,

[2] In chemistry, the normal concentration of sodium chloride is 0.5 mol of NaCl assuming complete dissociation. Physiological dissociation is approximately 1.7 ions per mole, so that a normal NaCl is 1/1.7 = 0.588 molar. This is approximately 4 times more concentrated than the medical term "normal saline" of 0.154 mol

makes this solution with a slightly higher osmolarity[3] than blood, on an average day the natremia[4] can range between 130-150 mEq / L (normonatremia) . However, the osmolarity of normal saline is very close to the osmolarity of the NaCl in the blood[22].

From Figure 22 and considering the dimensions of our saline cells, presents three values of resistivity, $\varrho = 2.65\Omega m$, $\varrho = 0.55\Omega m$ and $\varrho = 1.3\Omega m$, respectively.

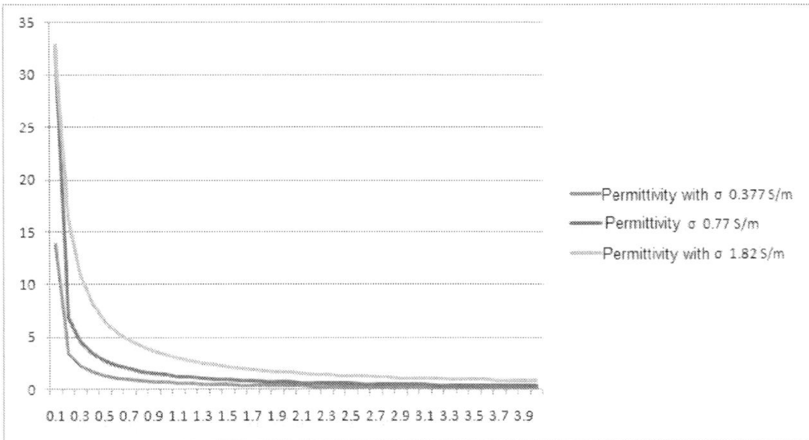

Figure 22. Measured frequency behavior of three electrolytic cells

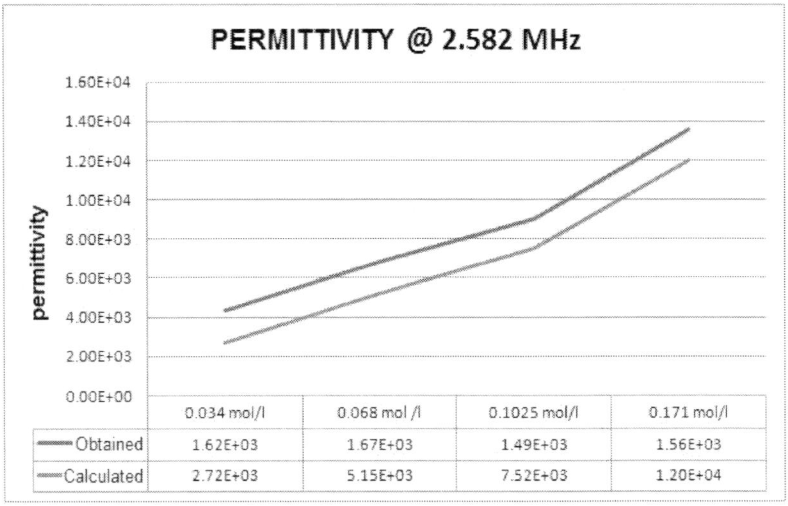

	0.034 mol/l	0.068 mol /l	0.1025 mol/l	0.171 mol/l
Obtained	1.62E+03	1.67E+03	1.49E+03	1.56E+03
Calculated	2.72E+03	5.15E+03	7.52E+03	1.20E+04

Figure 23. Saline solutions response at resonance frequency of the coil system

[3] Measuring solute concentration

[4] Concentration of sodium in the blood

Figure 23 presents a summary of values acquired with the network analyzer, Figure 24 represents the values obtained with the HP4192A impedance analyzer.

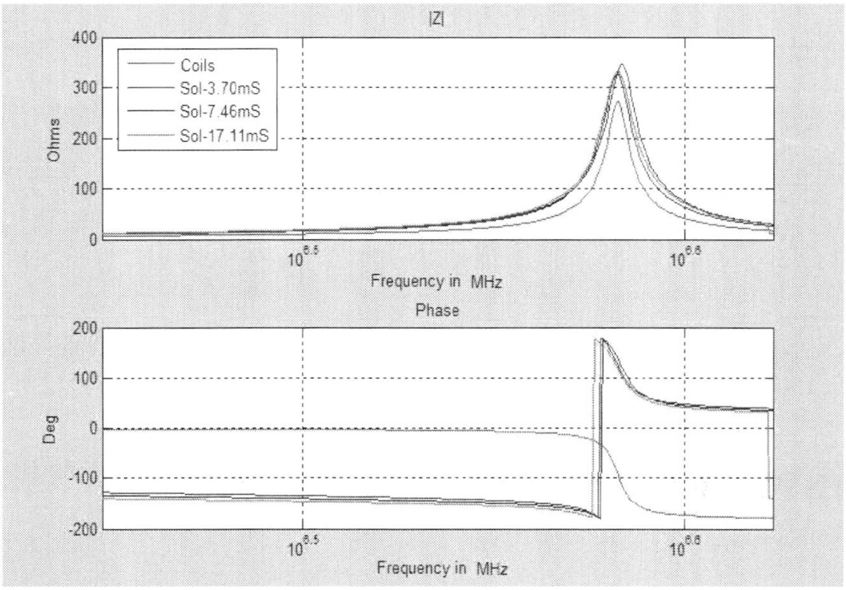

Figure 24. Frequency responses of Saline Solutions.

8. Biological suspensions

Typically, in vitro measures, is considered a sample of uniform section (A) and length (L), in the majority of cases it is more practical to consider the volume (V) in aqueous solutions. In this case the ratio between the conductivity and the permittivity of the sample with the resistance and reactance are:

$$C = \frac{\varepsilon\, \varepsilon_0\, V}{L^2}\ [F] \tag{22}$$

$$R = \frac{L^2}{\sigma\, V}\ [\Omega] \tag{23}$$

$$Z = \frac{L^2}{(\sigma + j\, \omega\, \varepsilon\, \varepsilon_0)\, V} \tag{24}$$

Following experiments with saline solutions, was considered to have the information needed to proceed to carry out experiments with biological solutions, hence the first use of coil system as a reference and according to the above expressions, and also with a 500 mL volume, and a tank length of 85 mm was obtained with a suspension of high concentration of biomass (80 g / l of yeast), a 8.305 dB attenuation, comparing this with a 13.71 dB attenuation of coil system, this means a representation of a 65% increase in impedance in the suspension.

It should be mentioned that the embodiment of the vast majority of measurements for suspension was made with the HP4192A impedance analyzer, considering the structure and existing coil system, the utilization ratio of the impedance analyzer is because the probes used to interconnect arrangement coils are constructed on purpose, greatly decreasing the amount of error attributable to the length of the coaxial cables.

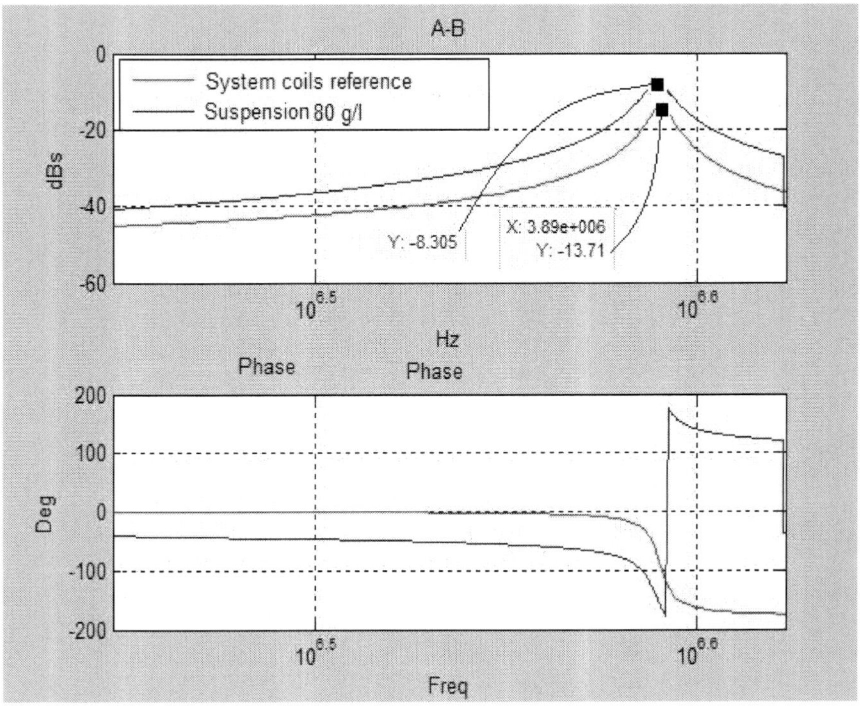

Figure 25. Impedance Representation of Biological Suspension, comparatively to the coil system.

After 35 minutes, the yeast was deposited at the bottom of the container; we proceeded to extract 350ml of water and recovered the same amount of water (350ml), but now with a salinity 8.78mS/cm.

The interesting thing in this experiment as shown in Figure 25, having increased conductivity, as a result of water replaced without electrolytes, with water with electrolytes (8.78mS/cm) decreased by 6% the impedance of the suspension by the lower proportion of yeast and major salinity, $Z_{lev} = 384.37\Omega$, $Z_{lev} + NaCl = 361.87\Omega$, $\Delta\Omega = 22.5$

9. Comparison of results

Measurements of saline cell and a biological suspension with HP impedance analyzer, allow us to characterize the results:

	Z (Ohms)	ε	μ (H/m)	σ (S/m)
2g NaCl	325.3	14.45	2.13E-04	0.00307409
4gNaCl	334.9	14.039	2.13E-04	0.00298597
10gNaCl	333.3	14.11	2.13E-04	0.0030003

Table 3. Characteristic values of a saline cell saline with varying degrees of salinity at resonance frequency of the coils system @ 3.825 MHz

	Z (Ohms)	ε	μ (H/m)	σ (S/m)
80g_yeast	395.64	11.68	2.16E-04	0.00252755
80g_yeast+6h	392.42	11.77	2.17E-04	0.00254829
80g_yeast+24h	385.08	11.99	2.17E-04	0.00259686

Table 4. Characteristic values of a biological suspension with various lengths of time at resonance frequency of the coils system @ 3.892 MHz

10. Conclusion

The presentation of results, by their very short nature, could be interpreted as an activity which is not time consuming, but the opposite is true, because from the moment of preparation of the experiments, there are always a number of imponderables, such as materials or materials that are needed to carry out the measurements do not have them, at least, operating conditions, climate, lighting, etc..

For the characterization of substances, suspensions and / or solutions must take into account, and in a very particular, maintain the same amounts, and make measurements in a repetitive manner, so as to place on record its findings to conditions geometric, physical and mechanical properties, different and contrast with conditions similar to those obtained. The measurements carried out with the Network Analyzer, which was mostly used equipment, are quite contrasting with measurements made with the HP impedance analyzer. As we can see with the first, which strongly influences the distance at which measurements were made, not with the impedance analyzer as it had a "fixture" on purpose.

Speaking of the actual material used and specific the experiment with saline cell, it was found that one of the factors that influence in the admittance at low frequencies is the electrical permittivity of water. The values depend on the frequency of measurements, and the frequency sweep, this due to the response of the dielectric constant of the solution, which varies considerably with frequency.

Using a saline solution with a high degree of salinity, more than 9 g / l of salt, we see that in the graphs of the results if ε'' / ε '<2, the ε'' accuracy degrades. This may be because the impedance of solutions with high salt is essentially resistive. In an opposite manner with a "window" at 999 KHz frequency on average, before the resonance frequency of the coil circuit, the impedance of the solution is essentially reactive, and since this type of impedance measurements of reflection / transmission were performed mainly with the network analyzer is a significant higher level of uncertainty in measuring the smaller capacitive component where ε'' is derived.

The HP impedance analyzer measurements, both saline solutions and suspensions, allowed us to compare results, specifically ε" values. In the suspensions could be evaluated especially an increase in the impedance, over time, possibly due to increased cell growth, hence an increase in the capacitive reactance of the suspension.

Considering the results obtained, we can consider the approach to the characterization as reasonably good since the relative permittivity is much greater than unity for the dielectric (both saline solutions and biological suspensions).

One of the properties of biological tissue, such as conductivity, is well reflected by the frequency dependence, particularly in the range of 10 KHz-10MHz (β dispersion range). The conductivity at low frequencies denotes the volume of extracellular fluid essence, the additional contribution of intracellular fluid volume with a significant increase in the applied frequency causes a significant increase in conductivity.

And finally and considering the substantial increase in interest in the development of magnetic induction spectroscopy (MIS) as a valid option for obtaining the conductivity of the human body without the need for direct contact with tissue (Korzhenevskii and Cherepenin 1997 Griffiths et al. 1999, Korjenevsky et al. 2000); In addition to other passive electrical properties of biological tissues (Hermann Scharfetter, Casañas Roberto and Javier Rosell, 2003).

Since MIS is based on measurements of small changes in magnetic fields, typically of the order of 1% or less at frequencies up to 10 MHz, and also because of their physical limitations is not recommended at frequencies below 10 kHz, this also represents a great challenge in electronic design, possibly one of its greatest disadvantages.

Author details

Jesús Rodarte Dávila, Jenaro C. Paz Gutierrez and Ricardo Perez Blanco
Department of Electrical and Computer Engineering, Juarez City Autonomous University, North Charro Avenue, Juarez City, Chihuahua, México

Acknowledgement

My eternal gratitude to Dr. Ramon Bragos Badia all his efforts and attention afforded to this humble servant, without which there would have been possible to achieve this work.

Ramon the Humans are more humans for humans like you, Thanks. Also I would like to thank to Petra Salazar for his typing work of this paper, without your help this work may be stay in the abstract.

11. References

[1] Grimnes S. Bioimpedance and Bioelectricity Basics, Academic Press ISBN 0-12-303260-1;p7

[2] Hoffer EC, Clifton KM, Simpson DC. Correlation of wholebody impedance with total body volume. J Appl Physiol 1969;27:531–4.

[3] Nyboer J. Electrical impedance plethysmograph, 2nd ed. Springfield, IL: CC Thomas; 1970.

[4] Boulier A, Fricker J, Thomasset A-L, Apfelbaum M. Fat-free mass estimation by the two-electrode impedance method. Am. J Clin. Nutr. 1990; 52:581–5.

[5] Harris A. Ductal epithelial cells cultured from human fetal epididymis and vas deferens: relevance to sterility in cystic fibrosis, Journal of Cell Science, Vol 92, Issue 4 687-690, Copyright © 1989 by Company of Biologists

[6] Asami K. Real-Time Monitoring of Yeast Cell Division by Dielectric Spectroscopy, Biophys J, June 1999, p. 3345-3348, Vol. 76, No. 6

[7] Ong K.G., Monitoring of bacteria growth using a wireless, remote query resonant-circuit sensor: application to environmental sensing. Copyright © 2001 Elsevier Science B.V. All rights reserved.

[8] Hofmann M. C., Transponder Based Sensor for Monitoring Electrical Properties of Biological Cell Solutions. Copyright © 2005 The Society for Biotechnology, Japan Published by Elsevier B.V.

[9] Scharfetter H., Casañas R., and Rosell J., "Biological Tissue Characterization by Magnetic Induction Spectroscopy (MIS): Requirements and Limitations,"IEEE Transactions on Biomedical Engineering, vol. 50, NO. 7, JULY 2003.

[10] http://www.ni.com/support/labview/lvtool.htm

[11] Service Manual R&S®ZVx Vector Network Analyzer Family: http://www2.rohde-schwarz.com/en/products/test_and_measurement/network_analysis/ZVx-%7C-Key_Facts-%7C-4-%7C-658.html

[12] Riedel C.H., Keppelen M., Nani S., Merges R.D. y Dössel. (2004) "Planar System for Magnetic Induction Conductivity Measurement Using a Sensor Matrix". Institute of Physics Publishing., Physiol. Meas. 25

[13] Rodarte J. (2008) Método Experimental de Medida de Impedancias Inalámbricamente Usando un Analizador de Redes. Instituto Tecnológico De Chihuahua., Electro 2008.

[14] Ong, K.G. and Grimes, C. A., "A resonant printed-circuit sensor for remote query monitoring of environmental parameters". Smart Mater. Struct., 9, 421–428 (2000).

[15] Edminister J., Nahvi M., "CIRCUITOS ELECTRICOS Y ELECTRONICOS", Cap. 13. Cuadripolos (circuitos de dos puertas), ISBN: 8448145437, Ed: McGrawHill, Serie Schaum

[16] Calculation of Formulae - Self-inductance ; www.ivorcatt.com/6_2.htm

[17] Donaldson N.deN., Perkins T.A., "Analysis of resonant coupled coils in the design of radio frequency transcutaneous links" , Med. & Biol. Eng. & Comput., 1983, 21, 612-627

[18] Commercial Software "PSpice" Ver 9.2 Copyright 1886-1999 by Cadence Design Systems

[19] Kenneth W. Misevich, Capacitive Humidity Transducer, IEEE Transactions on Industrial Electronics and Control Instrumentation, vol. IECI-16, NO. 1, JULY 1969.

[20] Mathcad, version 14.0.0.163 Copyright © 2007 Parametric Technology Corporation. All Rights Reserved.

[21] Pething, 1979 R. Pething. Dielectric and Electronic Properties of Biological Materials, Wiley (1979).

[22] Awad S., Allison S.P., Lobo D.N., "The history of 0.9% saline.", Clin. Nutr. 2008 Apr; 27 (2): 179-88. Epub 2008 Mar 3.

Chapter 5

Laser-Induced Breakdown Spectroscopy

Taesam Kim and Chhiu-Tsu Lin

Additional information is available at the end of the chapter

http://dx.doi.org/10.5772/48281

1. Introduction

Laser-induced breakdown spectroscopy (LIBS) is an atomic emission spectroscopy. Atoms are excited from the lower energy level to high energy level when they are in the high energy status. The conventional excitation energy source can be a hot flame, light or high temperature plasma. The excited energy that holds the atom at the higher energy level will be released and the atom returns to its ground state eventually. The released energy is well-defined for the specific excited atom, and this characteristic process utilizes emission spectroscopy for the analytical method. LIBS employs the laser pulse to atomize the sample and leads to atomic emission. Compared to the conventional flame emission spectroscopy, LIBS atomizes only the small portion of the sample by the focused laser pulse, which makes a tiny spark on the sample. Because of the short-life of the spark emission, capturing the instant light is a major skill to collect sufficient intensity of the emitting species. Three major parts of the LIBS system are a pulse laser, sample, and spectrometer. Control system is usually needed to manage timing and the spectrum capturing. Figure 1 illustrates those three major components and a computer in the conventional LIBS.

The LIBS has been used for the materials detection and analysis in various applications, such as steel and alloys[1-8], paints and coatings[9-15], wood pre-treatment[16], polymers [17], bacteria[18], molds, pollens, and proteins[19,20], and space exploration[21]. The great majority of LIBS results were consolidated in the reviews[22] and books[23, 24].

In spite of its advantage in analytical spectroscopy, LIBS application is still restricted within certain areas and propagation of the technology is not very wide. Many laboratory LIBS systems are built in schools, research labs and companies with discrete optical parts. Their pioneering approach in the new application seemed promising for a certain samples, but actual use in the application field is usually very limited. We can explain the situation with other analytical techniques, for example, Gas Chromatography (GC). The GC can separate the volatile species. However, one GC setup can work for a narrow range of species grouped in the sample. For different applications, the user must change the GC column, detector,

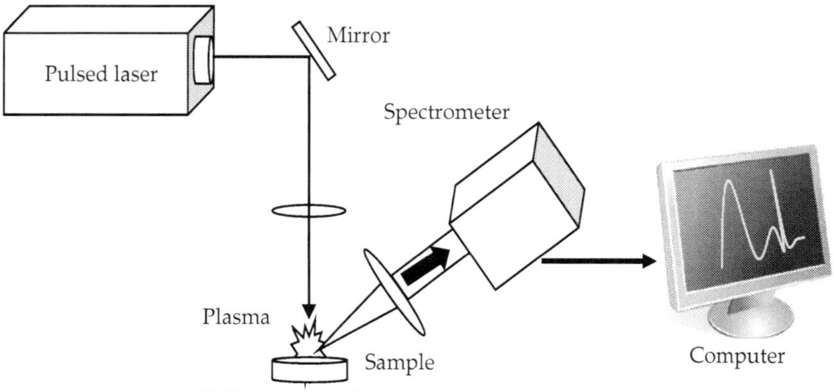

Figure 1. The conventional LIBS system configuration

carrier gas or at least use a new column temperature cycle. The application of LIBS also needs case-by-case adjustment. Many new applications start with looking at the advantages of LIBS and choosing a LIBS setup, and it still needs a detailed investigation for successful analysis.

This chapter describes how the LIBS system works and explains the major parts of LIBS to select specific functional requirements for its intended application. The three major parts: laser, sample and spectrometer are explained. The laser provides the breakdown energy and plasma generation. Analytical sample is the target of the laser shot and the source of emission species. The spectrometer comprises detection system with light detector and computer. Their disadvantages and limitations are discussed then suggesting how to select the equipment type and configuration to maximize the advantages of LIBS. This will provide a beginning inspiration of LIBS systems to install and apply the desired specific analytical purpose or application area.

2. LIBS system design with modern technology

a. Laser as a breakdown energy source

LIBS uses pulsed-laser light and focuses it onto the sample surface to make a plasma plume that contains the highly excited species of the sample composition. For generating plasma, there is a threshold value of the energy density. The threshold level will depend on the absorption coefficient of the sample surface of the laser wavelength, which is highly different by the sample phase. Gas and liquid need more energy to make breakdown. Solids with a dark color surface easily make a strong breakdown compared to clear or highly reflective solids. Figure 2 shows the effect of laser energy to make breakdown by the relation of laser power and focusing. Starting with a laser beam as 1 cm diameter, this light beam can be condensed by a convex lens. The focused beam density becomes 160 J/cm^2 as in Figure 2(a). Also, the laser is operating in the pulsed mode, assuming a10 nsec duration, total power per unit of time will be 16 GWatt/ cm^2 as

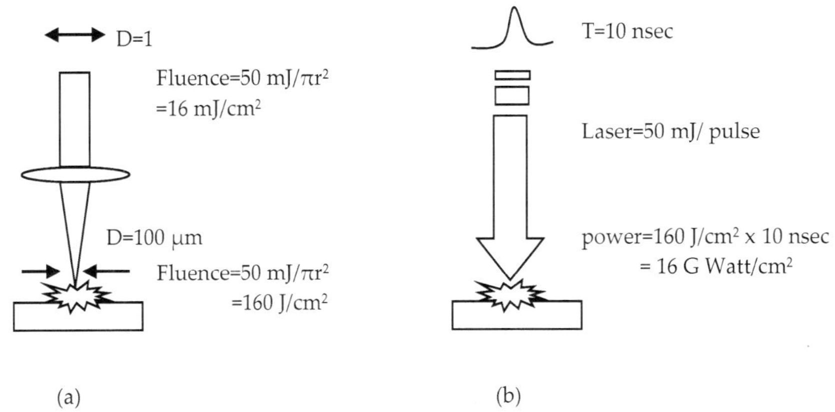

Figure 2. Laser energy delivery for breakdown condition. (a) focusing effect, (b) pulsing effect

in Figure. 2(b). Most of breakdown needs a few GW (10^6 Watt) of energy density, indicating that 50 mJ of laser energy is sufficient to make breakdown and evaporate most of material.

At the early stage of LIBS development, several types of pulse laser were used to make laser-induced breakdown plasma. An eximer laser was an important pulse laser especially for the UV light pulse. XeCl-eximer with 308 nm was used in the LIBS to measure elemental distribution on the paper coating[25]. The laser energy of 0.2 mJ was focused and made a crater of 30 μm diameter. This energy is corresponding to 10^8 W/cm^2. More than 90 % of ingredients in the paper coating are pigment, binder and other agents. The pigment's main component is usually aluminum oxide, silicon dioxide and calcium carbonate. The mass of coating material ablated by single laser pulse was estimated to be about 2 ng by a laser shot. A typical nitrogen laser has a wavelength 337.1 nm and a pulse duration of 10 nsec. Just like the eximer laser, the nitrogen laser LIBS configuration in Figure. 3 also includes discharge from the wide shape of the electrode. The laser beam is usually a few cm wide, so a tight focusing is needed. The surface of solar cell was measured by nitrogen laser breakdown and only a 40-nm-thick TiO_2 layer was detected[26]. The very popular pulse laser is Nd:YAG laser because it has a solid laser oscillator in a small size and light weight. The fundamental wavelength is 1064 nm with a pulse duration of 10 nsec typically. The Nd:YAG laser does not require any gas supply. The laser model for LIBS size usually has a closed loop water cooling that excludes external connection. A typical LIBS setup was shown in an earlier paper[27] as in Figure 4. A 50 mm focal length convex lens makes a simple optics configuration to make plasma on the sample.

b. Optical arrangement for laser–induced breakdown spectroscopy

When a laser shoots on the sample surface, a plasma plume arises from the inner to the outer surface. The actual size of plasma plume made by a 100 mJ laser pulse will be few millimeters. During the plasma propagation from the sample surface, the time profile features can be observed. The very initial emission is generated at the bottom of the plasma

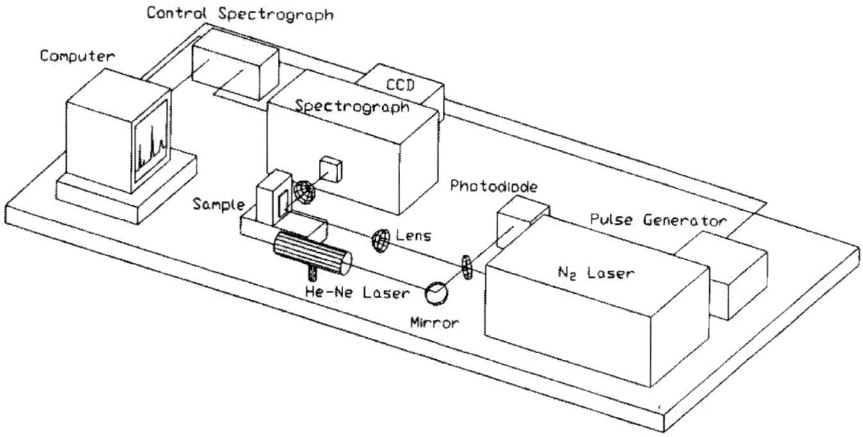

Figure 3. A LIBS setup with nitrogen laser.

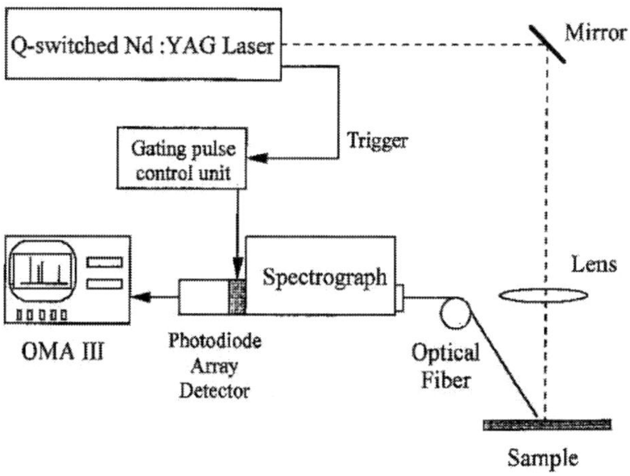

Figure 4. Schematic diagram of the LIBS setup with ND:YAG laser

plume, and then expanded to the outer plume. Depending on the light collecting optics, plasma propagation is captured at the different time. At the initial LIBS design uses a side-view emission collection as in the Figure. 5 (a). The angle between the laser light path and the collecting optic can be any angle, but is typically 30-45 degrees. Some experiments use 90 degrees, which is a complete side view of the plasma and will lose some portion of emission by the shadow of the sample itself. This configuration is occasionally used for plasma physics study. The collateral view design Figure. 5 (b) is a useful optical configuration for non-fixed sample distances. Laser light path shares emission collection optics. A selective wavelength reflector or prism can be used to separate laser light and emission through the light path. This design has several optical parts and needs complicated adjustments for optimum light measurement. The collateral configuration has many advantages. Collecting optics looks at the plasma in front of the plasma (or top of the plasma)at every point in the light axis and in the focus cone, which means they capture every light emitting species during plasma propagation to the space. Because some elements have different propagation profile than others, propagation height changes the signal significantly at the angled collection. The next advantage is that the optical part can be integrated in the compact enclosure, and it allows the operator to move the optics (detector head of LIBS) more freely. Remote monitoring LIBS, hand held design LIBS, should be compact and have a mostly collateral optics configuration.

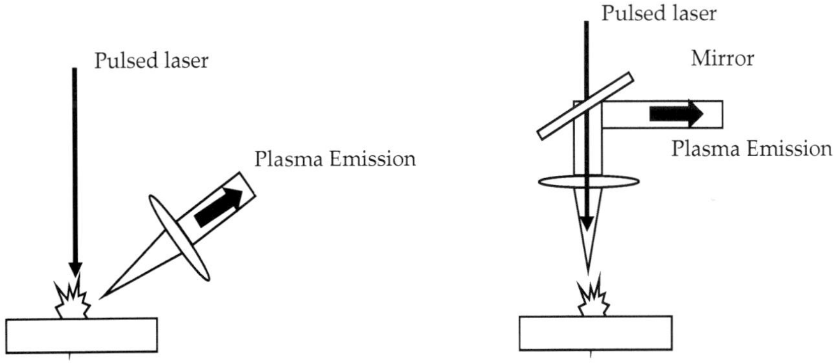

(a) (b)

Figure 5. Side collection and (b) collateral collection configuration of plasma emission

c. Sampling technique

The first mentioned advantage of LIBS has been no-sampling step. In the very beginning review in the *Encyclopedia of analytical Chemisstry*, Yueh, Singh and Zhang described it as "LIBS uses a very small amount of samples, and no sample preparation is necessary. It has the ability to perform real-time analysis because it prepares and excites the sample in one step". They then consecutively mentioned, "The disadvantage of LIBS is that the plasma conditions vary with the environmental conditions as well as the laser energy fluctuation."We can infer from the description of LIBS that no-sampling is both an advantage and a disadvantage. Most of analytical the techniques need a certain sampling procedure to bring the sample to the technique (or machine). During the sampling procedure, like the acid digestion in the flame analysis, the sample is homogenized and their matrices become concordant. However, if LIBS analyzed the sample without any pre-treatment process, then the irregular homogeneity is inevitable. As a result LIBS will include severe matrix effects at the real sample. It will mitigate the biggest advantage, i.e., no-sample process. In other words, if the sample is measured as it is, the species in same concentration do not make a consistent signal, the analytical result will be severely diverted.

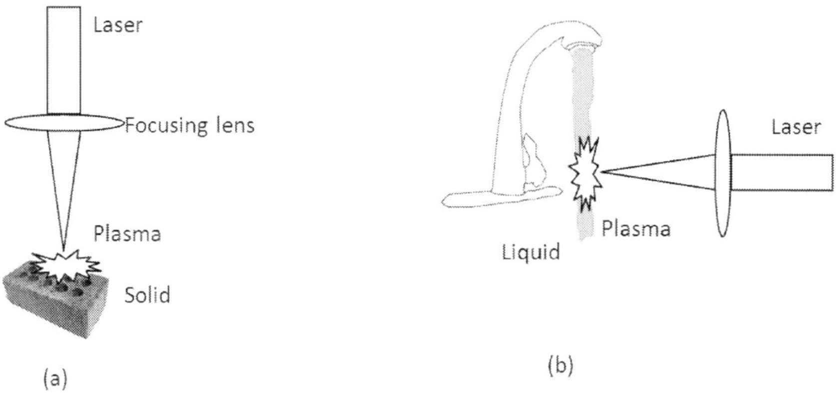

Figure 6. Solid sample and liquid sample under LIBS measurement

The fluctuation will be more serious because LIBS takes only a small amount of the sample, usually a micron sized spot. Two possible sample types are depicted in the Figure. 6. A solid sample is the most convenient sample type. Metal and ceramic samples include elements with strong atomic and ionic emission. Their emission spectra are measured at the range from UV to visible light, which is feasible by the most spectrometers. The spectra from many elements from the tool steel are shown by the Nd:YAG laser excitation[28]. In this research, the microscopic view of the ablated holes made on tool steel is about 10 microns in diameter. This resolution indicates that any inhomogeneity more than 10 microns will be clearly observed from each laser pulse measurement. The intensity of element-specific spectra provides a simple qualitative analysis. Their method was sufficient to characterize the nature of the defect by a simple estimation of the elemental composition between the basic material and the defect.

d. Capturing emission light

The LIBS signal is instantaneous and decays quickly. Temporal control of the detecting device is very important. In spite of the fact, overall emission can be captured by opening entire time of the spectrometer, most of LIBS measurement is controlled by time gate operation. Time control improves the signal-to-noise ratio by eliminating the continuum emission. A typical emission profile shown in the Figure 7, recorded at the different heights from the sand/ soil mixture sample[29]. As soon as the laser fires with the duration of a few nsec of pulse width, the plasma intensity is propagating outward from the sample surface. At about 0.5 μsec, plasma is observed at 0.3 mm away from the surface. At the propagating distance is 3 mm at 12 μsec, then plasma cool down with decreasing intensity until 20- 30 μsec. The plasma size will be much smaller and life is shorter when a weak laser power is used. The experiment uses aluminum[30] with a diode-pumped Nd:YAG laser, which can run at a faster repetition rate (kHz) with a laser energy of 80 μJ, was setup under the microscope excitation and detection optics. Like other flash lamp pumped lasers, the temporal profile of continuum emission is shown in Figure. 8 for aluminum atom (Al 396.1 nm) and aluminum ion (Al II at 358.6 nm) emission lines. The laser pulse was fired at the zero time of the x- axis. This profile indicates, the broad band continuum emission, which comes from high temperature heated plasma and regardless of the species in the plasma, has a lifetime of about 13 nsec. The ionic line from Al ion has shorter lifetime about 24 nsec. The neutral lines stay much longer, up to 80 nsec.

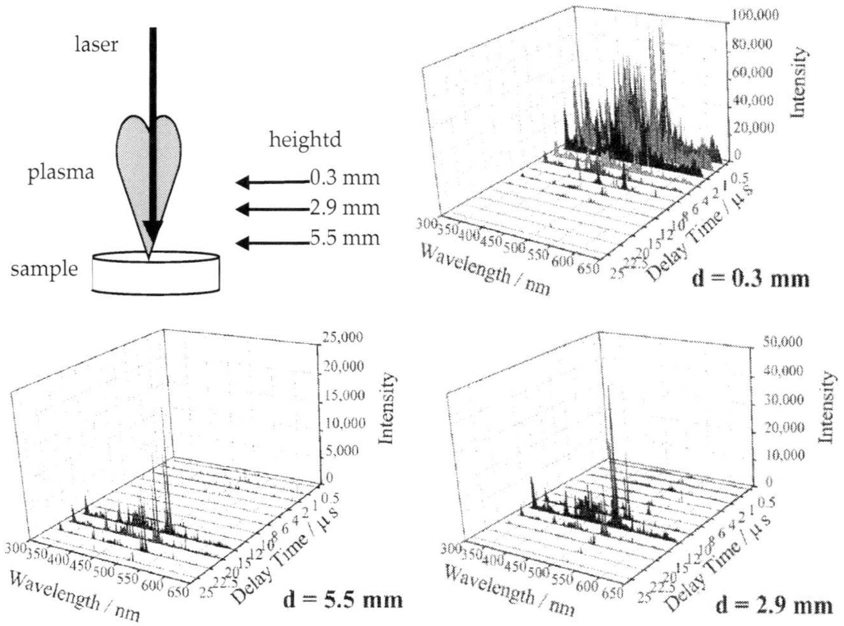

Figure 7. Spectra as a function of decay time measured at three observation distance from the sample surface. The original figure is rearranged to indicate observation height more clearly.

The lifetimes of laser-induced plasma are easily compared at various excitation energies from the silicon sample[25]. The time profile shows the plasma emission signal depends on the excitation pulse energy. Absolute intensity of the signal will increase by increasing laser pulse energy. The decay plot in the Figure 9 is normalized to a maximum intensity for comparison. This research explains the decay time dependence by the excitation energy that the probability of excitation to higher energy level is increased and more populated, leading to a longer decay time. Also, the upper state of the monitored transition receives population from this higher state at later times and lengthening of the rise time of the signal

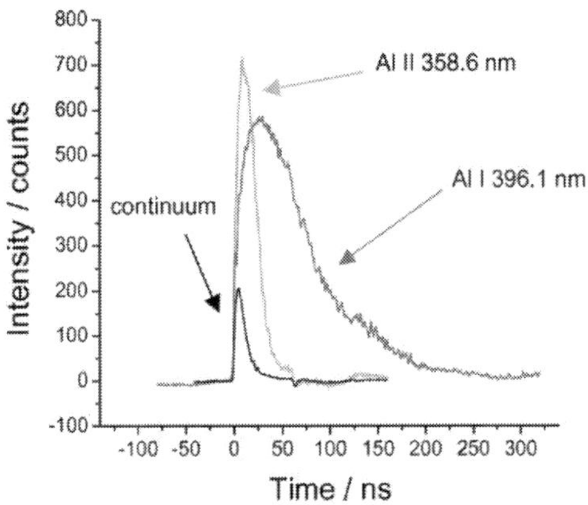

Figure 8. Temporal profile of continuum emission and aluminum (atom and ion) emission.

will result. As a result, the lifetime point of LIBS will be changed by the system setup, especially using laser power. Capturing time of emission signal should be determined empirically by looking at the profile, usually at peak intensity point.

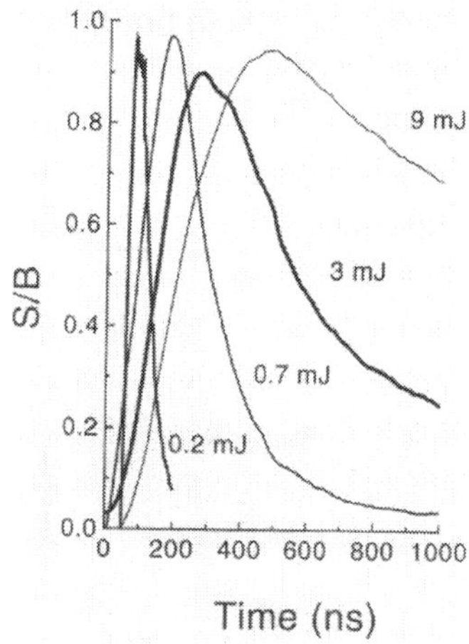

Figure 9. Time resolved signal-to-background ratio of the silicon line at 251 nm at various excitation energies.

e. Spectrometer and detector

Spectrometer completes the detecting part with a photo sensor and a manipulating computer. The spectrometer must have proper resolution and sensitivity. Also, in many cases the plasma emission needs to be separated from the continuum background signal, the detector has to be operated by timing control or gating operation. Various types of spectrometer with CCD array detector are available in the market. The wavelength range needed for LIBS is UV to visible range to have detection of most elements. If the dispersion of the spectrometer is 0.3 nm to measure 1 nm peak with three pixel, the pixel to pixel dispersion should be 0.3 nm. Total of 1024 pixel CCD array can have coverage 1024 x 0.3 = 307 nm, which can assign the range as 250 nm to 557 nm span. In many cases, the sample will have mixed elements and the emission lines will be overlapped and difficult to distinguish with 0.3 nm resolution. A conventional CCD array detector may not provide sufficient resolution and coverage to measure LIBS.

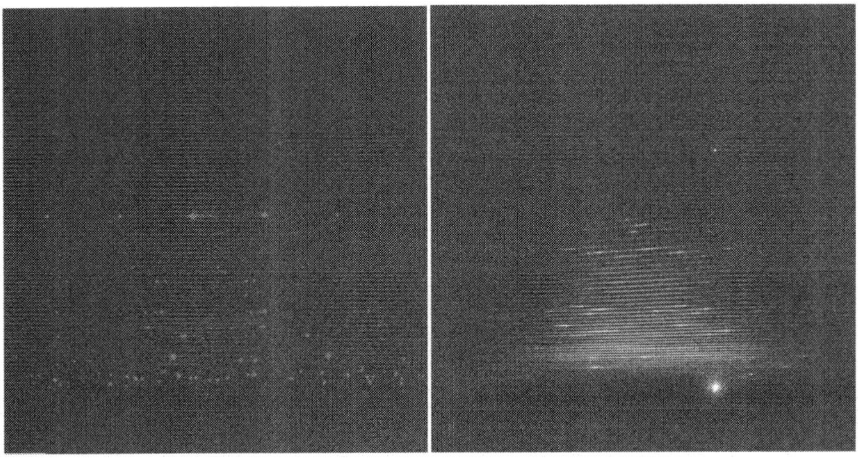

Figure 10. Echelle spectrometer dispersion image (a) Hg lamp, (b) LIBS spectrum of Sn metal

A correction of the array detector resolution is accomplished using multiple stacked spectrometers. For example, 5 spectrometers with 1000 array CCD stacks will cover a 500 nm span, in which each spectrometer covers a 100 nm range with 0.1 nm resolution. Echelle spectrometer uses very high orders of dispersion. One or two prisms are used to separate each diffraction order. As a result, the spectra are dispersed in two dimensional surfaces as shown Figure 10. The CCD detector in the Echelle spectrometer should be a two dimensional, the same as in the image camera. The continuum emission from the spark also engages in the Echelle spectrometer, so the detector must have gated operation. To satisfy those requirements, such as two dimensional, sensitive and gated operation, the cost of CCD detectors for the Echelle spectrometer is still significantly high.

3. Sample type and their application

This section illustrates various application examples that have performed from the authors' research group. As we have mentioned in the previous section, the LIBS technique needs individual verification for an application, because it does not need sampling. Three typical applications are explained in this section.

a. Paint and coating identification

Materials and techniques of paints and coatings require an appropriate verification process to achieve the desired property of the protective finishing. The organic coating involves a multi-step process in which the quality of the metal finish required for an industrial product would determine the number and the type of steps in a given process[31]. These multi-step coating processes include the selection and composition verification of substrates, surface cleaning, surface pre-treatment, primer, topcoat, and the application of paint curing methods. The paint formulation is a mixture of multi-ingredients, composing: resins, solvents, pigments, fillers, corrosion inhibitors, and other rheological additives. The organic coating in metal finishing practice is extremely complex. The complexities of paint compositions, paint types, and painting processes make their chemical analysis very difficult. In spite of some elemental analysis methods that have been well established for the general purpose in chemistry, the determination of metallic components in paint has been relied on the indirect analytical methods. For example, the metallic zinc dust in the Zn-rich epoxy primer was determined by differential scanning calorimetry (DSC)[32]. The DSC method measured the apparent heat of fusion of the paint sample, and compared this measured value to the standard value of pure zinc as an indirect measurement of zinc composition in paint. Infrared absorption spectroscopy is useful sometimes for the composition analysis if the paint ingredients contain any specific functional groups which are spectroscopically active[33], such as the isocyanate group in the urethane. The direct analysis of these functional groups may be possible only if the paint sample is uncured, and contained a relatively simple composition. In practice, there is no direct way for identifying a cured paint film. Once the paint, e.g. epoxy or urethane, is applied and fully cured, no more epoxy or isocyano functional group would remain in the paint film. Even though, the researchers have attempted to characterize the fully cured paint products by identifying the hydroxyl or amino groups, and use them for differentiating the epoxy paint or urethane paint. The results are generally inconclusive because the majority of other cured paints also have those functional groups as reaction products. The LIBS technique described in this section shows the capability of coating identification at the specimen surface. [34]

3.1. Materials of paints and coatings

The substrates selected for spectral fingerprinting by LIBS technique are: (i) aluminum alloys (2024-T3, 3003, 7075-T6 from Advanced Coating Technologies, Inc. (ACT), Hillsdale, MI) and pure aluminum foil (Aldrich Fine Chemicals), and (ii) cold-rolled steel (CRS from Q-PANEL, Cleveland, OH and Caterpillar's OEM facility) and pure iron (Aldrich). The surface pretreated substrates used for LIBS studies are: (i) Al 2024-T3/Clad, a ultra thin layer of pure aluminum is treated on 2024-T3 aluminum alloy, (ii) Al 2024-T3 Bare/Alodine 1200, the surface of 2024-T3 aluminum alloy is treated with Alodine 1200 solution which contains chromates (i.e., hexavalent chromium), (iii) phosphated (Bonderite 1000, or B-1000) and phosphated/chromated (B-1000/P-60) from ACT, and (iv) galvanized (electroplated and hot-dipped) and galvalume steel plates that have a treated surface layer of Zn and Zn/Al, respectively. Eleven heavy-machine OEM paint samples (four urethane, three epoxy, and four alkyd) were used for the spectral fingerprinting by LIBS technique. The paints were applied on 2 x 4 inch steel panels using a spray coating method and cured thermally or by air-dry as directed by the paint manufacturer.

3.2. LIBS: An *in situ* and quasi-nondestructive analytical technique

LIBS technique is capable of carrying out a depth profile analysis of successive surface layers by controlling and calibrating the working parameters of LIBS system. A Q-switched Nd-YAG laser (Continuum, Minilite II) operating at a wavelength of 1064 nm was employed as the excitation source. The pulse laser has a power of 50 mJ per pulse and a pulse width of 8 ns. The laser beam was focused onto the sample with a 5 cm focal length lens. A fiber optic cable collected the breakdown plasma emissions at the sample surface and directed them to a portable, miniature, CCD array fixed-grating spectrometer. Figure 11 shows the optical microscope images of some LIBS-measured sites on a painted steel panel: (a) paint surface before analysis, (b) one laser pulse applied, (c) two laser pulses applied, and (d) five laser pulses applied. The scale bars are 50 m in length. The first shot of the focused laser beam (70 mJ/pulse) made a burn pattern on the paint surface (Figure. 11b). The successive laser pulses penetrated into the coating layers and eventually reached the metal substrate (Figure. 11d). In principle, the LIBS spectrum recorded after each laser pulse, or for each layer of the multilayer paint samples, should generate the characteristic breakdown spectral peaks of the corresponding chemical compositions. The affected coating area by the laser pulse is limited to less than 100 μm in diameter (Figure. 11b, 11c and 11d). The layer thickness of materials that each laser pulse can penetrate is a function of laser fluent at the focal point, optical geometry and material type. It is important to mention that a well-established elemental analysis method, such as EDX can also perform a similar analysis. However, the sample used in EDX analysis must be cut into a few millimeter sizes for fitting inside the detection stage in a vacuum chamber. Also, the cut samples need to be covered with a conductive coating for EDX analysis because paints are the dielectric materials. This film deposition of conductive layer is again done under another vacuum facility. These complicated sampling processes are eliminated in the LIBS analysis.

3.3. LIBS characterization of substrates

A less trivial experiment was performed to determine whether the LIBS system could be used to distinguish between different alloys of the same main metal content or between the same metal alloys obtained from different manufacturing sources. Aluminum has many alloys in common use, and these alloys frequently need specific protective coatings for aerospace applications. The 2024-T3 Al alloy contains copper as the main dopant (i.e., 4.4% Cu, 0.6% Mn, and 1.5% Mg). The 7075-T6 Al alloy contains zinc as the main dopant (i.e., 5.6% Zn, 1.6% Cu, 2.5% Mg, and 0.23% Cr). The 3003 Al alloy contains no specific main dopant (0.0-0.6% Si, 0.0-0.7% Fe, 0.05-0.20% Cu, and 0.0-0.10% Zn). The Al alloys, 2024-T3 and 7075-T6 have high surface protection strength, whereas Al alloy 3003 displays a good pitting corrosion resistance. All three alloys should show aluminum peaks in LIBS spectra, and 7075-T6 Al should display zinc and magnesium peaks and 2024-T Al should display

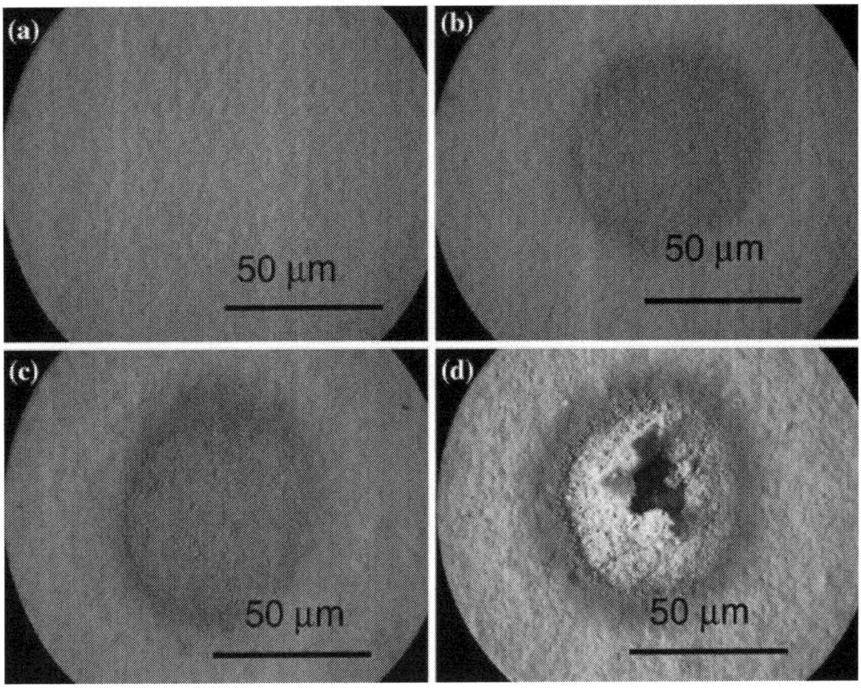

Figure 11. Microscopic images of laser burn patterns on paint film in LIBS experiment after (a) zero, (b) one, (c) two, and (d) five laser pulses

copper and manganese peaks in their breakdown spectra. Figure 12 compares the LIBS spectra recording from 250 nm to 450 nm for pure aluminum foil (spectrum a) and three Al alloys (spectra b, c, and d). As expected, spectrum 13a gives only aluminum peaks at 281.6 nm, 306.3 nm, 308.2/309.3 nm, 358.0 nm and 394.4/396.1 nm.

The spectrum of aluminum alloy shows, in addition to the aluminum peaks, three spectral peaks at 328.2 nm, 330.2 nm, and 334.8 nm are due to zinc (I) ionic states. The 7075-T6 Al alloy gives also the LIBS peaks at 278.6 nm, 285.2 nm and 383.5 nm for Mg and at 325.0 nm, 327.7 nm and 423.0 nm for Cu emission. The LIBS technique is not only able to identify the chemical compositions of alloys, but also capable of differentiating the possible contaminants in those alloys. For example, the contamination of Mn has been detected in Al 7075-T6 sample as illustrated in spectrum 13d. The contaminants of Mn and Mg are observed in spectrum 13b for Al 3003 sample.

The qualitative LIBS spectral assignment is also carried out for pure iron strip, cold-rolled steel, and industrial steel coupons used in the Caterpillar's OEM facility (referred to as CAT machine steel). The bare cold-rolled steel (CRS, SAE 1010) has a composition of 0.08-0.13 % C, 0.3-0.6% Mn, 0.04% P(max), and 0.05% S(max). The LIBS spectra recorded from 250 nm to

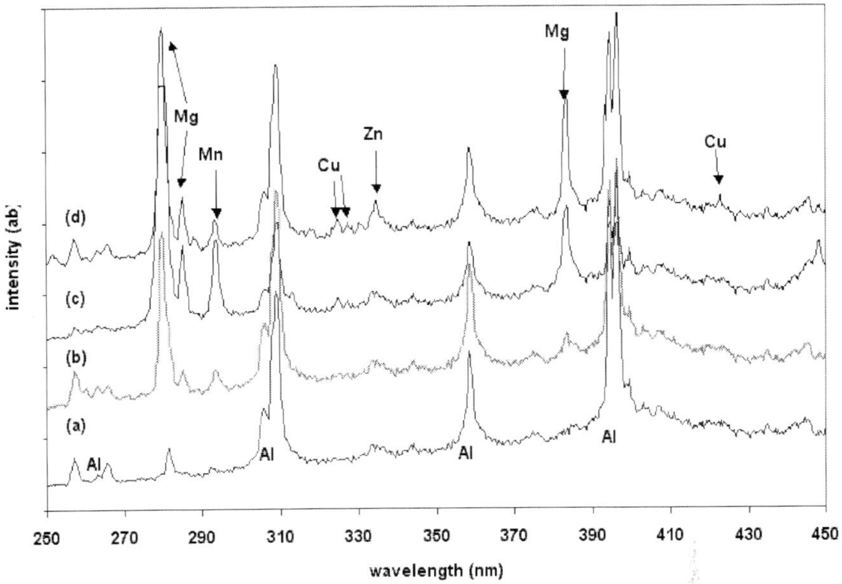

Figure 12. LIBS spectra of aluminum alloys, (a) pure Al foil, (b) 3003 alloy, (c) 2024-T3 alloy, and (d) 7075-T6 alloy

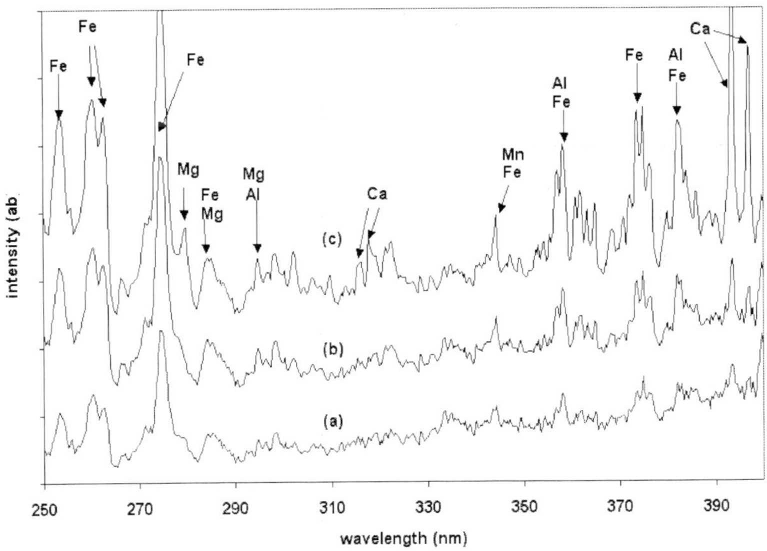

Figure 13. LIBS spectra of steel panels, (a) pure iron piece, (b) cold-rolled steel from Q-PANEL, and (c) CAT machine steel

400 nm are shown in Figure 13: (a) pure iron strip, (b) cold-rolled steel, and (c) CAT machine steel. Spectrum 13a shows LIBS peaks for the pure iron piece at 259.9 nm, 262.6 nm, 275.0 nm, 358.1 nm, 373.4 nm and 373.7 nm. The laser breakdown emission for CRS as shown in spectrum 13b is almost identical to that of the pure iron strip, except an additional peak at 344.3 nm which may be assigned to Mn as incorporated in the cold-rolled steel. The spectrum 13c indicates that CAT machine steel is not a pure iron piece or a standard CRS sample, but rather is a surface pretreated CRS. The surface layer of CAT machine steel contains Ca, Mg, Al, Mn, and P (at 589.1 nm), in addition to Fe. The results indicate that CAT machine steel is an iron phosphate treated CRS, containing a substantial quantity of Ca and Mg, and some small amount of Al, and Mn in the phosphating bath.

3.4. LIBS characterization of surface pretreatment layer on substrates

Another important part of this research is to establish the effectiveness of LIBS spectral fingerprinting technique for characterizing the composition of any metal surface pretreatment that may have been applied on the substrates. The common metal surface pretreatment used on aluminum alloys today is a chromium-based pretreatment (such as Alodine 1200 or Alodine 1000), which usually contains the chromates (i.e., the compounds contain hexavalent chromium). There are different processes used for surface pretreatment on aluminum alloys; some processes cause a color change of the metal surface to a yellowish color, and some cause no color change at all. In the latter case, it is almost impossible to tell, visually, whether the metal alloy has been pretreated. In this work, the different panels analyzed by LIBS are aluminum alloys of 2024-T3 bare, and 2024-T3 Clad (Clad: a thin layer of pure aluminum on 2024-T3 substrate). The surface pretreatment layer on 2024-T3 bare panel is Alodine 1200. The main active ingredient of Alodine solution is potassium dichromate or strontium chromate. Upon the deposition of a thin layer of pure aluminum on Al 2024-T3 bare, the LIBS spectrum of Al 2024-T3/Clad should give only the pure aluminum peaks which are the same as spectrum 12a for pure aluminum foil.

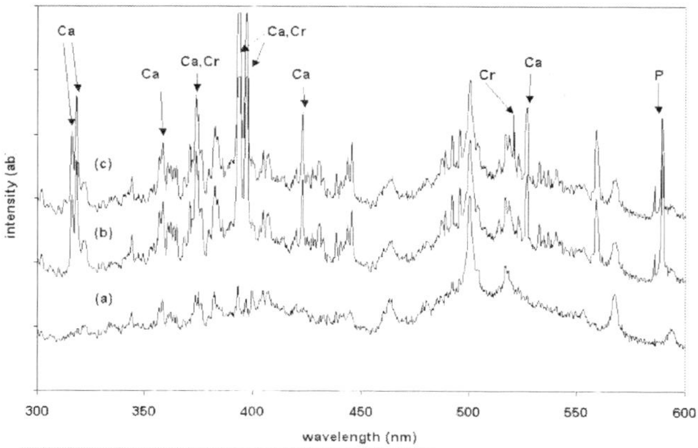

Figure 14. LIBS spectra of surface pretreated layers on CRS substrate, (a) untreated, (b) B-1000, and (c) B-1000/P60 panel

The surface pretreatment of metal prior to the application of a coating or adhesive is a conventional industrial practice to improve the coating adhesion and inhibit substrate corrosion. For cold-rolled steel, the phosphate conversion coating (e.g., Bonderite® B-1000) and phosphating/chromating (using parcolene 60) pretreatment (e.g., B-1000/P60) are commonly used. The LIBS technique is used to fingerprint the differences in chemical compositions of the surface pretreated layer. The LIBS spectra were taken at the first laser shot spot on (a) untreated CRS panel, (b) B-1000 CRS panel, and (c) B-1000/P60 CRS panel. The laser-induced breakdown spectra of untreated and different chemically treated CRS panels are clearly identifiable and their spectral assignments are marked in Figure 14. The LIBS peaks in spectrum 13a are assigned to Fe and Mn, and are similar to those in spectrum 14b. The phosphate treated B-1000 panel gives a few additional LIBS peaks in spectrum 14b, such as P at 589.1 nm and Ca at 315.8 nm, 318.2 nm, 358.3 nm, 394.0 nm, 423.0 nm, and 527.1 nm. In spectrum 16c, the additional P60 treatment on B-1000 CRS is evident by the appearance of chromium peaks at 373.9 nm, 396.8 nm, and 527.1 nm. When these LIBS spectra are compiled in the software system as a standard library file, they can be used to determine if the surface pretreatment processes (including composition, uniformity, and thickness) have been done in according to the products specification.

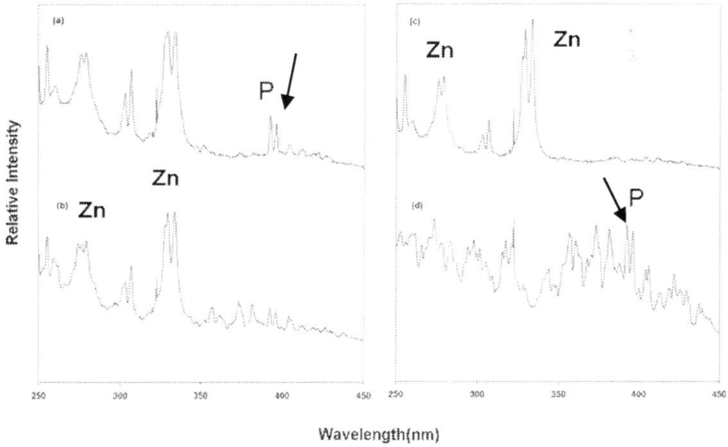

Figure 15. LIBS spectra of EZG panel, (a) from ACT Laboratories, Inc., (b) from China Steel Corp., Taiwan, (c) pure Zn metal piece, and (d) B-1000 CRS panel

Zinc-coated steel (such as Zn/B-1000) is known to inhibit iron corrosion, similar to the effect of zinc anodes. The addition of aluminum to zinc is highly beneficial in improving its corrosion resistance and has resulted in the development of coatings with aluminum contents between 5 and 55% (i.e., "galvalume" zinc-coated steel). Zinc coatings may be applied to steel panel by hot dipping (i.e., hot dipped galvanized steel, HDG) and electroplating (i.e., electrogalvanized steel, EZG). Due to the high degree of variations in the processing of EZG, HDG, and galvalume, it is critically important to have a versatile materials characterization technique, such as LIBS CoatID, to verify the manufacturing conditions of zinc-coated steel at the different factory sites. For a simple illustration, we use LIBS system to test two EZG panels (ACT Laboratories, Inc. vs. China Steel Corp., Taiwan), two HDG panels (ACT vs. Valspar Corp.), and two galvalume panels (Valspar Corp. vs. China Steel Corp.). Figure 15 compares the breakdown emission spectra (recorded from 250 nm to 450 nm) for (a) EZG panel from ACT, (b) EZG panel from China Steel Corp., (c) pure Zn metal piece, and (d) B-1000 CRS panel from ACT. The LIBS spectra of EZG panels (spectra 15a and 15b) should resemble those of the combined spectra of pure Zn (spectrum 15c) and B-1000 CRS (spectrum 15d), depending on the thickness of both phosphate layer on bare CRS and Zn-galvanized layer on B-1000 CRS panel. By comparing the EZG panels processed at ACT Laboratories, Inc. (spectrum 15a) and that processed at China Steel Corp. (spectrum 15b), it shows that both EZG panels have been subjected to the electrogalvanizing process as stated in their products data sheet. However, the LIBS was able to distinguish a thinner Zn-galvanized layer in the Taiwanese sample, because the steel plate was not covered fully by the Zn-layer and thus the B-1000 steel peaks are still quite visible as shown from 350 nm to 450 nm in spectrum 15b. On the other hand, both Zn-galvanized layer and B-1000 phosphate layer in ACT sample are thicker than those in the Taiwanese sample, as indicated by the appearance of a strong P emission doublet and also several intense Zn peaks. In ACT sample, the thicker Zn and phosphate layers give a higher coverage on the steel panel, and thus almost no steel peak is observed in Figure. 15a.

3.5. LIBS identification of paints and coating ingredients

Eleven paints from Caterpillar's OEM coating facility were selected for the identification test by LIBS technique, and listed in Table 1.It is noted that all paint samples have the same color (i.e., Caterpillar yellow) with only slightly different tint, the differences are hardly distinguishable with naked eyes. Samples 1 to 4 are two-pack urethane paints, 5 to 7 are two-pack epoxy paints, and 8 to 11 are one-pack alkyd paints. The paint samples 1 and 5-9 are primers, whereas those of 2-4 and 10-11 are topcoats. The processing methods used in coating applications, such as drying and thermal curing conditions are specified in the remark column of Table 1. The paint systems used in Caterpillar's OEM facility were specifically formulated by the paint manufacturers that have been successfully tested and verified for the required protection of heavy duty machines. Once the paint formulations were established, the manufacturer would strictly maintain the composition of paint ingredients in an effort to achieve a good quality control. This is the reason that LIBS technique may be effectively used for fingerprinting a specific brand of paint.

Figure. 16 displays the LIBS spectra for the eleven paint samples listed in Table 1. The topcoat paints (samples 2-4 and 10-11) display a relatively simpler LIBS spectrum than that of the primer paints (samples 1 and 5-9). In all spectra, the LIBS peaks grouped around 250 nm may be attributed to iron oxide as a dispersed pigment. The peaks originated from calcium at 393.3 nm and 396.8 nm are predominantly shown in the primer type paints. Calcium carbonate has been used at high levels for certain paints because of their low oil absorption. Calcium compound imparts some film structure to the wet paint by improving the stability to sedimentation of other heavier pigments in paint. It is not surprising that primer paint for CRS coating contains a rich calcium ingredient. The primer paints, samples 1, 6, 8, and 9, are shown to contain not only calcium carbonate but also magnesium silicate, as their corresponding LIBS peaks displayed at 279.8 nm and 383.5 nm.

In the previous section, the peak picking algorithm has been successfully used for characterizing substrates and surface pretreatment layers which contain only a few elements and have the well-characterized LIBS peaks. Since paint formulation contains a rather complex mixture of multi-ingredients, thus the decisions for paint identification could best be made by peak correlation algorithm. Any spectral pairs of identical samples must show a 100% correlation value. Due to the possible fluctuation in laser power density, the inhomogeneity of paint film compositions, and the variation in thickness of a paint film, the LIBS spectra for both testing and reference samples were measured at ten (10) different spots for each painted panel. A statistical average spectrum was made to achieve the reproducibility for the identification of a paint sample. The correlation values of identical samples show a 96-99% of reproducibility. On the other hand, the correlation values between two different types of paints, such as urethane and epoxy, show to be around 86.8 ± 0.7%. These correlation values give a clear discrimination between types of paints. Based on the correlation values, we can say that the test sample has a good match to the reference sample, if the correlation values are greater than 95%. We estimate from the use of peaks correlation algorithm, the LIBS system is capable of correlating the test paint samples to the standard paint films to give a 90-95% of perfect match. The remaining 5-10% near match or no match may due to the complex nature of paints and coatings, including the possible surface contaminations. In this case, a careful spectroscopic analysis is further required to achieve the proper paint sample identifications.

Sample No.	Resin	Type of paint	Remark
1	Urethane	Primer	2-part system, cured at 66 °C
2	Urethane	Top coat	2-part system, low temperature curing
3	Urethane	Top coat	2-part system, high temperature curing
4	Urethane	Top coat	2-part system, cured at 66 °C
5	Epoxy	Primer	Low temperature curing (54 °C)
6	Epoxy	Primer	Medium temperature curing (66 °C)
7	Epoxy	Primer	High temperature curing (82 °C)
8	Alkyd	Primer	Air-dry system
9	Alkyd	Primer	Baking system
10	Alkyd	Top coat	Air-dry system
11	Alkyd	Top coat	Baking system

Table 1. The sample paints obtained from Caterpillar Inc.

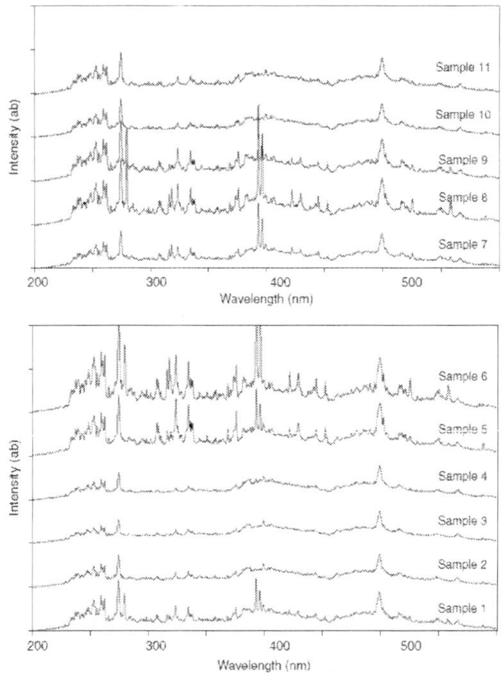

Figure 16. LIBS spectra for the eleven industrial paint samples obtained from Caterpillar Inc.

b. Organic and bio material screening

Biomaterial application has two areas depending on the analytical goal. The first goal is the analysis of metallic component in the biomaterial. The conventional elements like Na, K, Ca, Mg are included in plant, wood, grain, tissue and bio-remains. Their analysis is similar to other solid samples except those samples include high level of carbon compound. The second application of biomaterial is characterization of biomaterial itself. Breakdown spectrum from LIBS can have information of specific sample group. One of the researches has been made for classification of bacterial strains by major components analysis with LIBS[35]. A pulsed Nd:YAG laser (Continum, Powerlite8000, 10-ns pulse width) was focused on the sample solution by using a 20-cm-focal-length lens. The frequency-doubled laser output at 532 nm was used for plasma generation. The laser power used on the bacteria analysis was 50 mJ/pulse. A light collecting optical fiber was placed near the sample surface to detect the plasma emission which was sent to a spectrometer (Acton research, 1200 grooves/mm grating). The spectra were captured with a photodiode-array detector (OMA IV, EG&G, 1024 array) with a spectral resolution of 0.061 nm. The available spectral range is limited to about 50 nm from the full OMA coverage of 76 nm because of the shadow of optical components in the monochromator. The OMA output was processed and stored by using a personal computer.

Several bacterial strains have been classified depending on their major components analyzed by laser-induced breakdown spectroscopy (LIBS). The bacteria studied were *Bacillus megaterium, B. Subtillis, B. Thuringiensis, and Escherichia Coli.* Each strain was streaked on the cultivating plate and grown to prepare the colonies of vegetative or spore forms. The major inorganic components of the bacteria samples, including Ca, Mn, K, Na, carbon, and phosphorus, were clearly identified from the LIBS data. The vegetative forms of bacteria, beginning step of bacteria life, represent the similar quantities of analytical components between bacteria. After the bacteria have used up the available food supply the bacillus enter into their non-vegetative spore form. The bacteria spores accumulate a lot of calcium on the spore shell which showed strong emission of 393.7 nm and 396.9 nm in the LIBS spectrum. The diverse emission from phosphate at 588.1 nm and 588.7 nm provides a fingerprint of the bacteria. The relative change of inclusions of bacteria was clearly distinguished on the 2-dimensional chart of the bacterial components. This work demonstrates the potential of this method for the rapid and precise classification of bacteria with minimum sample preparation. The quick process of LIBS expected to be used in the real-time analysis of intentionally cultured bulk bacteria in the industrial or weaponized microorganism.

3.6. Preparation of microorganism

Five types of bacterial samples were prepared from the biology lab. Te laboratory stock stains used were *Bacillus Megaterium* QM B1551 (seven indigenous plasmid), *Bacillus Megaterium* PV361 (QM B1551 with all plasmid removed), *Bacillus Subtillis* 168M, *Bacillus Thuringiensis* T34, and *Escherichia Coli* carrying pHT315. QM and PV of *B. Meg* are closely related on their genetic origin and *B. thu.* is a divergence of *B.sub.* Only *E. Coli* is a gram-negative genus among them with antibiotic and enzyme resistive cell wall, and *E. Coli* does not make dormant spore on the contrary to other *bacillus*. All bacterial genera used are biosafety level 1 (non pathogenic). Each stain was streaked on *Luria-Bertani* (LB) plates (10.0 g tryptone, 5.0 g yeast extract and 5.0 g of NaCl, 15% agar, in 1 L double distilled H_2O) and grown overnight for both tests. LB plates were then spread with 0.1 mL of culture and grown for 24 hours for confluent plate test. The same set of vegetative bacteria were kept more than 5 days at room temperature to be spore forms after consuming nutrient and drying .

The series of bacteria cultured on the plastic dish are measured on the LIBS system without any pretreatment. The bacteria colonies on the top of the culturing medium (LB) are grown to roughly a 0.5 mm thickness in the wet condition. The areas of the colonies are wide enough for manual mounting on the sample stage and focusing into less than 50 micron diameter of breakdown diameter. The LIBS spectra from bacteria and culturing medium in the UV and visible spectral range have higher background level compare to solid metallic samples such as aluminum, copper and steel. The lack of light absorption on the sample requires a more intense laser for the breakdown. The threshold intensity of the laser pulse for stable and sufficient breakdown was 40 mJ. This value is bigger than 10-20 mJ of the solid LIBS application because the solid samples are not transparent and thus absorb more of the light.

3.7. LIBS spectra of bacterial strains

Although the entire spectrum, from the UV to visible range was initially scanned it was experimentally determined that the three ranges mentioned contained most of the peaks of possible interest. The presence of certain elements was investigated in the micro-organisms such as chlorine, sulfur, phosphorous, calcium, sodium, and potassium. Also of interest were the possible trace elements such as zinc, magnesium, manganese, cadmium, nickel, cobalt, and strontium. Listed in the table are peaks that could be candidates for these possible elements. Because of the atomic elemental nature of the LIBS we expect to match the major or minor peaks to the above mentioned elements certain series of peaks. The peaks were matched to the possible elements in microorganism using the library available in the NIST data base In our preliminary studies we were unable to match several trace elements such as Mg, Cd, Ni, Co, and Sr. The complexity of the iron emission can make it difficult to distinguish it from other elemental peaks of interest. Despite the different outward appearance and life cycle of the bacterial species they all shared a similar elemental composition. Strong peaks were found at 252.8 nm, 279.7 nm, 393.7 nm, 396.9 nm, 398.3 nm, 578.9 nm, 588.1 nm, and 588.7 nm. Using a spectrometer with 1200 gr/mm grating at 50 nm blocks of the spectrum four of the major peaks of interest were identified. The spectra shown in Figure 17 are emissions from bacteria around 400 nm. The peaks at 393.7 nm and 396.9 nm attributed to the calcium atomic transition $4s^2S_{0,1/2} - 4p\ ^2P_{1,1/2}$, and $4s^2S_{0,1/2} - 4p\ ^2P_{0,1/2}$, the strongest emissions from calcium used in many other atomic spectroscopy. These calcium emission were verified by using $CaCl_2$, $CaPO_4$ as spectral references.

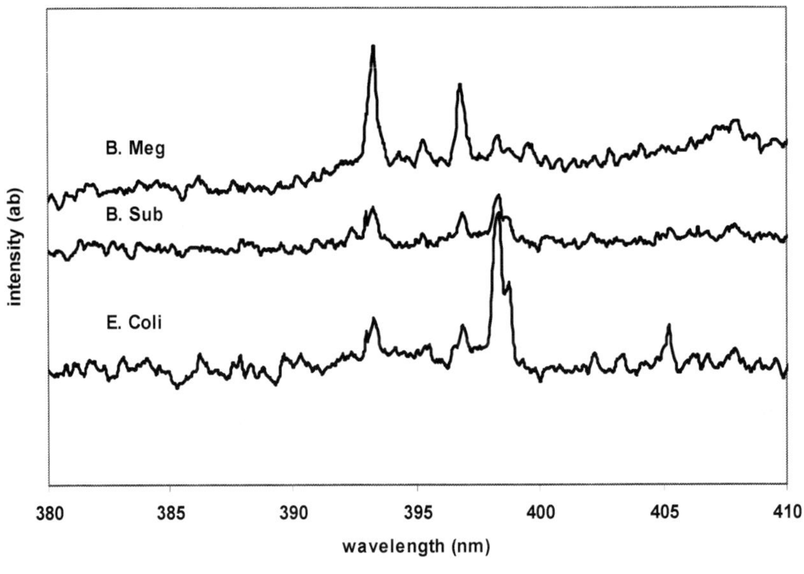

Figure 17. Spectrum after cell death and sporulation of the bacillus has occurred.

At the spore stage of the bacteria, after remaining for 5 days at room temperature to consume all of the nutrients, most of the surface water evaporated. It is known that the water content of spores is only about 10-30% of the water content of vegetative cells (active bacteria) to survive spores at levels of dehydration that would kill vegetative cells. The low water content also provides the spore with chemical resistance (to chemicals such as hydrogen peroxide) and it causes the remaining enzymes of the spore cell to become inactive. This inactivity makes the immunological detection hard to improve sufficient sensitivity. One chemical produced by spores that is thought to lend to their high resistance is dipicolinic acid. Dipicolinic acid interacts with calcium ions to form calcium dipicolinate, which is the main substance believed to lend spores their resistance and represents about 10% of the dry weight of a spore. The intensity of calcium is strong on the spore sample of *B. Meg, B. Sub.,* and *B. Thu.* colonies. *E. Coli* colonies have low calcium content and the composition does not change after aging. This is a proper result that because *E. coli* does not make spores. The spore cell also contains special spore proteins. These protective wall structures are highly resistant to heat desiccation, chemical disinfection and radiation. These functions are to protect DNA from harsh environments, but also disturb measurement chemical property on the conventional spectroscopy. Another component of spores that contributes to their resistance to chemical agents is the strong spore coat, which is composed of highly cross-linked keratin. Laser breakdown is strong enough to break the protective shell of the spores and take out inner component of the cell. The peak at 398.3 nm is overlapped to Mn emission from library data. The strong intensity in this wavelength could not be assigned as Mn because of relative intensity of Mn in other wavelengths. On the other hand, the peak on this wavelength is observed on the samples of organic compounds such as LB of culture medium, cellulose, and many organic polymers. This result leads to assigning this peak as carbon compound fragment.

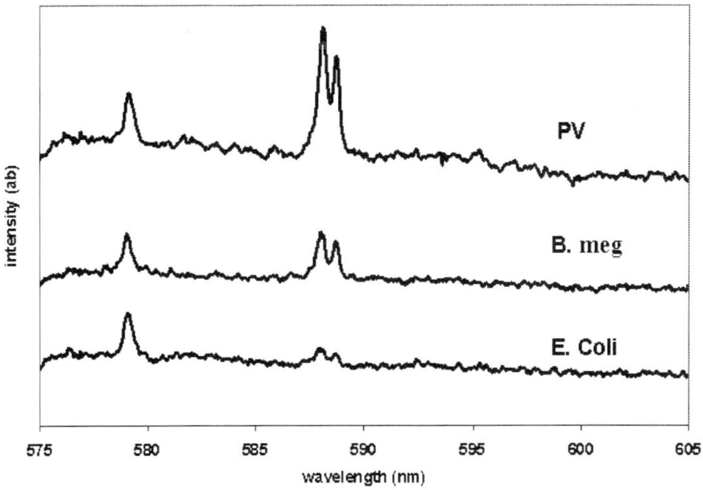

Figure 18. Spectrum to show the amount of phosphate compound in three different bacteria samples.

Figure 18 shows LIBS spectra of 3 bacteria colonies. The doublet peaks appeared at 588 nm is due to phosphate functional compound in the bacteria. In spite of phosphorus elemental emission library doesn't have significant emission on this wavelength, we assign this peak as phosphate because of strong peak observed from some other phosphate compound examination. *PV* shows strongest peak at the phosphate emission. *E. coli* shows weaker intensity. The other bacteria have intermediate peak height, like *B. meg.* in the middle. The amount of phosphate seems to be related to the strain of the bacteria not to the life progress of the microorganism. The phosphate intensities are always weak at the vegetative step and then increase at the spore. The increment of phosphate is biggest for *PV* bacteria. The peak at 578.9 nm is not assigned properly for a certain component of our culturing system. Every bacterial sample and culturing medium shows similar intense peak at this wavelength. With the result of LIBS experiments for several organic compounds, this component is identified as an organic functional group. The proper identification is still on the research.

Figure 19 shows distribution of bacteria on the spread chart of intensity ratio. The X axis is the intensity ratio to represent calcium amount on the bacteria samples. *B. meg* and its plasmid treated *PV* bacteria strains are at high amount of calcium. *B. sub* and *B. thur* are at a relatively lower than *B. meg* strains. *E. coli* does not store calcium whether vegetate or aged colonies. The outstanding feature of phosphate intensity is on the difference of *B. Meg* and *PV*. These two strains are genetically same organism but only *PV* have been removed their plasmid from *B. Meg* type bacteria. The modification of *PV* leads more storage of phosphate in the spore cells.

To identify the selective intake of certain element from culturing medium, we examine the components of the culture medium. The calcium in the medium was lower level than LIBS detectable. This result shows the sporation of bacteria absolutely needs calcium and they collect calcium very efficiently.

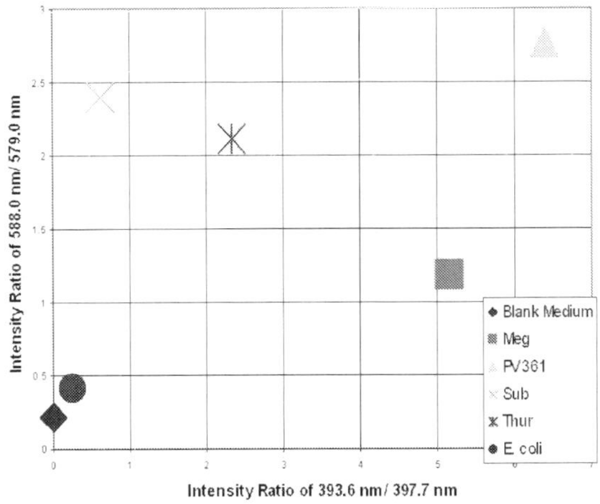

Figure 19. Intensity ratio of two selected wavelength range plotted against each other. The values for X axis are made the intensity of the samples at 393.6 nm divide by their intensity at 398.24 nm. The values for Y axis are made by intensity of the samples at 58

c. Application of LIBS for metallic component in aqueous solution

As mentioned at the LIBS property, solid samples are most convenient and strong LIBS signal. Liquidor gas samples need more specific optical arrangement to generate breakdown and emitting light collection. There are several ideas to overcome the sampling difficulty of gas and liquid samples. One of the publications here is a typical sample type conversion from liquid sample to solid with concentrating effect. Ion-exchange resins are conventional substances used to capture metal ions and hold them in the solid resin matrix. Chemically activated microporous membranes functionalized with polycarboxylic acid are typically employed[36]. The matrix encapsulation technique has been applied to collect trace metals from water, and converting metal ions to a solid form. The pre-concentration of analyte from a liquid sample into the ion exchange membrane was extensively studied for LIBS measurement by Schmidt and Goode[37]. The captured Cu on the small area of the ion exchange membrane has shown that the pre-concentration can provide a large volume of liquid filtration. Many of the test elements gave results in the sub mg/L range of detection limits by liquid filtration method. Total chromium elements were captured in the membrane and measured by LIBS[38] in the range of ng/mL detection limit, where Cr(VI) was chemically converted to Cr(III). A fast analysis technique with automated LIBS analyzer is configured for monitoring of metal ions in water[39]. The ion exchange membrane is used to develop Copper solutions were used to establish pre-concentration parameters. The chelating resin based filter membrane was used to capture Cu ions in water. A series of standard solutions were filtered through the ion exchange membrane by using a vacuum suction system to reduce the filtration time. The LIBS signals of copper absorbed on the layers of the membrane were investigated to determine parameters for practical analysis.

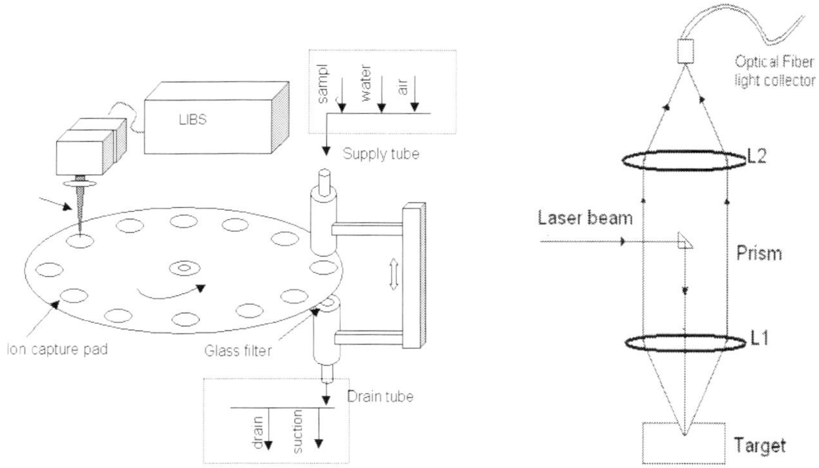

Figure 20. Experimental set up for ion-capture filtering and LIBS (top) and detail of collection optics (bottom).

3.8. Solution samples and instrumentation

The stock solutions were prepared using copper salts as the source of the copper ions from, cupric nitrate ($Cu(NO_3)_2 \cdot 3H_2O$, Fisher), cupric chloride and cupric sulfate, at a 500 mg/L concentration, which were then diluted to make a series of test solutions with well-defined concentrations. A copper sheet (99.9% Cu) was obtained. A commercially available extraction membrane was obtained from 3M filtration Products (St. Paul, MN). The original purpose of using the ion-exchange membrane was to extract multivalent metal ions for environmental analysis by chelating in the PTFE matrix. The ionic selectivity of the membrane is known to follow roughly along the EDTA complex formation constants. The membrane was affixed between the edges of two Teflon tubes with a 12 mm inner diameter. The upper tube was connected to the supply manifold valve for switching sample solutions, flushing water and drying air. The lower tube was depressurized for filtering and suction through a drain reservoir. The drainage tube held a round glass filter to support the membrane during solution filtering.

For this LIBS system, an ion-exchange membrane concentrator was assembled as shown in the Figure 20 for use in the experiment. A fiber optic cable collected the breakdown emissions through collection optics (L1 and L2) and directed them to a fixed-grating spectrometer. The spectrometer has a 3600-sensor array and covers a spectral range of 250 nm – 800 nm. The samples on the motorized stage were moved 0.5 mm to 1 mm stepwise to collect an averaged spectrum from a membrane sampler.

3.9. Operation conditions for membrane concentrator

A 10 mg/L copper solution prepared from $Cu(NO_3)_2$ was used to test the filtering conditions of the membrane concentrator. The LIBS spectra were obtained in the measurable intensity range. The inset in Figure 21 shows the major copper peak that was used for this test. The metal capturing function of the ion exchange membrane is known as that of the EDTA chelating process, therefore the chelating speed must be fast enough to capture all the copper ions during filtration. The estimated filtering time through a 0.2 mm thickness of membrane filter is 3 min when 20 mL of solution is passed through a 12 mm diameter filter at the vacuum suction pressure of 20 kPa. The amount of copper captured on the membrane filter should be proportional to the observed LIBS intensity. As shown in Figure 21, the average intensity for the LIBS peak at 324.75 nm is almost constant throughout the changes of suction pressure which control the filtering speed of test solution. The experimental results show that there is a large uncertainty in capturing Cu on membrane filter (a large error bar in Figure 21) when the sample solutions are filtered very slowly with low suction pressure. If the filtration process took more than 30 minutes at a pressure of less than 10 kPa, the error in measurement increased. It was also found that soaking for extended periods of time in test solution would lead to wrinkles on the membrane filter, due to membrane swelling. On a wrinkled membrane, the liquid filtration path is biased in certain areas of the filter paper resulting in uneven dispersion of captured ions on the membrane surface. On the other hand, when the test solution was filtered at a very high speed, i.e., the suction pressure is more than 80 kPa, the ions just pass through the membrane. A 20 mL solution takes only 10 sec to filter and ions begin to pass through without being captured which leads to a weaker intensity on the LIBS spectra. Based on this membrane operation test, a suction pressure of 30 kPa was maintained as the standard condition, so 20 mL of solution can be completely filtered within 2 minutes.

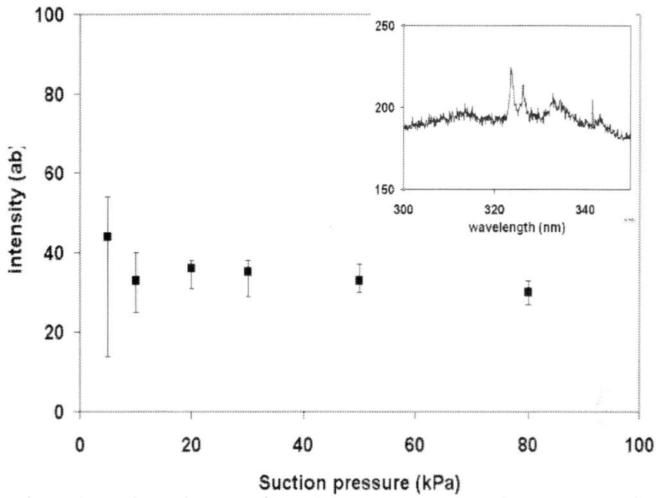

Figure 21. LIBS intensity of Cu at 324.7 nm sampled by filtering at different suction pressure. The inset shows the Cu emission that was concentrated from a 10 mg/L of Cu solution.

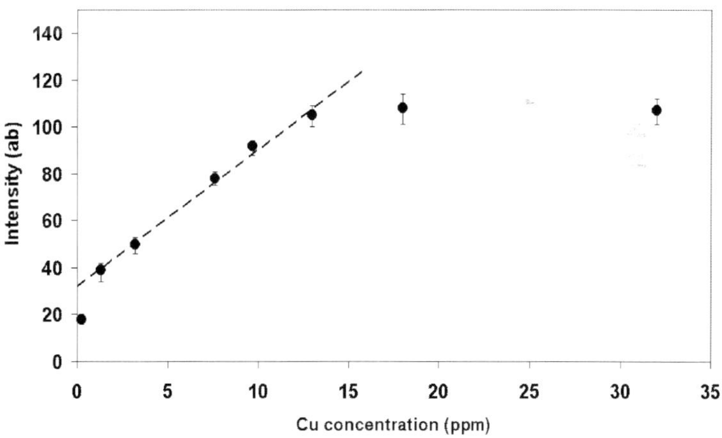

Figure 22. LIBS intensity measurements for low concentration samples. Each of the data points were obtained by averaging five individual measurements on a single filter surface at five different spots. The line was made using the data points from 0.5 mg/L to 15 mg

3.10. LIBS intensity from the top membrane surface

Figure 22 shows LIBS intensity of low concentration samples (a few mg/L Cu solutions) for the peaks at 324.75 nm. All data points were obtained by averaging five individual measurements at five different spots on a single membrane filter. Since the sample surface required for a single laser breakdown shot is less than 0.1mm in diameter, multiple measurements and their average values are easily obtained over the surface of the membrane filter. The LIBS intensity of copper displays a linear correlation over concentration ranges below 15 mg/L. The line in Figure 22 was made by using the data points from 0.5 mg/L to 15 mg/L of Cu solutions with the exclusion of higher concentration data. The results give a relatively too narrow dynamic range for general analytical use. However, the correlation of the line is R^2=0.9926 and it can be an acceptable analytical calibration concentration. The reason for the extremely limited dynamic range is investigated further in the next section.

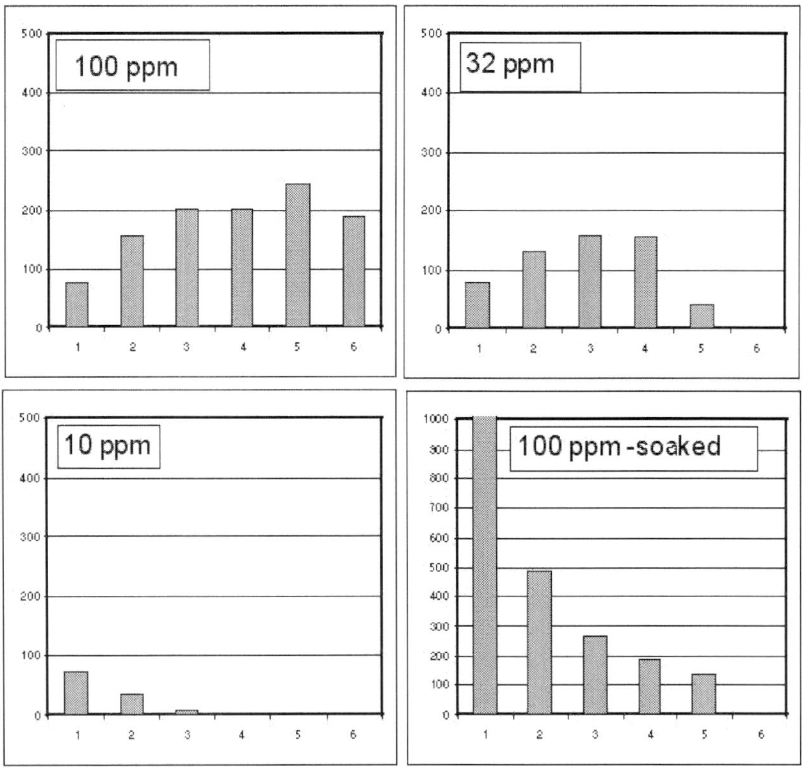

Figure 23. Intensity change by consecutive laser shots. The Cu concentration in the sample solutions for the membrane filtration is 100 mg/L, 32 mg/L and 10 mg/L. The membrane filter soaked for 24 hours in 100 mg/L solution (lower right) shows intensity greater

3.11. Depth effects on ion-capture membrane

In Figure 22, when the concentration of Cu solution is higher than 20 mg/L, the LIBS intensity of Cu at 324.75 nm seems to lose its proportionality relationship to the concentration and remain constantly extended to 35 mg/L. Initially, the retention capability of the membrane was suspected but this was shown not to be the case because the filtrate (drain) from high concentrations did not contain copper ions. To verify the effects of depth profiling on ion-capture membrane, further investigation was done using multiple laser shots at the same spot of membrane. Figure 23 shows the LIBS intensity change by consecutive laser shots at a single membrane point. The intensity of the first laser shots from the filtered samples of 100 mg/L, 32 mg/L and 10 mg/L solutions are unexpectedly similar, and then the next laser shots show intensity change relative to the solution concentration. The 100 mg/L sample makes strong LIBS intensities until the 10th laser shot (top-left graph in Figure 23, after the 6th shot is not shown). On the contrary, the LIBS intensity seems to disappear at the 3rd shot from 10 mg/L sample. A LIBS intensity comparison after the 10th pulse was not possible because the laser pulse had already penetrated through the membrane and a hole was generated. This observation can be explained by the thickness of membrane filter as resulted from the effects of depth profiling on the ion-capture membrane. The Cu ion in the sample solution is drawn inward on the ion-capture membrane during filtration and captured at a certain depth. It is clear that our laser power, 50 mJ/pulse, can ablate the Cu-membrane layer by layer. A well prepared calibration curve from other literature[37], using stronger laser power, also showed that the calibration began to taper off at around 10 mg/L (similar to Figure 22). The authors suggested a linearly regressed calibration curve, simply, for the entire concentration range. The effects of depth profiling on the ion-capture membrane is also proven by the sampling of passive extraction. For the passive extraction, the membrane filter was soaked for 24 hours in 100 mg/L solution and the LIBS intensity changes of Cu at 324.75 nm by the consecutive laser shots are shown in the graph of lower-right in Figure 23. The most intense spectrum was obtained from the first laser shot. None of the ions were drawn physically into membrane during the passive extraction, so they were mostly captured on the surface and gave the strongest LIBS intensity at the first laser shot. It is clear that the membrane captures ions at the deeper layer if the solution is drawn in by suction. As a result, the total ions through the entire thickness should be counted to get the proportional values to determine concentration. A modified calibration curved is made from the integration of 10 laser-shot intensities as in Figure 24 using a 2nd order equation.

3.12. Analysis of tap water using ion-capture membrane

Tap water was analyzed by using the ion-capture membrane concentrator and LIBS. The tap water to our lab at Northern Illinois University is supplied throughout the building by copper pipe. There were many studies which showed the copper contamination in the water supply from the pluming. The local government which supplies the tap water to this lab declared that the source of the city water is collected from the active public water supply wells. Inorganic contaminants, such as salt and metals, can be naturally occurring or

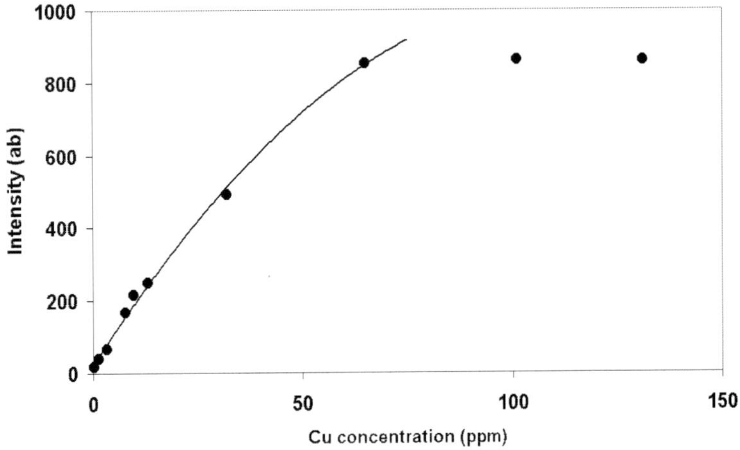

Figure 24. Modified calibration curve made from the integrated intensity of 10 laser shots and Cu concentration up to 70 mg/L. The obtained curve is y = -0.079x2 + 17.81x +23.46

resulted from urban storm water runoff, industrial, or domestic waste water discharges. The ions of Ca, Mg, Na, K, in the city water are known to be of reasonable concentrations within the EPA regulations. The annual water report from city shows that the concentration of Cu is 1.3 mg/L, and this amount is expected to maintain reasonably throughout the day and week. To make an illustration on the effects of pluming for copper contamination from the copper pipe, we took water samples during a specific time schedule, early Monday morning and Tuesday evening. Each sample contains 20 mL of water and filters through an ion-capture membrane. Figure 25 shows the LIBS spectrum of tab water sample captured by the membrane. Copper emissions were identified along with strong peaks of Ca. The concentrations are obtained from the integration of the same number of breakdown shots by using the calibration curve as in Figure 24. In case the integrated intensity is greater than 800 counts, which is around the saturation range, we simply dilute the original sample and measure again. The Tuesday sample shows Cu concentration in the range of 5- 10 mg/L, which is expected due to the large use of water during the active days in the building and the species in the water should be similar to the source water. However, the Monday morning samples show higher concentrations mostly around 100 mg/L or more. Some water samples show up to 270 mg/L of copper, especially for hot water pipe line. We can infer for the high concentration of copper on the Monday sample that water remained in the pipes during the weekend and did not move. This result shows that the contamination of copper from the pluming of building is significant and is shown to depend on the retention time of water in the building pipe.

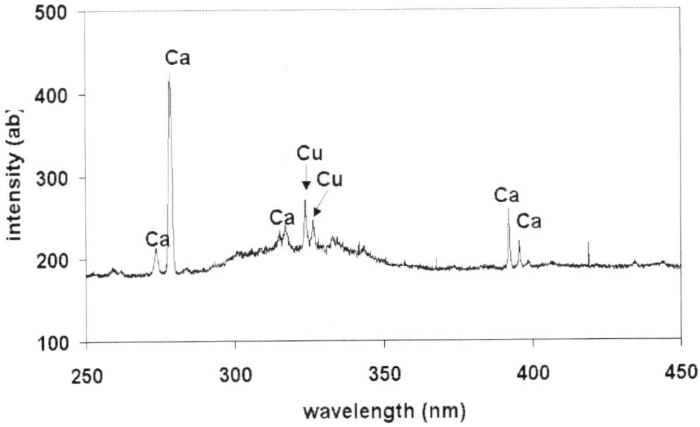

Figure 25. LIBS spectrum of ion captured membrane. The sample solutions are 20 mL of tap water filtered through membrane filter.

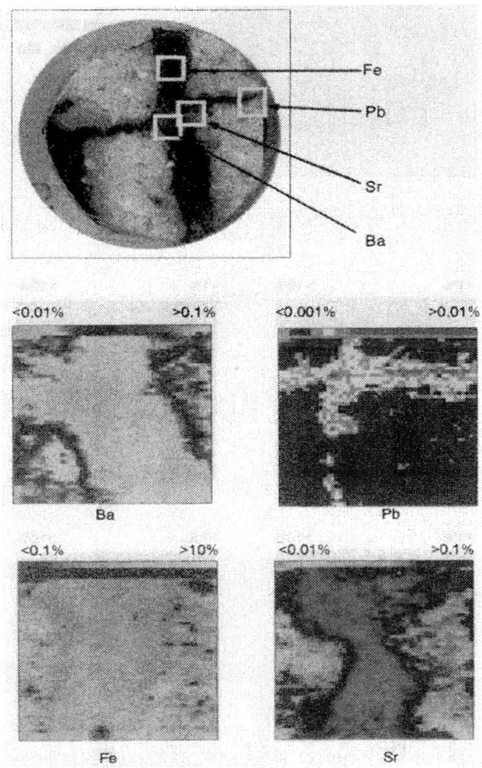

Figure 26. The elemental distribution of patterns for Ba, Pb, Sr, and Fe in a polished granite rock section.

d. Ceramic and geology sample application

Laser induced plasma spectroscopy has been applied to the analysis of element distribution mapping of polished rock sections[40]. The plasma was generated by focusing a frequency-doubled second harmonic 532 nm Nd:YAG laser on the target under atmospheric conditions. The experimental parameters, such as laser energy, atomic emission line and time profile of the plasma spectrum, were characterized to obtain optimum experimental conditions and estimate the element composition of the target surface. For the element mapping of samples, an X-Y stage was used to move the sample and an element image of 50 x 50 mm could be made in 30 min. Using this technique, the element concentration distribution of Ba, Cu, Fe, Mn, Pb, Si, and Sr in polished rock sections were obtained. Quantitative analysis was achieved by analyzing standard rock samples. Calibrated concentration versus plasma intensity was used for the color grading for the mapping of element concentration distribution. The elemental mapping analysis for a granite sample is illustrated in Figure 26. The ore vein within the existing sample was selected to identify the different compositions of ore and surface element distribution. The element distribution differences were represented by color grading, where the upper first line represents the color scale. The region where a lode was crossed during the analysis is rich in Pb and Sr but the Ba content is low. Iron does not show differences and is nearly uniformly distributed across the sample.

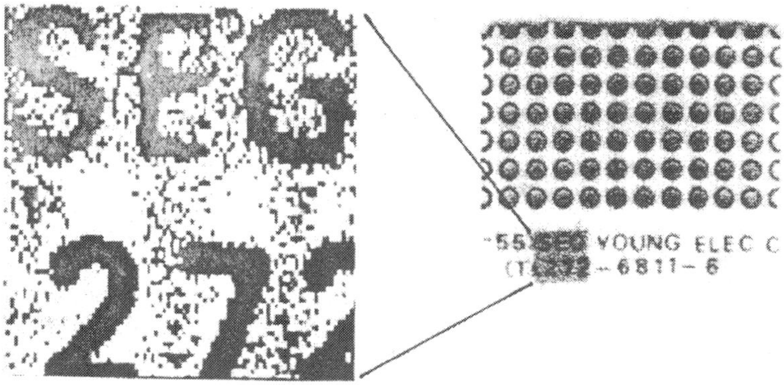

Figure 27. The mapping image of a commercial printed circuit board. Black circle and character on the right circuit board is copper layer for soldering electronic component. The measured values from LIBS constitute pixels on left map.

A sensitive optical technique for compositional mapping of solid surface using LIBS was described[41]. A pulsed Nd:YAG laser with second harmonic module was focused on the solid surface, giving a small ablation area, to produce plasma emission. Copper and magnesium emissions from a standard sample were carefully analyzed and assigned in the wavelength range 500-520 nm. The assigned spectral information was selected to construct an image of 100 x 100 pixels by mapping the measured emission intensity values from the analyzed points. The time required for image construction and image sharpness depends on the number of laser shots per point of analysis and the number of analyzed points per image. A clear image of a copper conductor pattern from a printed circuit board was generated. In addition, some copper contaminations around the conductor area are clearly visible in the scanning LIBS map. The contaminated copper salt probably resulted from the incomplete washing step during manufacturing that could cause a short circuit in an electronic device. A commercial printed circuit board is shown in the right portion of Figure 27. Circles on the top are patterned copper layer for soldering the electronic components. Characters on the bottom are the same copper layer for product identification. The laser-ablated area (gray shaded square) is 5 x 5 mm^2. The left portion of Figure 27 is a mapping image of copper corresponding to the gray shaded square. Each 100x100 pixels corresponds to a measured emission intensity value of LIBS on the ablated point.

4. Conclusion

The goal of LIBS development is to extend the analytical feasibility of LIBS for detecting organic, inorganic metals and ceramic material for the various applications. The sample types also not limited to the solid and expanded to liquid gas, aerosol, powder, bacteria, and industrial products. Several applications of LIBS were illustrated in this chapter. Detections for metallic components are usually accomplished easily from the measured spectral ranges from ultraviolet (230 nm) to visible (700 nm) of the plasma emission. Only a few seconds of measuring time is a great advantage and will be useful for screening and monitoring system for industry and security monitoring. Evaluation of analytical feasibility for detecting and identifying the sample should be decided by analyzing the LIBS spectra of specific components as well as matrix derived from the source of the samples.

Author details

Taesam Kim and Chhiu-Tsu Lin
Northern Illinois University, Illinois, USA

5. References

[1] Jurado-Lopez, A., Lque de Castro M.D. "Laser-Induced Breakdown Spectrometry in the Jewelry Industry. Part I. Determination of the Layer Thickness and Composition of Gold-Plated Pieces" *J. Anal.At.Spectrom.* 2002, 17, 544-547.

[2] Jurado-Lopez, A., Lque de Castro M.D. "Chemometric Approach to Laser-Induced Breakdown Analysis of Gold Alloys" *Appl. Spectrosc.* 2003, 57, 349-352.

[3] Sturm V., Peter. L., Noll, R. "Steel Analysis with Laser-Induced Breakdown Spectrometry in the Vacuum Ultraviolet" *Appl. Spectrosc.* 2000, 54, 1275-1278.

[4] Palanco S., Laserna, J. J. "Full automation of a laser-induced breakdown spectrometer for quality assessment in the steel industry with sample handling, surface preparation and quantitative analysis capabilities" *J. Anal. At.Spectrom.* 2000, 15, 1321-1327.

[5] Sattmann, R., Sturm, V., Noll, R. " Laser-induced breakdown spectroscopy of steel samples using multiple Q-switch Nd:YAG laser pulses" *J. Appl. Phys.*, 1995, 28, 2181-2187.

[6] Kraushaar, M., Noll, R., Schmitz, H. U." Slag Analysis with Laser-Induced Breakdown Spectrometry" *Appl. Spectrosc.*, 2003, 57, 1282-1287.

[7] Mateo, M. P., Cabalin, L. M., Laserna, J. J. "Automated Line-Focused Laser Ablation for Mapping of Inclusions in Stainless Steel" *Appl. Spectrosc.*, 2003, 57, 1461-1467.

[8] Thiem, T. L., Salter, R. H., Gardner, J. A., Lee, Y. I., Sneddon, J. "Quantitative Simultaneous Elemental Determinations in Alloys Using Laser-Induced Breakdown Spectroscopy (LIBS) in an Ultra-High Vacuum" *Appl. Spectrosc.*, 1994, 48, 58-64.

[9] Valdillo, J. M., Grcia, C. C., Palanco, S., Lasena, J. J. "Nanomertic range depth-resolved analysis of coated steels using laser-induced breakdown spectrometry with a 308 nm collimatied beam" *J. Anal. At.Spectrom.* 1998, 13, 793-797.

[10] Burgio, L., Clark, R. J., Stratoudaki, T, Doulgeridis, M., Anglos, D. "Pigment Identification in Painted Artworks: A Dual Analysis Approach Employing Laser-Induced Breakdown Spectroscopy and Raman Microscopy" *Appl. Spectrosc.* 2000, 54, 463-469.

[11] Anglos, D., Couris, S., Fotakis, C. "Laser Diagnostics of Painted Artworks: Laser-Induced Breakdown Spectroscopy in Pigment Identification" *Appl. Spectrosc.* 1997, 51, 1025-1030.

[12] Garcia C. C., Corral, M., Vadillo, J. M., Laserna, J. J. "Angle Resolved Laser-Induced Breakdown Spectrometry for Depth Profiling of Coated Materials" *Appl. Spectrosc.* 2000, 54, 1027-1031.

[13] Marquardt, B. J., Goode, S. R., Angel, S. M. "In Situ Determination of Lead in Paint by Laser-Induced Breakdown Spectroscopy Using a Fiber-Optic Probe" *Anal. Chem.*, 1996, 68, 977-981.

[14] Häkkänen, H. J., Korppi-Tommola, J. E. I. "UV-Laser Plasma Study of Elemental Distributions of Paper Coatings" *Appl. Spectrosc.*, 1995, 49, 1721-1728.

[15] Hidalgo, M., Martin, F., Lasema, J. J. " Laser-Induced Breakdown Spectrometry of Titanium Dioxide Antireflection Coatings in Photovoltaic Cells" *Anal. Chem.*, 1996, 68, 1095-1100.

[16] Moskal, T. M., Hahn, D. W. " On-Line Sorting of Wood Treated with Chromated Copper Arsenate Using Laser-Induced Breakdown Spectroscopy" *Appl. Spectrosc.*, 2002, 56, 1337-1344.

[17] Sattmann, R., Moüch, I., Krause, H., Noll, R., Souris, S., Hatziapostolou, A., Mavromanolakis, A., Fotakis, C., Larrauri, E., Miguel, R." Laser-Induced Breakdown Spectroscopy for Polymer Identification "*Appl. Spectrosc.*, 1998, 52, 456-461.

[18] Dixon, P. B., Hahn, D. W. " Feasibility of Detection and Identification of Individual Bioaerosols Using Laser-Induced Breakdown Spectroscopy" *Anal. Chem.*, 2005, 77, 631-638.

[19] Morel, S., Leone, N., Adam, P., Amouroux, J. " LIBS Applications - Detection of bacteria by Time-Resolved Laser-Induced Breakdown Spectroscopy" *Appl. Opt.*, 2003, 42, 6184-6191.

[20] Samuels, A. C., Delucia, F. C., McNesby, K. L., Miziolek, A. W. " LIBS Applications - Laser-induced breakdown spectroscopy of bacterial spores, molds, pollens, and protein: Initial studies of discrimination potential" *Appl. Opt.*, 2003, 42, 6205-6209.

[21] Knight, A. K., Scherbarth, N. L., Cremers, D. A., Ferris, M. J. " Characterization of Laser-Induced Breakdown Spectroscopy (LIBS) for Application to Space Exploration" *Appl. Spectrosc.*, 2000, 54, 331-340.

[22] Yueh, F., Singh, J. P., Zhang, H. "Laser induced Breakdown Spectroscopy, elemental analysis" in encyclopedia of Analytical Chemistry, John Wily & sons, 2000, pp2066 - 2087

[23] Sneddon, J., Thiem, T.L., Lee, Y. "Lasers in analytical Atomic Spectroscopy" 1996, VCH publish

[24] Miziolek, A. W., Palleschi, A., Schechter, I. "Laser induced Breakdown Spectroscopy"2006, Cambridge

[25] H.J. Hakkanen, J.E.I. Korppi-tommola, UV-Laser Plasma study of elemental distribution of paper coating. Appl. Spectrosc, v49,(12) 1995,p1721

[26] M. Hidalgo, F. Martin, J. J. Lasema, Laser induced Breakdown spectrometry of titanium dioxide antireflection coating sin photovoltaic cells. Anal. Chem. 1996, 68(7) pp1095-1100

[27] Demetriosanglos, StelionCouris, Costas Fotakis, Laser diagnostics of painted artworks: Laser-Induces breakdown spectroscopy in pigment identification, Appl. Spectrosco, 81(7),1997 1025-1030

[28] Klaus Loebe, Arnold Uhl, and HartmutLucht, Micro analysis of tool steel and glass with laser-induced breakdown spectroscopy, applie optics 2003, 42(30) 6166-6173

[29] Valery Bulatov, rivieKrasiker, and Israel Schechter, "Study of Matrix effect in laser plasma spectroscopy by combined multifiber spatial and temporal resolutions. Anal chem. 70(24) 1998, pp5302-11

[30] R Gabriele Cristoforetti, Stefano Legnaioli, Vincenzo Palleschi, Azenio Salvetti, Elisabetta Tognoni, Pier Alberto Benedetti, Franco Brioschi and Fabio Ferrario, Quantitative analysis of aluminium alloys by low-energy, high-repetitionrate laser-induced breakdown spectroscopy,J. Anal. At. Spectro. 2006,21 607-702

[31] Prefetti, B. M. *Metal Surface Characteristics Affecting Organic Coatings,* Federation Series on Coating Technology, FSCT, Blue Bell, PA, 1994.

[32] Weldon, D. G., Carl, B. M. "Determination of Metallic Zinc Content of Inorganic and Organic Zinc-Rich Primers by Differential Scanning Calorimetry" *J. coatings technology* 1997, 69, 45 – 49.

[33] Stamenkovic, J., Cakic, S., Konstantinovic, S., Stoilkovic, S. "Catalysis of the Isocyanate-Hydroxyl Reaction by Non-Tin Catalysts in Water borne Two Components Polyurethane Coatings", *Working and living environmental protection* 2004, 2, 243-250.

[34] Kim T, Nguyen B, Minassian V, Lin C.T Paints and coatings monitored by laser-induced breakdown spectroscopy. J. Coat.Technol. Res. 2007 4:242-255.

[35] Kim T, Specht Z, Vary P, Lin C. T. Spectral fingerprint of bacterial strains by laser-induced breakdown spectroscopy. J. Phys. Chem. 2004 108(17):5477-5482.

[36] Hestekin, J.; Sikdar, S.; Bhattachayya, D.; Bachas, L.; Cullen, L. *US Patent* 6139742, 2000.

[37] Schmidt, N.; Goode, S. *Appl. Spectrosco.*, 2002, 56, 370-374.

[38] Dockery, C.; Pender, J.; Goode, S. *Appl. Spectrosco.*, 2005, 59, pp. 252-257.

[39] Kim T, Ricchia M, Lin C. T. Analysis of copper in an aqueous solution by ion- exchange concentrator and laser-iduced breakdown spectroscopy. J. Chin. Chem. Soc. 2010 57(4B):829-835.

[40] Yoon,Y., Kim, T., Chung, K., Lee, K., and Lee, G., "Application of Laser induced Plasma Spectroscopy to the Analysis of Rock Samples", *Analyst* 1997, 22, pp. 1223-1227

[41] Kim, T., Lin, C., Yoon, Y., "Compositional Mapping by Laser-Induced Breakdown Spectroscopy, *J. Phys. Chem. B*, 1998, 102 ,pp. 4284-4287

Chapter 6

X-Ray Photoelectron Spectroscopy for Characterization of Engineered Elastomer Surfaces

Lidia Martínez, Elisa Román and Roman Nevshupa

Additional information is available at the end of the chapter

http://dx.doi.org/10.5772/48101

1. Introduction

Among different types of polymers, elastomers, also called rubbers, are of special interest for many industrial applications. This interest resides in the high yield strength of these materials that makes possible deforming them manifold their original length without permanent residual strain. However, elastomers can suffer from surface deterioration when subject to rubbing, contacting with aggressive media, ultraviolet light and other. Oxidation of elastomers can produce degradation of its chemical, physico-mechanical, rheological and surface properties. In tribological applications, the quality of the elastomer surfaces is also of special concern since significant degradation of mechanical and tribological behaviour is usually associated with small changes in the surface composition and properties [1]. Therefore, studying the mechanisms of surface degradation of elastomers is very important for comprehension of the failure modes of elastomer components and improving their durability.

For improving the performance of material surfaces, different surface modifications have been developed so far. Properties of elastomer surfaces depend, to a large degree, on the chemical constitution of molecules in the surface layer [2]. Therefore, tailoring polymer surfaces has attracted much interest of researchers in polymer chemistry [3]. Polymer surface modification allows obtaining good performance of components at lower costs than using expensive advanced bulk materials [4]. Presently, halogenation, etching, grafting, oxidation, and other surface modification techniques are intensively used. Another alternative is the application of coatings onto the elastomer surface, although the application of coatings on deformable substrates without occurrence of interfacial delamination is not straightforward. Among various coatings, amorphous diamond-like carbon (DLC) is considered by various authors as a good candidate for application on elastomer surfaces [5]. Such coatings have excellent tribological behaviour, i.e., low friction coefficient and wear rate [6].

Ethylene-propylene-diene elastomer (EPDM) is one of the most widely used elastomers in various outdoor and industrial applications, such as waterproof coatings, electrical insulation, pipes, and mounts. In general, it is employed in applications which demand a material with good mechanical properties and with a retained elastic nature [7]. World production of EPDM is estimated to be 41% of all elastomers [8-11]. Also it has a good resistance to degradation at elevated temperature, sunlight, in oxygen and, in particular, ozone [12]. Acrylonitrile-butadiene rubbers (NBR) and hydrogenated acrylonitrile-butadiene rubbers (HNBR) form another widely used family of elastomers. In hydrogenated rubber the double bonds of butadiene (CH_2=CH-C≡N) are saturated yielding rubber with much higher chemical inertness. These elastomers are extensively employed in automotive industry, especially for lip seals, due to their moderate cost, excellent resistance to oils, fuels and greases, processability and very good resistance to swelling by aliphatic hydrocarbons [13].

Our study is focused on characterization of surface chemical composition of different elastomers subject to rubbing, surface modification and application of coatings. The main technique used for this study was X-ray photoelectron spectroscopy (XPS), which is a very powerful technique for characterizing the chemical composition of very thin (few nm) surface layers. XPS is particularly useful when analysing elastomers, as it provides information about the chemical environment of the elements, i.e. type of bonds, chemical state, etc. Thus, XPS is well suited for investigation the changes in binding energy of chemical elements situated within the first tens of nanometres of the material surface [4]. Elastomers are typically composed of carbon, hydrogen, oxygen and nitrogen. Their surface and bulk properties depend on the way these elements are combined rather than on the presence of other chemical elements. XPS allows detection of new functional groups [4] and evaluation the variation in the amount of existing functional groups, e.g. C-O, as function of surface tailoring, ageing [1], or rubbing [14]. However, often it can be difficult to distinguish between different functional groups having similar binding energies. Therefore, in many cases some complementary techniques should be used to elucidate chemical features of elastomer surfaces. One of these complementary techniques consists in measuring of contact angles (CA) of sessile drops of various liquids placed on the elastomer surface. This very simple method provides valuable information on the types of surface groups [15]. In particular, by using water, presence of polar groups, e.g. C-O, can be determined. So, the degree of surface activation due to surface modification can be determined from measurements of surface hydrophobicity [16]. Then, more information on the surface chemistry and Surface Free Energy (SFE) can be obtained from measuring CA of various liquids with different characteristics. In the following sections we present some fundamental aspects of these techniques and case studies of elastomer surfaces.

2. XPS for characterization of elastomer surfaces

2.1. Introduction to the XPS technique

XPS is an analytical technique that has its fundamental origin in the photoelectric effect, which was first explained by Einstein in 1905 [17]. This effect has become a powerful tool for studying the composition and the electronic structure of the matter [18]. A schematic drawing of typical XPS measurement device is shown in Figure 1a. The measurements are performed in ultrahigh vacuum (UHV) in order to control the surface cleanliness and to reduce the electron scattering on gas molecules. To provide a beam of photons with given characteristics the device is equipped with an X-ray source focused on the sample surface. The photoelectrons emitted from the sample material at characteristic energies are analysed by a suitable electron analyser. The kinetic energy, E_k, at which electrons are emitted follows the fundamental energy conservation equation in photoemission:

$$h\nu = E_B + E_K + \varphi_a \,, \quad (1)$$

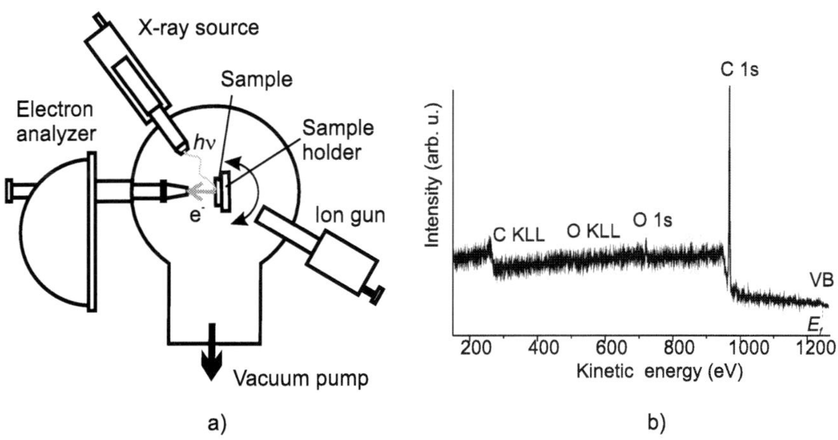

Figure 1. a) Schematic drawing of an experimental XPS system; b) typical photoemission spectra of an elastomer

in which h is Plank constant; ν is the photon frequency and the product $h\nu$ defines the energy of the incident photon; E_B is the binding energy of the electron in the atom. The origin of the binding energies is related with Fermi level, E_f, whereas the kinetic energies are referenced to the vacuum level. The difference between both levels corresponds to work function of the analyser, ϕ_a. By measuring the electron kinetic energies and knowing the spectrometer work function, it is possible to determine the binding energies of various inner levels (or core electrons), as well as those of the outer (or valence) electrons involved in chemical bonding. A typical photoemission (PE) spectrum, i.e. PE yield vs. kinetic energy of the emitted photoelectrons obtained from a photon-illuminated area, is shown in Figure 1b. The spectrum consists of a series of peaks on a background signal which generally increases at low kinetic energy due to secondary electrons, i.e. photoelectrons that are inelastically scattered in the way out of the sample. In summary, the XPS spectra consist of peaks at discrete kinetic energies corresponding to atomic core levels (CLs) and Auger transitions. Note that each element has a unique elemental spectrum. With the most commonly used excitation sources, the kinetic energy of photoelectrons is typically ranged between 0 and

1400 eV. Since inelastic mean free path of photoelectrons, λ, in solids is small [19], chemical information is obtained from the surface and few subsurface atomic layers. Quantitative information can be derived from the peaks areas, whereas chemical states can often be identified from the exact positions of the peaks and separations between them. The presence of chemical bonding causes binding energy shifts, which can be used to infer the chemical nature (such as atomic oxidation state) from the sample surface. Here, we limit ourselves to study elastomer samples. A complete description of XPS technique can be found in specialized literature [20, 21].

2.2. Advantages and shortcomings of XPS technique for characterization of elastomer surfaces. Operating conditions, measurements and semi-quantitative analysis

The standard XPS measurements are carried out under vacuum conditions by retarding-fields techniques. The most commonly used X-ray sources are Al Kα (1486.6 eV) and Mg Kα (1253.6 eV). The X-ray lines from these sources are narrow (less than 0.9 eV) and provide good energy resolution for many applications. Initially, a survey scan or wide energy range scan, typically from 1000 to 0 eV, should be obtained in order to identify the elements present on the surface. As each element emits electrons at characteristic energies, it is possible to identify all the elements present in the sample surface, except hydrogen and helium which are not detectable by this technique. Elastomers usually contain a small number of elements, of which the most common are C, O, N, F and Cl. Other elements like sulphur and zinc can be detected in small quantities. Sulphur is a typical curing agent, whereas zinc is usually employed as a curing activator [22]. In most of the cases, these elements will not be taken into account as they have no real influence on the surfaces properties. Normally, the elements are uniformly distributed in the bulk; however, under certain circumstances surface segregation may take place.

It should be stressed that XPS is a semi-quantitative technique. In order to quantify the amount of each element the integrated area of a particular peak should be divided by the corresponding relative sensitivity factor. The following is a generalized expression for determination of atom fraction, C_x, of a constituent x in a sample:

$$C_x = (I_x / S_x) / \left(\sum I_i / S_i\right), \tag{2}$$

where I_x is the peak area and S_x is the atomic sensitivity factor of the x-th element. The denominator corresponds to the atomic fraction of other elements in the sample. Assuming a homogeneous distribution of elements, a strong line for each element in the spectrum should be analyzed. In case the requirement of homogeneity is not fulfilled, the assumption of homogeneity can be used as a starting point for further calculations. Reference published data on elemental sensitivity factors could be used for determination of S, although the type of instrument and analysis conditions should be considered. With this technique it is also possible to identify chemical states of a given element by measuring the high resolution or core level peaks.

Depth distribution of elements can also be obtained using XPS in destructive or non-destructive modes. In the first one, ion sputtering is used to remove surface layers. Sputtering and XPS can be applied consecutively or simultaneously. In the non-destructive mode, depth profiling is obtained by varying the detection angle of the emitted electrons. In this case the probed depth is limited to 3λ. More detailed information about both methods can be found elsewhere [20].

Another important problem in XPS analysis is related with sample degradation due to X-ray radiation. In fact, this degradation comes from the secondary electrons emitted during the X-ray exposure [23, 24]. In most of the cases this degradation is slow enough as compared with time required for XPS analysis, thus the changes in composition due to X-ray can be neglected. Notwithstanding, this problem should be considered when analysing chemically unstable materials. In our work, no sample degradation due to X-ray radiation has been observed for all studied elastomers.

XPS measurements were performed in ultrahigh vacuum with base pressure of 2×10^{-10} mbar using a Phoibos 100 ESCA/Auger spectrometer with Mg Kα anode (1253.6 eV). To avoid X-ray damage on the samples low X-ray power of 150 W was used. The core level narrow spectra were recorded using pass energy of 15 eV. For the data analysis, the contributions of the Mg K_{α} satellite lines were subtracted and the spectra were subjected to a Shirley background subtraction formalism [25]. The binding energy, E_B, scale was calibrated with respect to the C 1s core level peak at 285 eV. The surface area subject to XPS analysis was around 5.6 mm^2 that is large enough to obtain an average surface chemical composition. When modified samples were analyzed, the surface area subject to XPS analysis was smaller than the treated area, thus the contribution from untreated surfaces was negligible. The shape of C 1s core level peak (HR C1s) measured with high energy resolution was analysed using peak fitting in order to identify functional groups. Depending on the chemical environment of the carbon atoms, important chemical shift of C 1s peak can be observed. Decomposition of the experimental peak in components allows identification of the contribution from each component. For the analysis of HR C1s, the spectrum recorded from the untreated sample was used as a reference. The HR C1s was fitted leaving the full width at half maximum (FWHM) of the C–C/C–H component to vary freely while the other components were forced to adopt the same value. The fit of the treated samples was performed using the same values of FWHM and the binding energies (with uncertainty of ± 0.1 eV) as for untreated elastomer. The only remaining free parameter in the fit procedure was the area of the peaks. By doing so, new carbon species derived from the treatment processes could be identified. An example of the analysis of HR C1s is presented in Figure 2 where some spectra of untreated and modified elastomers are compared. Presence of new carbon species (in this case C-O and C-F bonds) can be identified from the shape of the HR C1s. The contribution from these groups varied depending on the surface treatment. However, identification of chemical groups can be difficult when different species produce similar chemical shifts (see Table 1 for E_B of main carbon bonds identified in the present study). For example, C=C bond was included into the group of C-C/ C-H components since the shift between these two groups is only 0.3 eV [26] that is below the resolution limit of the experimental system used in this study. Detailed analysis of XPS spectra is presented in section 3.3.

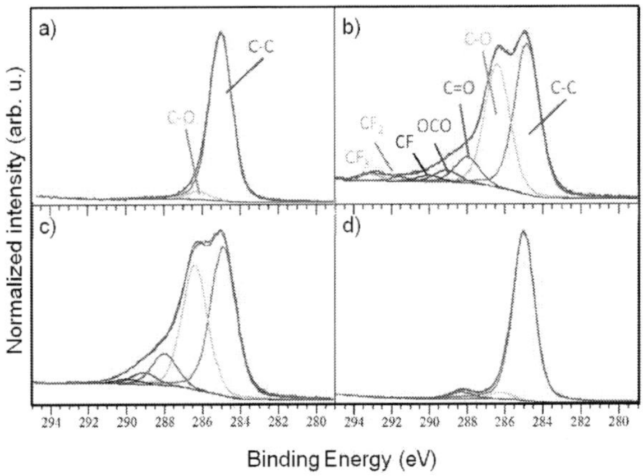

Figure 2. High resolution C1s core level spectra of elastomer samples: a) untreated EPDM, b) EPDM after fluorination with CF_4, c) EPDM after fluorination with SF_6, and d) HNBR after the same fluorination as c)

Bonds	E_B (±0.1 eV)	Ref.
C-C, C-H, C=C	285	[26, 27]
C-O, -CH-CF_2	286.3	[26-30]
C=O, -CH-CF	288.1	[26-30]
O-C=O, -CFH-CF_2	289	[26-30]
CF	289.8	[30, 31]
CF_2	291.8	[30, 31]
CF_3	293.3	[30-32]

Table 1. Components employed for the analysis of the C 1s core levels

2.3. Complementary techniques for the interpretation of the results of XPS

As we mentioned in the introduction section, surface and bulk properties of elastomers depend on the way the main constituents (C, H, O, N, etc.) are combined rather than on the presence of other chemical elements. Therefore, in some cases and depending on the light source employed for XPS analysis, it is difficult to distinguish between the presence of different functional groups, as occurs for HR C1s with C-O and C-N groups. Complementary information on surface chemistry of elastomers can be obtained from spectroscopy of inelastic scattering of light, e.g. Fourier Transformed Infra-Red spectroscopy (FTIR), Raman spectroscopy and others. Measurements of SFE of elastomer using sessile drop method is another very simple but powerful method which can provide valuable information on the type of the surface groups. The method is based on measuring the CA between a droplet of a certain liquid and an elastomer surface under well-controlled conditions. The CA is obtained from a balance of interfacial tensions between three phases: solid (S), liquid (L) and vapour (V) (Figure 3) and is defined from Young-Dupré equation:

$$\gamma_{SV} - \gamma_{SL} - \gamma_{LV} \cos\theta = 0 \,. \tag{3}$$

When a droplet contacts a rough surface, the measured or apparent contact angle, may differ from the intrinsic one, i.e. the CA of the same liquid on an ideally smooth surface of the same material. Wenzel [33] proposed to introduce a roughness factor, r, which is the real contact area divided by the geometrical, or projected, area. Homogeneous wetting regime of a liquid on a rough surface is described by:

$$r\cos\theta_a = \cos\theta \,. \tag{4}$$

The roughness factor can be determined numerically from 3D surface measurements obtained using appropriate technique, e.g. confocal microscopy, laser scanning profilometry, etc. [1]. Recently there were many criticisms on the Wenzel's approach. In [34] it was demonstrated that CA behaviour is determined by interactions of the liquid and the solid at the three-phase contact line alone and that the interfacial area within the perimeter is irrelevant. They suggested that Wenzel's equation is valid only to the extent that the structure of the contact area reflects the ground-state energies of contact lines and the transition states between them.

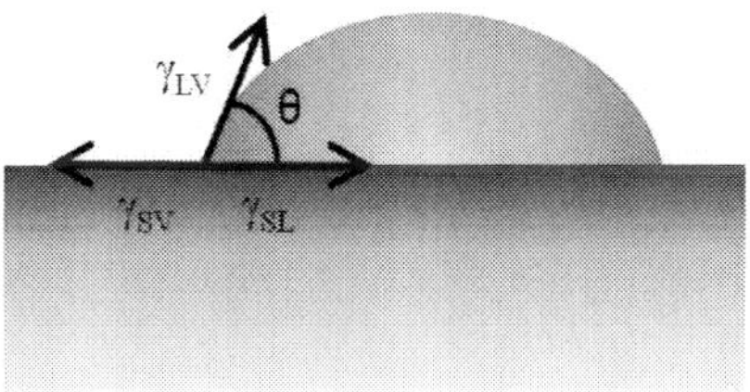

Figure 3. Schematic representation of contact angle

Depending on the specific method, the CA measurement allows determining total SFE, the polar and dispersive components of SFE (Fowke's approach), apolar Lifshitz – Van der Waals (LW) and polar acid - base components (van Oss's approach). According to van Oss's approach, the surface tension could be resolved into components due to dispersion, induction and dipole-dipole forces, and hydrogen bonding [35]. For non-metallic solid surfaces, in addition to apolar LW interactions, electron acceptor – electron donor interactions, or Lewis acid-base (AB) interactions may often occur. In this case the total surface tension is the sum of two components: γ^{LW} and γ^{AB} [36]. Unlike LW interactions, polar interactions are essentially asymmetrical. The polar component of the free energy of interaction between solid and liquid can be expressed as [35]:

$$\Delta F_{SL}^{AB} = -2\left(\sqrt{\gamma_S^+\gamma_L^-} + \sqrt{\gamma_S^-\gamma_L^+}\right), \tag{5}$$

where γ^+ is the electron acceptor, or Lewis acid component, and γ is the electron donor, or Lewis base component of the surface tension. Then, the expanded form of Young-Dupré equation can be obtained by combining (3) and (5):

$$0.5\gamma_L^t\left(1+\cos\theta\right) = \sqrt{\gamma_S^{LW}\gamma_L^{LW}} + \sqrt{\gamma_S^-\gamma_L^+} + \sqrt{\gamma_S^+\gamma_L^-}. \tag{6}$$

If for a given liquid the components of surface tension γ_L^t, γ_L^{LW}, γ_L^+, and γ_L^- are known, (6) is a linear function of three unknown parameters corresponding to the components of surface tension of the solid surface $\sqrt{\gamma_S^{LW}}$, $\sqrt{\gamma_S^-}$, and $\sqrt{\gamma_S^+}$. As this equation is underdetermined, the components of surface tension for the solid can be found by measuring CAs using at least three different liquids with known and different components of surface tension. If the values of the components of surface tension for the three liquids are close together, the calculated values for three parameters for the solid will be "unduly sensitive" [36] to small errors in the values of the parameters of surface tension of the liquids, and in the measured CAs. To overcome this problem, CA measurements should be performed with more than three liquids. These will constitute an overdetermined system of linear equations which can be solved by least-square method. In order to reduce the measurement error, each measurement of the CA should be repeated several times. Mean value, $\overline{\theta_i}$, and standard error of mean, seθ_i, should be determined for each liquid from these measurements.

The resulting set of simultaneous equations is the following:

$$0.5\gamma_{Li}^t\left(1+\cos\overline{\theta_i}\right) = \sqrt{\gamma_S^{LW}\gamma_{Li}^{LW}} + \sqrt{\gamma_S^-\gamma_{Li}^+} + \sqrt{\gamma_S^+\gamma_{Li}^-}, \tag{7}$$

where subscript i indicates the liquid. It can be written in the matrix form:

$$\mathbf{Y} = \mathbf{Ab}, \tag{8}$$

where $\mathbf{Y}$ is the matrix of independent variable (left side of eq. (7)), $\mathbf{A}$ is the (n×3) matrix of known coefficients, n is the number of liquids used for CA measurements, and $\mathbf{b}$ is the vector of unknown parameters:

$$\mathbf{Y} = 0.5\left[\left(1+\cos\overline{\theta_1}\right)\gamma_{L1}^t \cdots \left(1+\cos\overline{\theta_n}\right)\gamma_{Ln}^t\right]^T, \tag{9}$$

$$\mathbf{A} = \begin{pmatrix} \sqrt{\gamma_{L1}^{LW}} & \sqrt{\gamma_{L1}^+} & \sqrt{\gamma_{L1}^-} \\ \cdot & \cdot & \cdot \\ \sqrt{\gamma_{Ln}^{LW}} & \sqrt{\gamma_{Ln}^+} & \sqrt{\gamma_{Ln}^-} \end{pmatrix}, \tag{10}$$

$$\mathbf{b} = \left[\sqrt{\gamma_S^{LW}} \sqrt{\gamma_S^-} \sqrt{\gamma_S^+} \right]^T . \tag{11}$$

Then, mean values of the surface tension components can be determined from the matrix equation:

$$\mathbf{b} = \left(\mathbf{A}^T \mathbf{D}^{-1} \mathbf{A} \right)^{-1} \mathbf{A}^T \mathbf{D}^{-1} \mathbf{Y} , \tag{12}$$

where $\mathbf{D} = \begin{pmatrix} (se\theta_1)^2 & 0 & 0 \\ 0 & \ddots & 0 \\ 0 & 0 & (se\theta_n)^2 \end{pmatrix}$

is the covariance matrix of errors of CA measurements.

The standard error of mean of the unknown parameters can be found from the main diagonal of the covariance matrix:

$$\mathbf{K} = \left(\mathbf{A}^T \mathbf{D}^{-1} \mathbf{A} \right)^{-1} . \tag{13}$$

The calculated values of the parameters of the surface tension should be tested for statistical significance using t-test. In case some of the parameters are not statistically significant, it can be zero set and removed from **b**. Then, the calculation should be repeated using modified matrix **A**. By doing so, the standard error of the parameters of solid can be reduced.

Matrix method is also very useful for the analysis of surface tension variation in time, e.g. due to ageing. In this case, CA measurements are performed at different periods of time using a set of several liquids as described above. This constitutes a set of simultaneous equations at the selected points of time.

$$0.5\gamma_{Li}^t \left(1 + \cos\overline{\theta}_i(t_j) \right) = \sqrt{\gamma_S^{LW}(t_j)\gamma_{Li}^{LW}} + \sqrt{\gamma_S^-(t_j)\gamma_{Li}^+} + \sqrt{\gamma_S^+(t_j)\gamma_{Li}^-} . \tag{14}$$

Therefore, **Y** and **b** change to $(n \times p)$ matrixes, where p is the number of time points:

$$\mathbf{Y} = 0.5 \begin{pmatrix} (1+\cos\theta_{11})\gamma_{L1}^t & \cdots & (1+\cos\theta_{1j})\gamma_{L1}^t & \cdots & (1+\cos\theta_{1p})\gamma_{L1}^t \\ \vdots & \vdots & \vdots & \vdots & \vdots \\ (1+\cos\theta_{n1})\gamma_{Ln}^t & \cdots & (1+\cos\theta_{nj})\gamma_{Ln}^t & \cdots & (1+\cos\theta_{np})\gamma_{Ln}^t \end{pmatrix} , \tag{15}$$

$$\mathbf{b} = \begin{pmatrix} \sqrt{\gamma_{S1}^{LW}} & \cdots & \sqrt{\gamma_{Sj}^{LW}} & \cdots & \sqrt{\gamma_{Sp}^{LW}} \\ \sqrt{\gamma_{S1}^-} & \cdots & \sqrt{\gamma_{Sj}^-} & \cdots & \sqrt{\gamma_{Sp}^-} \\ \sqrt{\gamma_{S1}^+} & \cdots & \sqrt{\gamma_{Sj}^+} & \cdots & \sqrt{\gamma_{Sp}^+} \end{pmatrix} . \tag{16}$$

Assuming that all measurements have the same error, the matrix of parameters of solid surface can be found from the following equation:

$$\mathbf{b} = \left(\mathbf{A}^{\mathrm{T}}\mathbf{A}\right)^{-1}\mathbf{A}^{\mathrm{T}}\mathbf{Y}, \tag{17}$$

and standard errors of the unknown parameters can be found from the main diagonal of the covariance matrix:

$$\mathbf{K} = s^2\left(\mathbf{A}^{\mathrm{T}}\mathbf{A}\right)^{-1}, \tag{18}$$

where s^2 is the sample variance determined as:

$$s^2 = \frac{1}{n-p-1}\left(\mathbf{Y}-\mathbf{Ab}\right)^{\mathrm{T}}\left(\mathbf{Y}-\mathbf{Ab}\right). \tag{19}$$

Although different substances can be used as probe liquids, the following five liquids are the most widely used: water, glycerol, diiodomethane, formamide, and ethylene glycol [1, 29, 37, 38]. The values of the components of surface tension for these liquids are listed in Table 2.

Liquid	γ^t	γ^{LW}	γ^{AB}	γ^-	γ^+
Water	72.80	21.80	51.00	25.50	25.50
Glycerol	64.00	34.00	30.00	57.40	3.92
Formamide	58.00	39.00	19.00	39.60	2.28
Ethylene glycol	48.00	29.00	19.00	30.10	3.00
Diiodomethane	50.80	50.80	0.00	0.00	0.00

Table 2. The components of surface tension for different probe liquids (from [29, 37])

Additionally, the Fowke's model can be used to determine the polar and dispersive components of surface energy. The following is the set of simultaneous Young-Dupré equations corresponding to the measurements of the CA for p liquids at time t_j:

$$0.5\gamma_{Li}^{t}\left(1+\cos\overline{\theta}_i(t_j)\right) = \sqrt{\gamma_S^d(t_j)\gamma_{Li}^d} + \sqrt{\gamma_S^p(t_j)\gamma_{Li}^p}, i = 1\ldots p. \tag{20}$$

Since there are two unknown parameters in this model, the number of liquids used for the CA measurements can be smaller than for van Oss's model. After corresponding modification of **A** and **b**, solution of (20) can be found by the matrix method described above.

3. Case studies

3.1. Characterization of elastomer surface subject to ageing

Though synthetic elastomers like EPDM are very attractive to industry due to their high chemical stability and low permeability for water, they are sensitive to oxidation at elevated temperatures. Understanding of the chemical mechanisms of elastomer degradation is a key for designing advanced elastomers with higher resistance to oxidative degradation. Therefore, XPS and SFE analysis were employed to elucidate chemical changes produced on EPDM elastomer surfaces due to ageing.

XPS wide energy range scans were obtained for EPDM samples aged at 80 ºC and 120 ºC during up to 100 days. Surface chemical composition of the samples determined from these spectra as a function of ageing duration is shown in Table 3. The high carbon content in all samples arises from the contribution from the backbone structure of the elastomer. Oxygen, nitrogen, silicon and zinc are generally attributed to curing agents, amine-based accelerators and additives [1, 14].

With the increasing ageing duration, the O/C ratio also increases (last column of Table 3). In addition, for both temperatures there was certain increase in nitrogen and silicon concentrations for the 100 days ageing. As XPS is a superficial analysis technique, the variation of these elements present in small amounts on the surface could be related to diffusion processes during ageing and segregation of impurities on the surface. Being a thermally activated process, migration of additives is faster at higher temperatures, thus surface concentration of silicon after ageing at 120 ºC is higher than at 80 ºC.

Thermal ageing		Composition (% at.)					O/C ratio
T (ºC)	t_a (days)	C	O	N	Si	Zn	
As received		93	5	0	2	-*	0.054
	5	90	7	0	2	1	0.078
80	50	88	9	1	1	1	0.102
	100	82	14	3	1	-*	0.171
120	100	80	13	2	4	1	0.163

Table 3. Chemical composition of some EPDM samples obtained from XPS wide energy range scan (with permission from [1])
*traces

The results of curve fitting procedure of the HR C1s are shown in Figure 4. The broad carbon peak in the range of E_B from 283 eV to 289 eV can be attributed to different carbon-based surface functional groups. C 1s peak was fitted with four Gaussian/ Lorenzian components with the maximum intensity at E_B of 285 eV, 286.3 eV, 288.1 eV and 289 eV. According to the literature, these energies can be assigned to C-C or C-H, hydroxyl (C-O/ C-OH), carbonyl (C=O) and carboxyl (O-C=O), respectively (see Table 1). Most part of carbon was in form of C-C / C-H. For the samples aged at 80 ºC the amount of carbon bonded to oxygen, especially in form of hydroxyl, increased with the increase of ageing duration. After 100 days at 80 ºC, carbon-oxygen bonds were composed of hydroxyl (20% with respect to carbon), small portion of carbonyl (4%) and traces of carboxyl (1%). This effect was similar to the evolution of the oxygen content registered in the wide energy range scan (Table 3). Similar behaviour was also observed by [39] and [7]. When comparing the samples aged during 100 days at 80 ºC and 120 ºC the portion of carbon-oxygen functional groups was lower at higher temperature, though both samples had similar surface contents of oxygen. According to [39] C-OH bonds are the main product of EPDM ageing as inferred from the C 1s core level peak. However, the results obtained in our work suggested that ageing above 100 ºC could cause hydroxyl desorption. This could explain the lower C-O content registered after the treatment at higher temperature.

Variations in the components of the SFE for the elastomer as a function of the ageing parameters were determined using acid-base regression method with the five liquids listed in Table 2. The results are shown in Figure 5. For both ageing temperatures γ^+ was statistically insignificant, so this term was omitted from the model. Since the component γ^+ is null, EPDM surface is mainly γ_S^- monopolar. In the absence of a parameter of the opposite sign, energy parameters of a monopolar surface do not contribute to the total surface energy (energy of cohesion) since the polar component $\gamma^{AB} = 2\sqrt{\gamma^+\gamma^-} = 0$ [35]. Therefore, the total SFE is controlled solely by LW interaction. However, monopolar surfaces can strongly interact with bipolar liquids.

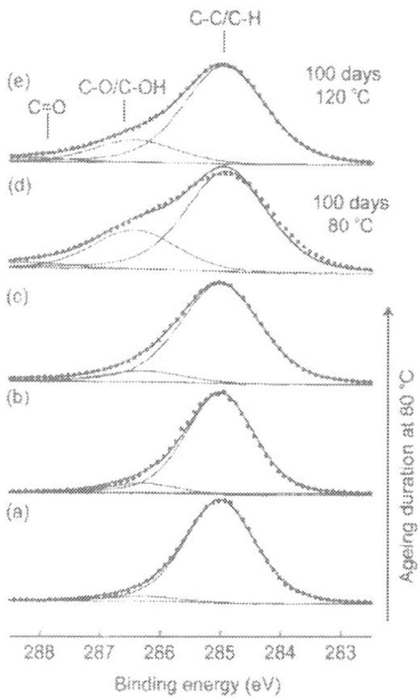

Figure 4. C 1s spectra of EPDM: a) as received, b) aged 5 days at 80 ºC, c) aged 50 days at 80 ºC, d) aged 100 days at 80 ºC, e) aged 100 days at 120 ºC. Dots – experimental data, solid lines – fitting (with permission from [1])

At 80 ºC, γ_S^{LW} increased exponentially with ageing duration reaching almost stable values after 60 days. Parameter γ_S^- had an induction period of approximately 5 days. These results agree with previous works in which the induction period during thermal oxidation of EPDM was determined as 130 h at 80 ºC [40] and 150 h at 150 ºC [7]. Variations of the induction period in different works can be due to differences in the EPDM composition, more specifically, in the carbon black and antioxidants content. After induction period, γ_S^- increased rapidly and reached the maximum in 30 days. Then, it remained almost constant with a slightly decreasing tendency, which, however, was within a standard error. The solid line connecting the filled circles in Figure 5a was obtained by fitting the experimental data with an exponential function having a time constant of 17.7±0.4 days.

At 120 °C (Figure 5b), γ_S^{LW} raised up at the beginning of ageing and then followed almost linear increasing behaviour with low rate. Surprisingly, after 100 days ageing at 120 ºC γ_S^{LW} was approximately 5% smaller than for ageing at 80 ºC. However, after 100 days γ_S^{LW} still maintained the linear growth, while at 80 ºC it stabilized.

The evolution of γ_S^- was similar to that of γ_S^{LW}, although the initial increase was not as steep as for γ_S^{LW} component. There is a large difference in the behaviour of γ_S^- for both ageing temperatures. On short ageing periods γ_S^- was notably smaller at the higher temperature, but this difference vanished on large ageing periods. In addition, at 120 °C the induction period was not observed. Probably, the induction period at higher temperature was less than one day, so it could not be measured in these tests. This finding is consistent with [40] who reported shortening of the induction period to 10 h at 120 ºC.

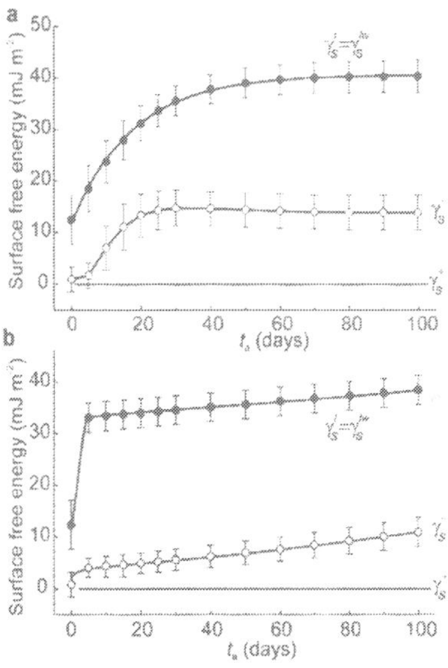

Figure 5. Components of SFE as function of ageing duration: a) ageing temperature 80 ºC, b) ageing temperature 120 ºC (with permission from [1])

The initial value of γ_S^t, which is equal to γ_S^{LW} in our case, is consistent with the findings of [39] for EPDM before weathering test at ambient temperature. They observed that γ_S^t first increased and then stabilized at 23.8 - 25.4 mJ m^{-2}. These values are almost two-fold smaller than in our thermal ageing experiments. This fact supports the hypothesis of a thermally activated nature of the processes responsible for the increase in the surface energy [1].

During ageing of EPDM, two competitive processes typically occur: (i) oxidation of the elastomer chains and (ii) crosslinking between the chains. The oxidation process resides in chain scission and recombination accompanied by formation of oxygen functional groups and radicals. Since double bonds are more chemically active due to the presence of a π-bond, cross-linking and oxidation at the initial stage of ageing mainly involves rupture of double bonds. Characteristic times for cross-linking of EPDM at 80 ºC and 120 ºC are 100 h and 12.5 h, respectively [40]. After these periods, the material is considered fully cross-linked (at given temperature) that implies significant reduction of the concentration of double bonds. Also, it is reasonable to expect that with the increasing temperature the degree of cross-linking increases and the residual concentration of double bonds decreases. During the induction period, cross-linking is the dominating process as can be inferred from the behaviour of γ_S^-, O/C ratio and very high activation energy for oxidation of EPDM, which ranges between 143.4 and 171.4 kJ mol^{-1} [41]. Further ageing of cross-linked elastomer is accompanied with slower oxidation of carbon chains. The higher reactivity of residual double bonds for EPDM aged at 80 °C can explain the steeper increase in SFE and higher concentration of oxygen after induction period. The evolution of SFE for ageing at 80° C is described by a first-order reaction with the activation energy between 63.5 and 83.7 kJ mol^{-1}[1]. These values are higher than those reported in [42], but similar to the activation energy for oxidation of long hydrocarbon chain alkanes and aromatics such as in heavy fuel oil [43]. For ageing at 120 ºC the linear increase in SFE is described by a zero-order reaction. Zero-order reaction was reported also for surface degradation of fully cross-linked EPDM under artificial weathering conditions [39].

In conclusion, oxygen functional groups, mainly hydroxyl, were identified on EPDM surface after ageing. The presence of these groups was more pronounced after the treatment at 80 ºC than at 120 ºC. Higher ageing temperatures lead to faster cross-linking processes. At lower temperature C=C bonds are not fully consumed due to cross-linking [3], hence the oxidation processes at lower temperature is more intensive than at higher temperature. In addition, ageing at long durations promotes changes in the surface chemical composition of EPDM. These changes can be attributed to migration of additives towards the surface as reflected by the increase in Si and N concentrations after 100 days ageing at both temperatures.

3.2. Characterization of surface chemical composition of elastomer surfaces subject to sliding friction

Degradation of elastomer surfaces can be accompanied by formation of specific surface texture like smearing or microfibrill formation [44-46]. In [45, 46] it was speculated that these effects could be due to tribochemical reactions and thermooxidative degradation, however no cogent experimental evidences have been presented so far. In order to provide deeper insight into the mechanisms of elastomer failure, surface chemical composition and SFE were studied before and after friction as a function of the amount of carbon black (CB) filler in EPDM [14]. Carbon black is one of the most widely used reinforcing fillers [47-49] that improves the stiffness and the toughness of rubbers, while maintaining high flexibility and good physical and mechanical properties at low manufacturing costs. The amount of CB varied between 0 and 60 parts per hundred rubber (phr). The samples were subjected to roller-on-plate (ROP) friction tests under conditions detailed in [46]. Friction coefficient was not influenced significantly by the CB content, whereas wear rate decreased with increasing the CB content [14].

Surface chemical composition (at. %) outside and inside the contact zones was determined from the analysis of wide energy XPS spectra (Figure 6). The dominating carbon contribution (95 % and higher) was due to the elastomer backbone structure. The atomic concentrations of other elements including O, Si, S, N and Zn remained below 5 %.

In case some thermooxidative processes and/or tribochemical reactions occur at the contact zone, one can expect certain increase not only in the oxygen concentration, but also in oxygen bonding to carbon atoms in the friction zone. However, the observed behaviour of surface chemical composition was more complex. More specifically, two different tendencies were observed as far as the amount of oxygen in the friction zone is concerned. For unfilled EPDM, the amount of oxygen on the surface of friction zone increased, whereas for filled EPDM it decreased. Notwithstanding these variations, on the surfaces not subjected to friction and for all CB contents, no changes in the binding energy of the oxygen and carbon were observed in high-resolution O 1s and C 1s spectra (Figure 7). The single contribution of C at 285 eV (Figure 7b) from the C-C / C-H component implies absence of oxygen-containing functional groups (see Table 1).

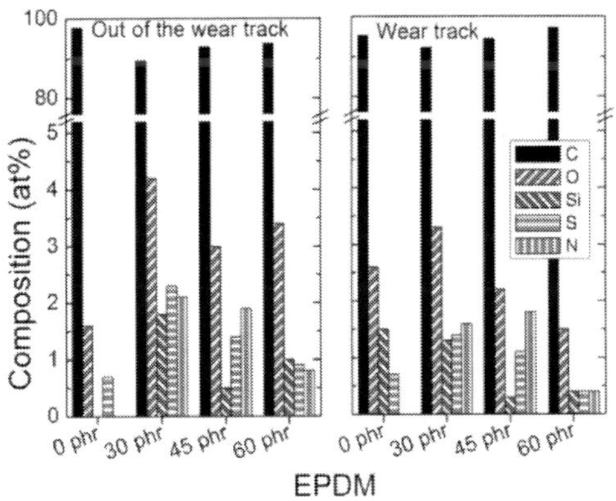

Figure 6. Surface composition of EPDM samples with different carbon black content determined from XPS analysis (with permission from [14])

Detailed analysis of the C 1s core level on the surface subjected to friction revealed that FWHM of the peaks were broader for EPDM 0 phr and 30 phr than for EPDM with higher

CB content (Figure 7b). The reason for this broadening could be initially attributed to higher surface roughness of these samples. However, an equivalent broadening did not occur for the O 1s peak. So, roughness could not explain the broadening of the C 1s core level peak. Moreover, the broadening was not completely symmetric and presented a shoulder at lower binding energies. The fitting of the spectra with a single component of the same FWHM for all samples evidenced this asymmetry (Fig. 7d). The shoulder corresponded to energies close to *sp*[2] carbon [26, 28, 29]. This finding suggests formation of carbon double bonds and/or graphitization in the elastomers with lower CB content. Since these elastomers have worse wear performance, peak broadening can be associated with higher wear rate and damage in the elastomer. A small mismatch between the fitting and the experimental data was observed at higher binding energies only for 0 phr EPDM. This mismatch could be attributed to the roughness effect since the contact surfaces of these samples were severely damaged. In the hypothetical case assuming that this mismatch was caused by C-O bonds, the amount of these species would be rather small. What is clear from these fittings is the absence of carbon-oxygen bonds in the wear track that could explain the different performance in response to friction of the EPDM samples.

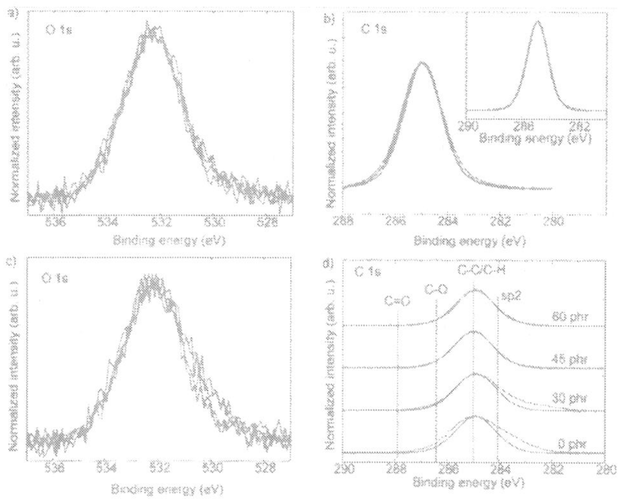

Figure 7. HR O 1s and C 1s core level spectra out of (a) and b) and in (c) and d) the wear track of the EPDM samples. Inset in b) represents the fitting of the EPDM 0 phr sample (with permission from [14])

Elastomer degradation is usually associated with bond scission and oxidation of the backbone structure [27, 39]. We argue that oxygen detected on the samples was not related with backbone structure oxidation since no C-O and C=O bonds were observed in XPS spectra. Migration of additives to the surface can be a plausible reason for the increase in oxygen. Actually, silicon was found at the characteristic binding energy of its oxide form (102 eV). Other authors have suggested that fracture of macromolecular chains is accompanied by generation of low molecular weight products as well as C=C structures [27, 50]. For the samples with larger amounts of CB (45 and 60 phr), the bond scission was smaller as can be inferred from their better wear performance, so the formation of C=C could not be appreciated with the given resolution of the XPS using a non-monochromatic light source.

On the wear track the amount of elements coming from additives (those different from carbon) as well as the amount of oxygen progressively decreased with the increasing amount of carbon black filler. Similarly to the unworn region, oxygen on the worn surface was associated mainly with silicon. Some changes in the sulphur spectrum also occurred. The as-received samples presented two peaks at about 162 eV and 168.5 eV. The first one is related to the S^{2-} sulphur state, while the second one is related to higher oxidation states. The peak at 168.5 eV significantly decreased after the removal of airborne contamination indicating superficial localisation of these oxides and the predominant S^{2-} state in the wear track. These findings suggested that the sulphur chemical state at the surface of the EPDM samples was altered in the ROP tests, and a part of the oxides located in the outer surface of the elastomer was removed.

Water CAs on worn, $\theta_{fr,a}$, and unworn, θ_{nfr}, surfaces were measured to study the changes in wettability caused by the presence of new superficial functional groups (mainly oxygen functional groups due to degradation). Subscript *a* denotes the apparent CA. The values of the roughness factor, *r*, and intrinsic CA for water, θ_{fr}, are shown in Table 4. For unworn samples, mean value of the CA was around 84º with no significant variations with different carbon black content. However, intrinsic CA was larger on the friction zone than on the unworn surfaces for all samples. The increase in the CA was statistically significant at the significance level 0.05. These findings imply that the worn surfaces were more hydrophobic than the initial ones. This behaviour is opposite to the tendency observed during ageing of a commercial EPDM with 52.6 phr of carbon black, when surface became more hydrophilic with 50% decrease in water CA [1]. The decrease in water CA in [1] was caused by thermooxidation of initially hydrophobic methyl-terminated surface of EPDM. This process was accompanied by an arrangement of polar oxygen functional groups (-C-OH, -C=O) on the outer surface layer [1, 2]. In case of frictional surfaces, no oxidation of the elastomer backbone could be found from XPS spectra. Furthermore, the increase in water CA could be associated with changes in the amount of additives present on the surface. Detailed analysis of the surface chemical composition, scanning electron microscopy and energy-dispersive X-ray spectroscopy revealed increase in zinc oxide and silica at the surface [14]. Both of these oxides have hydrophobic and superhydrophobic properties [51] that can explain the increase in water CA in the friction zone.

CB content (phr)	*r*	θ_{fr} (deg)	seθ_{fr} (deg)	θ_{nfr}(deg)	seθ_{nfr}(deg)
0	1.654	107	4.98	81	3.9
30	1.102	90.6	2.11	84	2.1
45	1.207	99.3	2.46	85	1.1
60	1.129	104	4.39	84	2.9

Table 4. Roughness factor (*r*), mean intrinsic contact angles on worn, θ_{fr}, and unworn, θ_{nfr}, surfaces with corresponding standard errors of mean (with permission from [14])

From the results of XPS and CA measurements we concluded that no thermooxidation processes were observed on friction zone under given experimental conditions for all EPDM samples and irrespective of the CB content. Chemical modification of the EPDM surface was due to mechanochemical effects rather than a thermooxidative effect [52-55]. Softer EPDMs with lower carbon black content were severely damaged during ROP test. The increase in C=C bonds for these samples can be attributed to bonds breaking accompanied by different radical reactions [56-58].

3.3. Characterization of surface chemical composition after atmospheric plasma treatments and thin coating of amorphous diamond-like carbon (DLC)

Surface modification is aimed at changing the characteristics of the surface and thin subsurface layer [59] or generation of active centres for further attachment of compounds [60]. By surface modification desired surface properties such as adhesion or wettability [61] can be obtained leaving the underlying bulk unchanged. By doing so, both the surface and bulk properties can be independently tailored and optimized. In this section we present some case studies with the purpose of demonstration the capabilities of XPS technique.

Plasma processing of materials is a crucial industrial technology in many areas including electronics, aerospace, automotive, and biomedical industries [62] due to its versatility [63]. Though nowadays plasma processing is performed mostly at low pressures, atmospheric plasma systems provide an appealing alternative to vacuum plasma systems because continuous processing can be performed at a lower cost [62, 64-66]. Operation under atmospheric pressure provides high flexibility and portability to this technique [32] and allows it expansion to processing of a larger number of materials [62]. One of the possibilities of atmospheric plasma treatments is the use of a plasma torch [67]. In this case, a reactive gas is added to the primary feed gas of the plasma torch in order to generate a flux of chemically active species, e.g. fluorine, toward treated surface. In this way, the surface being fluorinated should not be immersed in reagents and is not directly exposed to the plasma [62]. This treatment can be performed at room temperature and is faster than other fluorination methods [68]. All these advantages are important for industrial application. In our studies surface modification was carried out in two ways: using only ions of inert gases or using also chemically active gases. The reactive gases used for fluorination (commonly SF_6 and CF_4) should be thoroughly diluted with a carrier gas on order to create a stable plasma at atmospheric pressure. Nitrogen, argon or helium are typically used as a carrier gas as they can be easily ionized [65]. Due to the high reactivity of ionized fluorine-containing gases, surface reactions, etching, and plasma polymerization can occur simultaneously. The predominance of one or another will depend on the gas feed, the operating parameters and the chemical nature of the polymer substrates.

Table 5 shows the chemical composition of EPDM elastomers modified by atmospheric plasma using N_2, Ar and He carrier gases in combination with SF_6 and CF_4 fluorination precursors. The composition of untreated elastomers and those activated only by He plasma are included for comparison. The chemical composition was determined from XPS analysis.

All plasma-treated samples had fluorine (0.3 to 13 at. %), oxygen (13 to 22 at. %) and nitrogen (< 2 at. %) on their surfaces. It should be mentioned that plasma treatments promoted a significant increase in the oxygen content. This oxygen incorporation into the surface was similar when plasma activation only with He gas was used. Therefore, oxygen content on the treated surfaces does not significantly depend on the presence or absence of reactive gases during plasma treatments. Oxygen incorporation to the elastomer takes place through interactions between free radicals, O_2 and H_2O molecules from ambient air after the plasma treatment [64]. Traces of sulphur were detected on the EPDM surface after plasma processing using SF_6 [69]. The degree of fluorination for each treatment can be evaluated from O/C and F/C atomic ratios shown in Table 5. From these data it is evident that for fluorination purposes CF_4 is more effective than SF_6. Previously, these results were explained in terms of the dissociation products formed in the plasma [32, 66]. No significant differences in F/C ratio were found for the combination of inert carrier gases (Ar and He) and CF_4, whereas He was more effective for fluorination in combination with SF_6. At the same time, nitrogen carrier gas yielded lower F/C ratio. This is because radicals produced by dissociation of SF_6 and CF_4 could not react with N_2 due to the strong bond between nitrogen atoms. So far, only few information is available in the literature on these reactions [70, 71] to allow definite conclusion. The analysis of the HR C 1s (Figure 2 a, b) revealed the changes in the shape of the core level peak that evidenced not only significant increase in the C-O bonds after the fluorination treatments, but also the apparition of new functional groups involving carbon, oxygen and fluorine.

Process		Chemical composition (at. %)						
		C	O	N	F	S	F/C	O/C
Untreated		93	5	0	0	0	0	0.005
SF_6	N_2	80.4	16.3	1.5	1.3	0.5	0.02	0.20
	Ar	81.6	14.1	1.9	1.7	0.6	0.02	0.17
	He	71.1	19.6	1.6	6.8	0.9	0.09	0.27
CF_4	N_2	84.4	13.4	1.9	0.3	0	0.003	0.16
	Ar	65.9	19.3	1.8	13.0	0	0.20	0.29
	He	63.4	21.9	1.8	12.0	0	0.19	0.34
Activation	He	79	16	traces	-	-	-	0.20

Table 5. Surface composition of EPDM samples after atmospheric plasma treatments

	C1s core level components						
	C-C C-H	C-O -C-H-CF_2	C=O CH-CF	O-C=O CFH-CF_2	CF	CF_2	CF_3
Untreated	96	4	0	0	0	0	0
He activation	77	12	8	3	0	0	0
Fluorinated	45	36	7	5	3	3	1

Table 6. Analysis of the components of the C1s core level of EPDM elastomers: untreated, He activated and fluorinated

The presence of CF and CF_2 can be explained by H substitution and chain scissions [32]. CF_2- represents the main chain of the polymer,-CF- component could indicate cross-link sites [72] and -CF_3 component indicates end groups of polymer chains [72] and grafting [32]. These findings imply enhancing of cross-linking in the elastomer due to plasma processing with fluorine-containing gas.

Figure 2b and 2c show a comparison of HR C1s for elastomers treated with CF_4 and SF_6 using the same carrier gas. From the shape of the peaks one can observe again that CF_4 is more effective than SF_6 in fluorinating because of the presence of additional fluorine containing groups together with the increase in the total fluorine content already mentioned in Table 6. SF_6 molecules mainly dissociate into fluorine atoms and SF_5 radicals [66], whereas CF_4 produces fluorine, CF, CF_2 and CF_3 radicals that react with the elastomer surface leading to a substantial incorporation of fluorine [32]. Though fluorine atoms were considered as the main responsible of the fluorination process, CF_x radicals could contribute to formation of highly fluorinated components (in particular CF_3 groups) [65]. Thus, higher content of CF_x groups after treatment with CF_4 in comparison to SF_6 could be attributed to the larger number of fluorine-containing radicals. Our results demonstrated that the main effect of SF_6 resided in oxidation of the polymer surface, while CF_4, in addition to surface oxidation also induced incorporation of CF_x radicals enhancing the efficiency of the fluorination process [4]. SFE measurements of fluorinated samples revealed that increase in the surface free energy, γ, was noticeable only for the samples which chemical composition was significantly modified, especially by incorporation of oxygen and fluorine polar groups. The increase in SFE related mainly with the electron donor, γ^-, contribution of the polar component, γ^{AB}, [73].

It should be noted that the modification of elastomer surface is not always as evident as in the above examples. The extent of surface modification induced by a particular surface treatment is highly dependent on the type of the elastomer. For instance, when the same treatment is carried out for HNBR, the extent of the surface modification was much less as compared with EPDM (Figure 2d). Surface chemical composition of treated HNBR was almost the same as for untreated one with no presence of fluorine containing groups and no increase in oxygen content [15]. A saturated backbone structure of elastomer such as HNBR makes it less reactive to the plasma treatments.

From systematic studies of fluorinating process under different conditions we concluded that high concentration of fluorine on the elastomer surface, typically higher than 7%, is associated with formation of fluorine-containing functional groups in form of CF, CF_2 or CF_3.The presence of CF and CF_2 can be explained by H substitution and chain scissions [32], where -CF_2- represents the main chain of the polymer, -CF- component could indicate cross-link sites [72], and-CF_3 component indicates end groups of polymer chains [72] and grafting [32]. These findings imply enhancing of cross-linking in the elastomer due to plasma processing with fluorine-containing gas. On the contrary, in processing with lower fluorinating efficiency (< 7%) no evidence of C-F bonds was found [4].

The analysis of other core levels such as O 1s or F 1s can provide complementary information on surface chemical groups. From the analysis of O 1s core level peak of fluorinated samples having one symmetric peak at about 532.6 eV (not shown here, see [4]), we concluded that hydroxyl or ether species were predominant after all fluorinating treatments [62]. This finding is in agreement with the C 1s analysis presented above, as can be inferred from small contributions in the C 1s peak from carboxyl and carbonyl species at 288.1 and 289 eV, respectively [26]. On the other hand, F 1s core level spectra had certain differences when using SF_6 and CF_4 with the same carrier gas, e.g. He (Figure 8). When SF_6 was used, the peak was symmetric and centred at about 687.2 eV, whereas for CF_4 it was centred at 688.1 eV. These energies are close to those reported in the literature for fluorine covalently bonded to carbon [62, 64]. These findings clearly indicated an incorporation of fluorine in the polymer chains. The behaviour observed in the F 1s peaks presented a good correlation with the C 1s analysis given above. Higher fluorine concentration in the sample is associated with the presence of CF_2 bonds according to the C 1s analysis and displacement of the main F 1s peak towards higher binding energies [26].

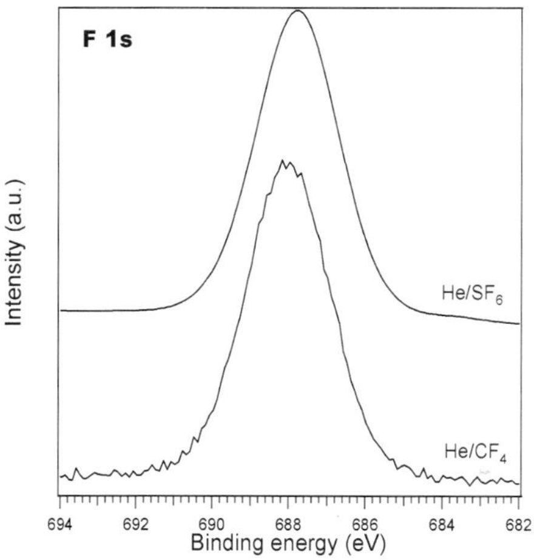

Figure 8. Comparison of the F 1s core level after fluorination treatments with different precursors

Another way to modify elastomer surfaces and improve their tribological properties consists in application of different coatings. Elastomer surfaces with low friction can be obtained by deposition of (3-aminopropyl)-triethoxysilane (APEO) or (3-glycidoxypropyl)trimethoxy-silane (GLYMO) coatings using siloxane precursors [74, 75] or polytetrafluoroethylene (PTFE). PTFE is commonly used as a coating on metallic substrates and it has been recently optimised for use on elastomers yielding low friction coefficient and enhancing other properties [75, 76]. Diamond like carbon (DLC) is another very promising candidate for coatings due to its excellent tribological properties and chemical inertness [6]. DLC coatings have been widely used on different substrates including metals, ceramics and other inorganic materials. Recently, elastic DLC coating on elastomers was developed [5]. Initially, the idea of applying a hard DLC film on soft elastomer materials with low elastic modulus received much scepticism because of the risk of loss of adherence and interfacial delamination. However, after successful demonstration of the efficiency of DLC coatings on elastomer substrates their application has been widely spread, especially in automotive industry. Further advances in application of DLC on elastomer surfaces can be made on the basis of deeper understanding of surface chemistry of coated systems. In this work we studied surface chemical composition of uncoated and DLC-coated NBR and HNBR elastomers using XPS and CA methods. HNBR is a hydrogenated elastomer with no double bonds in the elastomer backbone structure. Therefore, HNBR is less chemically reactive than NBR against the same treatment [4].

Surface chemical composition of elastomers obtained from the wide energy range scan XPS spectra are shown on Table 7. For all samples, carbon, oxygen and nitrogen were the main surface elements. Small amounts of other elements used in the elaboration processes of elastomers were also detected [22]. As we already mentioned before, these components have no important influence on the surface properties [13]. When DLC coating was applied, oxygen content slightly increased. In order to investigate changes in the main bonding of the elastomers, curve fitting procedure of XPS spectra of the C1s core level peaks was performed (see Figure 9 and Table 7). Four components: C-C/CH; C-O; C=O and O-C=O were used for fitting. These components were derived from the expected chemistry of the samples and taking into account natural oxidation process of the elastomers. The binding energies of these components are listed in Table 1. CN bond from the NBR structure was not considered due to the small contribution of nitrogen to the final composition. The results evidenced that C-C contribution corresponding to the backbone structure of the elastomers decreased for DLC-coated elastomers as a consequence of the formation of oxygen functional groups, mainly in form of C-O. HNBR presented larger variation of the carbon bond than NBR.

Despite the small variations observed in the surface chemical composition, DLC-coated elastomers presented better tribological performance reflected in a significant reduction in the coefficient of friction (COF) and friction noise [77]. Also, water CA increased for NBR after DLC deposition (Figure 10) indicating the increase in the hydrophobic character of NBR elastomer surfaces after DLC deposition. This finding is consistent with previous works where an increase in the hydrophobic properties of the DLC-coated elastomers was attributed to sp^2 and sp^3 hybridised carbon bonds in the DLC coating. One should bear in mind that higher hydrophobicity of the surface is usually related to a higher chemical stability. Actually, in our experiments SFE of NBR elastomers decreased by 9% after DLC deposition [77]. For hydrogenated HNBR elastomer, the variations in the hydrophobicity after DLC coatings were statistically insignificant. In contrast to NBR, SFE increased by 8% for DLC-coated HNBR. We suggested that hydrogenation of unsaturated bonds to form HNBR results in different reactivity of the elastomer towards the DLC coatings. Despite the fact that the same type of DLC coating was deposited on all the elastomers, the extent of the modifications was different depending on the substrate.

	C1s core level components						
	C	O	N	C-C C-H	C-O	C=O	O-C=O
NBR uncoated	98	1	0	97	3	0	0
NBR coated	98	1	0	93	7	0	0
HNBR uncoated	92	4	3	91	5	4	0
HNBR coated	89	6	3	86	11	2	1

Table 7. Surface composition and analysis of the components of the C1s core level of NBR and HNBR before and after DLC coating

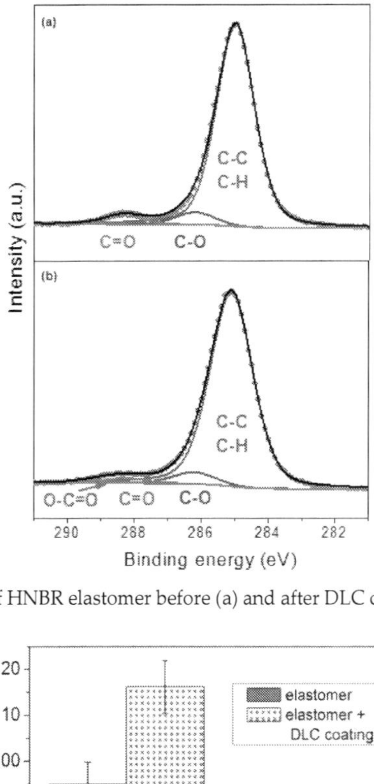

Figure 9. HRC 1s core level of HNBR elastomer before (a) and after DLC coating (b)

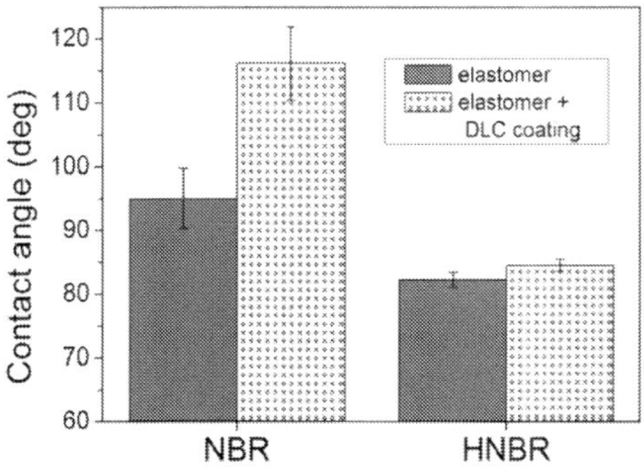

Figure 10. Water CA on elastomers before and after DLC coating

4. Concluding remarks

We have shown that XPS is a sensitive and versatile technique to characterize surface chemical composition of engineered elastomer surfaces. The combination of XPS with other techniques such as CA measurements or SFE calculations allows the evaluation of minute changes in surface chemical composition and structure of surface functional groups resulting from surface degradation or surface treatment. Therefore, factors like ageing processes, lubricant absorption or reaction of the elastomer chains under surface treatment have been analysed with this technique. As XPS binding energies are not only element-specific but also contain chemical information, it provides information about chemical states of a particular element. So, the degradation of the elastomer backbone structure during ageing and wear was evaluated in terms of C-O or C-OH bond formation. In general, the binding energy increases with increasing oxidation state and, for a fixed oxidation state, with the electronegativity of the ligands. The C 1s spectrum of fluorocarbon polymer is an example how the binding energy of carbon depends sensitively on the electronegativity of its neighbours. It was shown that an increase in the oxygen content is not necessarily related to oxidation of carbon bonds and, therefore, to degradation. The presence of new elements, e.g. fluorine due to plasma fluorination, is not always related to the formation of C-F bonds as well. This information cannot be obtained by many other analytical techniques, which makes XPS particularly interesting for this kind of studies. One limitation found was the lack of sensitivity to resolve the possible cross-linking after certain surface treatments. The energy gap between the C=C and C-C bonds are below 0.8 eV which is the range of resolution of nonmonocromated Mg Kα X-ray source. However, cross-linking effects were determined from the variation of SFE components and O/C and F/C ratios. When studying elastomer surfaces subjected to rubbing, no important oxidation indicative of thermochemical reactions was observed. Modifications of the elastomer surfaces were interpreted in terms of mechanochemical reactions and wear.

The final properties of elastomer components can be significantly modified even by small changes in chemical composition of thin surface layer. The extent of the surface modification is clearly influenced by the elastomer substrate and it is more significant for organic materials than other inorganic materials.

Author details

Lidia Martínez and Elisa Román
ICMM-CSIC, Dept. Surfaces and Coatings, Madrid, Spain

Roman Nevshupa*
CISDEM-CSIC, Spain

---

* Corresponding Author

Acknowledgement

Most of this study was developed under the framework of the EU project KRISTAL No. 515837-2 (6th FP). The authors acknowledge Dr. Yves Huttel for his fruitful discussion about the XPS results, IVW (Germany) and TRW (Spain) for the EPDM substrates, VITO (Belgium) for the plasma treatments, HEF (France) for the DLC coatings. The authors also acknowledge the Spanish National Research Council (CSIC) for grant PIE201160E085 and Spanish Ministry for Economy and Competitiveness for financial support provided in the projects MAT2011-29194-C02-02, BIA2011-25653 and "Ramón y Cajal" RYC-2009-04125.

5. References

[1] Nevshupa R, Martinez L, Alvarez L, Lopez MF, Huttel Y, Mendez J, et al. (2011) Influence of thermal ageing on surface degradation of ethylene-propylene-diene elastomer. Journal of Applied Polymer Science. 119:242-251.

[2] Laoharojanaphand P, Lin TJ, Stoffer JO (1990) Glow discharge polymerization of reactive functional silanes on poly(methyl methacrylate). Journal of Applied Polymer Science. 40:369-384.

[3] Desai SM, Bodas DS, Singh RP (2004) Fabrication of long-term hydrophilic elastomeric surfaces via plasma induced surface cross-linking of functional monomers. Surface and Coatings Technology. 184:6-12.

[4] Martínez L, Huttel Y, Verheyde B, Vanhulsel A, Román E (2010) Photoemission study of fluorination atmospheric pressure plasma processes on EPDM: Influence of the carrier and fluorinating gas. Applied Surface Science. 257:832-836.

[5] Nakahigashi T, Tanaka Y, Miyake K, Oohara H (2004) Properties of flexible DLC film deposited by amplitude-modulated rf p-cvd. Tribology International. 37:907-912.

[6] Lindholm P, Björklund S, Svahn F (2006) Method and surface roughness aspects for the design of DLC coatings. Wear. 261:107-111.

[7] Delor-Jestin F, Lacoste J, Barrois-Oudin N, Cardinet C, Lemaire J (2000) Photo-, thermal and natural ageing of ethylene–propylene–diene monomer (EPDM) rubber used in automotive applications. Influence of carbon black, crosslinking and stabilizing agents. Polymer Degradation and Stability. 67:469-477.

[8] Majumder PS, Bhowmick AK (1998) Friction behaviour of electron beam modified ethylene–propylene diene monomer rubber surface. Wear. 221:15-23.

[9] Lonkar SP, Kumar AP, Singh RP (2007) Photo-stabilization of EPDM–clay nanocomposites: Effect of antioxidant on the preparation and durability. Polymers for Advanced Technologies. 18:891-900.

[10] Farahani TD, Bakhshandeh GR (2005) The effect of curing on sorption and diffusion of a brake fluid in EPDM elastomer. e-Polymers. 47:1-10.

[11] Banik I, Bhowmick AK (2000) Electron beam modification of filled fluorocarbon rubber. Journal of Applied Polymer Science. 76:2016-2025.

[12] Tillier DL, Meuldijk J, Koning CE (2003) Production of colloidally stable latices from low molecular weight ethylene–propylene–diene copolymers. Polymer. 44:7883-7890.

[13] Degrange JM, Thomine M, Kapsa P, Pelletier JM, Chazeau L, Vigier G, et al. (2005) Influence of viscoelasticity on the tribological behaviour of carbon black filled nitrile rubber (NBR) for lip seal application. Wear. 259:684-692.

[14] Martinez L, Nevshupa R, Felhoes D, de Segovia JL, Roman E (2011) Influence of friction on the surface characteristics of EPDM elastomers with different carbon black contents. Tribology International. 44:996-1003.

[15] Martínez L, Álvarez L, Huttel Y, Méndez J, Román E, Vanhulsel A, et al. (2007) Surface analysis of NBR and HNBR elastomers modified with different plasma treatments. Vacuum. 81:1489-1492.

[16] Alisoy HZ, Baysar A, Alisoy GT (2005) Physicomathematical analysis of surface modification of polymers by glow discharge in medium. Physica A: Statistical Mechanics and its Applications. 351:347-357.

[17] Einstein A (1905) Über einen die erzeugung und verwandlung des lichtes betreffenden heuristischen gesichtspunkt. Annalen der Physik. 322:132-148.

[18] Hüfner S (2003) Photoelectron spectroscopy. Principles and applications, 3rd edition. Berlin: Springer, 662 p.

[19] Seah MP, Dench WA (1979) Quantitative electron spectroscopy of surfaces: A standard data base for electron inelastic mean free paths in solids. Surface and Interface Analysis. 1:2-11.

[20] Briggs D, Seah MP (1990) Practical surface analysis—auger and X-ray photoelectron spectroscopy. Chichester: Wiley, 657 p.

[21] Woodruff DP, Delchar TA (1986) Modern techniques of surface science. Cambridge: Cambridge University Press, 608 p.

[22] Mitra S, Ghanbari-Siahkali A, Kingshott P, Rehmeier HK, Abildgaard H, Almdal K (2006) Chemical degradation of crosslinked ethylene-propylene-diene rubber in an acidic environment. Part i. Effect on accelerated sulphur crosslinks. Polymer Degradation and Stability. 91:69-80.

[23] Akhter S, Allan K, Buchanan D, Cook JA, Campion A, White JM (1988) XPS and IR study of X-ray induced degradation of pva polymer film. Applied Surface Science. 35:241-258.

[24] Coullerez G, Chevolot Y, Léonard D, Xanthopoulos N, Mathieu HJ (1999) Degradation of polymers (PVC, PTFE, m-f) during X-ray photoelectron spectroscopy (ESCA) analysis. Journal of Surface Analysis. 5:235-239.

[25] Shirley DA (1972) High-resolution x-ray photoemission spectrum of the valence bands of gold. Physical Review B. 5:4709-4714.

[26] Beamson G, Briggs D (1992) High resolution XPS for organic polymers: The Scienta ESCA 300 database. Ney York: John Wiley & Sons, 306 p.

[27] Zhang SW (2004) Tribology of elastomers. Amsterdam: Elsevier, 282 p.

[28] Swaraj S, Oran U, Lippitz A, Friedrich JF, Unger WES (2005) Surface analysis of plasma-deposited polymer films, 6. Plasma Processes and Polymers. 2:572-580.

[29] Grythe KF, Hansen FK (2006) Surface modification of EPDM rubber by plasma treatment. Langmuir. 22:6109-6124.

[30] Nansé G, Papirer E, Fioux P, Moguet F, Tressaud A (1997) Fluorination of carbon blacks: An X-ray photoelectron spectroscopy study: I. A literature review of XPS studies of fluorinated carbons. XPS investigation of some reference compounds. Carbon. 35:175-194.

[31] Wen C-H, Chuang M-J, Hsiue G-H (2006) Plasma fluorination of polymers in glow discharge plasma with a continuous process. Thin Solid Films. 503:103-109.

[32] Hochart F, Levalois-Mitjaville J, De Jaeger R, Gengembre L, Grimblot J (1999) Plasma surface treatment of poly(acrylonitrile) films by fluorocarbon compounds. Applied Surface Science. 142:574-578.

[33] Wenzel RN (1936) Resistance of solid surfaces to wetting by water. Industrial & Engineering Chemistry. 28:988-994.

[34] Gao L, McCarthy TJ (2007) How Wenzel and Cassie were wrong. Langmuir. 23:3762-3765.

[35] Van Oss CJ, Chaudhury MK, Good RJ (1988) Interfacial Lifshitz-van der Waals and polar interactions in macroscopic systems. Chemical Reviews. 88:927-941.

[36] Wu W, Giese RF, Jr., van Oss CJ (1995) Evaluation of the Lifshitz-van der Waals/acid-base approach to determine surface tension components. Langmuir. 11:379-382.

[37] Navrátil Z, Buršíková V, St'ahel P, Šíra M, Zvěřina P (2004) On the analysis of surface free energy of DLC coatings deposited in low pressure rf discharge. Czechoslovak Journal of Physics. 54:C877-C882.

[38] Luner PE, Oh E (2001) Characterization of the surface free energy of cellulose ether films. Colloids and Surfaces A: Physicochemical and Engineering Aspects. 181:31-48.

[39] Zhao Q, Li X, Gao J (2008) Surface degradation of ethylene–propylene–diene monomer (EPDM) containing 5-ethylidene-2-norbornene (ENB) as diene in artificial weathering environment. Polymer Degradation and Stability. 93:692-699.

[40] Kumar A, Commereuc S, Verney V (2004) Ageing of elastomers: A molecular approach based on rheological characterization. Polymer Degradation and Stability. 85:751-757.

[41] Mason LR, Reynolds AB (1998) Comparison of oxidation induction time measurements with values derived from oxidation induction temperature measurements for EPDM and XLPE polymers. Polymer Engineering & Science. 38:1149-1153.

[42] Budrugeac P, Segal E (1994) On the kinetics of the thermal degradation of polymers with compensation effect and the dependence of activation energy on the degree of conversion. Polymer Degradation and Stability. 46:203-210.

[43] Ayala JA, Rincón ME (1981) The oxidation of fuel oil #6 studied by differential scanning calorimetry. ACS Fuel. 26:120-130.

[44] Schallamach A (1971) How does rubber slide? Wear. 17:301-312.

[45] Felhös D, Karger-Kocsis J (2008) Tribological testing of peroxide-cured EPDM rubbers with different carbon black contents under dry sliding conditions against steel. Tribology International. 41:404-415.

[46] Karger-Kocsis J, Mousa A, Major Z, Békési N (2008) Dry friction and sliding wear of EPDM rubbers against steel as a function of carbon black content. Wear. 264:359-367.

[47] Rathinasamy P, Balamurugan P, Balu S, Subrahmanian V (2004) Effect of adhesive-coated glass fiber in natural rubber (NR), acrylonitrile rubber (NBR), and ethylene–propylene–diene rubber (EPDM) formulations. I. Effect of adhesive-coated glass fiber on the curing and tensile properties of NR, NBR, and EPDM formulations. Journal of Applied Polymer Science. 91:1111-1123.

[48] Chou H-W, Huang J-S, Lin S-T (2007) Effects of thermal aging on fatigue of carbon black–reinforced EPDM rubber. Journal of Applied Polymer Science. 103:1244-1251.

[49] Wang M-J, Wolff S. Surface energy of carbon black. In: Donnet J-B, editor. Carbon black: Science and technology, second edition. Ney York: CRC Press; 1993. p. 289-355.

[50] Rizk RAM, Abdul-Kader AM, Ali ZI, Ali M (2009) Effect of ion bombardment on the optical properties of ldpe/EPDM polymer blends. Vacuum. 83:805-808.

[51] Zhang J, Huang W, Han Y (2006) Wettability of zinc oxide surfaces with controllable structures. Langmuir. 22:2946-2950.

[52] Nevshupa R, Roman E, de Segovia JL (2010) Model of the effect of local frictional heating on the tribodesorbed gases from metals in ultra-high vacuum. International Journal of Materials & Product Technology. 38:57-65.

[53] Nevshupa RA (2009) The role of athermal mechanisms in the activation of tribodesorption and triboluminisence in miniature and lightly loaded friction units. Journal of Friction and Wear. 30:118-126.

[54] Heinike G (1984) Tribochemistry. Munchen: Carl Hanser Verlag, 495 p.

[55] Kostetsky BI (1992) The structural-energetic concept in the theory of friction and wear (synergism and self-organization). Wear. 159:1-15.

[56] Malhotra M, Kumar S (1997) Thermal gas effusion from diamond-like carbon films. Diamond and Related Materials. 6:1830-1835.

[57] Butyagin PY (1971) Kinetics and nature of mechanochemical reactions. Russian Chemical Reviews. 40:901-915.

[58] Butyagin PY (1984) Structural disorder and mechanochemical reactions in solids. Russian Chemical Reviews. 53:1025-1038.

[59] Orellana LM, Pérez FJ, Gómez C (2005) The effect of nitrogen ion implantation on the corrosion behaviour of stainless steels in chloride media. Surface and Coatings Technology. 200:1609-1615.

[60] Goddard JM, Hotchkiss JH (2007) Polymer surface modification for the attachment of bioactive compounds. Progress in Polymer Science. 32:698-725.

[61] Minko S, Müller M, Motornov M, Nitschke M, Grundke K, Stamm M (2003) Two-level structured self-adaptive surfaces with reversibly tunable properties. Journal of the American Chemical Society. 125:3896-3900.

[62] Ho KKC, Lee AF, Bismarck A (2007) Fluorination of carbon fibres in atmospheric plasma. Carbon. 45:775-784.

[63] Mitra S, Ghanbari-Siahkali A, Kingshott P, Rehmeier HK, Abildgaard H, Almdal K (2006) Chemical degradation of crosslinked ethylene-propylene-diene rubber in an acidic environment. Part ii. Effect of peroxide crosslinking in the presence of a coagent. Polymer Degradation and Stability. 91:81-93.

[64] Felten A, Ghijsen J, Pireaux JJ, Johnson RL, Whelan CM, Liang D, et al. (2008) Photoemission study of CF_4 rf-plasma treated multi-wall carbon nanotubes. Carbon. 46:1271-1275.

[65] Fanelli F, Fracassi F, d'Agostino R (2008) Fluorination of polymers by means of he/CF_4-fed atmospheric pressure glow dielectric barrier discharges. Plasma Processes and Polymers. 5:424-432.

[66] Borisov S, Khotimsky VS, Rebrov AI, Rykov SV, Slovetsky DI, Pashunin YM (1997) Plasma fluorination of organosilicon polymeric films for gas separation applications. Journal of Membrane Science. 125:319-329.

[67] http://www.vitoplasma.com/en/30, last access on March 2012.

[68] Yasuda H, Hsu TS (1977) Some aspects of plasma polymerization of fluorine-containing organic compounds. Journal of Polymer Science: Polymer Chemistry Edition. 15:2411-2425.

[69] Barni R, Riccardi C, Selli E, Massafra MR, Marcandalli B, Orsini F, et al. (2005) Wettability and dyeability modulation of poly(ethylene terephthalate) fibers through cold SF_6 plasma treatment. Plasma Processes and Polymers. 2:64-72.

[70] Kumar SVK, Sathyamurthy N, Manogaran S, Mitra SK (1994) Possible reaction of atomic nitrogen with SF_x (x = 1–5) and CF_x (x = 1–3) fragments from N_2-SF_6 and N_2-CF_4 discharges. Chemical Physics Letters. 222:465-470.

[71] Radoiu MT (2003) Studies of 2.45 GHz microwave induced plasma abatement of CF_4. Environmental Science & Technology. 37:3985-3988.

[72] Tran ND, Dutta NK, Roy Choudhury N (2006) Weatherability and wear resistance characteristics of plasma fluoropolymer coatings deposited on an elastomer substrate. Polymer Degradation and Stability. 91:1052-1063.

[73] Schlögl S, Kramer R, Lenko D, Schröttner H, Schaller R, Holzner A, et al. (2011) Fluorination of elastomer materials. European Polymer Journal. 47:2321-2330.

[74] Verheyde B, Havermans D, Vanhulsel A (2011) Characterization and tribological behaviour of siloxane-based plasma coatings on HNBR rubber. Plasma Processes and Polymers. 8:755-762.

[75] Verheyde B, Rombouts M, Vanhulsel A, Havermans D, Meneve J, Wangenheim M (2009) Influence of surface treatment of elastomers on their frictional behaviour in sliding contact. Wear. 266:468-475.

[76] http://www.fluoroprecision.co.uk/ptfe-coated-elastomers-rubbers.html, last access on March 2012.

[77] Martinez L, Nevshupa R, Alvarez L, Huttel Y, Mendez J, Roman E, et al. (2009) Application of diamond-like carbon coatings to elastomers frictional surfaces. Tribology International. 42:584-590.

Chapter 7

Non-Destructive Surface Analysis by Low Energy Electron Loss Spectroscopy

Vitaliy Tinkov

Additional information is available at the end of the chapter

http://dx.doi.org/10.5772/48090

1. Introduction

The modern progress in such priority scientific directions as microelectronics, nanotechnology, material science, heterogeneous catalysis, etc., are impossible without obtaining quantitative information about physical–chemical properties from the nano-size near surface region of the materials.

It is known that the physical–chemical properties of the metallic alloy surfaces differ markedly from that of the bulk and, mainly, it is caused by segregation one of the alloy components on the surface [1]. It is related to the fact that the physical–chemical state of the surface substantially influences such surface processes as adsorption, catalysis, oxidation, friction and wear. Recently such phenomena as the thermo-induced surface segregation of alloy components as used for obtaining chemically-active surfaces have been widely used; being of great interest in terms of heterogeneous catalysis and the development of new nanotechnological processes. Study of the kinetics of surface segregation permits the determination of the bulk diffusion coefficients of the segregated elements; knowledge of which then permits the controlled change of surface structures under heat treatment and etc.

At the present for the investigation of the physical–chemical properties of the metallic alloy surfaces the nondestructive methods are widely used, such as an Rutherford Backscattering Spectrometry (RBS), X-ray Photoelectron Spectroscopy (XPS), Low Energy Ion Scattering (LEIS), Ultraviolet Photoelectron Spectroscopy (UPS) and other [2].

Physical phenomena such as secondary electron emission (SEE) can be used for investigation of the near surface region of a solid with a purpose to obtain quantitative information concerning its crystal structure, element composition and the electronic states of atoms [3,4]. On the Figure1 the total energy distribution of reflected SEE from a surface is shown which is irradiated by an electron beam of primary energy E_0. The shape is due to some types of interaction: elastic and inelastic scattering together with secondary electron emission. There are four ranges in N(E), in each if which one of these interactions predominates.

The elastic interaction produced a narrow peak on the right, where the electrons retain their energy E_0 and merely show altered momentum direction (Region I). The broadening is due to thermal spread in the beam energy and is also affected by the analyzer resolution. With standard equipment the broadening is usually 0.5-1 eV, so phonon excitations (energy loss 10-50 meV) can be detected only by special techniques involving highly monoenergetic primary beams and improved analyzer resolution, as realized in high resolution energy loss spectroscopy. With other methods, one can assume that the elastic peak is due to group electrons from the beam that have undergone elastic and quasi elastic interactions with the surface. Various methods are applied to the elastic backscattering, particularly diffraction ones such as Low Electron Energy Diffraction (LEED) and high energy diffraction with back scattering which have been applied to the spatial distributions of the backscattered beams. The methods have been applied to the atomic structures and dynamic characteristics in ordered surface layers.

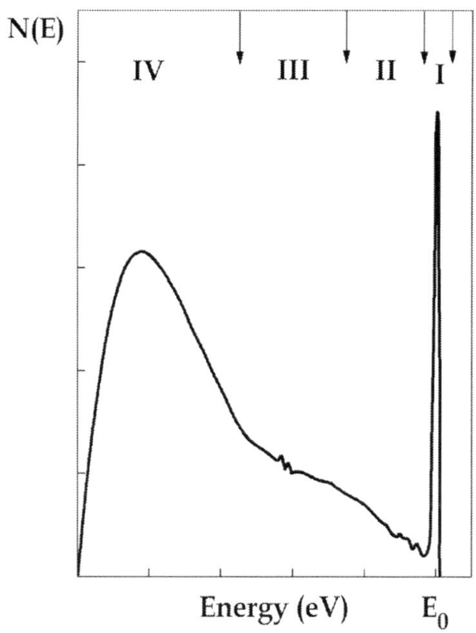

Figure 1. Total energy distribution of secondary electron emission from a surface which is irradiated by an electron beam of primary energy E_0

The broad low energy maximum (Region IV) is due to the true secondary electrons which have energies from zero up to some tens of eV and are formed by repeated inelastic electron-electron scattering in the cascade process. The true secondary electrons may constitute up to 70% of total energy distribution. Their energy distribution is related to the random filling of the final states and to the cascade multiplication.

In Region III there is also a fine structure due to electrons from the solid escaping in the vacuum by the Auger process (Auger Electron Spectroscopy (AES)). The Auger electron spectrum for a given element has a characteristic form and certain energies which have meant that AES is widely used in elemental analysis.

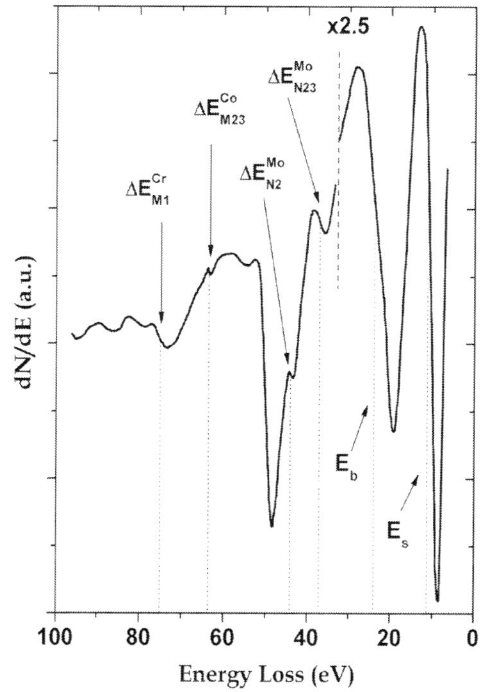

Figure 2. Example low EELS spectra obtained for the Co–Cr–Mo alloy surface at the primary energy E_0 = 350 eV with identification of energy losses

Region II is due manly to electrons that have lost some of their energy by inelastic scattering; directly by the elastic peak, one finds electrons that have suffered discrete energy losses from the excitation of inter- and intraband electronic transitions, surface and bulk plasmons, hybrid modes of plasmons and ionization losses (Ionization Spectroscopy). That range is usually 30-100 eV. Usually, the losses related to surface and bulk plasmon excitations are most intensive lines in the electron energy loss spectrum. The spectra of plasma oscillations are potential data carriers about composition and chemical state of elements on the surface of solid and in the adsorbed layers. The energy losses are called as characteristic losses because losses do not depend on the primary electron energy E_0 and its value is individual for the chemical element and compound. Region II is called as Electron Energy Loss Spectroscopy (EELS). At energy $E_0 < 1000$ eV it can be called as low EELS. On the Figure 2 really low EELS spectra is shown with interpretation of losses for the Co-Cr-Mo alloy surface which was measured at the primary electron beam energy $E_0 = 350$ eV in dN/dE mode [5].

Ionization Spectroscopy (IS) is a variant on EELS. Gerlach et al. [6,7] first applied this in terms of analysis of the surface composition analysis for V, Ni, Pd and Mo as impurities on the surface of polycrystalline metals without depth analysis. The IS method is based on measuring the energy spectra of electrons, which have lost a particular portion of the energy ΔE for the excitation of electrons from internal atomic levels into the empty states (conduction band) of the solid. Having lost energy ΔE, and after being inelastically scattered, the primary electrons escape into vacuum and are registered on the background secondary emission spectrum as individual monochromatic groups which form spectral lines. The advantages of IS as compared to other methods of electronic spectroscopy are (i) the position of ionization lines in the spectrum with respect to the lines of elastic scattered electrons is determined by the binding energy of electrons in the ground state and by the distribution of the density of empty states and does not depend on the value of the primary electron energy E_0, (this allows easy separation of IS lines from the AES lines) and, (ii) the possibility to vary the probing depth of the near-surface region because the change in primary energy E_0 induces a change in the mean free path λ of electrons. On the Figure 3 the different between Auger process and ionization process is shown which are generated by an electron beam of primary energy E_0.

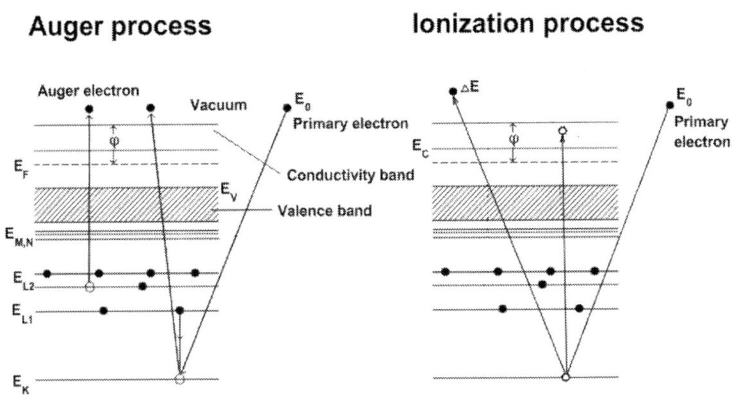

Figure 3. Example of different between Auger process and ionization process which are generated by an electron beam of primary energy E_0

The aim of the present chapter is to show the application of low Electron Energy Loss Spectroscopy as non-destructive method, namely Ionization Spectroscopy and surface and bulk plasmon excitations, at investigation of physical-chemical properties materials in the nano-size near surface region.

2. Low energy electron loss spectroscopy

2.1. Ionization spectroscopy

2.1.1. *Physical model*

Ionization Spectroscopy is based on the measurement of the energy spectra of electrons, which have lost a particular portion of the energy ΔE_β for the excitation of electronic transitions that are typical for a given kind of atom β. The position of an intensity line (IL) in the spectrum with respect to the primary electron energy E_0 is determined by the binding energy of electrons in the ground state and by the distribution of the density of empty states, but it does not depend on the value of E_0, on the work function or on the value of the surface charge.

The calculation of the contribution to the intensity of an IL by the electrons having lost an amount of energy ΔE_β at the depth Z from the sample surface by the ionization of the core states of the atoms β is simple when a traditional experimental configuration is used (an incident beam of the primary electrons is directed perpendicularly to the sample surface ($\theta_0 = 0$) and the secondary electrons are registered at the angle θ with respect to the normal). In this case calculations within the framework of a two-stage model allow us to obtain the following expression for the intensity of an IL [8]:

$$I_\beta\left(Z, E_{0j}\right) = K\sigma_\beta \tilde{r}_\beta n_\beta(Z)\exp(-Z/\Lambda_\beta), \tag{1}$$

where K is an instrumental factor, σ_β is the ionization cross-section of the core level, $n_\beta(Z)$ is concentration of atoms β at depth Z from surface, $\tilde{r}_\beta$ is the elastic scattering factor of electrons. Λ_β is the effective free-path of electrons in a sample with respect to inelastic collisions, which is determined by the equation

$$\Lambda_\beta^{-1} = \lambda_0^{-1} + \left(\lambda_\beta \cos\theta\right)^{-1}. \tag{2}$$

For the Pt-Me (Me: Fe, Co, Ni, Cu) alloys [9]

$$\lambda_0 = \frac{1194}{E_{0j}^2} + 0.429E_{0j}^{1/2}\ ;\ \lambda_\beta = \lambda(E_{0j} - \Delta E_\beta).$$

An effective probing depth in IS amounts to ~ $3\Lambda_\beta$ because the secondary electrons created in the near-surface region of this thickness contribute for 95% to the total intensity of an IL. An increase of the effective probing depth upon increasing the energy E_0 also results in an increased contribution from the deeper layers of the concentration profile into the IL intensity. This enables us to carry out a layer-by-layer reconstruction of the concentration profiles of the elements using the energy dependencies of the IL.

After integration of Eq. (1) with respect to depth and spatial angle of the four-grid energy analyzer, an expression for the total IL intensity has the following form

$$I_\beta(E_0) = 2\pi K\sigma_\beta \int_0^{\infty} \int_{\Theta_{min}}^{\Theta_{max}} \tilde{r}_\beta n_\beta(Z)\exp(-Z/\Lambda_\beta)\sin\Theta dZ d\Theta\,, \tag{3}$$

where $\Theta_{min} = 4^\circ$ and $\Theta_{max} = 70^\circ$ are respectively the minimum and maximum values of polar angle for the standard quasi-spherical four-grid energy analyzer.

As pointed out above, the offered method is essentially not sensitive to the type of the energy analyzer used. Only the values of Θ_{min} and Θ_{max} that correspond to the concrete conditions of an experiment should be substituted in Eq. (3). In the case of a binary A-B alloy, usually the ratio of intensities of A to B

$$R_A(E_{0j}) = \frac{I_A(E_{0j}, \Delta E_A)}{I_B(E_{0j}, \Delta E_B)} \tag{4}$$

is measured experimentally in order to eliminate the instrumental factor K, which is often unknown.

Let us consider as new variables the relative concentrations of the elements in a layer with number i:

$$N_\beta(i) = n_\beta(i)v_\beta,\ \beta = \overline{A,B}\,, \tag{5}$$

where v_β is the atomic volume of a pure component of an alloy. After replacing the integral in Eq. (3) by a summation over N and substituting the expression for I_β into formula (4), integration with respect to the width of isolated layer d leads to:

$$R_A(E_{0j}) = \frac{\sigma_A \lambda_i^A v_B}{\sigma_B \lambda_i^B v_A} \cdot \frac{\sum_{i=1}^{N} N_A(i)P_A(i,E_{0j})}{\sum_{i=1}^{N} N_B(i)P_B(i,E_{0j})}\,, \tag{6}$$

where

$$\begin{aligned} P_\beta(i,E_{0j}) &= \exp\left[-\frac{(i-1)d}{\lambda_0}\right] \int_{\Theta_{min}}^{\Theta_{max}} D_i(E_{0j},\Theta)\tilde{r}_\beta \left[1-\exp\left(-\frac{d}{\Lambda_\beta}\right)\right] d\Theta \\ D_i(E_{0j},\Theta) &= \exp\left[-\frac{(i-1)d}{\lambda_0\cos\Theta}\right] \frac{\cos\Theta\sin\Theta}{\lambda_0 + \lambda_i^\beta\cos\Theta}\,, \end{aligned} \tag{7}$$

i = 1, 2, ..., N - 1, and i = N

$$P_\beta(i,E_{0j}) = \exp\left[-\frac{(N-1)d}{\lambda_0}\right] \int_{\Theta_{min}}^{\Theta_{max}} D_N(E_{0j},\Theta)\tilde{r}_\beta d\Theta\,, \tag{8}$$

Following the approach offered in [10], the expression (6) is transformed into a system of linear equations (SLE) with respect to $N_A(i)$ using the relation $N_A(i) + N_B(i) = 1$. As a result, we obtain (j = 1, 2, ..., M)

$$\frac{\sum_{i=1}^{N} N_A(i)\left\{\frac{\sigma_A \lambda_i^A v_B}{\sigma_B \lambda_i^B v_A} P_A(i,E_{0j}) + R_A(E_{0j}) P_B(i,E_{0j})\right\}}{\sum_{i=1}^{N} P_B(i,E_{0j})} = R_A(E_{0j}), \tag{9}$$

where E_{0j} is the energy of primary electrons for which we have measured the ratio of IL intensities $R_A(E_{0j})$. Assuming that all interlayer distances in the near-surface region of a single-crystal alloy are identical and equal to d, for summations in Eqs. (6) and (9) the number of terms is selected that corresponds to the selection of N monolayers parallel to a free surface, using the relationship $(N-1)\cdot d = 3\Lambda_{max}$ (where $\Lambda_{max} = \left(\Lambda_A(E_0^{max}) + \Lambda_B(E_0^{max})\right)/2$). The system of linear equations (9) can be solved only when it is determined or overdetermined, i.e. if the inequality $M \geq N$ is true for this system.

In the following sections, methods are presented for building the solution of Eq. (9) and for the numerical calculation of the concentration profiles within the framework of the described model.

2.1.2. *Layer-by-layer reconstruction methods*

A system of equations, describing the deviations of the concentrations in a monolayer i ($N_A(i)$) from their bulk value N_A, can be represented in matrix form by the expression

$$\sum_{i=1}^{N} Q_{ji} \delta N_A(i) = \tilde{R}_A(E_{0j}), \tag{10}$$

where $\delta N_A(i) = N_A(i) - N_A$, $\tilde{R}_A(E_{0j}) = R_A(E_{0j}) - N_A \sum_{i=1}^{N} Q_{ji}$, and an explicit form of the matrix elements Q_{ji} is evident from the expression (9). However, the practical solution of the Eq. (10) presents particular difficulties because the matrix elements Q_{ji} correspond to close energy intervals that do not differ sufficiently. As a result, the determinant of the matrix Q is close to zero and the system (10) is ill-conditioned. As a consequence, the errors in the matrix elements Q_{ji} and in the $\tilde{R}_A$ values can result in an incorrect solution.

To construct a stable approximation for the solution of system (10), the condition-gradients projection method, the conjugate gradients projection method, the method of conjugate gradient projected on the Π^+–space and also the regularization method [11] were used in the present work. More detailed information on techniques for solving ill-posed problems can be found in [10,11]. The described regularization algorithm for the reconstruction of the elemental concentration profiles in a binary alloy on the basis of energy dependencies of the ratio of IL intensities is implemented in FORTRAN codes.

2.1.3. Results of the layer-by-layer reconstruction

Approbation of method of nondestructive layer-by-layer analysis was performed for the single crystal $Pt_{80}Co_{20}$ alloy with (100) and (111) surface orientations [12]. Initially, $Pt_{80}Co_{20}$(111) alloy surface was in a disordered state. First, we measured the spectra of the ionization losses for the clean (100) and (111) surfaces of $Pt_{80}Co_{20}$ alloy and polycrystals of platinum and cobalt in the dN/dE mode. The energy losses $\Delta E^{Pt}_{O_{2,3}} = 54eV$, $\Delta E^{Co}_{M_{2,3}} = 62eV$ and their IL were recorded in the range of primary electron energy E_0 = 200 - 500 eV.

In order to ignore in calculations an instrumental factor K (it is often unknown), the ionization cross-section σ_β and elastic scattering factor $\tilde{r}_\beta$ and the possible influence of matrix effects (for example, due to difference of atomic radii for platinum and cobalt r_{Pt}/r_{Co} = 1.104), usually the IL ratio of elements is measured, with normalization on standards by following equation:

$$R(E_0) = \frac{R_{alloy}(E_0)}{R_{st}(E_0)}, \quad (11)$$

where $R_{alloy}(E_0) = I^{alloy}_{Co}(E_0) / I^{alloy}_{Pt}(E_0)$; $R_{st}(E_0) = I^{st}_{Co}(E_0) / I^{st}_{Pt}(E_0)$; I^{alloy}_{Co}, I^{alloy}_{Pt} and I^{st}_{Co}, I^{st}_{Pt} are intensity lines of the ionization losses of the alloy components and standards (pure metals), respectively. Figure 4 shows the ratio of ionization peaks of Co to Pt as a function of the primary electron energy for (111) and (100) faces of $Pt_{80}Co_{20}$ alloy at room temperature before and after normalization on the standards.

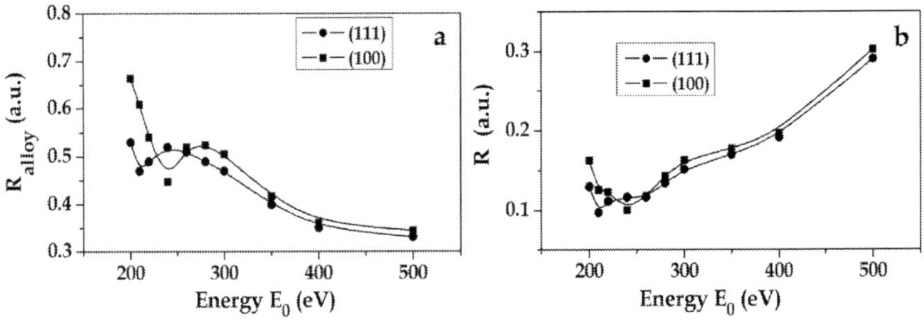

Figure 4. The ratio of ionization peaks of Co to Pt as a function of the primary electron energy for (111) and (100) faces of $Pt_{80}Co_{20}$ alloy at room temperature: (a) – before and (b) – after normalization on the standards

Based on experimental data $R(E_0)$, we calculated the layer-by-layer Pt concentration profiles for (100) and (111) faces of alloy $Pt_{80}Co_{20}$ by means of the condition-gradients projection method, the conjugate gradients projection method, the method of conjugate gradient projected on the Π^+-space with total level of experimental errors less than 3%. On Figure 5 the averaged Pt concentration are shows to all three methods by histograms.

It can be seen that the upper layer contains only platinum atoms for both faces and practically does not contain cobalt atoms. Moreover, there are strong orientation effects that affect on length of the platinum concentration oscillations for the (100) and (111) faces. The deeper oscillation is observed for a more "loose" (100) face, which affects the depth composition up to eight atomic layers. Whereas for the close-packed (111) face these changes are damped on the fifth level. The presented results are in good agreement with experimental data of concentration profiles which were obtained by means of LEED and LEIS [13,14].

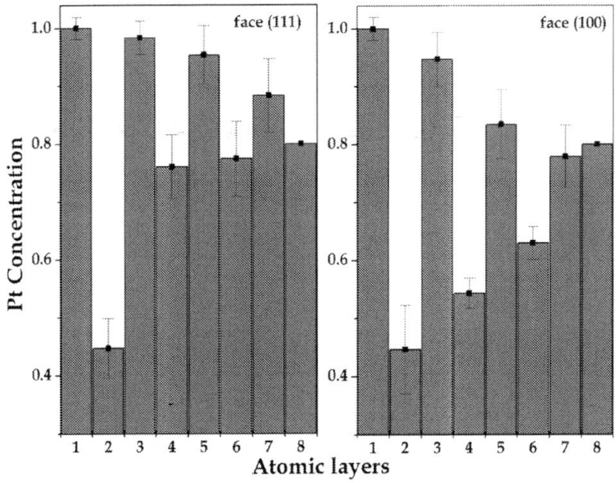

Figure 5. Layer-by-layer Pt concentration profiles reconstructed from the ionization spectra for (111) and (100) faces of $Pt_{80}Co_{20}$ alloy at room temperature

Non-destructive method of layer-by-layer analysis by IS can be effective at study of temperature concentration profiles. Authors [8] investigated influence of heating on concentration profile of $Pt_{80}Co_{20}(111)$ (see Figure 6). Heating the sample to 613 K leads to a depletion of Pt atoms in the 2nd layer ($C_{Pt}^{(2)}$ = 24%) and to an insignificant enrichment of Co atoms in layers 3-6 in comparison with the profile at room temperature. Increasing the temperature further to 673 K is accompanied by a negligible segregation of Co from the second layer ($C_{Pt}^{(2)}$ =31%) to the first ($C_{Pt}^{(1)}$ = 97%), while deeper layers remain practically unchanged. At 823K, a sandwich-like structure of the type Pt/Co/Pt was found in the first three atomic layers. As is obvious from Figure 6, heating the sample causes a smoothing of the oscillations in deeper layers towards the bulk concentration of the alloy. However, the first layer still consists of pure Pt up to 873 K. Further increasing the temperature gradually results in completely smoothed oscillations.

Consequently, the sample was slowly cooled during 10 hours from 1123 K to room temperature. As a result of this procedure, a chemically ordered alloy surface of the $L1_2$ type was obtained. LEED shows super-structural reflections in a diffraction pattern at E_0 = 112 eV. The result of the layer-by-layer reconstruction for the ordered state shows that the 1st

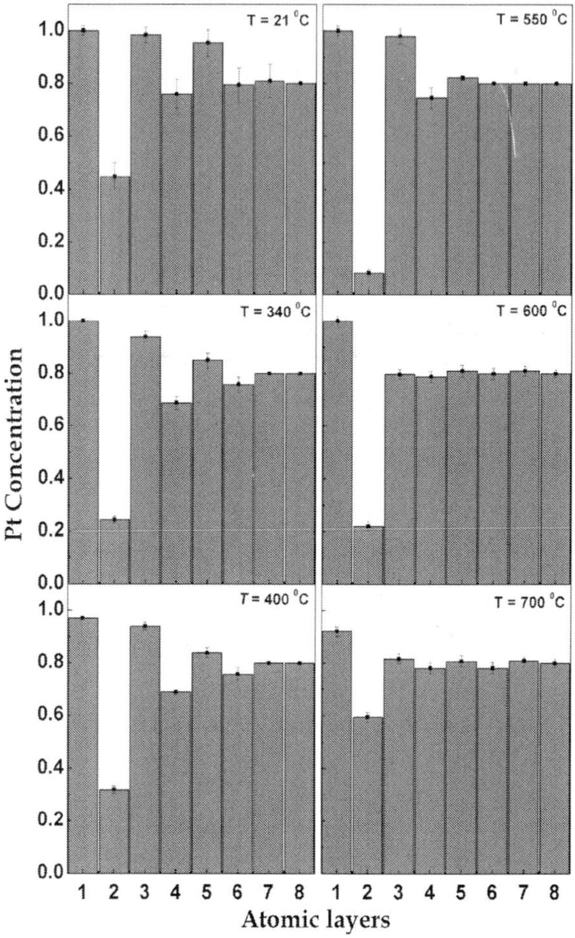

Figure 6. Layer-by-layer Pt concentration profiles reconstructed from the ionization spectra for the $Pt_{80}Co_{20}(111)$ alloy at the different heating

atomic layer consists of pure platinum, and that the other atomic layers have concentrations near the bulk value of the alloy. Probing the surface with primary electrons of 58 eV (corresponding to a probing depth of two atomic monolayers [9]), a p(2x2) structure was found. The appearance of these additional super-structural reflections in a diffraction pattern can be caused by two possible phenomena: chemical ordering at the surface of the alloy and/or a reconstruction of the surface [14].

In work [15] Electron Energy Loss Spectroscopy has been employed for investigation of the effect of 600 eV Ar^+-ion irradiation in the dose range $7\cdot10^{16}$–$4\cdot10^{17}$ ions/cm^2 on the atomic structure and surface composition of $Pt_{80}Co_{20}(111)$ alloy. Using the ionization energy loss spectra, a layer-by-layer concentration profile of the alloy components was reconstructed for different doses of ion irradiation of the surface. The Ar^+-ion bombardment of the alloy was found to result in the preferential sputtering of Co and in the enrichment of the near-surface region by Pt atoms with formation of an altered layer, which is characterized by a non-monotonic concentration profile dependent on the irradiation dose. The results obtained are discussed in the framework of the models of preferential sputtering and radiation-induced segregation.

Application of IS for the investigation of composition changes on the depth is not limited to the study by single crystal alloys. In references [8,16] IS was used to study the surface segregation in the ternary Co–Cr–Mo system. Since it was polycrystalline alloy, there can't be applied layer-by layer analysis with profile reconstruction. Nevertheless, the integral distribution of elements on the probing depth can be investigated by means of IS.

According to reference [16] the concentration of the Co–Cr–Mo alloy components on E_0 can be calculated by following expression

$$C_i = \frac{I_i(E_0) / I_i^{st}(E_0)}{\sum_{i=\text{Co,Cr,Mo}} I_i(E_0) / I_i^{st}(E_0)}, \tag{12}$$

where i = Co, Cr and Mo metals, $I_i(E_0)$ and $I_i^{st}(E_0)$ are intensity lines of the ionization losses of the alloy components and standards, respectively. For the thermodynamic equilibrium state the ionization spectra of the alloy components at different temperatures were measured. The condition of the thermodynamical steady–state of the alloy depended on the prolonged heating of the sample at every preset temperature for 15 hours. Figure 7 shows the concentration dependences $C_{Co,Cr,Mo}$ on E_0 for the polycrystalline alloy at a different heating temperatures.

At first, we estimated the thickness of the probing layers for Co-Cr-Mo alloy at change of E_0 from 200 eV to 800 eV. For estimation of the probing depth we used experimental data for the inelastic mean free path (IMFP) λ which are collected in reviews [9] for pure Co, Cr, Mo metals. After, these data was approximated by following equation

$$\lambda = kE_0^{\,n}, \tag{13}$$

where k, n are fitting parameters. As result k = 0.36 and n = 0.5 and variation $\lambda(E_0)$ is from 5Å to 10Å.

For the non–annealed Co–Cr–Mo alloy the Mo atoms showed preferred segregation in the outermost layers at a room temperature. Gradual increase of the probing depth by changing the primary electron energy E_0 to 600 eV shows that the Mo concentration in the near-surface region decreases and the Cr concentration greatly increases, while the Co concentration does not exceed 5–7 at.%. However, at the energy E_0 = 200 eV it was detected that $C_{Co} \approx 42$ at%, $C_{Cr} \approx 20$ at% and $C_{Mo} \approx 38$ at% were present in the near-surface layers.

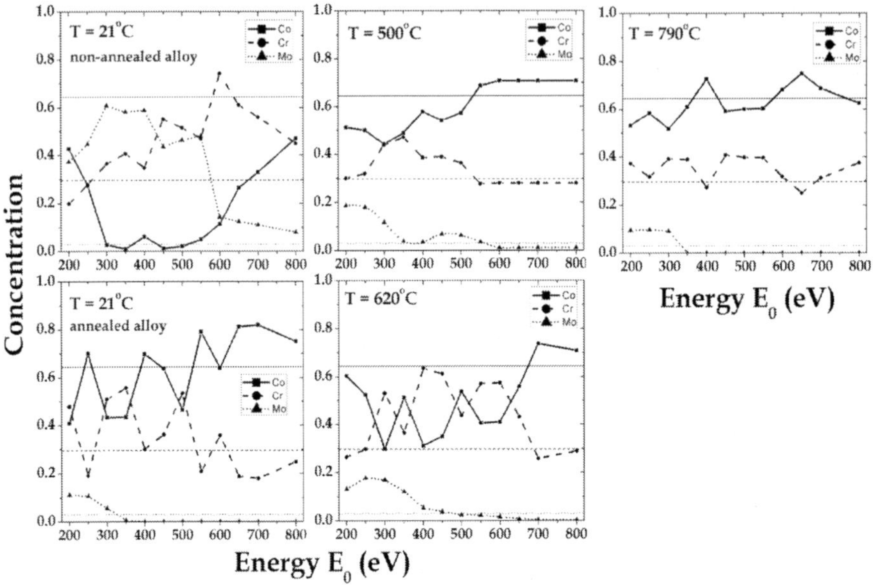

Figure 7. Concentration profiles of Co, Cr, Mo for the Co-Cr-Mo alloy at the different heating by means IS. Horizontal lines are bulk concentration for Co, Cr and Mo, respectively.

On the other hand, for an increase of probing depth at $E_0 > 250$ eV , the concentration of Co atoms sharply decreases to 1–6 at% in the below-surface region at $E_0 = 250 - 550$ eV. With the increase of primary electron energy $E_0 > 600$ eV the Co concentration rises but that of Cr and Mo atoms falls. Only an approximate tendency $C_{Co,Cr,Mo}$ towards the bulk concentration of the alloy is observed at $E_0 = 800$ eV. Heating of the alloy to temperature T = 500ºC essentially induced a change of the surface concentration in the Co–Cr–Mo alloy as compared to the surface concentration for the non–annealed state. Thus, the near surface layers contain $C_{Co} \approx$ 51 at%, $C_{Cr} \approx$ 30 at% and $C_{Mo} \approx$ 19 at% at the energy $E_0 = 200$ eV. With an increase of the probing depth the Mo concentration is lowered and the concentration of Co atoms is increased. Whereas in the interval of the energies $E_0 = 250 - 400$ eV the sharp growth of Cr concentration is observed and at $E_0 > 550$ eV the alloy composition is close to the bulk value. Further heating of alloy to T = 620ºC promotes an increase in concentration of Co atoms in the near surface region of Co–Cr–Mo alloy (at the $E_0 = 200$ eV). At the primary electron energy $E_0 = 250 - 650$ eV in the deeper layers growth in Cr concentration is detected as compared to the Cr bulk value and only at energy $E_0 > 700$ eV the composition of the alloy comes towards that of the bulk. Further increase of the alloy heating temperature to T = 790ºC is accompanied by smoothing of the alloy composition to the bulk. Nevertheless, an insignificant Mo segregation was still detected in the outermost layers of the alloy.

After prolonged annealing at T = 790ºC the ternary Co–Cr–Mo alloy was slowly cooled to room temperature over 12 hours. The concentration profile for the annealed alloy is shown on Figure 7. Also for the annealed state the preferred segregation of the Mo and Cr atoms is observed. At the energy E_0 = 200 eV the outermost layers contain $C_{Co} \approx 40$ at%, $C_{Cr} \approx 50$ at% and $C_{Mo} \approx 10$ at%. At increasing primary electron energy E_0 the Mo concentration sharply diminishes and at E_0 > 400 eV Mo atoms are not detected anymore, though insignificant oscillations of the composition for the Cr and Co atoms are found near to the bulk concentration at varying E_0. We suggest that the thermodynamic steady–state of the alloy corresponds to that of the annealed alloy at room temperature, but not for the non–annealed alloy.

2.2. Kinetics of surface segregation by IS

IS can be effective at investigation of kinetic processes in the thin layers of a solid. In work [17] for studying kinetics of surface segregation of the $Pt_{80}Co_{20}(111)$ alloy, the temperature interval T = 613 - 973K (T = 340 - 700 °C) is chosen at which the bulk alloy is in the ordered state. A special device allowed heating the sample to predetermined temperature, keeping it constant and changing with an accuracy ±2°C. Platinum - Pt alloy and 1%Rh thermo-couple was welded to the investigated sample for control temperature. Spectra of ionization losses were measured at every chosen temperature with a fixed time interval for platinum ($\Delta E^{Pt}_{O_{2,3}} = 54eV$) and cobalt ($\Delta E^{Co}_{M_{2,3}} = 62eV$). Primary electron beam with energy E_0 = 250 eV was taken for surface probing which by converting into monatomic layers corresponds to the 3rd monatomic layer over the depth [8]. Figure 8 shows the kinetics of segregation for Pt and Co atoms in the near-surface region at different temperatures of $Pt_{80}Co_{20}(111)$ alloy.

Note, that diffusion processes (internal diffusion) for single crystal alloys course mainly according to the vacancy mechanism [1]. Under heating up to 613K the kinetics of segregation atom $C_\beta(t)$ has a classical dependence which may be provisionally divided into two regions: I is the region of fast diffusion when strong segregation of Co atoms is observed; II is the saturation region when the steady-state equilibrium of segregating atoms is set in the near surface alloy region. The character of kinetic curve $C_\beta(t)$ dependence changes substantially at higher temperatures. Thus, when heating is up to 673K the fast diffusion region I has more gentle appearance and region II acquires two characteristic sites: IIa is the region of changing the direction of Pt and Co atoms segregation (temporary S-shaped fold), smoothly transient into IIb, which is region of steady-state equilibrium of segregating elements. We consider that such S-shaped fold is associated with eventual formation of the ordered phase in the near surface region. One of the reasons for nucleation of the composition close to the ordering is a decrease of interatomic interaction constants and as the consequence, an increase of the amplitude of thermal atomic oscillations. We suppose that, most probably, an ordered phase is formed between the 3-5th atomic layers. Since at probing of the alloy surface by electrons with the energy E_0 = 200 eV (1-2nd monolayer [8]), the integral concentration of Pt and Co atoms was C_{Pt} = 0.9 at.% and C_{Co}= 0.1 at.% within the whole time and temperature interval. This confirms the preferred segregation of Pt atoms in topmost layers. The further increase temperature for $Pt_{80}Co_{20}(111)$ alloy will lead to growth of the vacancies number due to thermal oscillation expansion of atoms and hence it will accelerate diffusion processes. Heating of alloy up to 823K, especially at 873K, leads to increase Co atoms enrichment in the near-surface alloy region as compared to other heating temperatures, and to decrease time which needed for the possible alloy ordering (IIa region). At heating of sample higher ordering temperature the rapid segregation of Co atoms was observed in region I which soon will be replaced by the segregation of atoms Pt (region II) and tend to bulk concentration. Such character of temporal diffusion of cobalt and subsequent segregation of platinum we suppose with redistribution of atoms in the near surface region of alloy.

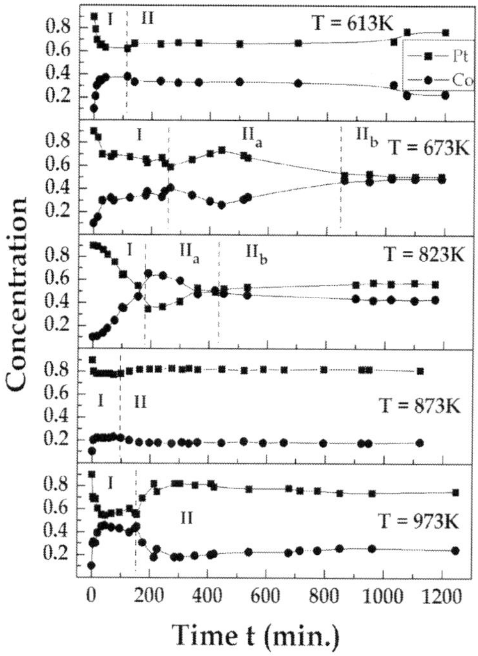

Figure 8. Kinetics of surface segregation in the near surface region for $Pt_{80}Co_{20}$(111) alloy at different heating

At studying kinetics of segregation of the binary alloys in the work [18] was established that the concentration C_x^t of segregating atoms to the surface from the bulk for the time t out of the depth x may be given by the following ratio:

$$C_x^t = C_\infty - C_\infty\left(1-\frac{1}{\alpha}\right)\exp\left(\frac{x}{\alpha d}+\frac{Dt}{\alpha^2 d^2}\right)erfc\left[\frac{x}{\sqrt{4Dt}}+\left(\frac{Dt}{\alpha^2 d^2}\right)^{1/2}\right], \quad (14)$$

where C_∞ is the bulk concentration of diffusing atoms; D is the diffusion coefficient; d is the thickness of the surface layer; α is the degree of surface enrichment defined as

$$C_s^t = \alpha C_0^t, \tag{15}$$

where C_s^t is the concentration of segregating atoms in the surface region; C_0^t is the atom concentration at the depth d at the initial time. By comparing formula (14) and (15) provided that $\alpha >> 1$, we get

$$C_x^t = \alpha C_\infty \left[1 - \exp\left(\frac{Dt}{\alpha^2 d^2} \right) erfc \left(\frac{Dt}{\alpha^2 d^2} \right)^{1/2} \right]. \tag{16}$$

Data approximation of the kinetics of segregation atoms cobalt by Eq. (16) allowed to determine the mean coefficient values of cobalt diffusion at different temperatures, the order of which corresponds to diffusion bulk values. According to these results, the temperature dependence of Co diffusion coefficient in $Pt_{80}Co_{20}(111)$ alloy was plotted (Figure 9), by which pre-exponential factor D_0= 5.1 m^2 s^{-1} and energy activation E = (327±22) kJ/mol were determined. The value of energy activation is close to sublimation heat of pure cobalt E = 309.73 kJ/mol.

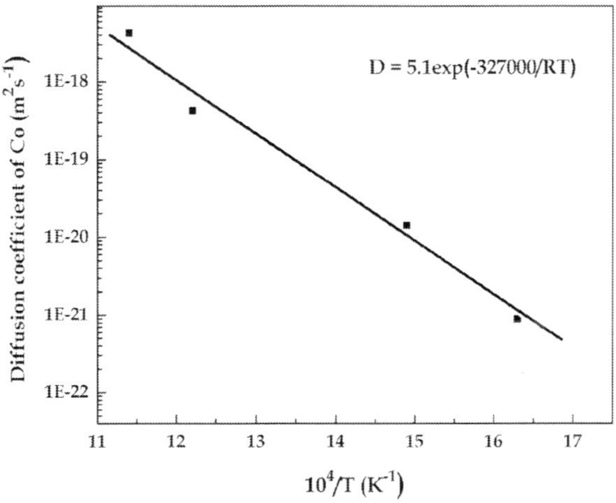

Figure 9. Diffusion coefficient of Co atoms in $Pt_{80}Co_{20}(111)$ alloy surface

2.3. Plasmon excitation

Plasmon excitations are potential data carriers about composition and chemical state of elements on the surface and bulk of solid and in the adsorbed layers.

2.3.1. Plasmon energy

A longitudinal plasma wave along the crystal produces long–range Coulomb forces between positive and negative charges and excites collective oscillations. These are called plasmons in the case of a free–electron gas model. The plasmon energy is obtained from the Fourier modes of the electron density $\rho(r) = \sum_k \rho_k e^{-ik \cdot r}$ and the ϱ_k are amplitudes of harmonic density fluctuations obeying [3]:

$$\ddot{\rho}_k + \omega_p \rho_k = 0 \tag{17}$$

in which ω_p is the Langmuir frequency. Then plasmon energy can be determined by the following expression:

$$E_b = \hbar\omega_b = \hbar\sqrt{\frac{e^2 N}{m\varepsilon_0}}, \tag{18}$$

where $\hbar$ is Plank's constant; ω_b is the cycle frequency of the bulk plasmon; e and m are the electronic charge and mass, respectively; n is the number of valence electrons per unit volume and ε_0 is the permittivity of the free space. The surface plasmon energy E_s is related to bulk plasmon energy by the following equation [19]:

$$E_s = E_b / \sqrt{1 + \varepsilon_s}, \tag{19}$$

where ε_s is the dielectric constant. In the framework of the model under consideration, $\varepsilon_s = 1$, i.e. $E_s = E_b / \sqrt{2}$.

In references [5, 20-24] the surface and bulk plasmon excitations were investigated for the $Pt_{80}Co_{20}(111)$ and $Cu_{75}Pd_{25}(100)$ single crystal alloys, ternary Co–Cr–Mo alloy and amorphous and crystalline $Fe_{73.6}Cu_1Nb_{2.4}Si_{15.8}B_{7.2}$ (FINEMET) alloy surface and their alloy components in range primary electron energy E_0 from 150 eV to 800 eV. It was found that the experimental values of plasma oscillation energy for all pure elements differ from the theoretical calculations but the data obtained in the given works are in good agreement with the results obtained by other authors.

Actually, the difference between experimental data and the free–electron gas model has been observed repeatedly for a lot of chemical elements. This may be a result of: (i) incomplete participation of valence electrons in the collective excitations; (ii) the involvement of filled d-band states and the appearance of inter– or intra-band transitions in characteristic spectra for the transition metals; (iii) cleanness and roughness of the surface region of specimens [3].

For example, on Figure 10 the bulk plasmon energies are shown for the range of primary electron beam energy 150 – 650 eV for pure Fe, Si, B, Nb, Cu and $Fe_{73.6}Cu_1Nb_{2.4}Si_{15.8}B_{7.2}$ alloy. It is known that, for silicon, the surface and bulk plasmon energy are 12 eV and 17 eV, respectively [4]. In our experimental data the plasmon energies are ~ 9 eV and ~15 eV. Most probably the shift of plasmon energies toward lower energy is related to the surface effects when comparing with other work because the probing depth is not deep and varied from 5.4 Å – 5 Å for silicon in the chosen range of E_0. Appearance of silicon oxides on the surface there can be eliminated since forming of oxides would lead to considerable increase in plasmon energy. For preparation of an atomically clean surface the amorphised silicon surface was first bombarded by argon ions and, subsequently, the sample was annealed at a high temperature. Consequently, both amorphous and crystalline phases can exist in the surface layers of Si. Also we cannot eliminate the fact that residual defects and implanted ions of argon may exist in the near-surface region, which is caused by ion irradiation. In the case of silicon we suggest that the total contribution of the above-mentioned surface effects will influence the shift of energy of plasma excitations to a lower energy. The experiments showed that, for the pure Fe, Si, B, Nb and Cu, the plasmon energy relation E_b/E_s exceeds the theoretical values and is equal to 1.79, 1.67, 2.13, 1.78 and 1.27, respectively. This discrepancy between theory and experiment has been observed repeatedly for many metals [3]. It should be noted that the theory supposes a perfectly flat surface in the vacuum–solid region and does not take into consideration the real physical–chemical state of the metallic surfaces.

According to reference [24] the experimental number of the valence electrons per unit volume n_{alloy} for the amorphous $Fe_{73.6}Cu_1Nb_{2.4}Si_{15.8}B_{7.2}$ alloy on E_0 can be calculated by the following expression

$$n_{\text{alloy}}(E_0) = \sum_{\text{j}} N_{\text{j}}\, n_{\text{j}}(E_0)\,, \tag{20}$$

where j = Fe, Si, B, Nb and Cu metals; N_j is number of the j–atoms per unit bulk (in our approach for the amorphous state, it is a bulk atomic concentration of the alloy components); $n_{\text{j}}(E_0)$ is the experimental number of the valence electrons per unit volume of the pure j–elements at fixed energy E_0. Substituting Eq.(20) into Eq.(18) and using experimental data we calculated the surface and bulk plasmon energy depending on primary electron energy E_0. The results of the calculations for the bulk plasmon energy of FINEMET are shown in Figure 10.

The obtained results are in good agreement with experimental data. For the surface plasmon the design function $E_s(E_0)$ is localized between values for the amorphous and crystalline alloy whereas for $E_b(E_0)$ there is a different situation. At low average primary electron energy $E_0 < 200$ eV the calculated function $E_b(E_0)$ is absolutely identical to energies of the bulk plasmon for the amorphous alloy and at energy $E_0 > 250$ eV the function $E_b(E_0)$ is close to the experimental data E_b for the crystalline state of the alloy. It was observed that, for the crystalline alloy, the energy of plasmon excitations is localized at lower loss energies as compared to those for the amorphous state.

It is known that the electronic state densities in the surface layers can be induced by such an effect as surface segregation. These phenomena are typical for major complex alloys when the composition in the near surface region differs from the bulk composition and it is caused by minimization of the free surface energy of the alloy [1]. Therefore the crystalline and phase structure and altered surface layers will also influence the dispersion of the surface and bulk plasmons when changing the primary E_0 or probing depth on amorphous or crystalline states of the alloy.

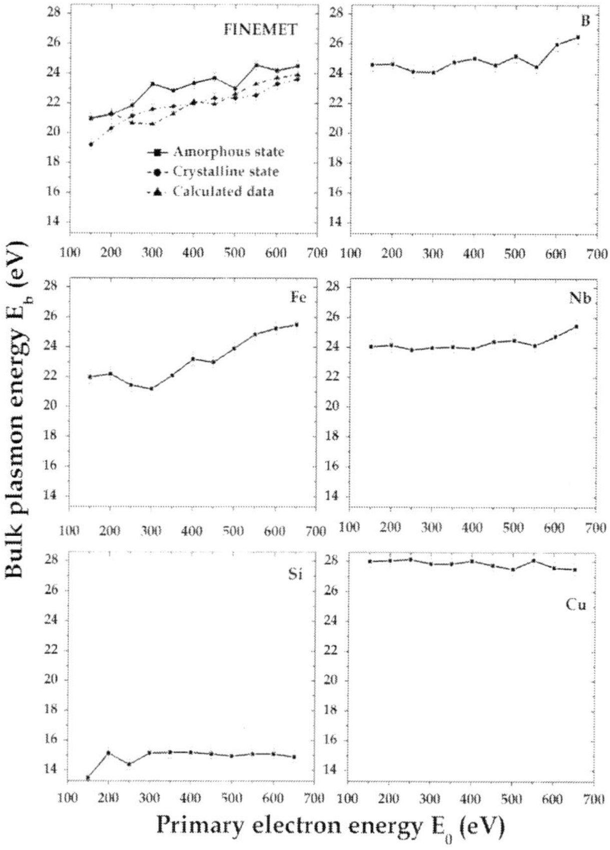

Figure 10. Dependence of the bulk plasmon energy E_b on the primary electron energy E_0 for the $Fe_{73.6}Cu_1Nb_{2.4}Si_{15.8}B_{7.2}$ alloy ribbons surface and pure alloy components

Similarly situation was observed at study plasmon energies for the disordered and ordered states of $Pt_{80}Co_{20}(111)$ alloy where are E_s = 10.57 eV, E_b = 22.17 eV and E_s = 15.8eV, E_b = 25.31 eV, respectively [21,22]. The plasma oscillations for the disordered state are localized at lower loss energies than it was established for ordered state. For the ordered alloy the bulk plasmon energy is 2–3eV more than that of the disordered alloy, whereas the difference for the surface plasmon energy is 4–7 eV in the whole range E_0. Probably it is related to changes of the DOS of valence electrons at the ordering alloy and surface segregation in the atomic layers.

Surface and bulk plasmon energy is sensitive not only to surface segregation, phase state etc but to heating too. EELS has been employed for investigation of the surface and bulk plasmon excitations versus heating in the Co–Cr–Mo alloy surface for the primary electron beam energies E_0 ranging from 150 to 800 eV (see Figure 11) [23].

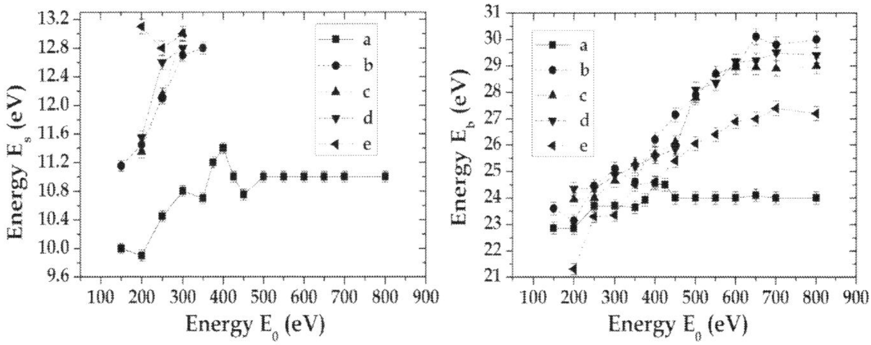

Figure 11. Dependence of surface and bulk plasmons energy from the primary electron energy E_0 for the Co-Cr-Mo alloy at different heating: (a) non-annealed state at T = 21 °C, (b) annealed state at T = 21 °C, (c) T = 500 °C, (d) T = 620 °C, (e) T = 790 °C.

As shown on Figure 11 for the annealed alloy the energies of surface plasmon E_s and bulk plasmon E_b are localized at greater energies than for the non–annealed alloy. In the range of the primary electron energy E_0 = 150 – 800 eV for the surface plasmon E_s this difference is 1 – 2 eV. For the (non–)annealed alloy at room temperature the surface plasmon energy E_s have a linearly growth with a increase of the probing depth of alloy. Significant changes of bulk plasmon energy was observed for the annealed alloy in the range of the primary electron energy E_0 = 150 – 800 eV. For the non–annealed alloy in this energy range of E_0 the bulk plasmon is varied in small region of energy E_b = 22.8–24.5 eV, whereas it strongly changes for E_b = 23.1–30.1 eV in case of annealed alloy. For the (non–)annealed alloy the plasmon energies E_b are close in the energy range E_0 = 150 – 200 eV. With an increase of primary electron energy E_0 for the annealed alloy the bulk plasmon energy linearly increases and remains unchanging at E_0 > 650 eV.

At heating the surface plasmon energy E_s is shifted with an increase of the energy, and than more temperature of sample the more shift of energy E_s. However, the energy shift of surface plasmon, which is induced by heating there strongly differs against to annealed and non–annealed states of alloy. In all region of heating of the ternary Co–Cr–Mo alloy the energy of bulk oscillation E_b increases linearly with an increase of the primary electron energy E_0. With respect to dependence E_b from E_0 for the annealed state the alloy heating to temperatures 500ºC and 620ºC it is accompanied by growth of plasma energy E_b in the range of the energy E_0 = 150 – 350 eV (the near surface region) and decrease this value at the E_0 > 400 eV. Further heating of alloy to T = 790ºC promotes to an insignificant shift of long wavelength plasmon oscillations E_b to sideways decrease of their energy in all region of E_0 as compared to other temperatures of Co–Cr–Mo alloy. Thus, for example, at the temperatures 500ºC and 620ºC the difference of bulk plasmon energy from 0.1 eV to 1.2 eV modulo with respect to annealed alloy, whereas at T = 790ºC it changes from 0 eV to 2.7 eV at corresponding energies E_0. In the range of the primary electron energy E_0 > 650 eV the bulk plasmon energy E_b has a linearly dependence in all temperature regions. We suppose that this value will correspond to the real bulk plasmon energy crystal at the given temperature of heating.

The authors [25–34] investigated the influence of heating on EEL spectra from the surface of pure elements: C, Al, Ni, Mo, Ta, Pb, Nb, W and Ag. It was observed that owing to heating the surface and bulk plasmon energy suffers shifts in the characteristic spectra. After leaded systematic analysis of this effect by means of Transmission Electron Microscopy with EELS detector, a method of definition of the linear expansion coefficient was proposed using data to thermo–induced shifts of long wavelength plasmons [25–29]. This approach was based on the supposition that at heating of metal in consequence of the expansion/compression of crystal lattice the conductive electron density will lower/raise as a result it must lead to decrease/increase plasmon energy. In this case, number of valence electrons per unit n(T) changes due to the thermal expansion of the crystal, Eq. (18) is rewritten using the linear thermal expansion coefficient α(T) of the crystal as follows:

$$\begin{aligned} E_b(T) &= E_b(T_0)\left\{1-\frac{2}{3}\int_{T_0}^{T}\alpha(T')dT'\right\}, \\ E_b(T_0) &= \hbar\sqrt{\frac{e^2 n(T_0)}{m\varepsilon_0}} \end{aligned} \quad . \tag{21}$$

The obtained results for the thermal expansion coefficient are in a good agreement with tabular data for the clean Al, Ag and Pb. Also angular–resolved high resolution EELS was applied to study the plasmon excitations in the spectra of poly and single crystals as a function of temperature T. For example, in Ref. [33] a particular attention was devoted to silver because of the presence of an extremely sharp surface plasmon as observed for thin films and for all low Miller index surfaces. It was established that energy displacement of surface plasmon depends on temperature because of thermal expansion of the solid. Though Jensen et al. [34] observed with EELS strong temperature effects on the surface plasmon energy on graphite, which have been explained as a consequence of the unusual semimetallic band structure. Therefore this approach does not give us an ambiguous explanation of the reason for the plasmon shift in the ternary Co–Cr–Mo alloy. As noted above, the heating of alloy to T = 620ºC promotes to an increase of the bulk plasmon energy with respect to the annealed alloy and only at T = 790ºC the bulk plasmon suffers shift with a decrease of energy. More over, this approach doesn't take into account changes of the surface plasmon energy and their coupling with bulk plasmon in the near surface region.

It is known that the electrons in metals, which are neutralized by the fixed positive ions tightly sufficiently coupled between themselves and disposed in the lattice site, it is possible to consider as the special type of plasma [3]. From the classical point view the plasma oscillations in metals are oscillations of valence electrons with respect to positive ions which formed the lattice. These oscillations are conditioned owing to long–range Coulomb forces. Therefore, besides of the crystal lattice parameter at heating of ternary Co–Cr–Mo alloy it is necessary to take into account the change of Coulomb interaction force between the plasmons and atomic core that can not be calculated within the framework of the classic approach. The chemical elements Co and Cr are metals for which either the valence electrons are strongly bond s, d–electrons and the core electrons are weakly bond. Probably that at change of heating of these metals and those alloys the shift of plasmon energy with an increase or decrease of energy also will be determined by the change of Coulomb interaction force between the valence electrons and the core. It can lead to change of free s, d–electron concentration and effective electronic mass m (in Eq.(18)), which do participate in plasma excitations and, as a result, to shift the plasmon loss line relative to initial state of Co–Cr–Mo alloy.

Fact, for the pure chemical elements the surface and bulk plasmon energy can substantially differ from the plasmon energies of their alloys or compounds. In reference [16] was observed that in the near surface region the profile concentration versus temperature is differs to bulk of ternary Co–Cr–Mo alloy. Therefore, it is necessary to expect a displacement of long wavelength plasmon oscillations in the range of the energies E_0 = 150 – 650 eV as a result of segregation of the alloy components. However as far as the changed composition of the near surface region of alloys can strongly influence on plasma excitations at different temperatures at the present time is not clean.

The experimental data obtained in Figure 11 are indicated about the complex nature of the plasmon shifts in the near surface region of ternary Co–Cr–Mo alloy. Although most authors meet an opinion, that the energy plasmon shift mainly can be related to lattice parameter of solids, we suppose that in case of the complicated Co–Cr–Mo system the shift of plasmon energy will be defined by the summary balance of above mentioned possible causes at heating.

2.3.2. *Intensity lines of plasmons*

The nature of the surface plasmon appearance in the EELS spectra is related to the physical and chemical state of the surface layer nanosize thickness. It is also known that probability of the surface plasmon excitation by primary electrons will be directly related to their probing depth of the solid. Growth of the primary electron energy will lead to the increasing of the bulk plasmon excitation probability and, on the other hand, to the decreasing of the surface plasmon excitation probability and to damping of the surface plasmon intensity line in the EELS spectra. Consequently, for every chemical element and their alloys it is possible to define the range of the primary electron energy, in which the line of the surface plasmon will be detected in the characteristic loss spectra. Based on this concept, in references [20, 21] it was proposed to determine of the ratio R_s (in a.u.) of IL surface and bulk plasmons from the energy E_0 by the following equation:

$$R_{s,b}(E_0) = \frac{I_{pl}^{s,b}(E_0)}{I_{pl}^{s}(E_0) + I_{pl}^{b}(E_0)}, \tag{22}$$

where $I_{pl}^{s,b}$ is IL of the surface and bulk plasmons from primary electron energy E_0.

In works [20 - 24] the changes of IL for surface and bulk plasmon were studied for the $Pt_{80}Co_{20}(111)$ and $Cu_{75}Pd_{25}(100)$ single crystal alloys, ternary Co–Cr–Mo alloy and amorphous and crystalline $Fe_{73.6}Cu_1Nb_{2.4}Si_{15.8}B_{7.2}$ (FINEMET) alloy surface and their alloy components in range primary electron energy E_0 from 150 eV to 800 eV. There was found that damping of the function $R_s(E_0)$ is different for all specimen and different value of the primary electron energy E_0 for which the intensity line of the surface and bulk plasmons are equal. In case of pure elements the damping of function $R_s(E_0)$ is related to a decrease in probability of their excitation dependant on respective probing depth of the near surface layer and contrariwise this increases the probability excitation of the bulk plasmon and with altered near surface layers. In case of alloy, there was advanced a assumption that decay of intensity line of surface plasmon relative to bulk plasmon can be associated with changing of surface composition on the depth for the alloys and it confirms an assumption as to possibility of establishing the range of primary electron energy E_0, at which the electron beam will probe only the near surface region for the different materials.

Good correlation between the damping of surface plasmon $R_s(E_0)$ and concentration profile was established for the $Cu_{75}Pd_{25}(100)$ alloy surface at room temperature [20] and for the $Pt_{80}Co_{20}(111)$ alloy surface and Co–Cr–Mo alloy at a different heating [8, 23].

The results of measurement for $Pt_{80}Co_{20}(111)$ alloy are shown on Figure 12 at different heating. For the disordered alloy the damping of surface plasmon R_s have a more prolonged dependence compared to the ordered alloy at room temperature. If we will estimate a probing depth at primary electron energies E_0 = 550 eV and E_0 = 350 eV when surface plasmon does not appear ($R_s \approx 0$) in the EELS spectra, then we founds approximately 6-7th and 2-3rd atomic layers (bulk concentration) for the disordered and ordered states of $Pt_{80}Co_{20}(111)$ alloy, respectively (see Figure 6).

As in case of thermo–induced shift of plasmon excitations the changes in IL of surface plasmon relative to bulk plasmon were observed. Heating of alloy induces decreasing intensity line of surface plasmon and then higher temperature that more damping of surface plasmon R_s at variation of the primary electron energy E_0. In case of the $Pt_{80}Co_{20}(111)$ alloy surface with increasing of heating the damping of oscillating concentration depth profile is decreases [8]. More over, there is observes correlation between damping of surface plasmon R_s relative to bulk plasmon and damping of oscillating concentration depth profile at every given temperature.

The results of measurement for the Co-Cr-Mo alloy are shown on Figure 13 at different heating. For the non–annealed alloy the damping of surface plasmon R_s has a prolonged dependence and only at E_0 > 800 eV the surface plasmon peak disappears in EELS spectra. For the annealed state of alloy with an increase of the energy E_0 the dependence R_s decays quickly compared to the non–annealed alloy and at energy E_0 > 350 – 400 eV the surface plasmon does not appears in EELS spectra. As in case of shifts of the surface and bulk plasmon energy at heating of alloy the essential changes on intensity lines of surface plasmon relative to bulk plasmon were observed.

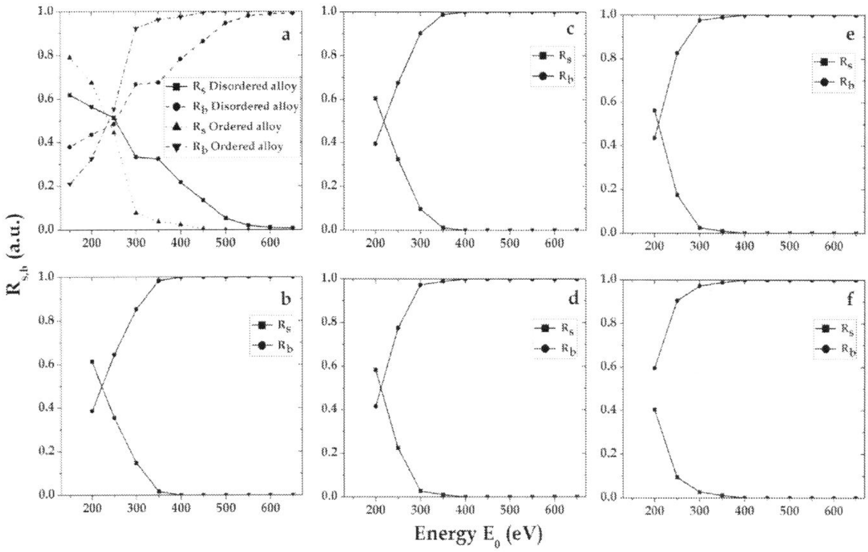

Figure 12. Dependence $R_{s,b}(E_0)$ from the primary electron energy E_0 for the $Pt_{80}Co_{20}(111)$ alloy at different heating: (a) T = 21 °C, (b) T = 340 °C, (c) T = 400 °C, (d) T = 550 °C, (e) T = 600 °C, (f) T = 700 °C

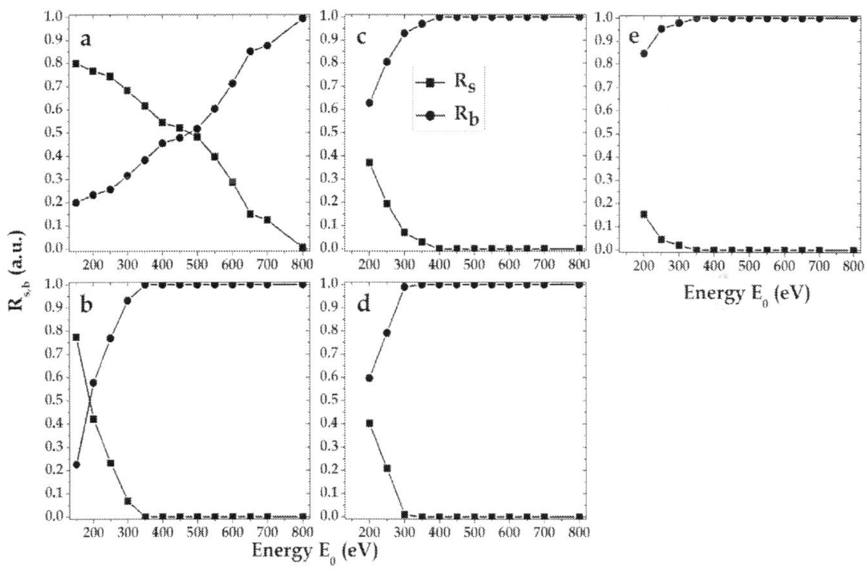

Figure 13. Dependence $R_{s,b}(E_0)$ from the primary electron energy E_0 for the Co-Cr-Mo alloy at different heating: (a) non-annealed state at T = 21 °C, (b) annealed state at T = 21 °C, (c) T = 500 °C, (d) T = 620 °C, (e) T = 790 °C.

Heating of Co-Cr-Mo alloy induces decreasing IL of surface plasmon as well as probability of their appearance in EELS spectra dependence on temperature at variation of the energy E_0. High–temperature heating of Co–Cr–Mo alloy promotes to increase the emission and background of secondary electrons in characteristic spectra, that did not allow us exactly to separate the peaks of plasma oscillations at small primary electron energy E_0 = 150 eV. As shown on Figure 13 for temperatures 500ºC and 620ºC the line of surface plasmon disappearances at E_0 > 350 eV and E_0 > 300 eV, respectively. At heating of alloy to T = 790ºC the surface plasmon is detected only in range of the energy E_0 = 200 – 350 eV, however dependence R_s from E_0 decays quickly compared to other temperatures. Besides increasing the intensity line of plasmons with the increase of heating of Co–Cr–Mo alloy the observed a broadening of bulk plasmon line. The correlation between the damping of function R_s from E_0 and formation of concentration profile in the near surface region of alloy was established for the non–annealed Co–Cr–Mo alloy (see Figure 7). For the annealed alloy the surface plasmon detects in EELS spectra in the range of E_0 = 150 – 350 eV. This range of energy E_0 corresponding to the near surface region where was observed the largest variation of alloy composition relative to bulk concentration. The similar situation occurs at heating to T = 790ºC for which the alloy concentration in the near surface region comes towards to the bulk at energy E_0 > 300 eV. Only the qualitative the correlation between the damping function R_s from E_0 and comes towards to the bulk concentration are observes for the Co–Cr–Mo alloy at the temperatures 500ºC and 620ºC.

3. Conclusion

Low Electron Energy Loss Spectroscopy can be used as effective non-destructive method at investigation of physical-chemical properties materials in the nano-size near surface region.

Ionization energy losses allows to investigation layer-by-layer concentration profile for the singe crystal alloys with monolayer resolution, element distribution on the depth for the polycrystalline alloys and study of kinetics of surface processes at thermo-induced treatment or after ion irradiation of the surface.

Plasmon excitations are very sensitive to structural and chemical state of surface and bulk and it can be used for study of electronic states of free electrons in the near surface region and influence of different kinetic processes on changing of electronic structure of materials.

Analysis of intensity line of surface and bulk plasmons depending on primary electron energy E_0 allows to define a surface-bulk interface when electron beam probes just near surface region with different physical-chemical properties as compared to the bulk material. These results have good correlation with data of surface composition on depth which obtained by IS and AES.

Author details

Vitaliy Tinkov
Department of the Surface Atomic Structure and Dynamic, Institute for Metal Physics of NAS of Ukraine, Ukraine

Acknowledgement

The author would like to acknowledge professor M.A. Vasylyev for his help in discussion this paper.

4. References

[1] Vasiliev, M A. Surface effects of ordering in binary alloys. Journal of Physics D: Applied Physics 1997;30(22) 3037-3070.

[2] O'Connor, D.J., Sexton, B.A. & Smart, R.St.C. Surface Analysis Methods in Materials Science, 2nd edit, Berlin: Springer; 2003.

[3] Raether, H. Excitation of Plasmon and Interband Transitions by Electrons. Springer Tracts Modern Physics 1980;88 1-196.

[4] Lüth, H. Surface and Interfaces of Solids. Springer Series in Surface Science 1993;15 1-356.

[5] Vasylyev, M.A., Tinkov, V.A. & Gurin, P.A.. Electron Energy Loss Spectroscopy study of the near surface region of dental Co–Cr–Mo alloy. Applied Surface Science 2008;254 4671–4680.

[6] Gerlach, R.L., Houston, J.E. & Park, R.L. Ionization spectroscopy of surfaces. Applied Physics Letters 1970;16(4) 147-188.

[7] Gerlach, R.L. Ionization Spectroscopy of Contaminated Metal Surfaces. Journal of Vacuum Science & Technology 1971;8 599-604.

[8] Vasylyev, M.A., Tinkov, V.A., Blaschuk, A.G., Luyten, J. & Creemers, C.. Thermo–stimulated surface segregation in the ordering alloy $Pt_{80}Co_{20}$(111): Experiment and Modeling. Applied Surface Science 2006;253 1081–1089.

[9] Seah, M.P. & Dench, W.A. Quantitative electron spectroscopy of surfaces: A standard data base for electron inelastic mean free paths in solids. Surface and Interface Analysis 1979;1(1) 2-11.

[10] Baschenko, O.A. & Nefedov, V.I. Depth profiling of elements in surface layers of solids based on angular resolved X-ray Photoelectron Spectroscopy. Journal of Electron Spectroscopy and Related Phenomena 1990;53 1 – 18.

[11] Cherkashin, G.Yu. Inverse problem: the concentration depth profile of elements from ARXPS data. Journal of Electron Spectroscopy and Related Phenomena 1995;74 67 – 75.

[12] Vasylyev, M.A., Blaschuk, A.G. & Tinkov, V.A. Reconstruction of concentration profiles in the surface region of $Pt_{80}Co_{20}$ alloy for (100) and (111) faces by means of ionization spectroscopy. Metal Physics and Advanced Technologies 2003;25(12) 1617–1632.

[13] Bardi, U., Atrei, A., Zanazzi, E., Rovida, G. & Ross, P.N. Study of the reconstructed (001) surface of the $Pt_{80}Co_{20}$ alloy. Vacuum 1990;41(1-3).

[14] Gauthier, Y., Baudoing-Savois, R., Bugnard, J.M., Bardi, U. & Atrei, A. Influence of the transition metal and of order on the composition profile of $Pt_{80}M_{20}$(111) (M = Ni, Co, Fe) alloy surfaces: LEED study of $Pt_{80}Co_{20}$(111). Surface Science 1992;276(1-3) 1 – 11.

[15] Vasylyev, M.A., Chenakin, S.P. & Tinkov, V.A. Electron Energy Loss Spectroscopy study of the effect of low–energy Ar^{+}–ion bombardment on the surface structure and composition of $Pt_{80}Co_{20}$(111) alloy. Vacuum 2005;78 19–26.

[16] Tinkov, V.A., Vasylyev, M.A. & Gurin, A.P. Investigation of the thermo–stimulated surface segregation in the ternary Co–Cr–Mo alloy by means of Ionization Spectroscopy. Vacuum 2009;83 1014–1017.

[17] Vasylyev, M.A., Tinkov, V.A., Sidorenko, S. & Voloshko, S. The temperature dependence of the atoms Co diffusion coefficient in $Pt_{80}Co_{20}$(111) alloy. Defect and Diffusion Forum 2007;265 19–23.

[18] Lea, C. & Seah, M.P. Kinetics of surface segregation. Philosophical Magazine 1977;35(1) 213-228.

[19] Maier, S.A. Plasmonics: Fundamentals and Applications, Springer Science+Business Media LLC; 2007.

[20] Vasylyev, M.A., Tinkov V.A. & Nieuwenhuys, B.E. Electron energy–loss spectroscopy of the metals Pd, Cu and the ordered $Cu_{75}Pd_{25}$(100) alloy. Journal of Electron Spectroscopy and Related Phenomena 2007;159 53–61.

[21] Vasylyev, M.A. & Tinkov, V.A. Low energy electron induced plasmon excitations in the ordering $Pt_{80}Co_{20}$(111) alloy surface. Surface Review and Letters 2008;15(5) 635–640.

[22] Tinkov, V.A. & Vasylyev, M.A. Thermo–induced shift of plasmon energy in electron loss spectra for the ordering $Pt_{80}Co_{20}$(111) alloy surface. Surface Review and Letters 2009;16(2) 249–258.

[23] Tinkov, V.A. & Vasylyev, M.A. Thermo–induced plasmon excitations in the near surface region of ternary Co–Cr–Mo alloy. Vacuum 2011;85(8) 787-791.

[24] Tinkov, V.A., Vasylyev, M.A. & Galstyan, G.G. Low energy electron induced characteristic losses in the $Fe_{73.6}Cu_1Nb_{2.4}Si_{15.8}B_{7.2}$ (FINEMET) alloy surface. Vacuum 2011;85(6) 677-686.

[25] Watanabe, H. Experimental Evidence for the Collective Nature of the Characteristic Energy Loss of Electrons in Solids – Studies on the Dispersion Relation of Plasma Frequency. Journal of the Physical Society of Japan 1956;11(2) 112-119.

[26] Abe, H., Terauchi, M., Kuzuo, R. & Tanaka, M. Temperature Dependence of the Volume-Plasmon Energy in Aluminum. Journal of Electron Microscopy 1992;41(6) 465-468.

[27] Abe, H., Terauchi, M. & Tanaka, M. Temperature Dependence of the Volume-plasmon Energy in Silver. Journal of Electron Microscopy 1995;44(1) 45-48.

[28] Leder, L.B. & Marton, L. Temperature Dependence of the Characteristic Energy Loss of Electrons in Aluminum. Physical Review 1958;122 341–343.

[29] Imbuch, A. & Niedrig, H. Temperature effect on energy loss spectrum of fast electrons in aluminium and lead foils between 3 K and 295 K. Physics Letters A 1970;32(6) 375-376.

[30] Apholte, H.R. & Ulmer, K. Temperaturabhängigkeit der charakterischen energieverluste in niob, molybdän, tantal und wolfram. Physics Letters 1966;22(5) 552-553.

[31] Heimann, B. & Hölzl, J. Variation of Characteristic Energy Losses in the Curie-Temperature Region of Ni (111). Physical Review Letters 1971;26 1573–1575.

[32] Korsukov, V.E. & Lukyanenko, A.S. The surface relaxation of Al as determined by electron energy loss spectroscopy on plasmons. Zeitschrift für Physik B Condensed Matter 1983;53(2) 143-150.

[33] Rocca, M. Low-energy EELS investigation of surface electronic excitations on metals. Surface Science Reports 1995;22(1–2) 1-71.

[34] Jensen, E.T., Palmer, E.E., Allison, W. & Annett, H.F. Temperature-dependent plasmon frequency and linewidth in a semimetal. Physical Review Letters 1991;66 492–495.

Chapter 8

From Micro– to Macro–Raman Spectroscopy: Solar Silicon for a Case Study

George Sarau, Arne Bochmann, Renata Lewandowska and Silke Christiansen

Additional information is available at the end of the chapter

http://dx.doi.org/10.5772/48160

1. Introduction

The phenomenon of inelastic scattering of light by matter is referred as Raman spectroscopy named after Sir Chandrasekhara Venkata Raman who first observed it experimentally in 1928 [1]. Because only one photon out of 10^6-10^{12} incident photons is inelastically or Raman scattered, it took some time until lasers with high enough light intensities for efficient Raman excitation and very sensitive detectors for measuring the still low intensity Raman light were developed. Another important step in advancing Raman instrumentation was the efficient rejection of the very intense elastic scattered light, known as Rayleigh light, through a double or triple monochromator or filters [2].

Nowadays, Raman spectroscopy is being successfully applied to both in- and ex-situ analyses of various processes and materials in different states of matter (solid, liquid, gas or plasma). Moreover, Raman spectrometers have become small, portable and easy to use even for nonspecialists. This technique is covering a very broad range of application fields, at scientific and industrial levels, including pharmaceuticals, biology, environment, forensics, geology, art, archaeology, catalysis, corrosion, materials and others. Giving the large amount of specific information related to each of the abovementioned areas, we refer the interested reader to [3,4].

In the field of semiconductors, Raman spectroscopy has shown to be a powerful analytic tool for investigating mechanical stress, crystallographic orientation, doping, composition, phase, and crystallinity of semiconductor materials in bulk, thin film and device form [4-7]. In particular, the use of Raman spectroscopy to study solar silicon materials in form of thin films on glass, wafers, and ribbons, which are then processed to solar cells used as a clean and sustainable energy source has gained new momentum in the context of climate change and energy security. The physics behind Raman scattering in semiconductors or crystals is based on the inelastic interaction of light with lattice vibrations or phonons that are sensitive to internal and external perturbations. A short but relevant theoretical introduction in the case of silicon will be given in Section 2. Back to early 70th, it was Anastassakis et. al. first reporting on the shift of the first-order silicon Raman peak under uniaxial *external* stress [8]. This work triggered the application of Raman spectroscopy in measuring *internal* stresses present in semiconductor materials and structures. Particularly important for the present contribution are the studies on *local* internal stresses in microelectronics devices such as silicon integrated circuits using confocal micro-Raman spectroscopy where the exciting laser light is focused onto the sample's surface through a microscope objective thus enabling investigations on the micrometer scale [4,5].

The first experimental part of this chapter (Sections 3.2 and 4.1) is mainly focused on the application of confocal *micro-Raman spectroscopy* to map the spatial distribution of internal stresses, their magnitude and sign in different solar silicon materials following the existing work in silicon microelectronics. Because internal stresses may decrease mechanical strength increasing the breakage rate and induce recombination active defects when combined with external stresses, their understanding and control will improve both process yields and solar cell efficiencies. In addition to mechanical information, other useful material properties can be obtained from the *same* first-order silicon Raman peak. We will show how internal stresses, defects, doping, and microstructure can be directly correlated with each other on the same map, enabling the basic understanding of their interactions. The micro-Raman measurements are supported and complemented at identical positions by other techniques such as EBSD, EBIC, and defect etching. Such a combination allows the correlation of internal stresses, recombination activity and microstructure on the micrometer scale.

In the second experimental part of this contribution (Sections 3.3 and 4.2), confocal *macro-Raman spectroscopy* is introduced and its application to solar silicon is demonstrated for the first time. Macro-Raman spectroscopy represents the state-of-the-art in fast, large area Raman mapping being initially developed to analyze the chemical homogeneity in pharmaceutical tablets. We will present a statistical analysis using Macro-Raman mapping of solar silicon, which is usually characterized by large spatial properties variations. The combination of the two mapping techniques offers insights into the interplay between solar silicon properties at different length scales. Finally, the potential use of macro-Raman spectroscopy for optimization and in-line quality check in a PV factory will be discussed.

Such detailed Raman studies are not limited to solar silicon materials but they can be performed on all Raman active materials. In this context, it is clear that today Raman spectroscopy is a versatile and mature characterization method, which can be applied both at micro- and macro-scale to learn about the interaction between materials properties and their optimization in relation to individual processing steps.

2. Theoretical background of Raman spectroscopy in silicon

The aim of this section is to provide the basic equations and their interpretation necessary to understand the experimental results shown throughout the present chapter. A rigorous mathematical derivation is extensively documented in many textbooks and papers published during the long history of the Raman effect [4,5,8-14].

In a Raman experiment described from a classical point of view, monochromatic light of frequency ω_i originating from a laser is incident on a crystal in a direction $\boldsymbol{k}_i$ with $\boldsymbol{E} = \boldsymbol{E}_0\exp[i(\boldsymbol{k}_i\cdot\boldsymbol{r} - \omega_i t)]$. The electric field of light will induce an electric moment $\boldsymbol{P} = \varepsilon_0\chi\boldsymbol{E}$, with the interaction between light and crystal at position $\boldsymbol{r}$ being mediated by lattice vibrations or phonons characterized by a wavevector $\boldsymbol{q}_j$ and a frequency ω_j with $Q_j = A_j\exp[\pm i(\boldsymbol{q}_j\cdot\boldsymbol{r} - \omega_j t)]$. It is the electrical susceptibility χ, which is changed by phonons. This means that the induced electric moment will emit besides the elastic scattered Rayleigh light of ω_i, Raman light of $\omega_i + \omega_j$ and $\omega_i - \omega_j$ resulting from anti-Stokes and Stokes Raman scattering, respectively [5]:

$$\boldsymbol{P} = \varepsilon_0\chi_0 \cdot \boldsymbol{E}_0 \exp[i(\boldsymbol{k}_i \cdot \boldsymbol{r} - \omega_i t)] + \varepsilon_0 \boldsymbol{E}_0 \left(\frac{\partial\chi}{\partial Q_j}\right)_0 A_j \times \exp[-i(\omega_i \pm \omega_j)t]\exp[i(\boldsymbol{k}_i \pm \boldsymbol{q}_j)\cdot \boldsymbol{r}]. \quad (1)$$

From a quantum mechanical point of view, a photon described by $\boldsymbol{k}_i$, ω_i produces an electron-hole pair. The electron is excited from the ground state to a higher energy state and interacts with a phonon characterized by $\boldsymbol{q}_j$, ω_j. As a result of this interaction, the electron gains or losses energy and trough the recombination of the electron-hole pair a photon $\boldsymbol{k}_s$, ω_s is emitted, where $\omega_s = \omega_i + \omega_j$ and $\omega_s = \omega_i - \omega_j$ for anti-Stokes and Stokes Raman scattering, respectively. In most cases, only the silicon Stokes Raman peak known also as the first-order silicon Raman peak occurring in the absence of internal and external perturbations at $\omega_0 \sim 520$ cm^{-1} is measured and examined. This peak is referred as the silicon Raman peak throughout the next sections. It corresponds to lower energy scattered photons λ_1 than the incident ones λ_0. The conversion formula from nm to cm^{-1} is written as:

$$\Delta\omega(cm^{-1}) = \left(\frac{1}{\lambda_0(nm)} - \frac{1}{\lambda_1(nm)}\right) \times 10^7. \quad (2)$$

2.1. Orientation evaluation

The Raman scattering efficiency or intensity depends on the polarization direction of the incident ($\boldsymbol{e}_i$) and backscattered ($\boldsymbol{e}_s$) light and on the three silicon Raman tensors $\boldsymbol{R}'_j$ which are proportional to $(\partial\chi/\partial Q_j)_0$ (see Equation 1), I_0 is a constant including all fixed experimental parameters [5,10,14]:

$$I(\boldsymbol{e}_i,\boldsymbol{e}_s) \approx I_0 \cdot \sum_{j=1}^{3} \left|\boldsymbol{e}_i \cdot \boldsymbol{R}'_j \cdot \boldsymbol{e}_s\right|^2. \quad (3)$$

Here the polarization directions are defined in the stage coordinate system, while the Raman tensors refer to the crystal coordinate system. The crystal - stage transformation is performed by means of a rotation matrix $\boldsymbol{T}(\alpha,\beta,\gamma)$ applied to the Raman tensors $\boldsymbol{R}'_j$, where α, β, and γ are the three Euler angles [10]:

$$I(\boldsymbol{e}_i,\boldsymbol{e}_s) \approx I_0 \cdot \sum_{j=1}^{3} \left|\boldsymbol{e}_i \cdot \left(\boldsymbol{T}^{-1}(\alpha,\beta,\gamma) \cdot \boldsymbol{R}'_j \cdot \boldsymbol{T}(\alpha,\beta,\gamma)\right) \cdot \boldsymbol{e}_s\right|^2. \quad (4)$$

By continuously rotating the polarization direction of the incident laser light θ with a $\lambda/2$ plate for two analyzer positions x and y, it is possible to obtain two experimental curves showing the intensity variations of the silicon Raman peak. The data fitting based on Equation (4) results in the numerical evaluation of the three Euler angles which are needed to describe the crystallographic orientation of a particular grain with respect to the stage (reference) coordinate system. Having the grain orientation, the intensity variations of the three optical phonons with polarization settings $I^{X}_{j}(\theta)$ and $I^{Y}_{j}(\theta)$ can be simulated separately. These six intensity variations can be transformed into six intensity ratio functions [10]:

$$W_1^{X,Y}(\theta)=\frac{I_1^{X,Y}(\theta)}{I_2^{X,Y}(\theta)+I_3^{X,Y}(\theta)},W_2^{X,Y}(\theta)=\frac{I_2^{X,Y}(\theta)}{I_1^{X,Y}(\theta)+I_3^{X,Y}(\theta)},W_3^{X,Y}(\theta)=\frac{I_3^{X,Y}(\theta)}{I_1^{X,Y}(\theta)+I_2^{X,Y}(\theta)} \quad (5)$$

It has been shown that for almost any arbitrary oriented grain, distinct polarization settings (θ, x or y) for which the intensity of one phonon prevails over the intensity sum of the other two phonons can be found [10]. Performing three Raman measurements on the same grain, one for every single-phonon polarization settings, several stress tensor components can be determined as experimentally shown in Section 4.1.1.

2.2. Stress evaluation

In the absence of stress (internal or external), the three Raman optical phonons of silicon are degenerate leading to a single Raman peak at $\omega_0 \sim 520$ cm^{-1}. Large mechanical stresses in the GPa range lift the degeneracy causing frequency shifts of the three optical phonons $\Delta\omega_j = \omega_j - \omega_0$, which appear as separate peaks in the Raman spectrum depending on the direction of the applied stress and measurement conditions [8,12,13]. When the stress level is below 1 GPa, these frequency shifts are too small to be resolved being masked by the natural width of the silicon Raman peak. In such cases, particular polarization settings for the incident and backscattered Raman light can be found that allow the excitation and probing of the three optical phonons almost separately and consequently their frequency shifts can be determined. These settings vary with the orientation of the investigated grain as discussed in the previous section. Next step consists in using the secular equation that relates the frequency shifts to the strain tensor components [5,9,10]:

$$\begin{vmatrix} p\cdot\varepsilon'_{xx}+q\cdot(\varepsilon'_{yy}+\varepsilon'_{zz})-\lambda & 2r\cdot\varepsilon'_{xy} & 2r\cdot\varepsilon'_{xz} \\ 2r\cdot\varepsilon'_{xy} & p\cdot\varepsilon'_{yy}+q\cdot(\varepsilon'_{xx}+\varepsilon'_{zz})-\lambda & 2r\cdot\varepsilon'_{yz} \\ 2r\cdot\varepsilon'_{xz} & 2r\cdot\varepsilon'_{yz} & p\cdot\varepsilon'_{zz}+q\cdot(\varepsilon'_{xx}+\varepsilon'_{yy})-\lambda \end{vmatrix}=0. \quad (6)$$

Here p, q, and r are material constants so-called phonon deformation potentials being the only three independent components for cubic symmetry crystals such as silicon, ε'_{ij} are the strain tensor components in the crystal coordinate system, while the eigenvalues λ_j (j = 1,2,3) are given by

$$\lambda_j=\omega_j^2-\omega_0^2=(\omega_j-\omega_0)\cdot(\omega_j+\omega_0)\approx\Delta\omega_j\cdot 2\omega_0. \quad (7)$$

The stress tensor components are finally obtained from the inverse Hooke's law $\varepsilon'_{ij} = S_{ij} \cdot \sigma'_{ij}$ where S_{ij} represents the elastic compliance tensor whose components are material constants. It is evident from Equation (6,7) that the three frequency shifts $\Delta\omega_j$ are not enough to determine the six independent stress tensor components σ'_{ij}. The probing depth in silicon ranges from a few hundreds of nm to a few μm for visible excitations, and from a few nm to a few tenths of nm for UV excitations. Thus, due to wavelength dependent absorption, only the stress state close to the sample's surface is measured. This implies a predominant planar stress state described by three stress components σ'_{xx}, σ'_{yy}, and τ'_{xy} that can be numerically evaluated using the three frequency shifts $\Delta\omega_j$. The residual stress components in the z direction given by σ'_{zz}, τ'_{xz}, and τ'_{yz} are included in Δ'_z which also serves as a correction parameter [10]:

$$\sigma' = \begin{pmatrix} \sigma'_{xx} & \tau'_{xy} & 0 \\ \tau'_{xy} & \sigma'_{yy} & 0 \\ 0 & 0 & \Delta'_z \end{pmatrix}. \tag{8}$$

The following equation can be used to transform the stress tensor components into average or von Misses stress:

$$\sigma'_{av} = \sqrt{\sigma_x'^2 + \sigma_y'^2 + \sigma_z'^2 - \sigma'_x\sigma'_y - \sigma'_x\sigma'_z - \sigma'_y\sigma'_z + 3(\tau_{xy}'^2 + \tau_{xz}'^2 + \tau_{yz}'^2)} \tag{9}$$

Another more straightforward way to relate measured Raman shifts to stress values is the use of a simple stress model illustrating the stress state in the sample. The classical example in the case of silicon is the presence of uniaxial stress σ along the [100] direction and the measurement of the backscattered Raman signal from the (001) surface [5]. In this configuration, only one stress tensor component is non zero:

$$\Delta\omega_3(cm^{-1}) = -2\cdot 10^{-9}\sigma(Pa) \quad or \quad \sigma(MPa) = -500\times\Delta\omega(cm^{-1}). \tag{10}$$

If biaxial stress in the x-y plane with stress components σ_{xx} and σ_{yy} (or $\sigma_{xx} = \sigma_{yy}$ for isotropic stress) describes the stress distribution in the sample:

$$\Delta\omega_3(cm^{-1}) = -4\cdot 10^{-9}\left(\frac{\sigma_{xx}+\sigma_{yy}}{2}\right)(Pa) \quad or \quad \sigma_{xx} = \sigma_{yy}(MPa) = -250\times\Delta\omega(cm^{-1}). \tag{11}$$

These two formulas (10,11) written in the stage coordinate system are commonly used in the community for a fast and reliable estimation of the *average stress* independently of the crystallographic orientation of grains in multicrystalline silicon. Thus, 1 cm^{-1} shift of the silicon Raman peak with respect to the stress free value of ~ 520 cm^{-1} corresponds to a uniaxial stress of 500 MPa or to a biaxial in-plane isotropic stress of 250 MPa. It can be seen that tensile stress shifts the silicon Raman peak to lower frequency, while compressive stress to higher frequency as sketched in Figure 1(a). Experimental examples of the stress state evaluation using both methods described above will be given in Section 4.1.

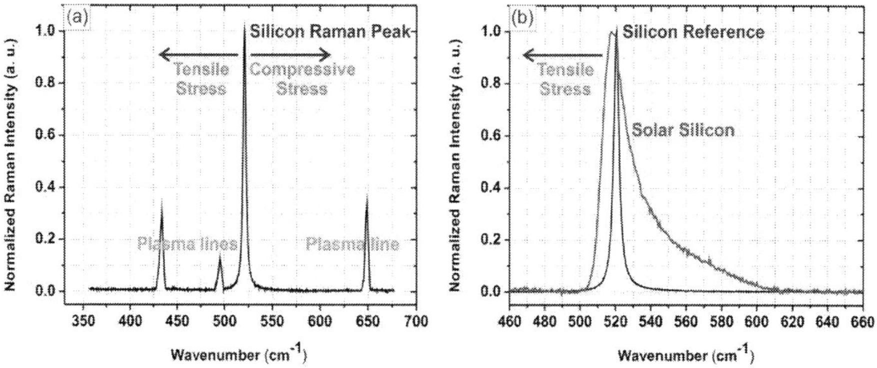

Figure 1. (a) Typical Raman spectrum of a silicon wafer used as reference. In the absence of stress, the three Raman optical phonons of silicon (1 x LO, 2 x TO) are degenerate resulting in a single Raman peak at ω_0 ~ 520 cm^{-1}. The plasma lines originating from an external reference lamp are used to correct the silicon peak position with respect to the thermal drift of the spectrometer grating. Tensile or compressive stresses (internal or external) below 1 GPa may shift the silicon Raman peak to lower or higher frequencies, respectively. (b) Comparison between Raman spectra of stress-free silicon reference and tensile stressed thin film solar silicon on glass. The peak asymmetry caused by high boron doping is clearly visible.

2.3. Doping evaluation

In the case of highly doped silicon, a resonant interaction occurs between the discrete optical phonon states (phonon Raman scattering) and the continuum of electronic states in the valence or conduction bands (electronic Raman scattering) because of electron-phonon coupling. This leads to Fano-type silicon Raman peak asymmetries, which can be observed as tails either on the right side (for p-type doping) or on the left side (for n-type doping) of the otherwise symmetric silicon Raman peak as shown in Figure 1(b) [11,14-17]. The function used to fit the intensity of the silicon Raman peak $I(\omega)$ is given by [11]

$$I(\omega,q,\Gamma,\omega_{\max}) = I_0 \frac{[q+2(\omega-\omega_{\max})/\Gamma]^2}{1+[2(\omega-\omega_{\max})/\Gamma]^2}. \tag{12}$$

Here I_0 is a scaling factor, q is so-called symmetry parameter, Γ is the linewidth of the peak, and ω_{max} is the peak position in the presence of Fano interaction. The symmetry parameter q describes the shape of the silicon Raman peak affected by Fano resonances. Large $q > 150$ values correspond to standard doping (< 10^{16} cm^{-3}) resulting in a nearly symmetric peak, while small $q < 50$ values correlate with high doping (> 10^{18} cm^{-3}) and pronounced peak asymmetry. For highly doped silicon, $1/q$ is approximately proportional to the free carrier concentration [11]. Thus, an accurate quantitative evaluation of doping on the micrometer scale is possible, provided a good calibration curve exist.

2.4. Qualitative defect density evaluation

As shown in the previous three sections, Raman spectroscopy can provide detailed information about semiconductor materials, in this contribution solar silicon, including crystal orientation, internal stresses and doping, which can be extracted from the intensity, position and asymmetry of the silicon Raman peak. In addition, the linewidth of the peak relates to the presence of extended crystal defects. In a perfect crystal, the phonon lifetimes are theoretically infinite in the harmonic approximation that neglects third- and higher-order derivatives of the crystalline potential resulting in narrow delta function-like linewidths of the optical phonon Raman spectra [18]. Defects act as anharmonic perturbations leading to finite phonon lifetimes that manifest themselves as a broadening of the peak described by its full-width at half maximum denoted FWHM or Γ. Therefore, the anharmonic lifetimes of phonons are defined as $1/\Gamma$ being evaluated using first principles calculations including both kinematic effects; i.e., the decay of phonons into vibrations of lower frequency and dynamic effects; i.e., the magnitude of the tensor of the third derivates of the crystal potential with respect to atomic displacements that describes the instability of one-phonon states. The calculated Γ values of the Raman-active optical phonons were found to agree well with those determined experimentally in the case of single crystalline semiconductors (defect free) such as diamond, Si, Ge, GaAs, GaP and InP [19,20].

Anharmonic effects and consequently broadening can also be induced by internal or external stresses approaching the stress-induced splitting limit of the peak as discussed in Section 2.2 or by large stress gradients within the probed volume [5]. Moreover, doping and/or impurities can either produce new Raman peaks through their own vibrational modes or alter/broaden the Raman spectrum of the host material through the change in mass and bond length (atomic effects) as well as through the resonant Fano interaction of free carriers (donors or acceptors) with the lattice (electronic effects). Because all these information originates from the silicon Raman peak, one can separate between the effects of stresses, doping and/or impurities and that of defects on the FWHM values. The Raman linewidths were also found to broaden in the case of relatively small grains (in the nm range) due to the phonon confinement effect, that is, the frequency distribution of the scattered light comes from a broader interval in k-space around the Γ-point in the Brillouin-zone since the $\Delta k = 0$ selection rule is partially lifted by the phonon scattering at grain boundaries [21]. Such a broadening does not occur in large grained solar silicon as presented here.

3. Experimental details

3.1. Sample preparation for Raman measurements

Raman spectroscopy investigates materials nondestructively, the appropriate excitation laser power to avoid damage being material dependent, without elaborated sample preparation. In the case of silicon thin film solar cells on glass, no sample preparation is needed because the as-grown material has low roughness. This is different for wafer- and ribbon-based solar cells for which a simple sample preparation procedure is necessary. Their surfaces have to be evened out by mechanical polishing prior to the Raman measurements to avoid artifacts induced by uncontrolled reflections at rough surface facets. The standard polishing procedure applied to small pieces consists in changing gradually from larger to smaller diamond particle sizes with the final polishing step removing most of the previously damage surface layer, thus leaving the samples in a negligible polishing-induced stress state. The cutting into small pieces leads to stress relaxation due to the creation of free surfaces as discussed in Section 4.1.2. After polishing, the samples are Secco-etched [22] for 5 seconds to make the grain boundaries and dislocations visible. This short defect etching step does not affect the Raman scattering or the other measurement techniques used herein.

3.2. Micro-Raman spectroscopy

The incident light needed for Raman excitation is provided by a laser with a main emission with narrow line width. An interferential filter is used to block the other emissions of the laser. After being reflected by an edge or a notch filter, the light is focused onto the sample's surface through a microscope objective, thus giving rise to the term micro-Raman spectroscopy. Depending on the objective (magnification, numerical aperture) as well as on the excitation wavelength, the diameter of the incident laser beam is different. For the 100x objective (numerical aperture 0.9) and 633 nm excitation employed in the micro-Raman measurements presented here, the probing diameter is ~ 1 μm. The laser power density can be quite high, thus a low laser power of ~ 2 mW at the silicon sample's surface should be used. In these conditions, no shift or increase in the FWHM of the silicon Raman peak due to the local heating of the sample by the laser beam were observed.

As already mentioned in Section 2.2, the probing depth is controlled by the material absorption, which is wavelength dependent. In crystalline silicon, an excitation wavelength of 633 nm results in a penetration depth of ~ 3 μm, while 457 nm gives ~ 300 nm. Since we use 633 nm, the entire thickness of the silicon thin films on glass is probed, while only the surface of the ~ 200 – 250 μm thick silicon wafers and ribbons is measured. The backscattered Raman light passes back through an edge or a notch filter which cuts most of the Rayleigh light, it is dispersed using a grating and then detected by a silicon CCD detector. All Raman measurements herein were performed at room temperature in the backscattering configuration using a LabRam HR800 spectrometer from Horiba Jobin Yvon. A schematic picture of our micro-Raman spectrometer is displayed in Figure 2.

In order to draw correct conclusions about materials with varying spatial properties, not only several but many Raman spectra are acquired while moving the microscope stage with the sample in x- and y-directions in steps equal or smaller than the diameter of the laser probing beam as shown in Figure 2. This results in a complete micro-Raman mapping of the investigated areas, which are usually in the range of a few tens of μm^2. The exposure time is typically up to 1 s per spectrum. These spectra are fitted with a Gauss-Lorentzian function and maps of the shift, FWHM and intensity of the silicon Raman peak corresponding to the spatial distributions of internal stress, defect density and grain orientation are obtained. A Fano-like fitting function as in Equation 12 is used for doping evaluation [11]. To ensure correct interpretation of the Raman data as well as to be able to visualize small mechanical stresses, the effect of the thermal drift of the spectrometer grating on the silicon peak position is corrected [5,14]. This is done by using one of the plasma lines visible in Figure 1(a) emitted by an external lamp located either close to the spectrometer's confocal hole or above the microscope [23,24].

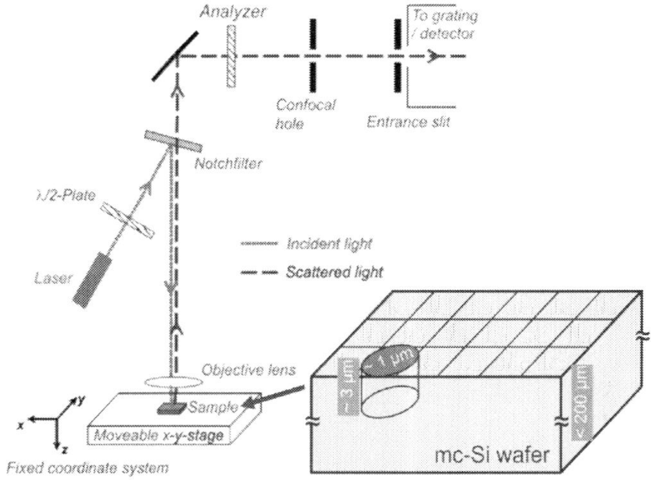

Figure 2. Schematic picture of the used micro-Raman spectrometer. The $\lambda/2$ plate and the analyzer adjust the polarization direction of the incident and backscattered light with respect to the stage (fixed) coordinate system. The polarized micro-Raman procedure enables the evaluation of the crystallographic orientation of arbitrary grains and of stress components as described in Sections 2.1 and 2.2. A graphical representation of micro-Raman mapping obtained by moving the microscope stage with the sample (multicrystalline silicon wafer) in x- and y-directions under a 633 nm exciting laser along with the Raman probing volume are also shown.

3.3. Macro-Raman spectroscopy

Macro-Raman spectroscopy enables fast, large area Raman mapping in the cm^2 range needed for statistical studies of materials properties and the correlations between them and with processing. In the context of PV, this technique can be used not only for fundamental studies in laboratory scientific research but also for optimization and in-line quality check in a PV factory. Macro-Raman mapping is possible through two add-ons that can be integrated in any existing micro-Raman spectrometer. The two new DuoScan™ (hardware) and SWIFT™ (software) Raman scanning modules developed by HORIBA Jobin Yvon provide significant reduction by orders of magnitude of the measurement times by means of large area probing beam (macro-beam) and high speed detector-stage coordination, respectively. Even faster Raman imaging is possible by combing these two technologies [7].

DuoScan™ Raman imaging technology extends the imaging capabilities of micro-Raman instruments from (sub-) micron to macro-scale mapping. The integration of the DuoScan unit to an existing micro-Raman spectrometer is shown in Figure 3(left). This mode is based on a combination of two orthogonally rotating piezo-mirrors that scan the laser beam across the sample following a user-defined pattern as displayed in Figure 3(middle). The size of the resulting macro-beam is adjustable being limited only by the opening of the used microscope objective. The maximum macro-beam sizes achievable with our 50x (NA 0.80) or 10x (NA 0.30) NIKON microscope objectives are 100 x 100 μm^2 or 1 x 1 mm^2. DuoScan allows to integrate the Raman signal over the macro-beam area giving an average spectrum, which contains the same spectral information as that obtained by averaging all micro-Raman spectra for the same area. The gain in acquisition time is evident, macro-Raman being orders of magnitude faster than conventional micro-Raman. For example, if an area of 30 x 30 μm^2 is entirely probed by macro-Raman in one second, micro-Raman with a spot-size of 1 x 1 μm^2 needs 900 seconds to cover the same area. The price one has to pay is the loss of lateral resolution.

Furthermore, DuoScan can be used in a step-by-step mode where the mapping takes place without moving the stage with the sample. A minimum step size of 50 nm is reached by deflecting the laser beam, which complements successfully the stepping capability of the stage specified to be ~ 500 nm. This mode applies for Raman imaging of nanoscale objects and features. The DuoScan mapping capabilities are summarized in Figure 3(right).

SWIFT™ Raman imaging technology enables ultra fast mapping without losing lateral resolution and thus image quality. In this mode, the time intervals needed for the stage to accelerate/decelerate as well as for the shutter in front of the detector to open/close for each measurement point are eliminated. Basically, these are dead times, which are not used for the acquisition of the Raman signal. The breakthrough consists in continuously moving the stage with the sample while keeping the shutter open and measuring continuously Raman spectra by means of high speed detector-stage coordination coupled with the high optical throughput of the Raman system. The SWIFT option can also be used for time resolved Raman imaging provided the investigated processes occur on the measurement time scale.

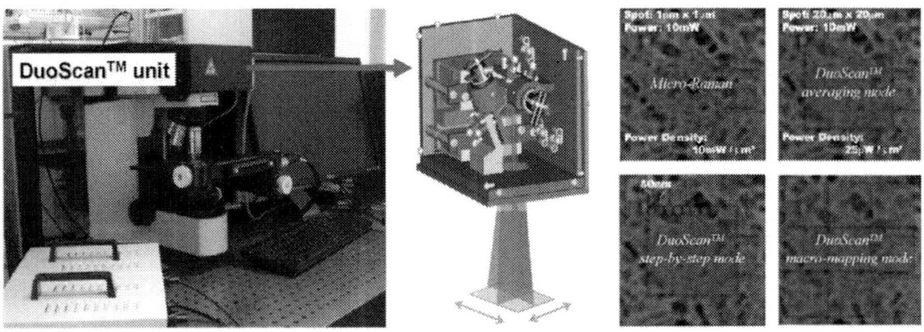

Figure 3. (left) DuoScan unit attached to a micro-Raman spectrometer. (middle) Schematic drawing illustrating the DuoScan working principle. The probing micro-beam is scanned by two orthogonally rotating piezo-mirrors resulting in a macro-beam, thus giving rise to the term macro-Raman spectroscopy. (right) Comparison between standard and DuoScan mapping modes described in the text. The left and middle pictures are taken from HORIBA's official webpage.

3.4. Complementary techniques to Raman spectroscopy for solar silicon studies

The Raman investigations on semiconductor materials represent an important step towards their fundamental understanding and the control of their properties for designing devices with specific functions. As already mentioned, Raman spectroscopy can provide detailed information regarding the spatial distributions of internal stress, defect density, doping, and grain orientation. We will show that even more insight into the interaction between different properties of silicon PV materials can be achieved when Raman measurements are supported and complemented at identical positions by other techniques such as EBSD, EBIC and defect etching.

EBSD measurements are performed using an EDAX system attached to a TESCAN LYRA XMU scanning electron microscope (SEM) to determine the grain orientations and grain boundary types. The crystal orientation is given in the {hkl}<uvw> representation where {hkl} is the crystal plane perpendicular to the sample normal direction (z axis) and <uvw> is the crystal direction aligned with the transverse direction of the sample (y axis). The inverse pole figure (unit triangle) shows the sample normal direction relative to the axes of the measured crystal. The misorientation between adjacent grains is given in the angle/axis notation, that is, the rotation angle about the axis common to both lattices to bring them into coincidence, and in terms of Σ-value which denotes the fraction of atoms in the GB plane coincident in both lattices [25].

EBIC measurements are done with an EVO 40 SEM at 20 keV beam energy both at room temperature and 80K to image most of the electrically active defects. In order to render the inhomogeneities of recombination clearly visible, a color scale is used for the EBIC maps. The maps represent the local EBIC signal normalized by the maximum EBIC signal. The lower the EBIC signal, the higher the recombination activity.

4. Results and discussions

4.1. Micro-Raman measurements

4.1.1. Silicon thin films on glass for solar cells

The unique characterization power of the Raman technique consists in the detailed mechanical and microstructural information that can be extracted from the silicon Raman peak: (1) the peak position map - the distribution, amount and sign of internal stresses, (2) the peak full-with at half maximum (FWHM) map - the distribution and qualitative comparison of defect densities, (3) the peak asymmetry map – the distribution and amount of doping, and (4) the peak intensity map – the grain orientation and the grain boundary pattern [7,10,11].

One representative example is shown in Figure 4 for the laser crystallized silicon seed layers of thin film solar cells on glass (in this example, 110 nm thick silicon seed film). The larger FWHM values in Figure 4(a) indicating a broadening of the Raman spectra are produced by a decrease in the phonon lifetimes, which in turn is mainly due to defects acting as anharmonic perturbations (see Section 2.3). Indeed, the dashed line contours in Figure 4(a) corresponds to low angle GBs indicated by arrows in the EBSD maps shown in Figure 4(e and f). It is well known that low angle GBs consist of dislocation networks/arrays. Their presence at these positions is further supported by the continuously changing crystallographic orientation within the studied grain as indicated by the gradual changing color in the intergranular misorientation gradient map by EBSD in Figure 4(f). Such intergranular misorientation is attributed to geometrically necessary dislocations forming low angle GBs [26].

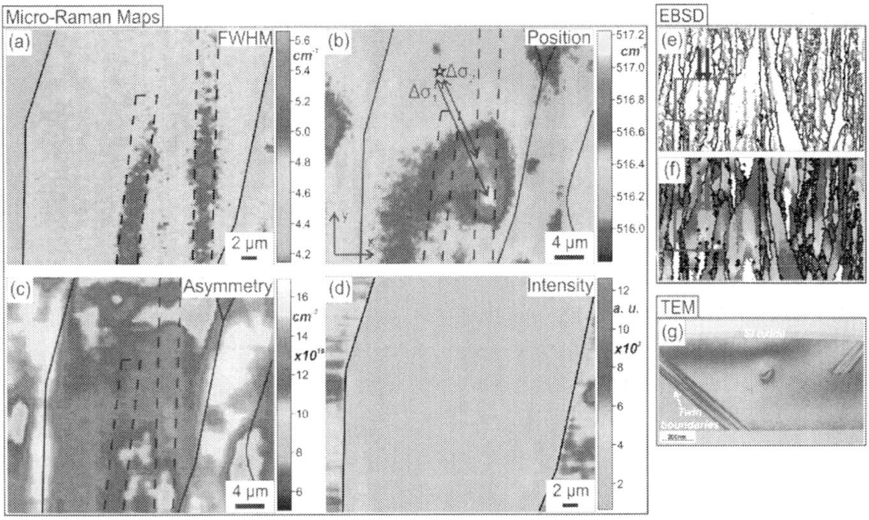

Figure 4. Micro-Raman maps of a laser crystallized silicon seed layer of a thin film solar cell on glass (110 nm thick, nominal boron doping of $2.1\text{x}10^{19}$ cm^{-3}) obtained from the fitting of the first-order Raman spectra of silicon: (a) peak FWHM – defect density map, (b) peak position – internal stress map with two lateral stress gradients $\Delta\sigma_1$=227, $\Delta\sigma_2$=197 ± 12 MPa (von Mises stresses), (c) peak asymmetry – doping map, (d) peak intensity – grain orientation map. EBSD maps: (e) grain boundary map including high angle GBs (black lines), low angle GBs (orange lines), Σ3 GBs (blue lines), Σ9 GBs (green lines), (f) Intergranular misorientation gradient map. The two vertical arrows indicate low angle GBs corresponding to the areas delineating by dashed lines in (a, b and c). (g) TEM cross-section image.

Dislocations are considered to be among the most detrimental type of defect controlling not only the mechanical but also the electrical properties of silicon and other materials. They are produced by the partial or total relaxation of thermally induced stresses during the crystallization and cooling processes as long as plastic deformation is allowed by temperature, this means above the brittle-ductile transition temperature of silicon. Below this temperature, the remaining thermally induced stresses are incorporated as thermally induced residual stresses in the silicon material. Once created, dislocations can move on glide planes leading to further plastic deformation through their multiplication until the lattice friction becomes larger than the effective stress needed for moving the existing dislocations [6].

By comparing the FWHM (defect density) and position (stress) maps displayed in Figure 4(a and b), it can be seen that the two patterns are unlike. Regions of similar FWHM values (similar defect densities) along the two lines of dislocations exhibit different Raman peak positions (different stress levels) showing virtually no influence of internal stresses on peak broadening. The fact that the two dislocation lines are only partly accompanied by stress can be explained by the locally different superposition of stress fields of dislocations and thermally induced residual stresses. Their interaction can lead to local configurations in which internal stresses get cancelled totally, partially or not at all [6,23,27]. Thus, we attribute the stress concentrations in Figure 4(b) to particular combinations of (1) defect configurations/structures, which do not necessary result in higher FWHM values and (2) thermally induced residual stresses. The stress map is corrected for the compressive contributions produced by the Fano effect and by the addition of boron by means of the lattice parameter using Vergard's law [7]. The presence of different types of defects and their nonuniform distribution inside the laser crystallized seed layer are supported by transmission electron microscopy (TEM) investigations shown in Figure 4(g).

Two lateral stress gradients inside the central grain in Figure 4(b) are evaluated in form of stress-tensors following the polarized micro-Raman procedure described in [10]. First, the crystallographic grain orientation is determined using the Raman intensity dependences on the polarization direction of the incident light (θ) and Raman backscattered light (x or y analyzer positions) as explained in Section 2.1 at the point marked by the star within this particular grain. The measured plots are displayed in Figure 5.

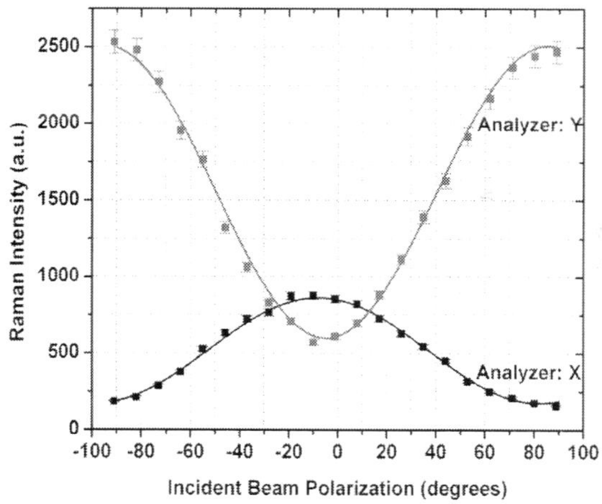

Figure 5. Raman intensity dependences on the polarization direction of the incident light for the X and Y analyzer positions measured at the point marked by a star in Figure 4(b). The error bars account for the ~ 3% intensity variations of the incident laser light. The continuous curves represent fit functions based on Equation 4 used to obtain the three Euler angles: $\alpha = 51° \pm 2°$, $\beta = 27° \pm 2°$, and $\gamma = -2° \pm 2°$ necessary to determine the grain orientation and the stress tensor components.

Their fitting by the Equation 4 gives the following three Euler angles: α = 51° ± 2°, β = 27° ± 2°, and γ = -2° ± 2°, which in turn provide the following rotation matrix to bring this arbitrary oriented grain in the stage (reference) coordinate system:

$$T(\alpha,\beta,\gamma)=\begin{pmatrix} 0.656 & -0.664 & 0.356 \\ 0.753 & 0.587 & -0.293 \\ -0.014 & 0.461 & 0.887 \end{pmatrix}.$$

Second, since the internal stresses are too small to produce a visible lifting of degeneracy of the three silicon optical phonon frequencies, the polarization settings for the incident and backscattered light for which the intensity of one of the three phonon modes dominates the other two are simulated using the previously determined Euler angles and Equation 5. The simulations of the six intensity ratio functions $W^{XY}{}_j(\theta)$ are shown in Figure 6. It can be seen that the polarization settings to measure separately the three phonon frequency shits $\Delta\omega_j$ for the given grain are: Phonon 1: 1°, X, Phonon 2: -22°, Y, and Phonon 3: 15°, Y.

Third, three Raman maps of the area in Figure 4(b) are measured for the three polarization settings above. Difference stress-tensors are calculated numerically from the three Raman frequency shifts $\Delta\omega_j$ with respect to the stage (reference) coordinate system as described in Section 2.2:

$$\Delta\sigma_1=\begin{pmatrix} -184\pm10 & -33\pm1 & 0 \\ -33\pm1 & -192\pm10 & 0 \\ 0 & 0 & -151\pm10 \end{pmatrix} MPa\,,$$

$$\Delta\sigma_2=\begin{pmatrix} -200\pm10 & -51\pm1 & 0 \\ -51\pm1 & -213\pm10 & 0 \\ 0 & 0 & -152\pm10 \end{pmatrix} MPa.$$

The two lateral stress gradients indicate compressive stresses at these positions with respect to the point marked by the star in Figure 4(b), while the shift towards lower frequencies (<~ 520 cm^{-1}) in the position of the Raman spectra implies tensile stress inside the silicon thin film. Their conversion into average or von Misses stresses using Equation 9 gives $\Delta\sigma_1$=227, $\Delta\sigma_2$=197 ± 12.5 MPa.

Figure 4(c) shows the asymmetry (doping) map obtained from the symmetry parameter q of the Raman spectra as defined in Section 2.3. For the quantitative doping evaluation, the free carrier concentration vs. q calibration curve in Figure 5 of Reference 11 was used. The free hole concentrations are found to be lower than the nominal boron doping of 2.1x10^{19} presumably due to the incomplete activation of dopants during laser crystallization and cooling [11]. Higher doping is observed both along GBs and inside grains. Regarding the influence of doping/impurities on the FWHM as discussed in Section 2.3, there is no correlation between them as seen by comparing the FWHM map with the asymmetry map displayed in Figure 4(a) and (c), respectively.

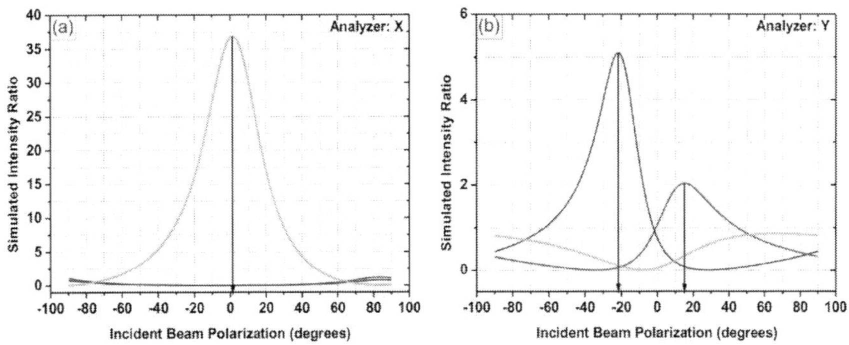

Figure 6. Simulation of the six intensity ratio functions $W^{XY}{}_j(\theta)$ for the central grain in Figure 4(b). The plot maxima marked by arrows indicate the polarization direction of the incident laser light for the two analyzer positions where the intensity of one of the three silicon phonons dominates over the sum of the other two phonons.

The different Raman scattering efficiencies caused by distinct crystallographic grain orientations and the polarization directions of the incident and backscattered laser light can be used to image the grains and to determine their orientations as shown in Section 2.1. This results in intensity maps such as displayed in Figure 4(d), which enable tracing of GBs represented as solid lines in all Raman maps of Figure 4. Thus, it is possible to relate grains and GBs to defect, stress and doping distributions, all data being provided by the same Raman mapping.

4.1.2. *Wafer and ribbon-based silicon solar cells*

Next examples illustrate the application of micro-Raman spectroscopy to block-cast and edge-defined film-feed (EFG) multicrystalline silicon materials, two industrial relevant materials with the former having the largest share (> 50%) in the PV market. Internal stresses are the result of the superposition between the thermally induced residual stresses that is the thermally induced stresses at the end of crystallization and cooling processes and the defect-related stresses. By cutting the silicon blocks and EFG tubes into wafers and then into small pieces for micrometer scale investigations, the thermally induced residual stresses are expected to relax to a large extent due to the creation of free surfaces. This is different in the case of silicon thin films on glass, which are measures as-prepared without any cutting. Thus, in the block-cast and EFG samples, the internal stresses produced mainly by defects are measured.

The resolution of micro-Raman can go down to single dislocation characterization as demonstrated in Figure 7(a) in the case of p-type block-cast mc-Si material taken from a PV factory production line. Here, the localized and quite symmetric stress distribution including both compressive (red area) and tensile (blue area) and its decay length resemble the stress field of an edge dislocation, which, in this case, is superimposed on the Σ27a GB [23,27]. The polarized micro-Raman stress measurements give the following difference stress tensors referring to the stage (reference) coordinate system shown in Figure 7(a). They have been evaluated between stressed positions close to the GB and positions at a distance from the GB, which are not affected by the dislocation stress field:

$$\Delta\sigma_1 = \begin{pmatrix} -40 \pm 10 & -14 \pm 3 & 0 \\ -14 \pm 3 & -38 \pm 10 & 0 \\ 0 & 0 & -25 \pm 10 \end{pmatrix} MPa,$$

$$\Delta\sigma_2 = \begin{pmatrix} 33 \pm 10 & -7 \pm 1 & 0 \\ -7 \pm 1 & 31 \pm 10 & 0 \\ 0 & 0 & 34 \pm 10 \end{pmatrix} MPa.$$

As shown in the previous example, it is worth to combine at the same position micro-Raman with other techniques not only to support the interpretation of the Raman results but also to get new insights into other material properties and their interplay. The EBIC images in Figure 7(b and c) show lower signal corresponding to reduced minority carrier lifetime of 79% at 300K and 63% at 80K in the region of the GB trajectory change where the edge dislocation is located as well as a signal variation along the Σ27a GB. By comparing the stress and EBIC images, it can be seen that the stressed area close to the change in the Σ27a GB trajectory denoted K and the stress-free area above and below it show similar EBIC signals, and thus similar recombination activities.

Figure 7(e) shows a Raman stress map of the same Σ27a GB at a distance of several millimeters from the position displayed in Figure 7(a). The compressive (red area) and tensile (blue area) stresses are more extended along the GB, less symmetric, and change positions with respect to the GB as compared with the stress map in Figure 7(a). This stress distribution is attributed to the stress field of an array of edge dislocations superimposing the GB. The band-like less compressed region on the right-hand side of the Σ27a GB can be explained by the presence of dislocations (edge, screw and/or mixed), in the grain and close to the GB, which have locally rearranged during crystal growth and cooling to reduce the strain energy and thus, the stresses in this region [23,27]. The following stress-tensor gradients referring to the stage (reference) coordinate system shown in Figure 7(e) have been determined by polarized micro-Raman:

$$\Delta\sigma_3 = \begin{pmatrix} 29 \pm 10 & -7 \pm 1 & 0 \\ -7 \pm 1 & 28 \pm 10 & 0 \\ 0 & 0 & 36 \pm 10 \end{pmatrix} MPa,$$

$$\Delta\sigma_4 = \begin{pmatrix} -34 \pm 10 & -3 \pm 1 & 0 \\ -3 \pm 1 & -37 \pm 10 & 0 \\ 0 & 0 & -37 \pm 10 \end{pmatrix} MPa.$$

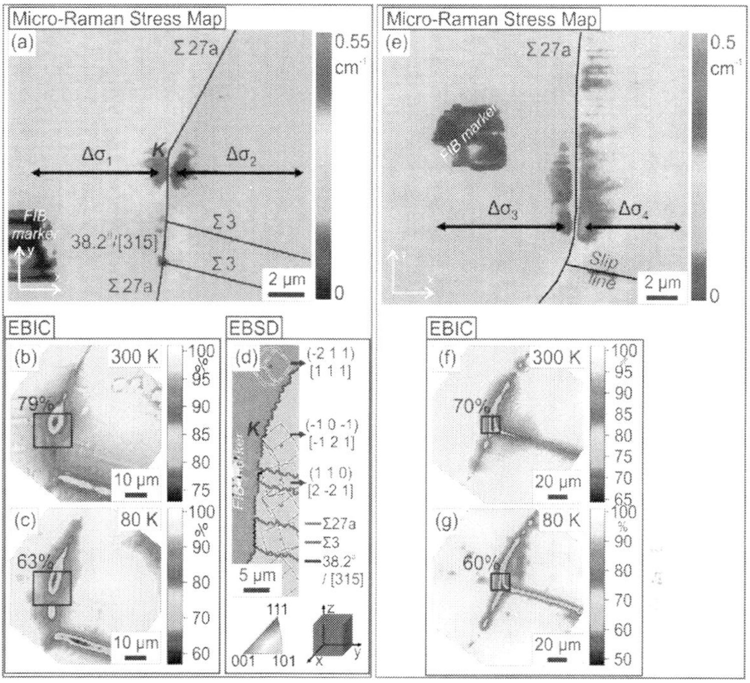

Figure 7. Micro-Raman, EBIC and EBSD studies of block-cast solar silicon at two positions along the same Σ27a GB. The Raman stress distributions are attributed to a single edge dislocation (a) and to an array of edge dislocations (e) superimposing the GB. The regions enclosed by rectangles in the EBIC images (b, c) and (f, g) correspond to the Raman mapped areas in (a) and (e), where the numbers indicate the maximum EBIC signal. The lower the EBIC signal, the higher the recombination activity. The focused ion beam (FIB) markers in (a, e) allows exact spatial correlation between different measurement techniques. (d) EBSD map showing the grain orientations and GB types along with the orientation triangle and the sample reference frame.

Like in the previous case the stressed and stress-free areas around the GB in Figure 7(e) are located in a region of similar (lower) EBIC signal of 70% at 300K and 60% at 80K as indicated in Figure 7(f) and (g), respectively and the recombination activity is inhomogeneous along the Σ27a GB. These two representative examples demonstrate the presence of spatial variations in mechanical and electrical properties of block-cast solar silicon on the micrometer scale.

Similar spatial properties variations are observed in the p-type EFG mc-Si material taken also from a PV factory production line. In order to illustrate the correlation between internal stresses, defect structure and electrical activity in the EFG material, we show here three positions along the same GB that contain representative examples of this correlation. Here, the internal stresses are evaluated using Equation 11 without employing the polarized micro-Raman procedure as in the case of silicon thin films on glass and block-cast mc-Si.

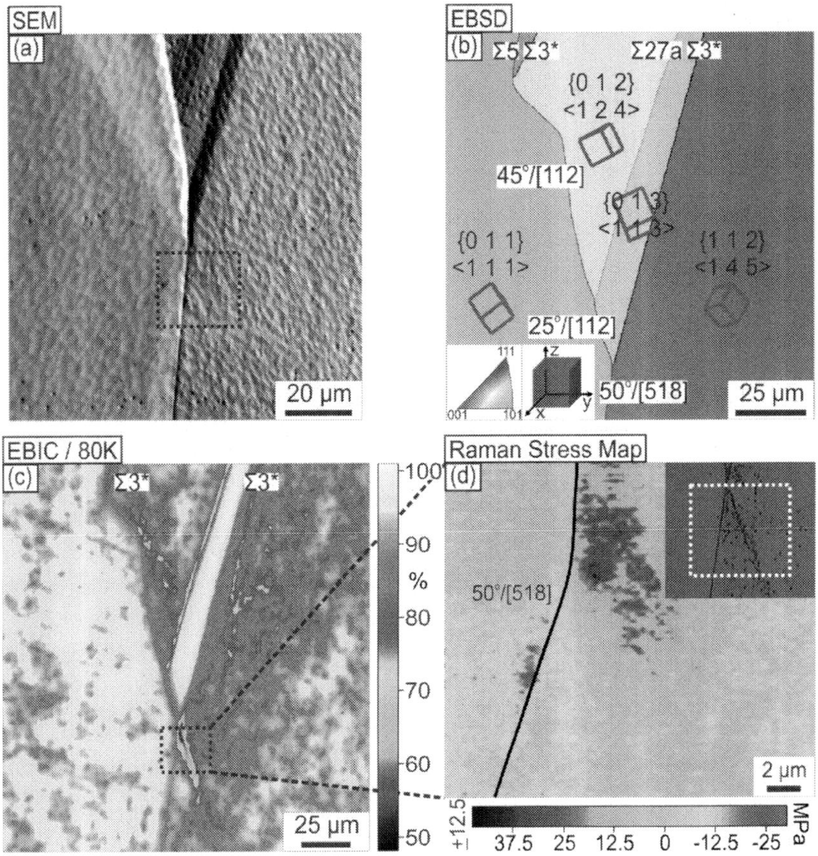

Figure 8. Position 1 (a) SEM image of the as-grown EFG wafer before mechanical polishing. (b) EBSD map showing the grain orientations and GB types along with the orientation triangle and the sample reference frame. (c) EBIC image taken at 80K where the inhomogeneous recombination activity inside grains and at GBs is mainly attributed to dislocations decorated with metallic impurities. (d) Not all dislocations visible in the defect etching image shown in the inset or measured by EBIC are accompanied by internal stresses as probed by micro-Raman. The dashed rectangle in the inset represents the Raman mapped area. At this position, the lowest EBIC current corresponds to the largest (tensile) stress.

The EBSD, EBIC, and micro-Raman measurements at the first position are displayed in Figure 8. The Raman stress map shows concentrated tensile (in blue) and compressive (in red) stresses close to a large-angle random GB described by a misorientation angle/axis of 50°/[518]. Except these areas, nearly no stresses are found neither along the GB nor inside the two adjacent grains of {011}<111> and {112}<145> orientations. By comparing the stress map with the corresponding EBIC map enclosed by the rectangle in Figure 8(c), it can be seen that not all recombination active dislocations visible at 80K are accompanied by stresses.

That is because dislocations interact with each other and tend to locally rearrange in configurations of minimum strain energy that can result in stresses or virtually no stresses. The EBIC image in Figure 8(c) shows an inhomogeneous electrical activity along different types of GBs as well as inside grains of different crystallographic orientations indicated in Figure 8(b). The recombination-active Σ3 GBs {60°/[111]} in Figures 8-10 are marked with an asterisk to distinguished them from the recombination-free Σ3 GBs in Figure 9.

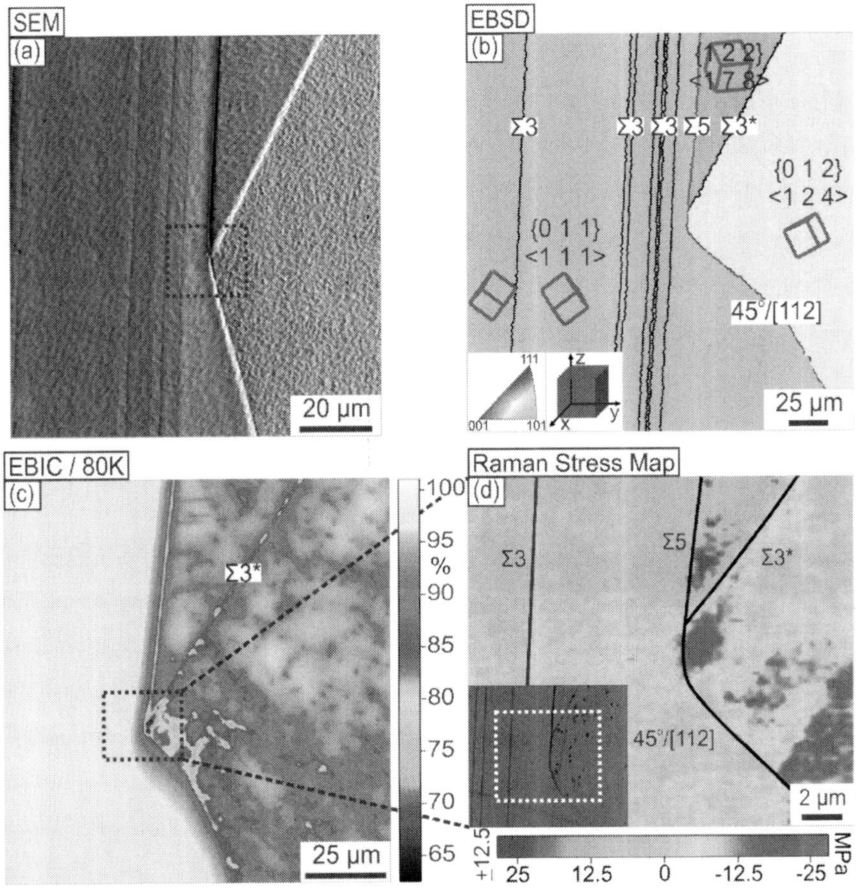

Figure 9. Position 2 (a) SEM image of the as-grown EFG wafer before mechanical polishing. (b) EBSD map. (c) EBIC image where the same left-hand side grain like in Figure 8 shows at this position no electrical activity. The Σ3 GBs are either recombination-free (Σ3) or recombination-active (Σ3*), while being virtually stress-free. (d) The defect etching image in the inset indicates that the presence of dislocation etch pits on Σ3* GBs leads to electrical activity provided the dislocations are decorated with metallic impurities. Here, the highest recombination activity corresponds to the largest (compressive) stress.

The EBSD, EBIC, and micro-Raman results obtained at the second position are shown in Figure 9. Like in the previous case, we did not find a one-to-one correspondence between electrically active dislocations and stresses, both exhibiting inhomogeneous spatial and magnitude distributions. These findings are similar to those on block-cast mc-Si displayed in Figure 7. Different at this position is the presence of tensile (in blue) and compressive (in red) stresses concentrated close to a GB triple point where a $\Sigma5$ GB {36.86°/[100]}, a $\Sigma3$ GB, and a large-angle random GB {45°/[112]} meet. It is worth noting that GBs of the same type, here $\Sigma3$ GBs, can be either recombination-free ($\Sigma3$) or recombination active ($\Sigma3^*$), while being both nearly stress-free. Essentially, independent of the GB type, such large differences in electrical activity originate mainly from the absence or presence of recombination-active dislocations on or very close to the GB. This point is confirmed by comparing the defect-etched optical image with the EBIC map: the $\Sigma3$ GBs decorated by dislocation etch pits (denoted $\Sigma3^*$) show increased electrical activity, while the $\Sigma3$ GBs without dislocation etch pits show no recombination activity. It can be seen that despite the same GB type assignment by the EBSD software, the $\Sigma3^*$ and $\Sigma3$ GBs are formed between adjacent grains of different crystallographic orientations. This fact suggests distinct kinematic conditions at these $\Sigma3$ GBs that can lead to dissimilar thermally induced stress levels and as a result to the generation or absence of dislocations. On the other hand, Raman measures only those configurations of dislocations (including the recombination-free dislocations) that lead to stresses. In contrast with the first (Figure 8) and third (Figure 10) positions, the {011}<111> left-hand side grain shows no reduction of the EBIC signal at the second position despite quite similar grain geometries at these three positions. This indicates that the thermally induced stresses present during the EFG growth relaxed not through the generation of dislocations but through the formation of twins found at the second position by EBSD.

Similar to the previous two cases, we observe non-uniform distributions of electrical activity and stresses along GBs and inside grains at the third position as displayed in Figure 10(c, d). However, we choose this position to show that the largest recombination activity is not always accompanied by the largest internal stresses as in the case of the first and second positions.

The local variations in the sign and values of the dislocation-related stresses as well as in the strength of the recombination activity are attributed to the cumulative effect of metallic impurity decoration, intrinsic structure, type, density, and distribution of dislocations inside grains and on GBs. This non-uniform distribution of dislocations originates from locally different mechanisms of nucleation, motion, multiplication and annihilation of dislocations controlled by the grain structure including the orientation, size and geometry of the grains, the kinematic constraints at GBs, and temperature. The presence of impurities is confirmed by EBIC measurements which show quite strong reduction of the EBIC signal at room temperature up to 70-80% which further reduces with decreasing temperature up to 50-65% at 80K. Such EBIC behavior corresponding to increasing recombination can be explained by the interaction of shallow levels related to the strain fields of dislocations with deep levels due to metallic impurity decoration and/or intrinsic core defects at dislocations. It is known that impurity accumulation in silicon can be enhanced due to the presence of stresses

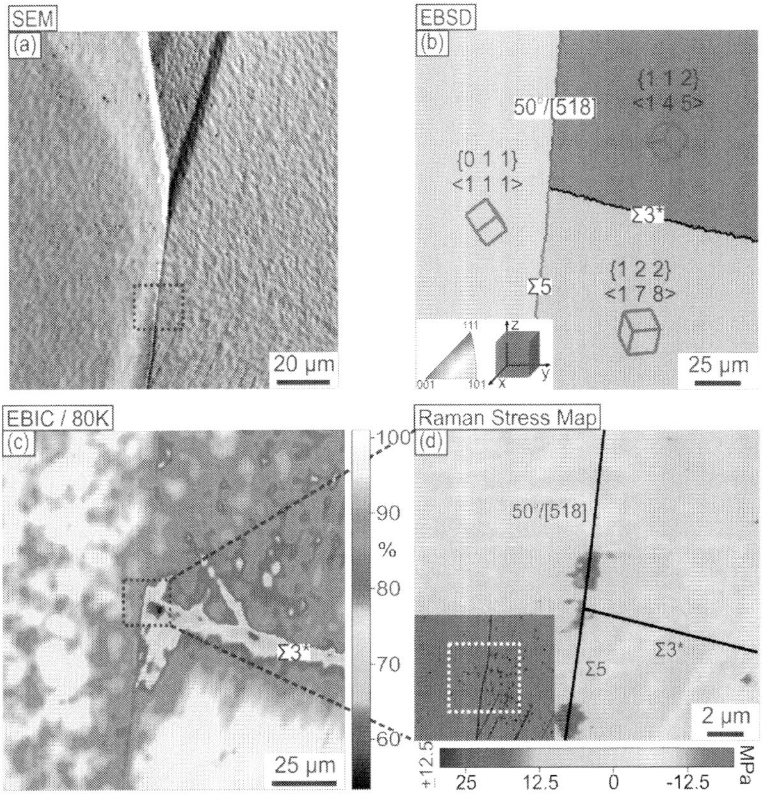

Figure 10. Position 3 (a) SEM image of the as-grown EFG wafer before mechanical polishing. (b) EBSD map. (c) EBIC image where the same left-hand side grain like in Figure 8 and 9 shows recombination activity. (d) The lowest EBIC current is not accompanied by stress at this position.

(thermally induced residual stresses and/or defect-related stresses) at temperatures where both impurities and dislocations are mobile. This can explain the increased electrical activity at regions of higher dislocation densities as at the first and second positions where dislocations are spatially distributed in such a way that their stress fields cancel partially or not at all so that an overall long-range stress field from these dislocations is measured by micro-Raman. The dislocations arranged in configurations in which their stress fields cancel totally (or below the detection limit of our Raman spectrometer of ± 12.5 MPa) are only visible by EBIC (when recombination-active) but not by micro-Raman, as at the third position. Point-by-point correlation of the micro-Raman and EBIC measurements indicates that internal stresses of several tens of MPa do not influence the minority carrier recombination in block-cast and EFG mc-Si. Comparably high stresses of up to 1.2 GPa are necessary in silicon in order to influence its electrical properties such as enhanced carrier mobility in the transistor channel through band structure modification and effective mass reduction [6,23].

4.2. Macro-Raman measurements

Representative macro-Raman mappings acquired using the DuoScan option described in Section 3.3 on the laser crystallized silicon seed layers of thin film solar cells on glass (in this example, 290 nm thick silicon seed film) are displayed in Figure 11. These measurements are performed at identical positions using probing macro-beams of 30 x 30 μm^2 and 100 x 100 μm^2 with the 50x and 10x NIKON microscope objectives. The distribution of internal stresses (a, c) and defect densities (b, d) obtained from the position and FWHM of the measured Raman spectra are quite similar when measuring with different DuoScan macro-beam sizes. The inhomogeneous stress patterns in (a, c) are the result of the interaction between defects through their own intrinsic stress fields and thermally induced residual stresses, while the line shape regions in (b, d) correlate with the laser traces where higher defect densities corresponding to larger FWHM values develop predominantly at adjacent laser scan lines where irradiated areas overlap. It can be seen that there is no correlation between the shift/position (stress) and FWHM (defect density) maps both at macro-scale (Figure 11(a-d)) as well as at micro-scale (Figure 4(a and b)). This further supports the argument used to explain the results in the previous sections, namely the locally different interaction between dislocations themselves and with thermally induced residual stresses.

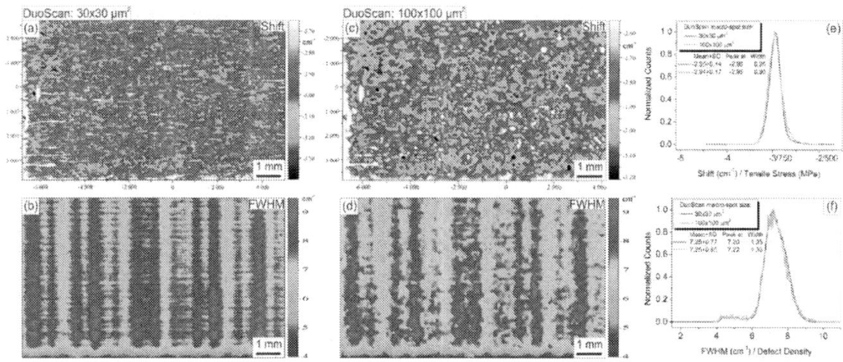

Figure 11. DuoScan Raman maps of the same area using probing macro-beams of 30 x 30 μm^2 (a, b) and 100 x 100 μm^2 (c, d), where the sharpness of the features decreases due the loss of lateral resolution. (a, c) The Raman peak position shifts with respect to a stress-free silicon reference are negative indicating the presence of tensile stresses inside the 290 nm thick laser crystallized silicon thin film on glass. (b, d) The FWHM maps show areas of different crystal quality related to different defect densities, which correlate with the laser traces as visible from the line shape character of the FWHM distributions. (e, f) Statistical evaluation using histograms for the two macro-beam sizes demonstrating that DuoScan can be used for large scale mappings without losing average information.

As expected, the sharpness of the features decreases with increasing the size of the probing macro-beam due the loss of lateral resolution. However, the spectral information is not altered by integrating the Raman signal over the 30 x 30 μm^2 or 100 x 100 μm^2 macro-beam areas. Indeed, the normalized histograms in Figure 11(e, f) calculated from the shift and

FWHM DuoScan maps show a good overlap demonstrating that DuoScan can be used for large-scale mapping of PV and other Raman active materials without losing average spectral information. The presence of thermally induced residual stresses is reflected in the position of the shift/stress histogram in Figure 11(e). Smaller thermally induced residual stresses corresponding to shift/stress histograms closer to or at zero prevent cracking or peeling off the silicon thin film solar cells or even substrate bending minimizing the breakage risk and processing/handling difficulties. When small, they also impede under external mechanical and thermal loads the occurrence of new stress-induced defects, which are commonly recombination active. The qualitative estimation of defects is apparent in the position of the FWHM/defect density histogram in Figure 11(f) where FWHM values closer to ~ 3 cm^{-1} corresponding to defect-free silicon indicate lower defect densities in the silicon thin film solar cells. Thus, macro-Raman can be used to evaluate statistically the materials properties and to see clearly the changes originating from different preparation conditions and processing.

5. Conclusions

The characterization power of the Raman technique at micro- and macro-scale in the case of multicrystalline solar silicon materials is demonstrated. Raman investigations at length scales ranging from μm^2 to cm^2 are possible through two new developed scanning modules, DuoScan™ (hardware) and SWIFT™ (software), which can be integrated in any standard micro-Raman spectrometer. The statistical evaluation of the large area Raman maps measured by macro-Raman spectroscopy shows that macro-scale Raman mapping integrates data over the macro-beam area giving an average spectrum that contains the full spectral information at the cost of decreasing lateral resolution. Moreover, Macro-Raman enables significant reduction by orders of magnitude of the acquisition time: if an area of 30 x 30 μm^2 is entirely probed by macro-Raman in one second, micro-Raman with a spot-size of 1 x 1 μm^2 needs 900 seconds to cover the same area. Deeper insights into the interplay between internal stresses, defects, doping, microstructure, and recombination activity with practical impact on the mechanical stability and conversion efficiency of solar cells have been obtained by combining Raman, EBSD, EBIC, TEM, and defect etching techniques. By tuning the crystallization process, the interaction between dislocations driven by the strain energy minimization can be used to reduce internal stresses resulting in mechanically stronger wafers and cells and to prevent metallic impurity precipitation at dislocations that should lead to improved energy conversion efficiencies.

Author details

George Sarau and Silke Christiansen
Max Planck Institute for the Science of Light, Erlangen, Germany
Institute of Photonic Technology, Jena, Germany

Arne Bochmann
Institute of Photonic Technology, Jena, Germany

Renata Lewandowska
Horiba Scientific, Villeneuve d'Ascq, France

Acknowledgement

This work was financially supported by (1) the German Federal Ministry for the Environment, Nature Conservation and Nuclear Safety and all the industry partners within the research cluster "SolarFocus", (2) the Max-Planck Society within the project "Nanostress", and (3) the European Commission within the FP7-Energy priority project "High-EF". Within the "High-EF" project, Horiba Scientific developed the DuoScan hardware and the SWIFT software that permit fast, large area Raman analyses. The first demonstrator was deployed to the Max Planck Institute for the Science of Light for further development of this technique with respect to applications in photovoltaics. The authors are thankful for this great research opportunity. We would also like to thank A. Gawlik of IPHT for preparing the laser crystallized samples used in this study as well as to M. Holla and W. Seifert of Joint Lab IHP/BTU for their help with the EBIC measurements and fruitful discussions. The content of this publication is the responsibility of the authors.

6. References

[1] Raman C.V, Krishnan K.S (1928) A New Type of Secondary Radiation. Nature 121: 501-502.

[2] Ferraro J.R, Nakamoto K, Brown C.W (2003) Introductory Raman Spectroscopy. Academic Press.

[3] Smith E, Dent G (2005) Modern Raman Spectroscopy - A Practical Approach. John Wiley & Sons.

[4] Pelletier M.J (Ed.) (1999) Analytical Applications of Raman Spectroscopy. Blackwell Publishing.

[5] Wolf I (1996) Micro-Raman Spectroscopy to Study Local Mechanical Stress in Silicon Integrated Circuits. Semicond. Sci. Technol. 11: 139-154.

[6] Sarau G, Christiansen S, Holla M, Seifert W (2011) Correlating Internal Stresses, Electrical Activity and Defect Structure on the Micrometer Scale in EFG Silicon Ribbons. Sol. Energy Mater. Sol. Cells 95: 2264–2271.

[7] Sarau G, Christiansen S, Lewandowska R, Roussel B (2010) Future of Raman in PV Development. Proc. 35th IEEE PVSC: 001770-001775.

[8] Anastassakis E, Pinczuk A, Burstein E, Pollak F.H, Cardona M (1970) Effect of Static Uniaxial Stress on the Raman Spectrum of Silicon. Solid State Commun. 8: 133-138.

[9] Wolf I, Maes H.E, Jones S.K (1996) Stress measurements in silicon devices through Raman spectroscopy: Bridging the gap between theory and experiment. J. Appl. Phys. 79: 7148-7156.

[10] Becker M, Scheel H, Christiansen S, Strunk H.P (2007) Grain orientation, texture, and internal stress optically evaluated by micro-Raman spectroscopy. J. Appl. Phys. 101: 063531 (1-10).

[11] Becker M, Gösele U, Hofmann A, Christiansen S (2009) Highly p-doped Regions in Silicon Solar Cells Quantitatively Analyzed by Small Angle Beveling and micro-Raman Spectroscopy. J. Appl. Phys. 106: 074515 (1-9).

[12] Anastassakis E (1999) Strain Characterization of Polycrystalline Diamond and Silicon Systems. J. Appl. Phys. 86: 249-258.

[13] Puech P, Pinel S, Jasinevicius R.G, Pizani P.S (2000) Mapping the three-dimensional strain field around a microindentation on silicon using polishing and Raman spectroscopy. J. Appl. Phys. 88: 4582-4585.

[14] Hanbücken M, Müller P, Wehrspohn R.B (Eds.) (2011) Mechanical Stress on the Nanoscale: Simulation, Material Systems and Characterization Techniques. Wiley-VCH.

[15] Fano U (1961) Effects of Configuration Interaction on Intensities and Phase Shifts. Phys. Rev. 124: 1866- 1878.

[16] Nickel N.H, Lengsfeld P, Sieber I (2000) Raman Spectroscopy of Heavily Doped Polycrystalline Silicon Thin Film. Phys. Rev. B 61: 15558-15561.

[17] Lengsfeld P, Brehme S, Brendel K, Genzel Ch, Nickel N.H (2003) Raman Spectroscopy of Heavily Doped Polycrystalline and Microcrystalline Silicon. phys. stat. sol. (b) 235: 170–178.

[18] Weber W.H, Merlin R (Eds.) (2000) Raman Scattering in Materials Science. Springer. pp. 56-64.

[19] Debernardi A, Baroni S, Molinari E (1995) Anharmonic Phonon Lifetimes in Semiconductors from Density-Functional Perturbation Theory, Phys. Rev. Lett. 75: 1819–1822.

[20] Debernardi A (1998) Phonon linewidth in III-V semiconductors from density-functional perturbation theory, Phys. Rev. B 57: 12847.

[21] Iqbal Z, Veprek S, Webb A.P, Capezzuto P (1981) Raman Scattering from Small Particle Size Polycrystalline Silicon, Solid State Commun. 37: 993-996.

[22] Secco D' Aragona F (1972) Dislocation Etch for (100) Planes in Silicon, J. Electrochem. Soc. 119: 948-951.

[23] Sarau G, Becker M, Christiansen S, Holla M, Seifert W (2009) Micro-Raman Mapping of Residual Stresses at Grain Boundaries in Multicrystalline Block-Cast Silicon Solar Cell Material: Their Relation to the Grain Boundary Microstructure and Recombination Activity, Proc. 24th European Photovoltaic Solar Energy Conference: 969-973.

[24] Sarau G, Becker M, Berger A, Schneider J, Christiansen S (2007) Stress Distribution in Polycrystalline Silicon Thin Film Solar Cells on Glass Measured by Micro-Raman Spectroscopy, Mater. Res. Soc. Symp. Proc. 1024E: 1024-A07-04.

[25] Grimmer H, Bollmann W, Warrington D.H (1974) Coincidence-Site Lattices and Complete Pattern-Shift in Cubic Crystals, Acta Cryst. A30: 197-207.

[26] Niederberger Ch, Michler J, Jacot A (2008) Origin of Intragranular Crystallographic Misorientations in Hot-Dip Al–Zn–Si Coating, Acta Materialia 56: 4002–4011.

[27] Hull D, Bacon D.J (2001) Introduction to Dislocations. Butterworth-Heinemann. Chapters 4 and 9.

Chapter 9

Stress Measurements in Si and SiGe by Liquid-Immersion Raman Spectroscopy

Daisuke Kosemura, Motohiro Tomita, Koji Usuda and Atsushi Ogura

Additional information is available at the end of the chapter

http://dx.doi.org/10.5772/47831

1. Introduction

Strained Si technology is important for engineering field-effect transistors (FETs) [1,2]. There are two types of the strained Si technologies. One is so-called global strained Si technology. Another is so-called local strained Si technology. The former is the technology of using a strained Si substrate which has a several-dozen-nanometers-thick strained Si layer at the top of the substrate [3-5]. The strained Si layer is obtained by growing Si on SiGe, therefore, large tensile strain with biaxial isotropy can be induced in Si. The isotropic biaxial tensile strain in Si allows for performance improvements for both of n- and p- type FETs. Homogeneous strain distribution can be obtained under the critical thickness of the strained Si layer [6].

In the latter case, the strain is induced only in the desired region, the channel region of FET [7,8]. A SiN film is used as the stressor that can induce tensile or compressive uniaxial stress in Si by changing the deposition conditions of the SiN film [9,10]. The uniaxial tensile strain enhances electron mobility, while the uniaxial compressive strain enhances hole mobility. Various kinds of the local strained Si techniques have so far been suggested by many researchers [11,12]. The combination of the global and local strained Si technologies is considered effective to induce extremely large strain in Si. Fin-type structures have been reported for high-performance FETs [13]. It is considered that the stress relaxation occurs during the fabrication of the fin-shaped strained Si layer. There are many other origins of strain fluctuations, e.g., shallow trench isolation (STI), metal gate electrodes, silicide, interconnections, and its layout. As a result, the stress states in the future generation FETs become complicated. The relationship between the electrical properties of FETs and the strain is also complicated. Therefore, to measure the complicated stress states in Si has great demand in order to improve the FET performance effectively.

Several kinds of strain or stress measurements have been studied, e.g., X-ray diffraction (XRD), transmission electron microscopy (TEM), electron backscattering diffraction (EBSD), and Raman spectroscopy [4,14-16]. Among them, Raman spectroscopy has the advantages such as high sensitive to local strain, submicron spatial resolution, nondestructive measurements, fast measurements, and ease of use. Consequently, Raman spectroscopy has been frequently used by many researchers to measure the strain in Si [3,7-9,17-21]. However, conventional Raman spectroscopy fails to measure the complicated stress states in Si. The reason is as follows. Backscattering geometry from a (001) Si substrate is generally used in Raman measurements of strained Si. In this geometry, only one of three optical phonon modes is Raman active, while two of three modes are Raman inactive. The limitation arises from the extremely high symmetry of the Si crystal. As a result, the weighted average value of the complicated stress state is obtained, that is, it is impossible to perform quantitative measurements of strain by conventional Raman spectroscopy.

Si has three optical phonon modes: one longitudinal optical (LO) phonon mode and two transverse optical (TO) phonon modes, the polarizations of which are parallel and perpendicular to the phonon wave vector, respectively. Recently, the forbidden optical phonon modes, the TO phonon modes, were excited even under the (001) Si backscattering geometry, using a high-numerical aperture (NA) liquid-immersion lens [22-24]. If all of the three optical phonon modes are detectable, the unknown three components of a stress tensor in Si can be obtained in theory [25-35]. The high-NA liquid-immersion Raman spectroscopy has great potential for measuring the complicated stress states in Si with high spatial resolution.

On the other hand, the number of stress tensor components is six. Therefore, the evaluation of nondiagonal stress components, shear stress components, is considered difficult even detecting the TO phonon modes. The shear stress is often generated at the discontinuous region, e.g., around STI and at the edge of a contact etch stop layer. The shear stress often produces dislocations in Si, which cause leakage current during transistor operation [36,37]. The shear stress measurements are desired for failure analysis. The induction of the stress with the nondiagonal components requires the transformation of the Raman tensors. Therefore, Raman spectroscopy is essentially sensitive to the shear stress.

In this study, anisotropic stress states in Si were measured by the high-NA liquid-immersion Raman spectroscopy. Strained SiGe was also measured by the same technique. SiGe has been suggested as the channel material of next generation FETs, because the both mobilities of electrons and holes in SiGe are higher than those in Si. Furthermore, the strain induction in SiGe is considered effective for improving electrical properties of SiGe FETs in the same way as strained Si [38-41]. The nondiagonal stress components, shear stress components, were measured by analyzing the dependence of Raman spectra on the relative polarization direction between sample orientation and electrical fields of incident and scattered light.

2. Experimental procedure

2.1. Excitation of TO phonon modes

The Raman intensity is calculated by the following equation [42];

$$I \propto \sum_j \left| e_s{}^T R_j e_i \right|^2 , \tag{1}$$

$$R_1 = \begin{pmatrix} 0 & 0 & 0 \\ 0 & 0 & d \\ 0 & d & 0 \end{pmatrix}, R_2 = \begin{pmatrix} 0 & 0 & d \\ 0 & 0 & 0 \\ d & 0 & 0 \end{pmatrix}, R_3 = \begin{pmatrix} 0 & d & 0 \\ d & 0 & 0 \\ 0 & 0 & 0 \end{pmatrix}, \tag{2}$$

where R_j is the Raman tensors of Si [43]. e_i and e_s are the electrical fields of incident and scattered light, respectively. The superscript T denotes transpose. From Eqs. (1) and (2), the TO phonon modes are not excited under the (001) Si backscattering geometry. This is because the component of z polarization of the incident light is reduced to almost zero in the case of the (001) Si backscattering configuration. The z polarization is thus needed to excite the TO phonon modes in Si. It is considered that oblique light relative to the (001) Si surface gives rise to the z polarization.

Fig. 1 shows the experimental set-up for oblique incident light configuration in this study. The glancing angles of the laser against the sample were 30° and 90°, as shown in Fig. 1. Fig. 2 shows the examples of calculations for the Raman intensities in the 90° and 30° configurations, using Eqs. (1) and (2). In the case of the 90° configuration, the (001) Si backscattering configuration, the TO phonon modes are Raman inactive and the LO phonon mode is Raman active, as mentioned above. On the other hand, in the case of the 30° configuration, the oblique incident light configuration, the TO phonon mode is Raman active. This fact arises because the z polarization of the incident light can be obtained in the oblique incident light configuration. In the rough approximations, the Raman intensities are considered to be the same for the 90° and 30° configurations, as shown in Fig. 2.

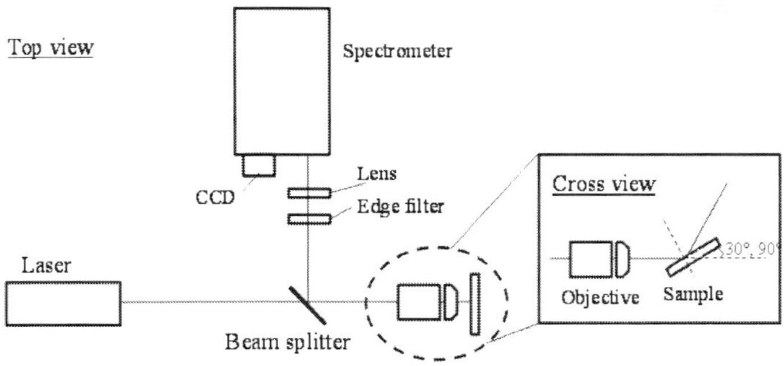

Figure 1. Experimental set-up for oblique incident light configuration

90° configuration

R_1: $$(1\ 0\ 0)\cdot\begin{pmatrix}0&0&0\\0&0&d\\0&d&0\end{pmatrix}\cdot\begin{pmatrix}0\\1\\0\end{pmatrix}=0$$

R_2: $$(1\ 0\ 0)\cdot\begin{pmatrix}0&0&d\\0&0&0\\d&0&0\end{pmatrix}\cdot\begin{pmatrix}0\\1\\0\end{pmatrix}=0$$

R_3: $$(1\ 0\ 0)\cdot\begin{pmatrix}0&d&0\\d&0&0\\0&0&0\end{pmatrix}\cdot\begin{pmatrix}0\\1\\0\end{pmatrix}=d$$

$$\boxed{I=d^2}$$

30° configuration

R_1: $$(\cos\theta\ \ 0\ \ \sin\theta)\cdot\begin{pmatrix}0&0&0\\0&0&d\\0&d&0\end{pmatrix}\cdot\begin{pmatrix}0\\1\\0\end{pmatrix}=d\sin\theta$$

R_2: $$(\cos\theta\ \ 0\ \ \sin\theta)\cdot\begin{pmatrix}0&0&d\\0&0&0\\d&0&0\end{pmatrix}\cdot\begin{pmatrix}0\\1\\0\end{pmatrix}=0$$

R_3: $$(\cos\theta\ \ 0\ \ \sin\theta)\cdot\begin{pmatrix}0&d&0\\d&0&0\\0&0&0\end{pmatrix}\cdot\begin{pmatrix}0\\1\\0\end{pmatrix}=d\cos\theta$$

$$\boxed{I=d^2}$$

Figure 2. Calculations of Raman intensities for 90° and 30° configurations.

The intensity ratio of the TO phonon mode to the LO phonon mode in the 30° configuration is calculated to be;

$$\frac{I_{TO}}{T_{LO}}=\frac{|d\sin\theta|^2}{|d\cos\theta|^2}\approx 0.046\,, \tag{3}$$

where θ is the aperture angle in Si. In the 30° configuration, θ is approximately 12.13° in the case of λ = 532 nm laser because Si has the large refraction index [44]. Therefore, the intensity of the TO phonon mode are much small, compared to that of the LO phonon mode even in the 30° configuration. The detection of the TO phonon modes is basically considered difficult. Moreover, for the oblique incident light configuration, high-resolution measurements cannot be achieved because it is difficult to use the high-NA lens and the beam spot becomes an ellipse. In this study, the high-NA liquid-immersion lens was used in order to obtain the oblique light relative to the (001) Si surface.

An aperture angle θ is calculated by NA = $n\times\sin\theta$ (where n is a refractive index). θ is equal to 44.4° in conventional Raman spectroscopy with the use of NA = 0.7 objective (n = 1.0). On the other hand, θ is equal to 69.0° in high-NA liquid-immersio Raman spectroscopy with the NA = 1.4 liquid-immersio lens (n = 1.5). However, the incident light widely refracts at the interface of the Si surface because Si has the large refractive index as mentioned above. The refractive index of Si for the λ = 364 nm light (where λ is wavelength) is approximately 6.5 [44]. Therefore, θ in Si results in 6.2° in conventional Raman spectroscopy (NA = 0.7). This configuration is almost under the (001) Si backscattering geometry, i.e., the component of the z polarization is reduced to almost zero. This fact causes that the TO phonon modes are Raman inactive in conventional Raman spectroscopy. On the other hand, θ in Si results in 12.4° in high-NA liquid-immersio Raman spectroscopy (NA = 1.4). It is considered that the value of θ in Si is still small to excite the TO phonon modes effectively, although the value is two times larger than that in conventional Raman spectroscopy. It is considered that the use of the UV light has the drawback for the excitation of the TO phonon modes. In the case of visible light (λ = 532 nm), θ in Si are calculated to be 9.8° and 19.8° in conventional and liquid-immersion Raman spectroscopy, respectively. The value for oil-immersion Raman spectroscopy with the use of the visible light is relatively large, therefore, the large component of the z polarization is obtained. Table 1 shows θs as a function of NA. θ_1 and θ_2 are the aperture angles in the medium and Si, respectively.

NA	With use of UV light (λ = 364 nm)			With use of visible light (λ = 532 nm)		
	θ_1	θ_2	TO/LO	θ_1	θ_2	TO/LO
0.7	44.4	6.2	0.06	44.4	9.8	0.25
1.1	57.8	9.7	0.2	57.8	15.5	1.0
1.2	67.4	10.6	0.3	67.4	16.9	1.3
1.4	69.0	12.4	0.5	69.0	19.8	2.0
1.7	-----	-----	-----	70.8	24.4	3.8

Table 1. θs with use of visible and UV light and intensity ratio of TO to LO phonon modes as a function of NA.

It is important to choose the appropriate NA and the wavelength for the excitation of the TO phonon modes. The intensity of the TO phonon mode excited by high-NA liquid-immersion Raman spectroscopy is estimated as follows. The TO phonon modes are excited mainly by the marginal ray of incident light. The Raman intensity can be calculated by the following equation [45]:

$$S_\Omega = A\sum_j \int_{\Omega_s}\int_{\Omega_i} \left(e_i R_j e_s\right)^2 d\Omega_i d\Omega_s, \tag{4}$$

where Ω_i and Ω_s are the solid angles of incident and scattered light, respectively. The intensity ratio of the TO to LO phonon modes is considered to be the intensity ratio of the z component of the marginal ray to the paraxial ray. Fig. 3 shows the intensity ratio of TO to LO phonon modes as a function of NA with the use of visible light. The aperture angle dependence on NA is also shown in Fig. 3.

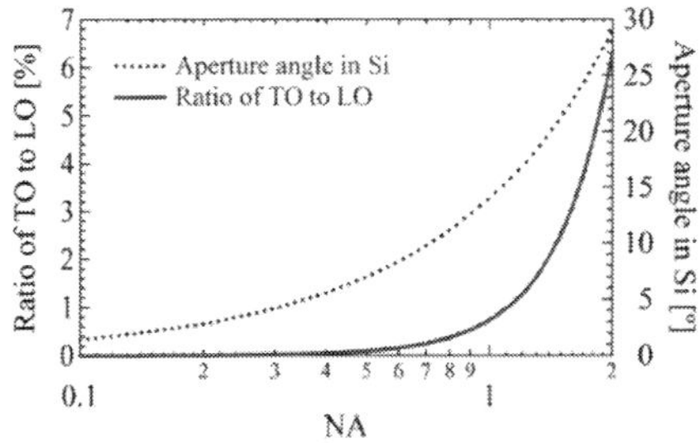

Figure 3. Intensity ratio of TO to LO phonon modes and aperture angle in Si vs. NA

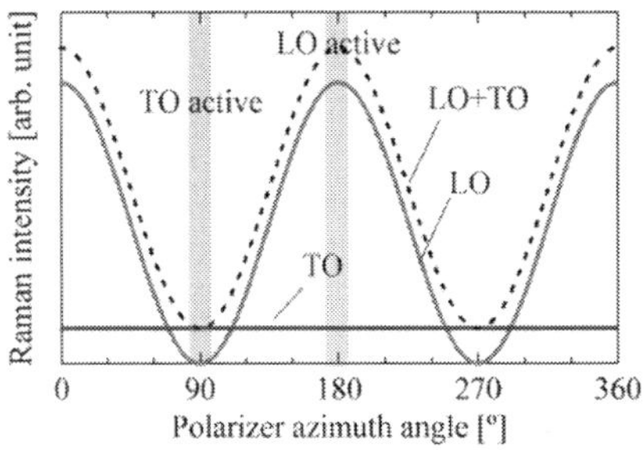

Figure 4. Raman intensities of LO and TO phonon modes vs polarizer azimuth angle

The intensity ratio of the TO to LO phonon modes as well as the aperture angle in Si increases with the increase in NA. Note that drastic increase is confirmed especially beyond NA = 1.0 for the intensity ratio of TO to LO. The value is approximately 2.0% for NA = 1.4 with the use of visible light, while the value is approximately 0.5% with the use of UV light, as shown in Table 1. From the estimations, the intensity of the TO phonon mode is very low, compared to that of the LO phonon mode. Actually, it is important to suppress the intensity of the LO phonon mode for the excitation of the TO phonon modes. This can be accomplished by the Raman polarization selection rules [42].

Fig. 4 shows the Si Raman intensities of the LO and TO phonon modes calculated by Eq. (1) as a function of a polarizer azimuth angle. For the LO phonon mode, the intensity changes in the period of 180°. On the other hand, for the TO phonon mode, the intensity is independent on the polarizer azimuth angle (the value is exaggerated for ease to view). This is because the component of the z polarization obtained by the oblique light remains constant all over the angles. As a result, the measurable Raman intensity profile is the sum of the intensities of the LO and TO phonon modes, which is shown by the dashed line in Fig. 4. From Fig. 4, the LO phonon mode can be detected at the polarizer azimuth angle of 0°, 180°, and 360°. These correspond to the LO active configurations. On the other hand, the TO phonon mode can be detected at the angle of 90° and 270°. These correspond to the TO active configurations. The TO and LO phonon modes can be separately detected by the Raman polarization selection rules. It is possible to evaluate complicated stress states in Si by analyzing multi-optical phonon modes.

2.2. Methodology of measurements for anisotropic biaxial stress states in Si

In this section, the methodology of measurements for anisotropic biaxial stress states in Si by liquid-immersion Raman spectroscopy is shown. In the previous section, it was shown that the z polarization can be created by the oblique light due to the high-NA liquid-immersion lens. Consequently, the TO phonon modes in Si can be excited by the z polarization even under the (001) Si backscattering geometry. The TO phonon modes allow for the measurements of the anisotropic biaxial stress states in Si.

The force constant of a Si crystal changes by the induction of strain. Consequently, the optical-phonon frequencies also change. The difference of the force constant ΔK is represented as a second-rank tensor. The eigenvalues of ΔK after the induction of the strain can be obtained by solving the secular equation [29]. The square roots of the eigenvalues correspond to the changes of the optical-phonon frequencies (the Raman wavenumber shifts). Three equations between the strain (stress) and the Raman wavenumber shifts are obtained because Si has three optical-phonon modes (two TOs and one LO).

Suppose that there is a linear relationship between ΔK and strain [26]. In a coordinate system of x: [100], y: [010], and z: [001], ΔK is represented as the following equation:

$$\Delta K = A\varepsilon , \tag{5}$$

$$A = \begin{pmatrix} p & q & q & 0 & 0 & 0 \\ q & p & q & 0 & 0 & 0 \\ q & q & p & 0 & 0 & 0 \\ 0 & 0 & 0 & 2r & 0 & 0 \\ 0 & 0 & 0 & 0 & 2r & 0 \\ 0 & 0 & 0 & 0 & 0 & 2r \end{pmatrix} , \tag{6}$$

where ε is a strain tensor. A is a fourth-rank tensor whose components are p, q, and r called phonon deformation potentials (PDPs). Generally, transistors are fabricated on (001) Si substrate in the direction of [110] Si. Therefore, the coordinate transformation makes analysis easy [46]. Second-rank and fourth-rank tensors are transformed in the coordinate system of x' = [110], y' = [−110], and z' = [001] by the following equations:

$$T_{ij} = a_{ik} a_{jl} T_{kl} , \tag{7}$$

$$T_{ijkl} = a_{im} a_{jn} a_{ko} a_{lp} T_{mnop} , \tag{8}$$

$$a = \begin{pmatrix} 1/\sqrt{2} & 1/\sqrt{2} & 0 \\ -1/\sqrt{2} & 1/\sqrt{2} & 0 \\ 0 & 0 & 1 \end{pmatrix} , \tag{9}$$

where T and a are a second- or fourth- rank tensor and a transformation matrix, respectively. Hence, Eq. (5) results in:

$$\Delta K' = A'\varepsilon' , \tag{10}$$

where the primes denote the components in the coordinate $x'y'z'$. The secular equation of $\Delta K'$ is below:

$$\begin{vmatrix} \Delta K_{xx}'-\lambda & \Delta K_{xy}' & \Delta K_{xz}' \\ \Delta K_{xy}' & \Delta K_{yy}'-\lambda & \Delta K_{yz}' \\ \Delta K_{xz}' & \Delta K_{yz}' & \Delta K_{zz}'-\lambda \end{vmatrix} = 0 , \tag{11}$$

where λ is the eigenvalues. An anisotropic biaxial stress state is represented as the following second-rank tensor:

$$\sigma' = \begin{pmatrix} \sigma_{xx}' & 0 & 0 \\ 0 & \sigma_{yy}' & 0 \\ 0 & 0 & 0 \end{pmatrix} , \tag{12}$$

where σ_{xx}' and σ_{yy}' are the stress components in the directions of [110] and [−110], respectively. Generally, stress induction changes not only optical-phonon frequencies but also Raman tensors. However, in the case of the stress tensors only with the diagonal components, there are no changes of the Raman tensors. Therefore, the Raman polarization selection rules after the induction of the biaxial stresses σ_{xx} and σ_{yy} remains as those of stress-free Si [46]. The Raman tensors in the coordinate $x'y'z'$ are as follows [46]:

$$R_1' = \frac{1}{\sqrt{2}} \begin{pmatrix} 0 & 0 & d \\ 0 & 0 & d \\ d & d & 0 \end{pmatrix} , \quad R_2' = \frac{1}{\sqrt{2}} \begin{pmatrix} 0 & 0 & d \\ 0 & 0 & -d \\ d & -d & 0 \end{pmatrix} , \quad R_3' = \begin{pmatrix} d & 0 & 0 \\ 0 & -d & 0 \\ 0 & 0 & 0 \end{pmatrix} \tag{13}$$

Stress tensors are transformed to strain tensors by Hooke's law:

$$\varepsilon' = S'\sigma' , \tag{14}$$

$$S = \begin{pmatrix} S_{11} & S_{12} & S_{12} & 0 & 0 & 0 \\ S_{12} & S_{11} & S_{12} & 0 & 0 & 0 \\ S_{12} & S_{12} & S_{11} & 0 & 0 & 0 \\ 0 & 0 & 0 & S_{44}/4 & 0 & 0 \\ 0 & 0 & 0 & 0 & S_{44}/4 & 0 \\ 0 & 0 & 0 & 0 & 0 & S_{44}/4 \end{pmatrix} , \tag{15}$$

where S expressed by Eq. (15) is the elastic compliance tensor. The components of S, S_{11}, S_{12}, and S_{44} are 7.68 × 10^{-12}, −2.14 × 10^{-12}, and 12.7 × 10^{-12} 1/Pa, respectively [17]. The transformation of the fourth-rank tensor S by Eq. (8) is needed. The strain tensor ε' expressed by Eq. (14) is substituted for Eq. (11) and then the eigenvalues λ_i are calculated. As a result, using Eq. (16), the relationship between the Raman wavenumber shifts $\Delta\omega$s and the anisotropic biaxial stresses σ_{xx}' and σ_{yy}' are obtained as follows [35]:

$$\lambda_i = \omega_i^2 - \omega_0^2 = (\omega_i + \omega_0)(\omega_i - \omega_0) \approx 2\omega_0(\omega_i - \omega_0) , \quad \Delta\omega_i \approx \frac{\lambda_i}{2\omega_0} . \tag{16}$$

$$\Delta\omega_1 = \frac{\lambda_1}{2\omega_0} = \frac{1}{2\omega_0}\left[\frac{1}{2}p\left(S_{11}+S_{12}\right)+\frac{1}{2}q\left(S_{11}+3S_{12}\right)+\frac{1}{2}rS_{44}\right]\times\sigma_{xx}' + \frac{1}{2\omega_0}\left[\frac{1}{2}p\left(S_{11}+S_{12}\right)+\frac{1}{2}q\left(S_{11}+3S_{12}\right)-\frac{1}{2}rS_{44}\right]\times\sigma_{yy}', \tag{17-1}$$

$$\Delta\omega_2 = \frac{\lambda_2}{2\omega_0} = \frac{1}{2\omega_0}\left[\frac{1}{2}p\left(S_{11}+S_{12}\right)+\frac{1}{2}q\left(S_{11}+3S_{12}\right)-\frac{1}{2}rS_{44}\right]\times\sigma_{xx}' + \frac{1}{2\omega_0}\left[\frac{1}{2}p\left(S_{11}+S_{12}\right)+\frac{1}{2}q\left(S_{11}+3S_{12}\right)+\frac{1}{2}rS_{44}\right]\times\sigma_{yy}', \tag{17-2}$$

$$\Delta\omega_3 = \frac{\lambda_3}{2\omega_0} = \frac{1}{2\omega_0}\left[pS_{12}+q\left(S_{11}+S_{12}\right)\right]\times\left(\sigma_{xx}'+\sigma_{yy}'\right), \tag{17-3}$$

$$\Delta\omega_3 = \frac{\lambda_3}{2\omega_0} = \frac{1}{\omega_0}\left[\frac{S_{12}}{S_{11}+S_{12}}p+q\right]\times\varepsilon_{biaxial} = b\times\varepsilon_{biaxial}, \tag{18}$$

where b is so-called the b coefficient which is used for the evaluation of isotropic biaxial strain $\varepsilon_{biaxial}$ in strained Si substrates using the Raman wavenumber shift of the LO phonon mode $\Delta\omega_3$ [21,47].

Authors (year)	Sample	p/ω_0^2, q/ω_0^2, r/ω_0^2	b cm^{-1}	Citation
Anastassakis *et al.* (1970)[a]	Si bar	−1.25, −1.87, −0.66	−721	4 (9%)
Chandrasekhar *et al.* (1978)[b]	Si bar	−1.43, −1.89, −0.59	−696	18 (40%)
Anastassakis *et al.* (1990)[c]	Si bar	−1.85, −2.31, −0.71	−830	15 (33%)
Nakashima *et al.* (2006)[d]	Strained Si substrate	----------	−723	1 (2%)
JEITA (2007)[e]	Strained Si substrate	----------	−769	1 (2%)
Others[f-j]	Strained Si substrate	----------	−1040 ~ −715	6 (13%)

[a]Reference 14, [b]Reference 15, [c]Reference 16, [d]Reference 12, [e]Reference 39, [f-j]Reference 46, 47, 71, 78, and 79.

Table 2. Various PDPs suggested so far and statistics of citations.

Various PDPs have so far been suggested by many researchers. The suggested PDPs and the citation count are shown in Table 2. Forty-five papers were confirmed. As shown in Table 2, the values of PDPs are fluctuated. Thirty-seven of forty-five papers, approximately eighty-two percent papers, referred PDPs suggested by the Cardona's group in 1970-1990. Nakashima *et al.* examined the b coefficient in detail using strained Si substrates by Raman spectroscopy and high-resolution XRD in 2006 [21]. Furthermore, the detailed investigation of the b coefficient was performed in the working group of Japan electronics and information technology industries association (JEITA) in 2007 [48]. Eight organizations attended the working group: three companies for XRD measurements, three companies and one University for Raman measurements, and one company for Rutherford back scattering (RBS) measurements. The b coefficient of −769 cm^{-1} was obtained [16]. Extreme care is needed to choose appropriate PDPs.

In this study, the validity of three sets of PDPs was evaluated by liquid-immersion Raman spectroscopy: first, $p/\omega_0^2 = -1.25$, $q/\omega_0^2 = -1.87$, and $r/\omega_0^2 = -0.66$ reported by Anastassakis *et al.* in 1970 [49], second, $p/\omega_0^2 = -1.43$, $q/\omega_0^2 = -1.89$, and $r/\omega_0^2 = -0.59$ reported by Chandrasekhar *et al.* in 1978 [50], and third, $p/\omega_0^2 = -1.85$, $q/\omega_0^2 = -2.31$, and $r/\omega_0^2 = -0.71$ reported by Anastassakis *et al.* 1990 [51]. The first set of PDPs was obtained from the first investigation. The second set appears to be the most commonly used, and the third set is the most recently reported among the three sets of PDPs.

The relationship between Raman wavenumber shifts $\Delta\omega$s and the anisotropic biaxial stresses σ_{xx}' and σ_{yy}' are obtained by substituting PDPs shown above for Eq. (17). When PDPs reported by Anastassakis *et al.* in 1970 are used,

$$\Delta\omega_1 = -2.30 \times \sigma_{xx}' - 0.12 \times \sigma_{yy}', \tag{19-1}$$

$$\Delta\omega_2 = -0.12 \times \sigma_{xx}' - 2.30 \times \sigma_{yy}', \tag{19-2}$$

$$\Delta\omega_3 = -2.00 \times \sigma_{xx}' - 2.00 \times \sigma_{yy}'. \tag{19-3}$$

When PDPs reported by Chandrasekhar *et al.* in 1978 are used,

$$\Delta\omega_1 = -2.31 \times \sigma_{xx}' - 0.37 \times \sigma_{yy}', \tag{20-1}$$

$$\Delta\omega_2 = -0.37 \times \sigma_{xx}' - 2.31 \times \sigma_{yy}', \tag{20-2}$$

$$\Delta\omega_3 = -1.93 \times \sigma_{xx}' - 1.93 \times \sigma_{yy}'. \tag{20-3}$$

When PDPs reported by Anastassakis *et al.* in 1990 are used,

$$\Delta\omega_1 = -2.88 \times \sigma_{xx}' - 0.54 \times \sigma_{yy}', \tag{21-1}$$

$$\Delta\omega_2 = -0.54 \times \sigma_{xx}' - 2.88 \times \sigma_{yy}', \tag{21-2}$$

$$\Delta\omega_3 = -2.30 \times \sigma_{xx}' - 2.30 \times \sigma_{yy}'. \tag{21-3}$$

2.3. Methodology of measurements for nondiagonal stress components

In the case of the induction of stress with the only diagonal stress components, strain-modified phonon eigenvectors ξ_i's which are obtained by solving the secular equation expressed by Eq. (11) coincide with the coordinate $x'y'z'$. In this case, the Raman tensors of Si remains in the same form expressed by Eq. (13). On the other hand, shear stress causes a deviation between the phonon wave vector and ξ_i's, i.e., in the case of the induction of stress with the nondiagonal stress components, ξ_i's no longer coincide with the coordinate $x'y'z'$ [52]. The difference between ξ_i' and the coordinate $x'y'z'$ requires a change of the Raman tensors. The new Raman tensors $R_i'^*$ is expressed by:

$$R_i'^* = (\xi_i'^* \cdot \xi_1')R_1' + (\xi_i'^* \cdot \xi_2')R_2' + (\xi_i'^* \cdot \xi_3')R_3', \tag{22}$$

where $\xi_i'^*$ and ξ_i' are the strain-modified eigenvectors for the introduction of stress with the nondiagonal and only diagonal stress components, respectively. Assuming the stress tensor shown by Eq. (12), the Raman tensors R_i' changes to $R_i'^*$:

$$R_1'^* = \begin{pmatrix} \times & 0 & \times \\ 0 & \times & \times \\ \times & \times & 0 \end{pmatrix},\ R_2'^* = \begin{pmatrix} 0 & 0 & \times \\ 0 & 0 & \times \\ \times & \times & 0 \end{pmatrix},\ R_3'^* = \begin{pmatrix} \times & 0 & \times \\ 0 & \times & \times \\ \times & \times & 0 \end{pmatrix}, \tag{23-1)-(23-3}$$

where × indicates nonzero components (each value is not always the same), some of which depend on the eigenvalues obtained by solving the secular equation of Eq. (11). $R_2'^*$ has the same form as the Raman tensor R_2' because of $\tau_{xy} = \tau_{yz} = 0$. Therefore, R_2' corresponds to the TO phonon mode with the eigenvector in the y' direction. On the other hand, $R_1'^*$ and $R_3'^*$ no longer correspond to purely transverse and longitudinal modes, respectively, because their eigenvectors do not coincide with the x' and z' axes, respectively. Consequently, the Raman intensity is changed by the nondiagonal stress components obeying the Raman polarization selection rules. The nondiagonal stress components can be evaluated by analyzing the dependence of the Raman spectra on the relative polarization direction between the sample orientation and the electrical fields of incident and scattered light. The methodology is described as follows.

The methodology for evaluating complicated stress states was reported by Ossikovski *et al* [33]. They employed an experimental configuration that used oblique incident light to observe the forbidden modes, i.e., the TO phonon modes. In our experiments, the high-NA liquid-immersion lens was used to observe the TO phonon modes. High spatial resolution was preserved in liquid-immersion Raman spectroscopy.

First, a stress tensor is considered, and then the strain tensor is calculated by Hooke's law by Eq. (14). The strain tensor is substituted for the secular equation of Eq. (11). Three phonon eigenfrequencies of Si have been determined so far. Si is a nonpolar cubic crystal, that is, there is no difference between the TO phonon modes and LO phonon mode [52]. As a result, the three determined phonon eigenfrequencies are independent of the phonon wave vector. Second, the Raman tensors are determined using Eq. (22). Their forms change when the nondiagonal stress components are nonzero. Subsequently, the Raman intensity of each phonon mode is calculated by the Raman polarization selection rules given by Eq. (1). Third, each Raman spectrum is considered to be a Lorentzian function $\Lambda_i(\omega)$ [35]:

$$\Lambda_i(\omega) = \frac{I_i \Gamma^2}{(\omega - \Omega_i)^2 + \Gamma^2}, \tag{24}$$

where ω, Ω_i, and Γ are the Raman shift, the phonon eigenfrequencies , and the half width at half maximum of the spectrum, respectively. It is considered difficult to analyze each spectrum of the TO and LO phonon modes. Therefore the effective phonon eigenfrequency Ω_{eff} is used as a representative value. The effective value is the weighted average of the phonon eigenfrequencies relative to their intensities, as expressed by the following equation [26]:

$$\Omega_{eff} = \sum_i \frac{I_i \Omega_i}{I_T}, \tag{25}$$

where I_T is the total intensity of the three phonon modes. Eq. (25) is valid because the strain-induced splitting between the TO and LO phonon modes is small, compared to Γ. An example of a spectrum with the effective phonon eigenfrequency and the spectra of the TO and LO phonon modes are shown in Fig. 5. A uniaxial stress σ_{xx}' of 1.0 GPa is assumed in the calculation. In Fig. 5, the Raman spectra with the eigenfrequencies of Ω_1, Ω_2, and Ω_3 appear, which correspond to the optical phonon modes with the eigenvectors x', y', and z', respectively. The dashed line shows the Raman spectrum with the weighted average eigenfrequency. The Raman signal of the TO phonon modes with the eigenvectors x' and y' are obtained by the z polarization due to the high-NA lens (the component of z polarization is enlarged for ease to view).

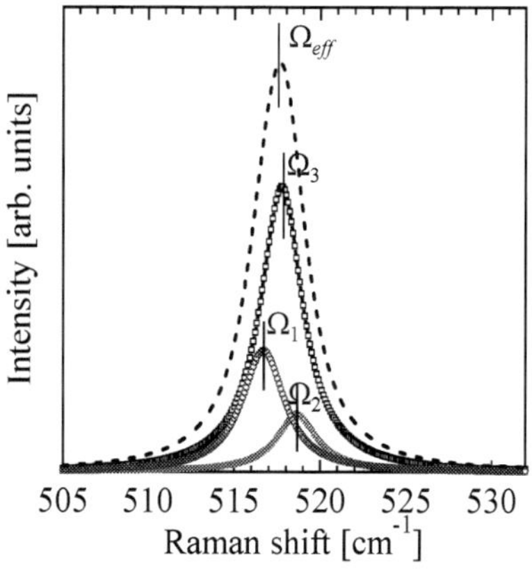

Figure 5. Raman spectrum with effective phonon frwuency and raman spectra with Ω_1, Ω_2, and Ω_3

Finally, the dependence of the Raman spectra on the polarization direction between the sample orientation and the electrical fields of incident and scattered light is obtained as follows. The electrical fields of incident and scattered light are fixed in the y' direction. Regarding e_i, because the high-NA liquid-immersion lens is used, the z polarization can be obtained:

$$e_i = \frac{1}{\sqrt{1+\alpha^2}} \begin{pmatrix} 0 \\ 1 \\ \alpha \end{pmatrix}, \tag{26}$$

where α is the component of the z polarization. For $\alpha = 0$, this is correct in the (001) Si backscattering geometry. α was experimentally determined. Eq. (1) shows that for the Raman intensity, the rotations of the polarization directions of incident and scattered light are equivalent to the rotation of the sample, although the period for the sample rotation is half compared to those for the polarization rotations. In the experiments, the sample was rotated from 0° to 180°, which is represented by the following equations:

$$R_i'^{*}{}_{rot} = T R_i'^{*} T, \tag{27}$$

$$T(\phi) = \begin{pmatrix} \cos\varphi & \sin\varphi & 0 \\ -\sin\varphi & \cos\varphi & 0 \\ 0 & 0 & 1 \end{pmatrix}, \tag{28}$$

where $R_i'^{*}{}_{rot}$ and T are the Raman tensors after rotation by φ and the transformation matrix, respectively.

Fig. 6 shows the dependence of the effective Raman shifts on the sample rotation angle for the various stress states including hydrostatic stress, uniaxial stress, biaxial stress, and stress with the nondiagonal components, which are represented by:

$$\sigma_{hydrostatic} = \begin{pmatrix} 0.33 & 0 & 0 \\ 0 & 0.33 & 0 \\ 0 & 0 & 0.33 \end{pmatrix}, \quad \sigma_{uniaxial} = \begin{pmatrix} 1.0 & 0 & 0 \\ 0 & 0 & 0 \\ 0 & 0 & 0 \end{pmatrix}, \tag{29) and (30}$$

$$\sigma_{biaxial} = \begin{pmatrix} 0.5 & 0 & 0 \\ 0 & 0 & 0 \\ 0 & 0 & 0.5 \end{pmatrix}, \quad \sigma_{shear} = \begin{pmatrix} 0.5 & 0 & 0.5 \\ 0 & 0 & 0 \\ 0.5 & 0 & 0.5 \end{pmatrix}. \tag{31) and (32}$$

The above stress states correspond to the load of 1.0 GPa. A unique profile can be obtained for each stress state, as shown in Fig. 6. The profile for the hydrostatic stress remains constant all over the sample rotation angles because the degeneracy of the long-wavelength optical phonons does not lift under the hydrostatic stress. It should be noted that the profile becomes asymmetric only for the stress state with the nondiagonal components. As a result, the shear stress in Si is considered to be detectable by analyzing the dependence of the effective Raman shifts on the sample rotation angle.

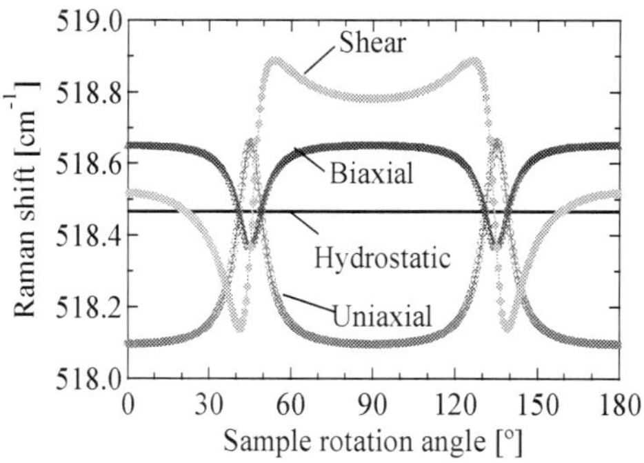

Figure 6. Effective Raman shift dependence on sample rotation angle for hydrostatic, uniaxial, biaxial, and shear stress.

2.4. Samples

(001)-oriented SSOI substrates were used as the samples [53,54]. Fig. 7(a) shows the cross sectional TEM image of the SSOI substrate. The structure of SSOI was strained Si layer/buried oxide (BOX) layer/Si substrate, which is the simplest structure among the strained Si substrates. The low-power consumption operation can be achieved due to the structure of Si on insulator (SOI) [55,56]. The thicknesses of the strained Si layers were 30, 50, and 70 nm. An isotropic biaxial tensile stress state exists in the strained Si layer.

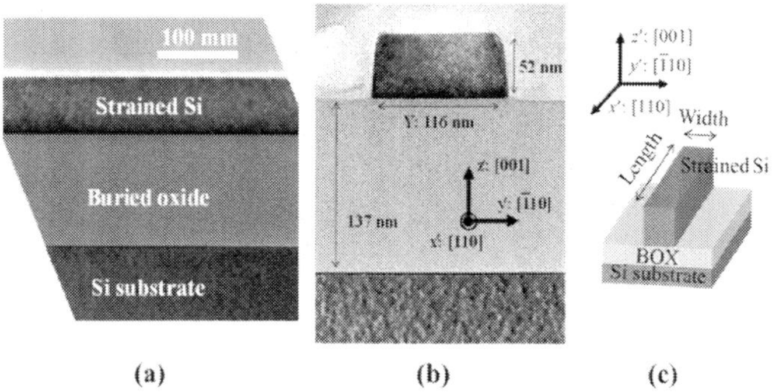

Figure 7. (a) Cross sectional TEM image of SSOI, (b) cross sectional TEM image of SSOI nanostructure, and (c) schematic of SSOI nanostructure

For Si, three long-wavelength optical phonon modes are degenerate at the center of the Brillouin zone. On the other hand, the degeneracy lifts after the induction of stress and the frequency of each mode individually shifts depending on the stress. For the isotropic biaxial tensile stress, the frequency of each mode shifts on the lower-frequency side and splits into singlet and doublet. In the case of (001) Si backscattering geometry, the singlet and doublet correspond to the LO and TO phonon modes, respectively. Fig. 8 shows the optical phonon frequencies for Si and SSOI. Generally, the LO phonon mode which is Raman active under the (001) Si backscattering geometry is measured and the isotropic biaxial stress in the strained Si layer is evaluated using the *b* coefficient shown in Table 2.

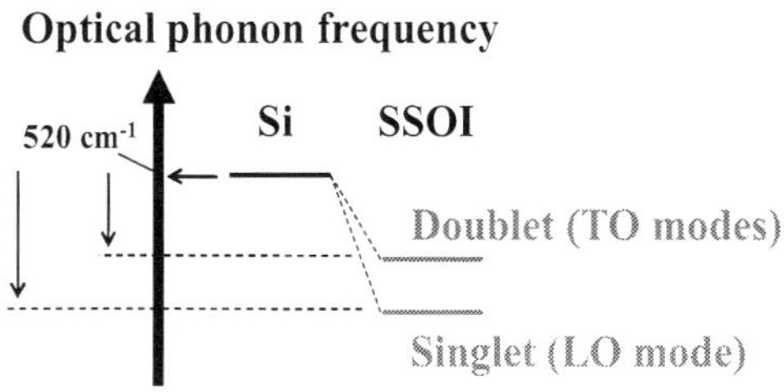

Figure 8. Optical phonon frequencies for Si and SSOI

SSOI nanostructures were fabricated with arbitrary forms by electron beam (EB) lithography and reactive ion etching (RIE). Fig. 7(b) and (c) show the cross sectional TEM image and the schematic of the SSOI nanostructure. The coordinate system in the experiments is also shown in Fig. 7(b) and (c). The SSOI lengths (*Ls*) were 5.0, 3.0, 2.0, 1.5, 1.0, 0.8, and 0.5 μm. The SSOI widths (*Ws*) were 1.0, 0.5, 0.2, 0.1, and 0.05 μm. The SSOI nanostructure shapes were anisotropic. Therefore, the stress states are also considered anisotropic. The stress component in the *z* direction is considered to be zero because of free-standing surface. As a result, the stress tensors in the SSOI nanostructures are considered to be expressed by Eq. (12).

SiGe nanostructures were fabricated as the same manner. SiGe with approximately 30% Ge concentration was epitaxially grown on a Si substrate. The thickness of the SiGe layer was approximately 35 nm. The *Ls* and *Ws* of the SiGe nanostructures were the same as those of the SSOI nanostructures. The cross sectional TEM image and the schematic of the SiGe nanostructure are shown in Fig. 9(a) and (b), respectively. As shown in the TEM image, overetching of the Si substrate is confirmed.

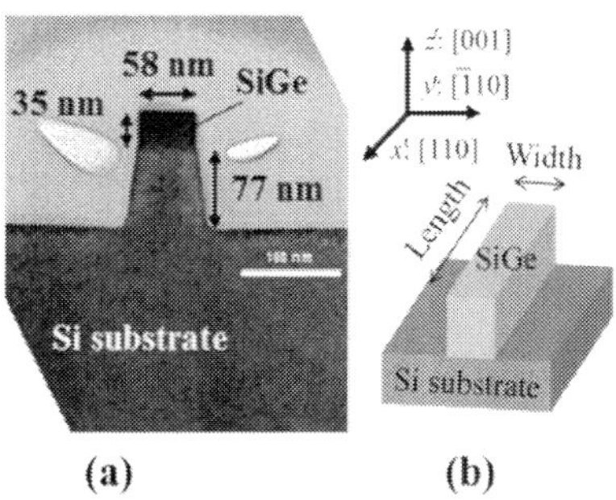

Figure 9. (a) Cross sectional TEM image of SiGe nanostructure and (b) schematic

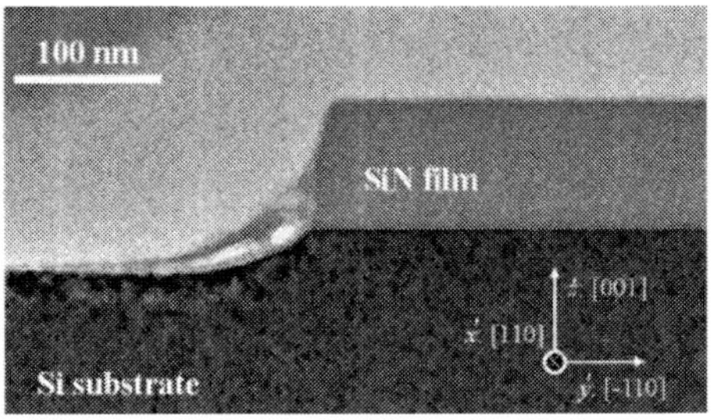

Figure 10. 10 TEM image of SiN film on Si substrate

A 80-nm-thick SiN film was deposited on a Si substrate by low-pressure vapor deposition. The inner stress of the SiN film was compressive due to its high density [10]. The compressive stress of approximately −1.0 GPa was observed by wafer bowing measurements [57]. Subsequently, the SiN film was etched to form an edge structure by EB and RIE. The cross-sectional TEM image of the sample is shown in Fig. 10. The stress distribution in Si around the SiN film edge was reproduced using the edge force model [58]. According to this model, the nondiagonal stress component, i.e., shear stress component, is induced in Si at the edge of the SiN film.

The stress distribution around the edge of the stress film is validated using the following equations of the edge force model [58]:

$$\sigma_{xx} = -\frac{2F_x}{\pi} \cdot \frac{x^3}{(x^2+z^2)^2}, \tag{33-1}$$

$$\sigma_{zz} = -\frac{2F_x}{\pi} \cdot \frac{xz^2}{(x^2+z^2)^2}, \tag{33-2}$$

$$\tau_{xz} = -\frac{2F_x}{\pi} \cdot \frac{x^2 z}{(x^2+z^2)^2}, \tag{33-3}$$

where σ_{xx} and σ_{zz} are the normal stress components in the direction of the x and z axes, respectively. τ_{xz} is the nondiagonal stress component (shear stress component). F_x is the tangential stress at the interface of the stress film and a substrate at the edge of the stress film, which is represented by $f \times t$, where f and t are the inner stress and the film thickness, respectively [58]. Each stress component is a function of x and z, which correspond to the lateral and depth directions of the substrate, respectively. The displacement along the y direction can be ignored because of the geometry. The plane strain assumption gives the stress components σ_{xx}, σ_{yy}, σ_{zz}, and τ_{xz}. Therefore, the stress tensor is represented by:

$$\sigma = \begin{pmatrix} \sigma_{xx} & 0 & \tau_{xz} \\ 0 & \sigma_{yy} & 0 \\ \tau_{xz} & 0 & \sigma_{zz} \end{pmatrix} \tag{34}$$

Fig. 11 shows the stress distribution in the substrate around the edge of the stress film, as calculated by Eq. 33. The inner stress of the film is assumed to be compressive (−1.0 GPa), and the film thickness is 80 nm. The positive and negative values indicate tensile and compressive stresses, respectively. First, large stress is induced around the edge of the stress film. The stress distribution around the edge is steep, especially for the stress components σ_{zz} and τ_{xz}. This fact indicates that high spatial resolution is needed to evaluate the nondiagonal stress component. Second, the opposite stress components σ_{xx} and σ_{zz} are confirmed between the region under the stress film and the space region; tensile stress appears in the region under the stress film, whereas compressive stress appears in the space region.

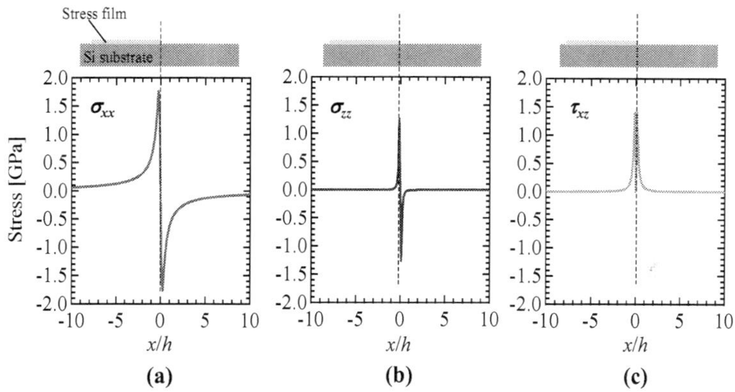

Figure 11. Stress distributions in Si calculated by (a) Eq. (33-1), (b) Eq. (33-2), and (c) Eq. (33-3).

2.5. Experimental configurations

We selectively obtained each optical phonon mode in Si by controlling incident and scattered electrical fields using polarizers and by sample rotation, which was based on the Raman polarization selection rules expressed by Eq. (1). Fig. 12 shows the various polarization configurations in liquid-immersion Raman spectroscopy. In the case of configuration (a), the LO phonon mode is Raman active. As shown in Fig. 12(a), the directions of the incident and scattered electrical fields are parallel to each other. The parallel-polarization configuration is generally applied in conventional Raman spectroscopy. On the other hand, the cross-polarization configuration by rotating the polarizer by 90° for the scattered light shown in Fig. 12(b) results in the fact that the TO phonon modes are Raman active. The TO phonon modes are excited by the z polarization due to the high-NA liquid-immersion lens. In this case, the peak separation of the two TO phonon modes is needed in the analysis. In the case of configuration (c), the sample is rotated by 45° in the parallel-polarization configuration. In this case, one of the two TO phonon modes is Raman active. In the experiments, we applied the configurations (a) and (c) to separately obtain the LO and TO phonon modes.

For the measurements of nondiagonal sress components in Si, the dependence of the Raman spectra from Si at the edge of the SiN film on the relative polarization direction between the sample orientation and the electrical fields of incident and scattered light was analyzed in detail. The experimental polarization configuration in liquid-immersion Raman spectroscopy is shown in Fig. 13. Both of the polarizations of the excitation laser and the scattered light were in the y'direction. The sample was rotated from 0° to 180°, as shown in Fig. 13.

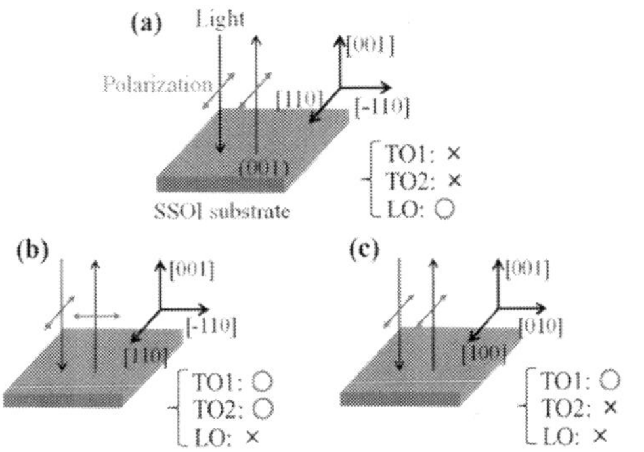

Figure 12. Polarization configurations in oil-immersion Raman spectroscopy: (a) LO active, (b) two TOs active, and (c) one of TOs active configuration

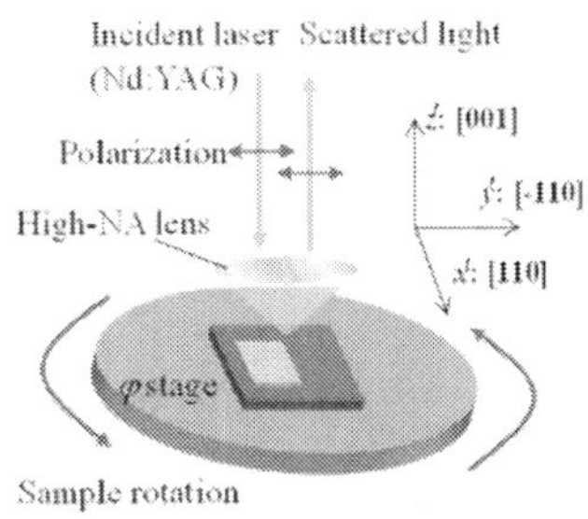

Figure 13. Polarization configuration for measurments of nondiagonal stress components in Si

Second harmonic generation of a neodymium-doped yttrium aluminum garnet (Nd:YAG) laser was used as the excitation source light in liquid-immersion Raman spectroscopy, the optical penetration depth of which is approximately 450 nm into Si [44]. The focal length of the spectroscope and the number of grating grooves were 2,000 mm and 1,800 mm^{-1}, respectively. Therefore, the high-wavenumber resolution of approximately 0.1 cm^{-1} was obtained. The detail explanations of the equipment are shown in Ref. 59. High-NA liquid-immersion lenses were used in this study. An oil-immersion lens with NA of 1.7 was used for the excitation of the TO phonon modes in the SSOI substrate with the 70-nm-thick strained Si layer. The refraction index n of the oil was 1.8. The oil-immersion lens with NA of 1.4 (n = 1.5) was used for the measurements of the anisotropic biaxial stress states in the SSOI nanostructures and the strained SiGe nanostructures. A water-immersion lens with NA of 1.2 (n = 1.3) was used for the measurements of the nondiagonal stress components. High spatial resolution was achieved owing to the high-NA liquid-immersion lens. The beam spot size was approximately 275, 334 and 390 nm for NA of 1.7, 1.4, and 1.2 liquid-immersion lenses, respectively, according to $0.88\times\lambda/\mathrm{NA}$ [60]. For the oblique incident light configuration as shown in Fig. 1, NA of the objective was 0.7. The glancing angles were 30° and 90°, as shown above.

2.6. Stress calculation

Stress calculations in the SSOI nanostructures were performed by finite element method (FEM). The results of FEM were compared with the values of the anisotropic biaxial stresses σ_{xx}' *and* σ_{yy}' obtained by oil-immersion Raman spectroscopy. The virtual biaxial thermal expansion of Si was used and the nodes between the interface of SSOI and BOX were fixed in the FEM calculations. The initial stress value of SSOI before the etching was defined as 1.1 GPa, which was equal to the value obtained by the Raman measurements. The number of meshes was constant for all the SSOI nanostructures: the number of nodes was 13,226 and the number of elements was 11,500. The averaged stress value in the circle area with a diameter of 334 nm corresponding to the beam spot size at the center of the SSOI nanostructure was compared with the measured data. For the depth direction, the stress values throughout the SSOI thickness were averaged because the optical penetration depth of the excitation light was large enough.

3. Results and discussion

3.1. Excitation of TO phonon modes in oil-immersion Raman spectroscopy

Fig. 14 shows the Raman spectra of the LO phonon modes from the SSOI substrate in the oblique incident light configurations with the glancing angles of 30° and 90°. Two peaks are seen in the Raman spectra because the excitation light (λ = 532 nm) penetrates the strained Si layer, the BOX layer, and reaches the Si substrate. Therefore, the wavenumber on the high-frequency side (defined to be 520 cm^{-1} in this study) originates from the Si substrate, and the wavenumber on the low-frequency side originates from the strained Si layer with the isotropic biaxial tensile stress state. The Raman intensities obtained in the 30° and 90° configurations are almost the same, which is consistent with the calculations shown in Fig. 2. Fig. 15 shows the Raman spectrum from the SSOI substrate in the TO active configuration. The glancing angle was 30°. The fitting curves for the strained Si layer and the Si substrate are also shown in Fig. 15. The explanation about the peak positions of the LO and TO phonon modes for SSOI are shown later.

From the results, the intensity ratio of the TO phonon mode from the strained Si layer obtained in the TO active configuration to the LO phonon mode from the strained Si layer obtained in the LO active configuration was calculated to be approximately 0.04. This value is almost the same as the theoretical value shown by Eq. (3). Using the oblique incident light configuration, the TO phonon modes were excited even for (001) Si. It is possible to completely eliminate the intensity of the LO phonon mode in the oblique incident light configuration in theory. Nevertheless, as shown in Fig. 15, the LO phonon mode was observed in the TO active configuration. This result is considered that there are misalignments of polarization between the incident/scattered light and the orientation of the Si substrate. It is considered that this behavior easily happens because the intensity of the LO phonon mode is much higher than that of the TO phonon mode.

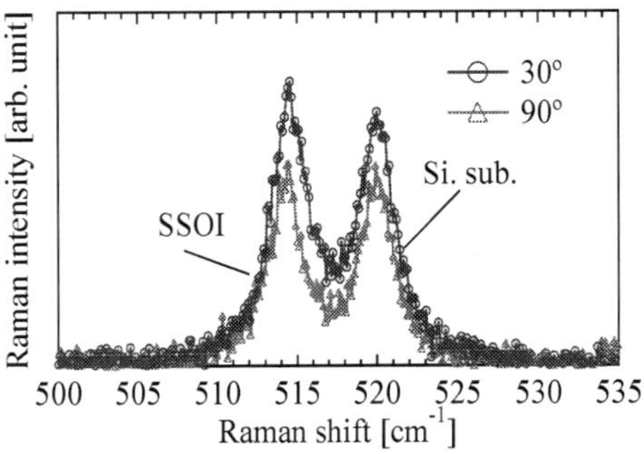

Figure 14. Raman spectra in 30° and 90° configurations

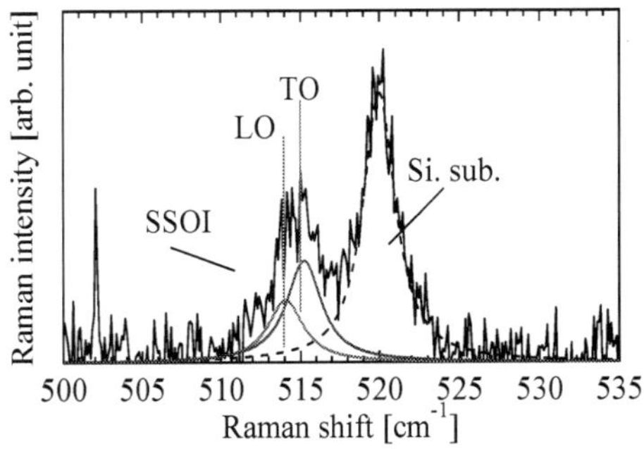

Figure 15. Raman spectrum in TO active configuration with fitting curves

Fig. 16 shows the Raman spectra from SSOI in the LO and TO active conditions, respectively, obtained by oil-immersion Raman spectroscopy. The light-exposure time is 5 s and 300 s for the excitations of the LO and TO phonons, respectively. The Raman intensity shown in Fig. 16 was normalized by the Raman signal from the strained Si layer. It should be noted that the low-frequency peak from strained Si in the LO active condition is lower than that in the TO active condition, although the peaks from the Si substrate in each condition are at the same wavenumber. The difference of the peak positions can be explained by Eq. (17); the LO phonon mode is more affected by biaxial stress than are the TO phonon modes.

Fig. 17 shows the result of fitting each peak. The LO phonon mode is detected irrespective of the TO active condition. This fact arises because the incident light with the polarization in the [010] Si direction generates even in configuration (c) shown in Fig. 12 due to depolarization effects [24]. It is considered difficult to avoid the depolarization effects for the SSOI substrate. On the other hand, it is reported that the contribution of the LO phonon modes can be decreased for the SSOI nanostructures because the depolarization effects relax due to the nanostructure [24]. It is considered that the peak separation of the LO and TO phonon modes is needed for the SSOI substrate to analyze the Raman spectrum obtained in the TO active configuration, while not necessary for the SSOI nanostructure.

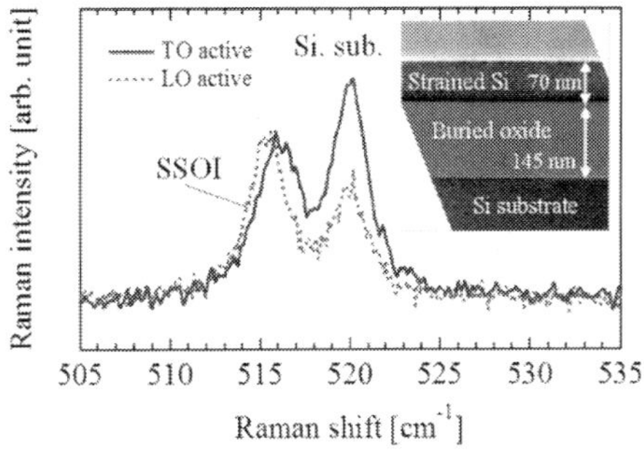

Figure 16. Raman spectra from SSOI in TO and LO active configurations

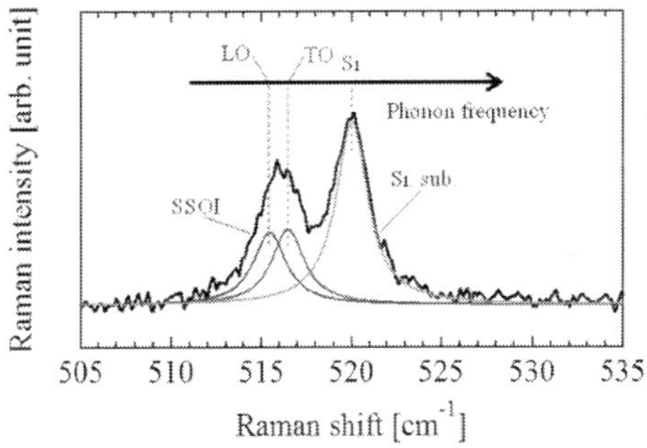

Figure 17. Raman spectrum in TO active configuration with fitting curves

The wavenumber shift of the LO phonon mode for the strained Si layer in the TO active configuration was −4.56 cm^{-1}, which is consistent with the value of −4.60 cm^{-1} for the Raman peak obtained in the LO active configuration. Furthermore, the Raman peak intensity from the Si substrate in the TO active condition is higher than that in the LO active configuration. This behavior indicates that the Raman peak that originates from the LO phonon mode is superimposed onto the Raman peak that originates from the TO phonon mode. We claim that the TO phonon mode was excited by using the high-NA oil-immersion lens.

Fig. 18 shows the Raman spectra from SSOI obtained by conventional Raman spectroscopy with the use of the NA = 0.7 objective. The dashed and solid lines denote the Raman spectra in the LO and TO active configurations, respectively. The light-exposure time of 3600 s for the TO active configuration is 400 times longer than that of 9 s (0.25% of 3600 s) for the LO active configuration. The intensity ratio of the TO to LO phonon modes are anticipated by the calculation shown in Table 2. In Fig. 18, the Raman intensity in each configuration is close to one another. Furthermore, the difference of the peak positions of the strained-Si layer in each configuration is confirmed, similarly to the results in oil-immersion Raman spectroscopy. These results indicate that the TO phonon mode was excited even in conventional Raman spectroscopy. However, the signal to noise ratio of the Raman intensity is bad. Moreover, the extremely long time measurements are necessary. In fact, it is difficult to perform mapping for obtaining biaxial-stress distributions in conventional Raman spectroscopy. We consider that it is important to use the high-NA liquid-immersion lens in order to excite TO phonon mode effectively and obtain biaxial-stress distributions in a realistic time.

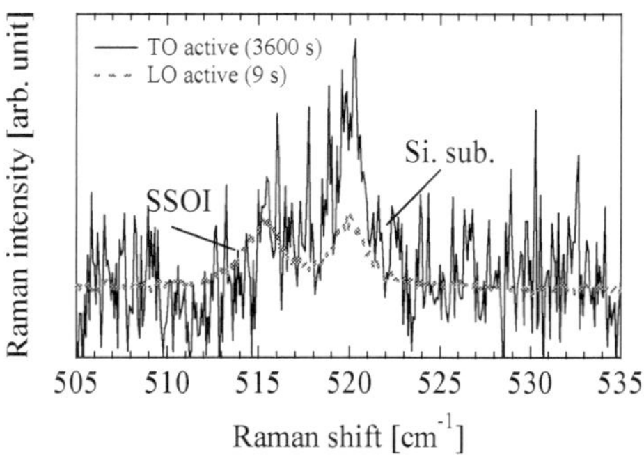

Figure 18. Raman spectra from SSOI with use of dry objective.

3.2. Measurements of anisotropic biaxial stress states in SSOI nanostructures

In-plane XRD measurements were performed to confirm strain in the strained Si layer. The diffraction from Si (220) and (–220) were measured (the results are not shown). As a result, ε_{xx} and ε_{yy} were 7.5 × 10^{-3} and 7.4 × 10^{-3}, respectively. These results indicate that the stress state in the strained Si layer is almost isotropic biaxial throughout the wide area which is equivalent to the footprint of the incident X-ray.

PDPs were evaluated by oil-immersion Raman spectroscopy. The calculated biaxial stresses σ_{xx} and σ_{yy} in the strained Si layer are summarized in Fig. 19. In the calculation of the biaxial stresses, PDPs reported in 1970, 1978, and 1990, as mentioned above, were used and Eqs. (19)-(21) were used for the stress calculations. In the oil-immersion Raman measurements, five points were obtained. The solid symbols indicate the average values. From the results, it appears that the biaxial stress values fluctuate in the range of 50–150 MPa, which is attributed to the dislocation conditions in the strained Si layer [61-63]. It should be noted that the apparent anisotropic natures of biaxial stress states were observed in the case of using PDPs reported in 1970 and 1978. The differences in the biaxial stresses are 530 and 170 MPa for PDPs reported in 1970 and 1978, respectively, which are inconsistent with the results of XRD. On the other hand, the isotropic nature was clearly observed in the case of using PDPs reported in 1990. As a result, PDPs of $p/\omega_0^2 = -1.85$, $q/\omega_0^2 = -2.31$, and $r/\omega_0^2 = -0.71$ reported by Anastassakis *et al.* in 1990 are considered the most accurate for evaluating stress in Si among the three sets of PDPs.

Figs. 20(a) and (b) show the Raman spectra from the SSOI nanostructures with W = 1.0 and 0.05 μm, respectively, in configuration (c). Ls of the nanostructures were both 5.0 μm. We subtracted the signal of the Si substrate fitting curves from the raw data in order to analyze the spectra from the SSOI nanostructures in detail. From Fig. 20(b), the signal from the SSOI nanostructure even with W = 50 nm can be clearly observed. This observation is attributed to the high spatial resolution in oil-immersion Raman spectroscopy.

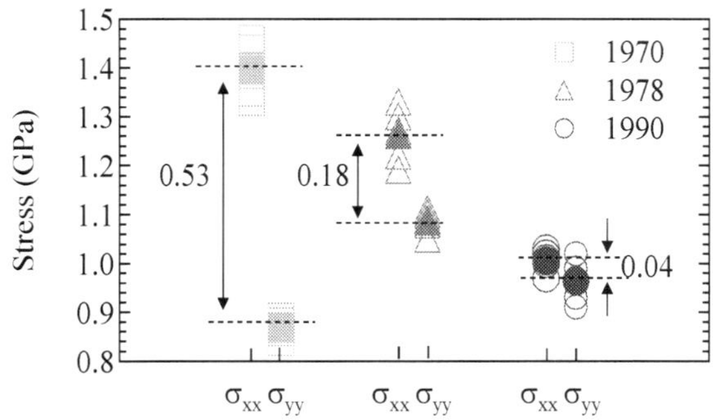

Figure 19. Biaxial stresses σ_{xx} and σ_{yy} in SSOI obtained by oil-immersion Raman spectroscopy using three sets of PDPs.

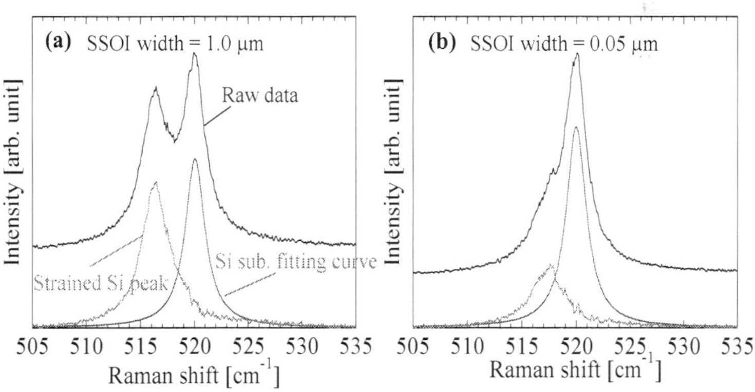

Figure 20. Rama spectra from SSOI nanostructures with Ws of (a) 1.0 and (b) 0.05 μm. L is both 5.0 μm.

The normalized Raman spectra as a function of L are shown in Fig. 21. W was fixed to 0.2 μm. The peaks from bulk Si are shown for comparison. The peak positions of strained Si gradually shift toward the high-frequency side with the decrease in L from 5.0 to 0.5 μm, as shown in Fig.21. Using Eq. (21), the anisotropic biaxial stresses σ_{xx}' and σ_{yy}' in the SSOI nanostructures were calculated, as shown in Fig. 22. Figs. 22(a) and (b) show the results for the SSOI nanostructures with the thickness of 50 and 30 nm, respectively. The results of the Raman measurements were compared with those of FEM. The example of the FEM calculations is shown in Fig. 23. Fig. 23 shows the three dimensional distribution of the σ_{xx}' component for the SSOI nanostructure with L = 1.0 and W = 0.2 μm. From the results of FEM, the stress relaxation is confirmed at the edge of the SSOI nanostructure, while large tensile stress remains in the center of the SSOI nanostructure and at the interface of the strained Si layer and the BOX layer. There is a good correlation between the results of oil-immersion Raman spectroscopy and FEM, as shown in Fig. 22. σ_{xx}' decreases with the decrease in L for the SSOI nanostructures with the thickness of 50 and 30 nm, while σ_{yy}' remains almost constant. Moreover, the values of σ_{yy}' for the SSOI nanostructures with the thickness of 30 nm are larger than those of 50 nm. Therefore, the thin SSOI nanostructures had immunity to the stress relaxation. Using oil-immersion Raman spectroscopy, the evaluation of the anisotropic biaxial stress states was accomplished for the SSOI nanostructures. It is considered that the results obtained in this study have important implications for the SSOI nanostructure fabrication.

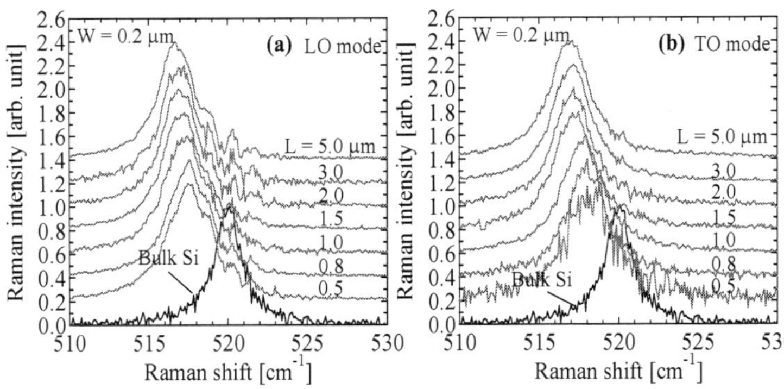

Figure 21. Normalized Raman spectra from SSOI nanostructures as a function of L in (a) LO and (b) TO active configurations.

3.3. Measurements of anisotropic biaxial stress states in strained SiGe nanostructures

In this section, the evaluation of the anisotropic biaxial stress states in the SiGe nanostructures is shown. For the SiGe nanostructures, large compressive stress is induced because the lattice constant of SiGe is larger than that of Si. The stress states in the SiGe nanostructures are considered to be expressed by Eq. (12) similar to the stress states in the

SSOI nanostructures. The crystal structure of SiGe remains diamond type, i.e., the methodology of evaluation for the anisotropic biaxial stress in Si shown above can be directly applied to strained SiGe.

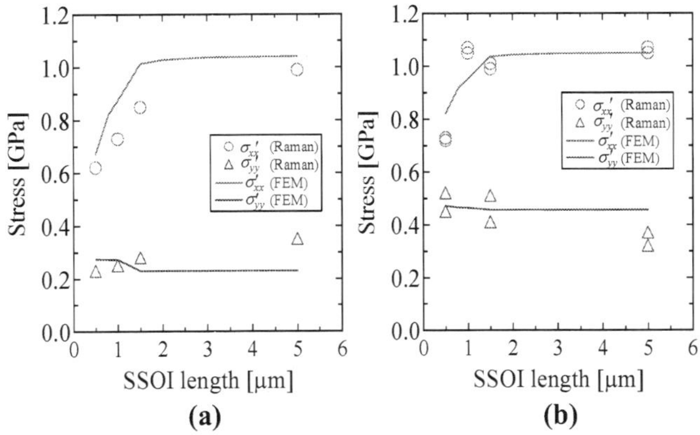

Figure 22. Biaxial stresses σ_{xx}' and σ_{yy}' as a function of *L* for (a) 50-nm-thick and (b) 30-nm-thick SSOI nanostructures.

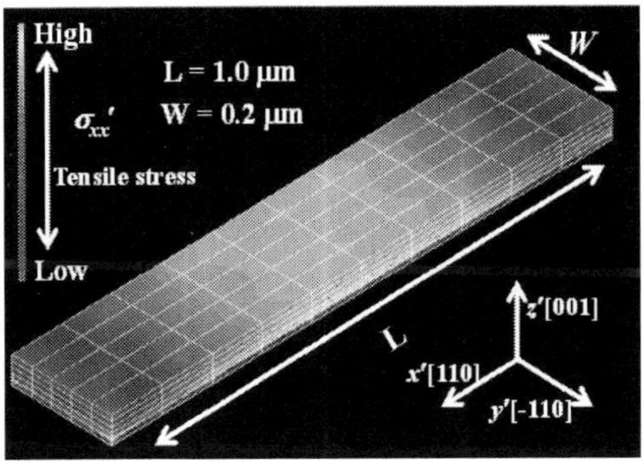

Figure 23. Three dimensional stress distribution obtained by FEM.

Fig. 24 shows the examples of Raman spectra from the SiGe nanostructures. As shown in Fig. 24, the intensity from Si-Si phonon mode in SiGe appeared to be weak, while the intensity of the Si substrate is very strong. This behavior makes the analysis difficult. Nevertheless, the LO and TO phonon modes can be separately obtained. The peak positions of the TO and LO phonon modes for the SiGe nanostructures with L = 5.0 and W = 1.0 μm are clearly different. Moreover, it should be noted that the difference decreases with the decrease in W. For W = 0.1 μm, there is little difference between the peak positions of the TO and LO phonon modes. This behavior indicates that the stress states in the SiGe nanostructures change from a biaxial state to a uniaxial state.

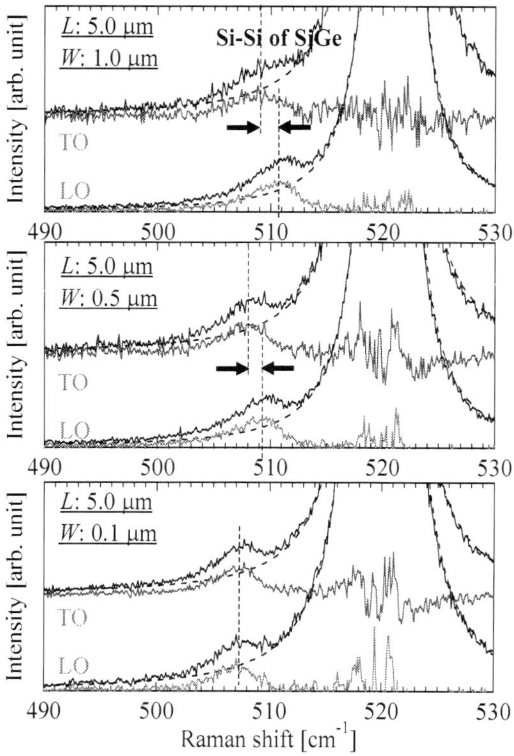

Figure 24. Raman spectra of TO and LO phonon modes from SiGe nanostructures.

Fig. 25 shows the Raman shifts of the TO and LO phonon modes for the SiGe nanostructures with W = 1.0, 0.5, and 0.1 μm as a function of L. First, the large compressive stress exists because the Raman shift of stress-free SiGe with the 30% Ge concentration is approximately 500 cm^{-1} [64,65]. From the results, the clear dependence of the Raman shifts on L and W were observed. It appears that the evaluation of the anisotropic biaxial stress states in strained SiGe is accomplished by oil-immersion Raman spectroscopy similar to evaluating strained Si. However, it is considered that the evaluation of strained SiGe is more difficult than that of Si. There are several unknown parameters for measuring stress in strained SiGe, e.g., PDPs of SiGe, precise Ge concentration, and peak position of stress-free SiGe. Various parameters have so far been suggested [64-70]. These problems are now under investigation.

3.4. Measurements of nondiagonal stress components

First, α (the contribution of z polarization) was determined from the Raman intensity ratio of the TO to LO phonon modes (the results are not shown). From the results, α was calculated to be 0.09 in the water-immersion Raman measurements. Fig. 26 shows the comparison between the measured and calculated data of the Raman shifts dependences on the sample rotation angle. The experimental results were obtained by high-NA water-immersion Raman spectroscopy. The measurement position was the region at the edge of the SiN film (the edge under the SiN film rather than in the space region). Features indicating the induction of stress with the nondiagonal component are clearly observed in the experimental results; the profiles of the Raman shift dependence on the sample rotation angle are asymmetric relative to 45° from 0° to 90° (or relative to 135° from 90° to 180°). As previously shown, this behavior indicates that the shear stress component τ_{xz} is induced in Si at the edge of the SiN film. From Fig. 26, there is a good correlation between the measured and calculated data. On the other hand, disagreements appear around sample rotation angles of 45° and 135°. One possible explanation is that the polarization of the electrical fields of the incident and scattered light is modified by the SiN film. This modification is not included in the calculation.

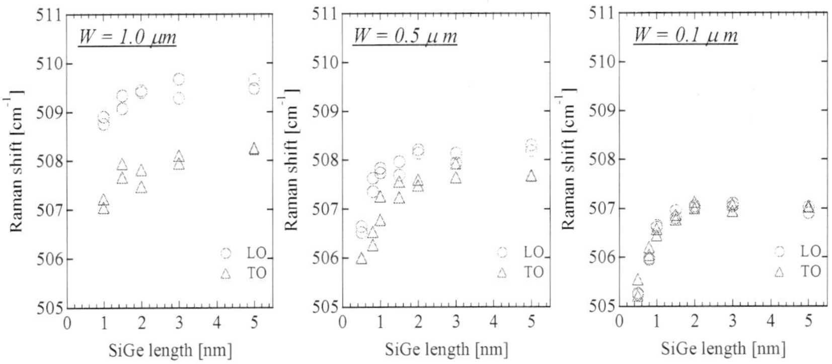

Figure 25. Raman shifts of TO and LO phonon modes for SiGe nanostructures as a function of *L*.

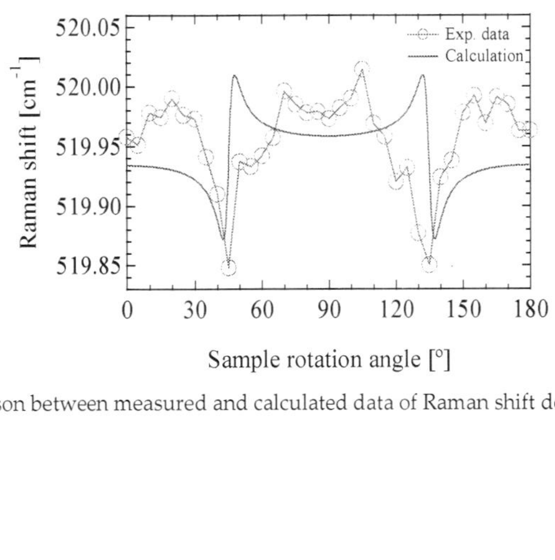

Figure 26. Comparison between measured and calculated data of Raman shift dependences on sample rotation angle.

The Raman shifts appear to be small. This is explained as follows. The stress induced at the edge of the SiN film is localized in width and depth. It is considered that the measured stress is averaged in the region of approximately 390 and 450 nm which are the spot size and optical penetration depth of the laser, respectively. From the water-immersion Raman measurements, τ_{xz}' was approximately 0.1 GPa. This value was consistent with the result obtained by the edge force model after the correction for the spot size and the optical penetration depth of the laser (the results are not shown) [46,58]. The methodology described here has the potential to measure complicated stress states in Si and SiGe even those with nondiagonal stress components.

4. Conclusion

We demonstrated the measurements of the complicated stress states in Si by high-NA liquid-immersion Raman spectroscopy. The z polarization was obtained due to the high-NA liquid-immersion lens, which allows for exciting the forbidden modes, the TO phonon modes, even under the (001) Si backscattering geometry. First, the TO phonon mode of the strained Si layer was observed in oil-immersion Raman spectroscopy. The peak positions of the strained Si layer were clearly separated in the TO and LO active configurations, although the Si substrate peaks remain at the same position in each configuration. This behavior indicates that the biaxial isotropic tensile stress state in the strained Si layer gives rise to the splitting of the optical phonon modes in Si. From the results, the LO phonon mode was more affected by biaxial stress than the TO phonon mode, which was consistent with the result obtained by solving the secular equations. Using the TO phonon mode as well as the LO phonon mode, the anisotropic biaxial stress states in the SSOI nanostructures were measured. As a result, the clear dependences of the biaxial stresses σ_{xx}' and σ_{yy}' on L and thickness were observed. σ_{xx}' decreased with the decrease in L, especially under L = 1.5 μm, while σ_{yy}' remains almost constant. The values of σ_{yy}' for the SSOI nanostructures with the thickness of 30 nm are larger than those of 50 nm, which indicates that the thin SSOI nanostructures had immunity to the stress relaxation. The results obtained by oil-immersion Raman spectroscopy were consistent with the FEM calculations. We also measured the anisotropic biaxial stress states in the strained SiGe nanostructures by the same technique. Consequently, the clear dependence of the Raman shifts on L and W were observed similarly to the results for the SSOI nanostructures. Furthermore, the stress with the nondiagonal component, the shear stress component, in Si was measured by water-immersion Raman spectroscopy. As a result, the asymmetric profile was obtained for the dependence of the Raman shifts on the sample rotation angle. This behavior indicates that the shear stress component τ_{xz} is induced in Si at the edge of the SiN film. There is a good correlation between the measured and calculated data. High-NA liquid-immersion Raman spectroscopy enabled us to measure the complicated stress states in strained Si and SiGe with high spatial resolution even those with the nondiagonal stress component.

Author details

Daisuke Kosemura, Motohiro Tomita and Atsushi Ogura
School of Science and Technology, Meiji University, Kawasaki, Japan

Koji Usuda
Green Nanoelectronics Collaborative Research Center, AIST, Tsukuba, Ibaraki, Japan

5. Acknowledgement

The authors thank Ryosuke Shimidzu of PHOTON Design Corporation for the fruitful discussion about the z polarization. The authors thank Dr. Kazuhiko Omote of Rigaku Corporation for his great help in high-resolution XRD measurements. This study was partially supported by the Semiconductor Technology Academic Research Center (STARC), the Japan Society for the Promotion of Science (JSPS) through the "Funding Program for World-Leading Innovative R&D on Science and Technology", "Scientific Research B"and the Japan Science and Technology Agency through the "Adaptable and Seamless Technology transfer Program (A-STEP) through target-driven R&D."

6. References

[1] S. Takagi, J. L. Hoyt, J. Welser, and J. F. Gibborns, J. Appl. Phys. 80, 1567 (1996).

[2] C. S. Smith. Phys. Rev. 94, 42 (1964).

[3] A. Ogura, D. Kosemura, K. Yamasaki, S. Tanaka, Y. Kakemura, A. Kitano, and I. Hirosawa, Solid-State Electronics 51, 219 (2007).

[4] K. Usuda, T. Irisawa, T. Numata, N. Hirashita, and S. Takagi, Semicond. Sci. Technol. 22, s227 (2007).

[5] K. Usuda, T. Mizuno, T. Tezuka, N. Sugiyama, Y. Moriyama, S. Nakaharai, S. Takagi, Appl. Surf. Sci. 224, 113 (2004).

[6] A. Ogura, T. Yoshida, D. Kosemura, Y. Kakemura, M. Takei, H. Saito, T. Shimura, T. Koganezawa, and H. Hirosawa, Solid-State Electronics 52, 1845 (2008).

[7] M. Takei, D. Kosemura, K. Nagata, H. Akamatsu, S. Mayuzumi, S. Yamakawa, H. Wakabayashi, and A. Ogura, J. Appl. Phys. 107, 124507 (2010).

[8] M. Takei, H. Hashiguchi, T. Yamaguchi, D. Kosemura, K. Nagata, and Atsushi Ogura, Jpn. J. Appl. Phys. 51, 04DA04 (2012).

[9] D. Kosemura, Y. Kakemura, T. Yoshida, A. Ogura, M. Kohno, T. Nishita, and T. Nakanishi, Jpn. J. Appl. Phys. 47, 2538 (2008).

[10] A. Ogura, H. Saitoh, D. Kosemura, Y. Kakemura, T. Yoshida, M. Takei, T. Koganezawa, I. Hirosawa, M. Kohno, T. Nishita, and T. Nakanishi, Electrochem. Solid-State Lett. 12, H117 (2009).

[11] S. Mayuzumi, S. Yamakawa, D. Kosemura, M. Takei, Y. Tateshita, H. Wakabayashi, M. Tsukamoto, T. Ono, A. Ogura, and N. Nagashima, IEEE Transactions on Electron Devices 56, 2778 (2009).

[12] T. Yamaguchi, Y. Kawasaki, T. Yamashita, N. Miura, M. Mizuo, J. Tsuchimoto, K. Eikyu, K. Maekawa, M. Fujisawa, and K. Asai, Jpn. J. Appl. Phys. 50, 04DA02 (2011).

[13] W. Xiong, C. R. Cleavelin, P. Kohli, C. Huffman, T. Schulz, K. Schruefer, G. Gebara, K. Mathews, P. Patruno, Y.-M. L. Vaillant, I. Cayrefourcq, M. Kennard, C. Mazure, K. Shin, and T.-J. K. Liu, IEEE Electron Device Lett. 27, 612 (2006).

[14] M. Tomita, D. Kosemura, M. Takei, K. Nagata, H. Akamatsu, and A. Ogura, Jpn. J. Appl. Phys. 50, 010111 (2011).

[15] M. D. Vaudin, Y. B. Gerbig, S. J. Stranick, and R. F. Cook, Appl. Phys. Lett. 93, 193116 (2008).

[16] K. Omote, J. Phys., Condens. Matter. 22, 474004 (2010).

[17] I. D. Wolf, Semicond. Sci. Technol. 11, 139 (1996).

[18] L. Zhu, C. Georgi, M. Hecker, J. Rinderknecht, A. Mai, Y. Ritz, and E. Zschech, J. Appl. Phys. 101, 104305 (2007).

[19] S. C. Jain, B. Dietrich, H. Richter, A. Atkinson, and A. H. Harker, Phys. Rev. B 52, 6247 (1995).

[20] T. Ito, H. Azuma, and S. Noda, Jpn. J. Appl. Phys. 33, 171 (1994).

[21] S. Nakashima, T. Mitani, M. Ninomiya, and K. Matsumoto, J. Appl. Phys. 99, 053512 (2006).

[22] D. Kosemura and A. Ogura, Appl. Phys. Lett. 96, 212106 (2010).

[23] V. Poborchii, T. Tada, and T. Kanayama, Appl. Phys. Lett. 97, 041915 (2010).

[24] A. Tarun, N. Hayazawa, H. Ishitobi, S. Kawata, M. Reiche, and O. Moutanabbir, Nano Lett. 11, 4780 (2011).

[25] E. Bonera, M. Fanciulli, and D. N. Batchelder, J. Appl. Phys. 94, 2729 (2003).

[26] S. Narayanan, S. R. Kalidindi, and L. S. Schadler, J. Appl. Phys. 82, 2595 (1997).

[27] M. Yoshikawa, M. Maegawa, G. Katagiri, and H. Ishida, J. Appl. Phys. 78, 941 (1995) 941.

[28] M. Becker and H. Scheel, J. Appl. Phys. 101, 063531 (2007).

[29] S. J. Harris, A. E. O'Neill, W. Yang, P. Gustafson, J. Boileau, W. H. Weber, B. Majumdar, and S. Ghosh, J. Appl. Phys. 96, 7195 (2004).

[30] D. Kosemura and A. Ogura, Jpn. J. Appl. Phys. 50, 04DA06 (2011).

[31] D. Kosemura, M. Tomita, K. Usuda, and A. Ogura, Jpn. J. Appl. Phys. 51, 02BA03 (2012).

[32] V. Poborchii, T. Tada, K. Usuda, and T. Kanayama, Appl. Phys. Lett. 99, 191911 (2011).

[33] R. Ossikovski, Q. Nguyen, G. Picardi, and J. Schreiber, J. Appl. Phys. 103, 093525 (2008).

[34] T. Tada, V. Poporchii, and T. Kanayama, J. Appl. Phys. 107, 113539 (2010).

[35] G. H. Loechelt, H. G. Cave, and J. Menendez, J. Appl. Phys. 86, 6164 (1999).

[36] P. Kumar, I. Dutta, and M. S. Bakir, J. Electronic Mater. 41, 322 (2011).

[37] M. Feron, Z. Zhang, and Z. Suo, J. Appl. Phys. 102, 023502 (2007).

[38] Y. Moriyama, Y. Kamimuta, Keiji Ikeda, Tsutomu Tezuka, Thin Solid Films 520, 3236 (2012).

[39] K. Ikeda, M. Oda, Y. Kamimuta, Y. Moriyama, and T. Tezuka, Appl. Phys. Exp. 3, 124201 (2010).

[40] T. Tezuka, E. Toyoda, T. Irisawa, N. Hirashita, Y. Moriyama, N. Sugiyama, K.Usuda, and S. Takagi, Appl. Phys. Lett. 94, 081910 (2009).
[41] T. Irisawa, T. Numata, T. Tezuka, K. Usuda, N. Hirashita, N. Sugiyama, E. Toyoda, and S. Takagi, IEEE Transactions on Electron Devices 53, 2809 (2006).
[42] E. Anastassakis, J. Appl. Phys. 82, 1582 (1997).
[43] R. Loudon, Adv. Phys. 13, 423 (1964).
[44] D. E. Aspnes and A. Studna, Phys. Rev. B. 27, 985 (1983).
[45] K. Mizoguchi and S. Nakashima, J. Appl. Phys. 65, 2583 (1989).
[46] I. D. Wolf, H. E. Maes, and S .K. Jones, J. Appl. Phys. 79, 7148 (1996).
[47] L. H. Wong, C. C. Wong, J. P. Liu, D. K. Sohn, L. Chan, L. C. Hsia, H. Zang, Z. H. Ni, and Z. X. Shen, Jpn. J. Appl. Phys. 44, 7922 (2005).
[48] Ogura A *et al* 2007 Report of JEITA Standard Strain MeasurementWorking Group JEITA.
[49] E. Anastassakis, A. Pinczuk, E. Burstein, F. H. Pollak, and M. Cardona, Solid State Commun. 8, 133 (1970).
[50] M. Chandrasekhar, J. B. Renucci, and M. Cardona, Phys. Rev. B 17, 1523 (1978).
[51] E. Anastassakis, A. Cantarero, and M. Cardona, Phys. Rev. B 41, 7529 (1990).
[52] E. Anastassakis, J. Appl. Phys. 81, 3046 (1997).
[53] O. Moutanabbir, M. Reiche, A. Ha¨hnel, M. Oehme, and E. Kasper, Appl. Phys. Lett. 97, 053105 (2010).
[54] T. S. Drake, C. Ni Chleirigh, M. L. Lee, A. J. Pitera, E. A. Fitzgerald, D. H. Anjum, J. Li, R. Hull, N. Klymko, and J. L. Hoyt, Appl. Phys. Lett. 83, 875 (2003).
[55] A. Ogura and O. Okabayashi, Thin Solid Films 488, 189 (2005).
[56] A. Ogura, T. Tatsumi, T. Hamajima, and H. Kikuchi, Appl. Phys. Lett. 69, 1367 (1996).
[57] M. Finot, I. A, Blech, S. Suresh, and H. Fujimoto, J. Appl. Phys. 81, 3457 (1997).
[58] S. M. Hu, J. Appl. Phys. 50, 4661 (1979).
[59] A. Ogura, K. Yamasaki, D. Kosemura, S. Tanaka, I, Chiba, and R. Shimidzu, Jpn. J. Appl. Phys. 45, 3007 (2006).
[60] I. D. Wolf, Theoretical and experimental study of the effects of the different optical parameters and lenses on the spatial resolution of the Raman system, STREAM consortium, Doc. No. IST-1999-10341, pp. 1-20.
[61] S. Nakashima, T. Yamamoto, A. Ogura, K. Uejima, and T. Yamamoto, Appl. Phys. Lett. 84, 2533 (2004).
[62] K. Kutsukake, N. Usami, T. Ujihara, K. Fujiwara, G. Sazaki, and K. Nakajima; Appl. Phys. Lett. 85, 1335 (2004).
[63] K. Sawano, S. Koh, Y. Shiraki, N. Usami, and K. Nakagawa, Appl. Phys. Lett. 83, 4339 (2003).
[64] J. C. Tsang, P. M. Mooney, F. Dacol, and J. O. Chu, J. Appl. Phys. 75, 8098 (1994).
[65] M. I. Alonso and K. Winer, Phys. Rev. B 39, 10056 (1989).
[66] D. J. Lockwood and J.-M. Baribeau, Phys. Rev. B 45, 8565 (1992).
[67] M. Holtz, W. M. Duncan, S. Zollner, and R. Liu, J. Appl. Phys. 88, 2523 (2000).
[68] F. Pezzoli, E. Bonera, E. Grilli, M. Guzzi, S. Sanguinetti, D. Chrastina, G. Isella, H. von Känel, E. Wintersberger, J. Stangl, and G. Bauer, J. Appl. Phys. 103, 093521 (2008).

[69] J. P. Dismukes, L. Ekstrom, and R. J. Paff, J. Phys. Chem. 68, 3021 (1964).
[70] F. Cerdeira, A. Pinczuk, J. C. Bean, B. Batlogg, and B. A. Wilson, Appl. Phys. Lett. 45, 1138 (1984).
[71] Y. Hoshi, A. Fukumoto, K. Sawano, I. Cayrefourcq, M. Yoshimi, and Y. Shiraki, Jpn. J. Appl. Phys. 46, 7294 (2007).

Chapter 10

HR-MAS NMR Spectroscopy in Material Science

Todd M. Alam and Janelle E. Jenkins

Additional information is available at the end of the chapter

http://dx.doi.org/10.5772/48340

1. Introduction

In the early to mid-90's, NMR studies were being published that recognized the power of magic angle spinning (MAS) to increase resolution in materials that were not strictly solids by averaging differences in magnetic susceptibility and residual dipolar coupling inherent in these samples. The method of utilizing MAS for non-solid materials to produce liquid-like NMR lines was termed High-Resolution Magic Angle Spinning (HR-MAS). A few of the first HR-MAS examples included investigation of resins for combinatorial chemistry,[1] solvent swollen polystyrene gels,[2] and lipid systems.[3] Then in 1996, Maas *et al.* [4] advanced the field of HR-MAS NMR by adding a magnetic field gradient along the magic angle (see Figure 1). Like high resolution solution NMR, this gradient improved sensitivity and resolution with the ability to more easily select coherence pathways and by reducing indirect dimension (t_1) noise.[4]

There are currently commercially available HR-MAS probes with magic angle gradients from companies like Bruker BioSpin Corporation (Billerica, MA),[5] Agilent Technologies (Santa Clara, CA),[6] JEOL USA, Inc. (Peabody, MA),[7] and Doty Scientific, Inc. (Columbia, SC).[8] In addition to magic angle gradients, many of these probes also have a deuterium (^{2}H) lock channel, allowing improved ease of shimming and long term stability. With the emergence of commercially available probes, HR-MAS NMR has become more popular in the last few years, especially in the biological and biomedical fields. This popularity is mainly due to the heterogeneous nature of tissues and cells that are well suited for HR-MAS. Multiple HR-MAS NMR studies involving different tissue biopsies, like brain, kidney, liver, and muscle tissues for metabonomics studies, as well as identification of abnormal tissues (*i.e.* cancerous tissues) have been reported.[9-11] HR-MAS NMR has also been applied to the characterization of foodstuffs, including the assignment of metabolites in tomatoes and apples, the study of biopolymers in fruit cuticles, quantification of *n*-3 fatty acids content in different fish species, and tracking the chemistry of coffee beans during the roasting process.[12, 13] These types of biological and foodstuff HR-MAS NMR investigations highlight the diverse range of information that can be obtained.

The application of HR-MAS NMR to material science was initially focused almost exclusively on the analysis of solid-phase (*i.e.* utilizing support resins) organic and peptide synthesis, or analysis of combinatorial solid-phase results. In these studies, the material was swollen in appropriate solvents such that the mobility of the attached ligands was increased, allowing high resolution NMR spectra to be obtained. The application of HR-MAS to solid state synthetic chemistry remains an active area of research, but will not be discussed in detail. The readers are encouraged to consult several very extensive reviews in this area.[14-16] In comparison to the numerous HR-MAS NMR studies on biological and solid-phase synthetic chemistry systems, there are fewer examples that focus on the use of this technique to material science. This chapter will review the application of HR-MAS NMR to a wide range of systems, including ceramics, zeolites, liquid crystals, ionic liquids, and surface modified nanoparticles.

1.1. How HR-MAS works

Materials that are crystalline or rigid solids (for example resins, ceramics, *etc.*) exhibit an extremely broad NMR signal due to extensive homo- and hetero-nuclear dipolar coupling, chemical shift anisotropy (CSA), and quadrupolar interactions. A variety of solid state NMR techniques have been developed to improve the resolution and sensitivity including the use of multiple pulse sequences, cross polarization, high power decoupling, multiple quantum NMR and fast MAS NMR. For example, solid state ^{1}H MAS NMR has found a number of applications in material characterization.[17] Even with these advances, solution-like NMR spectra are rarely realized for solid samples.

For samples that are in the liquid/solid classification, motional averaging will partially reduce or remove many of these broadening interactions. It is also possible to swell or plasticize a material to increase the local mobility, thereby reducing the magnitude of these interactions.[18] Even for liquid environments in a heterogeneous sample, differences in the magnetic susceptibility within the material can drastically reduce the observed resolution. For these types of dynamically averaged or susceptibility broadened systems, MAS even at moderate speeds will produce high resolution NMR spectra: this is the niche of HR-MAS NMR.

This improved resolution due to MAS arises because the Hamiltonians describing the dipolar, CSA and magnetic susceptibility interactions all contain an orientational component that scales as $(3\cos^2\theta-1)$, where θ is the angle between the rotor spinning axis and the magnetic field (Figure 1). When a sample is spun about an axis that is at the "magic angle" ($\theta = 54.7^{o}$) these interactions vanish. A simplistic schematic of a HR-MAS stator is pictured in Figure 1. As mentioned before, the main difference between a standard MAS stator and an HR-MAS stator is the magic angle gradient coil that can produce a gradient along the rotor spinning axis.

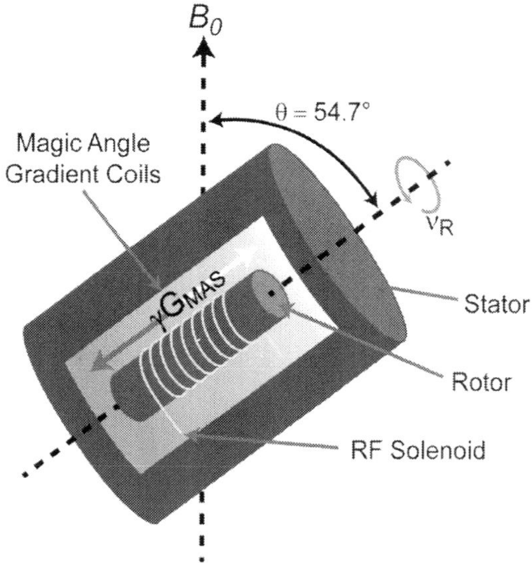

Figure 1. Schematic of a HR-MAS stator with a magic angle gradient along the rotor spinning axis.

For samples where residual dipolar interactions or differences in magnetic susceptibility are small, HR-MAS NMR reduces the observed line widths to be similar to those observed in solution high resolution NMR spectroscopy. For liquids adsorbed into/onto materials MAS speeds between 1 and 2 kHz may result in improved resolution, while for plasticized/swollen materials MAS speeds between 4 and 10 kHz may be necessary to obtain the desired resolution. Care must also be taken to ensure that the residual spinning sidebands (due to incomplete averaging) fall outside the spectral window of interest.

As an example of the enhanced resolution that can be achieved under HR-MAS two different swollen polymer systems are shown in Figure 2. These anion exchange membranes (AEM) operate in methanol solution, and readily adsorb significant amounts of water and methanol from the solution during use. Our group is interested in identifying the different chemical environments that these solvent molecules see within the membrane, as well as directly measuring the diffusion rates for these different species. Under static conditions, the water and methanol are not resolved in the NMR spectrum (Figure 2A, green). This lack of chemical shift resolution precludes any information about the local membrane environment being obtained. The broadening is caused by the magnetic susceptibility differences between the rigid AEM and the methanol solution. In this example, minimal MAS of ~750 Hz (Figure 2A, red) was sufficient to average the magnetic susceptibility allowing individual solvent environments to be easily observed. Increasing the spinning speed to 4 kHz removed the spinning side bands from the spectral region of interest, but did not dramatically increase the resolution of the solvent resonances. In this swollen AEM example, signals from the actual membrane are not observed at 4 kHz (only the swelling solvent is observed) indicating that the dipolar interactions present in the membrane remain much larger than the MAS speed. Since the protons from the membrane are not readily observable, it provides a unique opportunity to further study the solvent behavior in the membrane without interference. Additional discussion of this system is presented in Section 2.1.1. The second example, involves the swelling of a polyButadiene-AcryloNitrile (pBAN) polymer in $CDCl_3$. Inspection of Figure 2B (green) shows that under static conditions this rather soft material produces a relatively unresolved NMR spectrum. Even after extensive solvent swelling, the NMR spectrum is essentially unchanged (Figure 2B, red). With HR-MAS (4 kHz spinning) the resolution is dramatically enhanced, enabling a more detailed analysis of the pBAN ^1H NMR spectra.

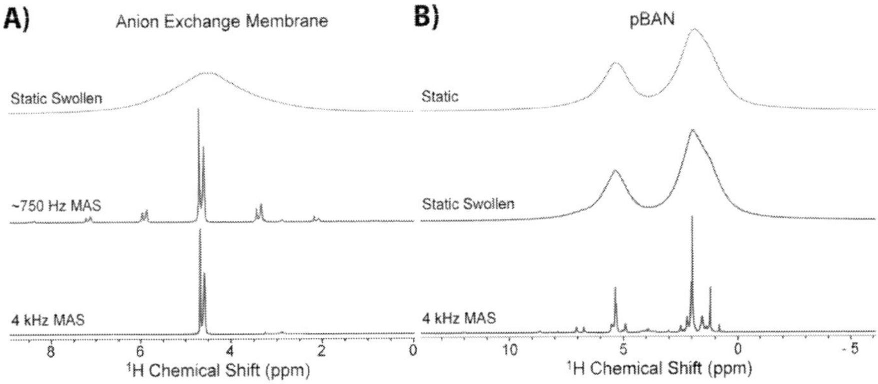

Figure 2. The improved resolution observed using ^1H HR-MAS NMR for the A) methanol swollen anion exchange membrane, and B) the $CDCl_3$ swollen pBAN (polyButadiene-AcryloNitrile) polymer.

With the dramatic increase in resolution observed utilizing HR-MAS, the arsenal of standard solution NMR techniques can be implemented for HR-MAS experiments, including solvent suppression, gradient-assisted sequences, and multiple-dimensional experiments. These can include INEPT (**I**nsensitive **N**uclei **E**nhanced by **P**olarization **T**ransfer), COSY (**CO**rrelation **S**pectroscop**Y**), NOESY (**N**uclear **O**verhauser **E**ffect **S**pectroscop**Y**), TOCSY (**TO**tal **C**orrelation **S**pectroscop**Y**), HETCOR (**HET**eronuclear **COR**relation), HMQC (**H**eteronuclear **M**ultiple **Q**uantum **C**oherence), and HMBC (**H**eteronuclear **M**ultiple **B**ond **C**orrelation) to name a few. As an example, a gradient-assisted 2D ^1H-^1H HR-MAS COSY NMR spectra for an ionic liquid adsorbed on an aluminum oxide membrane is shown in Figure 3. These types of correlation experiments could not be realized without the resolution afforded by HR-MAS. Additional discussion of this IL material is presented in Section 2.3.1.

It is also important to note that because anisotropic interactions (dipolar, CSA) are still present within these HR-MAS samples (though greatly reduced by molecular motion), it is also possible to apply some solid state NMR techniques during HR-MAS experiments. For example, to measure the residual homonuclear dipole-dipole interaction it is possible to incorporate a dipolar recoupling sequence to re-introduce this interaction under MAS. This allows the incoherent NOE exchange process to be replaced by a coherent polarization transfer process. These types of HR-MAS recoupling experiments have been demonstrated using DQ COSY for solid state synthesis samples,[19] while our group has incorporated a radio frequency dipolar recoupling (RFDR) sequence into the standard NOESY experiment for lipid membranes, including the extension of this dipolar recoupling into a mixing period during ^{1}H-^{13}C Heteronuclear correlation experiments.[20]

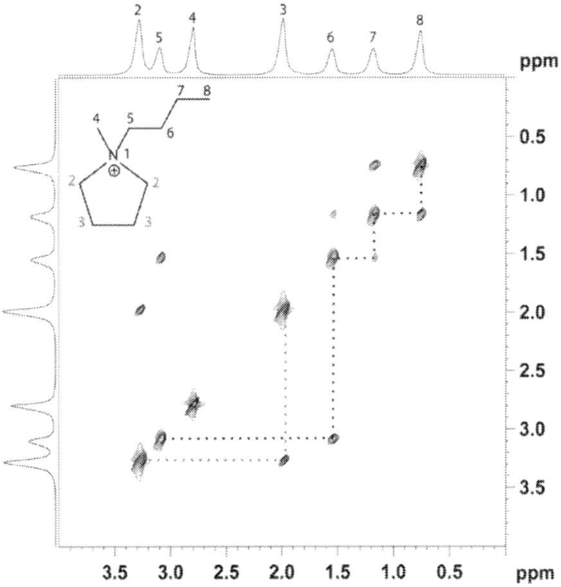

Figure 3. The gradient 2D ^{1}H HR-MAS NMR COSY spectrum for the ionic liquid $[MBPyrr]^{+}[TFSI]^{-}$ adsorbed into an inorganic aluminum oxide membrane. Even though the individual *J* couplings were not resolvable, these types of correlation experiments can still be realized under HR-MAS.

1.2. Experimental protocol and limitations

Many HR-MAS samples contain liquids or liquid-like materials. These samples require special care to prevent dehydration that can occur in a standard rotor with only a rotor cap. Various inserts are commercially available for packing of HR-MAS samples and can be seen in Figure 4. These inserts provide a tight seal to prevent dehydration or loss of solvent during MAS experiments. The inserts used generally depend on the sample size and need of the user. Examples of various insert options from Bruker BioSpin for a standard 4 mm rotor are shown in Figure 4. For ~12 μL samples, a Kel-F® bottom spacer (or a half drilled out rotor) and a top spacer with a seal screw can be utilized (Figure 4B, 4F and 4G). The top spacer insert (Figure 4F) contains a small hole in the top. When placed into the rotor at the properly gauged distance, any additional liquid will protrude through this hole. A Kimwipe® can then be utilized to remove this liquid, followed by using a small Kel-F®

screw to seal the insert (Figure 4B) and prevent dehydration. For ~50 μL volumes, a standard rotor can be utilized with the above top spacer and seal screw (Figure 4B, 4F, 4L). In addition to the spacers, a 30 μL disposable insert is also available (Figure 4I). The disposable Kel-F® inserts use a plug and screw cap to keep the sample well sealed (Figure 4B, 4H). The inserts are efficient if multiple samples need to be run, as samples can be packed into inserts and then easily placed into and removed from rotors, without having to wash the rotors between runs, and without having to own a large number of MAS rotors.

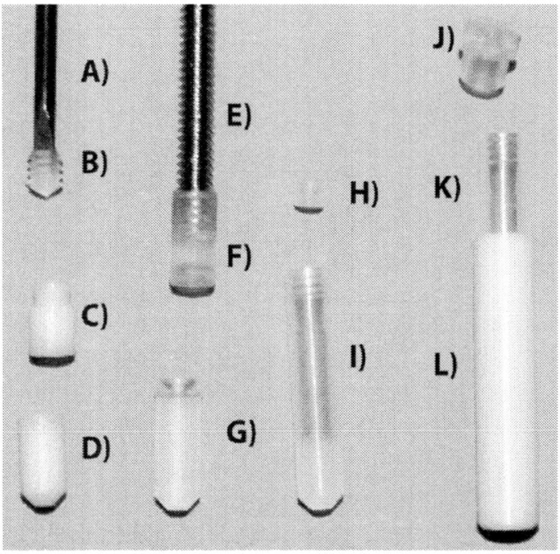

Figure 4. The tools and inserts used for HR-MAS NMR. These include A) the specialized tool for screw cap insertion, B) the sealing screw cap, C) the upper Teflon® insert, D) lower Teflon® insert for 30 μL volume, E) screw for insertion/extraction of top insert, F) top Kel-F® insert, G) bottom Kel-F® insert for 12 μL sample volume, H) plug for disposable insert, I) disposable 30 uL Kel-F® insert, J) 4 mm rotor cap, K) disposable inserted partially in a 4 mm rotor, L) 4 mm zirconia MAS rotor. All these parts are for the Bruker HR-MAS system, and may vary between vendors.

Temperature regulation is important in HR-MAS NMR for both sample purposes, and experimental reasons. The temperature of the bearing gas in many probes can be regulated to maintain a stable sample temperature; however, there is additional frictional heating that comes from spinning the rotor. To compensate for MAS related heating of the sample, the HR-MAS probe temperature should be calibrated at various MAS speeds and changes made to the temperature of the bearing gas to regulate the sample at the desired temperature. Temperature calibration of a HR-MAS probe has been discussed in literature.[21] This study utilized both methanol and glucose as NMR thermometers to determine the impact of MAS on temperature. Temperature changes measured at 2, 4, 6, and 8 kHz MAS where found to be 0.8 ± 0.1, 2.2 ± 0.1, 5.0 ± 0.1, and 7.9 ± 0.2 K in methanol and -0.3 ± 0.1, 1.2 ± 0.2, 3.0 ± 0.2, and 6.1 ± 0.1 K for glucose.[21]

Like all high resolution NMR techniques where resonances can be very narrow, it is important to be able to shim a sample to achieve the best possible resolution. Unlike standard high resolution NMR probes, the HR-MAS probes have the samples spinning at the magic angle, therefore the standard shimming protocols cannot be utilized. Instead, a combination of the standard shims must be used to address inhomogeneity of the B_0 field. By placing the probe so that the rotor is in the (x, z) plane in the laboratory frame, a combination of the laboratory shims can be used to express the Z shims along the magic angle. To obtain the equivalent response of a Z shim in standard shimming protocol, *i.e.* the Z shim along the magic angle (B_Z^{MAS}), the user needs to optimize a combination of the Z and X room temperature shims given by the linear relationship $B_Z^{MAS} = (1/\sqrt{3})B_Z^{LAB} - (\sqrt{2/3})B_X^{LAB}$. Likewise, the equivalent of a Z^2 shim is given by $B_{Z^2}^{MAS} = B_{(X^2-Y^2)}^{LAB} - 2\sqrt{2}B_{XZ}^{LAB}$ and the equivalent Z^3 shim is defined using $B_{Z^3}^{MAS} = -(2/3\sqrt{3})B_{Z^3}^{LAB} - (1/\sqrt{6})B_{XZ^2}^{LAB} + (5/\sqrt{3})B_{(X^2-Y^2)Z}^{LAB} - (5/3\sqrt{6})B_{X^3}^{LAB}$. The higher order Z shims map directly as $B_{Z^4}^{MAS} = -(7/18)B_{Z^4}^{LAB}$ and $B_{Z^5}^{MAS} = -(1/6\sqrt{3})B_{Z^5}^{LAB}$.[22] For the experimental examples discussed in this chapter, the authors typically only adjusted the shims up through $B_{Z^3}^{MAS}$. The Bruker Biospin Manual for High Resolution Magic Angle Spinning Spectroscopy points out that in theory shimming could be performed with just laboratory shims X, XZ,XZ^2, Z^4, and Z^5.[23] In reality, additional shims can be used to compensate for any inefficiency in the shim coil. A previous study has shown that once good shims are obtained, those shims can be utilized for samples with the same detection volume and position, for example the same sample size in the same type of rotor.[24] The shims should be independent of solvent because any susceptibility differences caused by solvent will be averaged out by MAS and does not affect the shims.[24] Piotto *et al.* also demonstrated shimming issues involved in systems containing water. In many cases, improved shims will be detected in an increase in lock signal. Deuterated water (HDO), however, has a strong chemical shift dependence on the temperature. When a sample spins, there can be a temperature gradient across the sample which can cause a portion of the HDO resonances to shift and broaden. If the probe is shimmed again based on the lock level from the water resonance, then any other signal in the spectra will most likely be deshimmed as those signals do not have as large of a chemical shift temperature dependence. With this in mind, Piotto *et al.* emphasize that shimming should be performed based on line shape, not lock level and attention should be taken when working with materials containing water.[24] More recently the MAGIC SHIMMING method has been described which utilizes a conventional homospoil gradient pulse to perform gradient shimming on MAS probes. This techniques does not require a gradient along the magic angle, and will reduce some of the trial and error presently inherent with manual shimming of HR-MAS probes.[25]

2. HR-MAS NMR spectroscopy in materials characterization

In this section a brief review of HR-MAS NMR studies involving the characterization of materials is presented. The majority of studies involved modified surfaces or surface interactions, and demonstrates the characterization need that HR-MAS NMR can fulfill.

2.1. Polymers

For rigid polymer materials the dipolar interactions are very large, leading to broadening of the NMR resonances, and can be characterized using standard solid state MAS NMR techniques. As noted before, these rigid polymers are not readily observed under HR-MAS conditions. For solvent swollen polymers the increased mobility of the polymer segments leads to a semi-solid regime where the moderate spinning speeds of HR-MAS are sufficient to produce high resolution NMR spectra. For example see the improved resolution observed for $CDCl_3$ swollen pBAN (Figure 2B). This ability to obtain spectra for mobile domains or components is the basis for the majority of HR-MAS polymer studies. A variety of different solvents are employed in combinatorial and solid phase synthesis, many of these with multiple ^1H resonances. In addition, there may be reasons to avoid having to dry the sample prior to introducing a deuterated solvent. In these cases, complicated solvent suppression techniques may be required.[26] It has been demonstrated that diffusion-filter HR-MAS spectra (see PFG discussion, Section 3.1) can provide suppression of solvent resonances based on differential diffusion behavior.[27]

One example of using HR-MAS NMR for characterizing polymer materials is the demonstration of the cyclic polyamide receptor threading onto the highly flexible polyethylene glycol (PEG) polymer chain attached to a polystyrene bead. Distinct PEG resonances for the threaded and non-threaded complexes could be easily resolved under HR-MAS conditions, while 2D NOESY spectra showed cross-peaks between the aromatic protons ($\delta \sim +8.87$ ppm) of the rotaxane and the methylene protons ($\delta \sim +3.55$ ppm) of the PEG polymer chain.[28] In another study, the complexation of Zn and Ru metalloporphyrins to beads functionalized with pyridyl ligands revealed that supermolecular interactions and changes in the dynamics are directly probed by HR-MAS techniques.[29]

The vulcanization of butadiene rubber (BR) with different curing systems has also been monitored by ^1H and ^{13}C HR-MAS NMR. This rubber is a very mobile system, with the increased resolution afforded by HR-MAS allowing the chemical identity of cross links to be determined, and revealing that α-substitution or addition depended on the disulfide cross-linkers employed.[30, 31] Other HR-MAS studies observed the impact of cross-linking in solvent swollen poly(amidoamine) polymers, the nature of water interactions in these same class of polymers, [32, 33] or the cross-linking performance in silicon-containing soybean-oil copolymers.[34] HR-MAS NMR has also been used to study the functionalization of poly hydroxyethyl methacrylate (HEMA) cryogels,[35] cyclomaltoheptaose polymers,[36] the synthesis of hyper-branched *bis* (hydroxymethyl) propionic acid (*bis*-MPA) polymers as a function of catalysts,[37] and polymers for nanoparticle stabilization.[38]

HR-MAS NMR has found use in investigation of porous polyalkylvinyl ether polymer particles being used for stationary phases in chromatographic applications. These measurements were performed using the same solvent as HPLC and allowed details about the polymer structure and mobility to be evaluated; these properties are expected to impact the chromatographic process.[39] Interactions between solution molecules and the polymer component of HPLC stationary phases (C18, C30 and PEAA)[40] or the interaction with molecularly imprinted polymers [41] have also been investigated using HR-MAS NMR. These studies used saturation transfer difference pulse sequences to identify those molecules that were involved in binding to the larger stationary phase.

The NMR spectroscopic characterization of polymer degradation is commonly directed towards analysis of the small degradation fragments that are solvent soluble. By using HR-MAS NMR the hydrolytic degradation of the biodegradable photo-initiated cross linked poly(DL-lactide)-dimethacrylate (PDLLA) polymer network was directly monitored. Swelling these polymers in the solvent dimethyl sulfoxide (DMSO) and a combination of 1D and 2D (COSY, NOESY) experiments allowed the different signals in the PDLLA chain to be identified. Degradation in a base was shown to occur through hydrolysis of the ester bonds within the poly(lactide) segment.[42] HR-MAS has also been used to study the role of partial hydrolysis in controlling the composition of hydrophobic polyacrylamide gels.[43]

2.1.1. *Example: HR-MAS NMR investigations of anion exchange membranes*

The increased resolution achieved with ^{1}H HR-MAS on the solvent swollen anion exchange membrane (AEM) was demonstrated in Figure 2. Due to differences in the magnetic susceptibility between the swelling solution and the polymer membrane, the water and methanol resonances were not resolved under static conditions. Under HR-MAS four distinct NMR resonances were observed for the solvent species; two water resonances and two methanol resonances. Based on line widths and PFG HR-MAS diffusion measurements (Section 3.2.1) the higher ppm shifted water and methanol were assigned to bulk (or free) methanol or water within the membrane, while the lower ppm shifted water and methanol resonances exhibit slower diffusion rates and were assigned to water or methanol associated (bound) within the membrane. The increased resolution obtained under HR-MAS was further exploited to explore the interchange of water and methanol between the different binding environments within the AEM using 2D HR-MAS NMR NOESY exchange experiments (Figure 5). At short mixing times (1 ms) no correlations were observed between any of the resonances. With increasing mixing time (> 10 ms), cross peaks were observed between free water and membrane-associated water, plus cross peaks between free methanol and membrane-associated methanol, resulting from physical exchange of water (red dashed lines) or methanol (red dashed lines) between the free and unbound environments. With long mixing times (> 200 ms), NOE correlation between methanol and water were observed (green dashed line), and result from through space NOE magnetization exchange (not physical spatial exchange). Interestingly, these NOE correlations were only observed between associated water and associated methanol, as well as free water and free methanol, supporting the assignment and indicating that the spatial contact of water and methanol is maintained in these associate membrane environments. Additional exchange studies are ongoing.

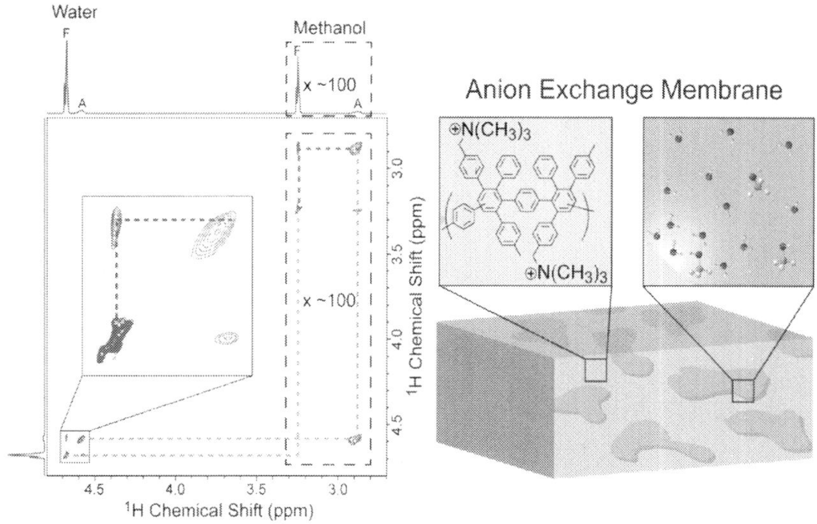

Figure 5. 2D ^{1}H HR-MAS NMR NOESY spectrum (500 ms mixing time) of an anion exchange membrane swollen in a 1N methanol solution. The increased spectral resolution obtained for the swelling solvent made it possible to detect exchange between free and associated water and between free and associated methanol (red dashed lines), as well as NOE magnetization exchange between free water and methanol and associated water and methanol (green dashed lines).

2.2. Ceramics, zeolites, catalysts surfaces and composites

Due to the rigid or crystalline nature of ceramics, zeolites, catalysts or inorganic/organic composites, solid state MAS NMR spectroscopy remains the dominant NMR characterization method, and can involve a range of nuclei and multi-pulse NMR techniques. In contrast, the limited number of HR-MAS NMR studies is directed towards the investigation of gas/solid adsorption or intercalation within these materials. For example, the degradation of the pesticide diethyl parathion and the chemical warfare agent (CWA) simulant diisopropyl fluorophosphates in functionalized montmorillonite clay using ^{31}P HR-MAS NMR has been reported. The HR-MAS removed susceptibility effects present from the clay material, and allowed a direct measure of the decomposition kinetics. The adsorption of organosilanes on the $MgCl_2$ in Ziegler-Natta catalysts has been investigated using ^{1}H HR-MAS, and demonstrates differential binding (differential mobility) to the Mg surface with changing surface coverage and as a function of the degree of saturation in the Mg coordination sites.[44]The polymer mobility in the condensed sol-gels produced from organosilanes and phenyl siloxanes was also evaluated.[45]

2.3. Ionic liquids and liquid crystals

HR-MAS NMR should be an ideal tool for the characterization of ionic liquids (ILs) and liquid crystals (LC) due to the high viscosities commonly encountered in ILs, and the thermal- or concentration-induced ordering transitions of LCs. Ionic liquids are interesting compounds, and continue to be used for a wide range of material science applications. ILs are used as liquid electrolytes in energy storage and production devices, solvents for CO_2 capture and green chemistry, solutions for biomass processing, nanostructured synthesis and solvents for catalysis reactions. For many of these applications room temperature ionic liquids (RTIL) are utilized as neat solutions without another solvent. Strong attractions between the IL cation and anion component may lead to the formation of important structural motifs that could be missed or changed with the introduction of a solvent. For NMR characterization, solvent-IL interactions may also mask subtle changes in the chemical shift produced by other molecular interactions of interest. In the case of ILs, it is now recognized that HR-MAS NMR provides a powerful tool for the characterization of these systems.

One of the earliest reports involving HR-MAS NMR of ILs studied silica-immobilized ILs suspended in DMSO,[46] which produced nearly liquid-like resolution. In this same study, they were also able to obtain 2D ^{1}H-^{13}C HR-MAS NMR HMQC spectra of these attached ILs. Another study explored the use of ILs as a solvent or chemical reaction media. The standard solution NMR of compounds dissolved in these ILs suffered from low resolution, produced by the high viscosities of the ILs, dynamic range issues due to the lack of deuterated ILs, and concerns about chemical shift referencing due to the high magnetic susceptibility of ILs. To demonstrate the capabilities of HR-MAS NMR, Rencurosi and co-workers[47] dissolved *para*-methoxy benzyl acetate and a glucopyranoside in a series of different ILs. Even at moderate spinning speeds there was significant improvement in resolution, and the chemical shift referencing became consistent with that observed in solution NMR using traditional deuterated solvents. In addition, they were able to follow acetylation of *para*-methoxybenzyl alcohol directly in the ILs. The increased resolution inherent in this technique was also important for investigations of CO_2 interactions with imidazolium based ILs using ^{13}C HR-MAS NMR.[48] ^{1}H HR-MAS NMR has also been used to probe the interaction of the RTIL *N*-methylimidazolium chloride ($[Hmim]^{+}Cl^{-}$) with the silica surface of Aerosil. This adsorption was monitored through changes in chemical shift with increased IL loading and Cl^{-} salt concentration. Based on the magnitude of the chemical shift variations, the interaction was found to preferentially involve the H(2) position of $Hmim^{+}$, and was a physisorption versus a chemisorption process.[49] In another study, the dynamics of IL's confined in monolithic silica ionogels were followed by ^{1}H HR-MAS NMR and relaxation experiments, and demonstrated that the IL maintained liquid like behavior with very little reduction in motions even as the pore size diminished from 12 nm to 1.5 nm.[50] Surprisingly, only a single ^{1}H HR-MAS NMR diffusion study of a non-biological LC material confined in nanopores has been reported.[51]

2.3.1. *Example: HR-MAS NMR investigation of ionic liquids on surfaces*

While many RTILs are "liquids" at room temperature, they can be highly viscous, making standard solution NMR analysis on the neat ILs challenging (as noted above). In our laboratory, a novel series of quaternary ammonium and cyclic pyrrolidinium RTIL have been studied using different NMR techniques. This includes pulse field gradient (PFG) NMR to measure diffusion and ^{14}N NMR relaxation experiments to determine IL molecular reorientation times.[52] More recently, we have explored using ^{1}H HR-MAS NMR to characterize the interactions between IL and metal oxide surfaces. Figure 6A shows the ^{1}H HR-MAS NMR spectra of the neat ionic liquid, *N*-methyl-*N*-(n-butyl) pyrrolidinium bis(trifluoromethanesufonyl) imide ([MBPyrr]$^{+}$[TFSI]$^{-}$). The changes in the NMR spectrum for the neat solution under static conditions (top) and under MAS conditions (bottom) demonstrate the simple resolution enhancing properties of HR-MAS (previously discussed in Section 1.1). The static NMR spectrum revealed distorted line shapes reflecting poor magnetic field homogeneity across the sample volume, along with broadening (FWHM ~ 115 Hz) due to differences in local magnetic susceptibility within the sample. The broad line width is commonly encountered when working with highly viscous RTIL that are not dissolved into solvents. Increasing the temperature (within the limitations of the instrumentation) will improve the resolution for these RTIL, but may not be desirable in some instances. Not surprisingly, even slow spinning (~1 kHz) of the neat RTIL provides an immediate improvement in the resolution. For the ([MBPyrr]$^{+}$[TFSI]$^{-}$) example, the line width are reduced to ~ 3 Hz (~40 fold reduction), such that small ^{1}H-^{1}H spin-spin *J* coupling are clearly resolved (Figure 6A, bottom).

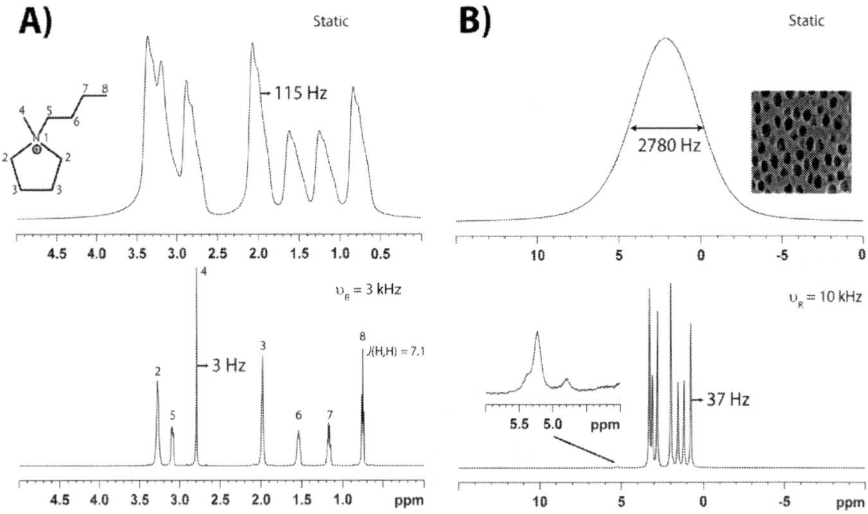

Figure 6. The ^{1}H HR-MAS NMR of the ionic liquid [MBPyrr]$^{+}$TFSI^{-} at 298K as a A) neat liquid and B) adsorbed (20 wt%) into an aluminum oxide ANOPORE inorganic membranes (200 nm pore diameter). The spectra in the top portion of each figure were obtained under static conditions, with the spectra in the bottom portions of each figure obtained under MAS conditions.

The loss of resolution is even more dramatic when $[MBPyrr]^+[TFSI]^-$ is adsorbed into the small pores (200 nm) of an inorganic aluminum oxide membrane. The ^1H NMR spectrum of the static sample is now a broad (~2780 Hz) featureless line providing no chemical shift resolution. Under HR-MAS the individual ^1H resonances become resolved, with a line width on the order of 37 Hz. While the individual J couplings are no longer resolved as they were in the neat IL case, the resolved chemical shifts allow one to compare directly with the neat IL. In the adsorbed $[MBPyrr]^+[TFSI]^-$, the *n*-butyl methyl group protons (H-8) and the ring methylene protons (H-3) show a very small chemical shift change ($\Delta\delta = 0.007$ ppm) due to adsorption with the ANOPORE membrane. In contrast the methyl group (H-4) and the *n*-butyl methylene protons (H-5) reveal an asymmetric line shape with chemical shift variations ranging from $\Delta\delta = +0.01$ to $+0.02$ ppm. This result suggests that while the adsorption process is weak, it occurs preferentially through the IL N^+ with the surface. The ability to resolve such small differences under HR-MAS NMR is important in identifying these surface interactions.

Even though the ^1H HR-MAS NMR spectra for the surface adsorbed $[MBPyrr]^+[TFSI]^-$ does not reveal resolvable J coupling, it is still possible to obtain a 2D COSY spectrum for this material, as shown in Figure 3. This correlation experiment immediately confirms the chemical shift assignments for the neat IL sample, and clarifies differences seen for the IL dissolved in a solvent. The 2D ^1H HR-MAS NOESY NMR spectra for the IL $[MBPyrr]^+[TFSI]^-$ adsorbed into the ANAPORE membrane is shown in Figure 7. Interestingly there are numerous cross peaks observed even at moderate mixing times, reminiscent of the liquid-ordered phase observed in lipid bilayers (For example see Ref. [20, 53, 54]). These through-space correlations are stronger than those observed in the neat IL (not shown), and suggest that an increase in the dipolar-dipolar interactions is occurring for the surface adsorbed species. This argues that local motions have become reduced for the adsorbed species. There are also some missing or weak correlations (dashed circle) implying distinct conformations present for the *n*-butane chain on the adsorbed species. Also of interest is the appearance of new chemical environments for the adsorbed IL. This includes a weak resonance at $\delta = +4.8$ ppm which is assigned to residual water, but shows no correlations with the IL. There are two additional ^1H environments at $\delta = +5.2$ and $+5.3$ ppm, which are attributed to additional surface water species on the aluminum oxide surface. The environment at $\delta = +5.2$ ppm shows a through space NOE correlation with a new resonance near 2 ppm. This small resonance is buried under the shoulder of the dominant methylene (H-2) resonance and has not been presently assigned. These types of correlation experiments demonstrate the capabilities and information that can be obtained from HR-MAS studies of ILs, and suggest future efforts along these lines are warranted.

2.4. Surface modified nanoparticles

HR-MAS NMR studies of surface modified nanoparticles (NP) are closely related to those of the polymer resin studies mentioned in the previous sections, with accurate characterization of the surface-attached ligands being the primary objective. For NPs the solvent plays the role of suspending or dispersing the material, in contrast to the role of swelling that the solvent plays in analysis of resins. For HR-MAS NMR studies of NPs, the optimal solvent provides both high solubility, and good dispersion (prevents aggregation) of the NPs.

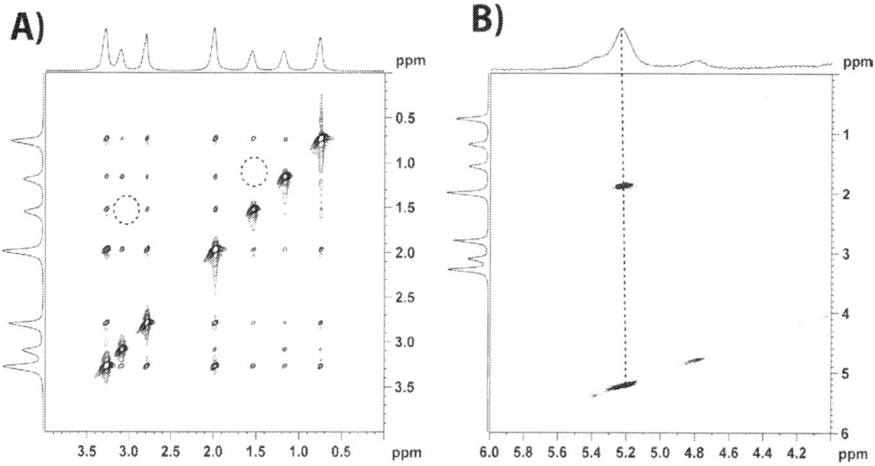

Figure 7. The 2D ^1H HR-MAS NOESY NMR correlation spectra for the ionic liquid $[MBPyrr]^+[TFSI]^-$ adsorbed into an inorganic aluminum oxide membrane (pore size 200 nm). The spectrum in A) shows numerous through space correlation between the protons of the IL, while B) shows the spectral expansion for the water and surface associate-water species and corresponding NOE correlations.

HR-MAS NMR studies of surface modified NPs include the 1D and 2D ^1H NMR investigation of modified gold (Au) NPs using a wide range of high resolution correlation experiments: COSY, TOCSY and HMQC.[55] This revealed that the relative signal sensitivity (intensity) depends on the distance between the detected ^1H and the surface of the Au NP. This distance correlation is either a function of reduced local dynamics, or spin-spin T_2 relaxation effects governed by the Au surface. Another study involved the active molecular compounds Aloin A and Aloesin extracted from the leaf of Cape Aloe, and was able to show how these compounds selectively stabilize Au NPs preferentially through the glucose component.[56] HR-MAS NMR has also been used to identify the binding motif of peptides on the Au NP surface.[57] While HR-MAS NMR is typically employed to overcome anisotropy of magnetic susceptibility or residual dipolar interactions, Poito and co-workers[58, 59] have reported an interesting set of experiments that demonstrated HR-MAS can be employed to overcome paramagnetic effects present in iron oxide NPs. Through a careful analysis of these iron oxide systems, the surface structure and ligand binding (chelation) configuration was determined, and in several cases were quite different from the standard single point attachment proposed by others.[58]

^{1}H HR-MAS NMR has also been used to look at the surface modification in polymer based NPs, including the monitoring of multiple synthetic steps employed during the peptide-surface modification of poly(vinylidene fluoride) (PVDF) nanoparticles,[60] or the surface modification of Dendron based NPs.[61] Polystyrene embedded silver clusters produced by a thermolysis reactions have also been characterized using HR-MAS NMR.[62]

2.5. Surface immobilized linkers and catalysts and chiral stationary phases

HR-MAS NMR spectroscopy has also been extensively used in investigating the mobility of linkers and catalysts attached to other materials, including ^{1}H and ^{31}P HR-MAS NMR studies of phosphine, bisphosphinoamine linkers and corresponding metal catalysts.[63-66] These investigations demonstrate very nicely that the surface mobility as probed by HR-MAS is strongly dependent on the swelling solvent employed in the studies.[67] The grafting of organotin catalysts to polystyrene for the tranesterification reaction between ethyl acetate and *n*-octanol or the ring-opening polymerization of ε-caprolactone has been followed by both 1D and 2D ^{1}H, ^{13}C and ^{119}Sn HR-MAS NMR.[68-70] While the catalytic activity of the attached tin was unchanged following numerous cycles, the NMR revealed there was actually a change in the liquid-solid interface with a reduction in the mobility of the undecyltin trichloride catalyst at the resin surface. For experiments with rapid catalytic turnover rate there was a reduction in performance. This reduction in activity was reversible with solvent extraction, and was confirmed by HR-MAS to involve the change of the Sn environment back to the original state.[69] The use of ^{119}Sn HR-MAS NMR has also been demonstrated to provide a quantitative measure of the tin loading in supporting catalysts.[71] This HR-MAS NMR study was also the first to incorporate the ERETIC (*e*lectronic *re*ference *t*o access *i*n vivo *c*oncentration) method.

HR-MAS NMR techniques have been used to characterize the surfaces of silica and polystyrene particles modified with chiral agents that allow identification of different enantiomers. In one study the NMR was able to probe regiochemistry of the surface chemistry, along with monitoring the stability against decomposition of the polystyrene resins.[72] Another investigation used 2D transfer NOESY NMR experiments to evaluate the stereoselective binding of molecules to the chiral stationary phase based on the negative cross-peaks associated with the transfer NOE effect.[73]

2.6. Soil and humic materials

While standard solid state MAS NMR techniques have become common in the investigations of humic, soil and coal materials, the implementation of HR-MAS techniques for these materials has been more limited. The ability of HR-MAS to identify different motional regimes has proven powerful in the elucidation of the chemistry at water/soil interfaces.[74-79] These NMR studies show that at this interface fatty acids, aliphatic esters and alcohols are the prominent species, and that aromatic functional groups are protected by hydrophobic regions and are not directly accessible to the penetrating water,[74] while surface polymethylene groups may control the sorption properties of organo-clay complexes.[76, 80] These types of studies have been extended to three-dimensional (3D) HMQC-TOCSY to further increase the resolution of the highly overlapping spectra from humic materials.[81]

3. PFG HR-MAS NMR to measure diffusion in materials characterization

Modern HR-MAS probes include a gradient coil that can produce a magnetic gradient along the long axis of the MAS rotor which is set at the magic angle (θ = 54.7°). Standard MAS probes have also been combined with micro-imaging gradient systems in which gradient coils wrapped around the stator were not employed, but instead rely on imaging gradients that were external to the probe. [82] This is not the common configuration, and has been replaced by significant development efforts from the instrumental vendors involving the integration of gradients directly into the HR-MAS probes. With the gradient coil along the magic angle pulsed field gradient (PFG) experiments can be performed under MAS conditions. Fortuitously, enhanced T_2 times are generally observed under MAS allowing PFG experiments with longer diffusion times to be implemented than would have been accessible with static conditions.

During the PFG diffusion experiments, the application of a gradient "tags" a spin with a phase that is related to its spatial position. Figure 8A provides a pictorial representation of the dephasing of spins around the magic angle caused by the magic angle gradient. If the position of the spin does not change during the diffusion period (Δ), this dephasing is refocused and the original signal intensity (S_0) is recovered. If on the other hand the spin changes spatial position (diffuses) during Δ, the dephasing for that spin is not refocused, and the signal intensity decreases. The loss in signal intensity with increasing gradient strength is related to the self-diffusion rate with the classic Stejskal-Tanner equation:

$$\frac{S}{S_0} = e^{-\gamma^2 g^2 \delta^2 \left(\Delta - \frac{\delta}{3}\right) D} \tag{1}$$

where S is the experimental amplitude of the signal, γ is the gyro magnetic ratio, g is the gradient strength, δ is the gradient pulse length, Δ is the diffusion time, and the D is the diffusion constant.[83] By fitting this decay the diffusion constant can be determined. Figure 8B shows an example of the signal intensity loss observed during diffusion experiments for the different two water environments present in swollen AEM with increasing gradient strength.

PFG diffusion HR-MAS NMR experiments can also be used to obtain diffusion-filtered NMR spectra through the separation of different motional regimes present in complex mixtures. This filtering is accomplished by selecting Δ times where the signal intensity (Eqn. 1) for the fast diffusing components has been highly attenuated, while the slow diffusing components have significant signal intensity remaining. Practical aspects of diffusion measurements using HR-MAS have been previously discussed by Viel *et al.* [84] A significant finding from this study was that sample volume played a role in the reliability of diffusion rates measured. It was shown that small volumes (~12μL) exhibit reproducible diffusion rates, while larger volumes (~50μL, a full 4 mm rotor) produced inconsistent or unreliable data.

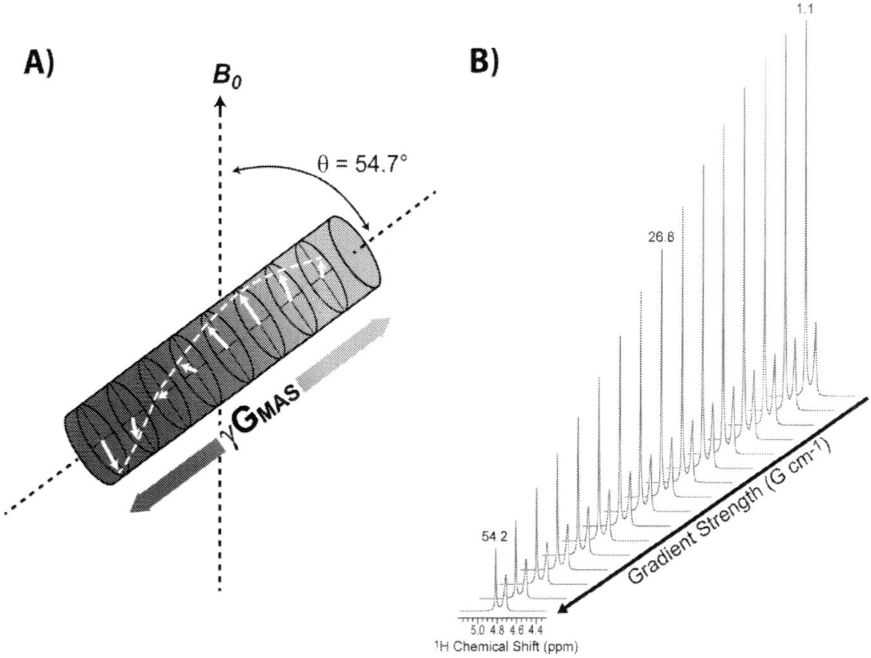

Figure 8. A) Pictorial representation of the gradient produced along the magic angle of the rotor. B) The decay of two different water signals found in a 1N methanol solution of an AEM membrane with increasing gradient strength. Gradient strength values (G/cm) are shown above the stack plot.

Three commonly used PFG diffusion pulse sequences are shown in Figure 9. The basic spin-echo diffusion sequences is depicted in Figure 9A, but is limited by loss of signal intensity due to spin-spin T_2 relaxation during the diffusion period Δ. Two variations of the stimulated echo (STE) sequence are shown in Figure 9B and 9C, and in this case spin-lattice T_1 relaxation is occurring during the diffusion Δ period. For most materials T_1 values are longer than T_1 making these STE sequences the preferred choice for material analysis. It should be noted that all of the gradient pulses in these sequences are trapezoidal shaped to compensate for the inability of instrumentation to generate perfect rectangular gradients. Shaped pulses, like sine or trapezoidal shapes, are used to produce experimentally reproducible gradient pulses. The PFG stimulated echo with dipolar gradients and spoil gradient, depicted in Figure 9B, is beneficial for the use with many HR-MAS samples which exhibit differences in magnetic susceptibility across the sample.[85, 86] The PFG stimulated echo in Figure 9C has an additional delay to the PFG stimulated echo in Figure 9B that is utilized to address eddy currents within the sample.[87]

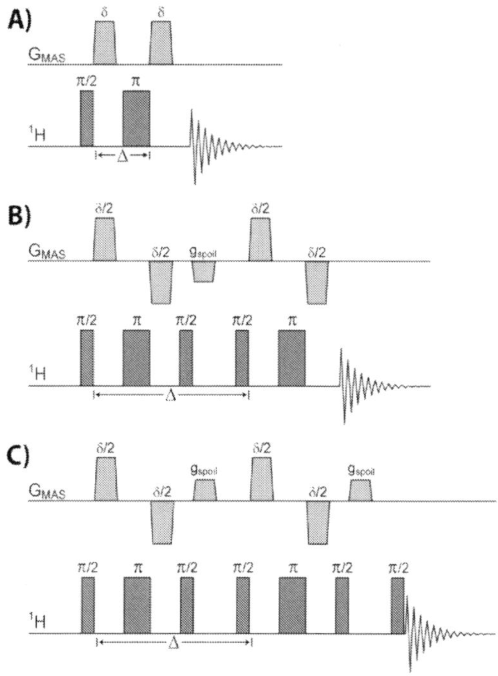

Figure 9. Diffusion pulse sequences. Pulse Field Gradient (PFG) A) Spin-Echo, B) PFG Stimulated Echo with dipolar gradients and spoil gradient based on Cotts *el al.* 13-interval sequence[85], and C) PFG Stimulated Echo with dipolar gradients and spoil gradient with an additional eddy current delay. G_{MAS} indicates that the gradient is applied along the magic angle.

3.1. Diffusion in zeolites, nanoparticles and liquid crystals

As noted for other material studies, heterogeneities in the magnetic susceptibility or restricted molecular motions within zeolite crystallites lead to broadening of the NMR signal, such that resolution of individual species in mixtures becomes difficult. HR-MAS NMR resolves this issue, and has led to the utilization of PFG NMR diffusion experiments on organic mixtures in zeolites. This technique has been used to study the diffusion of *n*-butane in silicalite-1,[88] ethane, water and benzene mixtures adsorbed to the zeolite NaX,[89] acetone-*n*-alkane (C_6 to C_9) mixtures in nanoporous silica,[90] or mixtures of *n*-butane and *iso*-butane adsorbed in MFI zeolite. The increased resolution afforded by these PFG HR-MAS studies on mixtures reveal the obstructive influence of the isopropyl molecules or bulky benzene on the diffusion of other molecular species, [89, 91] and that the creation of acetone-alkane complexes greatly impacts the observed diffusion properties.[90]

PFG HR-MAS NMR has also been used to obtain 2D DOSY (**D**iffusion **O**rdered **S**pectroscop**Y**) spectra of surface modified iron oxide NPs. These results were able to distinguish between NP-bound and free ligands in these materials.[58] These types of PFG

HR-MAS NMR experiments should prove useful in understanding the surface-ligand dynamics present in modified NPs.

Diffusion experiments using HR-MAS NMR has also been found useful in the analysis of transport properties in lipid membranes.[92-94] Due to the anisotropic molecular reorientation of these liquid crystalline (LC) systems, significant dipolar coupling remains, leading to broad lines and short relaxation times. However dipolar coupling can be reduced through the use of MAS. Surprisingly, the use of PFG HR-MAS NMR for non-biological LCs is more limited, with the single investigation of local molecular dynamics of the thermotropic LC 4′-pentyl-4-cyanobiphenyl (5CB) confined in Bioran glasses with pore diameters of 30 nm and 200 nm being reported. By utilizing PFG techniques it was possible to measure the diffusion constants as a function of temperature through the isotropization temperature of the liquid crystal, thus demonstrating that for this case there is only a minor reduction in the diffusion rates with molecular confinement.[51]

3.2. Diffusion in polymers

There is extensive literature on the use of PFG NMR to measure diffusion of polymer solutions and melts, along with PFG diffusion measurements of different species adsorbed into polymers, including gases, water, organic solvents and electrolytes. There have been a limited number of examples where the improved resolution afforded by HR-MAS NMR was coupled with PFG. This work includes the development of diffusion filtered HR-MAS NMR techniques to study the gelation process of super-molecular gels,[95] along with a combination diffusion-filtering and a spin-echo enhanced (T_2 –filtered) experiment on DMF-swollen resins.[96] These techniques allowed the identification of free and surface bound molecules, while eliminating the signal from the immobile bulk resin matrix. In complex mixtures there may be future avenues for HR-MAS to resolve subtle differences in the local chemical environments as demonstrated in the following example.

3.2.1. Example of HR-MAS diffusion in anion exchange membranes

As discussed in Section 2.1.1, four distinct resonances were observed in the 1D ^{1}H HR-MAS NMR spectra of AEM polymers swollen in a 1N methanol solution. In Figure 10A, NMR spectra for three AEM polymers with different ion exchange capacities (IEC) are shown. Two resonances were observed for both water and methanol, and were assigned to free (F) and membrane-associated (A) environments. From this HR-MAS data it is easy to see that there is a correlation between chemical shift of the associated species and the IEC of the membrane, with both the associated water and methanol resonance shifting to lower ppm with decreasing IEC. This decrease in chemical shift is most likely due to a change in hydrogen bonding between the solvent components and the membrane, reflecting how strongly the solvent molecules are associated with the membrane. Recall that the resolution of these individual environments was not observable in the static NMR spectra (Figure 2A). Using ^{1}H HR-MAS PFG diffusion experiments, the self-diffusion constants were obtained for each of these four different environments, which were not accessible from the static data.

Because the ^1H signal for the AEM membrane is not readily observable under HR-MAS conditions, the diffusion rates obtained for the resolved solvent resonances were not biased by the polymer membrane. Figure 10B shows the signal decay for the associate methanol environment as a function of gradient strength for the three different IEC levels. The magic angle gradients were used to perform diffusion measurements utilizing a PFG Stimulated Echo with dipolar gradients and spoil gradient with a Δ=100ms (Figure 9B). The signal decay shows that there is a correlation between diffusion rate and IEC, exhibiting a faster diffusion rate with increasing IEC values. A more detailed analysis of this work is forth coming, but this example demonstrates the power of combining HR-MAS and PFG diffusion experiments.

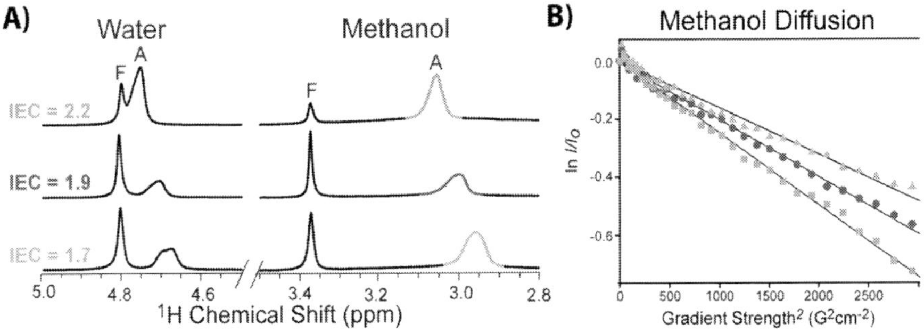

Figure 10. A) ^1H HR-MAS NMR spectra with the assigned free [F] and associated [A] water and methanol environments. B) The diffusion rates for the associated methanol in three different anion exchange membranes with varying ion exchange capacity (IEC) values. The colored peaks in the ^1H HR-MAS NMR spectra correlate to the colored symbols in the diffusion plot of the associated methanol peak of IEC = 2.2 (■), 1.9 (●), and 1.7 mequiv/g (▲).

4. Conclusions

The application of HR-MAS NMR to the characterization of materials or material interfaces that exist in the semi-solid range has been demonstrated. A wide variety of different material systems have been explored, showing that this technique can provide resolution and dynamic information where standard solution or solid state NMR techniques were unsuccessful. HR-MAS NMR is a powerful tool for the detailed characterization of modified surfaces and surface adsorbed species. This technique also provides a direct probe of differences in local mobility as reflected by line width variations. Through the combination of the enhanced resolution afforded by HR-MAS with pulse field gradient (PFG) capabilities, selective filtering and diffusion measurements of complex heterogeneous materials can also be realized. The ability to resolve and obtain diffusion rates for multiple environments in materials will prove beneficial for understanding the diffusion process in mixed chemical systems. While HR-MAS NMR is considered a mature, relatively routine technique, the application to the materials field is expected to continue being an active area of development. It is hoped that this review will encourage researchers to explore the application of HR-MAS NMR techniques to their different material systems.

Author details

Todd M. Alam* and Janelle E. Jenkins
Sandia National Laboratories, Department of Nanostructured and Electronic Materials, Albuquerque, NM, USA

Acknowledgement

Sandia National Laboratories is a multi-program laboratory managed and operated by Sandia Corporation, a wholly owned subsidiary of Lockheed Martin Corporation, for the U.S. Department of Energy's National Nuclear Security Administration under contract DE-AC04-94AL85000. The authors would like to acknowledge Michael Hibbs (Sandia) for providing the anion exchange membranes.

5. References

[1] Sarkar S. K., Garigipati R. S., Adams J. L., Keifer P. A. (1996) An NMR Method To Identify Nondestructively Chemical Compounds Bound to a Single Solid-Phase-Synthesis Bead for Combinitorial Chemistry Applications, *J. Am. Chem. Soc.* 118:2305-2306.

[2] Stöver H. D. H., Frèchet J. M. J. (1991) NMR Characterization of Cross-Linked Polystyrene Gels, *Macromolecules* 24:883-888.

[3] Gross J. D., Costa P. R., Dubacq J.-P., Warschawski D. E., Lirsac P.-N., Devaux P. F., Griffin R. G. (1995) Multidimensional NMR in Lipid Systems. Coherence Transfer Through *J* Couplings Under MAS, *J. Magn. Reson. Ser. B* 106:187-190.

[4] Maas W. E., Laukien F. H., Cory D. G. (1996) Gradient, High Resolution, Magic Angle Sample Spinning NMR, *J. Am. Chem. Soc.* 118:13085-13086.

[5] Bruker BioSpin, (2012) Biological Tissue Analysis. Available: http://www.bruker-biospin.com/probes_hrmas.html. Accessed 2012 April 10.

[6] Agilent Technologies, (2012) Nano Probes. Available: http://www.chem.agilent.com/en-US/Products/Instruments/magneticresonance/nmr/probes/liquids/nano/pages/default.aspx. Accessed 2012 April 10.

[7] JEOL, (2012) FGMAS. Available: http://www.jeol.cn/?p=1232. Accessed 2012 April 10.

[8] Doty Scientific, (2012) HR-MAS MAG Probe. Available: http://www.dotynmr.com/solids/HRMASMAG.htm. Accessed 2012 April 10.

[9] Lindon J. C., Beckonert O. P., Holmes E., Nicholson J. K. (2009) High-Resolution Magic Angle Spinning NMR Spectroscopy: Application to Biomedical Studies, *Prog. Nucl. Mag. Res. Sp.* 55:79-100.

[10] Zietkowski D., Davidson R. L., Eykyn T. R., De Silva S. S., deSouza N. M., Payne G. S. (2010) Detection of Cancer in Cervical Tissue Biopsies Using Mobile Lipid Resonances

* Corresponding Author

Measured with Diffusion-Weighted ^{1}H Magnetic Resonance Spectroscopy, *NMR Biomed.* 23:382-390.

[11] Beckonert O., Coen M., Keun H. C., Wang Y., Ebbels T. M. D., Holmes E., Lindon J. C., Nicholson J. K. (2010) High-Resolution Magic-Angle-Spinning NMR Spectroscopy for Metabolic Profiling of Intact Tissues, *Nat. Protoc.* 5:1019-1032.

[12] Valentini M., Ritota M., Cafiero C., Cozzolino S., Leita L., Sequi P. (2011) The HRMAS-NMR Tool in Foodstuff Characterization, *Magn. Reson. Chem.* 49:S121-S125.

[13] Vermathen M., Marzorati M., Baumgartner D., Good C., Vermathen P. (2011) Investigation of Different Apple Cultivars by High Resolution Magic Angle Spinning NMR. A Feasibility Study, *J. Agric. Food Chem.* 59:12784-12793.

[14] Shapiro M. J., Gounarides J. S. (2001) High Resolution MAS-NMR in Combinatorial Chemistry, *Biotechnology Bioengineering (Combinatorial Chemistry)* 71:130-148.

[15] Power W. P. (2003) High Resolution Magic Angle Spinning - Applications to Solid Phase Synthetic Systems and Other Semi-Solids, *Annual Reports NMR Spectroscopy* 51:261-295.

[16] Shapiro M. J., Gounarides J. S. (1999) NMR Methods Utilized in Combinatorial Chemistry Research, *Progress Nuclear Magnetic Resonance* 35:153-200.

[17] Brown S. P. (2012) Applications of High-Resolution ^{1}H Solid-State NMR, *Solid State Nuclear Magnetic Resonance* 41:1-27.

[18] Schröder H. (2003) High Resolution Magic Angle Spinning NMR for Analyzing Small Molecules Attached to Solid Support, *Comb. Chem. High T. Scr.* 6:741-753.

[19] Thieme K., Zech G., Kunz H., Spiess H. W., Schnell I. (2002) Dipolar Recoupling in NOESY-Type ^{1}H-^{1}H NMR Experiments Under HRMAS Conditions, *Organic Letters* 4:1559-1562.

[20] Alam T. M., Holland g. P. (2006) ^{1}H-^{13}C INEPT MAS NMR Correlation Experiments with ^{1}H-^{1}H Mediated Magnetization Exchange to Probe Organization in Lipid Biomembranes, *J. Magn. Reson.* 180:210-221.

[21] Nicholls A. W., Mortishire-Smith R. J. (2001) Temperature Calibration of a High-Resolution Magic-Angle Spinning NMR Probe for Analysis of Tissue Samples, *Magn. Reson. Chem.* 39:773-776.

[22] Sodickson A., Cory D. G. (1997) Shimming a High-Resolution MAS Probe, *J. Magn. Reson.* 128:87-91.

[23] Engelke F., Maas W. E. (1997) *High Resolution Magic Angle Spinning Spectroscopy User Manual Version 1.0*: Bruker Instruments, Inc. 50 p.

[24] Piotto M., Elbayed K., Wieruszeski J. M., Lippens G. (2005) Practical Aspects of Shimming a High Resolution Magic Angle Spinning Probe, *J. Magn. Reson.* 173:84-89.

[25] Nishiyama Y., Tsutsumi Y., Utsumi H. (2012) MAGIC SHIMMING: Gradient Shimming with Magic Angle Sample Shimming, *J. Magn. Reson.* 216:197-200.

[26] Smallcombe S. H., Patt S. L., Keifer P. A. (1995) WET Solvent Suppression and Its Applications to LC NMR and High-Resolution NMR Spectroscopy, *J. Magn. Reson., A* 117:295-303.

[27] Warrass R., Wieruszeski J.-M., Lippens G. (1999) Efficient Suppression of Solvent Resonances in HR-MAS of Resin-Supported Molecules, *J. Am. Chem. Soc.* 121:3787-3788.

[28] Ng Y.-F., Meillon J.-C., Ryan T., Dominey A. P., Davis A. P., Sanders J. K. M. (2001) Gel-Phase MAS NMR Spectroscopy of a Polymer-Supported Pseudorotaxane and Rotaxane: Receptor Binding to an "Inert" Polyethylene Glycol Spacer, *Angew. Chem. Int. Ed.* 40:1759-1760.

[29] de Miguel Y. R., Bampos N., de Silva K. M. N., Richards S. A., Sanders J. K. M. (1998) Gel Phase MAS ^{1}H NMR as a Probe for Supramolecular Interactionas the Solid-Liquid Interface, *Chem. Commun.* 2267-2268.

[30] Hulst R., Seyger R. M., van Duynhoven J. P. M., van der Does L., Noordermeer J. W. M., Bantjes A. (1999) Vulcanization of Butadiene Rubber by Means of Cyclic Disulfides. 3. A 2D Solid State HRMAS NMR Study on Accelerated Sulrfur Vulcanizates of BR Rubber, *Macromolecules* 32:7521-7529.

[31] Hulst R., Seyger R. M., Van Duynhoven J. P. M., van der Does L., Noordermeer J. W. M., Bantjes A. (1999) Vulcanization of Butadiene Rubber by Means of Cyclic Disulfides. 2. A 2D Solid State HRMAS NMR Study of Cross-Link Structures in BR Vulcanizates, *Macromolecules* 32:7509-7520.

[32] Calucci L., Forte C., Ranucci E. (2007) Water/Polymer Interactions in a Poly(amidoamine) Hydrogel Studied by NMR Spectroscopy, *Biomacromolecules* 8:2936-2942.

[33] Annunziata R., Tranchini J., Ranucci E., Ferruti P. (2007) Structural Characterisation of Poly(Amidoamine) Networks Via High-Resolution Magic Angle Spinning NMR, *Magn. Reson. Chem.* 45:51-58.

[34] Sacristán M., Ronda J. C., Cádiz V. (2009) Silicon-Containing Soybean-Oil-Based Copolymers. Synthesis and Properties, *Biomacromolecules* 10:2678-2685.

[35] Van Camp W., Dispinar T., Dervaux B., Du Prez F. E., Martins J. C., Frtizinger B. (2009) 'Click' Functionalization of Cryogels Conveniently Verified and Quantified Using High-Resolution MAS NMR Spectroscopy, *Macromolecular Rapid Communications* 30:1328-1333.

[36] Crini G., Bourdonneau M., Martel B., Piotto M., Morcellet M., Richert T., Vebrel J., Torri G., Morin N. (2000) Solid-State NMR Characterization of Cyclomaltoheptoase (β-Cyclodextrin) Polymers Using High-Resolution Magic Angle Spinning with Gradients, *J. Applied Polymer Science* 75:1288-1295.

[37] Komber H., Ziemer A., Voit B. (2002) Etherification as Side Reactions in the Hyperbranched Polycondensationof 2,2-Bis(hydroxymethyl)propionic Acid, *Macromolecules* 35:3514-3519.

[38] Favier I., Gómez M., Muller G., Picurelli D., Nowicki A., Roucoux A., Bou J. (2007) Synthesis of New Functionalized Polymers and their use as Stabilizers of Pd, Pt, and Rh Nanoparticles. Preliminary Catalytic Studies, *J. Applied Polymer Science* 105:2772-2782.

[39] Bachmann S., Hellriegel C., Wegmann J., Händel H., Albert K. (2000) Characterization of Polyalkylvinyl Ether Phases by Solid-State and Suspended-State Nuclear Magnetic Resonance Investigations, *Solid State Nuclear Magnetic Resonance* 17:39-51.

[40] Schauff S., Friebolin V., Grynbaum M. D., Meyer C., Albert K. (2007) Monitoring the Interactions of Tocopherol Homologues with Reversed-Phase Stationary HPLC Phases by ^{1}H Suspended-State Saturation Transfer Difference High-Resolution Magic Angle Spinning NMR Spectroscopy, *Anal. Chem.* 79:8323-8326.

[41] Courtois J., Fischer G., Schauff S., Albert K., Irgum K. (2006) Interactions of Bupivacaine with a Molecularly Imprinted Polymer in a Monolithic Format Studied by NMR, *Anal. Chem.* 78:580-584.

[42] Melchels F. P. W., Velders A. H., Feijen J., Grijpma D. W. (2010) Photo-Cross Linked Poly(DL-Lactide)-Based Networks. Structural Characterization by HR-MAS NMR Spectroscopy and Hydrolytic Degradation Behavior, *Macromolecules* 43:8570-8579.

[43] Feng Y., Billon L., Grassl B., Khoukh A., Francois J. (2002) Hydrophobically Associated Polyacrylamides and their Partially Hydrolyzed Derivatives Prepared by Post-Modification. 1. Synthesis and Characterization, *Polymer* 43:2055-2064.

[44] Busico V., Causà M., Cipullo R., Credendino R., Cutillo F., Friederichs N., Lamanna R., Serge A., Castelli V. V. A. (2008) Periodic DFT and High-Resolution Magic-Angle-Spinning (HR-MAS) ^{1}H NMR Investigation of the Active Surfaces of $MgCl_2$-Supported Ziegler-Natta Catalysts. the $MgCl_2$ Matrix., *J. Phys. Chem. C* 112:1081-1089.

[45] Linder E., Brugger S., Steinbrecher S., Plies E., Mayer H. A. (2001) Investigations on the Mobility of Novel Sol-Gel Processed Inorganic-Organic Hybrid Materials, *J. Mater. Chem.* 11:1393-1401.

[46] Brenna S., Posset T., Furrer J., Blümel J. (2006) ^{14}N NMR and Two-Dimensional Suspension ^{1}H and ^{13}C HRMAS NMR Spectrscopy of Ionic Liquids Immobilized on Silica, *Chem. Eur. J.* 12:2880-2888.

[47] Rencurosi A., Lay L., Russo G., Prosperi D., Poletti L., Caneva E. (2007) HRMAS NMR Analysis in Neat Ionic Liquids: A Powerful Tool to Investigate Complex Organic Molecules and Monitor Chemical Reactions, *Green Chemistry* 9:216-218.

[48] Carvalho P. J., Álvarez V. H., Schröder B., Gil A. M., Marrucho I. M., Aznar M., Santos L. M. N. B. F., Coutinho J. A. P. (2009) Specific Solvation Interaction of CO_2 on Acetate and Trifluoroacetate Imidazolium Based Ionic Liquids at High Pressures, *J. Phys. Chem. B* 113:6803-6812.

[49] Lungwitz R., Spange S. (2008) Structure and Polarity of the Phase Boundry of *N*-Methylimidazolium Chloride/Silica, *J. Phys. Chem. C* 112:19443-19448.

[50] Le Bideau J., Gaveau P., Bellayer S., Néouze M.-A., Vioux A. (2007) Effect of Confinement on Ionic Liquids Dynamics in Monolithic Silica Ionogels: ^{1}H NMR Study, *Phys. Chem. Chem. Phys.* 9:5419-5422.

[51] Romanova E. E., Grinberg F., Pampel A., Kärger J., Freude D. (2009) Diffusion Studies in Confined Nematic Liquid Crystals by MAS PFG NMR, *J. Magn. Reson.* 196:110-114.

[52] Alam T. M., Dreyer D. R., Bielwaski C. W., Ruoff R. S. (2011) Measuring Molecular Dynamics and Activation Energies for Quanternary Acyclic Ammonium and Cyclic Pyrrolidinium Ionic Liquids using ^{14}N NMR Spectroscopy, *J. Phys. Chem. A* 115:4307-4316.

[53] Huster D., Gawrisch K. (1999) NOESY NMR Crosspeaks Between Lipid Headgroups and Hydrocarbon Chains: Spin Diffusion or Molecular Disorder?, *J. Am. Chem. Soc.* 121:1992-1993.

[54] Huster D., Arnold K., Gawrisch K. (1999) Investigation of Lipid Organization in Biological Membranes by Two-Dimensional Nuclear Overhauser Enhancement Spectroscopy *J. Phys. Chem. B* 103:243-251.

[55] Zhou H., Du F., Li X., Zhang B., Li W., Yan B. (2008) Characterization of Organic Molecules Attached to Gold Nanoparticle Surface Using High Resolution Magic Angle Spinning ^{1}H NMR, *J. Phys. Chem. C* 112:19360-19366.

[56] Krpetic Z., Scarì G., Caneva E., Speranza G., Porta F. (2009) Gold Nanoparticles Prepared Using Cape Aloe Active Compounds, *Langmuir Lett.* 25:7217-7221.

[57] Krpetic Z., Nativo P., Porta F., Brust M. (2009) A Multidentate Peptide for Stabilization of Facile Bioconjugation of Gold Nanoparticles, *Bioconjugate Chem.* 20:619-624.

[58] Polito L., Colombo M., Monti D., Melato S., Caneva E., Prosperi D. (2008) Resolving the Structure of Ligands Bound to the Surface of Superparamagnetic Iron Oxide Nanoparticles by High-Resolution Magic-Angle Spinning NMR Spectroscopy, *J. Am. Chem. Soc.* 130:12712-12724.

[59] Polito L., Monti D., Caneva E., Delnevo E., Russo G., Prosperi D. (2008) One-Step Bioengineering of Magnetic Nanoparticles via a Surface Diazo Transfer/Azide–Alkyne Click Reaction Sequence, *Chem. Commun.* 621-623.

[60] Deshayes S., Maurizot V., Clochard M.-C., Berthelot T., Baudin C., Déléris G. (2010) Synthesis of Specific Nanoparticles for Targeting Tumor Angiogenesis Using Electron-Beam Irradiation, *Radiation Phys. Chem.* 79:208-213.

[61] Costantino L., Gandolfi F., Bossy-Nobs L., Tosi G., Gurny R., Rivasi F., Vandelli M. A., Forni F. (2006) Nanoparticulate Drug Carriers Based on Hybrid poly(D,L-Lactide-*co*-Glycolide)-Dendron Structures, *Biomaterials* 27:4635-4645.

[62] Conte P., Carotenuto G., Piccolo A., Perlo P., Nicolais L. (2007) NMR-Investigations of the Mechanism of Silver Mercaptide Themolysis in Amorphous Polystyrene, *J. Mater. Chem.* 17:201-205.

[63] Blümel J. (2008) Linkers and Catalysis Immobilized on Oxide Supports: New Insights by Solid-State NMR Spectroscopy, *Coordination Chemistry Reviews* 252:2410-2423.

[64] Posset T., Rominger F., Blümel J. (2005) Immobilization of Bisphosphinoamine Linkers on Silica: Identification of Previously Unrecongnized Byproducts vis ^{31}P CP/MAS and Suspension HR-MAS Studies, *Chem. Mater.* 17:586-595.

[65] Guenther J., Reibenspies J., Blümel J. (2011) Synthesis, Immobilization, MAS and HR-MAS NMR of a New Chelate Phosphine Linker System, and Catalysis by Rhodium Adducts Thereof, *Adv. Synth. Catal.* 353:443-460.

[66] Posset T., Guenther J., Pope J., Oeser T., Blümel J. (2011) Immobilized Sonogashira Catalyst Systems: New Insights by Multinuclear HRMAS NMR Studies, *Chem. Commun.* 2011:2059-2061.

[67] Blümel J. (2008) Linkers and Catalysts Immobilized in Oxide Supports: New Insights by Solid-State NMR Spectroscopy, *Coordination Chemistry Reviews* 252:2410-2423.

[68] Pinoie V., Poelmans K., Miltner H. E., Verbruggen I., Biesemans M., Van Assche G., Van Mele B., Martins J. C., Willem R. (2007) A Polystyrene-Supported Tin Trichloride Catalyst with a C11-Spacer. Catalysis Monitoring Using High-Resolution Magic Angle Spinning NMR, *Organometallics* 26:6718-6725.

[69] Poelmans K., Pinoie V., Verbruggen I., Biesemans M., Deshayes G., Duquesne E., Delcourt C., Degée P., Miltner H. E., Dubois P., Willem R. (2008) Undecyltin Trichloride Grafted onto Cross-Linked Polystyrene: An Effecient Catalyst for Ring-Opening Polymerization of ε-Caprolactone, *Organometallics* 27:1841-1849.

[70] Deshayes G., Poelmans K., Verbruggen I., Camacho-Camacho C., Degée P., Pinoie V., Martins J. C., Piotto M., Biesemans M., Willem R., Dubois P. (2005) Polystyrene-Supported Organotin Dichloride as a Recyclable Catalyst in Lactone Ring-Opening Polymerization: Assessment and Catalysis Monitoring by High-Resolution Magic-Angle-Spinning NMR Spectroscopy, *Chem. Eur. J.* 11:4552-4561.

[71] Pinoie V., Biesemans M., Willem R. (2008) Quantitative Tin Loading Determination of Supported Catalysts by ^{119}Sn HRMAS NMR using a Calibrated Internal Signal (ERETIC), *Organometallics* 27:3633-3634.

[72] Porto S., Seco J. M., Espinosa J. F., Quiñoá E., Riguera R. (2008) Resin-Bound Chiral Derivatizing Agents for Assignments of Configuration by NMR Spectroscopy, *J. Organic Chem.* 73:5714-5722.

[73] Hellriegel C., Skogsberg U., Albert K., Lämmerhofer M., Maier N. M., Linder W. (2004) Characterization of a Chiral Stationary Phase by HR-MAS NMR Spectroscopy and Investigation of Enatioselective Interaction with Chiral Ligates by Transferred NOE, *J. Am. Chem. Soc.* 126:3809-3816.

[74] Simpson A. J., Kingery W. L., Sahw D. R., Spraul M., Humpfer E., Dvorstak P. (2001) The Application of ^{1}H HR-MAS NMR Spectroscopy for the Study of Structures and Associations of Organic Components at the Solid-Aqueous Interface of a Whole Soil., *Environ. Sci. Technol.* 35:3321-3325.

[75] Simpson A. J., Simpson M., Smith E., Kelleher B. P. (2007) Microbially Derived Input to Soil Organic Matter; Are Current Estimates Too Low?, *Environ. Sci. Technol.* 41:8070-8076.

[76] Feng X., Simpson A. J., Simpson M. (2006) Investigating the Role of Mineral-Bound Humic Acid in Phenanthrene Sorption, *Environ. Sci. Technol.* 40:3260-3266.

[77] Shirzadi A., Simpson M. J., Kumar R., Baer A. J., Xu Y., Simpson A. J. (2008) Molecular Interactions of Pestisides at the Soil-Water Interface, *Environ. Sci. Technol.* 42:5514-5520.

[78] Combourieu B., Inacio J., Delort A.-M., Forando C. (2001) Differentation of Mobile and Immobile Pesticides on Anionic Clays by ^{1}H HR MAS NMR Spectroscopy, *Chem. Commun.* 2214-2215.

[79] Colnago L. A., Martin-Neto L., Pérez M. G., Daolio C., Ferreira A. G., Camargo O. A., Berton R., Bettiol W. (2003) Application of ^{1}H HR/MAS NMR to Soil Organic Matter Studies, *Ann. Magn. Reson.* 2:116-118.

[80] Simpson A. J., Simpson M., Kingery W. L., Lefebvre B. A., Moser A., Williams A. J., Kvasha M., Kelleher B. P. (2006) The Application of ^{1}H High-Resolution Magic-Angle

Spinning NMR for the Study of Clay-Organic Associations in Natural and Synthetic Complexes, *Langmuir* 22:4498-4503.

[81] Simpson A. J., Kingery W. L., Hatcher P. G. (2003) The Indentification of Plant Derived Structures in Humic Materials Using Three-Dimensional NMR Spectroscopy, *Environ. Sci. Technol.* 37:337-342.

[82] Pampel A., Zick K., Glauner H., Engelke F. (2004) Studying Lateral Diffusion in Lipid Bilayers by Combining a Magic Angle Spinning NMR Probe with a Microimaging Gradient System, *J. Am. Chem. Soc.* 126:9534-9535.

[83] Stejskal E. O., Tanner J. E. (1965) Spin Diffusion Measurements: Spin Echoes in the Presence of Time-Dependent Field Gradient, *J. Chem. Phys.* 42:288-292.

[84] Viel S., Ziarelli F., Pagès G., Carrara C., Caldarelli S. (2008) Pulsed Field Gradient Magic Angle Spinning NMR Self-Diffusion Measurements in Liquids, *J. Magn. Reson.* 190:113-123.

[85] Cotts R. M., Hoch M. J. R., Sun T., Marker J. T. (1989) Pulsed Field Gradient Stimulated Echo Methods for Improved NMR Diffusion Measurements in Heterogeneous Systems, *J. Magn. Reson.* 83:252-266.

[86] Tanner J. E. (1970) Use of the Stimulated Echo in NMR Diffusion Studies, *J. Chem. Phys.* 52:2523-2526.

[87] Gibbs S. J., Johnson Jr. C. S. (1991) A PFG NMR Experiment for Accurate Diffusion and Flow Studies in the Presence of Eddy Currents, *J. Magn. Reson.* 93:395-402.

[88] Pampel A., Fernandez M., Freude D., Kärger J. (2005) New Options for Measuring Molecular Diffusion in Zeolites by MAS PFG NMR, *Chem. Phys. Lett.* 407:53-57.

[89] Pampel A., Engelke F., Galvosas P., Krause C., Stallmach F., Michel D., Kärger J. (2006) Selective Multi-Component Diffusion Measurement in Zeolites by Pulsed Field Gradient NMR, *Micropor. Mesopor. Mat.* 90:271-277.

[90] Fernandez M., Pampel A., Takahashi R., Sato S., Freude D., Kärger J. (2008) Revealing Complex Formation in Acetone-*n*-Alkane Mixtures by MAS PFG NMR Diffusion Measurement in Nanoporous Hosts, *Phys. Chem. Chem. Phys.* 10:4165-4171.

[91] Fernandez M., Kärger J., Freude D., Pampel A., van Baten J. M., Krishna R. (2007) Mixture Diffusion in Zeolites Studied by MAS PFG NMR and Molecular Simulation, *Micropor. Mesopor. Mat.* 105:124-131.

[92] Gaede H. C., Gawrisch K. (2003) Lateral Diffusion Rates of Lipid, Water and a Hydrophobic Drug in a Multilamellar Liposome, *Biophys. J.* 85:1734-1740.

[93] Gaede H. C., Gawrisch K. (2004) Multi-Dimensional Pulse Field Gradient Magic Angle Spinning NMR Experiments on Membranes, *Magn. Reson. Chem.* 42: 115-122.

[94] Polozov I. V., Gawrisch K. (2004) Domains in Binary SOPC/POPE Lipid Mixtures Studied by Pulsed Field Gradient ^{1}H MAS NMR, *Biophys. J.* 87:1741-1751.

[95] Iqbal S., Rodríguez-Lansola F., Escuder B., Miravet J. F., Verbruggen I., Willem R. (2010) HRMAS ^{1}H NMR as a Tool for the Study of Supramolecular Gels, *Soft Matter* 6:1875-1878.

[96] Chin J. A., Chen A., Shapiro M. J. (2000) SPEEDY: Spin-Echo Enhanced Diffusion Filtered Spectroscopy. A New Tool for High Resolution MAS NMR, *J. Comb. Chem.* 2:293-296.

Section 3

Nano Spectroscopy

Spectroscopic Analyses of Nano-Dispersion Strengthened Transient Liquid Phase Bonds

Kavian Cooke

Additional information is available at the end of the chapter

http://dx.doi.org/10.5772/48086

1. Introduction

Transient liquid-phase (TLP) bonding is a material joining process that depends on the formation of a liquid at the faying surfaces by an interlayer that melts at a temperature lower than that of the substrate [1, 2, 3, 4, 5].

The TLP bonding is distinguished from brazing processes by the isothermal solidification of this liquid. This is accomplished since the interlayer is rich in a melting point depressant (MPD). Upon heating through the eutectic temperature, the interlayer will either melt or react with the base metal to form a liquid. During the hold time above the melting temperature, the melting point depressant (solute) into the base metal (solvent). This resulting solid/liquid interfacial motion via epitaxial growth of the substrate is termed "isothermal solidification." A homogeneous bond between the substrates is formed when the two solid/liquid interfaces meet at the joint centerline marking the end of the isothermal solidification stage.

TLP bonding provides an alternative to fusion welding and has been extensively studied for joining particle-reinforced Al-MMCs [1]. The advantage of using this process is that, reinforcing particles are incorporated into the bond region either by using a particle reinforced insert layer or by the melt-back due to a eutectic reaction between the interlayer and the aluminum alloy [6]. It has been shown in the scientific literature that melt-back can be controlled by using thin interlayer materials. It has also been suggested that heating rate, interlayer composition and thickness are most important in reducing melt-back during TLP bonding. These parameters also determine the width of the liquid phase, removal of surface oxide film and particulate redistribution in the bonded region [16, 7,5].

According to Bosco and Zok [8] there exists a critical interlayer thickness at which pore-free bonds are produced. The thickness of the interlayer must exceed that which is consumed through solid-state diffusion; otherwise, no liquid is formed at the bonding temperature. This sets a minimal requirement for the interlayer thickness, which can be determined using equation 1:

$$h_c = h_\eta C_\eta \left(\frac{\rho_\eta}{\rho_i} \right)^2 \tag{1}$$

Where, $2h_\eta$ is the total thickness of the reaction layer formed at the interlayer/MMC interface when the eutectic temperature is reached, C_η is the mass fraction of Ni in the reaction layer. The mass densities of the reaction layer and the Ni interlayer are ρ_η and ρ_{Ni} respectively. On the other hand, Li et al. [5] suggested that a maximum thickness exists that minimizes particle segregation and maximises joint strength. The composition of the interlayer has also been shown by Eagar and McDonald [9] to be equally as important as the interlayer thickness. In the published literature, it has been shown that pure metal interlayers such as nickel [10, 11], silver [12,13] and copper [13, 15] can react with the metal surface by means of eutectic or peritectic reaction to displace surface oxides and regulates the bonding temperature. In joints made using these pure interlayers, the existing problems can be summarized as: parent metal dissolution particulate segregation [13, 14,15], void formation in particle segregated region [12] and low strength micro-bonds at particle-metal interfaces resulting from the poor wettability of molten interlayer on ceramic reinforcement [12,16]. Particle segregation has been shown by Stefanescu et al [17, 18] to be promoted by the slow movement of the solid-liquid interface during isothermal solidification. Li et al. [5] showed that during TLP bonding of Al-MMCs containing particulate reinforcements with diameters less than 30 μm, particle segregation to the interface occurred when a copper foil thickness between 5 and 15 μm was used. Earlier research used alloyed interlayers such as; Zn–Al, Cu–Al and Cu/Ni/Cu systems to decrease bonding temperature in air and to prevent particulate segregation [19]. While composite interlayers such as Al–Si–W mixed powder and Al–Si–Ti–SiC mixed powder were used to improve the densification of a thick reaction layer which formed at the joint. This layer was reinforced with a metallic phase and a ceramic phase [20].

Recent studies into joining Al-MMCs have focused on decreasing bonding temperature by using Sn-based interlayers reinforced with silicon carbide. The results show that joint strength of Al6061 + 25% (Al_2O_3)p improved by approximately 100% when compared to unreinforced Sn-based joints formed by ultrasonic assisted soldering [21]. Yan et al. [22] developed a SiC particle reinforced Zn-based filler which was used to join SiCp/A356 composite. The results indicated that with the use of ultrasonic vibration suitable particle distribution and reduced void formation were achieved. Cooke et al. [23, 24] used an electrodeposited Ni-Al_2O_3 nano-composite coatings to join Al-6061 MMC. The results showed that the use of Ni-Al_2O_3 nano-composite coatings can be used successfully to increase joint strength when compared to TLP bond produced using pure Ni-coating. The results also showed that coating thickness of 5μm can be used to control particle segregation during TLP bonding of Al-6061 MMC

In order to study the kinetics of TLP bonding, techniques such as wavelength dispersive spectroscopy, energy dispersive spectroscopy and x-ray diffraction spectroscopy are normally used, since the compositional changes across the joint region is the mechanism by which the process progresses to completion. In previous studies EDS, WDS and XRD has been used extensive for studying materials due in part to the flexibility that the techniques afford. In most papers, a choice is made between the two processes depending on the information that is required. The difference between these processes are that, the energy-dispersive (ED) type records X-rays of all energies effectively simultaneously and produces an output in the form of a plot of intensity versus X-ray photon energy. The detector consists of one of several types of device producing output pulses proportional in height to the photon energy. Whereas the wavelength-dispersive (WD) type makes use of Bragg reflection by a crystal, and operates in 'serial' mode, the spectrometer being 'tuned' to only one wavelength at a time. Several crystals of different interplanar spacing are needed in order to cover the required wavelength range. Spectral resolution is better than for the ED type, but the latter is faster and more convenient to use. X-ray spectrometers attached to SEMs are usually of the ED type, though sometimes a single multi-crystal WD. This chapter examines the application of spectroscopic analyses such as EDS, WDS and XRD to the evaluation of nanostructure TLP bonds using Ni-Al_2O_3 nano-composite thin film as a filler material during TLP bonding of Al-6061MMC. The effects of process parameters on the mechanical and microstructural microstructural changes in the joint region will also be studied.

1.1. Spectroscopic analysis techniques

The characterization of materials using spectroscopic techniques such as energy dispersive spectroscopy or x-ray diffraction spectroscopy techniques is dependent on the generation of a beam of electrons which interacts with the sample to be analyzed. When electrons strike a anode with sufficient energy, X-rays are produced. This process is typically accomplished using a sealed x-ray tube, which consists of a metal target and a tungsten metal filament, which can be heated by passing a current through it resulting in the "boiling off" of electrons from the hot tungsten metal surface. These "hot" electrons are accelerated from the tungsten filament to the metal target by an applied voltage The collision between these energetic electrons and electrons in the target atoms results in electron from target atoms being excited out of their core-level orbitals, placing the atom in a short-lived excited state. The atom returns to its ground state by having electrons from lower binding energy levels make transitions to the empty core levels. The difference in energy between these lower and higher binding energy levels is radiated in the form of X-rays. This process results in the production of characteristic X-rays. X-rays are generated when the primary beam ejects an inner shell electron thus exciting the atom. As an electron from the outer shell drops in to fill the vacancy and de-excite the atom it must give off energy. This energy is specific to each individual element in the periodic table and is also specific to what particular electron dropped in to fill the vacancy. The conversion between energy, frequency, and wavelength is the well-known de Broglie relationship: $E = h\nu = hc/\lambda$, where ν is the frequency, h is Planck's constant (6.62×10^{-34} joule-second), c is the speed of light (2.998×10^{8}m/sec), and λ is the wavelength of the radiation (in m). Based on this relationship, two distinct types of x-ray detector systems are used. These two types of detector systems are called Energy-Dispersive x-ray Spectrometry (EDS) and Wavelength-Dispersive x-ray Spectrometry (WDS).

EDS spectrometer are most frequently attached to electron column instruments such as SEM or (EPMA). As the name implies is a method of x-ray spectroscopy by which x-rays emitted from a sample are sorted out and analyzed based on the difference in their energy level. An EDS system consists of a source of high-radiation; a sample, a solid-state detector (usually from lithium-drifted silicon (Si(Li)); and a signal processing electronics. When the sample atoms are ionized by a high-energy radiation, they emit characteristic x-rays. X-rays that enter the Si(Li) detector are converted into signals (charge pulses) that can be processed by the electronics into an x-ray energy histogram. This x-ray spectrum consists of a series of peaks representative of the type and relative amount of each element in the sample. The number of counts in each peak can be further converted to elemental weight concentration either by comparison standards or standardless calculations. In general, three principal types of data can be generated using an EDS detector: (i) x-ray dot maps or images of the sample using elemental distribution as a contrast mechanism, (ii) line scan data or elemental concentration variation across a given region, and (iii) overall chemical composition, both qualitative and quantitatively.

As the name implies, WDS is a detection system by which x-rays emitted from the sample are sorted out and analyzed based on differing wavelength (λ) in the WDS, or crystal spectrometry. As in EDS or imaging mode, the beam rasters the sample generating x-rays of which a small portion enters the spectrometer. As the fluorescent x-rays strike the analyzing crystal, they will either past through the crystal, be absorbed, be scattered, or be diffracted. Those which satisfy Bragg's Law; $n\lambda = 2d\,Sin\theta$.

(where n = an integer, d = the interplanar spacing of the crystal, θ = the angle of incidence, and λ = x-ray wavelength) will be diffracted and detected by a proportional counter. The signal from this detector is amplified, converted to standard pulse size in the single channel analyzer and counted with a scalar or displayed as rate vs time on rate meter. By varying the positioning crystal one changes the wavelength that will satisfy Bragg's Law. Therefore one can sequentially analyze different elements. By automating crystal movements one can dramatically speed up the analysis time. Typically the WDS analysis is used to gain the same type of information that the EDS is used for, qualitative and quantitative and quantitative information, line scan and dot maps for elemental distribution.

1.2. X-ray diffraction

X-ray diffraction (XRD) is another quantitative spectroscopic technique which reveals information about the crystal structure, chemical composition, and physical properties of materials and thin films. These techniques are based on observing the scattered intensity of an X-ray beam hitting a sample as a function of incident and scattered angle, polarization, and wavelength or energy. Similar to the EDS and WDS techniques discussed above X-ray diffraction is dependent on Bragg equation, which describes the condition for diffraction to occur in terms of the wavelength of the x-radiation (λ), the interplanar ("d") spacings of the crystal, and the angle of incidence of the radiation with respect to the crystal planes (θ). As the spacing between atoms is on the same order as X-ray wavelengths crystals can diffract the radiation when the diffracted beams are in-phase.

When the incident beam satisfies the Bragg condition, a set of planes forms a cone of diffracted radiation at an angle q to the sample. Since the cone of X-rays intersects the flat photographic filmstrip in two arcs equally spaced from the direct X-ray beam, two curved lines will be recorded on the photographic film. The distance of the lines from the center can be used to determine the angle q, which can then be used to determine the interplanar "d" spacing. X-ray powder diffractometers record all reflections using a scintillation detector (in counts per second of X-rays). The pattern of diffracted X-rays is unique for a particular structure type and can be used as a "fingerprint" to identify the structure type. Different minerals have different structure types, thus X-ray diffraction is an ideal tool for identifying different minerals.

2. Electrodeposition of the nano-composite Ni/Al_2O_3 coating

Composite coatings can be produced by co-deposition of fine inert particles into a metal matrix from an electrolytic or an electroless bath. This technique is getting interesting due to its ability to produce films with excellent mechanical properties such as wear resistance, corrosion resistance, and lubrication.

The preparation of a composite coating is done in two steps. First, an effective dispersion of fine inert particles is produced in the electrolyte. Next, the preparation of the composite coating is made by the manipulation of electrochemical conditions. An effective dispersion of inert particles in the electrolyte promotes the adsorption opportunity of inert particles on the cathode. It causes a higher volume content of inert particles in the composite coating. The mechanical properties of the composite coating are also promoted with the enhancement of the volume content of inert particles in the coating. In electrodeposited composites, the particles to be dispersed in the metal matrix are maintained in suspension in the bath by agitation and they become incorporated in the coating by a process known as electrophoresis.

A number of coating parameters were studied to determine the effect of each on the coating thickness: pH of the electrolyte, current density, electrolyte temperature, agitation frequency and deposition time [33]. Samples were coated for time intervals of 2, 4, 6, 8, 10, 15, 20 and 30 minutes. After electrodeposition, coating thickness was determined using a light microscope. Figure 1(b) shows the relationship between coating thickness and deposition time when; the current density was set to 1.0A/dm^2, pH 3.0, agitation frequency of 300 rpm and an electrolyte temperature of 50°C [25]. It was found that the coating thickness increased with increasing deposition time, in keeping with Faraday's law governing electrode reactions during electrolysis [26].

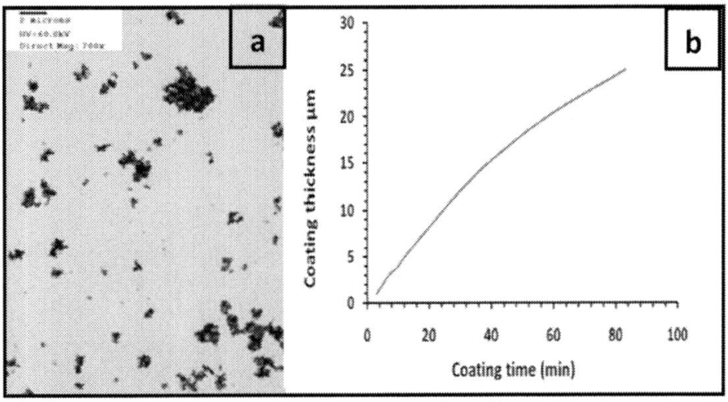

Figure 1. (a) TEM image of as-received nano-sized Al_2O_3 powder (b) Coating thickness as a function of electrodeposition time at 1.0 A/dm^2 and 50°C [33].

In this study, Al -6061/Al_2O_3p samples of dimensions 10 × 10× 5mm were prepared to 800 grit abrasive paper and polished to 1 μm diamond suspension after which they were cleaned in an acetone bath. Acid pickling took place in a solution of 15 wt.% HNO_3 and 2 wt.% HF at 50 °C for 2 minutes and then rinsed in distilled water. The plating solution was prepared by dissolving: 250 g/L $NiSO_4$ $6H_2O$, 45 g/L $NiCl_2.6H_2O$, 35 g/L H_3BO_3 and 1 g/L Saccharin in distilled water. The Ni- Al_2O_3 composite coating was produced by adding 50g/L of ceramic particles to a separately to the nickel bath. The particles were thoroughly mixed into the solution for two hours and kept in suspension in the bath with a magnetic stirrer rotating at 300 rpm. The coating solution was maintained at a temperature of 50°C and pH of 3.0 [33]. The thickness of Ni-Al_2O_3p coatings were controlled by the current density and plating time. The actual amount of Ni-Al_2O_3p electroplated onto a surface was determined by the weight gain after the plating process. The coating thickness was calculated by using the equations 2 and 3:

$$\rho_C = \frac{mass\ of\ coating}{Area\ x\ thickness} = \frac{m}{Axt} \tag{2}$$

The density of the composite coating (ρ_C) can be calculated by using Equation 3 (rule of mixtures) where x_v is the volume fraction of alumina particles in the Ni-Al_2O_3 coating.

$$\rho_C = \rho_{Al_2O_3} x_v + \rho_{Ni}(1 - x_v) \tag{3}$$

2.1. Spectroscopic analysis of the film structure and composition

Wavelength dispersive spectroscopic (WDS) and X-ray diffraction (XRD) spectroscopic analyses of the coatings deposited were used to evaluate the distribution of the dispersion particles. Figure 2 shows an SEM micrographs of the coating produced by the co-electrodeposition of Ni + 18vol% (nano-Al_2O_3)p. Examination of the coatings using a light microscope revealed the absence of surface defects and interfacial void, however Al_2O_3 particle clusters were present in the coating. This was attributed to particle clustering in the powder prior to the coating process as indicated by the TEM image of the as-received Al_2O_3 powder shown in Figure 1(a). The volume fraction of Al_2O_3 present in the coating was studied by digital x-ray mapping and these results are shown in Figure 3. From Figure 3b and 3c, the areas which contains a high concentration of Al_2O_3 are easily identify and corresponds to Al and O which would confirm the compound to be Al_2O_3 since the base coating is Ni.

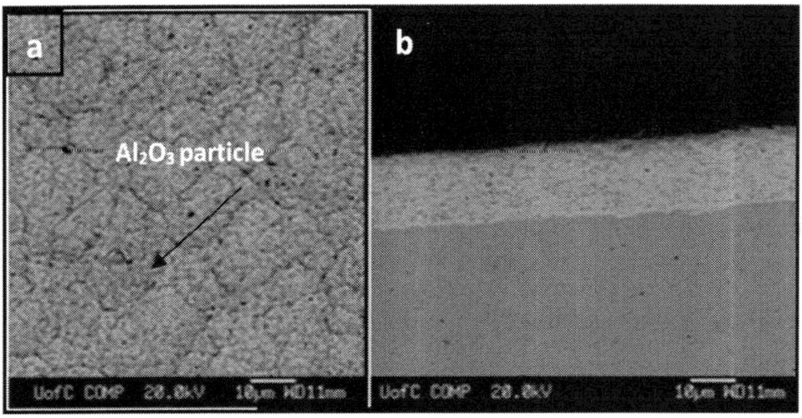

Figure 2. (a) Surface of the Ni-Al_2O_3 coating and (b) Cross-section of the Ni-Al_2O_3 coating produced by electrodeposition [33].

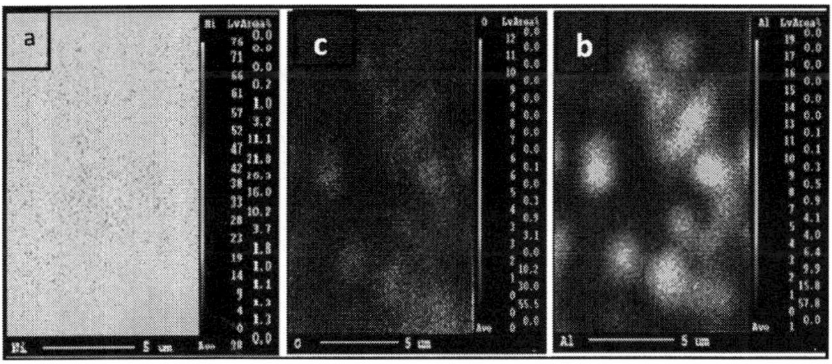

Figure 3. X-ray digital composition maps taken from the Ni-Al_2O_3 coating surface for; (a) Ni, (b) O and (c) Al [33].

3. Fundamentals of transient liquid phase bonding

TLP bonding requires that the base metal surfaces are brought into intimate contact with a thin interlayer placed between the bonding surfaces. The interlayer can be added in the form of a thin foil, powder or coating [27, 28] which is tailored to melt by eutectic or peritectic reaction with the base metal. The liquid filler metal wets the base metal surface and is then drawn into the joint by capillary action until the volume between components to be joined is completely filled.

The driving force of TLP bonding is diffusion. A process which can be described by Fick's first and second laws. The first law describes diffusion under steady-state conditions and is given by equation 4:

$$J = -D\frac{\partial C}{\partial x} \tag{4}$$

Fick's second law describes a non-steady state diffusion in which the concentration gradient changes with time and can be expressed as shown in equation 5:

$$\frac{\partial C}{\partial t} = D\frac{\partial^2 C}{\partial x^2} \tag{5}$$

Equation 6 shows a general solution for Equation-5 using separation of variables is [29]:

$$C(x,t) = \frac{1}{2\sqrt{D\pi t}}\int_{-\infty}^{+\infty} f(\xi)e^{\left(\frac{\xi - x)}{4Dt}\right.} d\xi \tag{6}$$

Where the error function solution for equation 6 is shown in equation 7:

$$C(x,t) = \frac{C_0}{2}\left[erf\left(\frac{x - x_1}{\sqrt{4Dt}}\right) - erf\left(\frac{x - x_2}{\sqrt{4Dt}}\right)\right] \tag{7}$$

If the following boundary conditions are applied to Equation-7, the concentration as a function of time can be calculated using Equation-8:

$$\begin{gathered} C(x,0) = \begin{cases} C_o & (x > 0) \\ 0 & (0 < x) \end{cases} \text{ and } C(x,0) = \begin{cases} C_o & x = +\infty \\ 0 & x = -\infty \end{cases} \\ C(x,t) = C_0\left(1 - erf\left(\frac{x}{\sqrt{4Dt}}\right) + erf\left(\frac{kx}{\sqrt{4Dt}}\right)\right) \end{gathered} \tag{8}$$

TLP bonding can be conducted by one of two distinct methods. Method-I employs a pure interlayer which forms a liquid through eutectic reaction with the base metal and Method-II employs an interlayer with a liquidus temperature near the bonding temperature [2]. Method-II is most commonly used as it reduces the overall process time by decreasing the volume of solute to be diffused from the interface before the liquid is formed. On the other hand, method-I can be considered to be more effective in TLP bonding as the eutectic reaction is able to displace surface oxides during bonding. TLP bonding process was first divided into five discrete stages by Duvall et al. [30]. These stages were: heating, melting, dissolution of the base-metal, isothermal solidification and homogenization of the excess solute at the bond-line. Zhou later condensed the TLP process to three stages: base-metal dissolution, isothermal solidification and homogenization [7]. In later work by MacDonald and Eager [9] the second and third stage described by Duvall were combined to give four stages; heating, melting and parent metal dissolution, isothermal solidification and homogenization. Zhou [7] reclassified his earlier work to include a heating stage. The second stage was also expanded to be dissolution and widening.

4. Effect of bonding variables on microstructural evolution

This section investigates the effects of bonding variables on the chemical and microstructural homogeneity across the joint region. Factors affecting TLP bonding are; bonding temperature, bonding time, contact pressure and interlayer thickness and composition. Contact pressure can also affect the strength of the bond produced and this has been reported extensive in the scientific literature. However in this work, when bonding pressure increased beyond 0.1 MPa the result is rapid creep failure during the bonding process. Therefore the bonding pressure was held constant at 0.01MPa. The effects of these parameters were evaluated using WDS and XRD to study the change in chemical composition across the joint region as a function of bonding variable.

4.1. Effect of bonding time on joint properties

Wavelength dispersive spectroscopic analyses of joints bonded at 600°C using a 5 μm thick Ni-Al_2O_3 coating as the interlayer as a function of bonding time as shown in Table 1. For joints bonded for 1 minute a large concentration of Ni remained at the interface after bonding. However, when the bonding time was increased to 30 minutes resulted in the elimination of the interface and an increase in the grain size within the joint. This disrupts the band of segregated particles at the interface (see Figure 4) and chemically homogenized the joint zone.

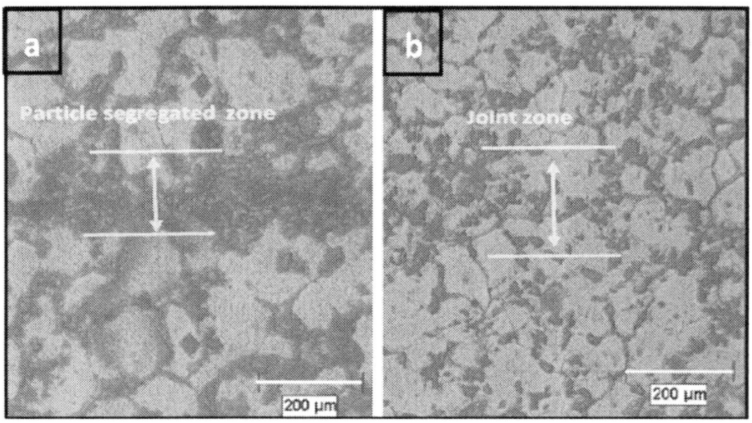

Figure 4. Light micrographs of joint bonded with 5μm thick Ni–Al_2O_3 coating for: (a) 1 min (b) 30 min [23].

The segregation of particles was accredited to the pushing of micro-Al_2O_3 particles by the solidifying liquid-solid interface. Stefanescu [18] showed that particle pushing can be assumed to be a steady-state condition under which the interface velocity can be assumed to be equal the rate of isothermal solidification. Li et al. [5] suggested that the segregation tendency is dependent on the relationship between the liquid film width produced at the bonding temperature, particle diameter and inter-particle spacing. When the liquid film width is large enough that sufficient particulate material is contained in the melt, particles will be pushed ahead of the solidifying liquid-solid interface resulting in particle segregation at the bond-line. However if the liquid film width is less than some critical value, segregation should not occur.

In the reported studies on transient liquid phase diffusion bonding of Al-MMCs it was shown that the width of the segregated zone at the joint center increased with increasing bonding time. The opposite of this relationship was seen when using the Ni-Al_2O_3 coating. As the bonding time increased, the width of the segregated region decrease. This can be attributed to the heterogeneous nucleation of grains within the joint zone during solidification and this lead to grain refining at the joint. A high resolution SEM micrograph shown in Figure 5 revealed the presence of a nano-sized alumina particle at the center of a grain. EDX spectra of the particle showed Al and O in high concentrations with traces of Mg [23, 24, 33]. Comparing Gibbs free energy of formation at the bonding temperature for MgO (-1195 kJ/mol) and Al_2O_3 (-985 kJ/mol) it is found that Al_2O_3 is unstable in the presence of Mg hence it is like that some of the nano-size Al_2O_3 will decomposed to form $MgAl_2O_4$ compound. WDS analysis across the joint zone as a function of bonding time indicated that the Ni volume at the joint center varied between 3.122 wt% after 1 minute and 0.37 wt% after 30 minutes bonding time (see Table 1).

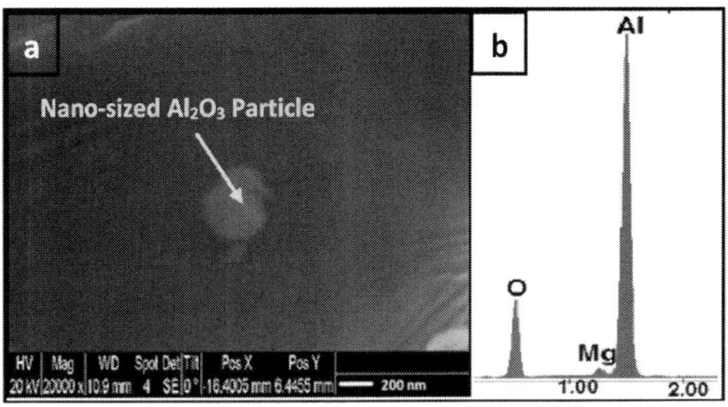

Figure 5. (a) SEM image of nano-particle present in the center of a grain (b) EDS analysis of nano-Al_2O_3 particle [23]

Time (min)	Fe / wt%	Ni / wt%	Si / wt%	Mg / wt%	Cu / wt%	Al / wt%
1	0.399	3.122	0.633	0.379	0.586	93.635
5	0.677	0.849	0.863	1.047	0.71	79.375
10	0.049	0.071	0.211	0.676	0.162	98.598
30	0.028	0.037	0.091	0.458	0.087	99.153

Table 1. WDS analysis of joints made at 600°C as a function of bonding time

4.2. Effect of temperature on joint properties

Wavelength dispersive spectroscopic analyses of the joint as a function of bonding temperature indicated that the Ni concentration at the interface decreased from 4.65 wt% to 0.19 wt% as the bonding temperature is increased from 570 to 620°C (see Table 2). This was attributed to an increase in the diffusivity of Ni from the interface into the base metal as the temperature increased. A review of the scientific literature shows that the diffusivity of Ni increased from D_{570}=4.69 x10^{-13} m^2/s to D_{620}= 1.58 x10^{-12} m^2/s when the bonding temperature was increased from 570 to 620°C [10, 31].

A study of the joint microstructure for a bond made at 570°C revealed the segregation of Al_2O_3 particles to the bond-line as shown in Figure 6(a). When the bonding temperature was increased to 590°C the width of the particle segregated zone within the joint decreased to approximately 150 μm as shown in Figure 6 (b). Further increase in bonding temperature to 600°C also resulted in a reduction of the width of the segregated zone. A similar result was obtained when the bonding temperature was increased to 620°C (see Figure 6d). This observation was consistent with earlier literature, which suggested that the use of thin interlayers during bonding can help to control the degree of particle segregation taking place within the joint [5]. The micrographs indicate that the width of the particle segregated zone decreases with increasing bonding temperature. This can be attributed to particle pushing by the primary α-phase during solidification as shown in section 4.1 [17, 18].

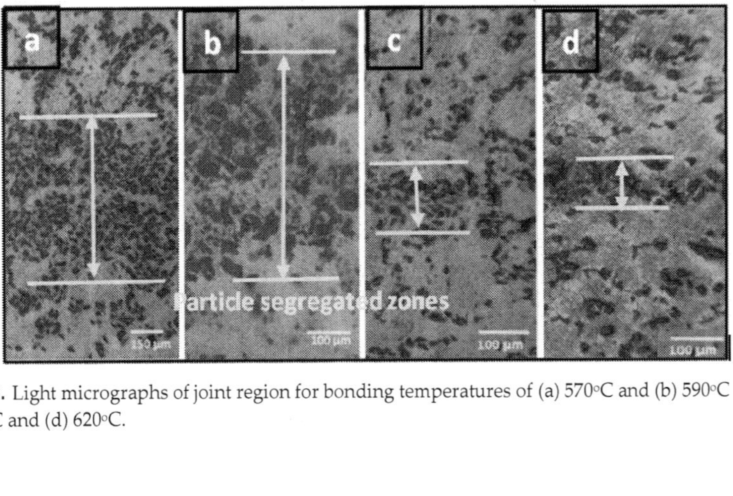

Figure 6. Light micrographs of joint region for bonding temperatures of (a) 570°C and (b) 590°C (c) 600°C and (d) 620°C.

Temperature	Mg/ wt%	Ni/ wt%	Si/ wt%	Fe/ wt%	Al/ wt%
570°C	2.53	4.65	0.72	0.35	91.75
590°C	1.73	1.69	0.52	0.28	95.78
600°C	1.52	0.35	0.41	0.21	97.51
620°C	0.94	0.19	0.21	0.19	98.47

Table 2. Wavelength dispersive spectroscopic analyses of joints made at 600°C for 10 minutes composition as a function of bonding temperature (wt %).

4.3. Effect interlayer thickness on joint properties

The effects of interlayer thickness on microstructural development across the joint region and subsequent effect on joint micro-hardness and shear strength were investigated. According to Bosco and Zok [8], there exists a critical interlayer thickness at which pore-free bonds are produced. This critical interlayer thickness should correspond to maximum joint strength. Therefore the objective in this section is to identify the critical interlayer thickness that maximizes joint strength. Joints made without the use of an interlayer resulted in the formation of a "planar interface" due to the presence of a layer of surface oxide, which prevents metal to metal contact (see Figure 7a). This was corroborated by studies on the solid-state diffusion bonding of Al-MMC [15]. The inability to achieve effective bonding in the solid-state highlights the need for low melting interlayers. When a 1μm thick Ni-Al_2O_3 coating was used as the interlayer a thin joint zone was achieved (see Figure 7b). However a WDS analysis of this region indicated the presence of a higher concentration of Al_2O_3 when compared to bonds made using a pure nickel coating of the same thickness. This was attributed to the presence of nano-sized Al_2O_3 particles in the joint zone and the presence of residual surface oxide.

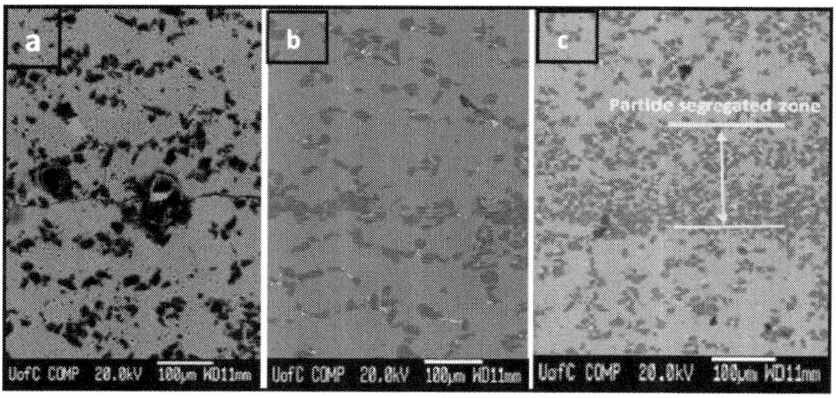

Figure 7. Microstructure of joints bonded at 600°C for 10 min using (a) no-interlayer used (b) 1 μm thick Ni–Al_2O_3 coating (c) 9μm thick Ni –Al_2O_3 coating [32].

Figures 8 show that the width of the segregated zone increased with increasing coating thickness. At a coating thickness of 2 μm a more chemically homogeneous joint was created however WDS analysis showed pockets of oxide with the following composition 65.56 wt% and 28.15 wt% were still present at the interface. When the coating thickness was increased to 4 μm, a 95 μm wide segregated zone was at the joint center and the concentration of nickel remaining at the interface after bonding increased from 0.47 wt% with 3μm thick coating to 0.58wt% when a 4μm thick coating (see Table 3). Further increase in coating thickness to 5 μm, resulted in the formation of a 110 μm wide segregated zone while the nickel increased to 0.97 wt%.

Interlayer thickness (μm)	**Mg** / wt%	**Ni** / wt%	**Si** / wt%	**Fe** / wt%	**Al** / wt%
1	0.46	0.37	0.53	0.22	98.42
2	0.83	0.49	0.62	0.19	97.87
3	0.98	0.47	0.67	0.21	97.67
4	0.97	0.58	0.72	0.27	97.46
5	0.84	0.97	0.53	0.18	97.48
7	0.86	0.88	0.71	0.38	97.17
9	1.05	1.01	0.69	0.24	97.01
11	1.13	1.45	0.65	0.52	96.25
13	0.99	1.63	0.67	0.97	95.74

Table 3. WDS analysis of joints made at 600°C for 10 minutes as a function of interlayer thickness

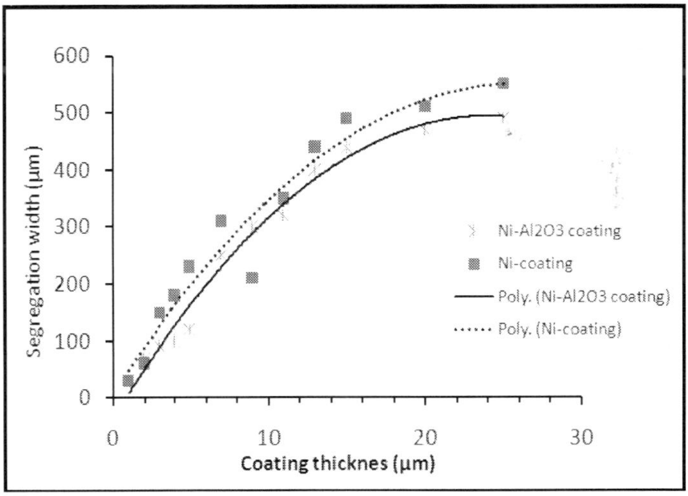

Figure 8. Width of the particle segregated zone formed at within the joint as a function of interlayer thickness for pure Ni coating and Ni–Al_2O_3 coating [32].

The increase in the width of the segregated zone was attributed to increased liquid formation with increasing coating thickness. As the width of the eutectic liquid increases more Al_2O_3 particles are immersed in the liquid phase. These particles are pushed by the solid/liquid interface during isothermal solidification [5].

The width of the particle segregated zone was significantly lower than that achieved when pure Ni-coatings are used as the interlayer. The difference in the width of the segregated zone between joint bonded using pure Ni coating and Ni-Al_2O_3 coating was attributed to the presence of nano-size Al_2O_3 particle in the joint center and a reduction in the concentration of Ni (81.6 wt%) present in the coating, when Ni-Al_2O_3 is used (see Figure 13b).

4.4. Effect of nano-sized particle on joint properties

The effect of nano-sized particles on the microstructural development across the joint region was studied using energy dispersive spectroscopy (EDS). Figure 9(a) shows the microstructure of a joint bonded using a 5μm thick Ni coating dispersed with 50 nm Al_2O_3 particles. From the micrograph a 50 μm wide particle segregated zone was seen within the joint center. Also present at the center of the joint are dark clumps, which EDS analysis suggested are oxide particles (Figure 10 a and b). The presence of the oxide clusters observed, are likely Al_2O_3 particles which agglomerated during the deposition process. When the coating particle size was increased to 500 nm Al_2O_3, a similar result was obtained (see Figure.9b). WDS analyses of the joints as a function of particle size indicated that the Ni concentration of 0.95 wt% and 0.79 wt% for samples bonded using 500 nm and 50 nm respectively. The lower concentration obtained when 50 nm particles are used suggest a faster diffusivity of Ni during the bonding process. This was attributed to greater surface contact between the uncoated Al-6061 sample and the Ni-Al_2O_3 coating surface. Analysis of the roughness using SEM indicated that the surface roughness increased from 0.1 μm for coatings containing 50 nm particles to 0.25 μm for coating containing 500 nm particles.

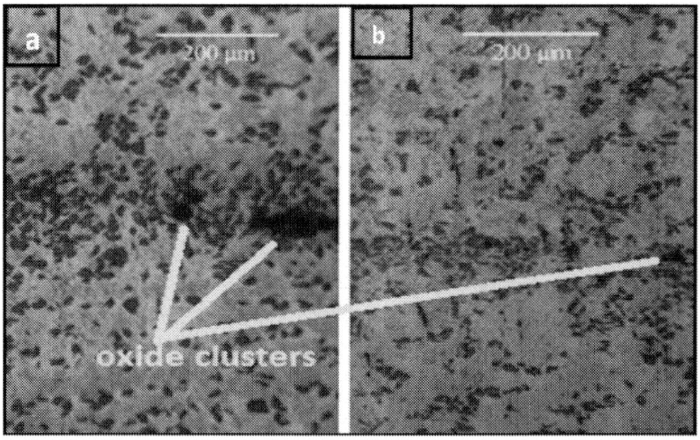

Figure 9. Microstructure of joints bonded at 600°C for 10 min using (a) 5μm thick Ni-(50nm) Al_2O_3 (b) 5μm thick Ni-(500nm) Al_2O_3 [32].

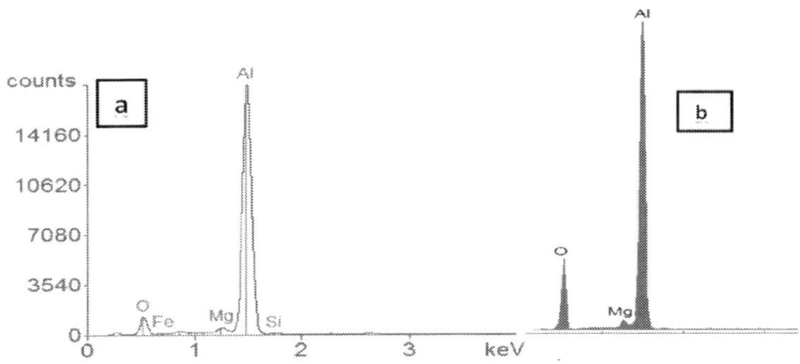

Figure 10. EDS analyses of joints bonded at 600°C for 10 min using (a) 5μm thick Ni-(500 nm) Al_2O_3 (b) 5μm thick Ni-(50 nm) Al_2O_3 [33]

The micrograph shown in Figure 9 a thin segregated zone was formed at the joint center in both cases. The hypothesis is that the difference in the width of the segregated zone obtained is dependent on the differences in particle size, surface roughness of the coating and the distribution of the nano-sized particles in the joint zone during bonding. The presence of nano-Al_2O_3 along the interface was confirmed by TEM analyses which indicated that the nano-particles are arranged along the grain boundary as shown in Figure 11 which would impart a pinning effect as described by Orowan [24].

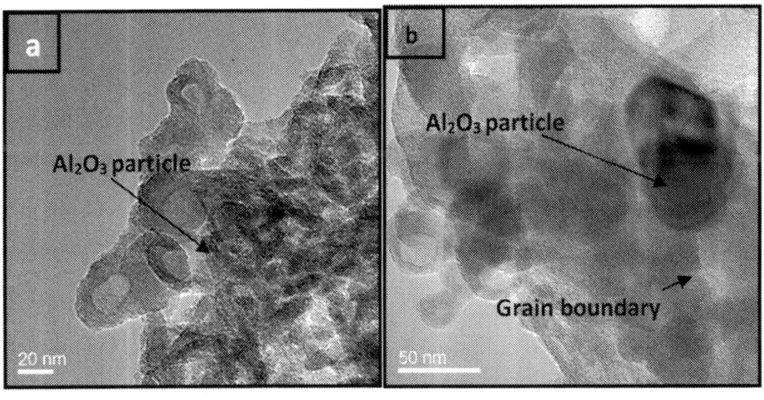

Figure 11. (a) TEM image of the bonded joint when nano-sized Al_2O_3 particles are used in the interlayer and (b) TEM image of a nano-Al_2O_3 particle located at a grain boundary [33].

5. Effect of bonding variable on the mechanical properties of the joint

5.1. Effect of bonding time the joint shear strength

The shear strength of joints made as a function of bonding time is shown in Figure 12 (a). The graph show that the shear strength increased with increasing bonding time from 68

MPa at 1 minute to 137 MPa at 30 minutes. When a Ni-Al_2O_3 coating was used as the interlayer for a bonding time of 10 minutes, shear strength of 136 MPa was recorded. However when a pure Ni coating was used under that same bonding conditions, shear strength of 117 MPa was achieved [23]. The differences in joint shear strengths obtained were attributed to the presence of a nano-sized dispersion of Al_2O_3 particles within the joint and the precipitation of nickel aluminide phases within the joint region. Shen et al. [34] showed that the increase in yield strength of the nano-particle reinforced aluminum alloy is related to particulate–dislocation interaction by means of the Orowan bowing mechanism. Orowan theory suggests nano-sized particles act as barriers to dislocation motion. This mechanism leads to dislocation pile-up and an increase in the joint shear strength [35].

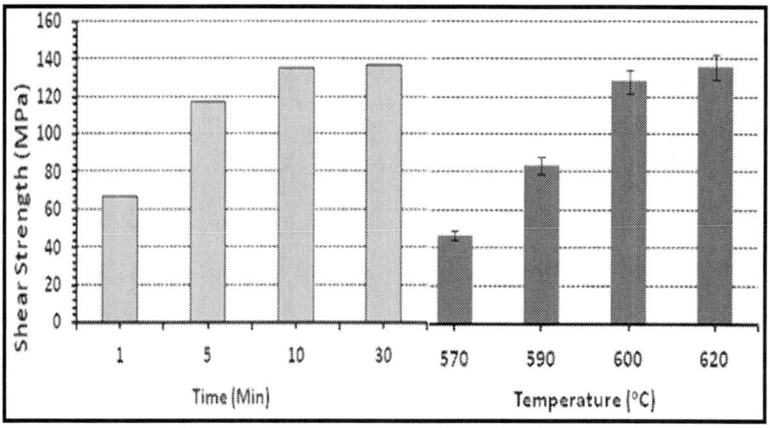

Figure 12. (a) Effect of bonding time on joint strength using 5μm thick Ni- Al_2O_3 coating at a bonding temperature of 600°C and (b) Effect of bonding temperature on joint shear strengths made using 5μm thick Ni-Al_2O_3 coatings for 10 minutes [23, 24].

The fractured surfaces of the shear tested joints were analyzed to identify the mechanism of joint failure. Figure 13(a) shows the fractured surface of a bond made for a bonding time of 1 minute. The fractograph shows an undulating surface containing shear plastic deformation with some cleavage facets indicating a mixed failure mode. The fracture appeared to have propagated through the bond-line. The result of the fractographic analyses suggests that the mechanism of failure transitioned from brittle to ductile as the bonding time increases. When the bonding time was increased to 10 minutes (see Figure 18b). The surface was characterized by an undulating appearance of plastic deformation indicative of ductile mode of failure. Additionally, fractured micro-Al_2O_3 particles were observed at the fractured surface, indicating a transgranular fracture through the bond-line.

XRD analyses of the fractured surfaces of bonds made at 1 and 10 minutes are shown in Figure 14. The results indicated the presence of peaks for phases such as $AlFe_6Si$ (2θ=38°), Al_9FeNi and AlFeSi. The literature showed that these ternary compounds forms readily in Al-Mg-Si-Fe-Ni systems through various peritectic reactions [17]. In addition, binary crystal phases such as Al_3Ni, Ni_3Si (2θ=78°). Al_3Si and Al_2O_3 compounds were also identified.

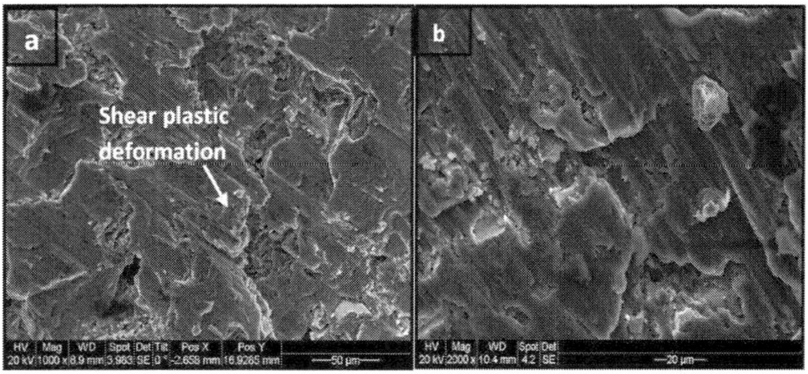

Figure 13. SEM micrograph the fractured surface of a bond made at 600°C with 5 μm thick Ni–Al_2O_3 for: (a) 1 minute and (b) 10 minutes

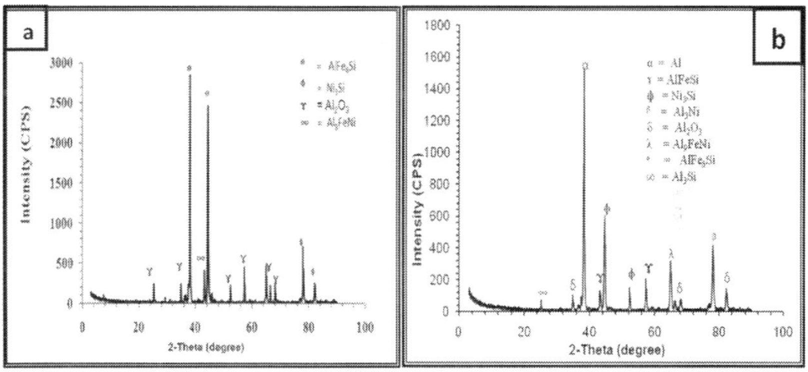

Figure 14. XRD analysis of the fractured surface of a bond made with 5 μm thick Ni–Al_2O_3 for (a) 1 minute and (b) 10 minutes and

5.2. Effect of temperature on joint shear strength

Joint shear strengths measured as a function of bonding temperature were obtained using a single lap shear test. A comparison of the joint shear strengths of bonds made at 570, 590,

600 and 620°C is shown in Figure 12(b). The test result show that the shear strength increased with increasing bonding temperature from 45 MPa at 570°C to 138 MPa 600°C. This increase in joint strength was attributed to the presence of nano-sized ceramic particles and the precipitation of intermetallic phases within the joint region. The formation of these nickel aluminide phases increased with increasing bonding temperature. Specimens bonded at 620°C gave the highest bond strength of 136 MPa. The effect of the nano-particles and the precipitated intermetallics on the composite was discussed by Zhang and Chan [45] and results in Orowan strengthening as discussed in the previous sections.

In order to compare the effect of bonding temperature on joint failure mechanisms the fractured surfaces were examined using SEM. The results collected suggested that the ductility of the joint increased with increasing bonding temperature. For a bonding temperature of 570°C a mix failure mode was observed with both shear rupture dimples and cleavage planes (Figure 15a). An XRD analysis of the fractured surface indicated a high concentration of Al_2O_3 particles (see Figure 16a). This indicated that at this temperature the matrix-particle (M-P) interface was the weakest point for crack propagation giving the lowest joint strength of (53 MPa). Fracture propagation was observed through the bond-line. When the bonding temperature was increased to between 590°C and 620°C XRD analyses of the fracture indicated that the amount of intermetallic formed within the joint increased to include the binary compounds Al_3Si, Al_3Ni and Ni_3Si which suggested that failure propagated through the bond-line. Additionally, peaks of the ternary phases AlFeSi (2θ=37 and 68°) and Al_9FeSi (2θ=78 and 84°). The presence of these compounds confirms the formation of peritectic reactions during bonding. At 620°C the fractured surface was characterized by shear ruptured dimples indicating a ductile failure mode which occurred in the parent metal adjacent to the bond-line.

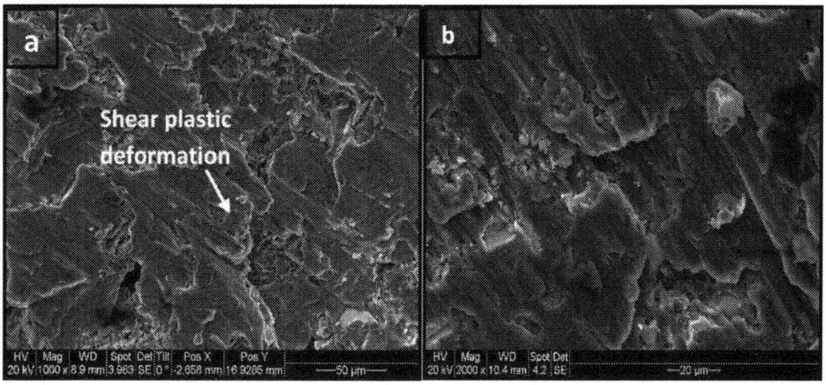

Figure 15. (a) SEM micrograph of the fractured surface of a joint made using a 5 µm thick Ni-Al_2O_3 coatings at 570°C and (b) 620°C.

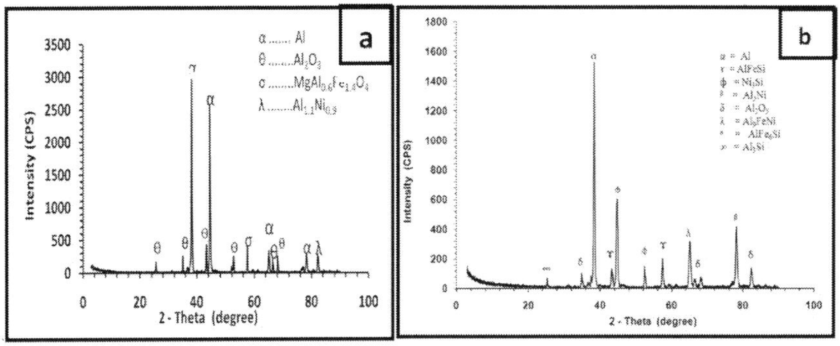

Figure 16. XRD spectrum of the fractured surface of a joint made using a 5 μm thick Ni-Al_2O_3 coatings at (a) 570°C and (b) 620°C.

5.3. Effect of coating thickness on shear strength measurements

Figure 17 (a) shows the variation in joint shear strength values as a function of coating thickness. The graph shows that the shear strength increased with increasing coating thickness from 53 MPa at 1 μm to 144 MPa at 11 μm. This increase in joint strength was attributed to three phenomena: the presence of nano-size Al_2O_3 particles in the joint center, the segregation of micro Al_2O_3 particle to the joint zone and the precipitation of intermetallic phases such as Al_3Si, Ni_3Si, Al_3Ni, and Al_9FeNi within the joint region. As discussed in the previous sections, for composites containing nano-sized particles, strengthening is often explained by the Orowan mechanism [36, 37]. Orowan bypassing theory shows that when smaller particle reinforcements are used the result is a more effective pinning of dislocation motion compared to when micro-particles are used. This mechanism leads to an increase in joint strength and hardness.

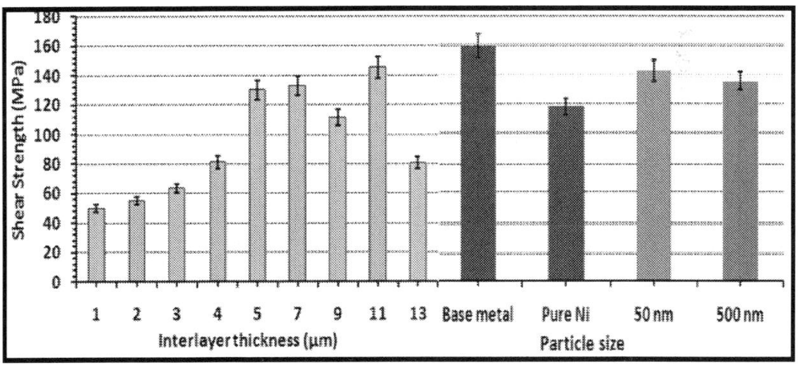

Figure 17. (a) Joint shear strengths as a function of particle size using 5μm thick coatings (b) Shear strength profile plotted as a function of Ni-Al_2O_3 coating thickness.

When the coating thickness was increased beyond 11 μm, a decline in the strength of the joints was observed. At a coating thickness of 13 μm, a joint strength of 80 MPa was recorded. This reduction in joint strength was attributed to the formation of densely packed micro-Al_2O_3 particle-rich regions along the bond interface and also due to an increase in the volume of intermetallics compounds such as $AlFe_3Si$ within the joint. The literature shows that the volume fraction of the micro-Al_2O_3 particles within the joint is inversely proportional to the joint ductility. Therefore, as the width of the particle segregated zone increased the ductility of the joint decreases. This leads to embrittlement of the joint region and causes a reduction in joint strength [38,39, 40, 41]. The findings published in the scientific literature supports the results collected in this study.

Fractured surfaces of the shear tested joints were analyzed to identify the mechanism of joint failure and the composition of the fractured surfaces. Figures 18 shows the micrographs of the typical fractured surfaces obtained for joints that were bonded using coating thickness ranging from 1 to 3μm, respectively. Fractographic analyses revealed that the fractured surface contained cleavage planes, which propagated through the bond-line. In addition, Al_2O_3 particles were visible at the fractured surface. Examination of the fractured surfaces revealed characteristics of a brittle fracture which suggest that insufficient eutectic liquid is formed when using coating thickness between 1 and 3μm. Composition analysis of the fractured surfaces using XRD indicated the presence of peaks for Al_2O_3, Ni_3Si and $Ni_{17}Al_{3.9}Si_{5.1}O_{48}$ compounds.

When an interlayer thickness of 11μm (see Figure 19a) was used, the fractured surface showed evidence of both shear plastic deformation and fractured micro- Al_2O_3 particles indicating a ductile transgranular fracture. Samples bonded at this condition had the highest shear strength. This indicated a critical combination of segregated micro- Al_2O_3 particles and nano-Al_2O_3. Crack propagation occurred in the base metal adjacent to the bond-line. The results suggest that within coating thickness range of 5 to 11μm, sufficient eutectic liquid is produced, which facilitate good particle to matrix bonding resulting increased joint strength. The XRD spectrum shown in Figure 19 (b) indicated the presence of Ni_3Si, $MgAl_2O_4$ and Al_2O_3 compound at the fractured surface. Increasing the coating thickness beyond 11μm resulted in the gradual transition of the fracture mode from ductile to brittle. At a coating thickness of 13μm thick Ni-Al_2O_3 coating (see Figure 20), the surface is characterized by dimples along the interparticle regions indicating a ductile failure through the particle-rich regions along the bond-line. XRD analyses of the fractured surfaces indicate the presence of peaks for Al_2O_3, $NiAl_2O_4$ and $AlFe_3Si$ compound at the fractured surface (see Figure 20b).

The results suggest that the ductility of the joint region increased with increasing coating thickness up to 9 μm. When the coating thickness was increased beyond 9 μm the joint region transitioned from ductile to brittle. These transitions were attributed to an increase in the volume of eutectic liquid that forms with increasing coating thickness leading to interparticle contact.

Figure 18. (a) SEM micrograph (b) cross-section (Mag. X10) and (c) XRD analysis of the fractured surface for a bond made with a 3 μm thick Ni–Al_2O_3 coating for 10 minutes at 600°C.

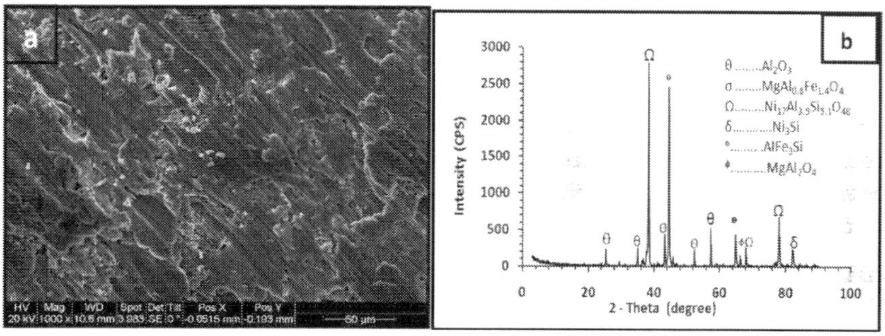

Figure 19. (a) SEM micrograph (b) cross-section (Mag. X10) and (c) XRD analysis of the fractured surface for a bond made with an 11 μm thick Ni–Al_2O_3 coating for 10 minutes at 600°C.

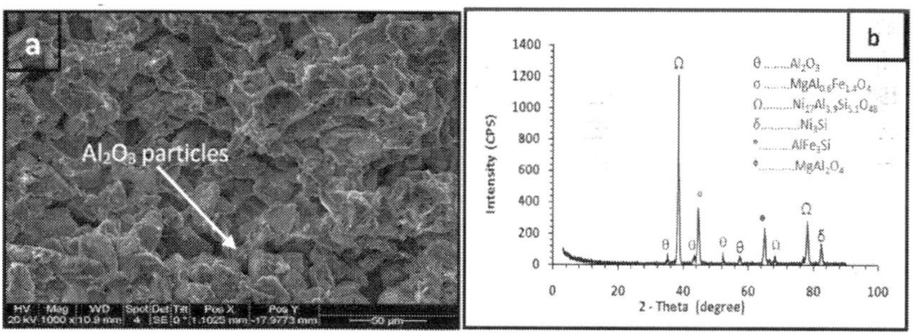

Figure 20. (a) SEM micrograph and (b) XRD analysis of the fractured surface for a bond made with a 13 μm thick Ni–Al_2O_3 coating for 10 minutes at 600°C.

5.4. Effect of interlayer particle size shear strength measurements

Figure 17(b) show the joint shear strength graph as a function of interlayer particle size. The result indicated that joint shear strength increased with decreasing particle size from 138

MPa with 500 nm to142 MPa with 50 nm. This increase in joint shear strength was attributed to better distribution of nano-sized particles within the interlayer when smaller particle sizes are used. In both cases higher shear strengths were obtained than when pure Ni coating is used (117 MPa) [23]. The results indicate that joint strength of up to 90% that of the base metal (BM) is achievable when using a 50 nm diameter nano-sized particle-reinforced interlayer. Tjong [42] showed that the nano-particle size has a strong effect on the yield strength. The author suggested that a particle size of 100 nm is a critical value for improving the yield strength of nano-composites. Below this critical value the yield strength increases significantly with decreasing particle size. Similar results were obtained by Gupta and coworkers [43, 44]. Zhang and Chen [45] showed that the Orowan stress plays a major role in strengthening the nano-composites.

Figure 21 shows the fractured surface for a bond made using a 5μm Ni- 50 nm Al_2O_3 particle. The fractograph showed shear plastic deformation, indicative of ductile fracture with a crack propagating primarily through the bond-line and a section of the base metal adjacent to the bond-line. A similar result was obtained when a dispersed particle size of 50 nm were used in the coating. XRD analyses of the fractured surfaces indicated the presence of peaks for Al_2O_3, $NiAl_2O_4$ and $Al_{11}Ni_9$ compound at the fractured surface.

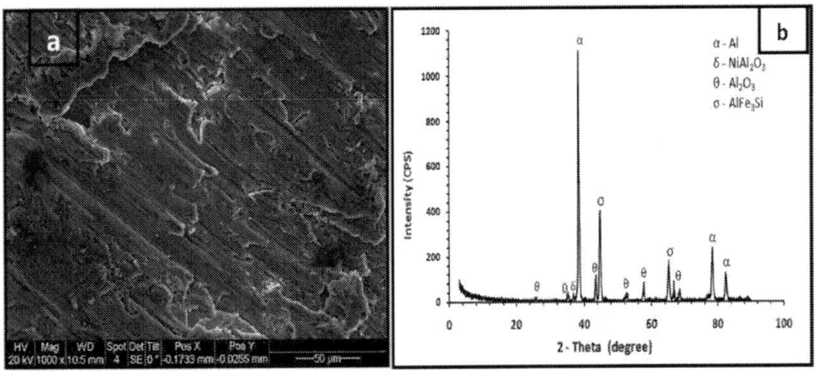

Figure 21. (a) SEM micrograph and (c) XRD analysis of the fractured surface for a bond made with 5 μm thick Ni–(50nm) Al_2O_3 for 10 minutes at 600°C.

6. Mechanism of joint formation

The mechanism of bond formation using composite Ni-Al_2O_3 coating is discussed in this section. The experimental results and the scientific literature show that the joint formation during transient liquid phase diffusion bonding is completed in five distinct stages: interfacial contact and solid-state diffusion, eutectic melting and base metal dissolution, and isothermal solidification. These stages will be discussed thoroughly with reference to the change in composition across the joint region. Mathematical models for predicting the parameter settings when nano-composite coatings are used in joining will also be presented. The sub-section of this topic are as follows:

6.1. Interfacial contact and Solid-state diffusion

The first stage of transient liquid phase bonding involved heating the sample to the bonding temperature. During this stage of bonding, two mechanisms are thought to occur simultaneously, firstly an increase in interfacial contact between the coating surface and the Al 6061 surface and secondly, solid-state diffusion along the coating/MMC contact interface. The initial contact area between the one side of the coating and the metal surfaces is only a small fraction of the theoretical area available, due to the presences of micro-asperities of the surface of both the metal sample and the coating. However, under the effects of heating and an external pressure an intimate contact can be establish at the bonding surfaces, as the micro-asperities suffer plastic deformation. As the temperature increases greater plastic flow is achieved at the interface and the percentage contact area increases. This increase in the contact area results in an increased diffusivity of the solute (Ni) into the base metal. During the heating stage Ni diffuses deep into the Al-MMCs resulting in the formation of complex intermetallic compounds.

Figure 22(a) shows an SEM micrograph of a joint bonded using a 15μm thick Ni-Al_2O_3 coating as the interlayer. The joint was made at a bonding temperature of 600°C for 1 minute and shows three distinct reaction layers at the interlayer/MMC interface. The nano- Al_2O_3 particles that were co-electrodeposited with Ni can are clearly shown in Figure 22.

The composition of the reaction layers was determined quantitatively by energy dispersive spectroscopy (EDS) analysis and is shown in Table 4. The formation of reaction layers (L_2 and L_3) shown in Figure 22, occurred as a result of the inter-diffusion of Ni and Al. EDS analysis showed that L_1 was composed of a Ni-Al layer dispersed with nano-sized Al_2O_3 particles after a bonding time of 1 minute (see Table 4). Reaction layer L_2 on the other hand appears to be a nickel-aluminide with compositions of 50.7 (at %) Ni and 46.70(at %) Al. The L_3 layer contains approximately 24.30 (at %) Ni and 77.0 (at %) Al, which is likely to be $NiAl_3$ intermetallic. This compound is believed to form due to the low solubility of Ni in Al. This has been reported to be approximately 2.9 at% [46]. The saturation of the aluminum interface through the inter-diffusion of Ni and Al leads to the precipitation of the nickel aluminide intermetallic $NiAl_3$.

The phase diagram of the Ni + Al system indicates the thermodynamic stability of the γ'-Ni_3Al phase when formed in the nickel concentration range of about (74 to 76) at.% [46]. Additionally, the phase diagram of the (Ni + Al) system proposed by Nash et al. [47] showed that for aluminum concentrations exceeding 40 mol%, there exist three coexistence fields: (Al + $NiAl_3$), ($NiAl_3$ + Ni_2Al_3) and the non-stoichiometric intermetallic β -NiAl, which is formed in the concentration range 43mol% to 59 mol% aluminum. Rog et al. [48] determined the Gibbs free energy of formation for various intermetallic compounds forming in the Ni + Al system. The results showed that within the temperature range of 570°C to 620°C (843K to 893K) the nickel aluminide compounds listed in Table 4 are formed.

Based on the scientific literature, the compound formed in the reaction layer L_3 is likely to be the $NiAl_3$. This compound also appears on the right side of Equation 9 and is believed to form due to low solubility of Ni in Al which has been reported to be approximately 2.9 at% [46]. The saturation of the aluminum interface through the diffusion of Ni can lead to the precipitation of the nickel aluminide intermetallic $NiAl_3$. The composition of L_2 indicates that the compound is likely NiAl.

Phase	ΔG_f^o (kJ.mol^{-1})
Ni +Al = NiAl	-133.0 +/- 1.0
$Ni+NiAl_3= Ni_2Al_3$	-144.1+/- 0.8
$2Ni +3Al = Ni_2Al_3$	-311.0+/- 1.7
$Ni + 3Al =NiAl_3$	-166.8+/- 0.9

Table 4. The standard Gibbs free energy values for the chemical reactions with nickel aluminides at 870K [48]

Layers	Al / wt%	Ni / wt%	Si / wt%	Mg / wt%	Compound
L_1	12.5	87.5	0	0	$Ni +Al_2O_3$
L_2	46.71	50.67	0.27	0	NiAl
L_3	72.84	24.31	0	0.55	$NiAl_3$
L_4	72.84	24.31	0	0.55	NiAl3

Table 5. Energy dispersive spectroscopic compositional analyses of the reaction layers developed during bonding (wt%)

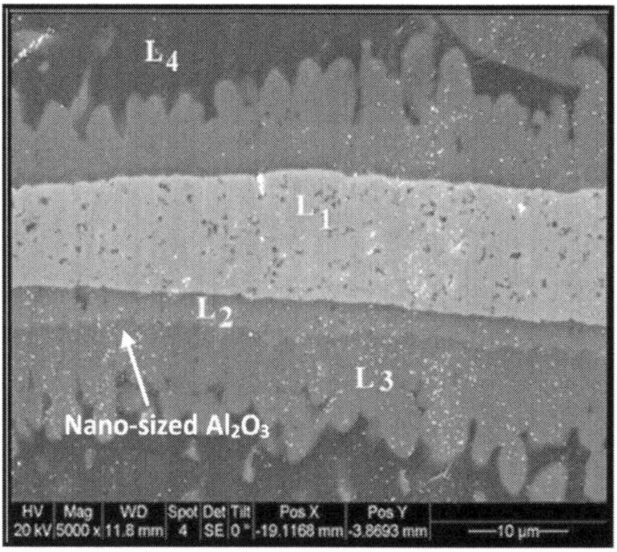

Figure 22. (a) SEM micrograph of joint bonded with a 15 μm $Ni-Al_2O_3$ coating for 1 min.

6.2. Liquid formation and base-metal dissolution

Immediately following the heating stage, eutectic melting ensues. According to Dmitry et al [50] there are two possible reactions which are capable of producing a liquid at the joint interface within this temperature range (570°C-620°C). Upon reaching a temperature of 565°C, the L_2 compound reacts with L_4 (88.4%Al and 2.08 % Si) to form a eutectic liquid (E1) along the bond interface as predicted by Equation 9 [46]. The compound formed in L_2 is consumed in the eutectic reaction and diffuses into the base metal. When the joint region was heated to 577°C a second eutectic liquid formed within L_4 as predicted by Equation 10 [49]. The formation of a eutectic liquid layer at the bond interface leads to faster inter-diffusion between Al and the Ni interlayer and this results in a gradual change in the composition of the joint region.

$$L_1 \xrightarrow{565\,^{o}C} (87\%)\,Al + (2\%)\,Si + (11\%)\,NiAl_3 \tag{9}$$

$$L_2 \xrightarrow{577\,^{o}C} (87.5\%)\,Al + (12.5\%)\,Si \tag{10}$$

Dmitry et al. [50] carried out thermodynamic calculations of (Al-Mg-Si-Fe-Ni) quinary systems formed in aluminum alloys. The results showed that in alloys containing Al-Mg-Si-Fe-Ni, numerous ternary and quaternary reactions can occur that has the potential of producing a liquid phase. Some of the phases that contribute to these reactions are shown in Table 6. The XRD analysis of the fractured surfaces shown in the previous also indicated that presence of $AlFe_3Si$. This phase is possibly a variant of the β-phase family listed in Table 6.

Phase	Designation	Composition (wt%)	Density (g/cm³)
Al_3Ni	ε	42Ni	3.95
Mg_2Si	M	63.2Mg; 36.8Si	4.34
Al_9FeNi	T	4.5-14Fe; 18-28Ni	3.40
Al_5FeSi	β	25-30Fe, 12-15 Si	3.45
$Al_8FeMg_3Si_6$	π	10.9Fe; 14.1Mg; 32.9Si	2.82

Table 6. Chemical composition and density of phases formed in Al-Mg-Si-Fe-Ni system [50]

6.3. Base metal dissolution

Base metal dissolution is also an important part of the second stage of bonding. Research work published [1,2] on the mechanisms of TLP bonding showed that dissolution of the interlayer and base metal occurs simultaneously at the bonding temperature. In this study, holding at a bonding temperature above 577°C resulted in increased diffusion of Ni into the base metal and causes the liquid phase to spread between the bonding surfaces due to the effects of pressure and capillary action. The application of pressure enhances spreading of the liquid phase between the bonding surfaces due to capillary action. This spreading increases the contact area and induces the diffusion of Ni and Al into the liquid phase. This results in more eutectic liquid formation and an increase in the width of the liquid phase at the joint, due to dissolution of a section of the base metal. Further increase in the bonding time to 5 minutes, resulted in the diffusion of the Ni into the base metal and away from the joint interface leading to the formation of eutectic or peritectic liquid along the grain boundary as indicated in the EDS analysis of the eutectic microstructure shown in Figure 23.

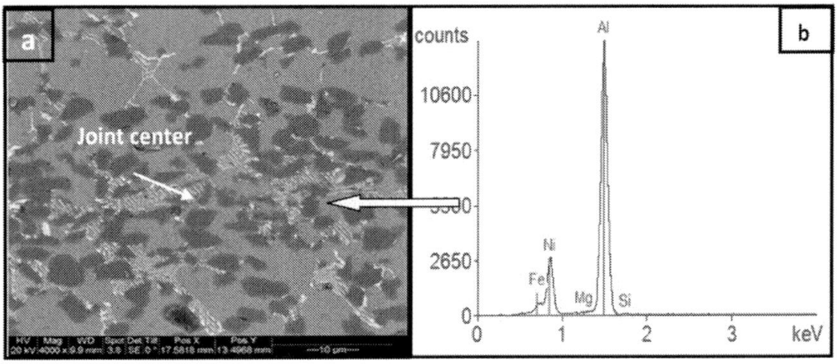

Figure 23. (a) SEM micrograph of joint bonded using a 25 μm Ni foil for 5 min and (b) EDS analysis of the joint region.

According to published studies [20,1] on the stages of TLP bonding, it is expected that dissolution of the interlayer and base metal occurs simultaneously at the bonding temperature followed by spreading of the liquid phase between the bonding surfaces. This spreading increased the bonded area and enhanced the diffusion of Ni and Al into the liquid phase. This continuing diffusion resulted in more liquid formation and an increase in the width of liquid phase. The maximum liquid width attained when using the Ni-Al_2O_3 composite coating as the interlayer was found by Cooke et al [24] to be $W_{max} = 20.6w_o$.

6.4. Isothermal solidification

In TLP bonding, prolonged hold time at the bonding temperature for 10 minutes allowed for the diffusion of Al into the eutectic liquid which caused the composition of the liquid phase to become aluminum rich, resulting in a change in the eutectic composition (see Table 6). The change in the joint composition initiates the isothermal solidification stage of TLP bonding as a function of bonding time since the temperature and interlayer thickness is constant. When the bonding time is increased to 10 minutes the interface is eliminated and the grain size within the joint increased. This disrupts the band of segregated particles at the interface and homogenized the joint zone. When the bonding time was increased to 30 minutes a corresponding increase was seen in grain size, resulting in more uniform distribution of micro-Al_2O_3 particles. In the reported studies on transient liquid phase diffusion bonding of Al-MMCs, it was shown that the width of the segregated zone at the joint center increased with increasing bonding time. The opposite of this relationship was seen when using the Ni-Al_2O_3 coating. As the bonding time increased, the width of the segregated region decrease. This can be attributed to the heterogeneous nucleation of grains within the joint zone during solidification and this lead to grain refining at the joint. The high resolution SEM micrograph shown in Figure 33 revealed the presence of a nano-Al_2O_3 particle at the center of a grain. EDX spectra of the particle showed Al and O in high concentrations with traces of Mg. Comparing Gibbs free energy of formation at the bonding temperature for MgO (-1195 kJ/mol) and Al_2O_3 (-985 kJ/mol) it is found that Al_2O_3 is unstable in the presence of Mg hence it is like that some of the nano-size Al_2O_3 will decomposed to form $MgAl_2O_4$ compound.

The segregation of particles during isothermal solidification was accredited to the pushing of strengthening particles by the solidifying liquid-solid interface. Stefanescu [17, 18] showed that particle pushing can be assumed to be a steady-state condition under which the interface velocity can be assumed to be equal the rate of isothermal solidification This rate can be calculated using a model proposed by Sinclair [51]. The constant, ξ, signifies the solidification rate of the system. Increasing ξ results in faster solid-liquid interface motion, and a shorter duration of the isothermal solidification stage. The rate of isothermal solidification can be calculated using Equation 11.

$$\xi = -2\,(k - 1)^{-1}\sqrt{\frac{D}{\pi}}\cdot\frac{\exp\left(\frac{-\xi^2}{4.D}\right)}{erfc\left(\frac{\xi}{2\sqrt{D}}\right)} \tag{11}$$

Where k is a partition coefficient given by $\frac{C_{L\alpha}}{C_{\alpha L}}$ and D is the diffusivity of Ni into Al. The diffusivity at 570°C is 4.69 x10^{-13} m^2/s, when the bonding temperature is increased to 620°C, the diffusivity also increased to 1.58 x10^{-12} m^2/s. This increase in diffusivity is reflected in a faster solid-liquid interface rate (ξ) and a shorter isothermal solidification stage.

The final concentration of Ni $\left(C_{L\alpha}\right)$ was taken from the Al-Ni-Si phase diagram [30] to be 4.9 wt % for the bonding temperature of 620°C. The diffusivity of Ni in Al at 620°C is D = 1.58 x 10^{-12} m^2/s [25, 31]. Using these values the predicted interface rate constant ξ = -0.395μm/s was calculated from Equation 8. This solidification rate is significantly less than the critical interface velocity (16 -400 μm/s) required to engulf dispersed particle during solidification [26]. Li et al. [21] suggested that particle segregation tendency is dependent on the relationship between the liquid film width produced at the bonding temperature, particle diameter and inter-particle spacing. When the liquid film width is large enough that sufficient particulate material is contained in the melt, particles will be pushed ahead of the solidifying liquid-solid interface resulting in particle segregation at the bond-line. However if the liquid film width is less than some critical value, segregation should not occur. WDS analysis across the joint zone as a function of bonding time indicated that the Ni volume at the joint center varied between 3.67 wt% after 1 minute and 0.15 vol.% after 30 minutes bonding time. The relationship between the width of segregated region and the maximum liquid width that formed during bonding was determined by using equation 12.

$$S_{sz} = W_{max}\left[\frac{\delta_p + \chi_2}{\delta_p + \chi_1}\right] \tag{12}$$

Where δ_p is the average particle size, χ_1 is the inter-particle spacing in the as received MMC and χ_2 is the inter-particle spacing after bonding at the joint. By substituting δ_p = 28 μm, $\chi_1 = 10\,\mu m$ and $\chi_2 = 0$ into Equation 6.3, the relationship between the width of the particle segregated zone and the maximum width of the eutectic liquid phase was found to be, $S_{sz} = 0.74\ W_{max}$. This means that the width of the segregated zone is approximately 74% of the width of the maximum width of the eutectic liquid phase formed during bonding.

7. Conclusion

Transient liquid phase diffusion bonding of particle reinforced Al-6061 MMC using Ni-Al_2O_3 interlayer was successfully achieved using nano-composite Ni-Al_2O_3 coating. The results obtained from the spectroscopic analyses using WDS and XRD showed that Ni-Al_2O_3 coating the bonding process can be characterized in four distinct stages: interfacial contact and solid-state diffusion, which resulted in the formation of three reaction layers promoted by the diffusion of nickel into the aluminum base metal. The second stage of the joining process was the formation of an Al-Ni-Si eutectic liquid at the bonding temperature. It is supported that the reaction layers formed within the joint melted to form a liquid phase, followed by dissolution of the base metal (third stage) as the liquid spread between the bonding surfaces through capillary action. The final stage of bonding involved isothermal solidification at the bonding temperature in which the diffusion of Ni into Al results in a change in the composition of the liquid phase leading to solidification.

The joint shear strength was studied as a function of bonding parameters, bonding time, bonding temperature, interlayer thickness and interlayer particle size. The results showed that the joint shear strength increased with increasing bonding time, bonding temperature and interlayer thickness. On the other hand the results showed that bond strength increased when the interlayer particle size was reduced from 500 nm to 50 nm. The increase joint shear strength seen when Ni-Al_2O_3 coating were used was attributed to the presence of highly-dispersed nano-sized reinforcement particles in the joint region act to strengthen the joint region by Orowan bowing mechanism.

The results showed that an optimum joint strength of 144 MPa can be achieved if the following bonding parameters are used: 30 minutes bonding time, 620°C bonding temperature, and 11 μm thick Ni–Al_2O_3 coating. Within the parameter ranges tested the bonding pressure had the lease effect on the joint shear strength of TLP bonded joints.

Author details

Kavian Cooke

Department of Mechanical Engineering, School of Engineering, University of Technology, Jamaica

8. References

[1] W. D. Macdonald and T. W. Eager 'Transient Liquid Phase bonding processes' Minerals Metals and Joining Society, 1992, 93-101

[2] I. Tuah-Poku, M. Dollar and T. B. Massalski, 'A study of the transient Liquid phase bonding process applied to a Ag/Cu/Ag sandwich joint' Metall. Trans. A, 1988, 19 A, 675 -686

[3] Y. Zhou, W.F. Gale, and T.H. North Modelling of Transient Liquid. Phase Bonding, International Material Review, (1995) Vol. 40, No.5, pp 181-196

[4] ASTM E3 -01 (2008) Standard preparation for metallographic examination, ASTMInternational 100 Barr Harbor drive United States

[5] Z. Li, Y. Zhou, T.H. North,'Counteraction of particulate segregation during transient liquid-phase bonding of aluminum-based MMC material' Mater. Sci. 1995, 30, 1075–1082

[6] A.A. Shirzadi, E.R. Wallach, (1997) New approaches for transient liquid phase diffusion bonding of Aluminum based metal matrix composites, Material Science Technology Vol.13, pg: 135–142.

[7] Y. Z hou W. F. Gale, and T. H. North, Modeling of transient liquid phase bonding, International Mater. Rev., 1995, Vol. 40, No.5, 181-196.

[8] N. S. Bosco, F. W. Zok, Critical interlayer thickness for transient liquid phase bonding in the Cu–Sn system, Acta Materialia, 2004, Vol. 52. pg: 2965–2972.

[9] W. D. Macdonald and T.W. Eager, 'Transient liquid phase bonding processes' Minerals Metals and Joining Society, 1992, 93-101.

[10] J.R. Askew, J.F. Wilde and T.I. Khan, Transient liquid phase bonding of 2124 aluminum metal matrix composite, Mater. Sci. Tech. 1998, 14 920–924.

[11] R.F. Chen, Y.H. Zhao, Z.X. Shen, L.G. Dai, X.L. Zhang, R. Zhu, Study on the joint strength of SiCp/ Al metal-matrix composite by magnetron sputtering method, Mater. Sci. 2009, pp 628 - 629

[12] A. Suzumura, and Y. Xing, Diffusion brazing of short Al2O3 fiber reinforced aluminum composite. Mater. Trans., 1996, Vol. 37, pg: 1109–1115.

[13] T. Enjo, K. Ikeuchi, Y. Murakami, N. Suzuki, Diffusion bonding of aluminum-magnesium-silicon series 6063 alloy reinforced with alumina short fibers, Trans. of JWRI, 1987, Vol.16, No. 2, pg: 285-92

[14] Y. Zhai, T.H. North and J. Serrato-Rrodrigues, Transient liquid-phase bonding of alumina and metal matrix composite base materials, Mater. Sc., 1997, Vol.32, pg;1393–1397.

[15] A. A. Shirzadi, E.R. Wallach, New approaches for transient liquid phase diffusion bonding of Aluminum based metal matrix composites, Mater. Sci. Tech., 1997, Vol.13, pg: 135–142.

[16] A. Urena, J.M. Gomez de Salazar, and M.D. Escalera, Diffusion bonding of and aluminum metal matrix composites Weld. J, 1997, Vol.76, Pg: 92–102.

[17] D.M Stefanescu, Science and engineering of casting solidification, Kluwer Academic/Plenum Publishers, New York (2002).

[18] D.M. Stefanescu, F.R. Juretzko, B.K. Dhindaw, A. Catalina, S. Sen, P.A. Curreri, "Particle Engulfment and Pushing by Solidifying Interfaces: Part II, Microgravity Experiments and Theoretical Analysis" Metall. Mater. Trans. A, Vol. 29 (1998), pp; 1697.

[19] Z.W. Xu, J.C. Yan, G.H. Wu, X.L. Kong, S.Q. Yang, Interface structure and strength of ultrasonic vibration liquid-phase bonded joints of Al2O3p/6061Al composites Scripta Materialia, 2005, Vol. 53, pg: 835–839

[20] J.H. Huang, Y.L. Dong, Y. Wan, J.G. Zhang, G.A. Zhou, Reactive diffusion bonding of SiCp/Al composites by insert layers of mixed powders, Mater. Sci. Tech., 2005, Vol. 21, pg: 1217–1221

[21] B. Wielage, I. Hoyer, S. Weis, Soldering aluminum matrix composites. Weld J 2007;86:67–70

[22] J. Yan, Z. Xu, L. Shi, X. Ma, S. Yang, Ultrasonic assisted fabrication of particle reinforced bonds joining aluminum metal matrix composites Mater Des. 32 (2011) 343–347

[23] K.O. Cooke, T.I. Khan, G.D. Oliver (2011) Nanostructure particle-reinforced transient liquid phase diffusion bonding; a comparative study, Metall. Mater. Trans. A, DOI: 10.1007/s11661-011-0663-6.

[24] K.O. Cooke, T.I. Khan, G.D. Oliver (2011) Transient liquid phase diffusion bonding Al-6061 using nano-dispersed Ni coatings, Mater. Des. DOI: 10.1016/j.matdes.2011.04.051.

[25] W. Deqing, S. Ziyuan and K. Tangshan, Composite plating of hard chromium on aluminum substrate, Surface and Coatings Technology, 2005, Vol.191, pp:324– 329

[26] D.R. Gabe, Principles of metal surface treatment and protection, Anti-Corrosion Methods and Materials, Published by Pargamon Press, Oxford,1972,

[27] W.F. Gale and D.A. Butts, Transient liquid phase bonding Science and Technology, Journal of Welding and Joining, 2004, Vol. 9, No. 4, pp: 283-300.

[28] W.D. Zhuang and T.W. Eagar, Transient liquid-phase bonding using coated metal powders Welding Journal, 1997, Vol. 76, Issue 12, Pages 157-167.

[29] M.L. Kuntz, Y. Zhou, and S.F. Corbin, A study of transient liquid-phase bonding of Ag-Cu using differential scanning calorimetry, Metallurgical and Materials transaction A, Vol. 37A, 2006, pp: 2493-2504.

[30] D.S. Duvall, W.A Owczarski, and D.F. Paulonis, TLP bonding: a new method of joining heat resisting alloys, Welding journal, Vol 53, No4. 1974, pp: 203-214

[31] W. F. Gale. and T. C. Totemeier 'Smithells Metals Reference Book', 8th Edn, 2004, Published by Elsevier

[32] K. Cooke, T. Khan and G. Oliver (2012) Critical interlayer thickness required for transient liquid phase bonding using nano-dispersed Ni coatings, submitted to Journal of Science and Technology of Welding and Joining Vol. 17 No. 1 page 22-31.

[33] K.O. Cooke, A Study of the Effect of Nanosized Particles on Transient Liquid Phase Diffusion Bonding Al6061 Metal–Matrix Composite (MMC) Using Ni/Al_2O_3 Nanocomposite Interlayer , Metallurgical and Materials Transaction B, (2012) (accepted article) DOI: 10.1007/s11663-012-9643-5

[34] Y.-L. Shen, E. Fishencord, N. Chawla, Correlating macrohardness and tensile behavior in discontinuously reinforced metal matrix composites Scripta. Materialia, Vol. 42, 2000, pp 427.

[35] Z.Y. Ma, Y.L. Li, Y. Liang, F. Zheng, J. Bi, S.C. Tjong, Nanometric Si3N4 particulate-reinforced aluminum composite, Journal of Material Science and Engineering A 219 (1996) 229.

[36] I. Shao, P.M. Vereecken, C.L. Chien, P.C. Searson and R.C. Cammarata, Synthesis and characterization of particle-reinforced J Mater Res 17 (2002), pp. 1412–1418.

[37] L. Thilly, M. Véron , O. Ludwig, F. Lecouturier Deformation mechanism in high strength Cu/Nb nano-composites Materials Science and Engineering A309–310 (2001) 510–513

[38] S.V. Kamat and M. Manoharan, Work hardening behaviour of alumina particulate reinforced 2024 aluminum alloy matrix composites, Journal of Composite Materials 1993, 27: 1714-1721 DOI: 10.1177/002199839302701801

[39] H. Nami, A. Halvaee and H. Adgi, Transient liquid phase diffusion bonding of Al/Mg2Si metal matrix composite, Materials and Design, 2011, doi:10.1016/j.matdes.2011.02.003

[40] G.A. Chadwick and P.J. Heath, Machining metal matrix composites Mater. 6/2 (1990), pp. 73–76.

[41] Z.Y. Ma, SC. Tjong, In situ ceramic particle-reinforced aluminum matrix composites fabricated by reaction pressing in the TiO2 (Ti)–Al–B (B2O3) systems Metallurgical Material Transaction 1997; 28(A):1931–42.

[42] S. C. Tjong, Novel nanoparticle-reinforced metal matrix composites with enhanced mechanical properties, Advanced Engineering Materials Volume 9, Issue 8, 2007

[43] S. F. Hassan, M. Gupta, "Development of high performance magnesium nano-composites using solidification processing route Journal of Material Science and Technology, 2004, 20, 1383.

[44] N. Srikanth, S. F. Hassan, M. Gupta, Energy dissipation studies of Mg-based nano-composites using an innovative circle-fit approach, Journal of Composite Material.2004, 38, 2037.

[45] Z. Zhang, D. L. Chen, Consideration of Orowan strengthening effect in particulate-reinforced metal matrix nano-composites: A model for predicting their yield strengthScr. Mater. 2006, 54, 1321.

[46] K. W. Richter, K. Chandrasekaran and H. Ipser, The Al–Ni–Si phase diagram. Part II: phase equilibria between 33.3 and 66.7 at.% Ni, Journal of Intermetallics, 12 (2004) 545–554

[47] P. Nash, M.F. Singleton, J.L. Murray, in: P. Nash (Ed.), Phase Diagrams of Binary Nickel Alloys, ASM International, Materials Park, OH, 1991, pp. 3–11.

[48] G. Rog , G. Borchardt, M. Wellen, W. Lose, determination of the activities in the (Ni + Al) alloys in the temperature range 870K to 920K by a solid-state galvanic cell using a CaF2 electrolyte J. Chem. Thermodynamics 35 (2003) 261–268

[49] V. Raghavan, Al-Ni-Si (Aluminum-Nickel-Silicon) Phase diagram evaluations, Journal of Phase Equilibria and Diffusion, 2005, Vol. 26, pp:262-267

[50] E. Dmitry, G. Belov, N.A. Aksenov, and A. Andrey, (2005) Multicomponent Phase Diagrams : Applications for Commercial Aluminum Publisher: Elsevier Science

[51] C.W. Sinclair, 'Modelling transient liquid phase bonding in multi-component systems Journal of Phase equilibria, 1998, 20, No. 4,

Chapter 12

Photo-Catalytic Degradation of Volatile Organic Compounds (VOCs) over Titanium Dioxide Thin Film

Wenjun Liang, Jian Li and Hong He

Additional information is available at the end of the chapter

http://dx.doi.org/10.5772/48099

1. Introduction

Volatile organic compounds (VOCs) are emitted as gases from certain solids or liquids. VOCs include a variety of chemicals, some of which may have short- and long-term adverse health effects. Concentrations of many VOCs are consistently higher indoors (up to ten times higher) than outdoors.

The control of VOCs in the atmosphere is a major environmental problem. The traditional methods of VOCs removal such as absorption, adsorption, or incineration, which are referred to the new environmental condition have many technical and economical disadvantages. So in recent years, some new technologies called advanced oxidation processes (AOPs), such as biological process, photo-catalysis process or plasma technology, are paid more and more attention.

Advanced oxidation processes (AOPs) are efficient novel methods useful to accelerate the non-selective oxidation and thus the destruction of a wide range of organic substances resistant to conventional technologies. AOPs are based on physicochemical processes that produce in situ powerful transitory species, principally hydroxyl radicals, by using chemical and/or other forms of energy, and have a high efficiency for organic matter oxidation.

Among AOPs, photocatalysis has demonstrated to be very effective to treat pollutants both in gas and in liquid phase. The photo-excitation of semiconductor particles (TiO_2) promotes an electron from the valence band to the conduction band thus leaving an electron deficiency or hole in the valence band; in this way, electron/hole pairs are generated. Both reductive and oxidative processes can occur at/or near the surface of the photo-excited semiconductor particle.

Photocatalytic degradation of VOCs on UV-illuminated titanium dioxide (TiO_2) is proposed as an alternative advanced oxidation process for the purification of water and air. Heterogeneous photo-catalysis using TiO_2 has several attractions: TiO_2 is relatively inexpensive, it dispenses with the use of other coadjutant reagents, it shows efficient destruction of toxic contaminants, it operates at ambient temperature and pressure, and the reaction products are usually CO_2 and H_2O, or HCl, in the case of chlorinated organic compounds. Decomposition path of VOCs with UV/TiO_2 or UV/TiO_2/doped ions is shown in Fig. 1. However, the formations of by-products, such as CO, carbonic acid and coke-like substances, were often observed. These by-product formations are due to low degradation rate of intermediate compounds that are formed by the partial oxidation of VOCs. In order to improve the VOC degradation rate, some authors reported on the enhancement of VOC degradation through the addition of anions (dopant = S, N, P, etc), cations (dopant = Pt, Cu, Mg, etc), polymers or co-doped with several ions on TiO_2, while the difference between doping agents has not been discussed yet.

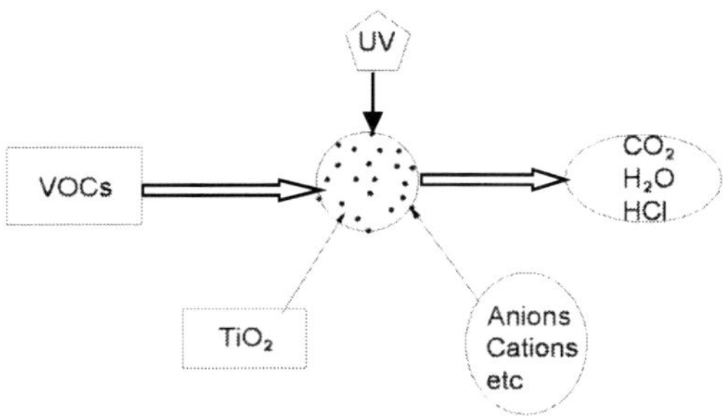

Figure 1. Decomposition path of VOCs with UV/TiO_2.

In this chapter, toluene, p-xylene, acetone and formaldehyde were chosen as the model VOCs because they were regarded as representative indoor VOCs for determining the effectiveness and capacity of gas-phase air filtration equipment for indoor air applications, the photo-catalytic degradation characters of them by TiO_2/UV, TiO_2/doped Ag/UV and TiO_2/doped Ce/UV was tested and compared. The effects of hydrogen peroxide, initial concentration, gas temperature, relative humidity of air stream, oxygen concentration, gas flow rate, UV light wavelength and photo-catalyst amount on decomposition of the pollutants by TiO_2/UV were analyzed simultaneously. Furthermore, the mechanism of titania-assisted photo-catalytic degradation was analyzed, and the end product of the reaction using GC-MS analysis was also performed.

2. Materials and methods

2.1. Chemicals and experimental set-up

Acetone, toluene, p-xylene and formaldehyde used in our experiment was analytical reagent. The TiO_2 photocatalyst was prepared with 100 % anatase using the sol-gel method, and immobilized as a film (thickness 0.2 mm) on glass springs. Ethanol, tetrabutyl orthotitanate, diethanolamine, N,N-dimethylformamide and polyethylene glycol used as raw materials for photocatalyst preparation were of analytical grade and utilized without further purification. $AgNO_3$, $Ce(NO_3)_3 \bullet 6H_2O$ were used as Ag or Ce source for modified TiO_2 samples. Deionized water was used throughout the study.

A schematic diagram of the experimental system for photo-oxidation is shown in Fig. 2. The experiments were performed in a cylindrical photo-catalytic reactor with inner diameter 18.0 mm. A germicidal lamp (wavelength range 200-300 nm) with the maximum light intensity at 254 nm was installed in the open central region. The desired amount of representative sample, that is acetone, toluene, p-xylene or formaldehyde, was injected into the obturator. Then, the photo-catalytic degradation was performed by transporting the gas across the photo-catalyst continuously when UV lamp was turned on. Glass spring coated by a TiO_2 thin film was filled around the lamp. In whole experiment, humidity was controlled and adjusted with vapour. In some experiments it was replaced with a 15 W black-light lamp with a maximum light intensity output at 365 nm. After a stabilized period of about 3 h, the pollutant concentrations in the outlet gas became the same as in the inlet gas, and the experiment was started by turning on the UV lamp. Relative humidity of the reactor was detected with humidity meter. Oxygen concentration was controlled with oxygen detector.

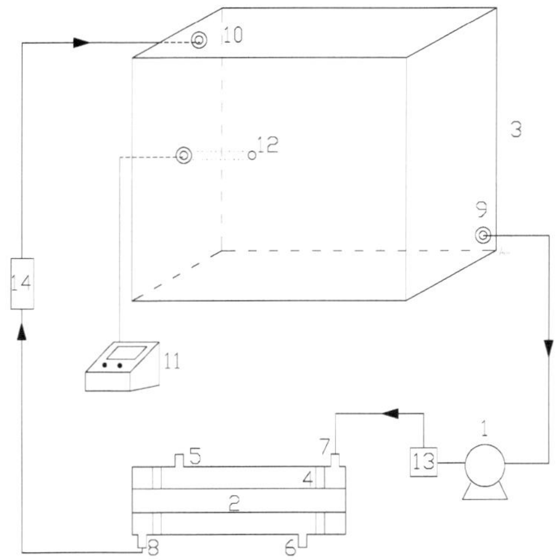

Figure 2. Schematic diagram of experimental set-up. 1- Minitype circulation pump; 2 - Germicidal lamp; 3 - Obturator (airproof tank, 125 L); 4 - Lacunaris clapboard; 5, 6 - Sampling spots; 7-10 – Inlet & Outlet; 11 - Temperature-humidity detector; 12 - Probe; 13 - Gas heated container; 14 - Humidity controller.

2.2. Photo-catalyst preparation

Fig. 3 shows the schematic flow-chart of the experimental procedure.TiO_2 precursor sols were prepared by adding tetrabutylorthotitanate (400 mL) into ethanol (960 mL) at room temperature. Then diethanolamine (69.1 mL) was added, and the mixture stirred for 2 hr. Subsequently, ethanol (120 mL), deionized water (25.2 g), 5 wt% $AgNO_3$ or $Ce(NO_3)_3$ were added dropwise to the solution. After stirring for 15 min, N,N-dimethylformamide (16.8 mL) was added. This reduced surface tension and made a smooth coating of the thin film on the carrier. The solution was then left to rest for 24 hr. Finally, polyethylene glycol (4.32 g) dissolved in ethanol (120 mL) at 50 °C was added dropwise to the solution. The final solution was left to sit for 12 hr, after which the TiO_2 gel had formed. The prepared mixture could remain stable for months at ambient temperature. Thin film TiO_2 photocatalyst was formed by dip-coating with a velocity of 5 cm/s. Glass springs were selected as the photocatalyst carrier due to their excellent transparency and long light diffusion distance. Fig. 4 was the sketch of glass spring structure. These were immersed in the TiO_2 gel mixture, and then dried at room temperature. This was followed by calcination at 500 °C in a muffle furnace for 2 hr. The glass springs were coated repeatedly (total of five times) using this method to form a thin TiO_2 photocatalyst film. The TiO_2 film was very stable and durable, and no loss was observed during its application.

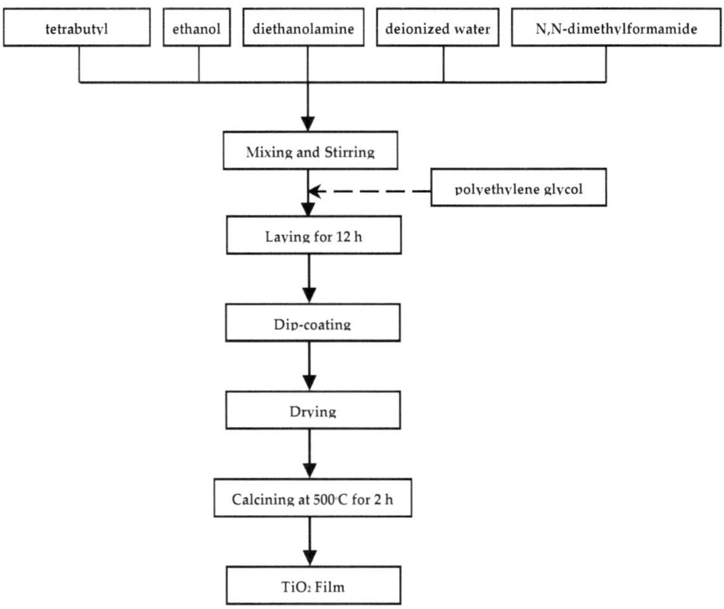

Figure 3. Flowchart of photo-catalyst preparation.

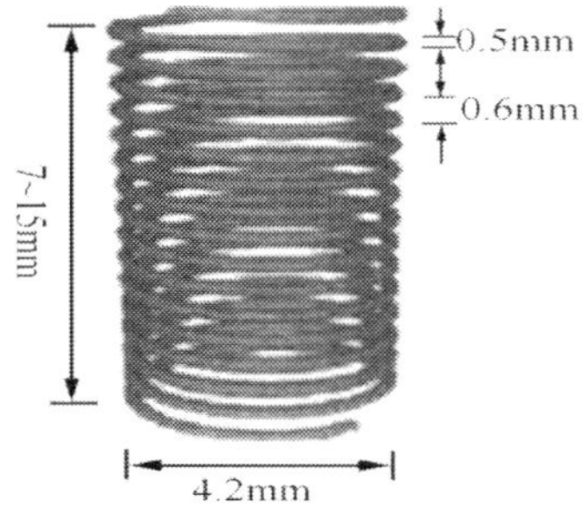

Figure 4. Sketch of glass spring structure.

2.3. Analytical methods

The concentrations of acetone, toluene, and p-xylene were analyzed by a gas chromatograph (Model GC-14C, Shimadzu, Japan) with a flame ionization detector (FID). The oven temperature was held at 60 °C and detector temperature maintained constant at 100 °C. The end products of the reaction were detected by GC-MS. GC-MS analysis was conducted using an HP 6890N GC and HP 5973i MSD. A HP-5 capillary column (30m×0.32mm ID) was used isothermally at 60 °C. The carrier gas (helium) flow-rate was 30 cm/s, and the injector and detector temperatures were 150°C and 280°C, respectively. Intermediate products analysis was done by EI mode and full scan. Formaldehyde concentration in gas stream was determined by acetylacetone spectrophotometric method. HCHO absorbed by deionized water in acetic acideammonium acetate solution would react with acetylacetone to form a steady yellow compound. HCHO concentration in the gas stream was then determined by measuring light absorbance at 413 nm with a spectrophotometer (UV/Vis 722). Temperature and humidity were measured with a temperature-humidity detector (Model LZB-10WB, Beijing Yijie Automatic Equipment Ltd., China).

The characteristics of the immobilized nano-structured TiO_2 thin film were analyzed by field-emission scanning electron microscopy (FE-SEM, Model JSM 6700F, JEOL, Japan) and X-ray diffractometry (XRD, Rigaku, D-max-γA XRD with Cu Kα radiation, λ = 1.54178 Å). The surface area of the TiO_2 film was also analyzed using gas adsorption principles (Detected by Micromeritics, American Quantachrome Co., NOVA 1000). The synthesized samples had a BET surface area of 56.3 m^2/g, compared with Degussa P25 TiO_2 with a surface area of 50.2 m^2/g.

The degradation rates (%) of acetone, toluene, p-xylene and formaldehyde were calculated as follows:

$$\eta = \frac{C_i - C_o}{C_i} \times 100\% \tag{1}$$

where C_i is the inlet concentration, and C_o is the outlet concentration at steady state.

3. Results and discussion

3.1. SEM and XRD Images of the Photocatalyst

FE-SEM analysis of the particle size and shape of the synthesized TiO_2 sample showed it consisted of uniform nano-particles (Fig. 5). However, some cracks were found on the surface. A major contributor to these cracks could be the greater surface tension resulting from the small diameter (0.5 mm) of the glass springs and the high-temperature sintering process. In further experiments, we decreased the temperature from 500 °C to 450 °C. At the lower temperature, there were fewer cracks on the surface but they were not eliminated completely.

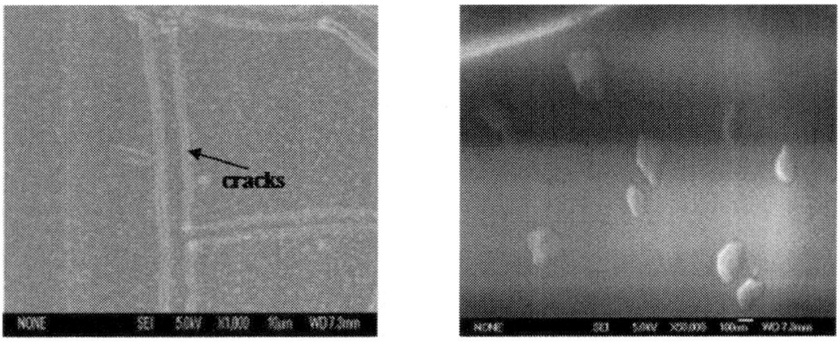

Figure 5. SEM photographs and macroscopic morphology of the TiO_2 thin film coated on a glass spring.

The left and right photographs were taken at 1000× and 50 000× magnification, respectively.

According to Scherrer's equation (Eq. 2) and XRD patterns, the particle size of TiO_2 (D) was calculated to be 35 nm.

$$D = k\lambda / \beta \cos\theta \tag{2}$$

The crystalline phase of the TiO_2 catalyst was analyzed by XRD (Fig. 6). All the diffraction peaks in the XRD pattern could be assigned to tetragonal anatase TiO_2, with lattice constants of a=0.3785 nm, b=0.3785 nm, and c=0.9514 nm.

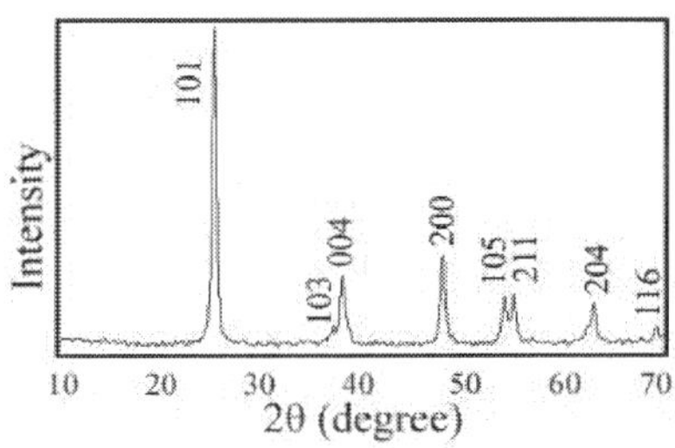

Figure 6. XRD spectrum of anatase crystalline phase of the TiO_2 catalyst

3.2. Effect of doped Ag/TiO_2 or Ce/TiO_2 on decomposition of VOCs

The characters of catalyst are important for the degradation of VOCs. Fig. 7 illustrates the degradation rates of acetone, toluene, and p-xylene (ATP) as functions of irradiation time when pure TiO_2, Ag-TiO_2 and Ce–TiO_2 were used. As controls, blank experiments in the absence of TiO_2 had been studied. The results corresponded to the flow-rate of 1 L/min, initial concentration of 0.1 mol/m^3 and relative humidity of 35%. It was found that all the conversions of ATP in the TiO_2/UV, Ag-TiO_2/UV and Ce-TiO_2/UV processes were increased with irradiation time. Table 1 shows the degradation rates for both catalysts after 8-h photo-catalytic reaction. It can be seen from Fig. 7 and Table 1 that the doping of silver or cerium ions could improve the photo-activity of TiO_2 effectively. Furthermore, the degradation character of the photo-catalyst was in the order Ce-TiO_2 > Ag-TiO_2 > TiO_2. Besides, the results of blank experiments in the absence of TiO_2 showed that the removal efficiency of ATP was very low. For example, the removal efficiency of acetone was merely 6.3% after 8 h and lower than 46.5% for pure TiO_2, which means that TiO_2 plays an important role in photo-catalytic reaction.

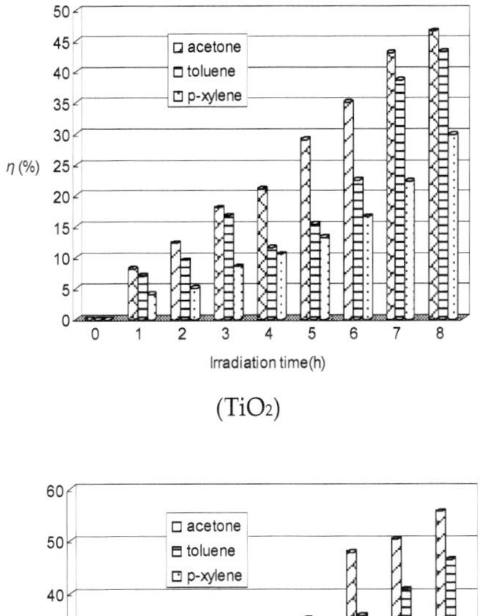

(TiO_2)

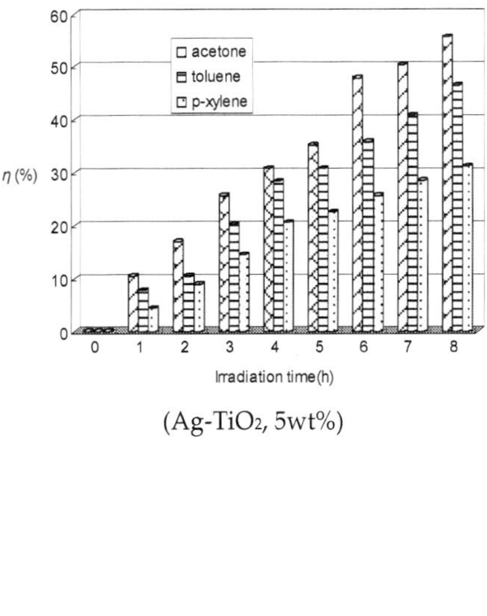

(Ag-TiO_2, 5wt%)

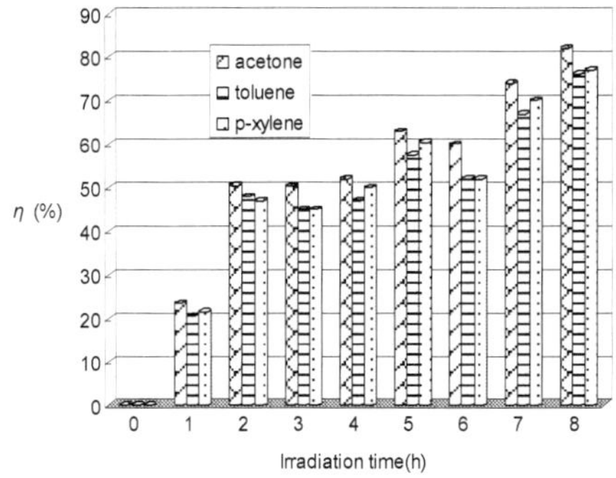

(Ce-TiO_2, 5wt%)

Figure 7. Effect of doped Ag/Ce/TiO_2 on decomposition of ATP.

Catalyst	TiO_2	Ag-TiO_2	Ce-TiO_2
η (acetone, %)	46.5	55.5	82.0
η(toluene, %)	43.2	46.4	76.2
η (*p*-xylene, %)	29.8	31.2	77.8

Table 1. ATP degradation rates for different catalysts after 8 hrs.

Fig. 8 illustrated the effect of doped Ag/Ce/TiO_2 on decomposition of HCHO. The conditions were as follows: flow-rate of 3 L/min, initial concentration of 0.1 mg/m^3, relative humidity of 35%. It was found that conversions of HCHO in the TiO_2/UV, Ag-TiO_2/UV and Ce-TiO_2/UV processes were increased with irradiation time. It could be seen that the doping of silver or cerium ions could improve the photo-activity of TiO_2 effectively. Furthermore, the degradation character of the photo-catalyst was in the order Ce-TiO_2 > Ag-TiO_2 > TiO_2.

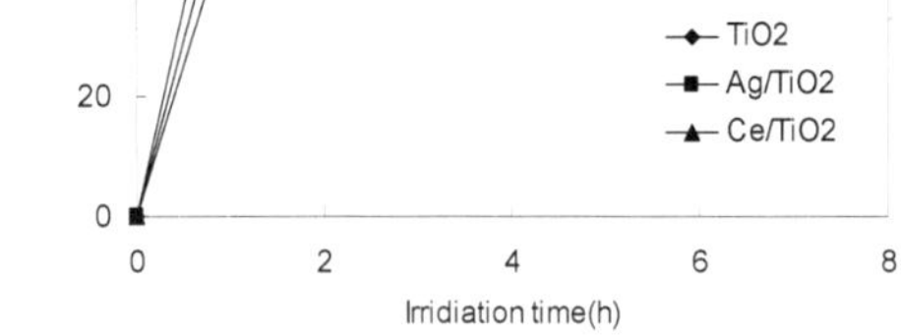

Figure 8. Effect of doped Ag/Ce/TiO_2 on decomposition of HCHO.

The reason was as follows: Ag/Ce doping could narrow the band gap. The narrower band gap will facilitate excitation of an electron from the valence band to the conduction band in the doped TiO_2, thus increasing the photo-catalytic activity of the material. At the same time, silver or cerium species could create a charge imbalance, vacancies and unsaturated chemical bonds on the catalyst surface. It will lead to the increase of chemisorbed oxygen on the surface. Surface chemisorbed oxygen has been reported to be the most active oxygen, and plays an important role in oxidation reaction. Herein, silver or cerium modified TiO_2 might have better activity for the oxidation of VOCs. Furthermore, samples after Ag/Ce doping treatment showed a slight change of colour from white to yellowish.

The photo-catalytic activity of Ce-TiO_2 in the oxidative degradation of VOCs being higher than that of Ag-TiO_2 may be explained as follows: Compared to Ag, Ce doping serves as an electron trap in the reaction because of its varied valences and special 4f level. For Ce^{3+}-TiO_2, the Ce 4 f level plays an important role in interfacial charge transfer and elimination of electron-hole recombination. So, Ce doping could enhance the electron-hole separation and the decomposition rate of VOCs could be elevated. Moreover, the valence electrons of TiO_2 catalyst are excited to the conduction band by UV light, and after various other events, electrons on the TiO_2 particle surface are scavenged by the molecular oxygen to produce reactive oxygen radicals. Furthermore, redox reactions between the pollutant molecules and reactive oxygen radicals happened, VOC molecules were turned into harmless inorganic compounds, such as CO_2 and H_2O at the end.

3.3. Effect of Hydrogen Peroxide

Hydrogen peroxide is considered to have two functions in the photo-catalytic degradation. It accepts a photo-generated conduction band electron, thus promoting the charge separation, and it also forms OH•. The addition of H_2O_2 increases the concentration of OH• radicals since it inhibits the electron-hole recombination.

Experiments were conducted to evaluate the effect of H_2O_2 on the toluene/p-xylene photo-degradation. The conditions were as follows: flow rate of 1 L/min, initial concentration of 0.1 mol/m^3, relative humidity of 35%, and photo-catalyst of pure TiO_2. As shown in Fig. 9, the removal efficiency of toluene or p-xylene increased with reaction time.

In the first 3 h, the degradation rate of toluene or p-xylene without H_2O_2 was higher because of the competitive adsorption between toluene or p-xylene molecules and hydrogen peroxide. Then, more reactants and/or radical molecules were produced during the photo-chemistry course, which led to the improvement of toluene or p-xylene decomposition. The final degradation rates of toluene and p-xylene with H_2O_2 were up to 97.1 and 95.4% after 8 h, respectively.

The degradation of acetone was studied with and without hydrogen peroxide (Fig. 10). Overall, the acetone removal efficiency increased with reaction time. Initially, the degradation rate of without H_2O_2 was higher than that of with H_2O_2 because of competitive adsorption between acetone and hydrogen peroxide after hydrogen peroxide addition to the

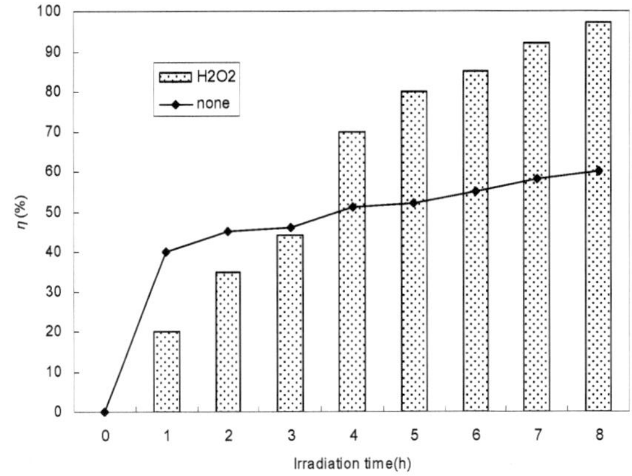

(Toluene)

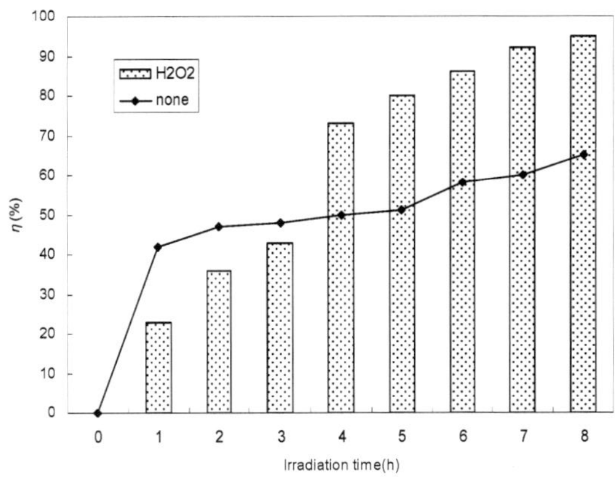

(*P*-xylene)

Figure 9. Effect of toluene/p-xylene degradation on hydrogen peroxide.

sample chamber (10 mL per 30 min, 30 % H_2O_2, RH 35 %). As the reactants and/or byproducts accumulated on the catalyst, and there was no new super-oxidation supplied, the catalyst deactivated and the degradation rate increased slowly after 2 hr. Hydroxyl radicals were produced due to the presence of hydrogen peroxide (Eq. 3). This decreased recombination of electron-hole pairs, and consequently the final acetone degradation rate was up to 91.8 % after 8 hr. Consumption of hydroxyl radical likely played an important role in deactivation of the catalyst. An appropriate volume of hydrogen peroxide could enhance the degradation rate, while too much could decrease the degradation rate (Eq. 4).

$$H_2O_2+e^- \rightarrow \cdot OH+OH^- \quad (3)$$

$$H_2O_2+ \cdot OH \rightarrow H_2O+HO_2 \cdot \quad (4)$$

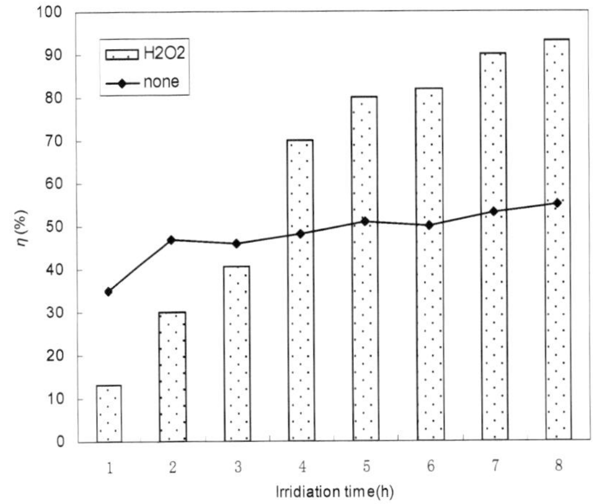

Figure 10. Effect of acetone degradation on hydrogen peroxide.

During deactivation the catalyst color in process without H_2O_2 changed from light white to khaki, while with H_2O_2 was light khaki. This suggests catalyst deactivation in process without H_2O_2 was more extensive than in with H_2O_2. Moreover, the change in catalyst color after the reaction indicates that the reaction occurred in the surface of the catalyst. After sintering at 390 °C for 1 h the catalyst recovered its original color. This again suggests that catalyst deactivation was due to the accumulation of reactants and by-products on the catalyst surface, which impeded degradation reactions. To compare the performance of H_2O_2/UV, we evaluated the potential of acetone degradation by H_2O_2 alone. The concentration of acetone remained almost the same over 8 h. Consequently, we concluded that H_2O_2 alone could not remove the VOCs.

3.4. Effect of initial concentration

In order to discuss the effect of VOCs initial concentration on photo-catalytic degradation rates, we investigated the removal efficiency of ATP and HCHO under different initial concentrations. The ATP concentrations in the experiment ranged between 0.05-0.3 mol/m^3. The conditions were as follows: gas flow-rate of 1 L/min, relative humidity of 35%, Ce-doped TiO_2 as photo-catalyst, and irradiation time of 8 hr. The results showed that the photo-catalytic degradation rates decreased with increasing ATP initial concentration, just illustrated in Fig. 11. Based on the Langmuir-Hinshelwood equation, the degradation rate decreased with increasing initial concentration while the absolute amount of degraded pollutants may increase. At higher initial concentration, the UV light might be absorbed by gaseous pollutants rather than the TiO_2 particles, which led to the reduction of the photo-degradation efficiency. Moreover, at different initial concentrations, acetone was easiest to be destructed, while p-xylene was difficult to be removed among ATP from gas flow.

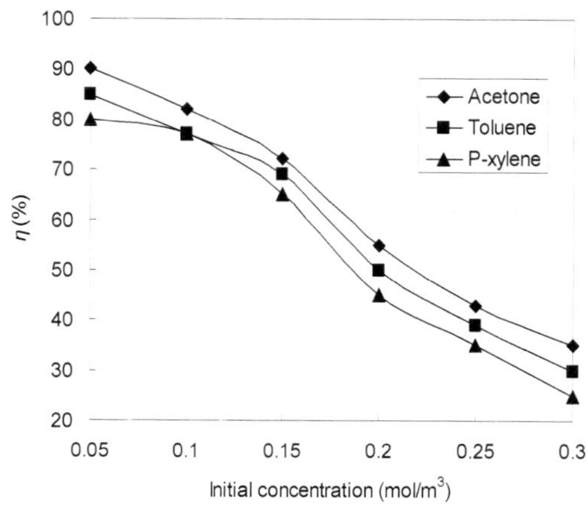

Figure 11. Effect of ATP initial concentration on the photo-catalysis of ATP by TiO_2.

As a main indoor pollutant, the indoor formaldehyde concentration is usually below 0.5 ppmv. It is worth discussing whether the low level of indoor HCHO can be decreased to a value below 0.1 mg/m^3 (specified in the indoor air quality standard of China). So in our experiment, the HCHO concentrations in the experiment ranged between 0.1-0.5 mg/m^3. The conditions were as follows: relative humidity of 35%, Ce-doped TiO_2 as photo-catalyst, and irradiation time of 120min. The results showed that the photo-catalytic degradation rates decreased with increasing HCHO initial concentration, just illustrated in Fig. 12.

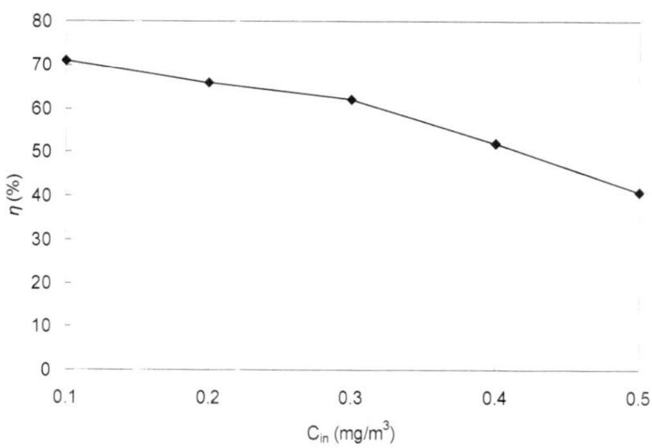

Figure 12. Effect of initial HCHO concentration on HCHO degradation by TiO_2.

In gas-phase photo-catalyst, collision frequency between radicals and HCHO affected the removal efficiency. When formaldehyde molecule reaches to the catalyst surface, the photo oxidation will occur. At higher initial concentration, the UV light might be absorbed by gaseous pollutants rather than the TiO_2 particles, which led to the reduction of the photo-degradation efficiency.

3.5. Effect of UV Light Wavelength

In order to investigate the influence of the UV intensity on the photo-catalytic efficiency, the experiments were performed using two lamp configurations (254 and 365 nm). The effect of UV light wavelength on the efficiency of HCHO degradation is shown in Fig. 13. Just shown in Fig.13, 254 nm UV light provided more effective HCHO photo-degradation than 365 nm UV light.

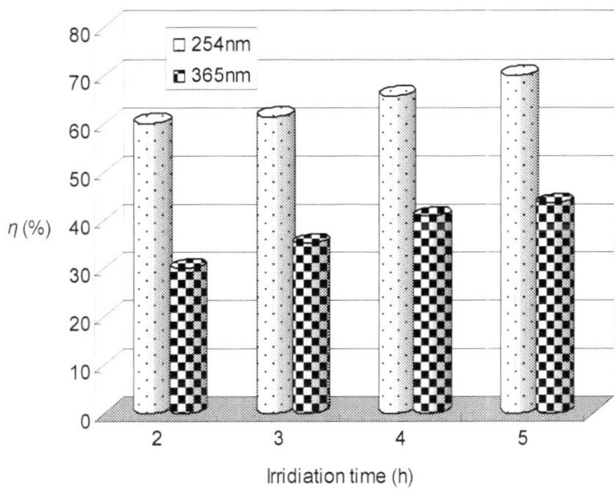

Figure 13. Effect of UV light wavelength on HCHO degradation.

The effect of UV light wavelength on the efficiency of ATP degradation is shown in Fig.14. 254 nm UV light provided more effective ATP photodegradation than 365 nm UV light. Degradation of ATP in the UV/TiO_2 process followed the same trend.

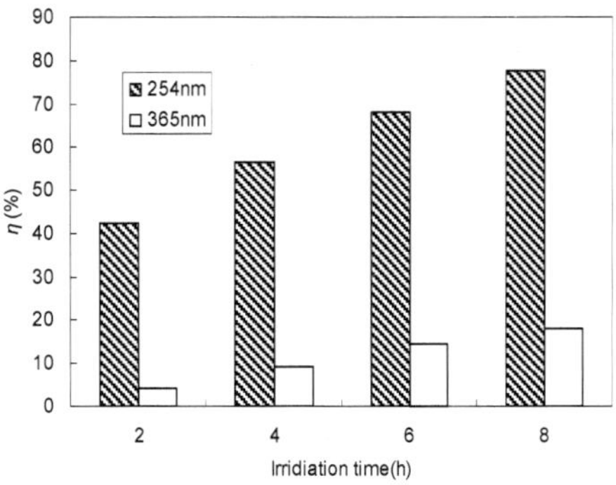

(a) Acetone

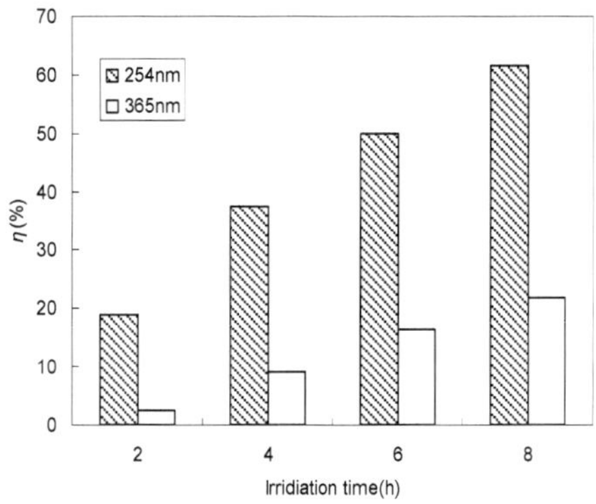

(b) Toluene

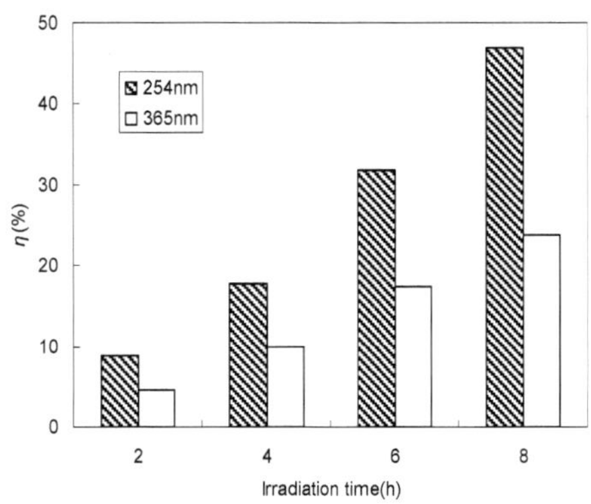

(c) P-xylene

Figure 14. Effect of UV wavelength on degradation of acetone, toluene, and p-xylene by TiO_2/UV processes.

The different results obtained with 254 and 365 nm UV lamps were mainly due to the stronger UV irradiation from the 254 nm lamp (about 58 W/m^2 on its surface) than that from the 365 nm lamp (30 W/m^2 on its surface). This illustrates that the 254 nm UV lamp irradiated photons with higher energy, which led to more efficient degradation with TiO_2/UV.

3.6. Effect of gas flow rate

The effect of gas flow rate on ATP degradation was studied at an initial concentration of 0.1 mol/m^3 and relative humidity of 35 %, just as shown in Fig. 15. When the flow rate was increased from 3–9 L/min, degradation of toluene and acetone decreased. With a flow rate >3 L/min the reactants have shorter residence time on the photocatalyst surface and consequently do not bind to the active sites. In general, an increase in gas flow rate results in two antagonistic effects. These are a decrease in residence time within the photocatalytic reactor, and an increase in the mass transfer rate. In our opinion, the decrease in degradation with increasing gas flow rate showed that the residence time of pollutant molecules with TiO_2 is an important factor. However, the degradation rate at 1 L/min was the lowest. This was due to adsorption of active species on the catalyst, which led to a decrease in the reaction between pollutant molecules and active species. For p-xylene, the degradation rate was the highest when the flow rate was 7 L/min. From these results it can be concluded that gas flow rate remarkably influences the degradation rate. While both toluene and p-xylene are aromatic hydrocarbons, toluene is an unsymmetrical molecule and p-xylene is symmetrical. Consequently, the adsorption and degradation of toluene were greater than for p-xylene under the same flow rate. The highest degradation rates for acetone, toluene, and p-xylene were 77.7, 61.9, and 55 %, respectively.

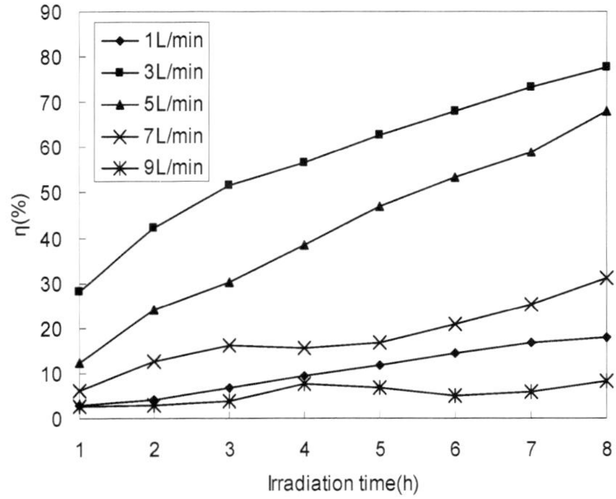

(a) Acetone

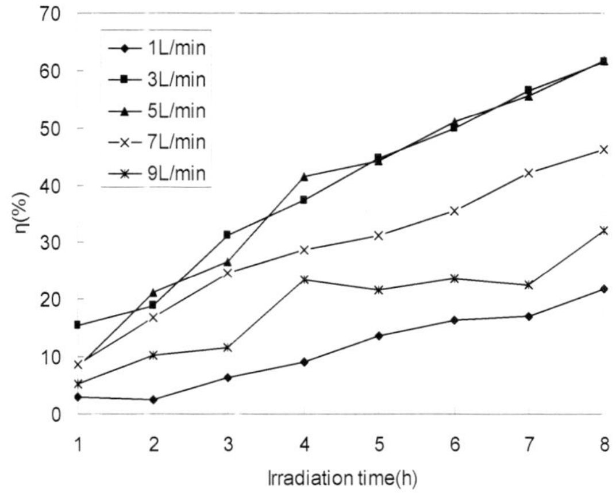

(b) Toluene

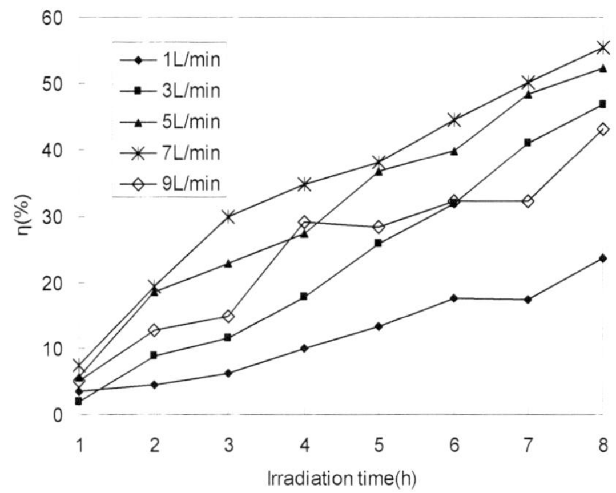

(c) P-xylene

Figure 15. Effect of flow rate on the degradation of acetone, toluene, and p-xylene by TiO_2/UV processes.

The Langmuir–Hinshelwood (L–H) rate expression has been widely used to describe the gas–solid phase reaction for heterogeneous photocatalysis. Assuming that mass transfer is not the limiting step, and that the effect of intermediate products is negligible, then the reaction rate in a plug-flow reactor can be expressed as:

$$r = -\frac{dC_{VOC}}{dt} = \frac{kKC}{1+KC} \tag{5}$$

where k and K are the L–H reaction rate constant and the L–H adsorption equilibrium constant, respectively; and t is the time taken for ATP molecules to pass through the reactor. After integration of Equation (5) the following linear expression can be obtained:

$$\frac{\ln(C_{in} / C_{out})}{(C_{in} - C_{out})} = \frac{kKT}{(C_{in} - C_{out})} - K \tag{6}$$

where C_{in} and C_{out} are the inlet and outlet concentrations of ATP, respectively; and T is the recurrent time of VOCs in the reactor.

If the L–H model is valid, a plot of $ln(C_{in}/C_{out})/(C_{in}-C_{out})$ versus $1/(C_{in}-C_{out})$ should be linear. This was the case with our data (Fig. 16), and the linearity correlation coefficients of acetone, toluene and p-xylene were 0.9989, 0.9995 and 0.9992, respectively. This result suggests that the reaction occurs on the photocatalyst surface through an L–H mechanism and not in the gas phase.

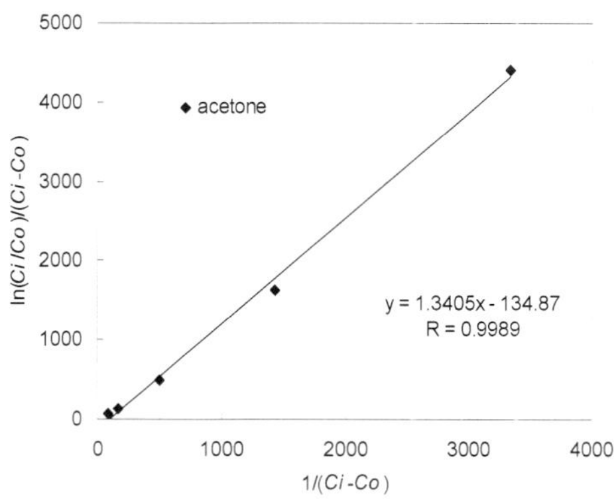

(a) Acetone

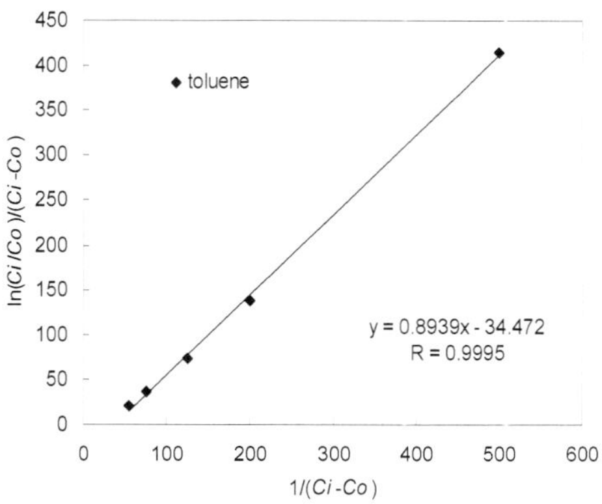

(b) Toluene

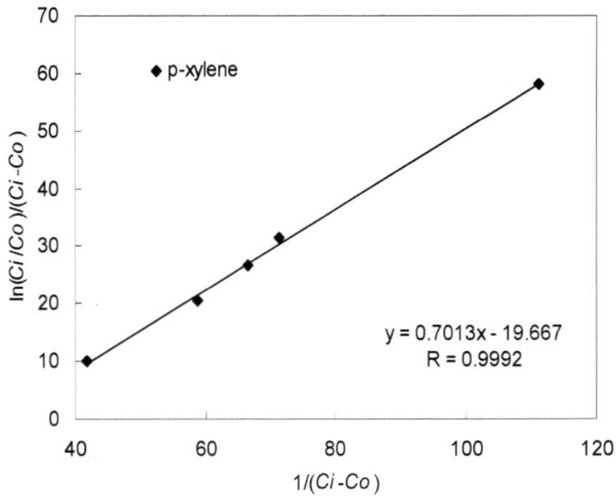

(c) P-xylene

Figure 16. Plot of $\ln(C_{in}/C_{out})/(C_{in}-C_{out})$ and $1/(C_{in}-C_{out})$.

3.7. Degradation of Pure Individual VOCs and Their Mixture

Gaseous-phase photo-degradation for pure individual VOCs (acetone, toluene, and p-xylene) and their mixture was carried out in the continuous flow reactor system. The gas stream passed through the reactor at a flow rate of 5 L/min and contained 0.1 mol/m^3 pure acetone, toluene, or p-xylene, or 0.3 mol/m^3 of their mixture. The gas residence time was 72 s in the reactor. The experiment was run for 8 hr, and samples were collected at hourly intervals.

Both acetone and p-xylene in the mixed gas degraded at much lower rates than their pure individual gases under the same conditions, just as shown in Fig. 17. However, the opposite trend was observed for toluene. Toluene has an unsymmetrical structure, which leads to instability and promotes adsorption and degradation of pollutant molecules on the catalyst surface according to the L-H mechanism. In addition, the byproducts of acetone and p-xylene produced in the reaction could promote toluene degradation. In contrast, degradation of acetone and p-xylene in the mixed gas was reduced by competitive adsorption and catalysis of toluene. Among the pure gases and the mixture, acetone had the highest degradation efficiency. Furthermore, the efficiency of pure toluene degradation was lower than that of pure p-xylene degradation due to structural stability.

3.8. Effect of gas temperature

The effect of gas temperature on photo-catalytic degradation of gaseous toluene was investigated in the range of 25-50 °C (Fig. 18). The conditions were as follows: gas flow-rate of 1 L/min, relative humidity of 35%, irradiation time of 8 h, photo-catalyst of Ce-doped TiO_2, and initial concentration of 0.1 mol/m^3. Degradation efficiency of toluene gradually

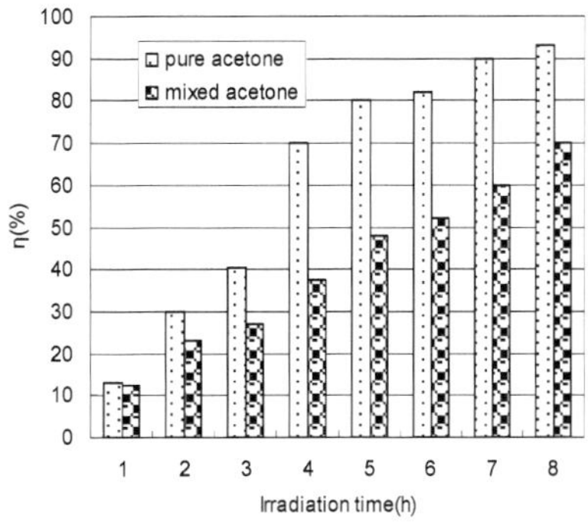

(a) Acetone

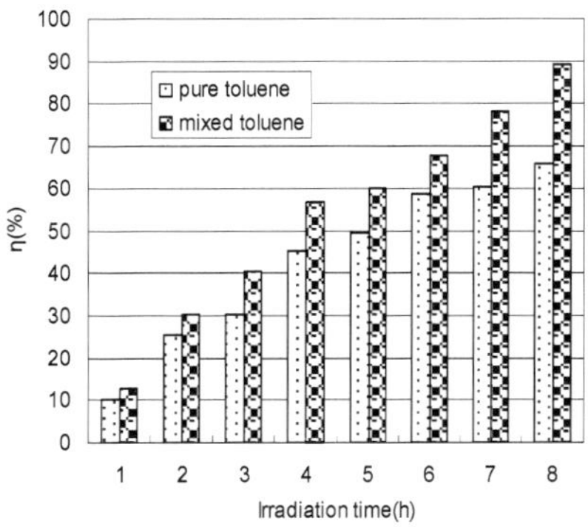

(b) Toluene

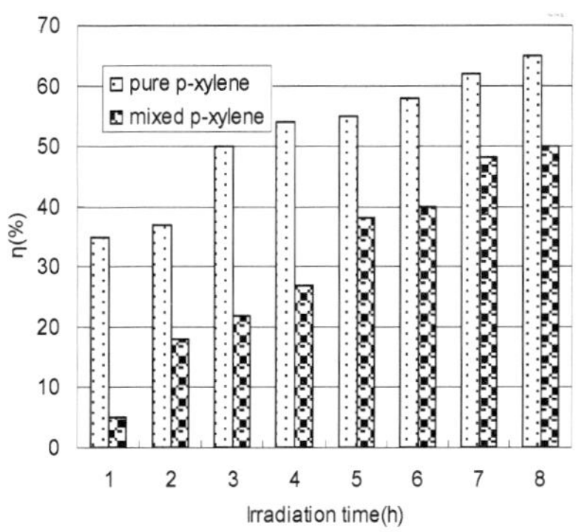

(c) P-xylene

Figure 17. Degradation with H_2O_2 of pure acetone, toluene, p-xylene, and their mixture.

increased when gas temperature was below 45 °C, but decreased at >45 °C. The increase in temperature would lead to the production of free radicals that could effectively collide with toluene molecules. Moreover, higher temperature may increase the oxidation rate of toluene at the interface. However, with increasing temperature, the adsorptive capacities of toluene on catalyst decreased, which led to the reduction of toluene removal efficiency.

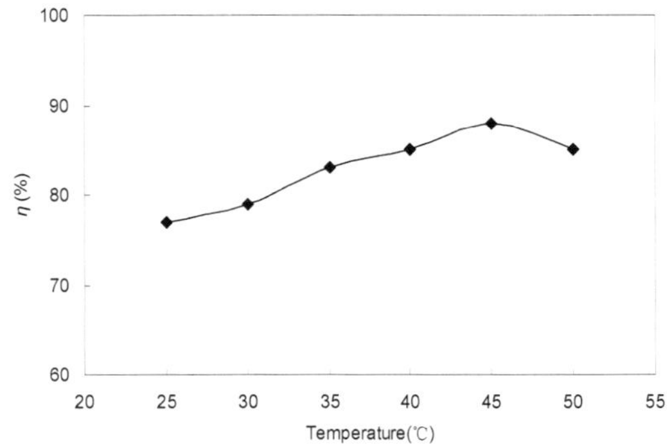

Figure 18. Effect of gas temperature on the photo-catalysis of toluene.

3.9. Effect of photo-catalyst amount

In photo-catalytic degradation of organic compounds, the optimal TiO_2 concentration depends mainly on both the nature of the compounds and the reactor geometry. In this work, the influence of TiO_2 amount on HCHO photo degradation was investigated. A set of gaseous experiments with different amount of TiO_2 from 0 to 100mg was carried out at the RH of 35% and the initial HCHO concentration of 0.1mg/m^3. The degradation rates of HCHO for different amount TiO_2 were presented in Fig. 19. The photo-catalytic degradation efficiency increased with increasing the amount of TiO_2 when TiO_2 amount was lower than 70mg. When the TiO_2 amount was more than 70mg, the photo-catalytic degradation efficiency was decreased. So 70mg of TiO_2 amount was the optional amount in our experiment. And the thickness of 70mg of TiO_2 amount was about 0.2mm.

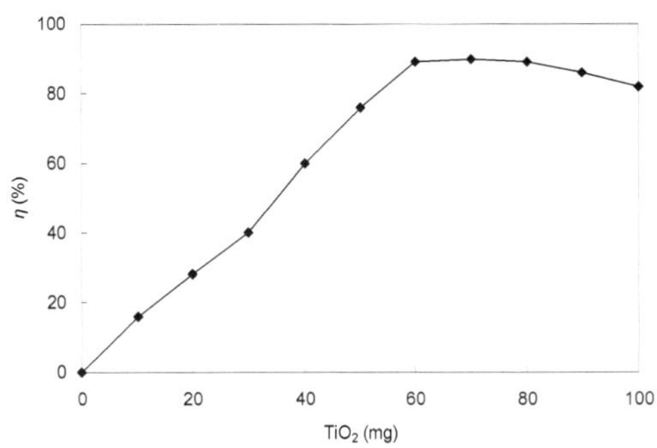

Figure 19. Effect of TiO_2 amount on HCHO degradation

At the same time, in our investigation, the effect of photo-catalyst concentration on the degradation of acetone in the gas flow was also analyzed in order to optimize the amount of TiO_2. Different concentrations (15-105 mg/L) of TiO_2 precursor sols were prepared by using different amounts of tetrabutyl orthotitanate. The conditions of the experiment were as follows: gas flow-rate of 1 L/min, relative humidity of 35%, Ce-doped TiO_2 as photo-catalyst, and irradiation time of 8 h. BET surface area of the synthesized samples was tested (see Table 2). The results showed that BET surface area increased with increasing photo-catalyst amount.

Sample concentration (mg/L)	15	30	45	60	75	90	105
BET (m^2/g)	50.2	66.2	68.2	72.2	78.3	88.8	88.9

Table 2. BET surface area for synthesized photo-catalyst.

Fig. 20 showed that the photo-catalytic degradation efficiency increased with increasing the amount of TiO_2. It was suggested that increasing efficiency was due to the increase of the surface area. It could be observed that the degradation efficiency increased with increasing the amount of the catalyst until it reached a plateau at 90-105 mg/L of TiO_2. This indicated that when the amount of TiO_2 was overdosed, the surface area was saturated, and then the intensity of UV was attenuated because of decreased light penetration and increased light scattering.

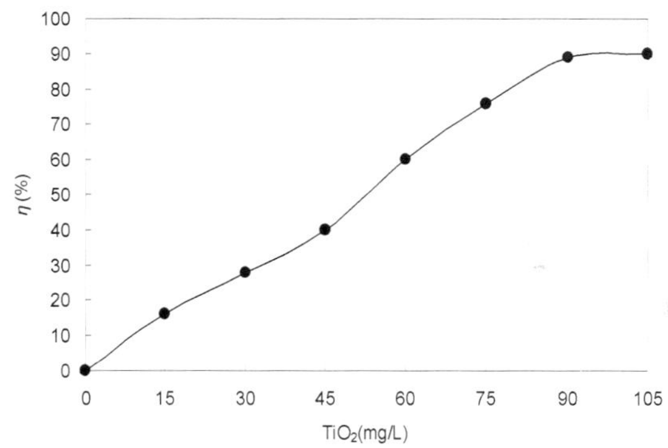

Figure 20. Effect of TiO_2 amount on the photo-catalysis of acetone.

3.10. Effect of relative humidity of air stream

The effect of relative humidity (0-60% RH) of air stream on HCHO decomposition was examined by adding water vapor to a fixed concentration of HCHO. TiO_2 photocatalyst was used in this experiment. Fig. 21 showed the experimental results at different relative humidity. The degradation rate increased with increasing relative humidity up to 35% and then started to decrease, which meant that 35% was the optimal humidity for photo-catalyst process under the experimental conditions. When the reaction time was 120min, the highest removal efficiency of HCHO was 60.2% when RH was 35%.

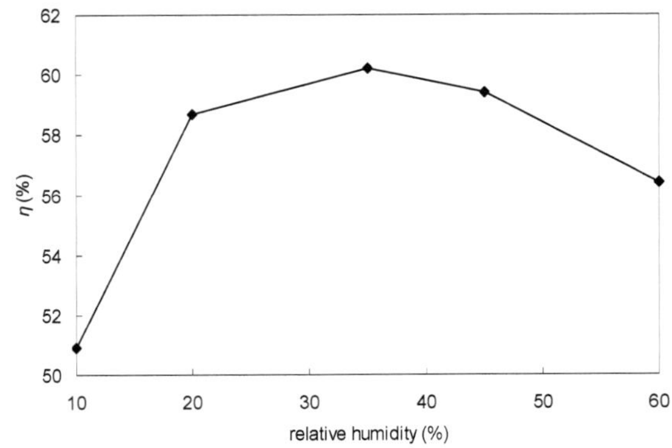

Figure 21. Effect of RH on decomposition of HCHO.

The enhancement of photo-catalytic reaction rate is frequently found in the presence of water vapor because hydroxyl groups or water molecules can behave as hole traps to form surface-adsorbed hydroxyl radicals. In photo-catalyst process, the hydroxyl radicals formed on the illuminated TiO_2 can not only directly attack HCHO molecules, but also suppress the electron-hole recombination. However, higher RH can be attributed to the competition for adsorption between HCHO and hydroxyl radicals, thus decrease the removal efficiency of HCHO.

3.11. Effect of oxygen concentration

The effect of oxygen concentration on HCHO degradation was presented in Fig. 22.

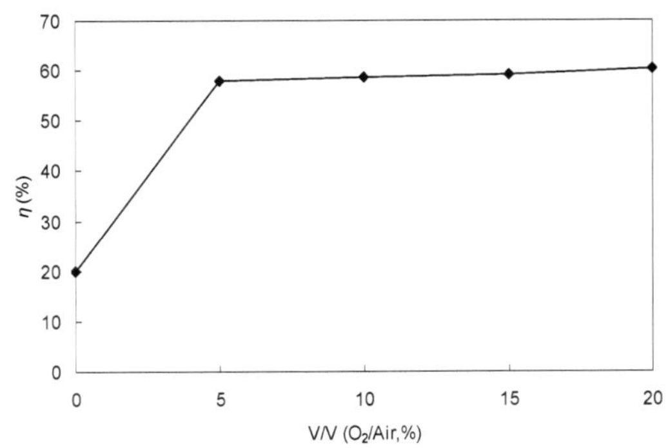

Figure 22. Effect of oxygen concentration on HCHO degradation.

The results corresponded to the initial concentration of 0.1mg/m^3, relative humidity of 35% and reaction time of 120min. It is obvious that oxidation rates for HCHO increased with increasing O_2 concentration under fixed conditions. As mentioned above, hydroxyl radical is an important factor to the HCHO photo-catalyst. At the same time, oxygen radical is also key factor for HCHO removal, which can react with HCHO on the TiO_2 surface and turn HCHO into CO_2 and H_2O.

3.12. Mechanism of photo-catalytic degradation of VOCs

The heterogeneous photo-catalytic process used in pollutant degradation involved the adsorption of pollutants on the surface sites, and the chemical reactions of converting pollutants into carbon dioxide and water. Activation of TiO_2 is achieved through the absorption of a photon (hv) with ultra-band energy from UV irradiation source. This results in the promotion of an electron (e−) from the valence band to the conduction band, with the generation of highly reactive positive holes (h+) in the valence band. This caused aggressive oxidation of the surface-adsorbed toxic organic pollutants and converts them into CO_2 and water.

In the degradation of toluene or p-xylene, the OH• radicals attack the phenyl ring of toluene or p-xylene, and some products, such as phenol, benzaldehyde or benzoic acid, may be produced during the reaction, and they were converted into CO_2 and H_2O at the end (Fig. 23). We could also observe that acetone was easily destructed to CO_2 and H_2O by photo-catalysts. By-products of toluene or p-xylene were detected by GC-MS, and involved phenol, benzaldehyde, aldehydes, alcohols, etc.

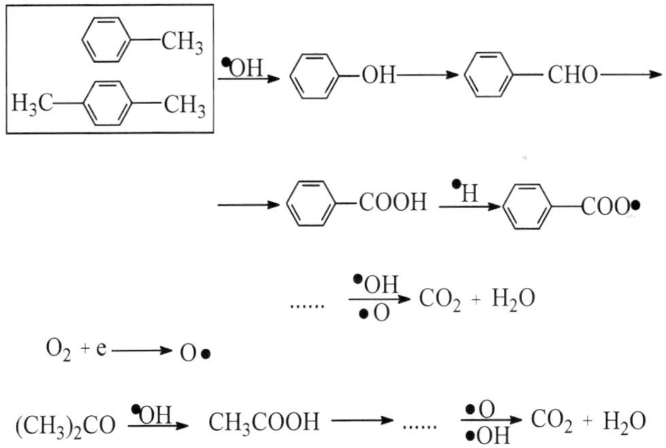

Figure 23. Suggested pathway for the photo-catalytic destruction of ATP.

The reaction rate constant (k) was chosen as the basic kinetic parameter for ATP since it was important in determination of VOCs photo-catalytic activity. The first order kinetic equation:

$$\ln\left(\frac{C_i}{C_o}\right) = k \times t + b \qquad (7)$$

was used to fit experimental data in Fig. 24

where C_o is the concentration of ATP remaining in the solution at t, and C_i is the initial concentration at t = 0.

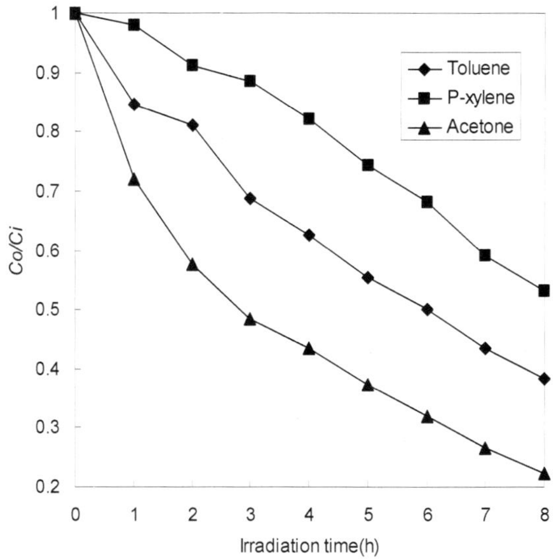

Figure 24. Kinetics of ATP degradation.

The variations in $\ln(C_i/C_o)$ as a function of irradiation time are given in Fig. 25. Reaction rate constant (k), linearity correlation coefficient (R) and intercept (b) data for the photo-catalytic destruction of ATP are exhibited in Table 3. The k of ATP could be ordered as follows: $k_{Acetone} > k_{Toluene} > k_{P\text{-}xylene}$, meaning that the decomposition capability of acetone was the best. The reason was probably due to molecular structure and molecular weight.

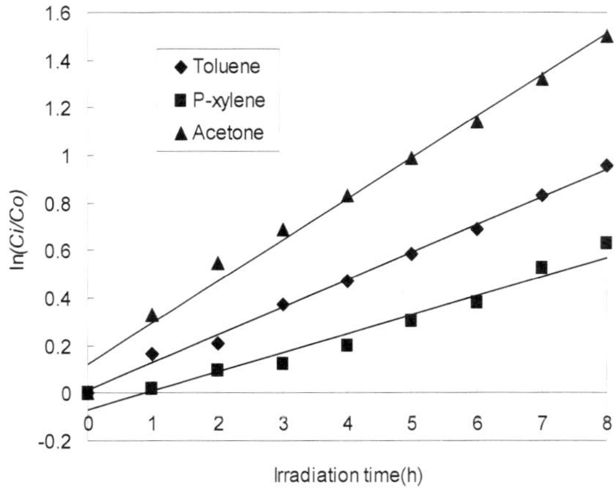

Figure 25. Relation between $\ln(C_i/C_o)$ and irradiation time, and linear fits for ATP.

	k (h^{-1})	R	b
Toluene	0.1165	0.998	0.0102
P-xylene	0.0797	0.980	-0.0668
Acetone	0.1742	0.993	0.12

Table 3. Values of k, R and b for the photo-catalytic destruction of ATP.

During the HCHO decomposition by photo-catalytic processing, formic acid was identified as the intermediate from the photo-degradation of formaldehyde. In our experiment, ion chromatography (IC) was used to determine the byproducts by sampling the gas products into distilled water. The result in this study showed that formic acid was also found. The probably pathway of HCHO destruction was shown in Fig. 26. The related reactions of HCHO destruction were shown with equations (8)-(21).

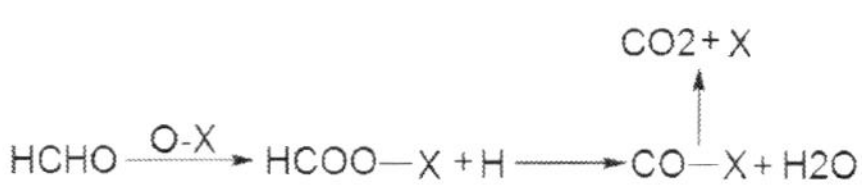

X : TiO_2, Ag/TiO_2 or Ce/TiO_2

Figure 26. Suggested pathway for the photo-catalytic destruction of HCHO

$$HCHO + e \longrightarrow H\bullet + CHO^{-} \tag{8}$$

$$HCHO + O\bullet \longrightarrow OH\bullet + CHO\bullet \tag{9}$$

$$HCHO + OH\bullet \longrightarrow H_2O + CHO\bullet \tag{10}$$

$$HCHO + OH\bullet \longrightarrow HCOOH + H\bullet \tag{11}$$

$$HCHO + H\bullet \longrightarrow H_2 + CHO\bullet \tag{12}$$

$$CHO\bullet + H\bullet \longrightarrow H_2 + CO \tag{13}$$

$$CHO\bullet + O_2 \longrightarrow CO_2 + OH\bullet \tag{14}$$

$$CHO\bullet + OH\bullet \longrightarrow H_2O + CO \tag{15}$$

$$CHO\bullet + HO_2\bullet \longrightarrow OH\bullet + H\bullet + CO_2 \tag{16}$$

$$CHO\bullet + O\bullet \longrightarrow OH\bullet + CO \tag{17}$$

$$CHO\bullet + O\bullet \longrightarrow H\bullet + CO_2 \tag{18}$$

$$HCOOH + OH\bullet \longrightarrow H_2O + H\bullet + CO_2 \tag{19}$$

$$CO + O\bullet \longrightarrow CO_2 \tag{20}$$

$$CO + OH\bullet \longrightarrow CO_2 + H\bullet \tag{21}$$

As mentioned in equation 7, k was the basic kinetic parameter for VOCs photo-catalytic activity. Fig. 27 showed the first order kinetic equation fitting the experimental data.

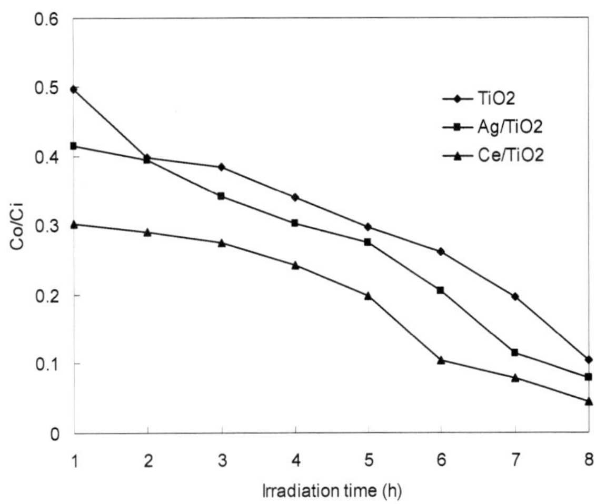

Figure 27. Kinetics of HCHO degradation.

The variations in $\ln(C_i/C_o)$ as a function of reaction time were given in Fig. 28. The reaction rate constant for TiO_2、Ag/TiO_2、Ce/TiO_2 were 0.1871、0.2302、0.2724, respectively, which meant that Ce/TiO_2 had the best photo-catalytic abilities among the catalysts.

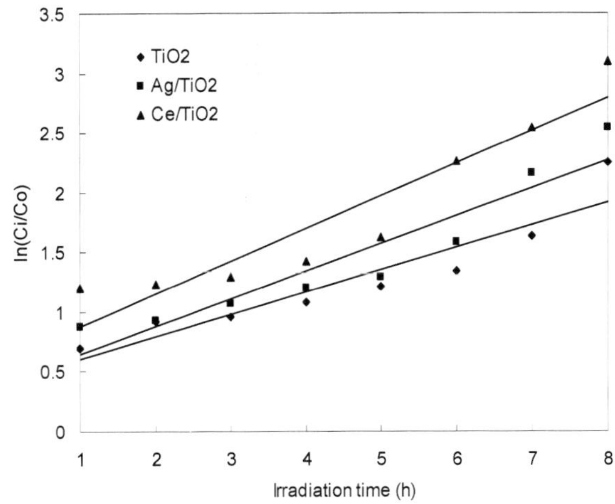

Figure 28. Relationship between $\ln(C_i/C_o)$ and reaction time.

4. Conclusion

In this chapter, nano-structured TiO_2, Ag-TiO_2 and Ce-TiO_2 thin films coated on glass springs were prepared by sol-gel method at room temperature. Toluene, p-xylene, acetone and formaldehyde were chosen as the model VOCs, the photo-catalytic degradation characters of them by TiO_2/UV, TiO_2/doped Ag/UV and TiO_2/doped Ce/UV was tested and compared. The effects of doped Ag/Ce ions, hydrogen peroxide, initial concentration, gas temperature, relative humidity of air stream, oxygen concentration, gas flow rate, UV light wavelength and photo-catalyst amount on decomposition of the pollutants by TiO_2/UV were analyzed simultaneously. Furthermore, the mechanism of titania-assisted photo-catalytic degradation was analyzed, and the end product of the reaction using GC-MS analysis was also performed.

Results were as follows: (1) Characterization of this film by SEM and XRD showed it consisted of nanoparticles, and the crystalline phase was anatase. (2) Doped Ag or Ce ions could enhance the photo-catalyst ability. The degradation character of the photo-catalyst was in the order Ce-TiO_2>Ag-TiO_2>TiO_2. (3) Hydrogen peroxide could promote the activation of catalyst, and toluene & p-xylene degradation rate with hydrogen peroxide was higher than that without it. The final degradation rates of toluene and p-xylene using H_2O_2 were up to 97.1 and 95.4% after 8 h, respectively. (4) The photo-catalytic degradation rates decreased with increasing VOCs initial concentration. Acetone was easiest to be destructed, while p-xylene was difficult to remove from gas flow. (5) The degradation efficiency gradually increased with gas temperature and 45 °C had the best removal efficiency. (6) 35% was the optimal humidity for photo-catalyst process under the experimental conditions. (7) Higher concentration of oxygen was better for HCHO removal. (8) The flow rate greatly influenced the degradation rate. For acetone and toluene, the degradation rate was highest with a flow rate of 3 L/min. For p-xylene, the degradation rate was highest when the flow rate was 7 L/min. The highest degradation rates for acetone, toluene and p-xylene were 77.7 %, 61.9 % and 55 %, respectively. (9) Illumination using a 254 nm light source was better than 365 nm. (10) The photo-catalytic degradation efficiency increased with increasing the amount of TiO_2 when TiO_2 amount was lower than 70mg. (11) In the gas mixture, acetone and p-xylene had much lower degradation rates than for their pure counterparts. The opposite trend was observed for toluene. Among acetone, toluene and p-xylene, the removal efficiency of acetone was highest both when pure and as a part of the gas mixture. (12) The photo-catalytic process used in pollutant degradation involved the adsorption of pollutants on the surface sites, and chemical reactions of converting pollutant into CO_2 and H_2O at the end. By-products of toluene or p-xylene were detected by GC-MS analysis, and involved phenol, benzaldehyde, aldehydes, alcohols, etc. The reaction rate constant (k) of ATP was sequenced $k_{Acetone}>k_{Toluene}>k_{P\text{-}xylene}$, meaning that the decomposition capability of acetone was the best, probably due to molecular structure and molecular weight. Formic acid was the main byproduct during the decomposition of HCHO. The reaction rate constant (k) of TiO_2、Ag/TiO_2、Ce/TiO_2 was sequenced $k_{Ce/TiO2}>k_{Ag/TiO2}>k_{TiO2}$, meaning that Ce/$TiO_2$ had the best photo-catalytic abilities among the catalysts.

Author details

Wenjun Liang, Jian Li and Hong He
College of Environmental and Energy Engineering,
Beijing University of Technology, Beijing, China

Acknowledgement

This work was supported by National high technology research and development program of China (2011AA03A406) and Project of Beijing Municipal Education Commission (KM201110005011).

5. References

Akira F., Tata N.R., Donald A. (2000). Titanium dioxide photocatalysis, Journal of Photochemistry and Photobiology C: Photochemistry Reviews, Vol. 1,pp.1–21.

Alberici, R.M., Jardim, W.E. (1997). Photocatalytic destruction of VOCs in the gas-phase using titanium dioxide. Appl. Catal. B-Environ., Vol.14, pp. 55-68.

Ao C.H., Lee S.C., Yu J.Z., Xu J.H. (2004). Photodegradation of formaldehyde by photocatalyst TiO2: effects on the presences of NO, SO2 and VOCs. Appl. Catal. B., Vol.54, pp.41–50.

Biard, P.F.; Bouzaza, A.; Wolbert, D. (2007). Photocatalytic Degradation of Two Volatile Fatty Acids in an Annular Plug-Flow Reactor; Kinetic Modeling and Contribution of Mass Transfer Rate. Environ. Sci. Technol., Vol.41, pp.2908-2914.

Boulamanti, A. K.; Philippopoulos, C. J. (2008). Photocatalytic degradation of methyl tert-butyl ether in the gas-phase: A kinetic study. J. Hazard Mater., Vol.160, pp. 83-87.

Boulamanti, A.K.; Korologos, C.A.; Philippopoulos, C.J. (2008). The rate of photocatalytic oxidation of aromatic volatile organic compounds in the gas-phase. Atmos. Environ., Vol.42, pp.7844-7850.

Bouzaza A., Vallet C., Laplanche A. (2006) Photocata-lytic degradation of some VOCs in the gas phase using an an-nular flow reactor. Determination of the contribution of mass transfer and chemical reaction steps in the photodegradation process. J. Photochem. Photobiol. A: Chem. Vol.177, pp.212-217.

Carp O., Huisman C.L., Reller A. (2004). Photoinduced reactivity of titanium dioxide. Prog. Solid State Chem., Vol.32, pp. 33-177.

Chatterjee D., Dasgupta S. (2005) Visible light induced photocatalytic degradation of organic pollutants. J. Photo-chem. Photobiol. C. Vol.6, pp.186-205.

Chen F., Pehkonen S.O., Ray M.B. (2002) Kinetics and mechanisms of UV-photodegradation of chlorinated organics in the gas phase. Water Res. Vol.36, pp.4203-4214.

Chen F., Yang Q., Pehkonen S.O., Ray M.B. (2004) Modeling of gas phase photodegradation of chlorinated VOCs. J. Air Waste Manage Assoc. Vol.54, pp.1281-1292.

Chen, X.B.; Mao, S.S. (2007). Titanium Dioxide Nanomaterials: Synthesis, Properties, Modifications, and Applications. Chem. Rev., Vol.107, pp. 2891-2959.

Collins J.J., Ness R., Tyl R.W., Krivanek N., Esmen N.A., Hall T.A. (2001). A Review of Adverse Pregnancy Outcomes and Formaldehyde Exposure in Human and Animal Studies. Regul. Toxicol. Pharm., Vol.34, pp.17-34.

Fujishima A., Zhang X.T. (2006). Titanium dioxide photocatalysis: present situation and future approaches, C. R. Chimie, Vol.9,pp.750–760.

Futamura S., Zhang A., Einaga H., Kabashima H. (2002) Involvement of catalyst materials in nonthermal plasma chemical processing of hazardous air pollutants. Catal. Today. Vol.72, pp.259-265.

González A. S., Martínez S. S. (2008) Study of the sono-photocatalytic degradation of basic blue 9 industrial textile dye over slurry titanium dioxide and influencing factors. Ultrason. Sonochem. Vol.15, pp. 1038-1042.

Hirakawa, T.; Koga, C.;Negishi, N.; Takeuchi, K.; Matsuzawa, S. (2009). An approach to elucidating photocatalytic reaction mechanisms by monitoring dissolved oxygen: Effect of H2O2 on photocatalysis, Appl Catal B: Environ., Vol.87, pp. 46-55.

Holzer F., Roland U., Kopinke F.D. (2002) Combination of non-thermal plasma and heterogeneous catalysis for oxidation of volatile organic compounds Part 1. Accessibility of the intra-particle volume. *Appl. Catal. B: Environ.* Vol.38, pp.163-181.

http://www.epa.gov/iaq/voc.html

Hu, Q.H.; Zhang, C.L.; Wang, Z.R.; Chen, Y.; Mao, K.H.; Zhang, X.Q.; Xiong, Y.L.; Zhu, M.J. (2008). Photodegradation of methyl tert-butyl ether (MTBE) by UV/H_2O_2 and UV/TiO_2. *J. Hazard Mater.*, Vol.154, pp.795-803.

Jacoby W.A., Blake D.M., Noble R.D., Koval C.A. (1995). Kinetics of the Oxidation of Trichloroethylene in Air via Heterogeneous Photocatalysis. *J. Catal.*, Vol.157, pp.87-96.

Jing L.Q., Xu Z.L., Sun X.J., Shang J., Cai W.M. (2001). The surface properties and photocatalytic activities of ZnO ultrafine particles. *Appl. Surf. Sci.*, Vol.180, pp.308-314.

Kecske's T., Rasko'J., Kiss J. (2004). FTIR and mass spectrometric studies on the interaction of formaldehyde with TiO_2 supported Pt and Au catalysts. *Appl. Catal. A*, Vol.273, pp.55-62.

Kim, S.B.; Hong, S.C. (2002). Kinetic study for photocatalytic degradation of volatile organic compounds in air using thin film TiO_2 photocatalyst. *Appl Catal B: Environ.*, Vol. 35, pp.305-315.

Li F.B., Li X.Z., Hou M.F., Cheah K.W., Choy W.C.H. (2005) Enhanced photocatalytic activity of Ce^{3+}-TiO_2 for 2-mercaptobenzothiazole degradation in aqueous suspension for odour control. *Appl. Catal. A: Gen.* Vol.285, pp.181-189.

Li, D.; Haneda, H.; Hishita, S.; Ohashi, N. (2005). Visible-light-driven nitrogen-doped TiO_2 photocatalysts: effect of nitrogen precursors on their photocatalysis for decomposition of gas-phase organic pollutants. *Mater. Sci. Eng. B.*, Vol.117, pp. 67-75.

Li, F.B.; Li, X.Z.; Ao, C.H.; Lee, S.C.; Hou, M.F. (2005).Enhanced photocatalytic degradation of VOCs using Ln^{3+}–TiO_2 catalysts for indoor air purification. *Chemosphere*, Vol.59, pp.787-800.

Lin L., Zheng R.Y., Xie J.L., Zhu Y.X., Xie Y.C. (2007). Synthesis and characterization of phosphor and nitrogen co-doped titania. *Appl. Catal. B.*, Vol.76, pp.196-202.

Liu H.M., Lian Z.W., Ye X.J., Shangguan W.F. (2005). Kinetic analysis of photocatalytic oxidation of gas-phase formaldehyde over titanium dioxide. *Chemosphere*, Vol.60, pp.630-635.

Liu T.X., Li F.B., Li X.Z. (2008). TiO_2 hydrosols with high activity for photocatalytic degradation of formaldehyde in a gaseous phase. *J. Hazard. Mater.*, Vol.152, pp.347-355.

Lu Y.W., Wang D.H., Ma C.F., Yang H.C. (2010). The effect of activated carbon adsorption on the photocatalytic removal of formaldehyde. *Building and Environment*, Vol.45, pp.615–621.

Mohseni, M. (2005). Gas phase trichloroethylene (TCE) photooxidation and byproduct formation: photolysis vs. titania/silica based photocatalysis. *Chemosphere*, Vol.59, pp. 335-342.

Obee, T.N. (1995). TiO_2 Photocatalysis for Indoor Air Applications: Effects of Humidity and Trace Contaminant Levels on the Oxidation Rates of Formaldehyde, Toluene, and 1, 3-Butadiene. *Environ. Sci. Technol.*, Vol.29, pp.1223-1231.

Ogata A., Einaga H., Kabashima H., Futamura S., Kushi-yama S., Kim H. H. (2003) Effective combination of nonthermal plasma and catalysts for decomposition of benzene in air. *Appl. Catal. B: Environ.* Vol.46, pp.87-95.

Ohno T., Tsubota T., Nakamura Y., Sayama K.(2005). Preparation of S, C cation-codoped $SrTiO_3$ and its photocata-lytic activity under visible light. *Appl. Catal. A.*, Vol.288, pp.74-79.

Ohtani B. (2010). Photocatalysis A to Z—What we know and what we do not know in a scientific sense, *Journal of Photochemistry and Photobiology C: Photochemistry Reviews*, Vol.11, pp.157–178.

Olmez T. (2008) Photocatalytic Treatment of phenol with visible light irradiation. *Fresen. Environ. Bull.* Vol.17, pp.1796-1802.

P′erez, M.; Torrades, F.; Garc′ıa-Hortal, J.A.; Dom`enech, X.; Peral, J. (2002). Removal of organic contaminants in paper pulp treatment effluents under Fenton and photo-Fenton conditions. *Appl. Catal. B: Environ.*, Vol.36, pp.63-74.

Parmar, G.R., Rao N.N. (2009). Emerging Control Technologies for Volatile Organic Compounds, *Critical Reviews in Environmental Science and Technology*, Vol.39, pp.41-78.

Sakthivel, S.; Kisch, H. (2003). Photocatalytic and photoelectrochemical properties of nitrogen-doped titanium dioxide. *Chem. Phys. Chem.*, Vol.4, pp. 487-490.

Sano T., Negishi N., Takeuchi K., Matsuzawa S. (2004). Degradation of toluene and acetaldehyde with Pt-loaded TiO_2 catalyst and parabolic trough concentrator. *Solar Energy*, Vol.77, pp.543-552.

Shan G.B., Yan S., Tyagi R.D., Surampalli R.Y., Zhang T.C. (2009). Applications of Nanomaterials in Environmental Science and Engineering: Review, *Practice Peri-odical of Hazardous, Toxic, and Radioactive Waste Management*, Vol.13, pp.110-119.

Shang J., Du Y.G., Xu Z.L. (2002). Photocatalytic oxidation of heptane in the gas-phase over TiO_2.*Chemosphere*, Vol.46, pp.93-99.

Shiraishi F., Yamaguchi S., Ohbuchi Y. (2003). A rapid treatment of formaldehyde in a highly tight room using a photocatalytic reactor combined with a continuous adsorption and desorption apparatus. *Chem. Eng. Sci.*, Vol.58, pp.929-934.

Sleiman, M.; Conchon, P.; Ferronato, C.; Chovelon, J.M. (2009). Photocatalytic oxidation of toluene at indoor air levels (ppbv): Towards a better assessment of conversion, reaction intermediates and mineralization. *Appl Catal B: Environ.*, Vol.86, pp.159-165.

Somekawa S., Kusumoto Y., Ikeda M., Ahmmad B., Horie Y. (2008). Fabrication of N-doped TiO_2 thin films by laser ablation method: Mechanism of N-doping and evaluation of the thin films. *Catal. Commun.*, Vol.9, pp.437-440.

Song L., Qiu R.L., Mo Y.Q., Zhang D.D., Wei H., Xiong Y. (2007). Photodegradation of phenol in a polymer-modified TiO2 semiconductor particulate system under the irradiation of visible light. *Catal. Commun.*, Vol.8, pp.429-433.

Sun R.B., Xi Z.G., Chao F.H., Zhang W., Zhang H.S., Yang D.F. (2007) Decomposition of low-concentration gas-phase toluene using plasma-driven photocatalyst reactor. *Atmos. Environ.* Vol.41, pp.6853-6859.

Tanada S., Kawasaki N., Nakamura T., Araki M., Isomura M. (1999). Removal of Formaldehyde by Activated Carbons Containing Amino Groups. *J. Colloid. Interf. Sci.*, Vol.214, pp.106-108.

Tokumura, M.; Nakajima, R.; Znad, H. T.; Kawase, Y. (2008). Chemical absorption process for degradation of VOC gas using heterogeneous gas–liquid photocatalytic oxidation: Toluene degradation by photo-Fenton reaction. *Chemosphere*, Vol.73, pp. 768-775.

Venkatachalam N., Palanichamy M., Arabindoo B., Mu-rugesan V. (2007). Alkaline earth metal doped nanoporous TiO_2 for enhanced photocatalytic mineralisation of bisphenol-A. *Catal. Commun.*, Vol.8, pp.1088-1093.

Vincenzo A., Marta L., Leonardo P., Javier S. (2006). The combination of heterogeneous photocatalysis with chemical and physical operations: A tool for improving the photoprocess performance, *Journal of Photochemistry and Photobiology C: Photochemistry Reviews*, Vol.7,pp.127–144.

Wang J.H., Ray M.B. (2000) Application of ultraviolet photooxidation to remove organic pollutants in the gas phase. *Sep. Purif. Technol.* Vol.19, pp.11-20.

Wang Y.Y., Zhou G.W., Li T.D., Qiao W.T., Li Y.J. (2009) Catalytic activity of mesoporous $TiO_{2-x}N_x$ photocatalysts for the decomposition of methyl orange under solar simulated light. *Catal. Commun.* Vol.10, pp.412–415.

Wang, X.Q.; Zhang, G.L.; Zhang, F.B.; Wang, Y. (2006). Study integration about Photocatalytic degradation of gaseous benzene in atmosphere by using TiO_2 photocatalyst. *Chemical Research and Application*, Vol.18, pp.344-348.

Wu C.D., Xu H., Li H.M., Chu J.Y., Yan Y.S., Li C.S. (2007) Photocatalytic decolorization of methylene blue via Ag-deposited BiVo(4) under UV-light irradiation. *Fresen. Environ. Bull.* Vol.16, pp. 242-246.

Xu Y.H., Chen H.R., Zeng Z.X., Lei B. (2006). Investigation on mechanism of photocatalytic activity enhancement of nanometer cerium-doped titania. *Appl. Surf. Sci.*, Vol.252, pp.8565–8570.

Yang S.X., Zhu W.P., Jiang Z.P., Chen Z.X., Wang J.B. (2006). The surface properties and the activities in catalytic wet air oxidation over CeO_2–TiO_2 catalysts. *Appl. Surf. Sci.*, Vol.252, pp.8499-8505.

Yang X., Xu L.L., Yu X.D., Guo Y.H. (2008) One-step preparation of silver and indium oxide co-doped TiO_2 photo-catalyst for the degradation of rhodamine B. *Catal. Commun.* Vol.9, pp.1224-1229.

Yi Z., Wei W., Lee S., Gao J.H. (2007). Photocatalytic performance of plasma sprayed Pt-modified TiO_2 coatings under visible light irradiation. *Catal. Commun.*, Vol.8, pp.906-912.

Yu J.C., Yu J.G., Ho W.K., Zhang L.Z. (2001). Preparation of highly photocatalytic active nano-sized TiO_2 particles via ultrasonic irradiation. *Chem. Commun.*, Vol.19, pp.1942-1943.

Yu J.G., Yu J.C., Cheng B., Hark S.K., Iu K. (2003). The effect of F−-doping and temperature on the structural and textural evolution of mesoporous TiO_2 powders. *J. Solid State Chem.*, Vol.174, pp.372-380.

Zhang C.B., He H. (2007). A comparative study of TiO_2 supported noble metal catalysts for the oxidation of formaldehyde at room temperature. *Catalysis Today*, Vol.126, pp.345–350.

Zhang C.B., He H., Tanaka K.I. (2005). Perfect catalytic oxidation of formaldehyde over a Pt/TiO_2 catalyst at room temperature. *Catal. Commun.*, Vol.6, pp.211-214.

Zhang C.B., He H., Tanaka K.I. (2006). Catalytic performance and mechanism of a Pt/TiO_2 catalyst for the oxidation of formaldehyde at room temperature. *Appl. Catal. B.*, Vol.65, pp.37–43.

Zhang, P.Y.; Liang, F.Y.; Yu, G.; Chen, Q.; Zhu, W.P. (2003). A comparative study on decomposition of gaseous toluene by O_3/UV, TiO_2/UV and O_3/TiO_2/UV. *J. Photoch. Photobio. A.*, Vol.156, pp.189-194.

Zou, L.D.; Luo, Y.G.; Hooper, M.; Hu, Eric. (2006). Removal of VOCs by photocatalysis process using adsorption enhanced TiO_2-SiO_2 catalyst. *Chem. Eng. Process.*, Vol.45, pp.959-964.

Zuo G.M., Cheng Z.X., Chen H., Li G.W., Miao T. (2006) Study on photocatalytic degradation of several volatile or-ganic compounds. *J. Hazard. Mater.* Vol.128, pp.158-163.

Chapter 13

Superparamagnetic Behaviour and Induced Ferrimagnetism of $LaFeO_3$ Nanoparticles Prepared by a Hot-Soap Technique

Tatsuo Fujii, Ikko Matsusue and Jun Takada

Additional information is available at the end of the chapter

http://dx.doi.org/10.5772/50031

1. Introduction

Lanthanum orthoferrite, $LaFeO_3$, is one of the most common perovskite-type oxides having an orthorhombic perovskite structure (space group Pbnm), where the distortion from the ideal cubic structure occurs to form the tilting of the FeO_6 octahedra. $LaFeO_3$ has much practical interest for electroceramic applications due to their attractive mixed conductivity displaying ionic and electronic defects [1, 2]. The mixed ionic-electronic conductivity of $LaFeO_3$ exhibits a linear response to oxygen pressure and provides oxygen sensor applications [3]. The excellent sensitivity and selectivity towards various toxic gases such as CO and NO_x are observed as well [4]. Moreover, $LaFeO_3$ nanoparticles exhibited good photocatalytic properties such as water decomposition and dye degradation under visible light irradiation [5, 6]. These properties are enhanced by the homogeneity and high surface area of the fabricated $LaFeO_3$ particles. Fine particles with diameter of less than 100 nm are potentially required for these purposes.

Besides, the orthoferrites are known to be prototype materials for magnetic bubble devices because of their large magnetic anisotropy with small magnetization [7]. $LaFeO_3$ is an interesting model system of orthoferrite antiferromagnets showing a weak ferromagnetism. The Néel temperature, T_N, of $LaFeO_3$ is 738 K, which is the highest temperature in the orthoferrite family [8]. The magnetic moments of Fe^{3+} ions are aligned antiferromagnetically along the orthorhombic a-axis. But they are slightly canted with respect to one another due to the presence of Dzyaloshinskii-Moriya interaction. A weak ferromagnetic component parallel to the c-axis appears. The magnetization of $LaFeO_3$ bulk crystals is considerably small, 0.044 μ_B/Fe [8]. However, magnetic structures of small particles are often different from those of bulk ones. For instance, antiferromagnetic nanoparticles exhibit increasing net magnetization due to the presence of uncompensated surface spins [9, 10]. If the ferromagnetic behavior is promoted in $LaFeO_3$, it should provide facile handling of their applications by using magnetic field. Magnetic properties of well-defined $LaFeO_3$ nanoparticles are worthy to investigate.

It is well known that the wet-chemical methods offer large advantages for low-temperature oxide formation with high surface area, small particle size, and exact cation-stoichiometry. Several methods such as co-precipitation technique [11, 12], polymerized complex method [13], combustion synthesis [14], and sol-gel technique [15] were reported to prepare $LaFeO_3$ nanoparticles. For instance the formation of a single phase of $LaFeO_3$ with the perovskite structure was observed at lower calcination temperatures of 300°C in [11, 12]. This temperature was much lower than that of conventional solid state reaction method. Recently we have successfully prepared $LaFeO_3$ nanoparticles by using the new chemical synthesis method, so-called "hot soap method" [16, 17]. It showed high controllability over the formation of nanoparticles with narrow size distribution, which was performed in the presence of surfactant molecules at high temperatures. The hot soap method is based on the thermal decomposition of reaction precursors of organometallic compounds in polyol solvent. But the presence of surfactant molecules in the solution prevents aggregation of precursors during growth. It was widely applied to prepare nanoparticles of compound semiconductors [18] and metallic alloys [19]. However there were few reports on preparing oxide nanoparticles [20].

In this paper we describe the details of our synthesis procedure of $LaFeO_3$ nanoparticles by using the hot soap method. The magnetic properties of the resultant particles were also discussed as a function of the particle sizes.

2. Experiment

$LaFeO_3$ nanoparticles were synthesized by the hot soap method. Their synthesis procedure is outlined in Figure 1. All chemicals used in this experiment were of reagent grade and used without any further purification. Iron acetylacetonate ($Fe(acac)_3$) and lanthanum acetate ($La(ac)_3 \cdot 1.5H_2O$) were preferred as iron and lanthanum sources, respectively, that were soluble in organic solvents such as polyethylene glycol (PEG 400). In a typical synthesis procedure, equal amounts of $Fe(acac)_3$ (1.2 mmol) and $La(ac)_3$ (1.2 mmol) were weighed out accurately and charged into a reaction flask with 20 mL of PEG 400. Coordinating organic protective agents of oleic acid (5 mmol) and oleylamine (5 mmol) were injected into the reaction mixture and the transparent brown solution was observed. Thereafter, the mixture solution was raised to 200°C and maintained for 3 h with stirring. Before cooling down to room temperature, 50 mL of ethanol was added to the reaction mixture, in order to precipitate the particles. The precipitated particles were rinsed with ethanol and dried at 100°C for 1 h. Some of the sample powders were heat-treated in air for 6 h at various temperatures between 300 and 500°C.

Obtained sample powders were characterized by x-ray powder diffraction (XRD) with monochromatic Cu Kα radiation, infrared spectroscopy (IR), and thermogravimetry and differential thermal analysis (TG-DTA). Powder morphologies of the products were observed by scanning electron microscopy (SEM, Hitachi S-4300) at 20 kV and transmission electron microscopy (TEM, Topcon EM-002B) at 200 kV. The BET surface areas were measured by using N_2 absorption at 77 K. The magnetic properties were investigated using a vibrating sample magnetometer with high-sensitivity SQUID sensor (MPMS SQUID-VSM) and conventional transmission Mössbauer spectroscopy with a 925 MBq $^{57}Co/Rh$ source. The velocity scale of Mössbauer spectra was calibrated with reference to α-Fe.

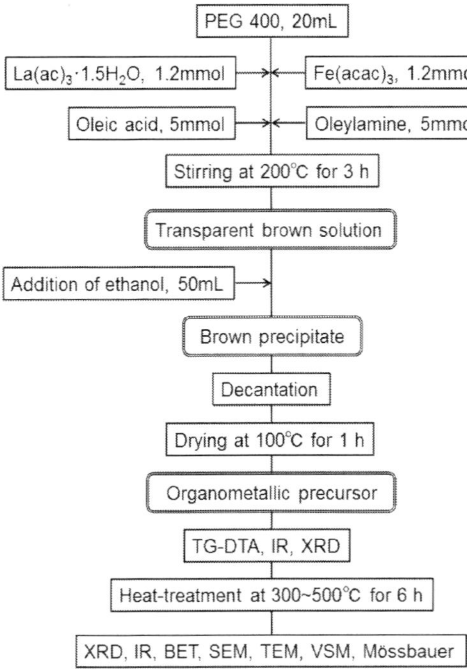

Figure 1. Flowchart of the procedure to prepare $LaFeO_3$ nanoparticles.

3. Results and discussion

3.1. Thermal decomposition of organometallic precursor

Figure 2 shows the TG-DTA curves of the organometallic precursor obtained by the hot soap method. In the TG curve, there are four temperature regions based on weight loss: (1) RT ~ 220°C, (2) 220 ~ 420°C, (3) 420 ~ 510°C, and (4) 510 ~ 600°C, in which the corresponding organic weight loss of 4%, 38%, 45% and 1% were observed, respectively. The small weight loss of the region (1) was ascribed to the evaporation of residual water and ethanol. While the large weight loss of the region (2), accompanied with faint endo- or exothermal peaks in the DTA curve, corresponded to the sublimation and the decomposition of excessive organic substances such as PEG, oleic acid and oleylamine. The temperature range of the region (2) was coincident with the boiling points of individual substances of 250°C (PEG), 223°C (oleic acid) and 350°C (oleylamine). The region (3) comprised the combustion reaction of the residual organics and carbonate components as suggested by the large exothermal peaks at 460°C and 500°C. The large weight loss was due to the decomposition of the most of the organics by oxidation and the release of NO_x, H_2O, CO and CO_2 gases, together with the formation of $LaFeO_3$ as discussed latter. Further heat-treatment in the region (4) gave no major weight loss anymore.

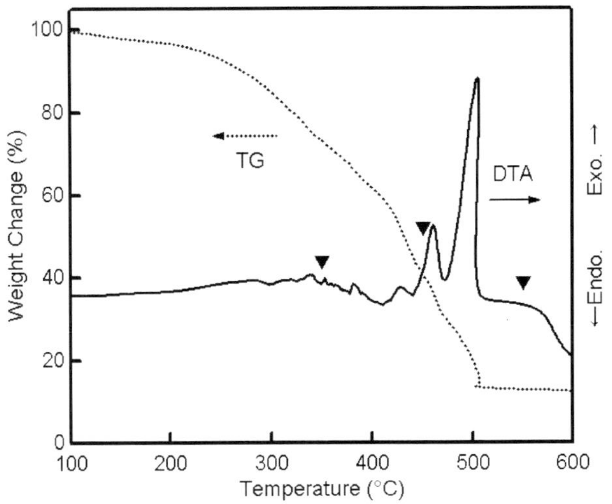

Figure 2. TG-DAT curves of the precursor. The solid triangles indicate the set temperatures for the subsequent XRD and IR observations.

In order to identify the structural changes of the resultant precursor after the heat-treatment, we measured XRD and IR spectra of heat-treated samples taken out from the TG-DTA furnace immediately after reaching to the set temperatures. Figure 3 shows the XRD patterns of the heat-treated samples at various set temperature by TG-DTA. The XRD pattern of the precursor powder had no sharp diffraction lines resulting from the formation of perovskite type oxides. The broad bands centered at around 2θ = 30° and 45° suggested the existence of disordered La_2O_3 phase [21]. Scarce changes in XRD patterns were observed up to the heating of 450 °C. However the XRD pattern of the specimen heated at 550 °C showed clear peaks attributed to $LaFeO_3$ perovskite phase. The pattern showed only the presence of the orthorhombic $LaFeO_3$ phase without the broad bands. Observed crystallization temperature between 450 and 550°C was very consistent with the thermal decomposition temperature of the precursor associated with the large exothermal peaks on the corresponding TG-DTA curves (Figure 2).

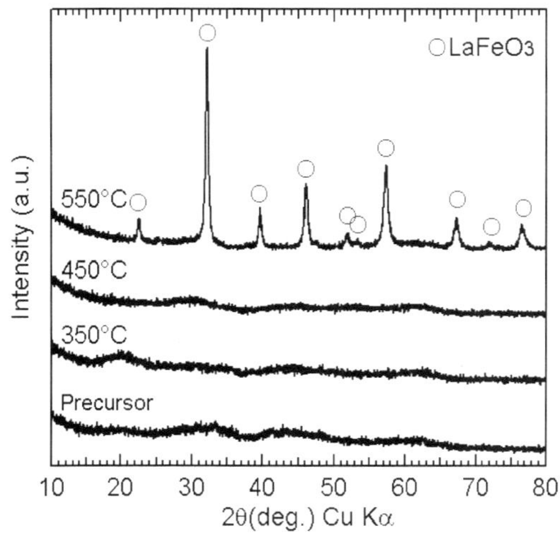

Figure 3. Typical XRD patterns of the obtained products heat-treated at various set temperatures in TG-DTA furnace with no holding time.

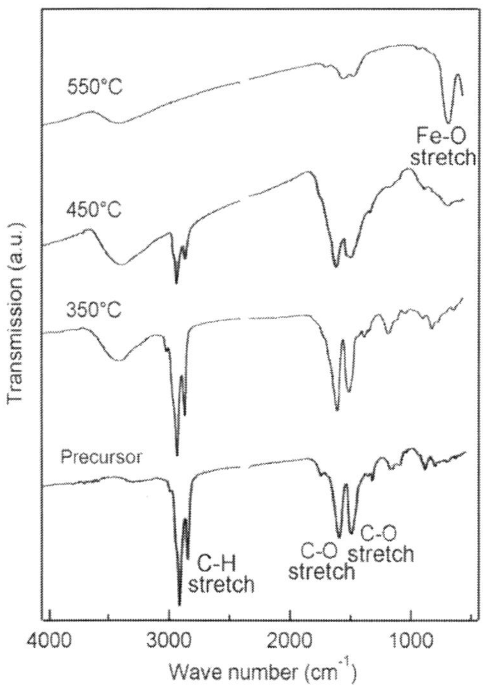

Figure 4. Corresponding IR spectra of the obtained products shown in Figure 3.

The crystallization of $LaFeO_3$ was fully accompanied with the decomposition of the organic substances. The corresponding IR spectra of the heat-treated samples at various set temperatures of TG-DTA are shown in Figure 4. The IR spectra clearly provided the evidence for the presence of organic substances in the precursor. The strong absorption bands at about 2900 cm^{-1} were assigned to the olefinic C-H stretching bands. While the bands appearing at 1400 ~ 1600 cm^{-1} were characteristics of the metal chelate compounds with carbonyl groups [22, 23]. Two strong peaks at 1550 and 1450 cm^{-1} were mainly assigned to the C=O asymmetric and symmetric stretching modes, respectively. The metal ions in the precipitate could be coordinated with large amount of organic molecules and the large weight loss (87%) at 500°C was confirmed after the thermal decomposition. With increasing the heat-treated temperature, the absorption bands assigned to the organic bonds were gradually decreased in intensity and nearly vanished at 550°C. Besides the strong new band appeared at 600 cm^{-1}, that can be attributed to the Fe-O stretching vibration band due to the formation of $LaFeO_3$ phase [24]. These results fully agreed with the XRD phase-analysis findings.

3.2. Formation of $LaFeO_3$ nanoparticles

As shown in Figure 2 in the previous section, a large weight loss of the precipitate started at 220°C, much lower than the temperature of the combustion of the residual organics and the crystallization of $LaFeO_3$. Moderate decomposition of organic substances in the precursor seemed to start prior to the combustion. The migration of metal ions should be facilitated even at the lower temperature. Therefore the long time heat-treatments below the combustion temperature were conducted to the precursors to prepare nanocrystalline particles. Figure 5 shows the XRD patterns of the samples heat-treated in air for 6 h at various temperatures below the combustion. The sample heated at 325°C showed the diffuse XRD pattern with no crystalline phases. While at 350°C, broad XRD peaks attributed to the $LaFeO_3$ perovskite phase started to be observed. This temperature was much lower than the crystallization temperature deduced by the TG-DTA curves. The average grain size, t, of $LaFeO_3$ particles estimated from the XRD peak broadening for the $LaFeO_3$(121) lines was summarized in Table 1 by using the Scherrer equation: $t = \frac{0.9\lambda}{B\cos\theta}$, where λ is the x-ray wavelength, B is the line broadening at half the maximum intensity (FWHM) in radians, and θ is the Bragg angle. The particle size of $LaFeO_3$ formed at 350°C was about 16 nm. With increasing the heat-treatment temperature, the XRD peaks became sharper because of the grain growth of the $LaFeO_3$ particles.

Heat-treatment temperature (°C)	Crystallite size (nm)
350	16
450	20
500	22

Table 1. Average crystallite size of $LaFeO_3$ nanoparticles as a function of the heat- treatment temperature.

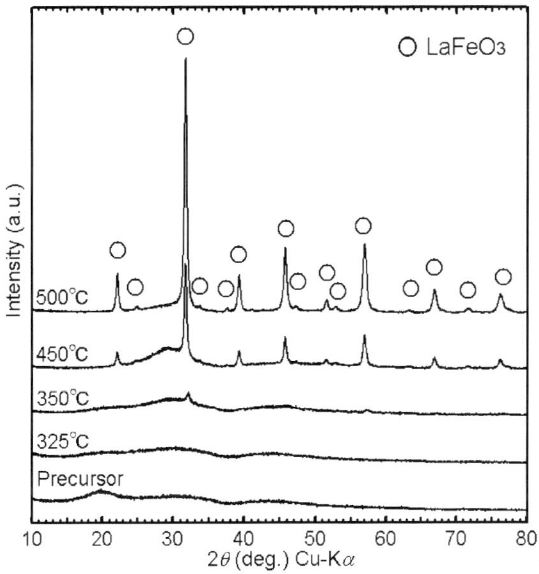

Figure 5. Typical XRD patterns of the obtained products after the long time heat-treatment (6 h) at various temperatures.

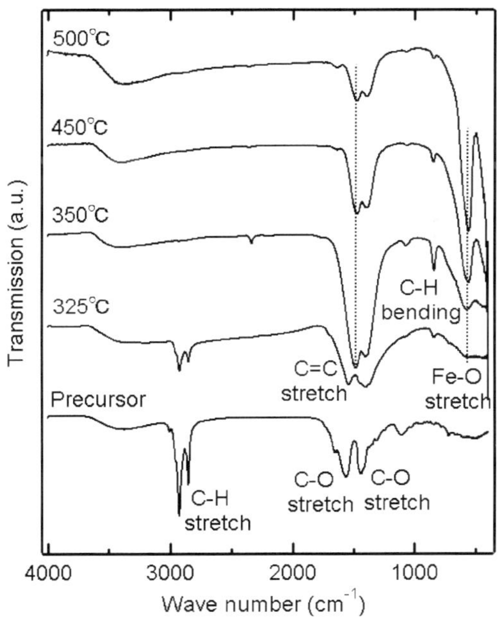

Figure 6. IR spectra of the obtained products after the long time heat-treatment (6 h) shown in Figure 5.

Decomposition of organic substances after the heat-treatment below the combustion temperature was examined by IR spectroscopy. Figure 6 shows the IR spectra of long time (6h) heat-treated samples at various temperatures. The precursor should consist of organometallic composites derived from starting materials. The IR spectrum indicated the clear bands assigned to C-H stretching (~2900 cm^{-1}) and C-O stretching (1400~1600 cm^{-1}) of organic substances [22, 23]. With increasing the heat-treatment temperature, the intensities of initial bands steeply decreased and a new band assigned to Fe-O stretching (~600 cm^{-1}) started to appear at 325°C. This temperature was slightly lower than the formation temperature of $LaFeO_3$ nanoparticles confirmed by XRD observations. And the Fe-O stretching band was intensified monotonically with increasing the temperature. By the way, subsequent to diminishing the intensity of olefinic C-H stretching bands, new strong band appeared in the spectra at 1475 cm^{-1} in addition to a weak band at 843 cm^{-1}. The former band could be assigned to aromatic C=C stretching vibration, while the latter was to aromatic C–H bending one [25]. This result suggested that the framework of organic matrix in precursor was considerably changed after the long time heat-treatment below the combustion temperature. The heat-treatment seemed to promote carbonization of organic substances as well as the crystallization of $LaFeO_3$ nanoparticles.

SEM images of the obtained powders heat-treated in air at various temperatures for 6 h are shown in Figure 7. The precursor particles with and without heat-treatment at 325°C revealed the smooth surface with squamous and wrinkled structures, respectively. These surface morphologies should be very consistent with the amorphous nature of the resultant organic matrix. With increasing the heat-treatment temperature above 350°C, the particles were crowned by squamous surfaces. The macrosized particles seemed to be disaggregated into fine particles after the crystallization. In spite of the nanometric crystal size determined from the XRD profiles, the BET surface area of the obtained $LaFeO_3$ particles at 350 and 500°C were only 3.9 and 3.1 m^2/g, respectively. The nanocrystalline fine particles formed in the precursors were strongly agglomerated with each other in present cases. The residual carbon matrix could also affect the morphology of the obtained powders.

The structure of the $LaFeO_3$ nanocrystallites was further analyzed by TEM micrograph. Figure 8 shows the TEM image and electron diffraction (ED) pattern of $LaFeO_3$ nanoparticles obtained by the long time (6 h) heat-treatment at 350°C. The particulates consisted of an agglomeration of numerous spherical primary particles loosely aggregated together, in contrast to the SEM observations. The average diameter of the primary particles estimated from the TEM image was about 15 nm, which was well consistent with that from XRD. The high-contract dotted rings in the ED patterns indicated the good crystallinity of the $LaFeO_3$ nanoparticles. These rings were indexed as the $LaFeO_3$ perovskite phase with random orientation. No diffraction spots attributed to the secondary phase were observed.

3.3. Magnetic properties of $LaFeO_3$ nanoparticles

Room temperature Mössbauer spectra of obtained powders after the long time heat-treatment (6h) at various temperatures are shown in Figure 9. The sample heated at 500°C, which had a larger crystallite size, showed a clear sextet pattern with a small amount of a

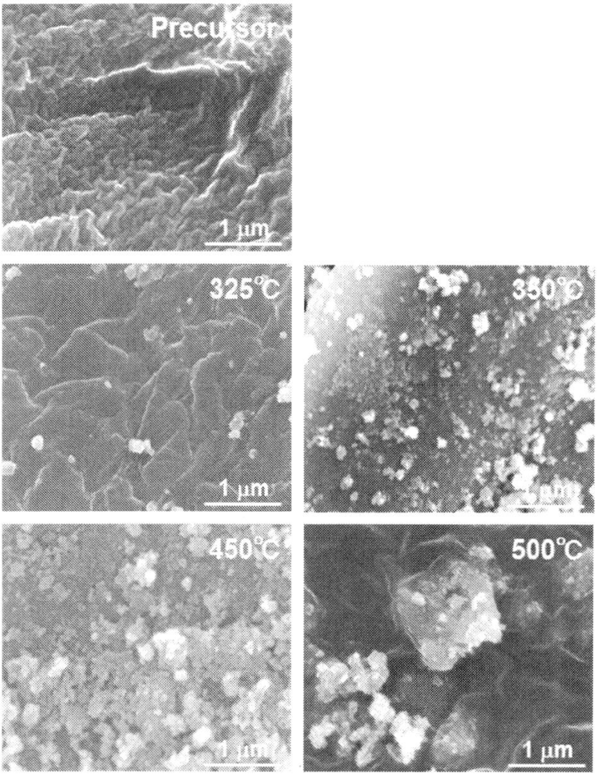

Figure 7. SEM images of the obtained products heat-treated at various temperatures.

doublet component. The observed Mössbauer parameters showing antiferromagnetic ordering were fully consistent with those of the $LaFeO_3$ bulk crystals [26]. The paramagnetic doublet patterns were dominant for other samples obtained at lower temperatures (350 and 450°C) in spite of the formation of $LaFeO_3$, as well as the non-crystalline samples at 300 and 325°C. This behavior was attributed to superparamagnetism because of the fine crystallite size of $LaFeO_3$ as discussed above. By the way, the doublet patterns showed asymmetric profiles. The asymmetric doublets were probably caused by the presence of iron ions in different environments such as surface or inner part of the crystals and/or the non-crystalline matrix. But it was difficult to decompose such complex spectra into unique components. Therefore the spectra were simply analyzed to assume one asymmetric doublet with different widths. The fitted parameters are listed in Table 2. The isomer shift values of ~0.3 mm/s for both doublet and sextet peaks indicated the Fe valence states in the all specimens were trivalent. No reduction occurred during the long time heat-treatment with organic substances. When looking in detail, the Mössbauer parameter of the doublet pattern indicated the systematic change depending on the heat-treatment temperature. The isomer shift gradually decreased while the quadrupole splitting gradually increased with increasing the heat-treatment temperature. This result suggested the formation of strong Fe-O bonds and ligand fields due to the crystallization of $LaFeO_3$ particles.

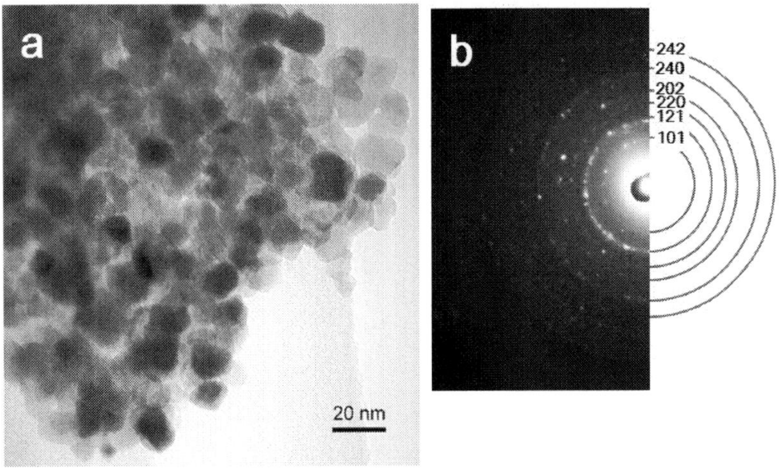

Figure 8. (a) TEM image and (b) its ED pattern of $LaFeO_3$ nanoparticles obtained at 350°C. Diffraction rings are indexed as their corresponding $LaFeO_3$ planes.

Heat-treatment temperature (°C)	Isomer shift (mm/s)	Quadupole splitting (mm/s)	Hyperfine filed (kOe)	Intensity (%)
500	0.369	-	517	65
	0.285	1.051	-	35
450	0.364	-	508	23
	0.292	1.008	-	77
350	0.355	-	512	21
	0.313	0.924	-	79
325	0.329	0.891	-	100
300	0.327	0.824	-	100

Table 2. Fitted Mössbauer parameters obtained from the spectra in Figure 9.

Magnetic measurements were performed on the heat-treated powders at various temperatures. The room-temperature magnetization curves of obtained samples are shown in Figure 10 as a function of the heat-treated temperature. $LaFeO_3$ is known to be antiferromagnetic with weak ferromagnetism [8]. The net magnetization of $LaFeO_3$ should be very small because of the antiferromagnetic order of the Fe^{3+} spins. Only a slight canting of the adjacent Fe^{3+} spins produced weak ferromagnetism. The spontaneous magnetization of $LaFeO_3$ bulk crystals was only ~0.1 emu/g at room temperature [27]. However all samples seemed to have large magnetization, especially for the sample heat-treated at 350°C. The shape of hysteresis loops showing the small spontaneous magnetization were characteristics of weak ferromagnetism. The maximum field applied of 60 kOe did not saturate the magnetization. This was mainly caused by antiferromagnetic ordering of the spins in the nanoparticles [27]. By the way, the sample prepared at 500°C showed the remarkable hysteresis gap than the other samples. Mössbauer spectra indicated that only the sample prepared at 500°C possessed predominant antiferromagnetic ordering at room temperature while the others were paramagnetic. The small hysteresis gaps of the samples obtained at lower temperatures (325~450°C) supported their superparamagnetic nature.

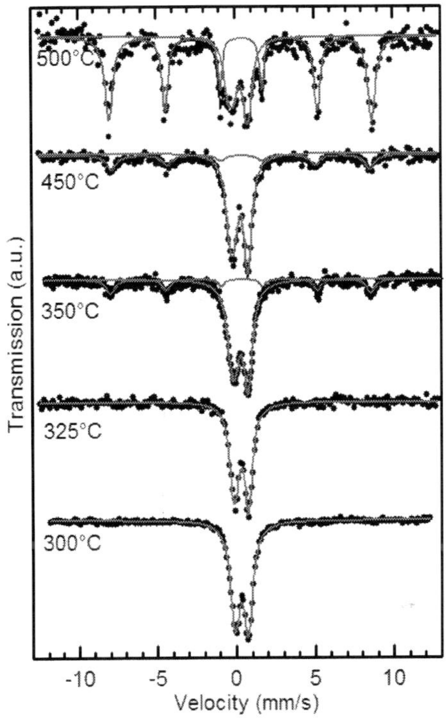

Figure 9. Room temperature Mössbauer spectra of the obtained products heat-treated at various temperatures.

The temperature dependence of magnetization behaviors under zero-field-cooling (ZFC) and field-cooling (FC) conditions measured in an external magnetic field of 200 Oe are shown in Figure 11. Magnetization of $LaFeO_3$ particles obtained at 500°C was gradually decreased with decreasing the temperature. This behavior was characteristic of the antiferromagnetic ordered states below the T_N. The higher T_N above the room temperature was confirmed by the Mössbauer spectroscopy. The difference of magnetization values between ZFC and FC curves was attributed to the residual paramagnetic components at room temperature. On the other hand, magnetization of other samples obtained at 300~450°C were smoothly increased

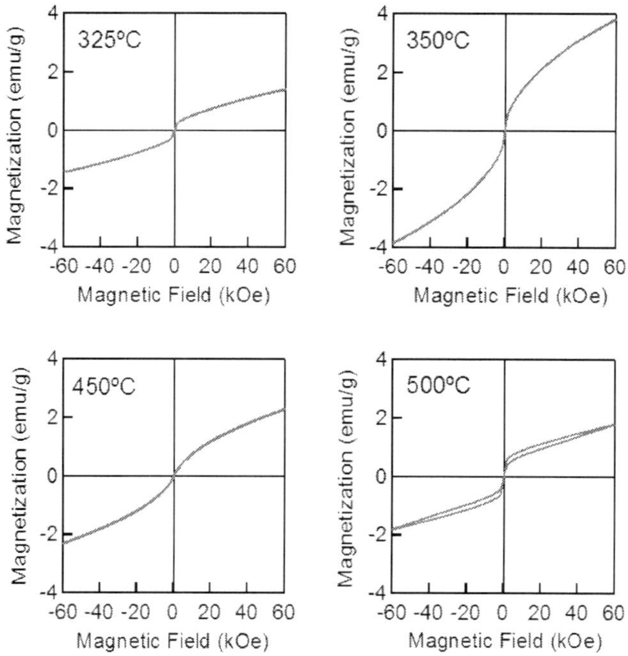

Figure 10. Room temperature magnetization curves of the obtained products heat-treated at various temperatures.

with the decreasing temperature. These behaviors were very consistent with the paramagnetic nature of the obtained samples at room temperature. Moreover, the sharp cusps in the ZFC curves were observed for the samples obtained at 325~450°C. These were considered to be the superparamagnetic blocking temperature (T_B), where the magnetization of $LaFeO_3$ nanoparticles begun to freeze [28]. The T_B detected by the ZFC curves was summarized in Table 3. It was decreased steeply with decreasing the heat-treatment temperature coincident with the decreasing particle sizes. Besides, the observed T_B values were very small, which indicated that the magnetic interactions between the obtained $LaFeO_3$ nanoparticles were very weak. The surface of $LaFeO_3$ nanoparticles was probably coated with the residual organic molecules to form the non-magnetic layers [29]. This assumption was fully supported by the IR spectra indicating the presence of carbonized components and the TEM micrograph showing the loosely aggregating $LaFeO_3$ nanoparticles. It should be mentioned here that the sample heat-treated at 325°C indicated the superparamagnetism, showing the small cusp in the ZFC curve at 7.7 K. The sample seemed to contain the $LaFeO_3$ nanoclusters, though it showed the non-crystalline XRD pattern. But the IR spectrum seemed to have good sensitivity to detect their formation as well as the magnetization measurement. By the way, in the case of conventional superparamagnetic behavior, the FC magnetization curves should increase continuously with decreasing the temperature below the T_B. But the FC curve of the heat-treated sample at 450°C showed a small cusp on cooling as well as the ZFC curve. This behavior suggested that the antiferromagnetic interactions between nanoparticles were present [28]. Long-range antiferromagnetic order was developed in associate with the grain growth of the $LaFeO_3$ nanoparticles.

Heat-treatment temperature (°C)	Blocking temperature (K)
325	7.7
350	27.6
450	36.9

Table 3. Blocking temperature (T_B) obtained from the ZFC curves shown in Figure 11.

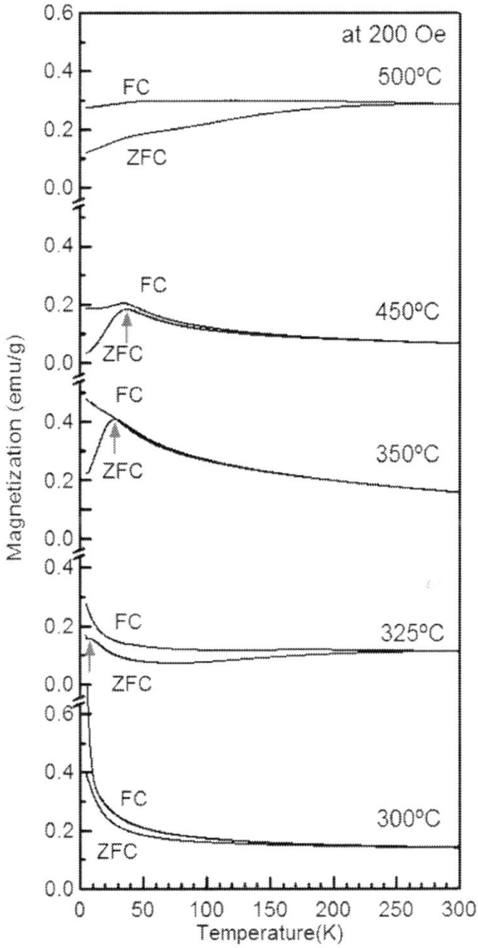

Figure 11. Temperature dependence of ZFC and FC magnetization curves at 200 Oe for the obtained products at various heat-treatment temperatures

In order to demonstrate the superparamagnetic behavior more directly, we measured the temperature dependence of their Mössbauer spectra down to 15 K. Figure 12(a) shows the typical law temperature Mössbauer spectra of the $LaFeO_3$ nanoparticles obtained at 350°C. The spectrum measured at room temperature (see Figure 9) was well consistent with that of the $LaFeO_3$ nanoparticles with average grain size of 16 nm reported in literature [26]. The spectrum was mainly composed of the super-paramagnetic doublet peak besides to the magnetically split sextet. The component showing the sharp sextet pattern could be attributed to the unexpected coarse $LaFeO_3$ particles which had a good crystallinity, while the most of the particles exhibited the paramagnetic doublet pattern at room temperature. With decreasing the temperature, the paramagnetic spectra corresponded to the enlarged sextets due to the presence of magnetic order. Therefore the spectra were decomposed into three components. One was the sharp sextet for the coarse particles; second and third were the broad sextet and paramagnetic doublet for the nanoparticles, respectively. Relative intensity ratio of each component was plotted in Figure 12(b) as a function of the temperature. The relative area of the paramagnetic component varies almost linearly with decreasing the temperature below 100 K. According to the Mössbauer measurements, we defined the

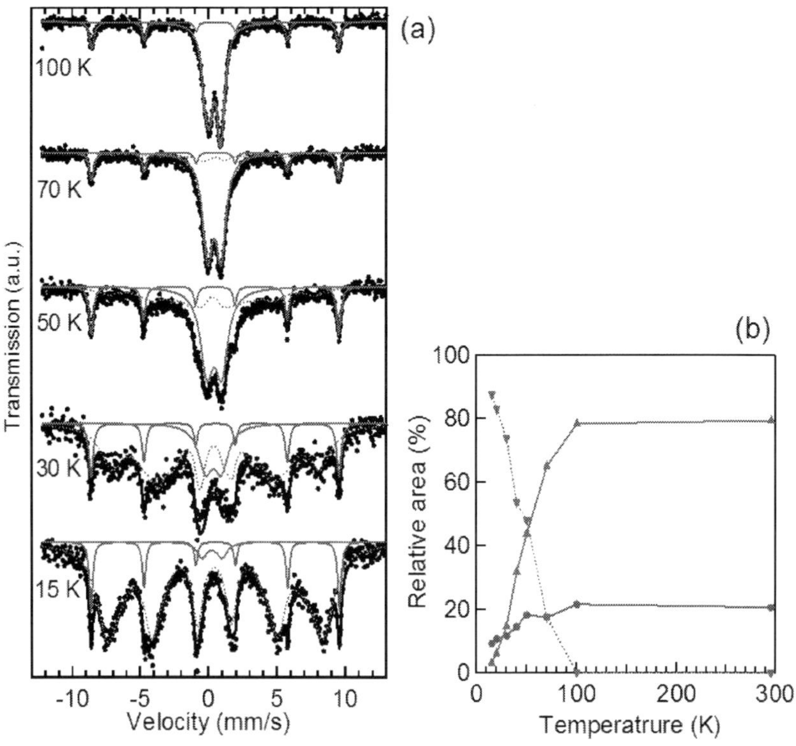

Figure 12. (a) Low temperature Mössbauer spectra of $LaFeO_3$ nanoparticles obtained at 350°C, and (b) relative area of each component as a function of the temperature. The details are described in the text.

median blocking temperature (T_{Bm}), as a temperature where the doublet collapsed to 50% of its initial value. This analysis yielded the T_{Bm} of ~50 K, which was much higher than the superparamagnetic T_B obtained by the magnetization measurement. It was well-known that the time scale of the superparamagnetic fluctuations becomes comparable to the time scale of the measurement [30]. The time scale of Mössbauer measurements (~10^{-9} s) was much faster than that of magnetization measurements (~10^{-1} s) and gave the higher T_{Bm} well above the T_B. This phenomenon evidenced the presence of superparamagnetic relaxation of the obtained samples.

Figure 13 shows the field dependent magnetization curves measured at lower temperature (5 K) for the $LaFeO_3$ nanoparticles prepared at various heat-treatment temperatures. The samples showed considerable magnetization at low temperature, though the net magnetization of $LaFeO_3$ bulk crystals should be very small because of the antiferromagnetic order of the Fe^{3+} spins. A spontaneous magnetization of the bulk crystals was evaluated to 0.15 emu/g at 5 K [16]. We have recently reported that the $LaFeO_3$ nanoparticles prepared at 350°C revealed an anomalous large spontaneous magnetization of 7.8 emu/g [16]. This phenomenon was almost reproducible. The spontaneous magnetization of the obtained powders exhibited the maximum value for the samples prepared at 350°C. The $LaFeO_3$ nanoparticles with the average diameter of ~15 nm seemed to have large ferromagnetism instead of antiferromagnetism. The induced ferromagnetism in $LaFeO_3$ nanoparticles was possibly associated with the uncompensated surface spins as reported in other antiferromagnetic nanocrystalline systems [9, 10]. It should be worth to mention that the induce magnetic moment per Fe^{3+} ion was evaluated to 0.34 μ_B for the sample obtained at 350°C. Taking into account that the moment of Fe^{3+} is 5 μ_B, a large fraction (~7%) of Fe^{3+} ions contribute to the magnetization of $LaFeO_3$ nanoparticles. Magnetic structure of $LaFeO_3$ nanoparticles should be strongly modified from the bulk one.

By the way, all magnetization curves revealed clear hysteresis gaps with small loop shifts at 5 K. The coercivity and the exchange bias field of each sample are summarized in Table 4 as well as the spontaneous magnetization. The exchange field was linearly increased with increasing the heat-treatment temperature. The exchange bias was generally attributed to the exchange coupling between the ferromagnetic shell and the antiferromagnetic core of the particles [31]. In the case of $LaFeO_3$ particles, the uncompensated surface spins and/or the enhanced spin cantings near the surface region should cause the considerable magnetization. With the increasing particle size, the reliable antiferromagnetic core was formed to produce the large exchange coupling.

Heat-treatment temperature (°C)	Spontaneous magnetization (emu/g)	Coercivity (Oe)	Exchange bias (Oe)
325	2.89	823	23
350	5.90	1304	25
450	2.94	2240	64
500	1.40	1275	208

Table 4. Magnetic parameters at 5 K of the obtained products at various heat-treatment temperatures.

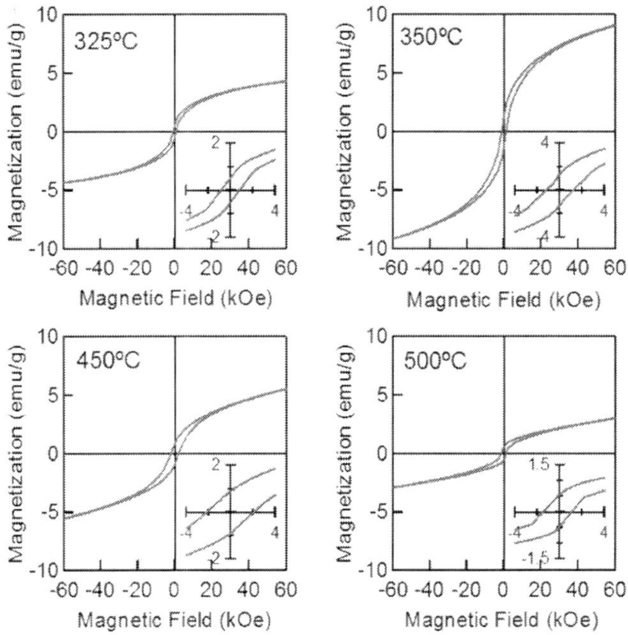

Figure 13. Low temperature (5 K) magnetization curves of the obtained products heat- treated at various temperatures. The insets are the enlarged view of the central region.

4. Conclusion

Nanocrystalline $LaFeO_3$ particles with an average diameter of 16 nm were prepared by using hot-soap technique. Crystalline $LaFeO_3$ were formed at a relatively low heat-treatment temperature of 350°C without any phase segregation. This temperature was much lower than the combustion point of organic substances. With increasing the heat-treatment temperature, the grain size of $LaFeO_3$ particles was increased. The $LaFeO_3$ nanoparticles exhibited superparamagnetic behavior and had anomalous large net magnetization. The spontaneous magnetization reached the maximum when the precursor was heat-treated at 350°C. Moreover the particles exhibited the exchange bias properties with increasing the grain size of $LaFeO_3$. The low blocking temperature observed by both magnetization and Mössbauer measurements indicated that the obtained nanoparticles at 350°C were magnetically isolated with each other.

Author details

Tatsuo Fujii*, Ikko Matsusue and Jun Takada
Department of Applied Chemistry, Okayama University, Japan

* Corresponding Author

Acknowledgement

The authors wish to thank their colleagues and students for their technical supports. In particular, the authors would like to acknowledge the contributions of Mr. Takuya Nonoyama for the sample preparations.

5. References

[1] Ming Q, Nersesyan MD, Wagner A, Ritchie J, Richardson JT, Luss D, Jacobson AJ, Yang YL (1999) Combustion synthesis and characterization of Sr and Ga doped $LaFeO_3$. Solid State Ionics 122: 113-121.

[2] Yoon JW, Grilli ML, Di Bartolomeo E , Polini R, Traversa E (2001) The NO_2 response of solid electrolyte sensors made using nano-sized $LaFeO_3$ electrodes. Sens Actuators B 76: 483-488.

[3] Hole I, Tybell T, Grepstad JK, Wærnhus I, Grande T, Wiik K (2003) High temperature transport kinetics in heteroepitaxial LaFeO3 thin films. Solid-State Electronics 47: 2279-2282.

[4] Toan NN, Saukko S, Lantto V (2003) Gas sensing with semiconducting perovskite oxide $LaFeO_3$. Physica B 327: 279–282.

[5] Parida KM, Reddy KH, Martha S, Das DP, Biswal N (2010) Fabrication of nanocrystalline $LaFeO_3$: An efficient solegel auto-combustion assisted visible light responsive photocatalyst for water decomposition. Int J Hyd Energy 35: 12161-12168.

[6] Su H, Jing L, Shi K, Yao C, Fu H (2010) Synthesis of large surface area $LaFeO_3$ nanoparticles by SBA-16 template method as high active visible photocatalysts. J Nanopart Res 12: 967-974.

[7] Szymczak R (1977) Temperature dependence of bubble domain structure in $YFeO_3$ and $DyFeO_3$ orthoferrites. Physica B+C 86-88: 1351-1353.

[8] Treves D (1965) Studies on Orthoferrites at the Weizmann Institute of Science. J Appl Phys 36: 1033-1039.

[9] Kodama RH, Berkowitz AE (1999) Atomic-scale magnetic modeling of oxide nanoparticles. Phys Rev B 59: 6321-6336.

[10] Lee YC, Parkhomov AB, Krishnan KM (2010) Size-driven magnetic transitions in monodisperse MnO nanocrystals. J Appl Phys 107: 09E124-1-3.

[11] Nakayama S (2001) LaFeO3 perovskite-type oxide prepared by oxide-mixing, co-precipitation and complex synthesis methods. J Mater Sci 36: 5643-5648.

[12] Li X, Zhang H, Zhao M (1994) Preparation of nanocrystalline $LaFeO_3$ using reverse drop coprecipitation with polyvinyl alcohol as protecting agent. Mater Chem Phys 37: 132-135.

[13] Popa M, Frantti J, Kakihana M (2002) Lanthanum ferrite $LaFeO_{3+d}$ nanopowders obtained by the polymerizable complex method. Solid State Ionics 154-155: 437-445.

[14] Wang JB, Liu QF, Xue DS, Li FS (2002) Synthesis and characterization of $LaFeO_3$ nano particles. J Matter Sci Lett 21: 1059-1062.

[15] Rajendran M, Bhattacharya AK (2006) Nanocrystalline orthoferrite powders: Synthesis and magnetic properties. J Eur Ceramic Soc 26: 3675-3679.

[16] Fujii T, Matsusue I, Nakatsuka D, Nakanishi M, Takada J (2011) Synthesis and anomalous magnetic properties of $LaFeO_3$ nanoparticles by hot soap method. Mater Chem Phys 129: 805-809.

[17] Fujii T, Matsusue I, Nakanishi M, Takada J (2012) Formation and superparamagnetic behaviors of $LaFeO_3$ nanoparticles. Hyperfine Interact 205: 97-100.

[18] Murray CB, Noms DJ, and Bawendi MG (1993) Synthesis and Characterization of Nearly Monodisperse CdE (E = S, Se, Te) Semiconductor Nanocrystallites. J Am Chem Soc 115: 8706-8715.

[19] Sun S, Murray CB, Weller D, Folks L, Moser A (2000) Monodisperse FePt Nanoparticles and Ferromagnetic FePt Nanocrystal Superlattices. Science 287: 1989-1992.

[20] Shouheng Sun S, Zeng H, Robinson DB, Raoux S, Rice PM, Wang SX, and Li G (2004) Monodisperse MFe_2O_4 (M= Fe, Co, Mn) Nanoparticles. J Am Chem Soc 126: 273-279.

[21] Sadaoka Y, Aono H, Traversa E, Sakamoto M (1998) Thermal evolution of nanosized $LaFeO_3$ powders from a heteronuclear complex, $La[Fe(CN)_6]\cdot nH_2O$. J Alloys Comp 278: 135-141.

[22] Tayyari SF, Bakhshi T, Ebrahimi M, Sammelson RE (2009) Structure and vibrational assignment of beryllium acetylacetonate. Spectrochim Acta A 73: 342-347.

[23] Nakamoto K, McCarthy PJ, Martell AE (1961) Infrared Spectra of Metal Chelate Compounds. III. Infrared Spectra of Acetylacetonates of Divalent Metals. J. Am. Chem. Soc. 83: 1272-1276

[24] Gosavi PV, Biniwale RB (2010) Pure phase $LaFeO_3$ perovskite with improved surface area synthesized using different routes and its characterization. Mater Chem Phys 119: 324-329.

[25] El-Hendawy AA (2006) Variation in the FTIR spectra of a biomass under impregnation, carbonization and oxidation conditions. J Anal Appl Pyrolysis 75: 159–166.

[26] Li X, Cui X, Liu X, Jin M, Xiao L, Zhao M (1991) Mössbauer spectroscopic study on nanocrystalline $LaFeO_3$ materials. Hyperfine Interact 69: 851-854.

[27] Shen H, Cheng G, Wu A, Xu J, Zhao J (2009) Combustion synthesis and characterization of nano-crystalline $LaFeO_3$ powder. Phys Status Solidi A 206: 1420-1424.

[28] Bedanta S, Kleemaann W (2009) Supermagnetism. J Phys D Appl Phys 42: 013001-1-27.

[29] Mørup S, Frandsen S, Bødker F, Klausen SN, Lefmann K, LindgÅrd PA, Hansen MF (2002) Magnetic Properties of Nanoparticles of Antiferromagnetic Materials. Hyperfine Interact 144/145: 347-357.

[30] Bødker F, Hansen MF, Koch CB, Lefmann K, Mørup S (2000) Magnetic properties of hematite nanoparticles. Phys Rev B 61: 6826-3838.

[31] Ahmadvand H, Salamati H, Kameli P, Poddar A, Acet M, Zakeri K (2010) Exchange bias in $LaFeO_3$ nanoparticles. J Phys D Appl Phys 43: 245002-1-5.

Section 4

Organic Spectroscopy

Chapter 14

Phenotiazinium Dyes as Photosensitizers (PS) in Photodynamic Therapy (PDT): Spectroscopic Properties and Photochemical Mechanisms

Leonardo M. Moreira, Juliana P. Lyon, Ana Paula Romani,
Divinomar Severino, Maira Regina Rodrigues and Hueder P. M. de Oliveira

Additional information is available at the end of the chapter

http://dx.doi.org/10.5772/48087

1. Introduction

Oscar Raab demonstrated, in 1900, that the light incidence on dyes can induce cell death [1]. A photosensitizer is a chemical compound that is activated by light of a specific wavelength that leads to tumor destruction [2]. Indeed, Photodynamic Therapy (PDT) is considered to have its origin in 1900 with the classical experiments by the german scientist Oscar Raab. Raab noticed that the exposure of *Paramecium caudatum* to acridine orange and later subjection to light resulted in death of this organism. Raab and his supervisor Hermann von Tappeiner later coined the term "photodynamic therapy" and applied PDT successfully for the treatment of cutaneous tumors using eosin. From that concept, photodynamic therapy (PDT) [3,4,5,6], as we known today, was founded. Since then, the development of other studies, culminating with those performed by Dougherty and co-workers resulted in a non-invasive technique for cancer treatment and other diseases [7,8]. In fact, precancerous cells, certain types of cancer cells and microbial infections can be treated this way.

Interesting data regarding the application of PDT against several diseases have been reported, since the employment of this therapy in different diseases has increased significantly. In fact, PDT has been used with phenotiazinium [methylene blue (MB) and toluidine blue] as photosensitizers against AIDS-related Kaposi's sarcoma, promoting complete sarcoma remission with excellent cosmetic results [9]. PDT with MB (and LED as light source), which is a very inexpensive system, has been applied against *Leishmania*, promoting significant reduction in the size of the lesions, diminishing the parasitic load in the draining lymph node and healing the lesions in hamsters experimentally infected with *L. amazonensis* [10]. This therapeutic alternative is very interesting due to the resistance of this organism to pentavalent antimonials (SbV), which constitutes the mainstay pharmacological alternative for leishmaniasis, due to emergence of drug resistance [11].

Tumor, which is also called neoplasm or blastoma, is the abnormal growth of tissues. Sick cells with genetic disturb develop more rapidly than the normal cells, which provokes the development of the tumor (that can be malign or non-malign cells). When the growth of the tumor is a very fast and chaotic process, with tendency to arrive in other organs, generally is a malign tumor [4]. Cancer is the general name of all malign tumors. This term cancer is originated from latin and means "crab". This name is due to the tendency of the tumor to be fixed in several biological tissues, which is correlated to the ability of the crab to be fixed in various surfaces [4].

Interestingly, the PDT procedure is easily performed in a physician's office or outpatient setting, which favors the application of this therapy in several environments, since PDT does not need great structural pre-requisites. In this context, it is important to notice that multicenter randomized controlled studies have demonstrated high efficacy and superior cosmetic outcome over standard therapies [12]. In fact, several cosmetic methodologies have been developed with PDT, such as resurfacing. For many non-oncologic dermatological diseases, such as *acne vulgaris*, viral warts and localized scleroderma, case reports and small series have confirmed the potential of PDT [12]. After the development of topical photosensitizers 5-aminolevulinic acid (ALA) or its methyl ester (MAL), PDT has gained worldwide popularity in dermatology, since these drugs do not induce prolonged phototoxicity as the systemic photosensitizing hematoporphyrin derivatives do [12]. PDT has essentially three steps. First, a light-sensitizing liquid, cream, or intravenous drug (photosensitizer) is applied or administered. Second, there is an incubation period of minutes to days. Finally, the target tissue is then exposed to a specific wavelength of light that activates the photosensitizing medication.

More than one million cases of skin cancer were diagnosed during 2008 in the U.S.A. and its worldwide incidence has risen throughout the last four decades. Squamous cell carcinoma (SCC) is the second most frequent skin cancer, only after basal cell carcinoma (BCC) [13]. In the 20th century, SCC was mainly linked to occupational sun exposure, whereas in the last decades the strongest link has been to ultraviolet (UV) radiation. On one hand, UVB exposure leads to direct DNA damage by pyrimidine dimer formation. On the other hand, UVA induces formation of reactive oxygen species which indirectly also cause DNA damage. Other factors such as the phototype, the genetic predisposition or the immune response are also involved in the carcinogenic process [13].

It is also important to notice that photoantimicrobial agents, that is, chemical compounds that exhibit increased inactivation of microorganisms when exposed to light, have been known also for over a century [14]. While there are several studies regarding the use of photosensitizers against bacterial and viral targets, the clinical use of photosensitizers in antimicrobial therapy has been developed very slowly through small scale trials. This is particularly a surprise considering the efficacy exhibited, especially by cationic photosensitizers, against pathogenic drug-resistant bacteria such as methicillin-resistant *Staphylococcus aureus* and vancomycinresistant *Enterococcus faecium* [14]. Furthermore, the exponentially increasing threat of microbial multidrug resistance has highlighted antimicrobial photodynamic inactivation (APDI) as a promising alternative treatment for localized infections [15]. APDI involves the direct application of the PS to the infected tissue rather than being injected intravenously, as the usual procedure for cancer treatment with PDT [15].

The photodynamic process involves photophysical and photochemical steps, which can be applied with several aims, such as therapies against cancer or infections. PDT light sources include laser, intense pulsed light, light-emitting diodes (LEDs), blue light, red light, and many other visible lights (including natural sunlight). Photosensitizer drugs may become activated by one or several types of light. The optimal light depends on the ideal wavelength for the particular drug used and target tissue.

Electron and energy transfer in the excited state govern the efficiency of a variety of photoinduced processes, including photosynthesis, light to energy conversion in semiconductor devices, cell damage induced by solar exposition and photodynamic action [16,17,18]. It is well reported that photophysical behavior of a dissolved dye depends on the nature of its environment, i.e., the solvent influences the spectra characteristics of the solute molecules [19]. Several factors influence the visible spectral behavior of dissolved dye molecules, especially the solvent's polarity and its hydrogen-bond donor/acceptor capacities [19]. The properties can be determined by the solvent dielectric constant, ε, and solvatochromic parameters. The strong solvatochromic behavior can be observed for dye molecules with large dipole moment changes during transitions between two electronic states. The solvent can differentially stabilize the ground and/or the excited state in polar and non-polar solvents [19].

The series of phenothiazine [thionine, methylene blue (MB), azure A (AZA) and azure B (AZB)] derivatives (Fig. 1) are positive dyes used as a model for phototherapeutic agent as well as for dye sensitized solar energy converter [20,21] due to their appropriate biological, chemical, photochemical and photophysical properties [22,23,24].

The intersystem cross quantum yield and the singlet oxygen formation for MB is 0.52 [25,26,27,28,29], the triplet lifetime is higher, approximately 3.0 μs, in air saturated aqueous solution, and up to 50 μs in nitrogen saturated aqueous solution. The singlet excited state has a lower lifetime, approximately 1,400 ps (Table 1), and it is due to the higher internal conversion and triplet formation, with a fluorescence quantum yield of 0.04 in methanol [30,31,32,33]. In addition, MB and and MB derivatives that have been used as photosensitizers in PDT showed a good biocompatibility (appropriate citotoxicity and phototoxicity) when used *in vitro* to attack key organelles in cells [14,21].

thionine
R1 = R2 = R3 = R4 = H

Methylene blue
R1 = R2 = R3= R4 = CH3

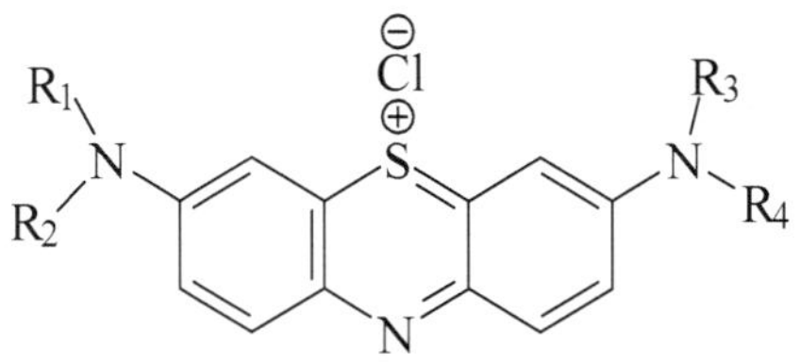

azure A
R1= R2 = CH3
R3 = R4 = H

azure B
R1 = R2 = R3 = CH3
R4=H

Figure 1. Thionine derivatives.

medium: water	
Dye	τ (ps)
Thionine	314,40
Nile blue	372,75
Azure A	421,28
Azure B	1268,56 (48,89%) e 306,00 (51,11%)
Toluidine blue	2179,65 (66,72%) e 358,03 (33,28%)
Methylene blue	328,84
medium: ethanol	
Dye	τ (ps)
Thionine	848,84
Nile blue	1170,03
Azure A	776,43
Azure B	724,78
Toluidine blue	643,89
Methylene blue	465,96

Table 1. Values of lifetime (τ) of some dyes at 25°C.

The Fluorescence decays of dyes were obtained by single-photo-counting technique. The excitation source was a Tsunami 3950 Spectra Physics titanium-sapphire laser, pumped by a Millenia X Spectra Physics solid state laser. The laser was tuned that a third harmonic generator BBO crystal (GWN-23PL Spectra Physics) gave the 292 nm excitation pulses that were directed to an Edinburgh FL900 spectrometer. The spectrometer was set in L-format configuration, the emission wavelength was selected by a monochromator (680 nm), and emitted photons were detected by a refrigerated Hamamatsu R3809U microchannel plate photomultiplier. The software provided by Edinburgh Instruments was used to analyze the individual decays. The quality of the fit was judged by the analysis of the statistical parameters reduced-χ^2 and Durbin-Watson, and by the inspection of the residuals distribution.

The dyes stock solutions were prepared in ethanol (6.0 x 10^{-5} M) and aliquots of these stock solutions were added, via a calibrated Hamilton microsyringe, to volumetric flasks containing water or ethanol, and the solutions were stirred for 30 minutes. The final concentrations of dyes were 1.0 x 10^{-6} M. All measurements were performed at 25°C using a cuvette with 0.2 cm of optical path.

The excited state lifetime depends on the solvent [34,35]. The dependence of the lifetime on the viscosity and solvent dielectric constant indicates that the dye excited state deactivation process is slow as the medium viscosity increases. This effect is related to the partial inhibition or the higher friction on the dye substitute groups rotation, such as $-CH_3$, $-NH_2$, $-N(CH_3)_2$ and $-N(CH_2CH_3)_2$ [36]. The lifetime values are in agreement with the results reported in the literature. Lee and Mills [37] showed the lifetime values for methylene blue aqueous solutions (358 ± 20 ps). Grofcsik et al [34] measured the lifetime of Nile blue excited state and oxazine 720 in different solvents at 20 °C. The thionine dye photophysics is well known [38]. In an aqueous solution, thionine has a fluorescence lifetime of 320 ± 60 ps when excited at 610 nm [37,39]. In organic solvent, the increase of the thionine fluorescence lifetime (450 ps in ethanol and 760 ps in terc-butilic alcochol) results in a increasing of the fluorescence quantum yield [38]. The thionine lifetime differences observed in an aqueous medium and ethanol is quite high, which shows the effect of microenvironment polarity on the excited state decaying [38]. In our experiments, in an aqueous medium thionine has a useful lifetime of 314.4012 ps, which is in agreement with the results presented in the literature.

The Nile blue lifetime in ethanol and water are 1420 and 418 ps, respectively. These results are higher than those that we found in our work. However, it should be taken into account that the temperatures used in our experiments are different from those whose results are different. It was shown that the lifetime of Nile blue depends on the temperature due to the intermolecular charge transfer [34,35,36]. This charge transfer process is facilitated by the presence of NH2 groups in molecule structure, such as on the Nile blue structure, which may change the lifetime values. Grofcsik et al [34] studied these probes in different solvents where it was observed that there is a relationship between the solvent permissivity and the excited state lifetime. It was shown that the lifetime is higher in nonpolar solvents, where protic solvents decreases the excited state lifetime. This behavior was observed in both dye molecules that were studied, which have a similar chemical structure.

Grofcsik et al [34,35] have shown that there is a relationship between the excited state lifetime of Nile blue and Oxazine 720 with the acidity of the medium. As the hydrogen ion concentration increases it is observed a decrease of the excited state lifetime [40]. It was also observed for methylene blue, azure A, azure B and azure C [41]. The reason for the rapid decay in acid medium is due to the formation of dications from the monocations reaction in the excited state with hydrogen ions. These results indicate that the reaction in the excited state the additional protons are located on the nitrogen atom of the ring and not on the terminal amine groups [40]. It is believed that for other compounds the results may be the same due to the similarity in the chemical structure of such molecules.

Dutt et al [42] studied the fluorescence lifetime of cresila violet, Nile blue, oxazine 720 and Nile red, using different solvents, such as alcohols, polyalcohols, amides and some aprotic solvents. The authors showed that the lifetime values for these dyes are approximately 3.5 ns for n-alcohols, which are higher than that for the Nile blue (1.62 ns in ethanol). This result is in agreement with our studies. When it is considered the behavior of bipolar solute in polar solvents, the hydrodynamic and dielectric contribution must be taken into account [42]. However, it is not well known how to measure these hydrodynamic and dielectric contributions individually. In the case of the four dyes, when in the presence of amides and aprotic solvents, as described above, the contributions are reasonably described by the hydrodynamic friction, where to describe the rotating relaxation in the presence of n-alcohols; the dielectric friction must be included.

Chen et al [43] studied the quantum yield of the methylene blue singlet oxygen as a function of the medium pH values. The authors showed that the protonated acid (3MBH2+) triplet state is similar to the base (3MB+) triplet state, and the quantum yield of the singlet oxygen formed is much higher in basic medium than that in acidic medium. The singlet oxygen formation increases as the pH of the medium is increased, while the singlet state lifetime decay the triplet state formation do not depend on the pH changes. It can be explained by the population decay rate of the singlet state due to the internal conversion to the fundamental state, and the intersystem crossing to the triplet state, which are much higher that the protonation rate [43,44]. Also [43] studied the behavior of methylene blue, 1,9-dimethyl-methylene blue and toluidine blue in aqueous medium and methanol. The triplet state formation and the singlet oxygen quantum yield in water were very similar to that for methylene blue and for 1,9-dimethyl-methylene blue. The kinetic studies results for the singlet state decay of methylene blue in water and in methanol were 0.37 and 0.62 ns, respectively, where for toluidine blue the results were 0.28 and 0.40 ns, respectively. In the case of methylene blue the decay useful life of the singlet excited state in methanol is approximately two times higher than in water. The authors showed that there is no influence of the solution concentration on the singlet state lifetime, where the differences on the lifetime decays that were observed in water and methanol are not related to the methylene blue dimerization in water. The methylene blue lifetime decay decreases with the increase of the dielectric constant of protic solvents due to the interaction of the methylene blue with the polar solvent [45]. In protic alcohols and in aqueous solutions the methylene blue excited state lifetime is higher than of the fundamental state. Therefore, the differences between the singlet and triplet states decrease as the relaxation rate is increased. In the presence of aprotic solvents, such as acetone, acetonitrile, and dimethyl sulfoxide, the dipole excited state is lower in the fundamental state, where the energy differences observed is higher and the relaxation lifetime is longer [46].

The use of these dyes as singlet oxygen photosensitizer in PDT, as well as tumor cells removal are being investigated [47,48,49,50]. It is known that under laser irradiation in the presence of photosensitizer dyes, the tumor cells undergo necrosis or apoptosis and the rate of tumor cell removal through apoptosis increases [51,52,53]. This behavior has been related to the presence of singlet oxygen in the tumor cells [54,55]. The increase of cell removal through apoptosis is of great importance in the PDT treatment [50,56]. There are no side effects in the cell removal through apoptosis because it is a controlled cell removal process, where there is no inflammation of the laser irradiated tissue. In some cases changes in the PDT mechanism has been observed, type I via free radical and type II via oxygen singlet, which could be related to the interaction among the dyes and the cellular system [57,58,59,60]. These changes involve the aggregation of two or more dye molecules in the same site [61].

2. Photodynamic Therapy (PDT): Mechanism of action

Selective tumor destruction without damaging surrounding healthy tissues can be reached by using PDT, which is treatment, activated by light, which requires the combination of three elements: a photosensitizer, visible or near-infrared light, and oxygen [62,63,64,65,66,67,68,69]. However, the precise mechanisms of PDT are not yet fully understood but two general mechanisms of photoinduced damage in biomolecules have been proposed: Type I and Type II [62,70,71]. Type I is the photodynamic mechanism in which the excited molecule induces radical formation that causes damage to biological targets (membranes, proteins and DNA), and an electron transfer event is the initial step [16]. In Type I mechanism, the photosensitizer in the excited state interacts directly with a neighbor molecules, preferentially O_2, producing radicals or radical ions through reactions of hydrogen or electron transfer [72]. Frequently, these radicals react immediately with the O_2 generating a complex mixture of reactive oxygen species (ROS), such as hydrogen peroxide, superoxide radical and hydroxyl radical, which are capable to promote oxidation a great number of biomolecules [62].

It is believed that $^1O_2$ produced through type II reaction is primarily responsible for cell death. It is known that several factors including the PS, the subcellular localization, the substrate and the presence of O_2 contribute to this process [71]. The lifetime of $^1O_2$ is very short (approximately 10-320 nanoseconds), limiting its diffusion to only approximately 10 nm to 55 nm in cells [73]. Type I photoreaction of some PSs are primarily responsible for sensitization through radical formation under hypoxic conditions. In the presence of oxygen, $^1O_2$ mediates photosensitization process, but the supplemental role of H_2O_2, $OH\bullet$ and $O_2\bullet$ must also to be considered. Only substrates situated very close to the places of ROS generation will be firstly affected by the photodynamic treatment because the half-life of $^1O_2$ in biological systems is under 0.04 μs and its action radius being lower than 0.02 μm [71]. This assumption is due to the fact that ROS are highly reactive and present a very short half-life. Type II is the photodynamic mechanism in which the photooxidation is mediated by singlet oxygen ($^1O_2$), where an energy transfer reaction from the photoexcited molecule to molecular oxygen is the initial step [16,62]. The process involves the excitation of the photosensitizer from a ground singlet state to an excited singlet state, where intersystem crossing to a longer-lived excited triplet state will occur. It is also important to point out that molecular oxygen is present in tissue with a ground triplet state. When the photosensitizer and an oxygen molecule are in proximity, an energy transfer can take place that allows the photosensitizer to relax to its ground singlet state, and create an excited singlet state oxygen molecule. Additionally, energy is transferred from triplet protoporphyrin IX to triplet oxygen, resulting in singlet ground state protoporphyrin IX and excited singlet oxygen, which reacts with biomolecules, which can damage some cells in the treatment area. Singlet oxygen is the usual name associated to the three possible excited electronic states immediately superior to the ground state of molecular oxygen (triplet oxygen) [3].

Due to the short half-life and diffusion distance of singlet oxygen in aqueous media, PDT can be considered a highly selective form of cancer treatment, as only the irradiated areas are affected, provided that the photosensitizer is nontoxic in the absence of light [74]. This combination of light/photosensitizer/oxygen as a mode of disease treatment has expanded from an initial focus on cancer tumors to include application in certain non-neoplastic diseases including age-related macular degeneration (AMD), coronary heart disease, periodontal diseases, and microbial infections [75].

Singlet oxygen is a very aggressive chemical species and will very rapidly react with any nearby biomolecules, being that the specific targets depend directly on the physical-chemistry properties of the photosensitizer used in the photodynamic process, which will result in no desired side effects, such as destructive reactions that will kill cells through apoptosis or necrosis. Therefore, depending on whether Type-I or Type-II mechanisms take place, the therapeutic efficiency of PDT may be completely altered. Therefore, the ratio of apoptotic versus necrotic cell death in tumors treated with PDT may depend on the competition between electron and energy transfer in the reaction site [16].

Oxidative stress generated by the photodynamic action occurs because in biological systems the singlet oxygen presents significantly low lifetimes, where the lifetimes of the singlet oxygen is lower than 0.04 μs, implying that its radius of action is also reduced, being usually lower than 0.02 μm [3]. Reactive oxygen species (e.g. hydroxyl radicals or superoxide) are their high reactivity and low specificity with a broad spectrum of organic substrates [76]. Various methods have been employed for the generations of hydroxyl radicals such as O_3/UV, H_2O_2/UV, TiO_2 photo-catalysis and photo assisted Fe(III)/H_2O_2 reaction.

3. Photosensitizers

3.1. Phenothiazinium dyes

The phenothiazinium dyes were first synthesized in the late 19th century—e.g. both Methylene Blue (Caro) and Thionin (Lauth) in 1876—during what might be considered to be a "gold rush" period of chemical experimentation after the discovery of the first aniline dyes [77]. Among photobactericidal compounds, the phenothiazinium photosensitizers methylene blue (MB) and toluidine blue O (TBO) have often been used as lead structures, being effective photosensitizers with singlet oxygen quantum yields of approximately 0.40

and exhibiting low toxicity levels in mammalian cells [14]. Members of the phenothiazine class are known to cross the blood-brain barrier and to be relatively nontoxic [78,79].

The biomedical use of phenothiazinium dyes has begun with specimen staining for microscopy by various medical scientists, among whom were famous scientists such as Romanovsky, Koch and Ehrlich. The idea of structure—activity relationships in stains developed in this era, particularly by Paul Ehrlich, laid the foundations for modern medicinal chemistry, and these principles should be followed by those attempting the properly organized photosensitizer synthesis [77]. Cellular uptake is determined by a combination of charge type/distribution and lipophilicity, both of which characteristics may be controlled by informed synthesis. Due to the expansion of PDT into the antimicrobial milieu, a far greater scope for photosensitizer design exists now. For example, in the field of blood product disinfection, an ideal candidate photosensitizer would be effective in the inactivation of bacteria, viruses, yeasts and protozoan, but would remain non-toxic and non-mutagenic in a human recipient. It is hardly surprising that none of the currently available agents fits all of these criteria [77].

Phenothiazinium dyes are cationic compounds with high redox potential that interacts with visible light inactivating several kinds of pathogenic agents in fresh plasma. Phenothiazinium dyes present great reactivity with the proteins and lipoproteins (cell membranes) and nucleic acids. These cationic compounds have limited capability to permeate the cell membrane as function of their elevated hydrophilic character [80]. Phenothiazinium dyes present significant action against encapsulated virus and some virus without capsule, such as parvovirus B19. As function of its genotoxic action, the employment of phenothiazinium dyes is prohibited in several countries, such as Germany[80]. On the other hand, the Methylene Blue is a highly hydrophobic compound with higher chemical affinity to the nucleic acids, which denotes its potential to application against virus.

Phenothiazinium dyes are photocytotoxic, and can cause photoinduced mutagenic effects [81]. In living systems, DNA acts as an important target for phenothiazinium dyes. It has been proved that these dyes can photosensitize biological damage. Azure B (AZB) is an easy available phenothiazinium dye, and has been widely employed both in metal determination and DNA staining detection. Owing positive charges on its molecular structure, AZB can bind to the DNA polyanion in living systems through electronic interactions. So, the study of the interaction of AZB with DNA in vitro is of importance.

Methylene Blue, MB (Figure 2) is a phenothiazinic dye current applied in PDT as therapeutic agent or photosensitizing compound. MB has a recognized antimicrobial effect in the dark (citotoxicity property) which can be increased, at oxygenated environment, by the incidence of light with a wavelength corresponding to its electronic absorption band [82,83].

Methylene Blue is a well-known photochemical oxidant. The photoreduction reaction of this dye by various types of electron donors has been studied quite often, and in most cases an electron transfer mechanism was proposed for explaining the observed results [84].

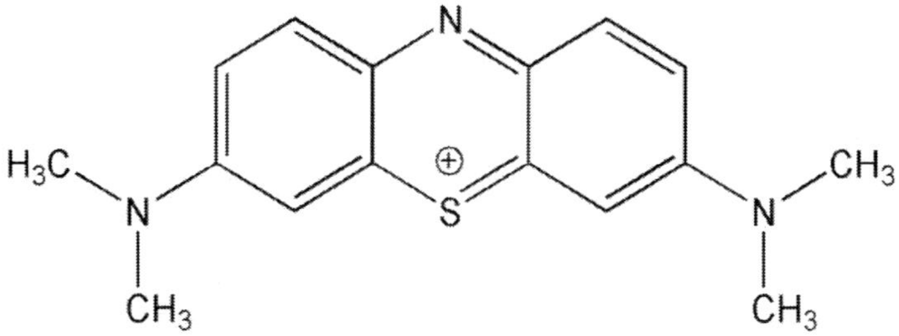

Figure 2. Structure of Methylene Blue

This molecule is particularly interesting for application in PDT due to its known physical chemical properties. For example, MB is a positive charged dye with three aromatic rings (6-members) very soluble in ethanol. It is already used clinically in humans for the treatment of metahemoglobinemia, without significant side effects. Besides these characteristics, MB presents a quantum yield of singlet oxygen formation around 0.5, with a low reduction potential, intense light absorption in the region of 664 nm in water (within the phototherapeutic window). Also, it displays a high photodynamic efficiency causing apoptosis of cancer cells, by mono or polychromatic light excitation). Currently, MB is used by several european agencies for disinfection of blood plasma, due to its efficiency in photodynamic inactivation of microorganisms such as viruses [85], including HIV, hepatitis B and C [86,87].

MB has been clinically used as a photosensitizer drug for PDT in the treatment of different types of tumors [88]. Phototherapeutic application examples include treatment of bladder cancer, inoperable esophagus tumors, skin virulence, psoriasis and adenocarcinomas [89]. Additionally, an important point to be considered is the extremely low cost of a treatment based on this dye compared with other available photo-drugs.

Although MB possesses a positive charge and the planar structure with delocalized charge, it has a tendency to form dimers, trimers or type H aggregated systems in the presence of certain additives, cell organelles or solvents, for example, water [90,91,92,93]. The development of self-aggregates compromises its photodynamic activity, impairing the production of singlet oxygen, principal phototoxic species in PDT. In self-aggregated states auto-quenching processes occur where the excited monomers have the energy suppressed by collisions with other monomers that constitute the aggregate [94,95,96,97,98].

Often, treatment protocols require unusual preparation methods, or conditions that may have many distinct characteristics of the most ideal conditions. One example is that the MB in diluted aqueous solution, with concentration around $2x10^{-5}$ mol L^{-1}, is found in monomeric form. However, its uses in topical treatments require concentrations higher than $6x10^{-2}$ mol L^{-1}, where self-aggregation and its consequences are significant [82].

Therefore, it is important to investigate the phenomena of MB self-aggregation present in solvent mixtures and / or interaction with biomolecules [90]. This study aims to investigate changes in MB spectroscopic properties caused by self-aggregate formation induced by solvent mixtures.

The MB is an oxazinic dye soluble in water or alcohol. It presents a quantum yield of oxygen singlet formation of about 0.5 and low reduction potential [25]. It is a dye with low toxicity, which absorb in the UV-visible light ($\lambda_{máx}$ = 664 nm; solvent: water) and shows good photodynamic efficiency to kill cancer cells, which can be excited by monochromatic and polychromatic light within the therapeutic window [82]. It is a hydrophobic dye, which forms aggregates when in the presence of aggregation agents such as polyelectrolyte, or when in the presence of solvents that induces the aggregation process. The aggregate formation changes photosensitization efficiency, decreasing the amount of singlet oxygen produced by light stimulation. The most important application of methylene blue (MB) is its use in PDT as a photosensitizer agent, in oncology and potentially in the treatment of other diseases, such as Leishmaniosis.

Teichert et al.[99] used *Candida albicans* strains that are resistant to the conventional treatment of *Candida* infections, which were collected from HIV-positive patients. These strains were inoculated in the oral cavity of rats that, subsequently, were submitted to the topic application of 1 mL of Methylene Blue at concentrations of 250, 275, 300, 350, 400, 450 and 500 μg mL^{-1}. After 10 minutes of dye application, the authors employed the diode laser with wavelength of 664 nm with potency of 400 mW (687.5 seconds), resulting in an energetic density of 275 J/cm^2 [100]. After one unique application, it was realized microbial culture exam of the respective samples and the individuals were sacrificed to the histological analysis of the tongue. The results obtained in this procedure demonstrated a complete elimination of the microorganisms, when the dye concentrations of 450 e 500 μg mL^{-1}were employed. In the histological analysis, the rats that were treated with PDT had no inflammatory signals. The tongues of the control group rats presented high level of infection by *Candida* which was located in the keratin layers [100]. The respective authors concluded that the PDT is a potential alternative to the treatment of the fungi infection, emphasizing, as advantages of this technique, its topic character, simple methodology and, mainly, the unspecific characteristic of PDT, i.e., the possibility of to be applied to a great number of microorganisms. Moreover, PDT can be applied several times without risk of selection of resistant yeasts [100].

Azures A, B and C, are examples of photosensitizer agents, which have the cationic derivatives, such as the Azure Bf4. The organic ions can interfere through fluorescent radiation absorption that is emitted by excited molecules, resulting in a photobactericidal effect on the *Staphylococcus aureus* and *Enterococcus faecium* colonies. This behavior is related to the light stimulation wavelength because the organic compounds present in the system absorb electromagnetic radiation. However, only organic compounds that present double bond conjugated system, such as azure A, B or C, are capable to absorb the visible light radiation.

It was observed that red visible light (600-700 nm) and nearinfrared are the wavelengths that can penetrate the human skin. The phenothiazinic dyes, such as Azures, absorb light in such wavelengths with high intensity. They show the formation of aggregates due to the presence of aggregation agents such as polyelectrolytes, or due to the presence of solvents that favors the aggregate formation, such as water. The aggregate formation changes the photosensitization process efficiency, decreasing the amount of singlet oxygen produced by the light stimulus. The self-aggregation phenomenon can be minimized by adding charged groups in the dye structure, which results in an electrostatic repulsion interaction, increasing the hydrophilic behavior of the dye, such as Azure B and Azure BF_4.

Azures are phenotiazine compounds. This class of dye has low toxicity in the dark, constant composition, being synthesized with high yield. Azures present great selectivity to the tumor cells and significant photo stability, being not maintained in the body for long interval of time. These dyes can be applied through endovenous and topic ways. Azures present high bactericide ability, being very auspicious compounds to be applied as photosensitizes in PDT, especially due to their favorable photodynamic properties and low cost [[101],[102]]. Azure dyes (including Azure B) are recalcitrant compounds used in the textile industry. For instance, Azure B has been used in a selective assay for detecting lignin peroxidase, the oxidative enzyme with the highest redox potential produced by white-rot fungi [[103],[104]].

Azure B is a very sensitive dye and extremely susceptible to detect slight alterations in its chemical environment, presenting significant solvatochromic processes. Physico-chemical properties of Azure B have motivated the employment of Azure B as a chromogenic reagent for the spectrophotometric determination of several compounds, which are relevant to biological and environmental chemistry such as periodate [[105]]. This cited method is simple and rapid, offering advantages of sensitivity and wide range of determinations, without involvement of any stringent reaction conditions, being successfully applied to the determination of periodate in solution and in several river water samples. In its time, Azure-C (AZC), and related phenothiazine compounds has been widely used for accelerating the oxidation of NADH, but not in connection to the NAD+ reduction process.

Thionine has been a subject of many studies, as for example in a photochemical and electrochemical biosensor [[106],[107],[108],[109],[110]] and in photovoltaic cells [[111]]. Thionine is a positively charged tricyclic heteroaromatic molecule, which has been investigated for its photoinduced mutagenic actions [[112],[113]], toxic effects, damage on binding to DNA [[114]] and photoinduced inactivation of viruses [[115]]. Thionins consist of 45–47 residues bound by three to four disulfide bonds, which includes α1-purothionin, βPTH, and β-hordothionin (βHTH) [[116],[117],[118]].

It has the ability to immobilize proteins and DNA and act as molecular adhesive [[119]]. Biophysical and calorimetric studies with three natural DNAs of varying base compositions, have shown the intercalative binding and high affinity of thionine to GC rich DNAs [[120]]. Thionine presented a high preference to the alternating GC sequences followed by the homo GC sequences contained in different synthetic polynucleotides [[121]]. AT polynucleotides presented a lower binding affinities but the alternating AT sequences had higher affinity compared to the homo stretches. The intercalation and the sequence of specific intercalative binding of thionine were shown by fluorescence, viscosity experiments and circular dichroic studies, respectively [121].

Studies based on absorbance, fluorescence, circular dichroic spectroscopy, viscosity, thermal melting and calorimetric techniques were used to understand the binding of thionine, with deoxyribonucleic acids of varying base composition, where strong binding of thionine to the DNAs were shown. Strong hypochromic and bathochromic effects and quenching of fluorescence were observed that showed strong binding of thionine to the DNAs [97]. The binding process is exothermic, which is associated to a large positive entropy changes and a negative enthalpy, and it showed that nonelectrostatic contributions are very important for the association of thionine to DNA. Studies on the interaction of thionine with sodium dodecylsulfate (SDS) micelles have shown that thionine binding affinity to SDS micelles was decreased with increasing temperature due to the thermal agitation [122].

The spectroscopic characteristics of thionine aggregates have shown that it depends on the concentration of thionine and on the chemical nature of the solvent [123]. Two peaks can be observed, at 597 nm and at the lower wavelength side of the 597 nm peak, and they related to the monomeric species and to the aggregate formation, respectively [124]. The understanding of the thionine aggregation process is very important for some application, such as in photovoltaic cells, where the reverse homogeneous redox reaction can be inhibited due to the presence of a surfactant in the system. The presence of a surfactant interferes in the thionine aggregation and polymerization process [125].

Several works about sensors have shown that the changing of spectroscopic and electrochemical properties of organic molecules, such as Toluidine Blue O (TBO), a phenothiazine dye, may indicate that there are some interaction with mediators and biological molecules [126,127,128]. Photochemical and electrochemical properties of TBO have been used to develop new photovoltaic devices for energy conversion and storage [129]. The aggregation behavior of such dyes in phase solution can be studied by using several optical rotation and circular dichroism techniques, as it can be seen in some studies of the interaction of TBO molecules on the DNA surface [130]. It was suggested that both intercalative and electrostatic interactions of TBO with DNA, where it was pointed out that the electrostatic interaction play an important role on the formation of the bridged structure of TBO with DNA [131].

TB can also be used as an oxygen radical inactivation, biological sensitizer and complexing agent in biological systems avoiding pathological changes [132]. Due to its low toxicity and high water solubility in salt form, which has an intense absorption peak in the visible region [133], it has been used in pharmaceutical formulations for cancer treatment [134]. Studies on the micellar solutions have shown that the aggregation properties and distribution behavior of toluidine blue in the presence of surfactant depend on the electrostatic interaction. In the case of surfactin, a natural surfactant, TB molecules can be located in the palisade of surfactin micelle [135].

Nile blue (NB) belongs to a class of molecules whose basic framework is that of a benzophenoxazine, a class which also includes Nile red, a phenoxazinone, here termed red

Nile blue (RNB) and Meldola's Blue. It has been found to be localized selectively in animal tumors [136] and can retard tumor growth [137,138]. NB has been used as a photosensitizer for oxygen in PDT applications [139,140], in processes that depend on solvent polarity [141,142], as a stain for Escherichia coli in flow cytometry [143], as a DNA probe [144] and many other applications [145,146,147,148]. Due to their high fluorescence quantum yield together with their solvatochromism, they have been used as stains and imaging agents. These dyes present relatively low solubility in aqueous medium as well as their fluorescence is reduced significantly in in the presence of polar medium, which opens up new possibilities to develop aqueous analogues of these benzophenoxazines [149]. Together with the increase of the solubility in water, it is believed that the self-assembly process to form aggregates can be disrupted resulting in an enhancement of the fluorescence intensity [150].

NB shows thermochromic and solvatochromic behavior in its ultraviolet/visible spectra [151]. The variation in the absorption spectrum is due to the equilibrium between the monocation and the neutral molecule, where the monocationic form is the more stable in most solvents. In strong basic conditions the neutral form is observed, where in strong acidic conditions the dicationic and tricationic forms can be observed [152]. The fast decay processes study can be used to get information on the effect of medium condition, basic and acidic, on determining the excited state lifetime on the picosecond scale. It was shown that the reason for the faster decay in acidic conditions results from the formation of dications by reaction of excited state monocations with hydrogen ions [153].

Despite the photophysics of NB in pure solvents is well characterized in literature, the NB interaction with microheterogeneous systems, such as micelles, reverse micelles (RMs) and DNA is still not well understood. Electrochemical studies have shown that NB-DNA duplexes modified microelectrode can be used as a rapid and sensitive method to detect TATA binding to DNA in the presence of other proteins [154]. However, there are no many works done on its interaction with DNA [155,156,157,158]. In a work done on the interaction of NB with biomimicking self-organized assemblies (SDS micelles and AOT reverse micelles) and a genomic DNA (extracted from salmon sperm) (SS DNA), it has been shown that there are two different binding modes of NB with genomic DNA, electrostatic and intercalative modes [144]. There was no explanation for the mechanism related to these interaction modes. The electrostatic mode is believed to be responsible for electron transfer between the probe and DNA, which may result in a quenching process of the NB fluorescence emission intensity when in the presence of low concentration of DNA. The intercalative mode is believed to be the subsequent release of quenching due to the intercalation of the dye in DNA base pairs. In another study, it was shown that binding affinity of the probe is higher with SDS micelles than with the DNAs within its structural integrity in presence of the micelles. The complex rigidity of NB with various DNAs and its fluorescence quenching with DNAs has shown a strong recognition mechanism between NB and DNA [159].

NB was immobilized in two different surfaces, a nonreactive surface (SiO_2), with its conduction band at much higher energies, and a reactive surface (SiO_2), with a conduction band situated at lower energies. The former is used to directly probe the excited-state dynamics of the dye undisturbed by other competing processes. The latter is used to study the charge injection process from the excited dye into the semiconductor nanocrystallites, acting as an electron acceptor. The transient absorption measurements of NB adsorbed on SiO_2 colloids (inert support) show that the NB aggregates have a relatively short-lived excitonic singlet state (τ = 40 ps) (Table 1). The lifetime of the excited singlet of the monomer in aqueous solution is ~ 390 ps. NB aggregates that were immobilized on reactive surface also inject electrons into SnO_2, resulting in the formation of the the cation radical, $(NB)_2^{\cdot+}$, of the NB aggregates and by the trapping of electrons the in SnO_2 nanocrystallites. The monophotonic dependence of the formation of $(NB)_2^{\cdot+}$ on SnO_2 surface supports the charge transfer from NB aggregates to SnO_2. The rate constant for this heterogeneous electron transfer process is ~ $3.3.10^8$ s^{-1} [[160]].

4. Aggregation of photosensitizers and its influence in PDT

Most of these dyes form aggregates in the ground state [[161],[162],[163]], even when the dye concentration is low (approximately 10^{-6} M) and in the presence of salts and aggregation inducing agents, such as anionic micelles, heparin, polyelectrolyte, liposome and vesicles. The planar structure of such dyes is a key factor that contributes to the approaching and dimerization of the dyes [[164],[165]].

The presence of hydrophobic ligands in the dye structure facilitates the aggregate formation in polar medium. The effects of the planar structure of the dye, hydrophobicity and the interaction with cell membranes were observed in photosynthetic systems II of plants [[166]] and other systems [[167]]. Some studies have shown that the interaction among phenothiazines and cyclodextrins results in the aggregate formation with different sizes depending on the cyclodextrins cavity size [162,[168]].

Studies that were conducted previously have shown that methylene blue molecules form aggregates and the photophysical behavior changes depending on the ground state aggregation. It results in a decreasing of the fluorescence intensity and on the singlet oxygen formation [49]. These studies have shown that the interaction with micelles is responsible for the dimerization process and not the interaction with monomer of the surfactant, as it has been postulated in some works [[169]]. In this stage of the work it is important to study the nature of the aggregates formed in different negatives interfaces and in biological systems, more specifically in micelles, vesicles and mitochondria.

It is well known that dimerization and medium composition effects changes the energy transfer process among triplets species and molecular oxygen and other triplet suppressors [[170],[171],[172],[173],[174],[175],[176],[177],[178]]. Some studies carried out using thionine and MB have shown some of these effects [[179]]. Azure A, azure B, thionine e MB are dimerized with different dimerization constants.

The aggregation of ionic dyes cannot be assigned to a specific type of chemical interaction. There is a significant contribution of several influences, such as van der Waals interactions, intermolecular hydrogen bounds and pi-electrons interactions, being that, frequently, it is not trivial to evaluate the specific contribution of each one of these interactions [[180]].

The quantum behavior of extended aggregates of atomic and molecular monomers, containing from a just a few up to thousands of subunits, is attracting increasing attention in chemistry and physics, being that proeminent examples are aggregates of large dye molecules, chromophore assemblies describing the photosynthetic unit of assemblies of ultra-cold atoms [181].

According to their structure, dyes, such as phenotiazinium, exhibit J- or H-aggregates, which present very typical J- or H-absorption bands [182]. The aggregate absorption band is red-shifted in relation to the monomer absorption. These are the J-aggregates showing a very narrow band whose position is well-predicted by a theory ignoring intramolecular vibrations. By contrast, other dyes showed a shift towards the blue (i.e. higher absorption energies) and were termed H-aggregates (hypsochromic shift). Unlike the J-band, the line shape of the H-band generally shows a rich vibrational structure and has a width of the order of the monomeric band [183]. The J-band is polarized parallel to the rods, while the H-band is polarized perpendicularly to the rod long-axis [184].

Self-organized J-aggregates of dye molecules, known for over 60 years, are emerging as remarkably versatile quantum systems with applications in photography, opto-electronics, solar cells, photobiology and as supramolecular fibres [185].

5. Future perspectives

Photodynamic Therapy has been used in clinical applications with significant success. Several studies have focused on the suitable conditions to improve the clinical results, such as the optimization of the incident light intensity. Indeed, the importance of irradiance is a determinant of PDT-induced pain. The increased use of low irradiance PDT may have a considerable impact on pain, which currently is the main limiting factor to successful delivery of PDT in some patients [186].

Another area of improvement of the PDT application is focused on the increase of the aqueous solubility of the photosensitizers. In fact, the photosensitizers require being suitable to several types of administration in biological medium. In this context, interesting photosensitizers that present low water solubility, such as C_{60}, constitute an area of scientific efforts. C60 can be accumulated selectively in the target point. However, the biological application of C60 is limited due to its poor solubility in water [187]. To improve the solubility of C60 in water, several water-soluble derivatives have been synthesized. Furthermore, other solubilization methods for C60 have been explored using cyclodextins, calixerenes, micelles, liposomes, and poly(N-vinyl-2-pyrroridone) (PNVP). In general, core-shell polymer micelles can be formed spontaneously by amphiphilic diblock copolymers due to association between hydrophobic blocks in water. The hydrophobic drugs can be incorporated into the hydrophobic core of the polymer micelle, and thus, the drugs can be solubilized in water. Nanosized water-soluble core-shell type polymer micelles can allow long circulation in the blood stream avoiding reticuloendothelial systems (RESs) and can be utilized for their enhanced permeability and retention (EPR) effect at solid tumor sites.

The production of ROS can be affected by factor, such as the aggregation and photobleaching of the photosensitizer. In fact, photosensitizers such as, for example, magnesium protoporphyrin (MgPpIX), have demonstrated that the aggregation and photobleaching reduce the photodynamic efficiency [188].

Low-level laser therapy has been used to speed up healing process of pressure ulcers due to its antiinflammatory, analgesic, anti-edematous, and scarring effects, since there is no consensus on its effect on infected ulcers [189]. It is an interesting topic to be evaluated in novel studies.

It is known that Gram positive bacteria are more sensitive to PDT as compared to Gram negative species. However, the use of cationic photosensitizers or agents that increase the permeability of the outer membrane allows the effective killing of Gram negative organisms [190]. Some photosensitizers have an innate positive charge, but some approaches are focused on to link photosensitizers to a cationic molecular vehicle, such as poly-L-lysine [190].

Photodynamic therapy has been also applied in dentistry, in endodontic treatments, with auspicious results regarding the control of microbial infections associated to this type of odontologic therapy [191].

The increasing application of PDT has motivated the development of other therapeutic techniques, with similar principles. We can mention the case of Sonodynamic Therapy (SDT). In 1989, Umemura and co-workers first pioneered the development of non-thermal ultrasound activating a group of photosensitizers for treating tumor, which is called Sonodynamic Therapy (SDT) [192]. They reported that the photosensitive compounds activated by ultrasound can kill cancerous cells and suppress the growth of tumor. Otherwise, they also thought highly of that the ultrasound could reach deep-seated tumor and maintain the focus energy in a small volume because of exceedingly strong penetration ability and mature focusing technology [193]. Particularly, SDT was developed from the well-known PDT but only put up low phototoxicity. Therefore, in recent years, along with the lucubration the SDT has attracted considerable attention and has been considered as a promising tumor treatment method [192].

Regarding the development of photosensitizers, it is important to register the relevant role of phthalocyanines. Phthalocyanines (Pcs) are highly delocalized p-conjugated organic systems and exhibit wide variety of roles in a various high technological areas such as semiconductor devices, liquid crystals, sensors, catalysts, non-linear optics, photovoltaic solar cells and PDT [194,195]. They are among the most important promising chemical compounds by advantage of their stability, photophysical, photochemical, redox and coordination properties. The properties of Pcs depend on their molecular composition with the number; position and nature of substituents and type of central metal play an important role in controlling their properties [194].

Indeed, Pcs have many considerable physical and chemical features, which have motivated the interest of several investigators because of their physico-chemical properties [195]. The presence of different substituents on the Pc ring also leads to increased solubility and supramolecular organizations with improved physicochemical characteristics, depending of the interest in terms of application [194]. In fact, phthalocyanines are very versatile chemical systems, which allow great variability of adjustment of properties in the process of chemical synthesis. This great number of structural possibilities has been utilized in many fields, since the different phthalocyanines can be applied in quite different areas, such as gas sensors, semiconductor materials, photovoltaic cells, liquid crystals, optical limiting devices, molecular electronics, non-linear optical applications, Langmuir-Blodgett films, fibrous assemblies and PDT [195].

6. Conclusions

The selection of photosensitizers that are more able to generate an efficient photodynamic action is one of the main questions involving PDT in the present days. The novel generations of photosensitizers is aiming to obtain the maximum quantum yield through the therapeutic window, avoiding spectral ranges that are absorbed by endogenous dyes, such as hemoglobin and melanin. In this way, the previous knowledge regarding the spectroscopic behavior of the new prototypes of photosensitizers is a relevant pre-requisite to the advancement of the types of applications and its respective repercussions, since the efficacy of the methodology depends on the capability of generations of reactive oxygen species (ROS) and reactive nitrogen species (RNS), which are intrinsically related to the optical profile of the photosensitizers.

Acknowledgment

Hueder Paulo Moisés de Oliveira thanks to the financial support propitiated by FAPESP (Project of research support 06/56701-3) and to CNPq to the research grants (479655/2008-1). Thanks to Msc Sandra Cruz dos Santos and Luiza Rosimeri Romano Santin to the revisions. Thanks also FFCLRP-USP, Prof. Amando Siuiti Ito's laboratory.

The author Máira Regina Rodrigues thanks CNPq and FAPESP for the financial support.

Author details

Leonardo M. Moreira
Departamento de Zootecnia (DEZOO), Universidade Federal de São João Del Rei (UFSJ), São João Del Rei, MG, Brazil

Juliana P. Lyon
Departamento de Ciências Naturais (DCNAT), Universidade Federal de São João Del Rei (UFSJ), São João Del Rei, MG, Brazil

Ana Paula Romani
Departamento de Química - Instituto de Ciências Exatas e Biológicas, Universidade Federal de Ouro Preto, Campus Morro do Cruzeiro, Ouro Preto , MG, Brazil

Divinomar Severino
Instituto de Química, Universidade de São Paulo, São Paulo, SP, Brazil

Maira Regina Rodrigues
Universidade Federal Fluminense, Polo Universitário de Rio das Ostras, Rio das Ostras, RJ, Brazil

Hueder P. M. de Oliveira*
Centro de Ciências Químicas, Farmacêuticas e de Alimentos,
Universidade Federal de Pelotas, Pelotas, RS, Brazil.

7. References

[1] Raab O (1900) Uber die Wirkung, fluorescirender Stoffe auf infusorien. Z. Biol. 39: 524-546.
[2] Manoto S L, Abrahamse H (2011) Effect of a newly synthesized Zn sulfophthalocyanine derivative on cell morphology, viability, proliferation, and cytotoxicity in a human lung cancer cell line (A549). Lasers Med Sci 26: 523–530.
[3] Machado A E H (2000) Terapia Fotodinâmica: Princípios, potencial de aplicação e perspectivas. Quim. Nova 23: 237-243.
[4] Simplicio F I S, Maionchi F, Hioka N (2002) Photodynamic Therapy: Pharmacological aspects, applications and news from medications development. Quim. Nova 25: 801-807.
[5] Almeida R D, Manadas B J, Carvalho A P, Duarte C B (2004) Intracellular signaling mechanisms in photodynamic therapy. Biochim. Biophys. Acta 1704: 59-86.
[6] Agostinis P, Buytaert E, Breyssens H, Hendrickx N (2004) Regulatory pathways in photodynamic therapy induced apoptosis. Photochem. Photobiol. Sci. 3: 721-729.
[7] Mitton D, Ackroyd R (2008) A brief overview of photodynamic therapy in Europe. Photodiagn. Photodyn. Ther. 5: 103-111.
[8] Dougherty T J, Gomer C J, Henderson B W, Jori G, Kessel D, Korbelik M, Moan J, Peng Q (1998) Photodynamic therapy. J. Natl. Cancer Inst. 90: 889-905.
[9] Tardivo J P, Giglio A D, Paschoal L H, Baptista M S (2006) New photodynamic therapy protocol to treat AIDS-related Kaposi`s sarcoma. Photomed. Laser Sug. 24: 528-531.
[10] Peloi L S, Biondo C E G, Kimura E, Politi M J, Lonardoni M V C, Aristides S M A, Dorea R G C, Hioka N, Silveira T G V (2011) Photodynamic therapy for American cutaneous leishmaniasis: The efficacy of methylene blue in hamsters experimentally infected with Leishmania (Leishmania) amazonensis. Experimental Parasitology 128: 353-356.
[11] Biyani N, Singh A K, Mandal S, Chawla B, Madhubala R (2011) Differential expression of proteins in antimony-susceptible and -resistant isolates of Leishmania donovani. Mol. Biochem. Parasitology 179: 91-99.
[12] Torezan L, Niwa A B M, Neto C F (2009) Photodynamic therapy in dermatology: basic principles and clinical use. An. Bras. Dermatol. 84: 445-459.

* Corresponding Author

[13] Bagazgoitia L, Santos J C, Juarranz A, Jaen P (2011) Photodynamic therapy reduces the histological features of actinic damage and the expression of early oncogenic markers. British Association of Dermatologists 165: 144–151.

[14] Wainwright M, (2007) Phenothiazinium photosensitisers: V. Photobactericidal activities of chromophore-methylated phenothiazinium salts. Dyes and Pigments 73: 7-12.

[15] Prates R A, Kato I T, Ribeiro M S, Tegos G P, Hamblin M R (2011) Influence of multidrug efflux systems on methylene blue-mediated photodynamic inactivation of Candida albicans. J Antimicrob Chemother 66: 1525–1532.

[16] Severino D, Junqueira H C, Gugliotti M, Gabrielli D S (2003) Baptista M S Influence of Negatively Charged Interfaces on the Ground and Excited State Properties of Methylene Blue. Photochem. Photobiol. 77: 459-468.

[17] Balzani V, Scandola F (1991) Supramolecular Photochemistry. Ellis Horwood, West Sussex, UK pp. 89–190.

[18] Kalyanasundaran K. (1987) Photochemistry in Microheterogeneous Systems. Academic Press, Orlando, FL pp. 1–151.

[19] Ghanadzadeh A, Zeini A, Kashef A, Ghanadzadeh A, Zeini A, Kashef A (2007) Environment effect on the electronic absorption spectra of crystal Violet. J. Mol. Liq. 133: 61–67.

[20] Danziger R M, Bareli K H, Weiss K (1967) Laser photolysis of methylene blue. J. Phys. Chem. 71: 2633-2640.

[21] Mellish K J, Cox R D, Vernon D I, Griffiths J, Brown S B (2002) In vitro photodynamic activity of a series of methylene blue analogues. Photochem. Photobiol. 75: 392-397.

[22] Kobayashi M, Maeda Y, Hoshi T, Okubo J, Tanizaki Y (1989) Analysis of the electronic absorption-Spectrum of adsorbed layers of methylene-blue. J. Soc. Dyers Colourists 105: 362-368.

[23] Alarcon E, Edwards A M, Aspee A, Moran F E, Borsarelli C D, Lissi E A, Nilo D G, Poblete H, Scaiano J C (2010) Photophysics and photochemistry of dyes bound to human serum albumin are determined by the dye localization. Photochem. Photobiol. Sci. 9: 93-102.

[24] Jacobs K Y, Schoonheydt R A (1999) Spectroscopy of methylene blue-smectite suspensions. J. Coll. Interf. Sci. 220: 103-111.

[25] Brenneisen P, Wenk J, Redmond R, Wlaschek M, Kochevar I E, Scharffetter-Kochanek K (1999) Requirement for FRAP and P70 ribosomal S6 kinase in the DNA-damage dependent signaling leading to induction of collagenase/MMP-1 and stromelysin-1/MMP-3 after UVB irradiation of dermal fibroblasts. Photochem. Photobiol. 69: 88S-88S.

[26] Schafer H, Stahn R, Schmidt W (1979) Solvent effects on fluorescence quantum yields of thionine and methylene-blue. Zeitschrift Fur Physikalische Chemie-Leipzig 260: 862-874.

[27] Kagan J, Prakash I, Dhawan S N, Jaworski J A (1984) The comparison of several butadiene and thiophene derivatives to 8-methoxypsoralen and methylene-blue as singlet oxygen sensitizers. Photobiochemistry and Photobiophysics 8: 25-33.

[28] Berkoff B, Hogan M, Legrange J, Austin R (1986) Dependence of oxygen quenching of intercalated methylene-blue triplet lifetime on DNA base-pair composition. Biopolymers 25: 307-316.

[29] Gak V Y, Nadtochenko V A, Kiwi J (1998) Triplet-excited dye molecules (eosine and methylene blue) quenching by H_2O_2 in aqueous solutions. J. Photochem. Photobiol. A. Chem 116: 57-62.

[30] Wilkinson F, Helman W P, Ross A B (1993) Quantum yields for the photosensitized formation of the lowest electronically excited singlet-state of molecular-oxygen in solution. J. Phys. Chem. Ref. Data 22: 113-262.

[31] Wainwright M, Phoenix D A, Rice L, Burrow S M, Waring J (1997) Increased cytotoxicity and phototoxity in the methylene blue series via chromophore methylation. J. Photochem. Photobiol. 40: 233-239.

[32] Kamat P V, Lichtin N N (1981) Electron-transfer in the quenching of protonated triplet methylene-blue by ground-state molecules of the dye. J. Phys. Chem. 85: 814-818.

[33] Nilsson R, Kearns D R, Merkel P B (1972) Kinetic properties of triplet-states of methylene-blue and other photosensitizing dyes. Photochem. Photobiol. 16: 109-115.

[34] Grofcsik A, Kubinyil M, Jones W J (1995) Fluorescence decay dynamics of organic dye molecules in solution. J. Mol. Struct. 348: 197-200.

[35] Grofcsik A, Jones W. J. (1992) Stimulated emission cross-sections in fluorescent dye solutions: gain spectra and excited-state lifetimes of Nile blue A and oxazine 720 J. Chem. Sot. Faraday Trans. 88: 1101-1106.

[36] Oliveira H P M, Junior A M, Legendre A O, Gehlen M H (2003) Transferência de energia entre corantes catiônicos em sistemas homogêneos. Quim. Nova 26: 564-569.

[37] Lee S-K, Mills A (2003) Luminescence of Leuco-Thiazine Dyes. J. Fluor. 13: 375-377.

[38] Viswanathan K, Natarajan P (1996) Photophysical properties of thionine and phenosafranine dyes covalently bound to macromolecules. J. Photochem. Photobiol. A. Chem. 95: 245-253.

[39] Tuite E, Kelly J M, Beddard G S, Reid G S (1994) Femtosecond deactivation of thionine singlet states by mononucleotides and polynucleotides. Chem. Phys. Lett. 226: 517-524.

[40] Grofcsik A, Kubinyi M, Jones W J (1996) Intermolecular photoinduced proton transfer in nile blue and oxazine 720. Chem. Phys. Lett. 250: 261-265.

[41] Havelcová M, Kubát P, Nemcová I (2000) Photophysical properties of thiazine dyes in aqueous solution and in micelles. Dyes and Pigments 44: 49-54.

[42] Dutt G B, Doraiswamy S, Periasamy N, Venkataraman B (1990) Rotational reorientation dynamics of polar dye molecular probes by picosecond laser spectroscopic technique. J. Chem. Phys. 93: 8498-8513.

[43] Chen J, Cesario T C, Rentzepis P M (2011) Effect of pH on Methylene Blue Transient States and Kinetics and Bacteria Photoinactivation J. Phys. Chem. A 115: 2702–2707.

[44] Sun H, Hoffman M Z (1993) Protonation of the excited states of ruthenium(II) complexes containing 2,2'-bipyridine, 2,2'-bipyrazine, and 2,2'-bipyrimidine ligands in aqueous solution. J. Phys. Chem. 97: 5014–5018.

[45] Chen J, Cesario T C, Rentzepis P M (2010) Time resolved spectroscopic studies of methylene blue and phenothiazine derivatives used for bacteria inactivation. Chem. Phys. Lett. 498: 81–85.

[46] Acemioglu A, Arık M, Efeoglu H, Onganer Y (2001) Solvent effect on the ground and excited state dipole moments of fluorescein. J. Mol. Struct.: Theochem 548: 165–171.

[47] Svanberg K, Anderson T, Killander D, Wang I, Stenram U, Engels, S A, Berg R, Johansson J, Svanberg S (1994) Photodynamic therapy of nonmelanoma malignant-tumors of the skin using topical delta-amino levulinic acid sensitization and laser irradiation. Br. J. Dermatol 130: 743-751.

[48] Tannock I F, Hill R P (1992) The basic science of oncology, 2nd ed. Mc Graw-Hill, New York.

[49] Dougherty T J, Gomer C J, Henderson B W, Jori G, Kessel D, Korbelik M, Moan J, Peng Q (1998) Photodynamic therapy. J. Natl. Cancer Inst. 90: 889-905.

[50] Ochsner M (1997) Photophysical and photobiological processes in the photodynamic therapy of tumours. J. Photochem. Photobiol. B: Biol, 39: 1-18.

[51] Kochevar I E, Lynch M C, Zhuang S G, Lambert C R (2000) Singlet oxygen, but not oxidizing radicals, induces apoptosis in HL-60 cells. Photochem. Photobiol. 72: 548-553.

[52] Fu Y C, Jin X P, Wei S M, Lin H F, Kacew S (2000) Ultraviolet radiation and reactive oxygen generation as inducers of keratinocyte apoptosis: Protective role of tea polyphenols. J. Toxicol. Environ. Health, Part A 61: 177-188.

[53] Lin C P, Lynch M C, Kochevar I E (2000) Reactive oxidizing species produced near the plasma membrane induce apoptosis in bovine aorta endothelial cells. Exp. Cell Res. 259: 351-359.

[54] Jori G, Fabris C (1998) Relative contributions of apoptosis and random necrosis in tumour response to photodynamic therapy: effect of the chemical structure of Zn(II)-phthalocyanines. J. Photochem. Photobiol. B: Biol. 43: 181-185.

[55] Reddi E, Jori G (1988) Steady-state and time-resolved spectroscopic studies of photodynamic sensitizers - porphyrins and phthalocyanines. Rev. Chem. Interm. 10: 241-268.

[56] Schuitmaker J J, Baas P, Leengoed H L L M v, Meulen F W V, Star W M, Zandwijk N V (1996) Photodynamic therapy: A promising new modality for the treatment of cancer. J. Photochem. Photobiol. B: Biol 34: 3-12.

[57] Lewis L M, Indig G L (2000) Solvent effects on the spectroscopic properties of triarylmethane dyes. Dyes Pigm. 46: 145-154.

[58] Baptista M S, Indig G L (1998) Effect of BSA binding on photophysical and photochemical properties of triarylmethane dyes. J. Phys. Chem B 102: 4678-4688.

[59] Amin K, Baptista M S, Indig G L (1998) Mechanisms of photoinactivation of enzymes mediated by triarylmethane dyes. Biophys. J. 74: 367-367

[60] Indig G L, Bartlett J A, Lewis L M (1999) Effect of self-association and protein finding on the photoreactivity of triarylmethane dyes. Photochem. Photobiol. 69: 78S-78S.

[61] Junqueira H C, Severino D, Dias L G, Gugliotti M S, Baptista M S (2002) Modulation of methylene blue photochemical properties based on adsorption at aqueous micelle interfaces. Phys. Chem. Chem. Phys. 4: 2320-2328.

[62] Dougherty T J, Gomer C J, Henderson B W, Jori G, Kessel D, Korbelik M, Moan J, Peng Q (1998) Photodynamic therapy. J. Natl. Cancer Inst. 90: 889–905.

[63] MacDonald J, Dougherty T J (2001) Basic principles of photodynamic therapy. J. Porphyrins Phthalocyanines 5: 105–129.

[64] Bonnett R. (2000) Chemical Aspects of Photodynamic Therapy. Gordon & Breach, Amsterdam. pp. 1-324.

[65] Dolmans D E, Fukumura D, Jain R K (2003) Photodynamic therapy for cancer. Nat. Rev. Cancer 3: 380–387.

[66] Sharman W M, Allen C M, van Lier J E (1999) Photodynamic therapeutics: basic principles and clinical applications. Drug Discov. Today 4: 507–517.

[67] Phillips D (2010) Light relief: photochemistry and medicine. Photochem. Photobiol. Sci. 9: 1589–1596.

[68] Agostinis P, Berg K, Cengel K A, Foster T H, Girotti AW, Golinick S O, Hahn S M, Hamblin M R, Juzeniene A, Kessel D, Korbelik M, Moan J, Mroz P, Nowis D, Piette J, Wilson B C, Golab J (2011) Photodynamic therapy of cancer: an update. CA Cancer J. Clin. 61: 250–281.

[69] Mitsunaga M, Ogawa M, Kosaka N, Rosenblum L T, Choyke P L, Kobayashi H (2011) Cancer cell-selective in vivo near infrared photoimmunotherapy targeting specific membrane molecules. Nat. Med. 17: 1685–1691.

[70] Gomer C J, Ferrario A, Luna M, Rucker N, Wong S (2006) Photodynamic therapy: combined modality approaches targeting the tumor microenvironment. Lasers Surg Med 38: 516–521.

[71] Moan J, Berg K (1991) The photodegradation of porphyrins in cells can be used to estimate the lifetime of singlet oxygen. Photochem. Photobiol. 53: 549–553.

[72] Ribeiro J N, Jorge R A, Silva A R, Flores A V, Ronchi L M, Tedesco A C (2007) Avaliação da atividade fotodinâmica de porfirinas para uso em terapia fotodinâmica através da fotoxidação de triptofano. Ecl. Quím. 32: 7-14.

[73] Dysart J S, Patterson M S (2005) Characterization of Photofrin photobleaching for singlet oxygen dose estimation during photodynamic therapy of MLL cells in vitro. Phys Med Biol. 50: 2597-2616.

[74] Gorman A, Killoran J, Shea C O, Kenna T, Gallagher W M, Shea D F O (2004) In Vitro Demonstration of the Heavy-Atom Effect for Photodynamic Therapy. J. Am. Chem. Soc. 126: 10619-10631.

[75] Ochsner M (1997) Photophysical and photobiological processes in the photodynamic therapy of tumours. J. Photochem. Photobiol. B: Biol, 39: 1-18.

[76] Verma P, Baldrian P, Gabriel J, Trnka T, Nerud F (2004) Copper–ligand complex for the decolorization of synthetic dyes. Chemosphere 57: 1207–1211.

[77] Wainwright M, Giddens R M (2003) Phenothiazinium photosensitisers: choices in synthesis and application. Dyes Pigm. 57: 245-257.

[78] Tanigushi S, Suzuki N, Masuda M, Hisanaga S, Iwatsubo T, Goedert M, Hasegawa M (2005) Inhibition of heparin-induced tau filament formation by phenothiazines, polyphenols, and porphyrins. J. Biol. Chem. 280: 7614-7623.

[79] Prusiner S B, May B C H, Cohen F E (2004) in: Prion Biology and Diseases (Prusiner, S. B., ed) Cold Spring Harbor LaboratoryPress, Cold Spring Harbor, NY pp. 961–1014.
[80] Rojo J, Picker S M, García J J G, Gathof B S (2006) Inactivación de patógenos en productos sanguíneos. Rev. Med. Hosp. Gen. Mex. 69: 99-107.
[81] Li Y F, Huang C Z, Li M (2002) Study of the interaction of Azur B with DNA and the determination of DNA based on resonance light scattering measurements. Anal. Chim. Acta 452: 285-294.
[82] Tardivo J P, Giglio A D, Paschoal L H C, Ito A S, Baptista M S (2004) Photodiagn. Photodyn. Ther. 1: 345-350.
[83] Peloi L S, Soares R R S, Biondo C E G, Souza V R, Hioka N, Kimura E (2008) Photodynamic effect of light-emitting diode light on cell growth inhibition induced by methylene blue. J. Biosci. 33: 231-237.
[84] Bertolotti S G, Previtali C M (1999) The excited states quenching of phenothiazine dyes by p-benzoquinones in polar solvents. Dyes Pigm. 41: 55-61.
[85] Huang Q, Fu W L, Chen B, Huang J F, Zhang X, Xue Q (2004) Inactivation of dengue virus by methylene blue/narrow bandwidth light system. J. Photochem. Photobiol. B 77: 39-43.
[86] Floyd R A, Schneider J E, Dittme D P (2004) Methylene blue photoinactivation of RNA viruses. Antivir. Res. 61: 141-151.
[87] Wainwright M (2000) Methylene blue derivatives - suitable photoantimicrobials for blood product disinfection? Int. J. Antimicrob. Agents 16: 381-394.
[88] Tardivo J P, Giglio A D, Oliveira C S, Gabrieli D S, Junqueira H C, Tada D B, Severino D, Turchiello R F, Baptista M S (2005) Photodiagnosis Photodyn Ther. 2: 175-191.
[89] J. R. Perussi (2007) Photodynamic inactivation of microorganisms. Quim. Nova, 30, 988 - 994.
[90] Gabrielli D S, Belisle E, Severino D, Kowaltowski A J, Baptista M S (2004) Binding, aggregation and photochemical properties of methylene blue in mitochondrial suspensions. Photochem. Photobiol. 79: 227-232.
[91] Zhao Z, Malinowski E R (1999) Window factor analysis of methylene blue in water. J. Chemom. 13: 83-94.
[92] Heger D, Jirkovsky J, Klán P (2005) Aggregation of Methylene Blue in Frozen Aqueous Solutions Studied by Absorption Spectroscopy. J. Phys. Chem. A *109:* 6702-6709.
[93] Bergmann K, O'Konski C T (1963) A spectroscopic study of methylene blue monomer, dimer, and complexes with montmorillonite J. Phys. Chem. 67: 2169-2177.
[94] Moreira L M, Lima A, Soares R R S, Batistela V R, Gerola A P, Hioka N, Bonacin J A, Severino D., Baptista M S, Machado A E H, Rodrigues M R, Codognoto L, Oliveira H P M (2009) Metalloclorophyls of Magnesium, Copper and Zinc: Evaluation of the Influence of the First Coordination Sphere on their Solvatochromism and Aggregation Properties. J. Braz. Chem. Soc. 20: 1653-1658.
[95] Delmarre D, Hioka N, Boch R, Sternberg E, Dolphin D (2001) Aggregation studies of benzoporphyrin derivative. Can. J. Chem. 79: 1068-1074.
[96] Simplicio F I, Maionchi F, Santin O, Hioka N (2004) Small aggregates of benzoporphyrin molecules observed in water-organic solvent mixtures. J. Phys. Org. Chem. 17: 325-331.

[97] Hioka N, Chowdhary R K, Chansarkar N, Delmarre D, Sternberg E, Dolphin D (2002) Studies of a benzoporphyrin derivative with pluronics. Can. J. Chem. 80: 1321-1326.
[98] Tessaro A L, Batistela V R, Gracetto A C, Oliveira H P M, R. Sernaglia R L, Souza V R, Caetano W, Hioka N (2011) Stability of benzoporphyrin photosensitizers in water/ethanol mixtures: pK(a) determination and self-aggregation processes. J. Phys. Org. Chem. 24: 155-161.
[99] Teichert M C, Jones J W, Usacheva M N, Biel M A (2002) Treatment of oral candidiasis with methylene blue-mediated photodynamic therapy in an immunodeficient murine model. Oral Surg Oral Med Oral Pathol Oral Radiol Endod. 93: 155-160.
[100] Almeida J M, Garcia V G, Theodoro L H, Bosco Á F, Nagata M J H, Macarini V C (2006) Photodynamic therapy: an option in periodontal therapy. Arquivos em Odontologia 42: 199-210.
[101] Varma R S, Singh A P (1990) Nucleophilic Addition-Elimination Reactions of 2-Hydrazinobenzothiazoles with Indolin-2,3-diones. J. Indian Chem. Soc. 67: 518-520.
[102] Hashiba I, Ando Y, Kawakami I, Sakota R, Nagano K, Mori T (1979) Jpn. Kokai Tokkyo Koho 79 73,771 1979. (CA 91:P193174v).
[103] Aguiar A, Ferraz A (2007) Fe^{3+}- and Cu^{2+}-reduction by phenol derivatives associated with Azure B degradation in Fenton-like reactions. Chemosphere 66: 947–954.
[104] Archibald F S, (1992) A new assay for lignin-type peroxidases employing the dye Azure B. Appl. Environ. Microbiol. 58: 3110–3116.
[105] Narayana B, Cherian T (2005) A Facile Spectrophotometric Method for the Determination of Periodate Using Azure B. J. Braz. Chem. Soc. 16: 978-981.
[106] Ou C, Yuan R, Chai Y, Tang M, Chai R, He X (2007) A novel amperometric immunosensor based on layer-by-layer assembly of gold nanoparticles–multi-walled carbon nanotubes-thionine multilayer films on polyelectrolyte surface. Anal. Chim. Acta 603: 205–213.
[107] Yang M, Yang Y, Yang Y, Shen G, Yu R (2004) Bienzymatic amperometric biosensor for choline based on mediator thionine in situ electropolymerized within a carbon paste electrode. Anal. Biochem. 334: 127–134.
[108] Huang M, Jiang H, Qu X, Xu Z, Wang Y, Dong S (2005) Small molecules as cross-linkers: fabrication of carbon nanotubes/thionine self-assembled multilayers on amino functionalized surfaces. Chem. Commun. 44: 5560–5562.
[109] Xu Y, Yang L, Ye X, He P, Fang Y (2006) Impedance-Based DNA Biosensor Employing Molecular Beacon DNA as Probe and Thionine as Charge Neutralizer. Electroanalysis 18: 873–881.
[110] Deng L, Wang Y, Shang L, Wen D, Wang F, Dong S (2008) A sensitive NADH and glucose biosensor tuned by visible light based on thionine bridged carbon nanotubes and gold nanoparticles multilayer. Biosens. Bioelectron. 24: 951–957.
[111] Nicotra V E, Mora M F, Iglesias R A, Baruzzi A M (2008) Spectroscopic characterization of thionine species in different media. Dyes Pigm. 76: 315-318.
[112] Muller W, Crothers D M (1975) Interactions of heteroaromatic compounds with nucleic acids. 1. The influence of heteroatoms and polarizability on the base specificity of intercalating ligands. Eur. J. Biochem. 54: 267–277.

[113] Tuite E, Kelly J M (1995) The interaction of methylene blue, azure B, and thionine with DNA: formation of complexes with polynucleotides and mononucleotides as model systems. Biopolymers 35: 419–433.

[114] Long X, Bi S, Tao X, Wang Y, Zhao H (2004) Resonance Rayleigh scattering study of the reaction of nucleic acids with thionine and its analytical application. Spectrochim. Acta Part A: Mol. Biomol. Spectrosc. 60: 455–462.

[115] Jockusch S, Lee D, Turro N J, Leonard E F (1996) Photo-induced inactivation of viruses: adsorption of methylene blue, thionine, and thiopyronine on Qbeta bacteriophage. Proc. Natl. Acad. Sci. U.S.A. 93: 7446–7451.

[116] Rao U, Stec B, Teeter M (1995) Refinement of purothionins reveals solute particles important for latice formation and toxicity. 1. a1-purothionin revisited,. Acta Crystallogr. D. Biol. Crystallogr. D 51: 904–913.

[117] Stec B, Rao U, Teeter M M (1995) Refinement of purothionins reveals solute particles important for lattice formation and toxicity. Part 2: structure of beta-purothionin at 1.7 angstroms resolution. Acta Crystallogr., D Biol. Crystallogr. 51: 914–924.

[118] Johnson K A, Kim E, Teeter M M, Suh S W, Stec B (2005) Crystal structure of alpha-hordothionin at 1.9 Angstrom resolution. FEBS Lett. 579: 2301–2306.

[119] Huang H Y, Wang C M (2010) Phenothiazine: An effective molecular adhesive for protein immobilization. J. Phys. Chem. B 114: 3560–3567.

[120] Paul P, Hossain M, Yadav R C, Suresh Kumar G (2010) Biophysical studies on the base specificity and energetics of the DNA interaction of photoactive dye thionine: spectroscopic and calorimetric approach. Biophys. Chem. 148: 93–103.

[121] Paul P, Kumar G S (2010) Toxic interaction of thionine to deoxyribonucleic acids: Elucidation of the sequence specificity of binding with polynucleotides. J. Hazard. Mater. 184: 620–626.

[122] Göktürk S, Talman R Y (2008) Effect of temperature on the binding and distribution characteristics of thionine in sodium dodecylsulfate micelles. J. Solution Chem. 37: 1709–1723.

[123] Nicotra V E, Mora M F, Iglesias R A, Baruzzi A M (2008) Spectroscopic characterization of thionine species in different media. Dyes Pigm. 76: 315-318.

[124] Lai W C, Dixit N S, Mackay R A (1984) Formation of H aggregates of thionine dye in water. J Phys Chem 88: 5364-5368.

[125] Mackay R A, Gratzel M (1985) The photoreduction of thionine by iron(II) in anionic micelles and microemulsions. Ber Bunsenges Phys Chem 89: 526-530.

[126] Chen S, Yuan R, Chai Y, Xu L, Wang N, Li X, Zhang L (2006) Amperometric hydrogen peroxide biosensor based on the immobilization of horseradish peroxidase (HRP) on the layer-by-layer assembly films of gold colloidal nanoparticles and toluidine blue. Electroanalysis 18: 471–477.

[127] Jiao K, Li Q J, Sun W, Wang Z (2005) Voltammetric detection of the DNA interaction with toluidine blue. Electroanalysis 17: 997–1002.

[128] Tian F, Zhu G (2004) Toluidine blue modified self-assembled silica gel coated gold electrode as biosensor for NADH. Sens. Actuators B 97: 103–108.

[129] Jana A K (2000) Solar cells based on dyes. J. Photochem. Photobiol. A 132: 1–17.

[130] Prento P (2001) A contribution to the theory of biological staining based on the principles for structural organization of biological macromolecules. Biotech. Histochem. 76: 137–161.

[131] Ilanchelian M, Ramaraj R (2011) Binding Interactions of Toluidine Blue O with Escherichia Coli DNA: Formation of Bridged Structure. J. Fluoresc. 21: 1439–1453.

[132] Fei D, Wang X M, Li H B, Ding L S, Hu Y M, Zhang H, Zhao S L (2008) Spectroscopy Studies of Interaction between Methylene Blue and Herring Sperm DNA. Acta Chim. Sin. 66: 443-448.

[133] Arikan B, Tunçay M (2005) Micellar effects and reactant incorporation in reduction of toluidine blue by ascorbic acid. Dyes Pigm. 64: 1-8.

[134] Tuite E, Norden B (1994) Sequence-Specific Interactions of Methylene Blue with Polynucleotides and DNA: A Spectroscopic Study. J. Am. Chem. Soc. 116: 7548-7556.

[135] Liu J, Zou A, Mu B (2010) Toluidine blue: Aggregation properties and distribution behavior in surfactin micelle solution. Colloids Surf., B 75: 496–500.

[136] Staveren H J, Speelman O C, Witjes M J H, Cincotta L, Star W M (2001) Fluorescence imaging and spectroscopy of ethyl Nile blue A in animal models of (Pre)malignancies. Photochem. Photobiol. 73: 32–38.

[137] Morgan J, Potter W R, Oseroff A R (2000) Comparison of photodynamic targets in a carcinoma cell line and its mitochondrial DNA-deficient derivative. Photochem. Photobiol. 71: 747–757.

[138] Singh G, Espiritu M, Shen X Y, Hanlon J G, Rainbow A J (2001) In vitro induction of PDT resistance in HT29, HT1376 and SKN-MC cells by various photosensitizers. Photochem. Photobiol. 73: 651–656.

[139] Lin C W, Shulok J R, Wong Y K, Schambacher C F, Cincotta L, Foley J W (1991) Photosensitization, uptake, and retention of phenoxazine Nile Blue derivatives in human bladder carcinoma cells. Cancer Res. 51: 1109-1116.

[140] Lin C W, Shulok J R (1994) Enhancement of Nile Blue derivative-induced photocytotoxicity by nigericin and low cytoplasmic pH. Photochem. Photobiol. 60: 143-146.

[141] Lee S H, Suh J K, Li M (2003) Determination of bovine serum albumin by its enhancement effect of Nile Blue fluorescence. Bull. Korean Chem. Soc. 24: 45-48.

[142] Krihak M, Murtagh M T, Shahriari M R (1997) A spectroscopic study of the effects of various solvents and sol-gel hosts on the chemical and photochemical properties of Thionin and Nile Blue A. J. Sol-Gel Sci. Technol. 10: 153-163.

[143] Betscheider D, Jose J (2009) Nile Blue A for staining Escherichia coli in flow cytometer experiments. Anal. Biochem. 384: 194-196.

[144] Mitra R K, Sinha S S, Pal S K (2008). Interactions of Nile Blue with Micelles, Reverse Micelles and a Genomic DNA. J. Fluoresc. 18: 423–432.

[145] Lee M H, Lee S W, Kim S H, Kang C (2009) Nanomolar Hg(II) detection using Nile Blue chemodosimeter in biological media. Org. Lett. 11: 2101-2104.

[146] Maliwal B P, Kusba J, Lakowicz J R (1995) Fluorescence energy transfer in one dimension: frequency-domain fluorescence study of DNA-fluorophore complexes. Biopolymers 35: 245-255.

[147] Lakowicz J R, Piszczek G, Kang J S (2001) On the possibility of long-wavelength longlifetime high-quantum-yield luminophores. Anal. Biochem. 288: 62-75.

[148] Tajalli H, Ghanadzadeh Gilani A, Zakerhamidi M S, Tajalli P (2008) The photophysical properties of Nile red and Nile blue in ordered anisotropic media. Dyes Pigm. 78: 15-24.

[149] Jose J, Burgess K (2006) Benzophenoxazine-based fluorescent dyes for labeling biomolecules. Tetrahedron 62: 11021-11037.

[150] Pal M K (1965) Effects of differently hydrophobic solvents on the aggregation of cationic dyes as measured by quenching of fluorescence and/or metachromasia of the dyes. Histochimie 5: 24-31.

[151] Rauf M A, Zaman M Z (1987) Spectral properties of oxazines in various solvents. Spectrochim. Acta A 43: 1171-1172.

[152] Gvishi R, Reisfeld R, Eisen M (1989) Structures, spectra and ground and excited state equilibria of polycations of oxazine-170. Chem. Phys. Letters 161: 455-460.

[153] Grofcsik A, Kubinyi M, Jeremy Jones W (1996) Intermolecular photoinduced proton transfer in nile blue and oxazine 720. Chem. Phys. Letters 250: 261-265.

[154] Gorodetsky A A, Ebrahim A, Barton J K (2008) Electrical detection of TATA binding protein at DNA modified microelectrodes. J. Am. Chem. Soc. 130: 2924–2925.

[155] Chen Q, Li D, Yang H, Zhu Q, Xu J, Zhao Y (1999) Interaction of a novel red-region fluorescent probe, Nile Blue, with DNA and its application to nucleic acids assay. Analyst 124: 901–907.

[156] Huang CZ, Li Y F, Zhang D J, Ao X P (1999) Spectrophotometric study on the supramolecular interactions of nile blue sulphate with nucleic acids. Talanta 49: 495–503.

[157] Yang Y, Hong H Y, Lee I S, Bai D G, Yoo G S, Choi J K (2000) Detection of DNA using a visible dye, Nile Blue, in electrophoresed gels. Anal. Biochem. 280: 322–324.

[158] Ju H, Ye Y, Zhu Y (2005) Interaction between nile blue and immobilized single- or double-stranded DNA and its application in electrochemical recognition. Electrochim. Acta 50: 1361–1367.

[159] Mitra R K, Sinha S S, Maiti S, Pal S K (2009) Sequence dependent ultrafast electron transfer of Nile blue in oligonucleotides. J. Fluoresc. 19: 353–361.

[160] Nasr C, Hotchandani S (2000) Excited-state behavior of Nile blue H-aggregates bound to SiO_2 and SnO_2 colloids. Chem. Mater. 12: 1529-1535.

[161] Bayoumi, A. M. E., Kasha, M. (1959). Exciton-type splitting of eletronics states in hydrogen-bounded molecular dimers of N-heterocyclics. Spectrochim. Acta. 15: 759-760.

[162] Lee C, Sung Y W, Park J W (1999) Multiple equilibria of phenothiazine dyes in aqueous cyclodextrin solutions. J. Phys. Chem. B. 103: 893-898.

[163] Patil K, Pawar R, Talap P (2000) Self-aggregation of methylene blue in aqueous medium and aqueous solutions of Bu_4NBr and urea. Phys. Chem. Chem. Phys. 2: 4313-4317.

[164] Ohline S M, Lee S, Willians S, Chang C (2001) Quantification of methylene blue aggregation on a fused silica surface and resolution of individual absorbance spectra. Chem. Phys. Lett. 346: 9-15.

[165] Zoratti M, Szabò I (1995) The mitochondrial permeability transition. Biochim. Biophys. Acta. 1241: 139-176.

[166] Misran M, Matheus D, Valente P, Hope A (1994) Photochemical electron transfer between methylene blue and quinones. Aust. J. Chem. 47: 209-216.

[167] Collings P J, Gibbs E J, Starr T E, Vafek O, Yee C, Pomerance L A, Pasternack R F (1999) Resonance light scattering and its application in determining the size, shape, and aggregation number for supramolecular assemblies of chromophores. J. Phys. Chem. B 103: 8474-8481.

[168] Liu D, Kamat P V (1996) Dye-capped semiconductor nanoclusters. One-electron reduction and oxidation of thionine and cresyl violet H-aggregates electrostatically bound to SnO_2 colloids. Langmuir 12: 2190-2195.

[169] Carroll M K, Unger M A, Leach A M, Morris M J, Ingersoll C M, Bright F V (1999) Interactions between methylene blue and sodium dodecyl sulfate in aqueous solution studied by molecular spectroscopy. Appl. Spect. 53: 780-784.

[170] Sakellarioufargues R, Maurette M T, Oliveros E, Riviere M, Lattes A (1982) Chemical and photochemical reactivity in micellar media and micro-emulsions. 4. Concentration effects on isophorone dimerization. J. Photochem. 18: 101-107.

[171] Sakellarioufargues R, Maurette M T, Oliveros E, Riviere M, Lattes A (1984) Chemical and photochemical reactivity in micellar media and microemulsions. 7. Effect of the interface on the reactivity of excited-states. Tetrahedron 40: 2381-2384.

[172] Reddi E, Jori G, Rodgers M A J, Spikes J D (1983) Flash-photolysis studies of hemato-porphyrins and copro-porphyrins in homogeneous and microheterogeneous aqueous dispersions. Photochem. Photobiol. 38: 639-645.

[173] Oliveros E, Pheulpin P, Braun A M (1987) Comparative-study of the sensitized photooxidation of N-methyl phenothiazine in homogeneous and microheterogeneous media. Tetrahedron 43: 1713-1723.

[174] Daraio M E, Aramendía P F, San Román E A, Braslavsky S E (1991) Carboxylated zinc phthalocyanines. 2. dimerization and singlet molecular-oxygen sensitization in hexadecyltrimethylammonium bromide micelles. Photochem. Photobiol. 54: 367-373.

[175] Kikteva T, Star D, Zhao Z, Baisley T L, Leach G W (1999) Molecular orientation, aggregation, and order in rhodamine films at the fused silica/air interface. J. Phys. Chem. B 103: 1124-1133.

[176] Monte F (1999) Identification of oblique and coplanar inclined fluorescent J-dimers in rhodamine 110 doped sol-gel-glasses. J. Phys. Chem. B 103: 8080-8086.

[177] Monte F, Mackenzie J D, Levy D (2000) Rhodamine fluorescent dimers adsorbed on the porous surface silica gel. Langmuir 16: 7377-7382.

[178] Borba E B, Amaral C L C, Politi M J, Villalobos R, Baptista M S (2000) Photophysical and photochemical properties of pyranine/methyl viologen complexes in solution and in supramolecular aggregates: A switchable complex. Langmuir 16: 5900-5907.

[179] Das S, Kamat P V (1999) Can H-aggregates serve as eight-harvesting antennae? Triplet-triplet energy transfer between excited aggregates and monomer thionine in aerosol-OT solutions. J. Phys. Chem. B. 103: 209-215.

[180] Neumann M G, Gessner F, Cione A P P, Sartori R A, Cavalheiro C C S (2000) Interaction between dyes and clays in aqueous suspension. Quim. Nova 23: 818-824.
[181] Eisfeld A, Schulz G, Briggs J (2011) The influence of geometry on the vibronic spectra of quantum aggregates. J. Lumin. 131: 2555-2564.
[182] Roden J, Eisfeld A, Briggs J S (2008) The J- and H- bands of dye aggregate spectra: Analysis of the coherent exciton scattering (CES) approximation, Chem. Physics 352: 258-266.
[183] Eisfeld A, Briggs J S (2006) The J- and H-bands of organic dyes aggregates, Chem. Physics 324: 376-384.
[184] Eisfeld A, Briggs J S (2007) The Shape of the J-band of pseudoisocyanine, Chem. Phys. Let. 446: 354-358.
[185] Eisfeld A, Briggs J S (2002) The J-band of organic dyes: lineshape and coherence length, Chem. Physics 281: 61-70.
[186] Ibbotson S H (2011) Irradiance is an important determinant of pain experienced during topical photodynamic therapy, J. Am. Acad. Dermatol. 65: 201-202.
[187] Yusa S, Awa S, Ito M, Kawase T, Takada T, Nakashima K, Liu D, Yamago S, Morishima Y (2011) Solubilization of C60 by Micellization with a Thermoresponsive Block Copolymer in Water: Characterization, Singlet Oxygen Generation, and DNA Photocleavage. J. Polym. Sci., Part A: Polym. Chem. 49: 2761–2770.
[188] Ronchi L M, Ribeiro A V F N, da Silva A R, de Sena G L, Jorge R A, Ribeiro J N (2007) The influence of aggregation and photobleaching in the photodynamic activity of magnesium protoporphyrin. Revista Capixaba de Ciência e Tecnologia 2: 5-12.
[189] Benvindo R G, Braun G, de Carvalho A R, Bertolini G R F (2008) Effects of photodynamic therapy and of a sole low-power laser irradiation on bacteria in vitro. Fisioterapia e Pesquisa 15: 53-7.
[190] Demidova T N, Hamblin M R (2004) Photodynamic Therapy targeted to pathogens. International Journal of Immunophatology and Pharmacology 17: 245-254.
[191] Amaral R R, Amorim J C F, Nunes E, Soares J A, Silveira F F (2010) Photodynamic therapy in endodontics - review of literature. RFO, 15: 207-211.
[192] Wang J, Guo Y, Gao J, Jin X, Wang Z, Wang B, Li K, Li Y (2011) Detection and comparison of reactive oxygen species (ROS) generated by chlorophyllin metal (Fe, Mg and Cu) complexes under ultrasonic and visible-light irradiation. Ultrason. Sonochem. 18: 1028–1034.
[193] Tangkuaram T, Wang J, Rodriguez M C, Laocharoensuk R, Veerasai W (2007) Highly stable amplified low-potential electrocatalytic detection of NAD+ at azure-chitosan modified carbon electrodes. Sens. Actuators, B 121: 277–281.
[194] Yuksel F, Durmus M, Ahsen V (2011) Photophysical, photochemical and liquid-crystalline properties of novel gallium(III) phthalocyanines. Dyes Pigm. 90: 191-200.
[195] Dilber G, Durmus M, Kantekin H, Çakır V (2011) Synthesis and characterization of a new soluble metal-free and metallophthalocyanines bearing biphenyl-4-yl methoxy groups. J. Organomet. Chem. 696: 2805-2814.

Chapter 15

Novel Fischer's Base Analogous of Leuco-TAM and TAM$^+$ Dyes – Synthesis and Spectroscopic Characterization

Sam-Rok Keum, So-Young Ma and Se-Jung Roh

Additional information is available at the end of the chapter

http://dx.doi.org/10.5772/48119

1. Introduction

Triarylmethane (TAM) dyes are organic compounds containing triphenylmethane backbones. TAM compounds are sometimes called leuco-TAMs (LTAMs) or leuco-bases. (Nair et al., 2006) LTAM molecules are the precursors of TAM$^+$ dyes since TAM$^+$ dyes are the oxidized form of LTAM molecules. Backbones of TAM molecules are also known to be an important group in intermediates in the synthesis of various organic functional compounds, including the preparation of polymers and supramolecules. (Bartholome & Klemm, 2006)

TAM$^+$ dyes potentially have numerous applications in the chemical, pharmaceutical, and life science industries, including as staining agents, ink dyes, thermal imaging materials, carbonless copying materials, drugs, leather, ceramics, cotton, and as a cytochemical staining agent. (Balko & Allison, 2000) A number of TAM$^+$ dye molecules are well known, such as malachite green (MG), brilliant green, crystal violet, and pararosanilin, *etc.* The chemical structures of some of these well-known TAM$^+$ dyes are shown in Fig. 1.

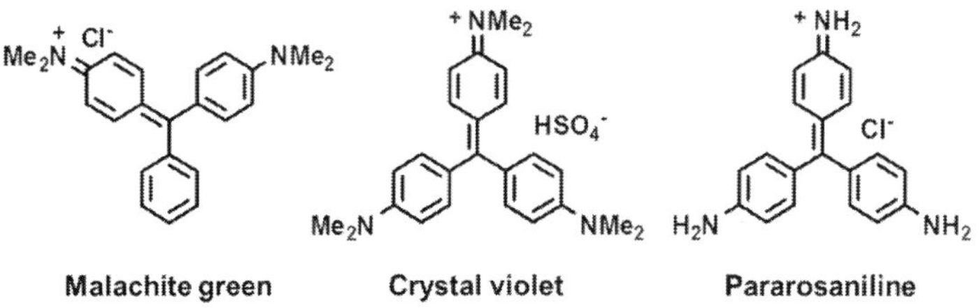

Figure 1. Chemical structures of well-known TAM$^+$ dyes.

Among them, MG is one of the most commonly used chemicals in dye chemistry. MG is a common green dye but it is absorbed into the human body in its carbinol and leuco forms (see the section on UV-Vis spectroscopy). MG is very active with the fungus *Saprolegnia*, which infects fish eggs in commercial aquaculture, and is known to be good for *Ichthyophthirius* in fresh water aquaria. (Indig et al., 2000) It has been known, however, for MG to be highly toxic to mammalian cells, even at low concentrations.(Plakas et al., 1999; Cho et al., 2003) Because of its low cost, effectiveness as an antifungal agent for commercial fish hatcheries, and ready availability, many people can be exposed to this dye through the consumption of treated fish. Since MG is similar in structure to carcinogenic triphenylmethane dyes, it may be a potential human health hazard.

In addition, in their oxidized form, TAM^+ dyes are highly absorbing fluorophores with extinction coefficients of ~2.0×10^5 $mol^{-1}cm^{-1}$ and high quantum yields. Their absorption maxima can easily be matched with the laser lines by simply changing the length of the conjugated chain and/or the heterocyclic moiety. Thus, TAM^+ dyes can be employed as fluorescence labels and sensors of biomolecules *in vivo* because their spectra reach the near-infrared region. Özer (2002) reported efficient non-photochemical bleaching of a TAM^+ dye by chicken ovalbumin and human serum albumin, showing that dye–protein adducts can also form and suggesting that proteins may be primary, rather than indirect, targets of TAM^+ action. However, the use of this substance has been banned in many countries because of its toxicity and possible carcinogenicity. Substitutive materials for MG compounds have thus been in considerable demand. Numbers of researchers (Gessner & Mayer, 2005) have been interested in developing TAM^+ molecules. We developed Fischer's base (FB) analogs of LTAM molecules (Keum et al., 2008, 2009, 2010, 2011, 2012), whose chemical structures contain a couple of heterocyclic FB rings and a substituted phenyl group on a central carbon of the molecules. The structures and numbering system for the Fischer's base (FB) analogs of LTAMs are shown in Fig. 2.

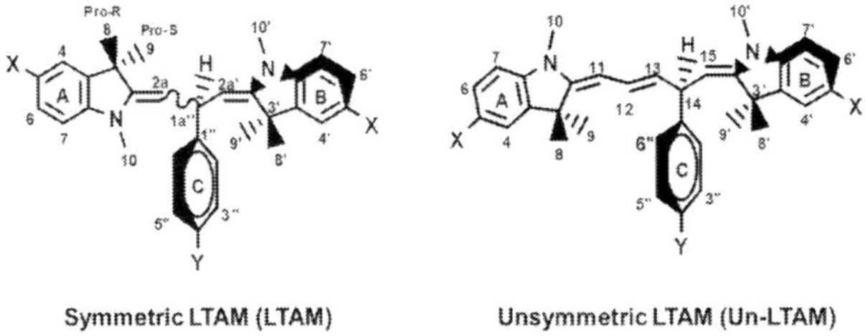

Figure 2. Structures and numbering systems for the FB analogs of symmetric and unsymmetric LTAM molecules.

In this chapter, abbreviations (LTAM and Un-LTAM) will be used to designate the Fischer's base analogs of symmetric and unsymmetric LTAMs, respectively. Note that "the general LTAM" in Section 2.2 denotes the LTAM/TAM^+ dyes that contain no FB moieties.

2. Preparation

2.1. General leuco-TAM and TAM+ molecules

Generally, the Friedel-Crafts-type catalytic alkylation of aromatic rings with aromatic aldehydes is an effective method for TAM+ formation. (Li et al., 2008; Kraus et al., 2008) Several mild and efficient triaryl- and triheteroarylmethanes formations using $[Ir(COD)Cl]_2$-$SnCl_4$, $AuCl_3$, $Cu(OTf)_2$, and $Sc(OTf)_3$ as catalysts have also been reported. Grignard reagents or n-butyl lithium compounds have also been used for their preparation. A brief summary of the preparation methods of general LTAM/TAM+ molecules is shown in Fig. 3.

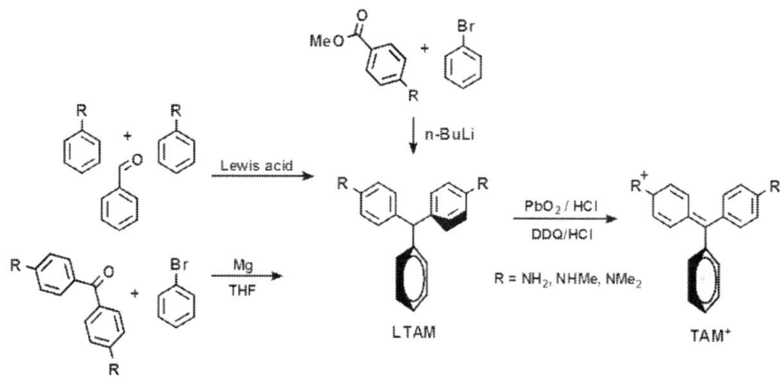

Figure 3. The synthetic procedures for the commercially well-known LTAM molecules.

Although a number of methods are available for the synthesis of triarylmethanes, most are multistep processes and/or require harsh reaction conditions.

2.2. Fischer's base analogs of Leuco-TAM

Fischer's base analogs of LTAM molecules can be obtained from a reaction of a molar excess of Fischer's base and substituted aryl aldehydes. The prepared LTAM dyes consist of two FB rings on the central carbon, where a substituted phenyl ring is located. The LTAM molecules can be symmetric or unsymmetric, depending on the identity of the two FB rings. They are the precursors of the TAM+ dyes, which are structurally close to the polymethine dyes (*e.g.*, Cy3, Cy5, *etc.*). (Ernst et al., 1989)

2.2.1. Symmetric LTAM FB analogs

The FB analogs of symmetric leuco-TAM molecules {2,2'-(2-phenylpropane-1,3-diylidene)bis(1,3,3-trimethylindoline)} derivatives were obtained from the reaction of 5-substituted benzaldehyde and excess (2- to 3-fold) FB in ethanol at room temperature for 2–4 h, as shown in Fig. 4. The white precipitate was filtered from the reaction mixture and washed thoroughly with cold ethyl alcohol. Purification was carried out through precipitation from acetone. TAM^+ dyes were then obtained from a reaction of LTAM molecule with 2,3-dichloro-5,6-dicyano-1,4-benzoquinone (DDQ) in the presence of HCl, followed by separation of the deep blue form from the product mixtures by column chromatography in MC/MeOH (9:1).

Melting points, yields, and other characteristic data of the prepared LTAM molecules are summarized in Table 1.

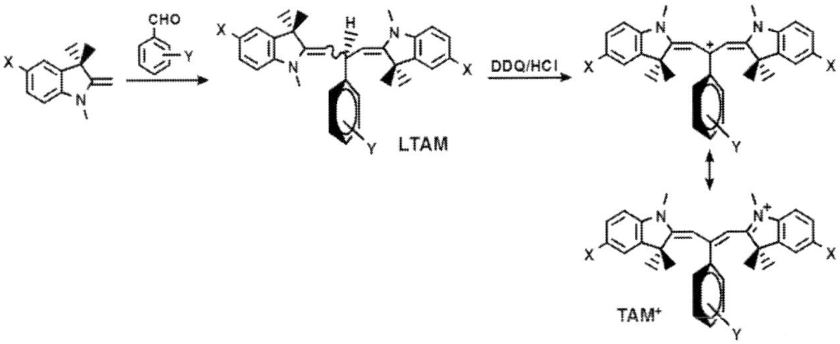

Figure 4. Synthetic scheme for symmetrical LTAM molecules.

2.2.2. *Unsymmetric LTAM FB analogs*

Unsymmetric LTAMs (Un-LTAM) were obtained from a reaction of excess Fischer's base with the substituted cinnamaldehydes, as shown in Fig. 5. The Un-LTAMs have two different FB skeletons on the central carbon, 1,3,3-trimethyl-2-methyleneindoline and 2-allylidene-1,3,3-trimethylindoline groups. The symmetric TAM^+ dyes with styryl-ring pendants are expected to possess elongated conjugation from the N^+ center of the FB ring to the phenyl ring. However, these LTAM molecules were not successfully obtained from the reactions of Fischer's base with the substituted cinnamaldehydes.

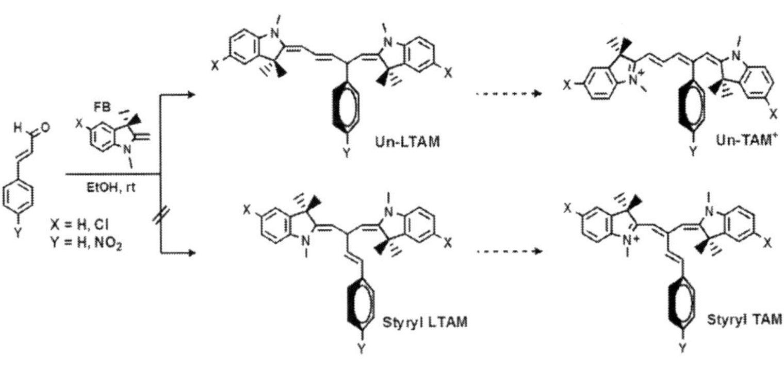

Figure 5. Synthetic scheme for unsymmetrical LTAM molecules.

Experimentally, Un-LTAM molecules were formed as the sole product and no symmetrical LTAM dyes were formed. This suggests that the Michael-type addition of the second molecule of FB occurs on the δ-carbon and not on the β-carbon of the extended α,β-unsaturated iminium salts that were formed from the reaction of FB and cinnamaldehydes. The mechanistic processes for the formation of Un-LTAM molecules are shown in Fig. 6.

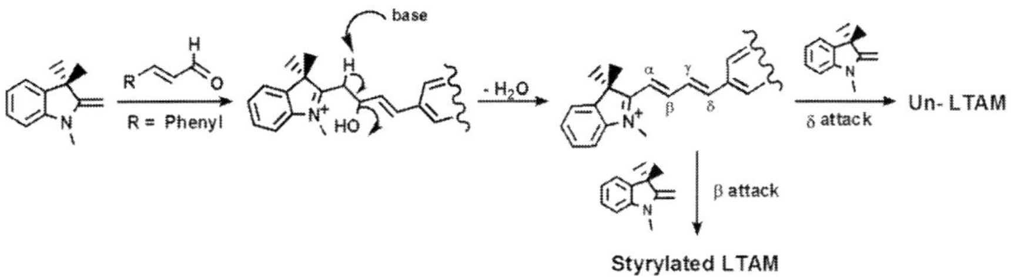

Figure 6. Mechanistic processes of the Michael-type addition of a FB molecule to the β and δ carbon of the α, β, γ, and δ-unsaturated iminium salts to form symmetrical and unsymmetrical LTAM dyes, respectively.

Melting points, yields, and other characteristic data of the prepared LTAM and Un-LTAM molecules are summarized in Table 1.

Compound	LTAM moecule X	Y	M.p. (°C)	Yield (%)	Colour
LTAM **1**	H	H	146	75	white
LTAM **2**	H	*p*-Cl	168	69	pink
LTAM **3**	H	*p*-NO_2	162(dec.)	67	reddish
LTAM **4**	Cl	H	189	82	white
LTAM **5**	Cl	*p*-NO_2	182	89	orange
LTAM **6**	Cl	*m*-NO_2	118(dec.)	72	reddish
LTAM **7**	Cl	2-Cl, 5-NO_2	204	65	orange
LTAM **8**	Cl	*p*-(N)	218-219	89	white
LTAM **9**	Cl	*m*-(N)	189-190	69	white
LTAM **10**	Cl	*p*-OMe	146	57	white
LTAM **11**	Cl	*p*-CHO	157	80	pale orange
LTAM **12**	Benzo[e]	H	181-182	21	pale green
LTAM **13**	Benzo[e]	*m*-(N)	152	48	pale lime
Un-LTAM **1**	H	H	191	42	brown
Un-LTAM **2**	H	*p*-NO_2	186	46	brown
Un-LTAM **3**	Cl	H	182	53	orange
Un-LTAM **4**	Cl	*p*-NO_2	177	55	orange

Table 1. Melting points (M.p.), yields, and colours of the prepared LTAM and Un-LTAM molecules.

3. Spectroscopic Characterization

Newly synthesized FB analogs of LTAM molecules are not expected to be fully planar because of steric crowding, rather they can be viewed as a screw or helical, and thus may possess propeller structures. Subsequently, they can adopt a conformation where all three rings are twisted in the same direction, making a right- or left-handed propeller. As an analogy to a common screw or bolt, right- and left-handed screws are nominated as *M* and *P*, respectively, as shown in the upper line of Fig. 7. The red arrows denote the direction of bond rotation, not the helical direction (blue arrows). Two rings rotate through a perpendicular conformation while one moves in the opposite direction.

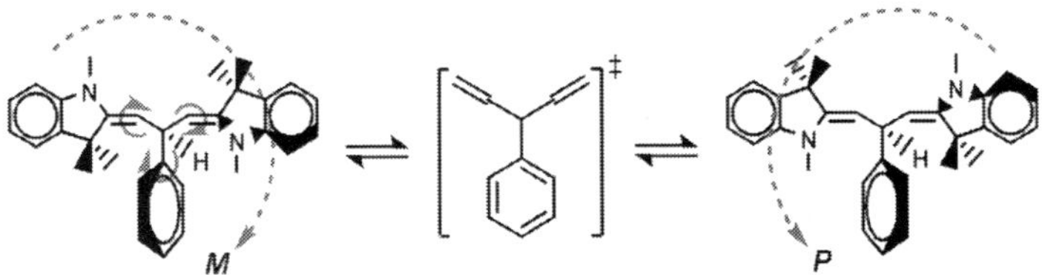

Figure 7. A new diastereomer can be formed via bond rotation (red arrows) of one diastereomer (blue arrows denote the helical direction between *M* and *P* isomers).

Theoretically, three configurational isomers, *ZE*, *EE*, and *ZZ*, can be proposed for these dyes. Since the *ZE* isomers can have *M* and *P* conformations, *ZE*-LTAM molecules can be obtained as a racemic mixture from the synthetic reaction described previously.

3.1. Diastereomeric identification for the prepared LTAM molecules by 1D ^{1}H NMR and 2D NMR experiments

The ^{1}H NMR spectra of LTAM molecules display characteristic signals (three groups) in the aliphatic region, namely a triplet and two doublets (group A) in the range of 4.20–5.40 ppm, two singlets (group B) between ~2.90 and ~3.30 ppm, and four identical singlets (group C) at 1.20–1.80 ppm. As a representative example, the ^{1}H NMR spectrum of LTAM **4** in $CDCl_3$ is shown in Fig. 8.

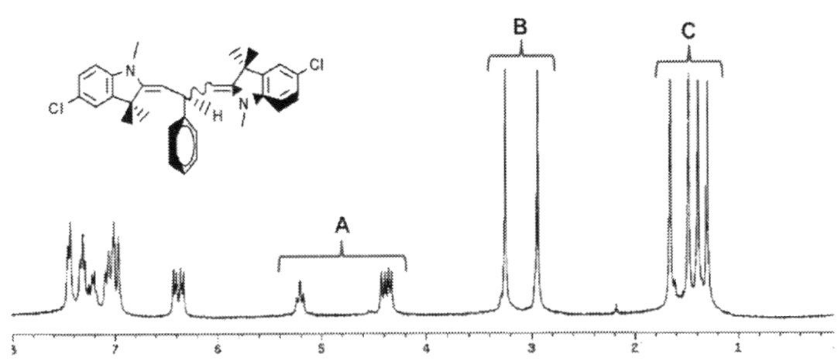

Figure 8. ^{1}H NMR spectroscopy of LTAM **4** as a representative example.

Judging from the chemical shifts of the signals in Fig. 8, group A may belong to sp^2 protons, H2a and H2a' and allylic proton H1a'', and groups B and C may belong to *N*-Me (H10) and *gem*-dimethyl groups (H8 and H9), respectively. Interestingly, the *gem*-dimethyl groups show four well-separated singlets, indicating that these *gem*-dimethyl groups are diastereotopic. Thus, the *gem*-dimethyl groups are not identical and they have different chemical shifts in the NMR spectra. The most common instance of diastereotopic groups is when two similar groups are substituents on a carbon adjacent to a stereogenic center. These LTAM molecules may not be fully planar because of steric crowding and would thus be expected to exhibit chirality, having no stereogenic centers. No detailed discussions are not given for the resoncances in the range of 6 to 8 ppm, since the resonance in those ranges are of the general aromatic protons.

The ^{1}H and ^{13}C resonances of the LTAM molecules were assigned using COSY and one-bond ^{1}H–^{13}C correlations obtained by both direct-detection HETCOR and indirect detection HSQC experiments. COSY was used to identify peaks from the A and B rings. The HETCOR and HSQC identified the shifts of the proton-containing carbons. HMBC was used to differentiate between the two A and B rings of one half of the molecules because these rings were not identical.

The HETCOR experiment identified the carbon shifts of those carbons with protons attached through one-bond coupling between ^{1}H and ^{13}C. Correlations between the protons such as H2a, H8, H9, H10 and H1a'', and their corresponding carbons are particularly useful for structure determination, as indicated in Fig. 9.

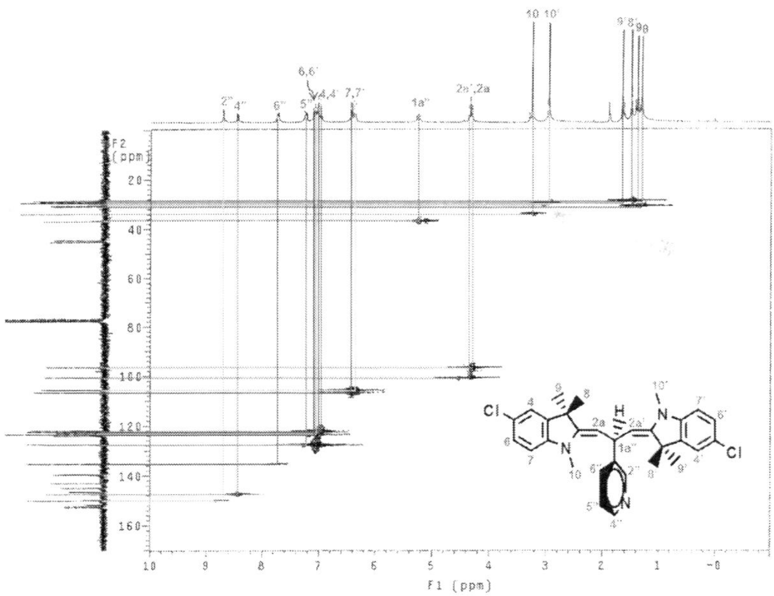

Figure 9. HETCOR of LTAM **9** in the range of of 0-170 ppm.

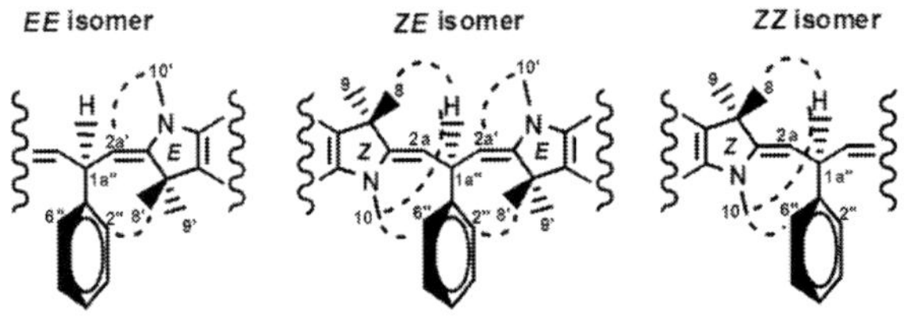

Figure 10. Structures and NOE correlations of the *ZE, EE,* and *ZZ* isomers of LTAM molecules.

The major chemical shift assignments within FB rings, A and B, of the molecules were mostly made using HMBC and NOE data. Since A and B rings are not identical, HMBC was needed to differentiate between the two groups of FB molecules. HMBC experiments have thus correlated C2 to H1a'', H2a, H8, H9, and H10 of ring A and correlated C2' to H1a'', H2a', H8', H9', and H10' in ring B. For the geometrical identification around the double bonds of the enamine moiety of the A and B (FB) rings, NOE experiments were carried out. Structures and NOE correlations of the *ZE, EE,* and *ZZ* isomers of the LTAM molecules are shown in Fig. 10. Determination of the configuration of the double bonds at positions 2–2a and 2'–2a' of the enamine moiety of the A and B (FB) rings was carried out by an NOE experiment. Strong NOE correlation was observed between the protons H10' (*N*-Me of B) at 2.97 ppm and H2a' (the same subunit) at 4.42 ppm. In addition, H10 (*N*-Me of A) has NOE with H2''/H6'', whereas the *gem*-dimethyl, H8 and H9, exhibits NOE with H2a. These observations are compatible with a *Z* arrangement around the double bond of ring A.

Compound	H or C	*Z* Ring		*E* Ring		Others[a]	
		δ (H)	δ (C)	δ (H)	δ (C)	δ (H)	δ (C)
LTAM **4**	2/2'	-	151.6	-	151.3	-	-
	2a/2a'	4.36(d)	97.3	4.42(d)	101.5	-	-
	3/3'	-	44.8	-	44.2	-	-
	3a/3a'	-	139.4	-	139.6	-	-
	8/8'	1.32(s)	30.6	1.50(s)	28.6	-	-
	9/9'	1.41(s)	30.8	1.68(s)	29.0	-	-
	10/10'	3.26(s)	33.5	2.95(s)	29.2	-	-
	1a''	-	-	-	-	5.22(dd)	38.7
	2''/6''	-	-	-	-	7.46(d)	127.6
LTAM **5**[b]	2/2'	-	-	-	155.8		
	2a/2a'	-	-	4.42	100.5		
	3/3'	-	-	-	45.16		
	3a/3a'	-	-	-	139.6		
	8/8'	-	-	1.39	30.6		
	9/9'	-	-	1.60	28.6		
	10/10'	-	-	2.97	29.0		
	1a''	-	-	-	-	5.19(t)	38.8
	2''/6''	-	-	-	-	7.53(d)	128.8

Compound	H or C	*Z* Ring		*E* Ring		Others[a]	
		δ (H)	δ (C)	δ (H)	δ (C)	δ (H)	δ (C)
LTAM **8**[b]	2/2'	-	-	-	152.0	-	-
	2a/2a'	-	-	4.41(d)	100.1	-	-
	3/3'	-	-	-	44.7	-	-
	3a/3a'	-	-	-	139.6	-	-
	8/8'	-	-	1.41(s)	29.3	-	-
	9/9'	-	-	1.51(s)	28.4	-	-
	10/10'	-	-	2.97(s)	28.6	-	-
	1a''	-	-	-	-	5.07(t)	38.3
	2''/6''	-	-	-	-	7.31(d)	122.9
LTAM **10**	2/2'	-	151.6	-	151.2	-	-
	2a/2a'	4.30(d)	97.9	4.37(d)	102.3	-	-
	3/3'	-	44.9	-	44.3	-	-
	3a/3a'	-	139.6	-	139.9	-	-
	8/8'	1.30(s)	30.8	1.47(s)	29.4	-	-
	9/9'	1.39(s)	30.7	1.65(s)	28.6	-	-
	10/10'	3.25(s)	33.7	2.94(s)	28.9	-	-
	1a''	-	-	-	-	5.15(dd)	37.8
	2''/6''	-	-	-	-	7.35(d)	127.4

[a]Others denote the phenyl ring and connecting groups of the LTAM molecules.

[b]Compound that has the *EE* configuration as the major isomer.

Table 2. ^{1}H and ^{13}C NMR spectral data for LTAM molecules in $CDCl_3$ (500 and 125 MHz, respectively).

Similarly, H2a' and H9' have NOE correlations with H10' and H2''/H6'', respectively. These NOE phenomena indicate a *ZE* geometry around the double bonds of the enamine moieties of A and B (FB) rings, respectively. Selected ^{1}H and ^{13}C NMR spectral data for the major diastereomer of various LTAM molecules in $CDCl_3$ (500 and 125 MHz, respectively) are listed in Table 2.

3.2. Thermal diastereomerization

3.2.1. *Diastereomeric mixtures in equilibrium state*

The ^{1}H NMR spectra of LTAM **4** became complicated upon thermal treatment. After approximately 2 h they exhibited three sets of signals corresponding to two other forms (a & b) in addition to the original major set, in $CDCl_3$, as shown in Fig. 11. Two of the double bonds in the LTAM molecules can exist as *ZE*, *EE*, and *ZZ* isomers, which may account for the complexity of the spectra. Their existence is most likely due to geometrical isomerism with respect to restricted rotation around the C=C double bond of the FB moiety and the C-CH(FB)(Ph) single bonds.

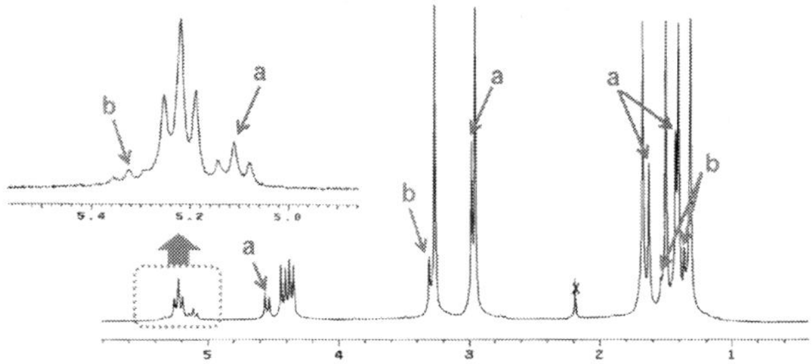

Figure 11. ¹H NMR spectra of LTAM **4** at the thermal equilibrium state, showing three sets of analog peaks (major, a and b groups).

Detailed analysis of the ¹H NMR spectra of the LTAM compounds in the thermal equilibrium state is important for determining the presence of a mixture of *ZE* and *EE* or *ZZ* isomers. After thermal equilibrium of LTAM **4** in $CDCl_3$, 2 h after sampling in an NMR tube at room temperature, three sets of complex signals were observed, namely triplet peaks at 5.0–5.4 ppm, three doublets at 4.3–4.6 ppm, four singlets at 2.8–3.4 ppm, and eight singlets at 1.3–1.7 ppm. Among these peaks, those signals assigned to the *ZE* isomers were a triplet at 5.22 ppm, two doublets at 4.36 and 4.42 ppm, two singlets at 2.95 and 3.26 ppm, and four singlets at 1.3–1.7 ppm, as discussed previously. The residual peaks might belong to the *EE* and/or *ZZ* isomers.

For identification of each of the diastereomers of LTAM **4** in organic solvents, 2D NMR experiments such as COSY, HMBC, and NOESY, were used at the equilibrium state. As an example, a COSY of diastereomeric mixtures of LTAM **4** in the range of 4.25–5.25 ppm in $CDCl_3$ is given in Fig. 12.

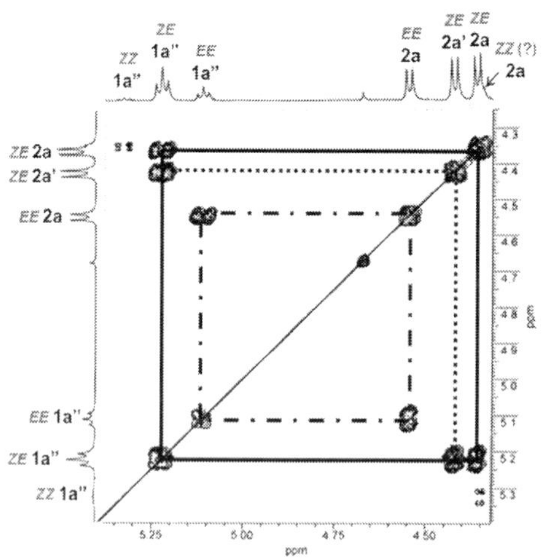

Figure 12. COSY of diastereomeric mixtures of LTAM **4** in the range of 4.25–5.25 ppm in $CDCl_3$.

^{1}H-^{1}H COSY in the range of 4.25–5.25 ppm showed two individual sets of H1a'' and H2a/2a' protons for the diastereomeric structures *ZE* (solid and dot) and *EE* (dash-dot) isomers of LTAM **4**, respectively. The methylene doublets of the *ZZ* isomers for this compound could not be detected due to their low concentration in the equilibrium state. Three sets of *N*-Me (H10 or H10') in the range of 2.8–3.4 ppm and the germinal methyl group (H8 and H9, and H8' and H9') could be easily distinguished through the visual peak ratios of the ^{1}H NMR.

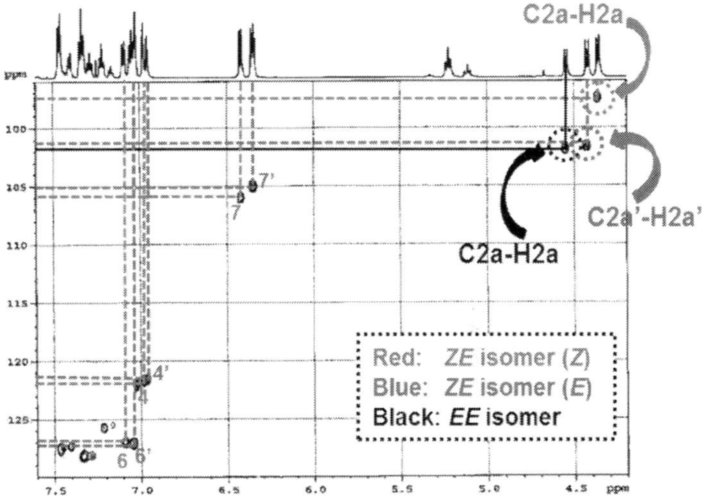

Figure 13. HSQC of diastereomeric mixtures of LTAM **4**, showing one-bond correlation of C2a-H2a, C2a'-H2a' for the *ZE* isomer and C2a-H2a for the *EE* isomer in the range of 95-130ppm.

^{1}H–^{13}C correlations were obtained by both direct-detection HETCOR and indirect detection HSQC experiments. Fig. 13 shows some of the one-bond ^{1}H–^{13}C correlations. HSQC identifies the shifts of the carbons bearing protons of the major *ZE* and minor *EE* isomers.

More particularly, HMBC can identify which protons belong to which unit. For the major *ZE* isomers, HMBC experiments have correlated C2 to H1a'', H2a, H8, H9, and H10 of ring A and correlated C2' to H1a'', H2a', H8', H9', and H10' in ring B. The *gem*-dimethyls (1.50 and 1.68 ppm) are correlated with C3' at 44.2 ppm. The H2a' at 4.42 ppm is correlated to the same C3', which allow us to assign it to the same subunit (B ring). Similarly, the *N*-Me at 2.95 ppm, correlated to the same C3', is also in the same subunit (B ring). The other *gem*-dimethyl groups (1.32 and 1.41 ppm) are correlated with C3 at 44.8 ppm. The H2a at 4.36 ppm and the *N*-Me at 3.26 ppm are also correlated to the same C3, indicating that they are in the second subunit (A ring). The low-field HMBC of the diastereomeric mixtures of LTAM **4** is given in Fig. 14 and the high-field HMBC (< 95 ppm) are not given here.

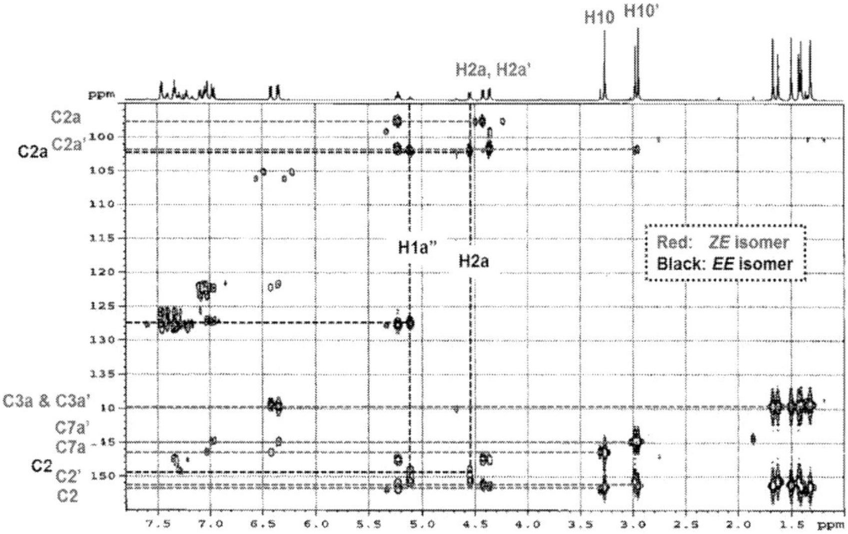

Figure 14. HMBC of the diastereomeric mixtures of LTAM **4** in the range of 95–155 ppm.

For the minor concentration *EE* isomers, the HMBC experiments correlated C2 at 149 ppm to H1a'', H2a, H8, H9, and H10. Similar correlations of C3 were made for H2a, H4, H8, and H9. HMBC could further correlate C3a to H7, H8, and H9. Similar to the extremely minor *ZZ* isomers, HMBC correlated C2 to H1a'', H2a, H8, H9, and H10.

Similar correlations of C3 were made to H2a, H4, H8, and H9. HMBC further correlated C3a to H7, H8, and H9. These correlations provided a clear distinction between diastereomeric isomers. H2a and *N*-Me were also coupled to the same C3, which confirms that all of the protons belong to the same isomer. As H9 or *N*-Me is coupled to C2''/C6'', the protons of the aromatic ring could be identified as a substructure of each isomer.

2D NOESY showed spatial correlations for each of the diastereomeric mixtures for LTAM **4** after reaching thermal equilibrium, as in Fig. 15. Namely, the spatial correlations labeled a–f in red were detected for the *ZE* isomer, and those labeled b', d', and f' in blue are detected for the *EE* isomer. In addition, one correlation, e' in green, was detected for the *ZZ* isomer.

Unfortunately, a few of the ^{1}H resonance peaks for the *ZZ* isomers were able to be detected, such as an *N*-methyl singlet and, very rarely, two *gem*-dimethyl peaks. This result indicates that the LTAM molecules equilibrate in a time-dependent manner, yielding a mixture of the *ZE*/*EE*/*ZZ* isomers in organic solvent ($CDCl_3$).

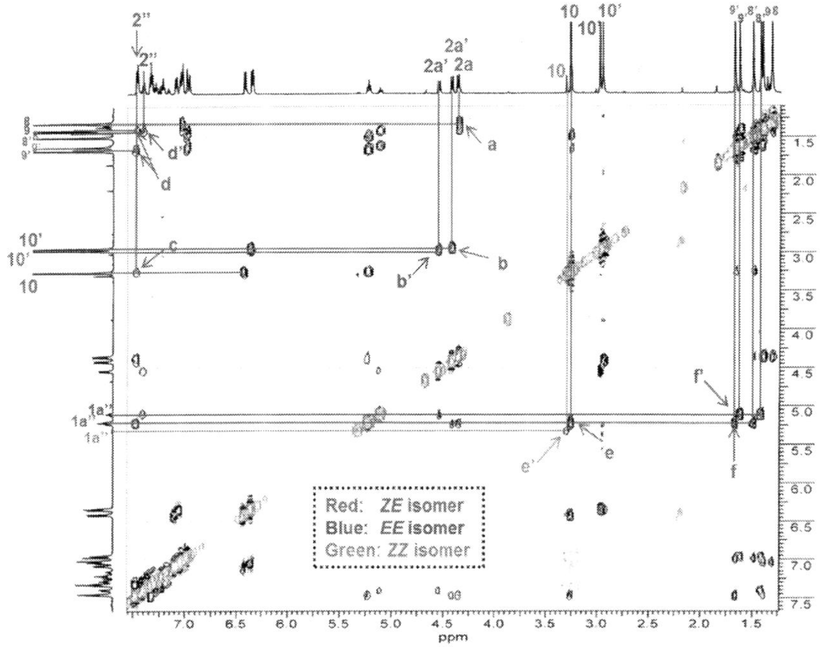

Figure 15. 2D NOESY of the diastereomeric mixtures of LTAM **4** after reaching thermal equilibrium in $CDCl_3$.

This NOE phenomenon indicates a *ZE* geometry around the double bonds of the enamine moieties of A and B (FB) rings of the major isomer. However, the minor isomer contains two symmetric Fischer's base units and 2D NOESY only showed the correlations of H2a'-H10' (blue mark, b'), H8'-H2''/H6'' (blue mark, d') but no correlation of H10'-H2''/H6''. This suggests that the *ZE* geometry around the double bonds of the enamine moieties does not exist for the minor isomer. Although the spatial correlations of H2a-H9, H10-H2''/H6'' and H1a''-H10 were expected for the extremely minor *ZZ* isomer, the proton peaks H2a and H2a' of the *ZZ* isomer were too small to be correlated with other protons in the 2D NOESY experiment. One spatial correlation of H1a''-H10 (green mark, e') was detected. This NOE phenomenon indicates that three diastereomers, such as *ZE*, *EE*, and *ZZ*, are in equilibrium with various ratios among the diastereomeric isomers, depending on the NMR solvent used.

The NMR data for the *Z* or *E* ring of the *ZE* isomer suggest that the signals for the geminal dimethyl group for the *EE* isomer should be shifted downfield compared to those of the *ZZ* isomer, whereas the signals of the *N*-methyl groups should be shifted upfield. The geminal dimethyl group signals for the *EE* isomer were shifted upfield compared to those of the *ZE* isomer, whereas the signals of the *N*-methyl and methylene groups were shifted downfield.

Compound	Ring	proton	Diastereomer (ppm)		
			ZE	*EE*	*ZZ*
LTAM **1**	*Z*	8-Me	1.31	-	1.37
		9-Me	1.41	-	1.57
		N-Me	3.28	-	3.32
		H2a	4.33	-	N/A
	E	8'-Me	1.51	1.44	-
		9'-Me	1.68	1.64	-
		N'-Me	2.97	3.00	-
		H2a'	4.41	4.53	-
LTAM **2**	*Z*	8-Me	1.33	-	1.38
		9-Me	1.42	-	1.55
		N-Me	3.29	-	3.33
		H2a	4.31	-	N/A
	E	8'-Me	1.52	1.46	-
		9'-Me	1.68	1.64	-
		N'-Me	2.99	3.01	-
		H2a'	4.36	4.47	-
LTAM **4**	*Z*	8-Me	1.32	-	N/A
		9-Me	1.41	-	N/A
		N-Me	3.26	-	N/A
		H2a	4.36	-	N/A
	E	8'-Me	1.50	1.43	-
		9'-Me	1.68	1.63	-
		N'-Me	2.95	2.98	-
		H2a'	4.42	4.55	-
LTAM **5**	*Z*	8-Me	1.30	-	N/A
		9-Me	1.39	-	N/A
		N-Me	3.22	-	3.28
		H2a	4.31	-	N/A
	E	8'-Me	1.41	1.39	-
		9'-Me	1.65	1.60	-
		N'-Me	2.94	2.97	-
		H2a'	4.32	4.42	-
LTAM **10**	*Z*	8-Me	1.30	-	N/A
		9-Me	1.39	-	N/A
		N-Me	3.25	-	3.28
		H2a	4.30	-	N/A
	E	8'-Me	1.47	1.41	-
		9'-Me	1.65	1.60	-
		N'-Me	2.94	2.96	-
		H2a'	4.37	4.49	-
LTAM **12**	*Z*	8-Me	1.67	-	N/A
		9-Me	1.77	-	N/A
		N-Me	3.44	-	N/A
		H2a	4.47	-	N/A
	E	8'-Me	1.90	1.83	-
		9'-Me	2.06	2.04	-
		N'-Me	3.08	3.11	-
		H2a'	4.52	4.60	-

Table 3. Selected ^{1}H resonances for some of the diastereomeric LTAMs.

Deshielding of the *gem*-dimethyl proton of the *EE* and the *N*-methyl proton of the *ZZ* isomer may be due to their relative proximity to the benzene ring, as indicated in the *ZE* isomers. Selected ^{1}H resonances for the diastereomeric LTAMs are listed in Table 3.

3.2.2. *Dynamic behavior of LTAM molecules*

Interestingly, the stability of these molecules in solution depends upon the solvent media. Namely, they are inert in polar organic solvents such as acetone and DMSO, but they are unstable in nonpolar solvents such as benzene, THF, and chloroform. They equilibrate time-dependently into a mixture of *ZE/EE* or *ZE/EE/ZZ* isomers, depending on the solvents used, as shown in Fig. 16.

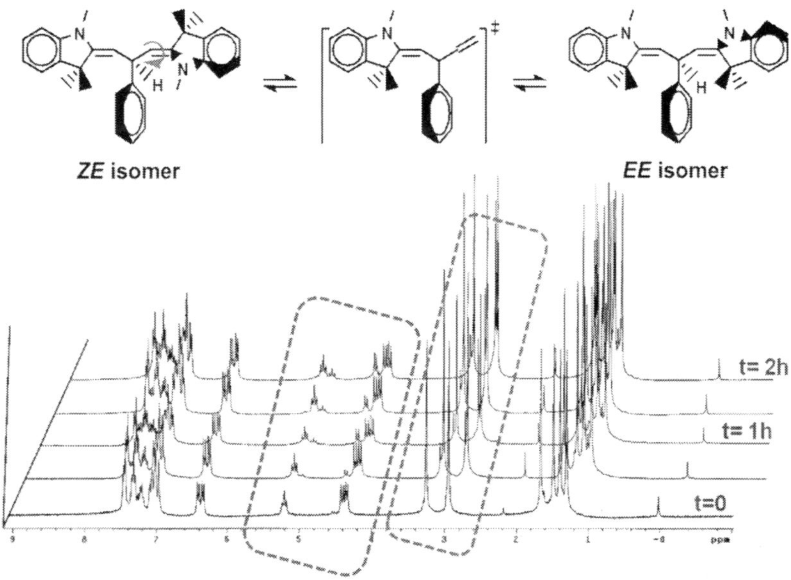

Figure 16. Dynamic behavior of LTAM **1** in $CDCl_3$, showing diastereomeric isomerization.

It has been reported (Keum et.al. 2008) that FB-analogs of LTAM molecules have very characteristic ^{1}H NMR resonance patterns in the range of 1.0–5.4 ppm as a result of three consecutive protons (H2a, H2a', and H1a''), two *N*-methyl, and four diastereotopic *gem*-dimethyl (8- and 9-Me) groups. Therefore, these characteristic peaks can be used to discriminate each of the diastereomers. (Ma et al., 2012) For example, ^{1}H NMR data of the 3-pyridinyl LTAM **9** showed the expected features (A → B) of resonances, *viz*. one triplet at 5.25 ppm, two doublets at 4.31 and 4.34 ppm, two singlets at 2.95 and 3.25 ppm, and four singlets at 1.30–1.65 ppm. In contrast, the spectra of 4-pyridinyl LTAM **8** showed very interesting features in the range of 2.90–5.40 ppm. This compound showed one triplet at 5.07 ppm, a doublet at 4.41 ppm, and a singlet at 2.97 ppm, as shown in C of Fig. 17.

The isomerization pattern of LTAM **8** is quite surprising because no FB-analog of MG showed a LTAM **8**-like feature (C → B) in the range of 2.90–5.40 ppm. Based upon quantum mechanical calculations (Keum et al., 2010), *ZE* would be expected to predominate over *ZZ* and *EE* in all media. Experimentally, the *ZE* isomers of LTAM compounds generally predominant in all organic media examined. The relative energy differences between the minor *EE* and extremely minor *ZZ* were 0.08 and 0.26 kcal/mol in $CDCl_3$ and DMSO-d_6, respectively. In addition, both of the spectra, A and C, converged to that displayed by B approximately 2–3 h after mixing the LTAM molecules with $CDCl_3$ in the NMR tube.

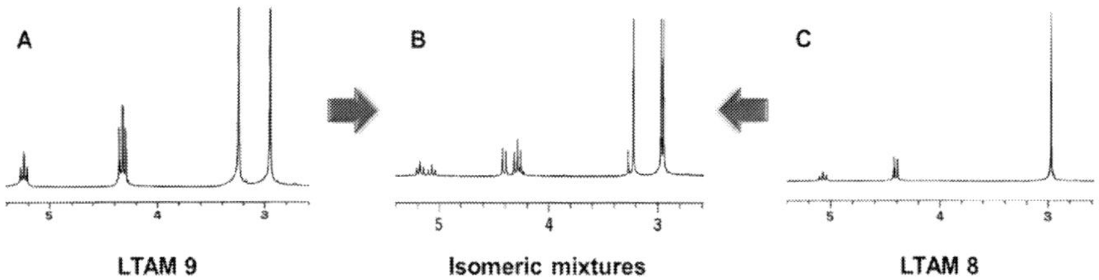

Figure 17. Characteristic proton resonance of the LTAM molecules in the range of 2.80~5.40 ppm in $CDCl_3$ (A: *ZE*, B: mixture after reaching thermal equilibrium, and C: *EE* diastereomer); arrows show the directions of isomerization.

3.2.3. Determination of diastereomeric ratios at equilibrium state

Since the characteristic peaks can be used to discriminate each of the diastereomers, the equilibrium ratios among these diastereomeric isomers in various organic solvents can be determined by ^{1}H NMR spectra. This is based on the intensities of either the *N*-methyl or *gem*-dimethyl signals corresponding to the three diastereomeric isomers at the equilibrium state. In some cases, the intensity of the H1a'' proton of the central carbon can be used. Since the *N*-methyl peaks show a well-separated singlet, it is more convenient to measure the ratio of the isomers. The cause of the isomerization of the LTAM compounds at room temperature is unclear. They belong to the group of conjugated enamine compounds. Enamine-imine tautomerism (C=C-NH and CH-C=N) may regulate *ZE* isomerization.

For most of the LTAM molecules examined that are listed in Table 1, the *ZE* isomers are the most stable and they at the equilibrium state for almost 100% of the time in polar solvents ($E_T(30) > 42$) and 60–80% in non-polar solvents ($E_T(30) < 42$) at room temperature. The minor *EE* minor and extremely minor *ZZ* isomers at the equilibrium state were 18–44% and 0–11% in nonpolar solvents, respectively, depending on the molecules examined. The percent ratios among the diastereomeric isomers of LTAM molecules in the thermal equilibrium states vary according to the molecules examined and solvents used.

However, some LTAM molecules, such as LTAMs **3**, **5**, and **8**, are exceptional. Surprisingly, the pure *EE* isomers are obtained, unlike for the LTAM molecules described previously. These exceptional compounds contain a resonance-electron withdrawing (-R) substituent, particularly on the para-position of the phenyl ring. This indicates that substituents such as *p*-NO_2, *p*-CHO, or *p*-(N) on the phenyl ring make the *EE* isomer more stable than the *ZE* isomer, which is predicted to be more stable theoretically. These are summarized in Table 4.

Further detailed studies are needed to determine how the isomerization occurs and what causes the unusual stability of a certain diastereomer.

Compound	Solvent	Percent ratio (%)			K_{eq}[a]	Note[b]
		ZE	*EE*	*ZZ*		
	$CDCl_3$	60.2	28.3	11.4	1.52	
LTAM **4**	Aceton-d_6	68.5	25.9	5.60	2.17	*ZE*
	DMSO- d_6	91.5	8.5	-	10.76	
	$CDCl_3$	61.0	34.1	4.90	0.52	
LTAM **5**	Aceton-d_6	59.9	34.8	5.30	0.53	*EE*
	DMSO- d_6	33.2	68.8	-	2.07	
	$CDCl_3$	63.3	31.3	5.40	0.46	
LTAM **8**	Aceton-d_6	61.3	32.8	5.90	0.49	*EE*
	DMSO- d_6	36.4	63.6	-	1.75	
	$CDCl_3$	64.0	31.1	4.80	1.78	
LTAM **9**	Aceton-d_6	62.4	32.0	5.60	1.66	*ZE*
	DMSO- d_6	59.2	40.8	-	1.45	
	$CDCl_3$	67.1	27.4	5.50	2.03	
LTAM **10**	Aceton-d_6	65.8	27.7	6.50	1.92	*ZE*
	DMSO- d_6	80.7	13.0	6.30	4.18	

[a]K_{eq} is the ratio of [*ZE*]/[*EE*+*ZZ*] or [*EE*]/[*ZE*+*ZZ*], depending on the identity of the major diastereomer.

[b]The major isomer in the solid state.

Table 4. Percent ratios among the diastereomeric isomers of LTAM molecules at thermal equilibrium states.

3.2.4. *Free energy change of activation, $\Delta G^{\ddagger}$*

The rate constants for the formation of the *EE* (and *ZZ*) isomers were measured from a plot of ln(A−A^0) versus time (in min), according to the peak-intensity of the central proton at ~5.15 ppm. As an example, excellent linearity was obtained with r = 0.999 and n = 6. The k_{obs} and half-life ($t_{1/2}$) for the isomerization of LTAM **4** were 5.95×10^{-4} s^{-1} and 19.4 min, respectively. The first-order rate constant is a sum of the rate constants for the backward and reverse reactions. From the rate constant obtained at room temperature, the obtained one-temperature $\Delta G^{\ddagger}$ value (Dougherty et al., 2006) for the *ZE* → *EE* isomerization of LTAM **4** in $CDCl_3$ was found to be 21.8 kcal/mol. Similarly, the rate constants for the diastereomeric isomerization of LTAM molecules were measured and the one-temperature $\Delta G^{\ddagger}$ values of all of them were obtained, using the equation 1 (Dougherty et.al. 2006) given below:

$$\Delta G^{\ddagger} = 4.576\ [10.319 + \log (T/k)]\ \text{kcal/mol} \quad (1)$$

These are summarized in Table 5.

Compound	k_{obs} x10^{-4} s^{-1}	$t_{1/2}$ (min)	Linearity		$\Delta G^{\ddagger}_{ZE \rightarrow EE}$	$\Delta G^{\ddagger}_{EE \rightarrow ZE}$
			r	n		
LTAM **4**	5.95	19.4	0.999	6	21.8	-
LTAM **5**	0.43	297	0.999	3	-	23.4
LTAM **8**	2.72	42.5	0.999	6	-	22.3
LTAM **9**	4.66	24.8	0.994	4	22.0	-
LTAM **10**	2.48	46.5	0.998	5	22.0	-
LTAM **11**	4.72	24.5	0.998	5	-	22.4
LTAM **13**	17.9	6.47	0.998	3	21.2	-

Table 5. Rate constants and $\Delta G^{\ddagger}$ values for the *ZE/EE* isomerization of LTAM molecules in $CDCl_3$.

It has been previously reported that the *ZE* isomerization of imines and their tautomeric isomers, enamines, has a very high energy barrier ($\Delta G^{\ddagger}$ = 23 kcal/mol), unless the process is strongly accelerated by either acid/base catalysts or by push-pull substituents. (Liao & Collum, 2003). The isomerization rate was found to be slow on the NMR time-scale.

3.3. UV-Vis spectroscopy of various forms of LTAM molecules

FB analogs of TAM^+ dyes were obtained from the reaction of FB analogs of LTAM molecules with DDQ in the presence of HCl, followed by separation of the deep blue form from the product mixtures by column chromatography in MC/MeOH (7:1). A reaction of TAM^+ with an inorganic base such as NaOH gives the carbinol form of the LTAM molecule. Only the TAM^+ cation shows deep coloration, in contrast to the LTAM and carbinol derivatives. This difference arises because only the cationic form has extended π-delocalization, which allows the molecule to absorb visible light.

The colored forms, TAM^+, of the prepared LTAM and Un-LTAM molecules have absorption maxima at 580–705 and 350–420 nm in ethanol for the x- and y-band, respectively. The carbinol form was detected at 325–385 nm in basic media. The leuco form of these molecules decomposed in HCl-saturated EtOH to form conjugated molecules observed at 385–435 nm. UV-Vis spectral data in $CDCl_3$ of the colored and decomposed forms LTAM **4**, and Un-LTAM **4**, as representative examples, are shown in Fig. 18.

UV-Vis spectral data for various forms of LTAM and Un-LTAM molecules, compared to those of commercial TAM^+ dyes, are summarized in Table 6.

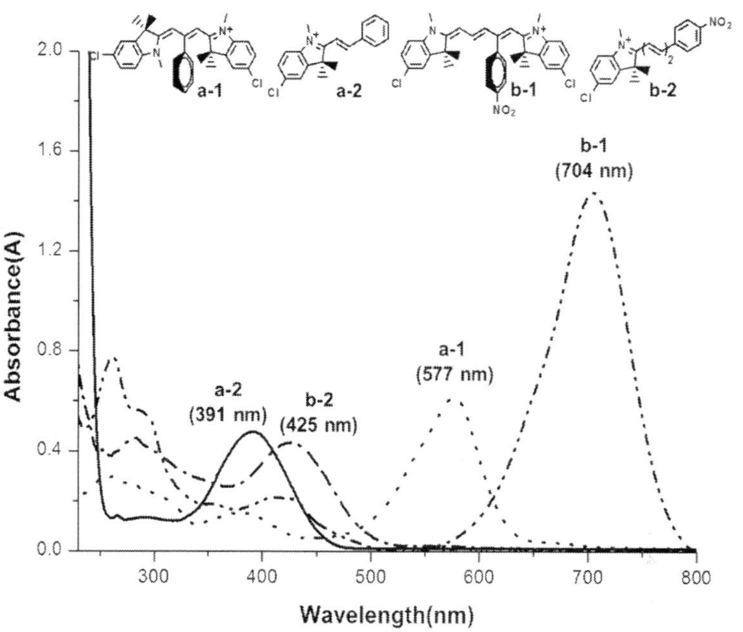

Figure 18. UV-Vis spectral data of LTAM **4** (a) and Un-LTAM **4** (b) in EtOH, showing the various forms such as the TAM+ (a-1 and b-1) and decomposed forms (a-2 and b-2).

Compound[a]	Various structural forms (λ_{max})				
	Leuco- (a)	Carbinol- (b)[b]	TAM+- (c)[c]		Decomposed dye (d)
			x-band	y-band	
Crystal Violet[d]	265	266	585	-	-
Malachite Green[e]	265	265	620	430	-
LTAM **1**	284	343	609	426	385
LTAM **4**	296	327	578	370	391
LTAM **5**	298	-	588	380	390
LTAM **8**	302	363	595	377	318
LTAM **9**	295	-	556	368	382
LTAM **11**	298	-	591	-	403
Un-LTAM **1**	322	381	693	412	420
Un-LTAM **2**	326	383	686	414	415
Un-LTAM **3**	324	368	686	353	435
Un-LTAM **4**	326	384	704	417	425

[a]Names of compounds are the same as in Table 1. [b]The carbinol denoted a hydroxylated TAM+ dye.

[c]Symbols (x-, y-band) are adopted from Ref. (Ernest et al., 1989) [d,e]Data for acetonitrile.

Table 6. UV-Vis spectral data for various forms of LTAM and Un-LTAM compounds.

In the UV-Vis spectral data of Table 6, MG and crystal violet dyes show absorption maxima at 620 and 430 nm for the x- and y-band, respectively, whereas the absorption maxima of the vinyl-log of MG are red-shifted for both the x- and y-bands, *i.e.*, 651 and 488 nm, respectively. This suggests that the vinyl effects of a vinyl unit may, to a large extent, behave like extended conjugation for both the x- and y-bands. Chemical skeletons for the N~N$^+$ and C(phenyl)~N$^+$ responsible for the x- and y-band, respectively, in the absorption spectra of various TAM$^+$ dyes are shown in Fig. 19.

Structurally, the FB analogs of symmetric and unsymmetric TAM$^+$ dyes in this work can be characterized as Cy3 and Cy5 dyes, respectively, as closed-chain cyanines.(Ernst et al., 1989) It was reported that Cy3 is maximally excited at 550 nm and maximally emits at 570 nm in the orange-red part of the spectrum, whereas Cy5 is maximally excited at 649 nm and maximally emits at 670 nm, which is in the red part of the spectrum. Therefore, the x-band of the Un-TAM$^+$ are expected to be higher than 650 nm and 550 nm, for the y- and x-band, respectively.

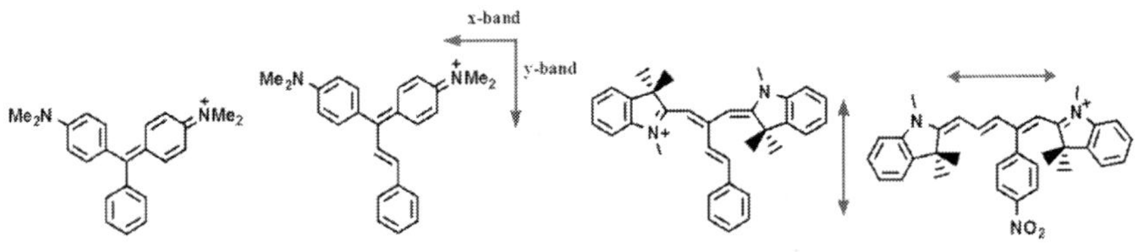

Figure 19. Chemical skeleton for the N~N$^+$ and C(phenyl)~N$^+$ responsible for the x- and y-band, respectively, in the absorption spectra of various TAM$^+$ dyes.

From the reaction of Un-LTAM **4** with $HClO_4$, the decomposed product {5-chloro-1,3,3-trimethyl-2-((1E,3E)-4-(4-nitrophenyl)buta-1,3-dienyl)indolium perchlorate} was isolated, brown, yield 57%, M.p.= 257–258 °C, IR (KBr) 3072, 2984, 2934, 1707, 1596, 1340, and 1086 cm^{-1}, ^{1}H NMR (DMSO-d_6) δ 1.76 (6H, s), 4.03 (3H, s), 7.37 (1H, d, *J* = 15.3 Hz), 7.66 (1H, dd, *J* = 10.2, 15.3 Hz), 7.73 (1H, d, *J* = 9.0 Hz), 7.79 (1H, d, *J* = 15.3 Hz), 7.92 (2H, d, *J* = 6.9 Hz), 7.95 (1H, d, *J* = 9.0 Hz), 8.09 (1H, s), 8.33(1H, dd, (*J* = 10.2, 15.3 Hz), and 8.33(2H, d, *J* = 6.9 Hz).

4. Solid state structure

4.1. LTAM molecules

The X-ray crystal structure of LTAM **12**, as a representative example, displays an orthorhombic crystal system with space group *Pna2₁*, with a residual factor of R_1 = *0.0517*. ORTEP diagrams of LTAM **12**, showing atom numbering, are provided in Fig. 20(a).

For LTAM **12**, the C7-C8 and C7-C24 distances are 1.512 and 1.513 Å, respectively, *i.e.*, typical lengths for C-C single bonds, and the enamine C8-C9 and C24-C25 bonds are 1.335 and 1.335 Å, respectively, which are typical lengths for C=C bonds. The LTAM **12** molecules possess three-bladed propeller conformations, similar to earlier reports for various non-hetaryl LTAM dyes. The inter-plane angles between the aromatic rings A-B, A-C, and B-C in LTAM **12** are 81.29, 87.79 and 86.02°, respectively, as shown in Fig. 20(b).

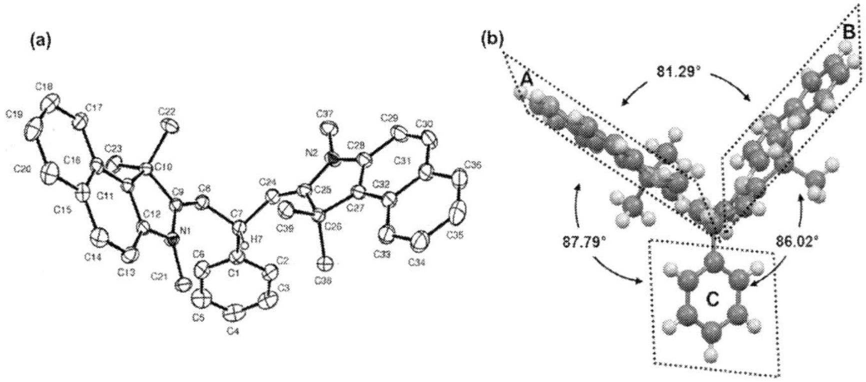

Figure 20. ORTEP diagrams with atom numbering scheme (a) and the propeller shape (b) of LTAM **12**, showing the inter-plane angles.

Selected bond lengths and bond angles for the LTAM molecules are listed in Table 7.

Ring	Bond length (Å) and Bond angle (°)	LTAM molecule			
		4	**5**	**9**	**12**
A	C7-C8	1.513	1.511	1.515	1.512
	C8-C9	1.330	1.341	1.334	1.335
	C9-N1	1.410	1.398	1.400	1.405
	C8-C9-N1	129.42	122.8	129.72	127.47
B	C7-C24	1.507	1.509	1.507	1.513
	C24-C25	1.332	1.334	1.330	1.335
	C25-N2	1.409	1.416	1.407	1.396
	C24-C25-N2	123.06	123.0	123.20	122.94
Others	C7-C8-C9	130.99	128.3	131.06	129.21
	C7-C8-H8	114.5	115.8	114.5	115.4
	C8-C7-H7	107.6	108.6	107.7	112.10
	C7-C24-C25	127.42	129.7	127.68	128.50
	C7-C24-H24	116.3	115.2	116.2	115.8
	C24-C7-H7	107.6	108.6	107.7	107.2

Table 7. Selected bond lengths and bond angles of LTAM molecules.

The dihedral angles H8-C8-C7-H7 and H24-C24-C7-H7 in LTAM **12** are 172.06° (θ_1) and 176.41° (θ_2), respectively. The inter-plane angles and dihedral angles for the LTAM molecules are given in Table 8.

Compound	Interplane angles[a] (º)			Dihedral angles[b] (º)	
	ring A-B	ring B-C	ring A-C	θ_1	θ_2
LTAM **1**	79.8	74.9	84.8	164.1	149.8
LTAM **4**	60.1	75.7	83.2	178.9	158.9
LTAM **7**	77.9	85.6	80.2	152.9	139.8
LTAM **8**	135.4	72.8	72.8	163.2	163.2
LTAM **9**	60.0	77.3	83.5	156.1	179.0
LTAM **11**	44.83	74.32	72.92	174.62	151.68
LTAM **12**	81.29	86.02	87.79	172.06	176.41

[a]Symbols (A-B) and numbering systems are as indicated in Fig. 20(b).

[b]Symbols (θ_1 and θ_2) are the dihedral angles H(8)-C(8)-C(7)-H(7) and H(24)-C(24)-C(7)-H(7), respectively.

Table 8. Inter-plane angles and dihedral angles for LTAM molecules in the solid state.

The C(7)=C(8) double bonds of the two 5-chloro Fischer's base moieties have *EE* configurations in **1**. In contrast, in **2**, the C(14)=C(15) and C(2)=C(3) double bonds of the two 5-chloro Fischer's base moieties belong to the *ZE* configuration. The *EE* (for **1**) and *ZE* (for **2**) isomers formed as the sole product in each case, despite the fact that three isomers, namely *ZE*, *EE*, and *ZZ*, are possible for these dyes which result from the reaction of excess 5-chloro Fischer's base and 4- and 3-pyridine carboxaldehyde.

Compound **12** is stacked so that a dimer is formed in the unit cell of the crystal. The packing in the unit cell of LTAM **12** is distinct, as can be seen in Fig. 21.

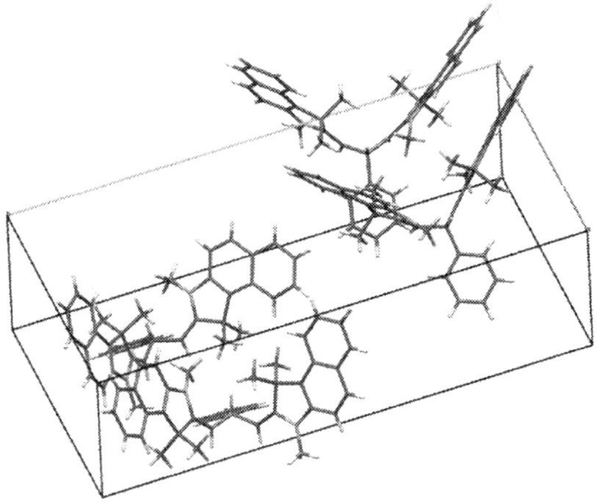

Figure 21. Molecular packing of LTAM **12**, showing the formation of a dimer.

4.2. Un-LTAM molecules

The Un-LTAM **4** was only successfully crystallized from acetone. Unfortunately, crystal growth was unsuccessful for the remainder of the Un-LTAM molecules. Selected bond lengths and bond angles are listed in Table 9.

Ring	Bond length (Å)		Bond angle (°)	
	N1-C2	1.397	C2-N1-C7a	111.7
A	N1-C10	1.439	C2-N1-C10	123.9
	C2-C7''	1.345	C2-C7''-H7''	117.1
	N2-C2'	1.408	C2'-N2-C7a'	111.3
B	N2-C10'	1.445	C7a'-N2-C10'	124.5
	C2'-C11''	1.324	C2'-N2-C10'	123.1
	C1''-C10''	1.516	C9''-C10''-C11''	110.92
	C9''-C10''	1.516	C1''-C10''-C11''	110.01
Conneting group	C10''-C11''	1.509	C9''-C10''-C11''	110.92
	C8''-C9''	1.328	C8''-C9''-C10''	127.29
	C7''-C8''	1.443	C11''-C2'-N2	122.42

Table 9. Selected bond lengths and bond angles of Un-LTAM **4**.

The X-ray crystal structure of Un-LTAM **4** shows a triclinic crystal system with space group *P-1*. An ORTEP diagram of Un-LTAM **4**, including the atom-numbering scheme, is shown in Fig. 22(a).

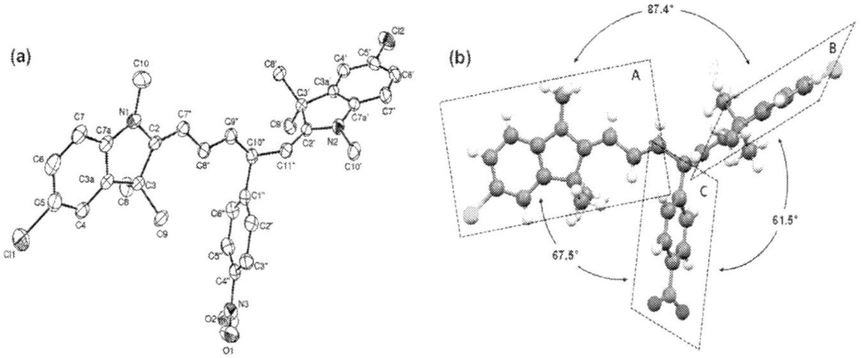

Figure 22. ORTEP diagrams with atom numbering scheme (a) and the propeller shape (b) of Un-LTAM **4**, showing the inter-plane angles.

The C9''-C10'' and C10''-C11''' distances are 1.516 and 1.509 Å, respectively, typical for C–C single bonds. The C7''-C8'' single bond distance, however, was 1.443 Å, which is shorter than a typical single bond and longer than a typical double bond. This is perhaps due to conjugation since the length (1.47 Å) of the central single bond of 1,3-butadiene is approximately 6 ppm shorter than that of the analogous single bond (1.53 Å) in butane. The two enamine C2=C7'' and C2'=C11'', and C8''=C9'' double bonds were 1.354, 1.324, and 1.328 Å, respectively, which are typical C=C bond lengths. In the crystal, the three aromatic rings of **4** are linked to three different layers, *viz.* a vinyl FB, a FB, and a phenyl group. The unsymmetrical molecule is a distorted version of the well-known three-bladed propeller conformation. (Keum et al., 2011). The inter-plane angles of the aromatic rings A–B, A–C, and B–C in Un-LTAM **4** are 87.4°, 67.5°, and 61.5°, respectively (Fig. 22(b)).

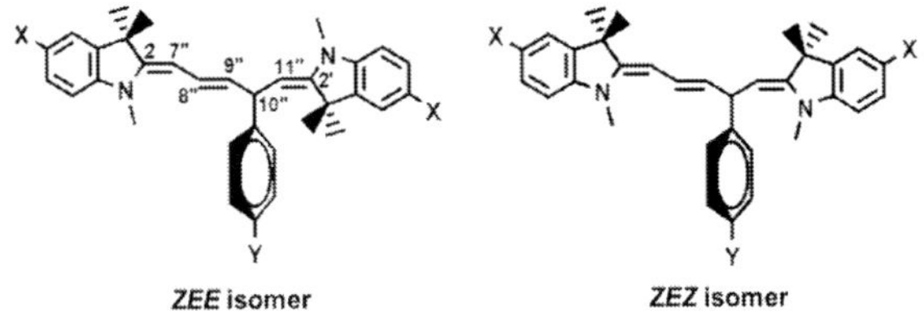

Figure 23. Chemical structures of Un-LTAM , with *ZEE* and *ZEZ* configurations.

The double bonds C2=C7", C8"=C9", and C2'=C11" of Un-LTAM **4** have *EEE* configurations. The *EEE* isomers of these LTAM dyes are formed as the sole product in all cases, even though there are three possible isomers, the two other diastereomers being the *ZEE* and *ZEZ* isomers, as shown in Fig. 23. Generally, the central carbon-carbon double bond of these LTAM dyes is expected to have an *E* configuration.

Although the presence of the *ZEE* and *ZEZ* diastereomers was generally expected to be found in organic solvents, none of these isomers were detected, unlike for the LTAM molecules examined previously.

Fig. 24 shows the molecular packing diagram of Un-LTAM **4**, showing the formation of the dimer, which is stacked in an alternating fashion in the unit cell of the crystal. The intermolecular distances in the dimer are 8.53 and 9.40 Å, for the FB and phenyl rings, respectively.

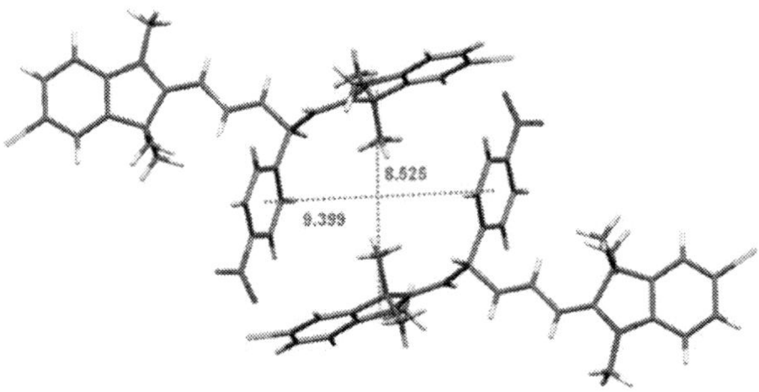

Figure 24. Molecular packing diagram of Un-LTAM **4**, showing formation of the dimer.

5. Conclusion

Novel Fischer's base analogs of LTAM and Un-LTAM molecules and their corresponding TAM^{+} dyes have been successfully developed. ^{1}H and ^{13}C NMR assignments for the prepared LTAM molecules have been completed by 1D and 2D NMR experiments, including DEPT, COSY, HSQC, HMBC, and NOESY. The geometry of the double bond was *ZE* in most cases, as measured directly by NOESY. The *EE* and *ZZ* isomers have C_2 symmetry, and hence, the two FB rings of these isomers are identical. Therefore, the ^{1}H NMR spectra of the *EE* and *ZZ* isomers are expected to be relatively simple compared to those of the *ZE* isomer. The novel LTAM molecules exist as a single isomer (*ZE* or *EE*) in the solid phase and they are equilibrated with other isomers in organic solvents. The percent ratios among the diastereomeric isomers of LTAM derivatives in the thermal equilibrium states vary according to the molecules examined and solvents used.

UV-Vis spectral data shows various structural forms of the LTAM and Un-LTAM molecules, such as (a) leuco-, (b) colored TAM$^+$, (c) carbinol-, and (d) decomposed-forms, similar to the commercially known TAM$^+$ dyes, such as MG, crystal violet, etc. Particularly, UV-Vis spectroscopic data for the Un-TAM$^+$ dyes showed absorptions in the near-IR region.

X-ray crystal analysis showed that the *ZE* isomers were generally formed with a so-called three-bladed propeller conformation. These isomers stacked to form a dimer or double dimer. However, the *EE* isomers were also formed specifically for the LTAMs **3**, **5**, **8**, and **11**, which have a resonance-electron withdrawing (-R) group at the para-position of the phenyl ring. Further analysis of a variety of substituted LTAM molecules is required to determine what makes the diastereomer structures change in the solid state.

Author details

Sam-Rok Keum, So-Young Ma and Se-Jung Roh
Korea University at Sejong Campus, South Korea

Acknowledgement

This research was supported by Basic Science Research Program through the National Research Foundation of Korea (NRF) funded by the Ministry of Education, Science and Technology (No.2012003244) and partly by the Brain Korea 21 project.

6. References

Anslyn, E.V. & Dougherty, D.A. (2006). *Modern Physical Organic Chemistry*, University Science Books, U.S. pp. 365-367.

Balko L, Allison J. The direct detection and identification of staining dyes from security inks in the presence of other colorants, on currency and fabrics, by laser desorption mass spectrometry. *J. Foren. Sci.* 2003, *48*: 1172–8.

Bartholome, D. & Klemm, E. (2006). Novel Polyarylene−Triarylmethane Dye Copolymers, *Macromolecules*, Vol. 39 pp. 5646-5651.

Cho BP, Yang T, Lonnie R, Blankenship L, Moody JD, Churchwell M, Beland FA. Culp S. *J. Chem. Res. Toxicol.* 2003, *19*, 285.

Ernst, L. A.; Gupta, R. K,; Mujumdar, R. B. & Waggoner, A. S. (1989). Cyanine dye labeling reagents for sulfhydryl groups, *Cytometry* Vol. 10, pp. 3-10.

Fengling, S. F. & Xiaojun, P. X. (2005). Heptamethine Cyanine Dyes with a Large Stokes Shift and Strong Fluorescence: A Paradigm Excited-State Intramolecular Charge Transfer. 2. *J. Am. Chem. Soc.* Vol. 127, pp. 4170-4171.

Gessner, T. & Mayer, U. (2005). Triarylmethane and Diarylmethane Dyes, *Ullmann's Encyclopedia of Industrial Chemistry*, Weinheim: Wiley-VCH, doi:10.1002/14356007.a27-179

Indig GL, Anderson GS, Nichols MG, Bartlett JA, Mellon WS, Sieber F. Effect of molecular structure on the performance of triarylmethane dyes as therapeutic agents for photochemical purging of autologous bone marrow grafts from residual tumor cells. *J. Pharm. Sci.* 2000, *89*: 88–99.

Kraus, G. A.; Jeon, I.; Nilsen-Hamilton, M.; Awad, A. M.; Banerjee J. & Parvin, B. (2008). Fluorinated Analogs of Malachite Green: Synthesis and Toxicity, *Molecules,* Vol. 13, No. 4, pp. 986-994; doi:10.3390/molecules13040986

Keum, S. R.; Roh, S. J,; Lee, M. H.; Saurial, F. & Buncel, E. (2008). ^{1}H and ^{13}C NMR assignments for new heterocyclic TAM leuco dyes, (2*Z*,2′*E*)-2,2′-(2-phenyl propane-1,3-diylidene)bis(1,3,3-trimethylindoline) derivatives. Part II. *Magn. Reson. Chem*. Vol. 46, pp. 872–877.

Keum, S. R.; Roh, S. J.; Kim, Y.N.; Im, D.H. & Ma, S. Y. (2009). X-ray crystal structure of hetaryl leuco-TAM dyes, (2Z,2′*E*)-2,2′-(2-phenylpropane-1,3-diylidene) bis(1,3,3-trimethyl indoline) derivatives. *Bull. Korean Chem. Soc.* Vol. 30, pp. 2608–2612.

Keum, S. R.; Roh, S. J,; Ma, S. Y.; Kim, D. K. & Cho, A. E. (2010). Diastereomeric isomerization of hetaryl leuco-TAM dyes, (2Z, 2′E)-2,2′-(2-phenyl propane-1,3-diylidene) bis(1,3,3-trimethylindoline) derivatives in various organic solvents. *Tetrahedron*, Vol. 66, pp. 8101–8107.

Keum, S. R.; Lee, M. H.; Ma, S. Y.; Kim, D. K. & Roh, S. J. (2011). Novel unsymmetrical leuco-TAM, (2*E*, 2′*E*)-2,2′-{(E)-4-phenylpent-2-ene-1,5-diylidene}bis(1,3,3-trimethyl indoline) derivatives: synthesis and structural elucidation. *Dyes and Pigments*, Vol. 90, pp. 233–238.

Liao, S. & Collum, D. B. (2003). Lithium Diisopropylamide-Mediated Lithiations of Imines: Insights into Highly Structure-Dependent Rates and Selectivities, *J. Am. Chem. Soc.* Vol. 125, pp. 15114-15127.

Li, Z.; Duan, Z.; Kang, J.; Wang, H.; Yu, L. & Wu, Y. (2008). A simple access to triarylmethane derivatives from aromatic aldehydes and electron-rich arenes catalyzed by $FeCl_3$, *Tetrahedron* Vol. 64, pp. 1924-1930.

Ma, S. Y.; Kim, D. K.; Lim, H. Y.; Roh, S. J. & Keum, S. R. (2012). Unusual Stability of Diastereomers of the Isomeric Pyridine-based Leuco-TAM Dyes 2,2'-(2-(Pyridin-4 or 3-yl) propane-1,3-diylidene)bis(5-chloro-1,3,3-trimethylindoline), *Bull. Korean Chem. Soc.* Vol. 33, pp. 681-684.

Nair, V.; Thomas, S.; Mathew, S. C. & Abhilash, K. G. (2006). Recent advances in the chemistry of triaryl- and triheteroarylmethanes. *Tetrahedron,* Vol. 62, pp. 6731-6747.

Özer, I. & Çaglar, A. (2002). Protein-mediated nonphotochemical bleaching of malachite green in aqueous solution. *Dyes & Pigments,* Vol. 54, pp. 11-16.

Plakas, S. M.; Doerge, D. R.; Turnipseed, S. B. In Xenobiotics in Fish; Kluwer Academic and Plenum Publisher: NY. 1999; pp. 149.

Keum, S. R.; Ma, S. Y.; Kim, D. K.; Lim, H. W. & Roh, S. J. (2012). Novel dimeric leuco-TAM dyes, 1,4-bis{(1*E*,3*Z*)-1,3-bis(1,3,3-trimethylindolin-2-ylidene)propan-2-yl}benzene derivatives: Structure and spectroscopic charactierization. *Journal of Molecular structure,* Vol. 1014, pp. 25-32.

Keum, S. R.; Ma, S. Y.; Kim, D. K.; Lim, H. W. & Roh, S. J. (2012). Unsymmetric leuco-TAM dyes, (2*E*, 2′*E*)-2,2′-{(E)-4-phenylpent-2-ene-1,5-diylidene}bis(1,3,3-trimethylindoline) derivatives. Part II: X-ray crystal structure☆. *Dyes and Pigments*, Vol. 94, pp. 490–495.

Section 5

Physical Spectroscopy

Chapter 16

Atomic and Molecular Low-n Rydberg States in Near Critical Point Fluids*

Luxi Li, Xianbo Shi, Cherice M. Evans and Gary L. Findley

Additional information is available at the end of the chapter

http://dx.doi.org/10.5772/48089

1. Introduction

Since electronic excited states are sensitive to the local fluid environment, dopant electronic transitions are an appropriate probe to study the structure of near critical point fluids (i.e., perturbers). In comparison to valence states, Rydberg states are more sensitive to their environment [1]. However, high-n Rydberg states are usually too sensitive to perturber density fluctuations, which makes these individual dopant states impossible to investigate. (Nevertheless, under the assumption that high-n Rydberg state energies behave similarly to the ionization threshold of the dopant, dopant high-n Rydberg state behavior in supercritical fluids can be probed indirectly by studying the energy of the quasi-free electron, through photoinjection [2–11] and field ionization [12–19].) Low-n Rydberg states, on the other hand, are excellent spectroscopic probes to investigate excited state/fluid interactions.

The study of low-n Rydberg states in dense fluids began with the photoabsorption of alkali metals in rare gas fluids [20, 21]. Later researchers expanded the investigation into rare gas dopants in supercritical rare gas fluids [22–26], and molecular dopants in atomic and molecular perturbers [27–36]. However, none of these previous groups studied dilute solutions near the critical point of the solvent. (The single theoretical study of a low-n Rydberg state in a near critical point fluid was performed by Larrégaray, *et al.* [35]; this investigation predicted a change in the line shape and in the perturber induced shift of the Rydberg transition.) These results from previous experimental and theoretical investigations of low-n Rydberg states in dense fluids are reviewed in Section 2.

In this Chapter, we present a systematic investigation of the photoabsorption of atomic and molecular dopant low-n Rydberg transitions in near critical point atomic fluids [37–40]. The individual systems probed allowed us to study (dopant/perturber) atomic/atomic interactions (i.e., Xe/Ar) and molecular/atomic interactions (i.e., CH_3I/Ar, CH_3I/Kr, CH_3I/Xe) near the perturber critical point. The experimental techniques and theoretical

*This work is adapted from that originally submitted by Luxi Li to the faculty of the Graduate Center of the City University of New York in partial fulfillment of the requirements for the degree of Doctor of Philosophy.

methodology for this extended study of dopant/perturber interactions are discussed in Section 3. Section 4 presents a review of our results for low-n atomic and molecular Rydberg states in atomic supercritical fluids. The accuracy of a semi-classical statistical line shape analysis is demonstrated, and the results are then used to obtain the perturber-induced energy shifts of the primary low-n Rydberg transitions. A striking critical point effect in this energy shift was observed for all of the dopant/perturber systems presented here. A discussion of the ways in which the dopant/perturber interactions influence the perturber-induced energy shift is also presented in Section 4. We conclude with an explanation of the importance of the inclusion of three-body interactions in the line-shape analysis, and with a discussion of how this model changes when confronted with non-spherical perturbers and polar fluids.

2. Perturber effects on low-n Rydberg states

2.1. Supercritical fluids

A supercritical fluid (SCF) exists at a temperature higher than the critical temperature and, therefore, has properties of both a liquid and a gas [41]. For example, SCFs have the large compressibility characteristic of gases, but the potential to solvate materials like a liquid. Moreover, near the critical point the correlation length of perturber molecules becomes unbounded, which induces an increase in local fluid inhomogeneities [41].

The local density $\rho(r)$ of a perturbing fluid around a central species (either the dopant or a single perturber) is defined by [42, 43]

$$\rho(r) = g(r)\,\rho\,,$$

where $g(r)$ is the radial distribution function and ρ is the bulk density. For neat fluids and for most dilute dopant/perturber systems, the interactions between the species in the system are attractive in nature. Thus, the perturber forms at least one solvent shell around the dopant (or a central perturber). As the dilute dopant/perturber system approaches the critical density and temperature of the perturber, the intermolecular interactions increase. This increase leads to a change in the behavior of the local density as a function of the bulk density [41]. Therefore, the changes in fluid properties near the critical point are due to the higher correlation between the species in the fluid. These intermolecular correlations are usually probed spectroscopically by dissolving a dopant molecule in the SCF. Since electronic excited states are incredibly sensitive to local fluid environment, fluctuations in the fluid environment can be investigated by monitoring variations in the absorption or emission of the dopant.

Fig. 1 gives three example spectroscopic studies of a dopant in near critical point SCFs. Fig 1a shows the energy position of an anthracene emission line for anthracene doped into near critical point carbon dioxide [44]. Unfortunately, no emission data on non-critical isotherms were measured in this fluorescence emission study. However, the maximum emission position shows a striking critical effect in comparison to a calculated baseline. A more complete investigation of temperature effects on the local density of SCFs via UV photoabsorption [45] of ethyl p-(N,N-dimethylamino)benzoate (DMAEB) in supercritical CHF_3 is presented in Fig. 1b for three isotherms near the critical isotherm. The isotherms shown are at the reduced temperature T_r [where $T_r \equiv T/T_c$ with T_c being the critical temperature of the fluid] of 1.01, 1.06 and 1.11. Although the three isotherms are evenly spaced, the photoabsorption shifts are very similar on the $T_r = 1.06$ and $T_r = 1.11$ isotherms, while the shift on the

$T_r = 1.01$ isotherm is significantly different. Therefore, the perturber induced shift is only temperature sensitive near the critical point of the perturber. Urdahl, *et al.* [46, 47] studied the $W(CO)_6$ T_{1u} asymmetric CO stretching mode doped into supercritical carbon dioxide, ethane and trifluoromethane by IR photoabsorption. All three systems show similar behavior, an example of which is presented in Fig. 1c. Again, the absorption band position changes significantly along the critical isotherm near the critical density. However, extracting the data presented in Fig. 1 is difficult. Moreover, most of the previous work [41] has used dopant valence transitions as probes, and these states tend to be less sensitive to local environments [1]. In the present work, we investigate near critical point SCFs using low-n Rydberg state transitions as the probe. Since Rydberg states are hydrogen-like, it should be possible to model these states within the statistical mechanical theory of spectral line-broadening.

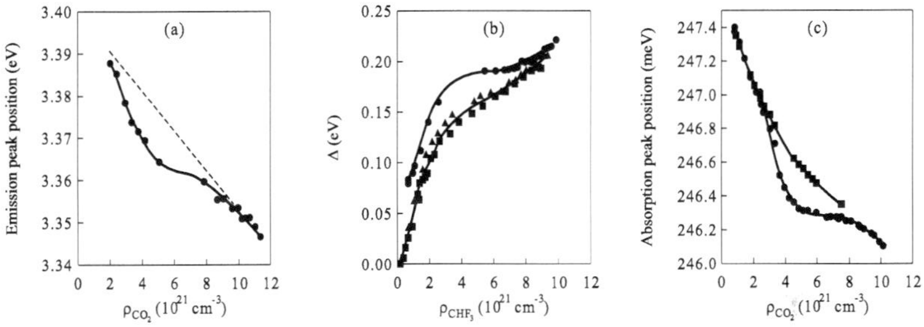

Figure 1. (a) The energy of fluorescence emission for (•) anthracene doped into supercritical carbon dioxide at a reduced temperature $T_r \simeq 1.01$ plotted as a function of carbon dioxide number density ρ_{CO_2}. The dashed line (- - -) is a reference line calculated using perturber bulk densities. Adapted from [44]. (b) The perturber induced energy shifts of the photoabsorption maximum for DMAEB doped into supercritical CHF_3 at (•) 30.0°C [i.e., $T_r \simeq 1.01$], (▲) 44.7°C, and (■) 59.6°C, plotted as a function of CHF_3 number density ρ_{CHF_3}. Adapted from [45]. (c) The energy position of infrared absorption for the $W(CO)_6$ CO stretching mode in supercritical carbon dioxide at (•) 33°C [i.e., $T_r \simeq 1.01$] and at (■) 50°C, plotted as a function of ρ_{CO_2}. Adapted from [46, 47]. The solid lines in (a) - (c) are provided as a visual aid.

2.2. Theoretical model

Due to the hydrogenic properties of Rydberg states, the optical electron is in general insensitive to the structure of the cationic core. Therefore, both atomic and molecular Rydberg transitions can be modeled within the same theory. In a very dilute dopant/perturber system, assuming that the dopant Rydberg transition is at high energy [i.e. $\beta\,(E_g - E_e) \gg 1$, $\beta = 1/(k\,T)$], the Schrödinger equation is

$$H|\Psi\rangle = E|\Psi\rangle\,, \tag{1}$$

where the eigenfunction $|\Psi\rangle$ is a product of the dopant electronic wavefunction $|\alpha\rangle$ and the individual perturber wavefunctions $|\psi_i;\,\alpha\rangle$. The Hamiltonian H in eq. (1) is the sum of several individual Hamiltonians, namely the Hamiltonian for the free dopant, the Hamiltonian for the free perturber, and the Hamiltonian for the dopant/perturber intermolecular correlation. The Hamiltonian H_{FD} for the free dopant is [48]

$$H_{FD} = \sum_{\alpha} E_\alpha |\alpha\rangle\langle\alpha|\,, \tag{2}$$

where $\alpha = e, g$ with e and g representing the excited state and ground state of the free dopant. The Hamiltonian H_{FP} for the free non-interacting perturber is given by [48]

$$H_{FP} = -\sum_{i}^{N} \frac{\hbar^2}{2m} \nabla_i^2 , \tag{3}$$

where N is the total number of perturbing atoms in the range of the Rydberg optical electron. Finally, the Hamiltonian H_{PD} for the intermolecular interaction is [48]

$$H_{PD} = \sum_{\alpha} \sum_{i} \left[V_{\alpha}(r_i) + V'(r_i)\right] |\Psi\rangle\langle\Psi| , \tag{4}$$

where $V(r)$ is the dopant/perturber intermolecular potential, and $V'(r)$ is the perturber/perturber intermolecular potential. Therefore, the total Hamiltonian H is [48]

$$H = H_{FD} + H_{FP} + H_{PD} , \tag{5}$$

which can be rewritten as a ground state Hamiltonian expectation value H_g and an excited state Hamiltonian expectation value H_e.

The absorption coefficient function is given as the Fourier transform [48]

$$\mathfrak{L}(\omega) \equiv \frac{1}{2\pi} \int_{-\infty}^{\infty} dt \, e^{-i\omega t} \langle \vec{\mu}(0) \cdot \vec{\mu}(t) \rangle , \tag{6}$$

where the autocorrelation function (i.e., $\langle \cdots \rangle$) is the thermal average of the scalar product of the dipole moment operator (i.e., $\vec{\mu}$) of the dopant at two different times. This autocorrelation function can be resolved within the Liouville operator formalism to give [48]

$$\langle \vec{\mu}(0) \cdot \vec{\mu}(t) \rangle \equiv \exp\left[\rho_P \langle e^{i L_g t} - 1 \rangle_g \right] , \tag{7}$$

where the two-body Liouville operator L_g is defined by

$$L_g \Omega = H_e \Omega - \Omega H_g = [H_g, \Omega] + (E_e - E_g)\Omega , \tag{8}$$

where Ω is an arbitrary operator. However, if lifetime broadening is neglected, only the dopant/perturber interaction and the dopant electronic energy change during the transition. Therefore, the autocorrelation function can be rewritten as [48]

$$\langle \vec{\mu}(0) \cdot \vec{\mu}(t) \rangle \equiv e^{i\omega_0 t} \exp\left[\rho_P \langle e^{i\Delta V t} - 1 \rangle_g \right] , \tag{9}$$

where ω_0 is the transition frequency of the neat dopant, ρ_P is the perturber density, and $\Delta V = V_e - V_g$, with V_e and V_g being the excited-state dopant/ground-state perturber and ground-state dopant/ground-state perturber intermolecular potentials, respectively.

In semi-classical line shape theory, the line shape function for an allowed transition is given by [20],

$$\mathfrak{L}(\omega) = \frac{1}{2\pi} \int_{-\infty}^{\infty} dt \, e^{-i[\omega(\mathbf{R}) - \omega_0] t} \, \frac{Z(\beta V_g + i t \Delta V)}{Z(\beta V_g)} , \tag{10}$$

where Z is the partition function and $\mathbf{R}$ denotes the collection of all dopant/perturber distances. Under the classical fluid approximation of Percus [49–51], the autocorrelation function $\Phi(t)$ is given by a density expansion [20]

$$\Phi(t) = \ln Z\left(\beta V_g + it\Delta V\right) - \ln Z\left(\beta V_g\right) = A_1(t) + A_2(t) + \cdots, \tag{11}$$

where [20, 25]

$$\begin{aligned} A_n(t) &= \frac{1}{n!}\int \cdots \int \prod_{j=1}^{n} d^3R_j\, \mathfrak{F}(\mathbf{R}_1,\ldots,\mathbf{R}_n) \\ &\times \prod_{j=1}^{n}\left[\exp(-i\,\Delta V(\mathbf{R}_j)\,t) - 1\right] . \end{aligned} \tag{12}$$

In eq. (12), $\mathfrak{F}(\mathbf{R}_1,\ldots,\mathbf{R}_n)$ is the Ursell distribution function [25, 26, 49–51], and $\Delta V(\mathbf{R}) = V_e(\mathbf{R}) - V_g(\mathbf{R})$. The Ursell distribution function for two body interactions [25, 26] is $\mathfrak{F}(\mathbf{R}) = \rho_{\mathrm{P}}\, g_{\mathrm{PD}}(\mathbf{R})$, where $g_{\mathrm{PD}}(\mathbf{R})$ is the perturber/dopant radial distribution function. The three body Ursell distribution function is estimated using the Kirkwood superposition approximation [52] to be

$$\mathfrak{F}(\mathbf{R}_1,\mathbf{R}_2) = \rho_{\mathrm{P}}^2\left[g_{\mathrm{PP}}(|\mathbf{R}_1 - \mathbf{R}_2|) - 1\right] g_{\mathrm{PD}}(\mathbf{R}_1)\, g_{\mathrm{PD}}(\mathbf{R}_2), \tag{13}$$

where $g_{\mathrm{PP}}(\mathbf{R})$ is the perturber/perturber radial distribution function. The density expansion terms are multibody interactions between dopant and perturber over all space. $A_1(t)$ is the dopant/perturber two-body interaction, $A_2(t)$ is the dopant/perturber three-body interaction, and $A_n(t)$ is the dopant/perturber $n+1$ body interaction. Substitution of these Ursell distribution functions into eq. (12) under the assumption of spherically symmetric potentials, gives [25, 26, 37]

$$A_1(t) = 4\pi\rho_{\mathrm{P}}\int_0^{\infty} dr\, r^2\, g_{\mathrm{PD}}(r)\left[\mathrm{e}^{-it\Delta V(r)} - 1\right], \tag{14}$$

and

$$\begin{aligned} A_2(t) &= 4\pi\rho_{\mathrm{P}}^2\int_0^{\infty} dr_1\, r_1^2\, g_{\mathrm{PD}}(r_1)\left[\mathrm{e}^{-it\Delta V(r_1)} - 1\right] \\ &\times \int_0^{\infty} dr_2\, r_2^2\, g_{\mathrm{PD}}(r_2)\left[\mathrm{e}^{-it\Delta V(r_2)} - 1\right] \\ &\times \frac{1}{r_1 r_2}\int_{|r_1-r_2|}^{|r_1+r_2|} s\left[g_{\mathrm{PP}}(s) - 1\right] ds . \end{aligned} \tag{15}$$

Since the strength of the interaction decreases as the number of bodies involved increases, and since higher order interactions are more difficult to model, most line shape simulations are truncated at the second term $A_2(t)$.

The autocorrelation function can be written as a power series expansion at $t = 0$, namely [21, 48, 53, 54]

$$\Phi(t) = \sum_{n=1}^{\infty}\frac{i^n}{n!}\, m_n\, t^n, \tag{16}$$

where the expansion coefficients m_i are given by

$$m_n = \frac{1}{(\sqrt{-1})^n}\,\frac{d^n}{dt^n}\Phi(t)\bigg|_{t=0} . \tag{17}$$

Using eq. (11) with $A_1(t)$ and $A_2(t)$ from eqs. (14) and (15), the first two expansion coefficients become [25]

$$m_1 = -4\pi\rho_P \int_0^\infty dr\, r^2\, g_{PD}(r)\, \Delta V(r)\,, \tag{18}$$

and

$$\begin{aligned} m_2 &= m_1^2 + 4\pi\rho_P \int_0^\infty dr\, r^2\, g_{PD}(r)\, \Delta V(r)^2 \\ &+ 8\pi^2 \rho_P{}^2 \int_0^\infty \int_0^\infty dr_1\, dr_2\, r_1 r_2\, g_{PD}(r_1)\, g_{PD}(r_2) \\ &\times \Delta V(r_1)\, \Delta V(r_2) \int_{|r_1 - r_2|}^{|r_1 + r_2|} s\, [\, g_{PP}(s) - 1\,]\, ds\,. \end{aligned} \tag{19}$$

The above expansion coefficients are equivalent to the moments of the optical coefficient, which are defined as [25]

$$m_n = \int \mathfrak{L}(E)\, E^n\, dE\,. \tag{20}$$

Thus, the perturber induced shift $\Delta(\rho_P)$ in the energy position of the optical coefficient maximum is [25, 26]

$$\Delta(\rho_P) = M_1 = \frac{m_1}{m_0} = \int \mathfrak{L}(E)\, E\, dE \Big/ \int \mathfrak{L}(E)\, dE\,, \tag{21}$$

while the full-width-half-maximum of the experimental absorption spectrum is proportional to

$$M_2 = \left[\frac{m_2}{m_0} - M_1^2\right]^{1/2} = \left[\left(\int \mathfrak{L}(E)\, E^2\, dE \Big/ \int \mathfrak{L}(E)\, dE\right) - M_1^2\right]^{1/2}, \tag{22}$$

where $\mathfrak{L}(E)$ is the absorption band for the transition and $E = \hbar(\omega - \omega_0)$ [$\hbar \equiv$ reduced Planck constant].

2.3. Previous studies

The interaction of dopant low-n Rydberg states with dense fluids has previously been the subject of some interest both experimentally and theoretically [1, 22–36]. The first detailed investigation of these interactions was a study of Xe low-n Rydberg states doped into the dense rare gas fluids (i.e., argon, neon and helium) by Messing, *et al.* [25, 26]. In the same year, Messing, *et al.* [27, 28] presented their studies of molecular low-n Rydberg states in dense Ar and Kr. A decade later, Morikawa, *et al.* [34] probed the NO valence and low-n Rydberg state transitions in dense argon and krypton. All of these experiments [25–28, 34] used a basic moment analysis of absorption bands to determine the energy shifts of these bands as a function of perturber number density. The photoabsorption energy shifts were then simulated [25–28, 34] using eq. (18) for various assumptions of intermolecular potentials and radial distribution functions. Messing, *et al.* also performed line shape simulations under the assumption of a Gaussian line shape [25–28] for selected perturber number densities. As molecular dynamics developed, research groups [30, 31, 33, 35, 36] returned to absorption line shape simulations in an attempt to match the asymmetric broadening observed experimentally.

2.3.1. Xe in Ar, Ne and He

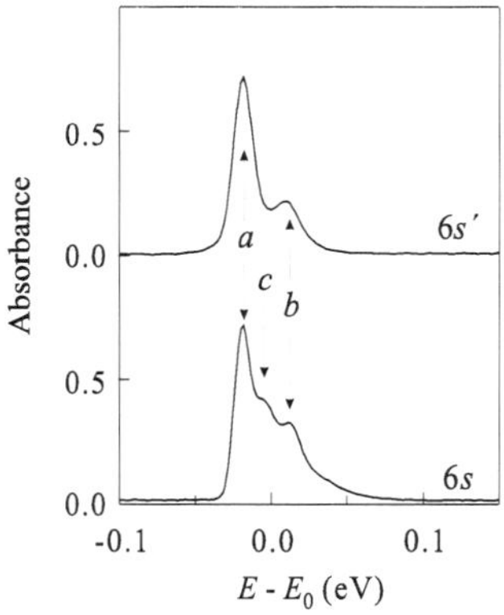

Figure 2. Photoabsorption spectra (relative units) of the Xe 6*s* and 6*s*′ transitions doped into Ar at an argon number density of $\rho_{Ar} = 1.47 \times 10^{21}$ cm^{-3} and a temperature of 23.6°C. *a* corresponds to eq. (23), *b* to eq. (24), and *c* to eq. (25). For the 6*s* Rydberg state $E_0 = 8.437$ eV and for the 6*s*′ Rydberg state $E_0 = 9.570$ eV. Data from the present work.

When Xe 6*s* and 6*s*′ Rydberg state transitions are excited in the presence of low density argon or krypton, satellite bands appear on the higher energy wing of the absorption or emission spectra [22–24]. These blue satellite bands increase with increasing perturber density [22–24] and with decreasing temperature. Therefore, Castex, *et al.* [22–24] concluded that these blue satellites are caused by the formation of dopant/perturber ground state and excited state dimers. An example of the Xe 6*s* and 6*s*′ Rydberg transitions in the presence of argon is shown in Fig. 2. The primary Xe absorption transition, or [37]

$$\mathrm{Xe} + h\nu \xrightarrow{\mathrm{Ar}} \mathrm{Xe}^* , \tag{23}$$

is indicated in Fig. 2 as *a*. The XeAr dimer transitions that yield the blue satellite bands are [22–24]

$$\mathrm{XeAr} + h\nu \xrightarrow{\mathrm{Ar}} \mathrm{Xe}^* + \mathrm{Ar} , \tag{24}$$

and

$$\mathrm{XeAr} + h\nu \xrightarrow{\mathrm{Ar}} \mathrm{Xe}^*\mathrm{Ar} . \tag{25}$$

These transitions are indicated in Fig. 2 as *b* and *c*, respectively.

Detailed investigations [25] of the Xe 6*s* Rydberg state doped in supercritical argon indicated that the energy position at the photoabsorption peak maximum shifted slightly to the red at low argon density and then strongly to the blue (cf. Fig. 3a). Similar results, which are shown in Fig. 3b, were then observed for the Xe 6*s*′ Rydberg states in supercritical argon [26]. These studies [25, 26] concluded that both the perturber-induced energy shift and the line shape broadening were temperature independent. However, since the blue satellite bands grow and broaden as a function of perturber number density, the energy position of the maximum absorption (or the first moment from a moment analysis) is not an accurate energy position for the primary Xe Rydberg transition. Thus, modeling the experimental first moment M_1 and

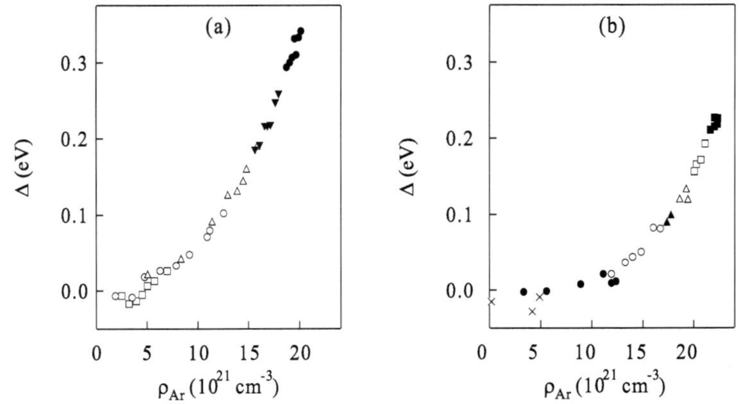

Figure 3. The perturber induced shift Δ of (a) the Xe 6*s* [25] and (b) the Xe 6*s*′ [26] absorption maximum plotted as a function of argon number density ρ_{Ar} at different temperatures. In (a), the markers are (□) 25°C; (○) −83°C; (△) −118°C; (▼) −138°C; and (•) −163°C. In (b), the markers are for temperature ranges of (×) −23°C to 27°C; (•) −93°C to −63°C; (○) −123°C to −113°C; (▲) −138°C to −128°C; (△) −158°C to −148°C; (□) −173°C to −163°C; and (■) −186°C to −178°C.

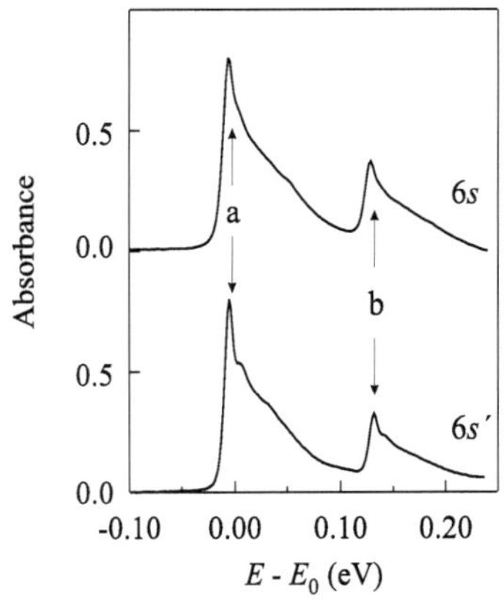

Figure 4. Photoabsorption spectra (relative units) of the CH_3I 6*s* and 6*s*′ transitions doped into Ar at an argon number density of $\rho_{Ar} = 1.89 \times 10^{21}$ cm^{-3} and a temperature of −79.8°C. *a* corresponds to eq. (26) and *b* to eq. (27). For the 6*s* Rydberg state $E_0 = 6.154$ eV and $E_0 = 6.767$ eV for the 6*s*′ Rydberg state. Data from the present work.

second moment M_2 using eqs. (19) and (20) required three groups of intermolecular potential parameters for different perturber density ranges.

The Xe 6*s* and 6*s*′ Rydberg transitions [26] in supercritical helium and neon show a similar perturber density dependence as that shown in Fig. 3. However, these two systems do not form ground state or excited state dimers and, therefore, do not have blue satellite bands. Thus, the moment analysis of the photoabsorption spectra presented a more accurate perturber induced shift as a function of perturber number density. Because of the

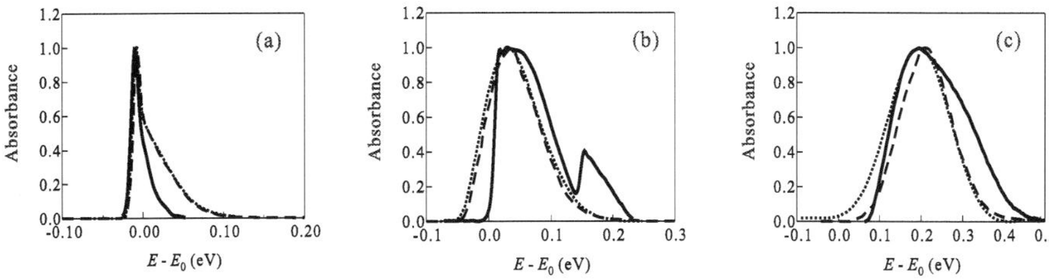

Figure 5. The line shape simulation of the CH_3I 6s transition doped into supercritical argon using (- - -) a semi-classical line shape function data [33] and using (· · ·) molecular dynamics [32] in comparison to (—) the photoabsorption spectra at (a) $\rho_{Ar} = 2.0 \times 10^{21}$ cm^{-3}, (b) $\rho_{Ar} = 7.6 \times 10^{21}$ cm^{-3}, and (c) $\rho_{Ar} = 2.0 \times 10^{22}$ cm^{-3}. Experimental data are from the present work.

simplicity of the absorption bands, Messing, *et al.* [26] simulated the line shapes of the Xe 6s and 6s' transitions in both helium and neon at a single perturber number density using eqs. (10) and (11). These simulations indicated that an accurate line shape could be obtained without blue satellite bands [26]. Unfortunately, no temperature dependence was reported in these papers [25, 26].

2.3.2. CH_3I in Ar and Kr

Since CH_3I is a molecular dopant, vibrational transitions as well as the adiabatic transition appear in the photoabsorption spectra. The adiabatic transition is given by [37]

$$CH_3I + h\nu \longrightarrow CH_3I^* , \tag{26}$$

denoted a in Fig. 4, as well as one quantum of the CH_3 deformation vibrational transition ν_2 in the excited state, or

$$CH_3I + h\nu \longrightarrow CH_3I^* \, (\nu_2) , \tag{27}$$

denoted b, in Fig. 4.

Although vibrational transitions are apparent, CH_3I [27–29, 32] has been investigated extensively because of the "atomic" like nature of the adiabatic 6s and 6s' Rydberg transitions. Messing, *et al.* [27, 28] extracted the perturber dependent shift $\Delta(\rho_P)$ of the CH_3I 6s and 6s' Rydberg states by performing a moment analysis on the photoabsorption bands using eq. (21). This analysis indicated that $\Delta(\rho_P)$ tended first to lower energy and then to higher energy as ρ_P increased from low density to the density of the triple-point liquid, similar to the trends shown in Fig. 3 for Xe in Ar. However, Messing, *et al.* [27, 28] did not explore critical temperature effects on $\Delta(\rho_P)$, nor did they correctly account for the vibrational bands on the blue side of the adiabatic Rydberg transition. Messing, *et al.* [28] did attempt to model the CH_3I 6s Rydberg transition in argon using a semi-classical statistical line shape function under the assumption that the adiabatic and vibrational transitions have exactly the same line shapes, although no comparison between the experimental spectra and the simulated line shapes was provided. Later researchers [30–33] concentrated on the simulation of the CH_3I 6s Rydberg state doped into argon using both molecular dynamics and semi-classical integral methods. Egorov, *et al.* [33] showed that the semi-classical integral method can yield results comparable to the molecular dynamics calculations of Ziegler, *et al.* [30–32]. Comparing the semi-classical method and molecular dynamics simulation to experimental spectra (cf. Fig. 5) shows that

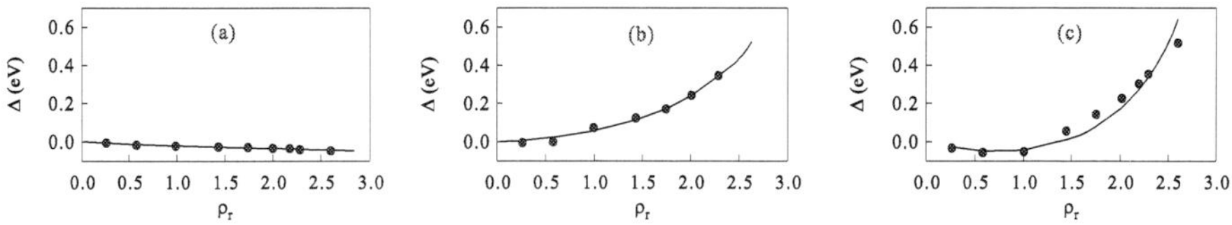

Figure 6. Experimental (•) and calculated (—) energy shift of the NO (a) $B'^2\Delta \leftarrow X^2\Pi$ ($v' = 7, 0$) valence transition, (b) $A^2\Sigma^+ \leftarrow X^2\Pi$ ($v' = 0, 0$) Rydberg transition and (c) $C^2\Pi^+ \leftarrow X^2\Pi$ ($v' = 0, 0$) Rydberg transition plotted versus the reduced argon number density [34].

both methods can be used to simulate the experimental spectra with an appropriate choice of intermolecular potentials.

2.3.3. NO in Ar

Morikawa, *et al.* [34] investigated valence and low-n Rydberg transitions doped into supercritical argon. They used eq. (21) to determine the perturber induced shift $\Delta(\rho_P)$ for several low-n Rydberg transitions as well as a valence transition. Under the assumption of spherically symmetric potentials for the ground and excited states of the NO/Ar systems, an accurate moment analysis using eq. (18) was performed (cf. Fig. 6). In The NO valence state transition (cf. Fig. 6a) shows a slight perturber-induced red shift, which differs significantly from the obvious blue shift of the low-n NO Rydberg state transitions shown in Figs. 6b and c. Fig. 6 also shows that low-n Rydberg states make a more sensitive probe to perturber effects than valence transitions. Later groups [35, 36] did line shape simulations to model the experimental spectra. Larrégaray, *et al.* [35] measured the NO 3sσ transition doped into supercritical argon at selected argon densities. Then, using molecular dynamics they successfully modeled the line shape of the transition. Once the intermolecular potentials and boundary conditions for the molecular dynamics simulations had been set against experimental data, Larrégaray, *et al.* [35] calculated the line shape for NO in Ar at the critical density and temperature. These calculations predicted that the photoabsorption peak maximum position would shift to the blue when argon was near its critical point. Later, Egorov, *et al.* [36] showed that similar results could be obtained using the semi-classical approximation. Therefore, line shape simulations using molecular dynamics and semi-classical line shape theory show comparable results.

3. Experiment techniques and theoretical methodology

3.1. Experimental techniques

All of the photoabsorption measurements presented in Sections 4 and 5 were obtained using the one-meter aluminum Seya-Namioka (Al-SEYA) beamline on bending magnet 8 of the Aladdin storage ring at the University of Wisconsin Synchrotron Radiation Center (SRC) in Stoughton, WI. This beamline, which was decommissioned during Winter 2007, produced monochromatic synchrotron radiation having a resolution of $\sim$ 8 meV in the energy range of 6 - 11 eV. The monochromatic synchrotron radiation enters the vacuum chamber which is equipped with a sample cell (cf. Fig. 7). The photoabsorption signal is detected by a photomultiplier that is connected to the data collection computer via a Keithley 486 picoammeter. The pressure in the vacuum chamber is maintained at low 10^{-8} to high 10^{-9}

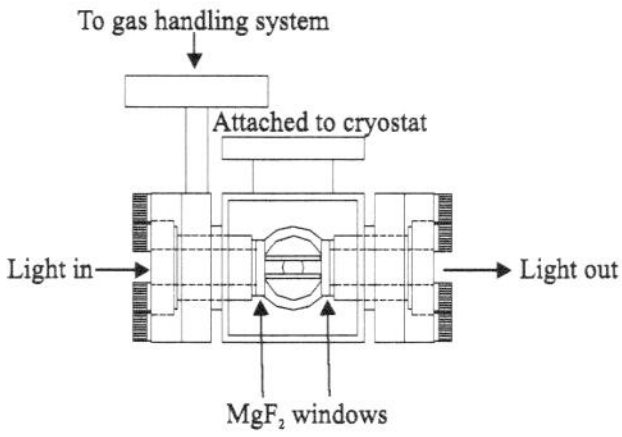

Figure 7. Schematic of the copper experimental cell.

Torr by a Perkin-Elmer ion pump. The experimental copper cell is equipped with entrance and exit MgF_2 windows (with an energy cut-off of 10.9 eV) that are capable of withstanding pressures of up to 100 bar and temperatures of up to 85°C. This cell, which has a path length of 1.0 cm, is connected to an open flow liquid nitrogen cryostat and resistive heater system allowing the temperature to be controlled to within ± 0.5°C with a Lakeshore 330 Autotuning Temperature controller. The gas sample is added through a gas handling system (GHS), which consists of 316-stainless steel components connected by copper gasket sealed flanges. The initial pressure for the GHS and sample cell is in the low 10^{-8} Torr range.

During the initial bakeout, the GHS and the vacuum chamber were heated to 100°C under vacuum for several days to remove any water adsorbed onto the surface of the stainless steel. The initial bake was stopped when the base pressure of the GHS and sample chamber is in the low 10^{-7} or high 10^{-8} Torr range, so that upon cooling the final GHS base pressure was $10^{-8} - 10^{-9}$ Torr. Anytime a system was changed (either the dopant or the perturber), the GHS was again baked in order to return the system to near the starting base pressure. This prevented cross-contamination between dopant/perturber systems.

The intensity of the synchrotron radiation exiting the monochromator was monitored by recording the beam current of the storage ring as well as the photoemission from a nickel mesh situated prior to the sample cell. The light then entered the experimental cell through a MgF_2 window, traveled through the sample and then a second MgF_2 window (cf. Fig. 7) before striking a thin layer of sodium salicylate powder on the inside of a glass window that preceded the photomultiplier tube. An empty cell (acquired for a base pressure of 10^{-7} or 10^{-8} Torr) spectrum was used to correct the dopant absorption spectra for the monochromator flux, for absorption by the MgF_2 windows, and for any fluctuations in the quantum efficiency of the sodium salicylate window.

Two dopants (i.e. methyl iodide and xenon) and five perturbers (i.e. argon, krypton, xenon, carbon tetrafluoride and methane) were investigated. All dopants and perturbers were used without further purification: methyl iodide (Aldrich, 99.45%), argon (Matheson Gas Products, 99.9999%), krypton (Matheson Gas Products, 99.998%), and xenon (Matheson Gas Products, 99.995%). When CH_3I was the dopant, the CH_3I was degassed with three freeze/pump/thaw cycles prior to use. Photoabsorption spectra for each neat dopant and each neat perturber were measured to verify the absence of impurities in the spectral range of interest. The atomic perturber number densities were calculated from the Strobridge equation of state [55] with the parameters obtained from [56] for argon, [57] for krypton, and [58] for xenon. All densities and temperatures were selected to maintain a single phase system in the sample cell.

At temperatures below T_c, a change in density required a change in temperature, since the isotherms are steeply sloped.

The sensitivity of the absorption spectra to local density required that the quality of a data set be monitored by performing basic data analysis during measurements. Any anomalies were corrected by immediately re-measuring the photoabsorption spectrum for the problem density/temperature/pressure after allowing additional time for increased density stabilization. Once a data set was obtained for non-critical temperatures, the photoabsorption data for perturber densities on an isotherm near the critical isotherm were then measured. For the near critical data set we selected a temperature that was $+0.5°$C above the critical temperature (chosen to prevent liquid formation in the cell during temperature stabilization near the critical density and to minimize critical opalescence during data acquisition). Near the critical density, the consistency of the density step is dependent on the slope of the critical isotherm. If the critical isotherm has a small slope in this region, it becomes difficult to acquire samples at a constant density step size due to our inability to vary the perturber pressure practically by less than 0.01 bar and to the difficulties encountered in maintaining temperature stability. For instance, near the critical density of xenon, a 1 mbar change in pressure or a $0.001°$C change in temperature causes a density change of 2.0×10^{21} cm^{-3}. Maintaining the necessary temperature stability (i.e., $\pm\, 0.2°$C) during the acquisition of data along the critical isotherm is difficult with an open flow liquid nitrogen cryostat system and usually required constant monitoring with manual adjustment of the nitrogen flow.

3.2. Theoretical methodology

3.2.1. *Line shape function*

The experimental line shapes were simulated using the semi-classical statistical line shape function given in eq. (10). Rewriting eq. (10) in terms of the autocorrelation function allows eq. (10) to be given as a Fourier transform, namely [25, 26, 33, 37]

$$\mathfrak{L}(\omega) \;=\; \frac{1}{2\pi}\,\mathrm{Re}\int_{-\infty}^{\infty} dt\; e^{-i\omega t}\,\langle e^{i\omega(\mathbf{R})\,t}\rangle\,, \tag{28}$$

where $\omega = \omega(\mathbf{R}) - \omega_0$, with ω_0 being the transition frequency for the neat dopant. Eq. (28) neglects lifetime broadening and assumes that the transition dipole moment is independent of **R**. In the substitution of the exponential density expansion [i.e., eq. (11)] for the autocorrelation function, the general term A_n represents a (n+1)-body interaction [20]. However, since the strength of the interaction decreases as the number of bodies involved increases, and since the higher order interactions are more difficult to model, our line shape simulations are truncated at the second term $A_2(t)$, or three body interactions. Within this approximation, eq. (28) becomes

$$\mathfrak{L}(\omega) \;=\; \frac{1}{2\pi}\,\mathrm{Re}\int_{-\infty}^{\infty} dt\; e^{-i\omega t}\,\exp\left[A_1(t) + A_2(t)\right]\;, \tag{29}$$

where the two terms are recalled from eq. (14),

$$A_1(t) \;=\; 4\pi\rho_P \int_0^{\infty} dr\, r^2\, g_{PD}(r)\left[\, e^{-it\Delta V(r)} - 1\right]\;,$$

and eq. (15),

$$
\begin{aligned}
A_2(t) &= 4\pi\rho_P^2 \int_0^\infty dr_1\, r_1^2\, g_{PD}(r_1) \left[e^{-it\Delta V(r_1)} - 1 \right] \\
&\times \int_0^\infty dr_2\, r_2^2\, g_{PD}(r_2) \left[e^{-it\Delta V(r_2)} - 1 \right] \\
&\times \frac{1}{r_1 r_2} \int_{|r_1 - r_2|}^{|r_1 + r_2|} s \left[g_{PP}(s) - 1 \right] ds\,.
\end{aligned}
$$

The required radial distribution functions (i.e., g_{PP} and g_{PD}) were obtained from the analytical solution of the Ornstein-Zernike equation for a binary system within the Percus-Yevick (PY) closure [59], while the Fourier transform for eq. (29) was performed using a standard fast Fourier transform algorithm [60]. The line shape obtained from the transform of eq. (29) was convoluted with a standard Gaussian slit function to account for the finite resolution (~ 8 meV) of the monochromator. More detailed discussions are given below.

3.2.2. Fast Fourier transform

A Fourier transform has the general form [60]

$$F(\omega) = \frac{1}{2\pi} \int_{-\infty}^{\infty} f(t)\, e^{-i\omega t}\, dt\,. \tag{30}$$

Since the line shape function is calculated numerically, the integration limits for eq. (30) must be finite and, therefore, an appropriate integration range must be determined. For any Fourier transformation, the integration limit and the total number of steps are related through [60]

$$\delta t \times \delta\omega = \frac{2\pi}{N}\,, \tag{31}$$

where δ stands for the sampling interval (i.e., the step size) of the corresponding variable and N is the total number of discrete points. Fourier transforms rely on the fact that data are usually obtained in discrete steps and the generating functions $f(t)$ and $F(\omega)$ can be represented by the set of points

$$
\begin{aligned}
f_k &\equiv f(t_k)\,, \quad t_k = k\,\delta t\,, \quad k = 1, \ldots, N\,, \\
F_n &\equiv F(\omega_n)\,, \quad \omega_n = n\,\delta\omega\,, \quad n = -\frac{N}{2}, \ldots, \frac{N}{2} - 1\,.
\end{aligned} \tag{32}
$$

Therefore, the function $F(\omega)$ is determined point-wise using

$$
\begin{aligned}
F(\omega_n) &= \frac{1}{2\pi} \int_{-\infty}^{\infty} f(t)\, e^{-i\omega_n t}\, dt = \frac{1}{2\pi} \sum_{k=1}^{N} f_k\, e^{-i\omega_n t_k}\, \delta t \\
&= \frac{\delta t}{2\pi} \sum_{k=1}^{N} f_k\, e^{-i 2\pi n k/N}\,.
\end{aligned} \tag{33}
$$

For simplicity, we will define the discrete Fourier transform from time to angular frequency as eq. (33). When computing the Fourier transform from eq. (33), the quickest method is known as a fast Fourier transform (FFT) and requires that the number of steps N be a power of 2. In our calculations, we use a Cooley-Tukey FFT algorithm [60] with $N = 1024$. The requirements for calculating eq. (30) within this FFT algorithm are, therefore, a complex array of the calculated values of the time dependent autocorrelation function truncated to the second term.

Rewriting eq. (29) using Euler's relation yields [38]

$$\langle e^{i\,\omega(\mathbf{R})\,t} \rangle = \mathrm{Re}\,\langle e^{i\,\omega(\mathbf{R})\,t} \rangle + \mathrm{Im}\,\langle e^{i\,\omega(\mathbf{R})\,t} \rangle\,, \tag{34}$$

where the real and the imaginary parts are given by

$$\begin{aligned} \mathrm{Re}\,\langle e^{i\,\omega(\mathbf{R})\,t} \rangle &= \exp\left[\mathrm{Re}\,(A_1(t) + A_2(t))\right]\,\cos\left[\mathrm{Im}\,(A_1(t) + A_2(t))\right]\,, \\ \mathrm{Im}\,\langle e^{i\,\omega(\mathbf{R})\,t} \rangle &= \exp\left[\mathrm{Re}\,(A_1(t) + A_2(t))\right]\,\sin\left[\mathrm{Im}\,(A_1(t) + A_2(t))\right]\,. \end{aligned} \tag{35}$$

In eq. (35),

$$\begin{aligned} &\mathrm{Re}[A_1(t) + A_2(t)] \\ &\quad = 4\,\pi\,\rho_{\mathrm{P}} \int_0^\infty dr\, r^2\, g_{\mathrm{PD}}(r)\, \left[\cos(\Delta V(r)\,t) - 1\right] \\ &\quad + 4\,\pi\,\rho_{\mathrm{P}}^2 \int_0^\infty \int_0^\infty dr_1\, dr_2\, h(r_1,r_2)\, \left[\cos(\Delta V(r_1)\,t)\, \cos(\Delta V(r_2)\,t)\right. \\ &\qquad\qquad + 1 - \sin(\Delta V(r_1)\,t)\, \sin(\Delta V(r_2)\,t) \\ &\qquad\qquad\qquad \left. - \cos(\Delta V(r_1)\,t) - \cos(\Delta V(r_2)\,t)\right]\,, \end{aligned} \tag{36}$$

and

$$\begin{aligned} &\mathrm{Im}[A_1(t) + A_2(t)] \\ &\quad = -\,4\,\pi\,\rho_{\mathrm{P}} \int_0^\infty dr\, r^2\, g_{\mathrm{PD}}(r)\, \left[\sin(\Delta V(r)\,t)\right] \\ &\quad - 4\,\pi\,\rho_{\mathrm{P}}^2 \int_0^\infty \int_0^\infty dr_1\, dr_2\, h(r_1,r_2)\, \left[\sin(\Delta V(r_1)\,t)\, \cos(\Delta V(r_2)\,t)\right. \\ &\qquad\qquad + \cos(\Delta V(r_1)\,t)\, \sin(\Delta V(r_2)\,t) \\ &\qquad\qquad\qquad \left. - \sin(\Delta V(r_1)\,t) - \sin(\Delta V(r_2)\,t)\right]\,, \end{aligned} \tag{37}$$

with

$$h(r_1,r_2) = r_1\, g_{\mathrm{PD}}(r_1)\, r_2\, g_{\mathrm{PD}}(r_2) \int_{|r_1 - r_2|}^{|r_1 + r_2|} s\,\left[g_{\mathrm{PP}}(s) - 1\right]\, ds\,.$$

The output of the FFT is a complex function of frequency. The real portion of this complex function is obtained and then convoluted with a standard Gaussian slit function. The final output is the simulated line shape function. Since eqs. (29), (36) and (37) depend on both the radial distribution functions and the ground-state and excited-state intermolecular potentials, these are discussed in more detail below.

3.2.3. Radial distribution function

After significant investigation, we found that the most stable calculation technique for obtaining radial distribution functions for this problem was the analytical solution of the Ornstein-Zernike relation within the Percus-Yevick (PY) closure [59]. Although this solution for a binary system yields four coupled integro-differential equations, dilute solutions (i.e., $\rho_D \ll \rho_P$) allows these equations to be reduced to the calculation of the perturber/dopant radial distribution function $g_{PD}(r)$ and the perturber/perturber radial distribution function $g_{PP}(r)$. This solution is given by [12, 59]

$$
\begin{aligned}
g_{PD}(r) &= r^{-1}\, e^{-\beta V_g(r)}\, Y_{PD}(r) \\
g_{PP}(r) &= r^{-1}\, e^{-\beta V_g'(r)}\, Y_{PP}(r)\,,
\end{aligned}
\tag{38}
$$

where

$$
\begin{aligned}
Y_{PD}(r) &= \int_0^r dt\, \frac{dY_{PD}(t)}{dt}\,, \\
Y_{PP}(r) &= \int_0^r dt\, \frac{dY_{PP}(t)}{dt}\,,
\end{aligned}
\tag{39}
$$

with

$$
\begin{aligned}
\frac{d}{dr} Y_{PD}(r) &= 1 + 2\pi\rho_P \int_0^\infty dt\, \left(e^{-\beta V_g(t)} - 1\right) Y_{PD}(t) \\
&\times \Big[e^{-\beta V_g'(r+t)}\, Y_{PP}(r+t) \\
&\qquad - \frac{r-t}{|r-t|}\, e^{-\beta V_g'(|r-t|)}\, Y_{PP}(|r-t|) - 2t \Big]\,, \\
\frac{d}{dr} Y_{PP}(r) &= 1 + 2\pi\rho_P \int_0^\infty dt\, \left(e^{-\beta V_g'(t)} - 1\right) Y_{PP}(t) \\
&\times \Big[e^{-\beta V_g'(r+t)}\, Y_{PP}(r+t) \\
&\qquad - \frac{r-t}{|r-t|}\, e^{-\beta V_g'(|r-t|)}\, Y_{PP}(|r-t|) - 2t \Big]\,,
\end{aligned}
\tag{40}
$$

and with V_g and V_g' being the ground state perturber/dopant and ground state perturber/perturber intermolecular potentials, respectively.

3.2.4. Intermolecular potentials

Eqs. (29), (34) - (40) are explicitly dependent on the excited-state and ground-state perturber/dopant intermolecular potentials through $\Delta V(r)$, and are implicitly dependent on the perturber/perturber and perturber/dopant ground-state intermolecular potential via $g_{PP}(r)$ and $g_{PD}(r)$. Thus, these simulations require one to develop a single set of ground-state and excited-state intermolecular potential parameters for each system. A standard Lennard-Jones 6-12 potential, or

$$V(r) = 4\varepsilon\left[\left(\frac{\sigma}{r}\right)^{12} - \left(\frac{\sigma}{r}\right)^{6}\right], \tag{41}$$

was chosen for the atomic perturber/perturber ground-state intermolecular interactions, and non-polar dopant/perturber ground-state intermolecular interactions. The ground-state molecular perturber/perturber intermolecular potential was a two-Yukawa potential, or

$$V(r) = -\frac{\kappa_0\,\varepsilon}{r}\left[\mathrm{e}^{-z_1(r-\sigma)} - \mathrm{e}^{-z_2(r-\sigma)}\right]. \tag{42}$$

The ground-state polar dopant/perturber intermolecular interactions were modeled using a modified Stockmeyer potential

$$V(r) = 4\varepsilon'\left[\left(\frac{\sigma'}{r}\right)^{12} - \left(\frac{\sigma'}{r}\right)^{6}\right] - \frac{1}{r^6}\alpha_P{\mu_D}^2, \tag{43}$$

which can be rewritten in standard Lennard-Jones 6-12 potential form [12], with

$$\varepsilon = \varepsilon'\left[1 + \frac{\alpha_P\,{\mu_D}^2}{4\,\varepsilon'\,\sigma'^6}\right]^2,$$

$$\sigma = \sigma'\left[1 + \frac{\alpha_P\,{\mu_D}^2}{4\,\varepsilon'\,\sigma'^6}\right]^{-1/6}.$$

(The modified Stockmeyer potential includes orientational effects via an angle average that presumes the free rotation of the polar dopant molecule.) An exponential-6 potential, given by

$$V(r) = \frac{\varepsilon}{1-(6/\gamma)}\left\{\frac{6}{\gamma}e^{\gamma(1-\chi)} - \chi^{-6}\right\}, \tag{44}$$

was chosen for the excited-state dopant/ground-state perturber interactions. In eqs. (41) - (44), ε is the well depth, σ is the collision parameter, α_P is the perturber polarizability, μ_D is dopant dipole moment, $\chi \equiv r/r_e$ (where r_e is the equilibrium distance), and γ is the potential steepness.

The Lennard-Jones parameters for the atomic fluids and the modified Stockmeyer parameters for the CH_3I/Ar and CH_3I/Kr interactions were identical to the parameters used to model accurately the perturber-induced shift of the dopant ionization energy for methyl iodide in argon [12–14, 61], krypton [12, 15, 61] and xenon [16, 61]. The parameters κ_0, z_1, z_2, ε and σ in eq. (42) were adjusted to give the best fit to the phase diagram of the perturber [61].The parameters ε, σ, χ and γ in eq. (44) were adjusted by hand to give the "best" fit to the experimental absorption spectra of the dopant low-n Rydberg states in each of the fluids investigated here.

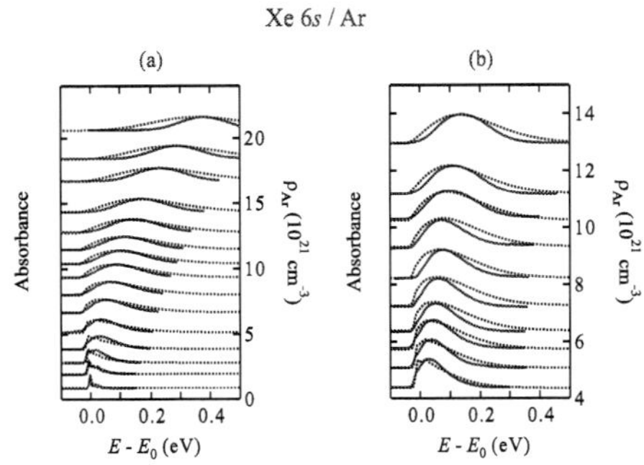

Figure 8. Selected photoabsorption spectra (–, relative scale) and simulated line shapes (· · ·) for the Xe 6s Rydberg transitions at (a) non-critical temperatures and (b) on an isotherm (i.e., -121.8°C) near the critical isotherm. The data are offset vertically by the argon number density ρ_{Ar}. The transition energy is $E_0 = 8.424$ eV for the unperturbed Xe 6s Rydberg transition.

4. Atomic perturbers

4.1. Xe low-*n* Rydberg states in Ar

4.1.1. Xe absorption

The Xe 6s and 6s′ Rydberg states (where *s* and *s*′ denote the $J = 3/2$ and $J = 1/2$ angular momentum core state, respectively) were experimentally measured in dense argon. As is discussed in Section 2.4.1, when Xe interacts with Ar, ground and excited state dimers form. These dimers are evidenced by blue satellite bands that arise on the higher energy side of the primary Rydberg transition. The Xe 6s Rydberg transition has two such blue satellite bands corresponding to eqs. (24)-(25), whereas the Xe 6s′ Rydberg transition has a single blue satellite band corresponding to eq. (24) (cf. Fig. 2) [22–24]. The absence of the ground state XeAr dimer to XeAr eximer transition [i.e., eq. (25)] for the Xe 6s′ Rydberg state may be caused by an extremely short lifetime preventing our ability to detect the transition or by the XeAr eximer decomposing during the excitation.

The solid lines in Fig. 8 present selected experimental Xe 6s Rydberg transitions doped into supercrtical argon at non-critical temperatures and along an isotherm near the critical isotherm offset by the argon number density. (Similar data for the Xe 6s′ Rydberg transitions are not shown for brevity.) It can be clearly seen that the Rydberg transitions broaden as a function of the argon number density. The maximum of the absorption band also shifts first slightly to the red and then strongly to the blue, similar to the original observations of Messing, *et al.* [25, 26]. Since the ground state interaction between Xe and Ar (or XeAr and Ar) is attractive, the ground states are stabilized by the argon solvent shell. The slight red shift observed at low argon number densities indicates that the xenon excited states (either Xe* or Xe*Ar) are also stabilized by the argon solvent shell. As the density increases, however, argon begins to shield the optical electron from the xenon cationic core, thereby decreasing the binding energy of the optical electron . Thus, as the density of argon increases the energy of the excited state also increases, leading to a blue shift in the transition energy at higher densities. Although not shown, the overall blue shift of the 6s Rydberg transition band is much larger than that of the 6s′ band at the triple point liquid density of argon. This difference in overall shift is caused by the difference in the core state of the cation, since the $J = 1/2$ core state has a permanent quadrupole moment. This permanent quadrupole moment increases the interaction of the cationic core and the optical electron, thereby implying that the optical electron is less perturbed by the argon solvent shell.

However, since the blue satellite bands also broaden and shift with increasing argon density, the primary Xe transition becomes indistinguishable at medium to large argon number densities. Thus, the argon-induced energy shift of the primary Xe transition cannot be investigated directly using these data. Therefore, to probe perturber critical effects on the dopant excited states, we must first accurately simulate the absorption spectra over the entire argon density range at non-critical temperatures and on an isotherm near the critical isotherm.

4.1.2. Discussion

In order to simulate accurately the absorption spectra at high density, any line shape simulation has to include the primary transition, denoted a in Fig. 2 and given by eq. (23), as well as the two XeAr dimer transitions that yield the blue satellite bands, denoted b and c in Fig. 2 and given by eqs. (24) and (25), respectively. For the simulation of Xe in Ar, we chose to use eq. (41) for the ground state Ar/Ar, Xe/Ar, and XeAr/Ar interactions and eq. (44) for the Xe^*/Ar and Xe^*Ar/Ar interactions. We also required that the simulation use a single set of intermolecular potential parameters for the entire argon density range at non-critical temperatures and along the critical isotherm. All intermolecular potential parameters except the Ar/Ar ground state parameters were adjusted by hand to give the best simulated line shape in comparison to the experimental data. The values of these parameters are given here in Appendix A [37, 40].

The relative intensities of the simulated bands were set by comparison to the absorption spectra of Xe doped into argon at argon number densities where all bands could be clearly identified. Experimentally, at low argon number densities, the ratio of heights between the b band and the primary transition is 0.2 for both the Xe $6s$ and $6s'$ Rydberg states in Ar. For the Xe $6s$ Rydberg state in Ar, the ratio of heights between the c band and the primary transition is 0.45. Although for concentrated Xe systems, the ratio of heights for the blue satellite bands to the primary transition would increase with decreasing temperature or increasing perturber number density, this is not the case for the very dilute Xe/Ar system investigated here (i.e., [Xe] < 10 ppm for all argon number densities). Therefore, we can assume that the intensity ratio of the blue satellite bands to the primary transitions stays constant at different temperatures and different argon densities.

The dotted lines in Fig. 8 are the simulated line shapes for the Xe $6s$ transition at non-critical temperatures (cf. Fig. 8a) and on an isotherm ($-121.8°$C) near the critical isotherm (cf. Fig. 8b). A similar figure for the Xe $6s'$ transition is not shown for brevity. Clearly, the simulated spectra closely match the experimental spectra for all densities. Both the simulated and experimental line shapes show a slight red shift at low argon number densities, followed by a strong blue shift at high argon number densities. With these accurate line shape simulations, moment analyses can be performed on the primary transition in order to investigate perturber critical point effects, as well as to discuss trends in solvation of different dopant electronic transitions in the same simple atomic fluid.

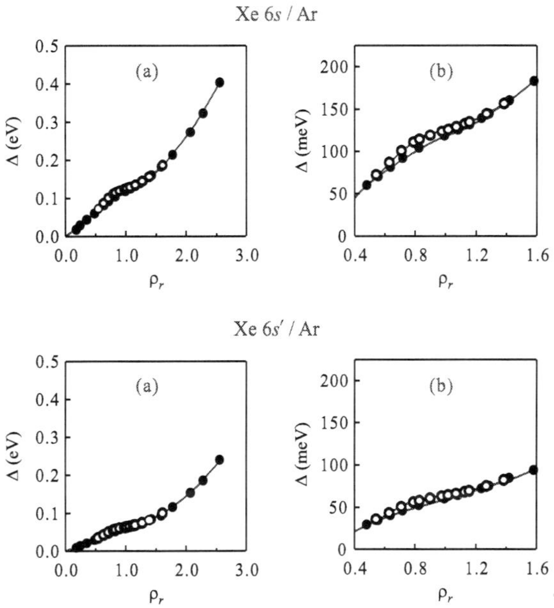

Figure 9. (a) The argon induced shift $\Delta(\rho_{Ar})$, as approximated by eq. (21), of the primary transition for the Xe 6*s* and 6*s*′ Rydberg states as a function of argon number density ρ_{Ar} at (•) non-critical temperatures and (○) along an isotherm near the critical isotherm. (b) An expanded view of $\Delta(\rho_{Ar})$The solid lines are a visual aid. See text for discussion.

A line shape analysis was performed on the accurate simulations of the primary Xe 6*s* and 6*s*′ Rydberg transitions in order to determine the average argon induced shift $\Delta(\rho_{Ar})$ of the primary transition, as approximated from the first moment [i.e., eq. (21)]. This moment analysis is shown in Fig 9 as a function of reduced argon number density ρ_r, where $\rho_r = \rho_{Ar}/\rho_c$ with $\rho_c \equiv 8.076 \times 10^{21}$ cm^{-3} [56]. Fig 9b shows an enhanced view of the perturber critical region with a critical effect in $\Delta(\rho_{Ar})$ clearly apparent. The absence of the red shift observed by Messing, *et al.* [25, 26] (cf. Fig. 3) results from our performance of a moment analysis on a blue degraded band, instead of a direct non-linear least square analysis using a Gaussian fit function on the primary transition. In other words, while the peak of the primary transition red shifts slightly at low argon densities, the first moment of the band does not, due to the perturber induced broadening.

General trends emerged in the behavior of the simulated line shape as a function of the intermolecular potential parameters. For instance, we observed that the strength of the asymmetric blue broadening of a band increases with increasing $\Delta r_e \equiv r_e^{(g)} - r_e^{(e)}$ [where $r_e^{(i)}$ is the equilibrium dopant/perturber distance for either the ground state dopant ($i = g$) or the excited state dopant ($i = e$)]. However, the overall perturber-induced energy shift of the band depended on the ground state intermolecular potential well depth $\varepsilon^{(g)}$ as well as Δr_e. The slight red shift at low perturber number densities, however, was controlled by the excited state intermolecular potential well depth $\varepsilon^{(e)}$. Comparison of the Xe 6*s* and 6*s*′ transition in Ar shows that the 6s Rydberg state broadens and shifts to higher energies more quickly than does the 6s′ state. Since both transitions are excited from the same ground state (implying that the ground state intermolecular potential parameters remain unchanged), Δr_e must decrease and $\varepsilon^{(e)}$ increase in order to simulate the Xe 6s′ Rydberg state in argon correctly. These general trends proved helpful when determining the intermolecular potential parameters for new systems.

Messing, *et al.* [25, 26] concluded that the argon induced energy shift is density dependent and temperature independent. However, both our experimental absorption spectra and the line shape simulations show a distinct temperature dependence near the argon critical point. To test the sensitivity of the perturber critical point effect, we extracted the perturber dependent shift $\Delta(\rho_{Ar})$ of the simulated primary Xe 6s transition in supercritical argon near the critical density along three different isotherms (i.e., T_r = 1.01, 1.06 and 1.11, where $T_r = T/T_c$ with $T_c = -122.3°C$). These data are shown in Fig. 10a and clearly indicate that the critical effect is extremely sensitive to temperature and can be easily missed if the temperature of the system is not maintained close to the critical isotherm.

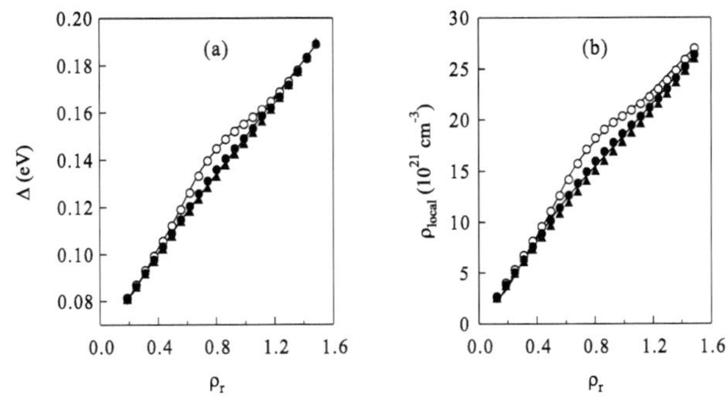

Figure 10. (a) The calculated argon induced shift $\Delta(\rho_{Ar})$ doped into supercritical argon plotted as a function of reduced argon number density at a reduced temperature $T_r = 1.01$ (○), 1.06 (●) and 1.11 (▲). (b) The local densities ($\rho_{\text{local}} = g_{\max}\,\rho_{\text{bulk}}$) of the first argon solvent shell around a central Xe atom plotted as a function of reduced argon number density at a reduced temperature $T_r \simeq 1.01$ (○), 1.06 (●) and 1.11 (▲). The solid lines are provided as visual aid.

If we return to the line shape equation [i.e., eq. (29)], we observe that the two-body interaction term $A_1(t)$ and the three-body interaction term $A_2(t)$ depend on the difference between the excited state and ground state intermolecular potentials and on the perturber/dopant radial distribution function. Since the potential difference will not depend dramatically on temperature, the critical point effect must be dominated by changes in the perturber/dopant radial distribution function $g_{PD}(r)$. In Fig. 10b, we plot the local density of the first solvent shell as a function of the bulk reduced argon number density on the same three isotherms. The $T_r = 1.01$ isotherm shows a much larger density deviation near the critical density in comparison to the other two isotherms. Thus, the argon induced blue shift is caused by the first perturber shell shielding the cationic core from the optical election. This increase in shielding decreases the binding energy of the electron, thereby increasing the excitation energy.

4.2. CH_3I low-n Rydberg states in Ar, Kr and Xe

4.2.1. CH_3I absorption

The CH_3I 6s and 6s' Rydberg states doped into supercritical argon, krypton and xenon were investigated both experimentally and theoretically [38, 40] from low perturber number density to the density of the triple point liquid, at both non-critical temperatures and on an isotherm near (i.e., $+0.5°$C) the critical isotherm of the perturber. The CH_3I 6s and 6s' Rydberg states show perturber-induced energy shifts and broadening similar to that observed for the Xe low-n Rydberg states in supercritical argon. The peak positions of the absorption spectra shift to the red slightly and then strongly to the blue as a function of perturber number densities. This is similar to the behavior for CH_3I in dense rare gases observed by Messing, *et al.* [27, 28]. Unlike Xe, which forms heterogenous dimers in argon, the CH_3I/perturber interactions are weaker. Thus, CH_3I does not possess blue satellite bands caused by dimer or excimer formation. However, CH_3I does possess a strong vibrational transition on the blue side of the adiabatic transition. Fig. 4 shows the absorption of both the 6s and 6s' Rydberg states of CH_3I and clearly illustrates the vibrational state, which represents the CH_3 group deformation vibrational band ν_2. The solid lines in Figs. 11 - 13 represent selected photoabsorption spectra for the CH_3I 6s Rydberg transition doped into supercrtical argon, krypton and xenon, while similar plots for the CH_3I 6s' transition are not shown for brevity. Experimental spectra of CH_3I in Xe at number densities between 5.0×10^{21} cm^{-3} and 7.0×10^{21} cm^{-3} could not be obtained, because of the large density deviation induced by small temperature fluctuations ($\approx 2.0 \times 10^{21}$ cm^{-3} for a 0.001°C temperature change) in this density region.

The experimental absorption of CH_3I low-n Rydberg transitions shows that as the perturber number density increases, the ν_2 vibrational band broadens and shifts until it merges with the adiabatic transition. Therefore, determining the perturber induced shift $\Delta(\rho_P)$ of the adiabatic transition from a simple moment analysis of the spectra presented in Figs. 11 - 13 is not possible, and we must perform an accurate line shape analysis of these data in order to extract $\Delta(\rho_P)$ and investigate the perturber critical effect. However, some qualitative information can be gleaned from Figs. 11 - 13. First, the rate of the broadening and the rate of shift for both the adiabatic transition band and the ν_2 vibrational transition band differ dramatically for different perturbers. However, although not shown, the CH_3I 6s and 6s' transitions have almost the same perturber induced shift, which differs from the behavior observed for the Xe in Ar system previously presented.

4.2.2. Discussion

Although CH_3I in the rare gases does not form dimers or excimers, the accurate simulation of the low-n Rydberg transitions must include both the adiabatic transition, given by eq. (26) and denoted a in Fig. 4, as well as one quantum of the CH_3 deformation vibrational transition ν_2 in the excited state, given by eq. (27) and denoted b in Fig. 4. For all of the simulations presented here, we again chose eq. (41) for the ground-state perturber/perturber intermolecular interactions. All of the ground-state dopant/perturber interactions, on the other hand, were approximated with eq. (43). The excited-state dopant/ground state perturber interactions were again modeled using eq. (44). All intermolecular potential parameters except the Ar/Ar, Kr/Kr, Xe/Xe, CH_3I/Ar, and CH_3I/Kr ground state potential parameters were adjusted by hand to give the best simulated line shape in comparison to our experimental absorption spectra. (The Ar/Ar, Kr/Kr, Xe/Xe, CH_3I/Ar, and CH_3I/Kr ground-state potential parameters used are in accord with those employed in our earlier studies of the quasi-free electron energy in rare gas perturbers [12].) Appendix A gives the values for all intermolecular potential parameters used in the line shape simulations presented here. The relative intensities of the simulated bands were fixed by comparison to the absorption spectra of CH_3I at perturber number densities where all bands (i.e., the adiabatic and vibrational transitions) could be clearly identified. Experimentally, at low perturber number densities the ratio of the vibrational band intensity to the adiabatic transition intensity is 0.22 for both the CH_3I 6*s* and 6*s*′ Rydberg states in all three perturbers.

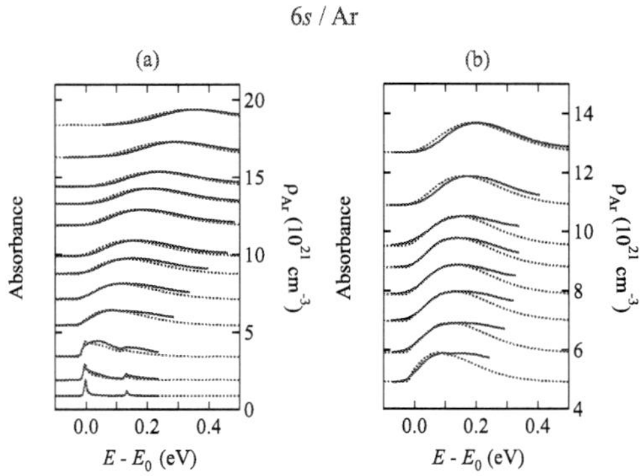

Figure 11. Selected photoabsorption spectra (—, relative units) and simulated line shapes (···) for the CH_3I 6*s* Rydberg transition in argon at (a) non-critical temperatures and (b) on an isotherm (−121.8°C) near the critical isotherm. The data are offset vertically by the argon number density ρ_{Ar}. The transition energy is $E_0 = 6.154$ eV for the unperturbed CH_3I 6*s* Rydberg transition. The variation between experiment and simulation is caused by other vibrational transitions and by perturber-dependent lifetime broadening not modeled here.

The dotted lines in Figs.11 - 13 present the simulated line shapes (dotted lines) of the low-*n* CH_3I Rydberg transitions in the atomic perturbers at non-critical temperatures and on an isotherm near the critical isotherm of the perturber. As was true for Xe in Ar, the simulated spectra closely match the experimental spectra for all densities. Both the simulated and experimental line shapes show a slight red shift at low perturber number densities, followed by a strong blue shift at high perturber densities. Given the accuracy of the simulated line shapes, simulated spectra for CH_3I in Xe in the region where experimental data were unobtainable are also presented in Fig. 13. We should note here that we were able to model the CH_3I 6*s* and 6*s*′ Rydberg states in Ar using the same set of intermolecular potential parameters for both states. This behavior was also observed for the CH_3I 6*s* and 6*s*′ Rydberg states in Kr. With identical potential parameters, the perturber induced shift $\Delta(\rho_P)$ will be the same for the 6*s* and 6*s*′ states. The independence of $\Delta(\rho_P)$ on the dopant cationic core state is different from that observed for Xe low-*n* Rydberg states in Ar and will be discussed in more detail below. The accurate line shape simulations allow $\Delta(\rho_P)$ for the adiabatic transitions to be extracted using eq. (21).

As with Xe in Ar, the accurate line shape simulations allow a moment analysis to be performed on the CH_3I low-*n* adiabatic Rydberg transition to obtain the perturber induced shift $\Delta(\rho_P)$

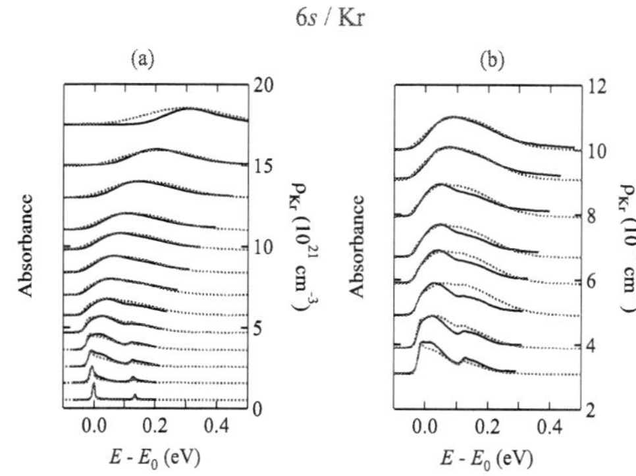

Figure 12. Selected photoabsorption spectra (—, relative units) and simulated line shapes (· · ·) for the CH_3I 6*s* Rydberg transition in krypton at (a) non-critical temperatures and (b) on an isotherm (−63.3°C) near the critical isotherm. The data are offset vertically by the krypton number density ρ_{Kr}. The transition energy is $E_0 = 6.154$ eV for the unperturbed CH_3I 6*s* Rydberg transition. The variation between experiment and simulation is caused by other vibrational transitions and by perturber-dependent lifetime broadening not modeled here.

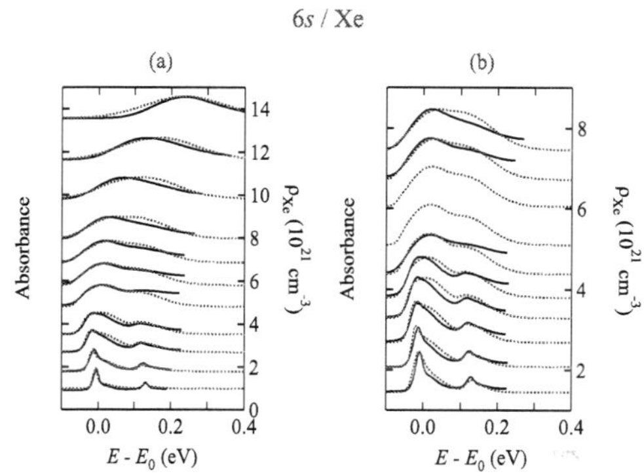

Figure 13. Selected photoabsorption spectra (—, relative units) and simulated line shapes (· · ·) for the CH_3I 6*s* Rydberg transition in xenon at (a) non-critical temperatures and (b) on an isotherm (17.0°C) near the critical isotherm. The data are offset vertically by the xenon number density ρ_{Xe}. The transition energy is $E_0 = 6.154$ eV for the unperturbed CH_3I 6*s* Rydberg transition. The variation between experiment and simulation is caused by other vibrational transitions and by perturber-dependent lifetime broadening not modeled here.

from eq.(21). The first moment of the simulated CH_3I 6*s* adiabatic transition is plotted as a function of the reduced perturber number density ρ_r in Fig. 14 for the 6*s* transition. (A similar figure for the 6*s*′ transition is not shown for brevity.) The first moment of the simulated adiabatic band does not red shift at low perturber density, as was originally stated by Messing, *et al.* [27, 28]. This absence of a red shift is again caused by the blue degradation of the adiabatic transition, which places the average energy (i.e., the first moment) of the band to the high energy side of the absorption maximum. The ground state interaction between CH_3I and the perturber is attractive, and therefore the ground state of the dopant is stabilized by the perturber solvent shell. The slight red shift of the absorption maximum observed at low perturber number densities is indicative of the stabilization of the CH_3I excited states by the perturber solvent shell. As the density increases, however, perturber molecules begin to shield the optical electron from the CH_3I cationic core, thereby increasing the excitation energy of the optical electron. Thus, as the perturber density increases, the energy of the excited state also increases, leading to a blue shift at higher perturber densities.

The 6*s* and 6*s*′ Rydberg states correspond to an optical electron in the same Rydberg orbital, but with the cation in a different core state: $J = 3/2$ for s and $J = 1/2$ for s′, where J is the total angular momentum of the core. In our investigation of $\Delta(\rho_P)$ for Xe in Ar, we found that $\Delta(\rho_P)$ of the 6*s* transition is 0.2 eV larger than that for the 6*s*′ transition, indicating that the change in the core quadrupole moment affects the dopant/perturber interactions in a dense perturbing medium. However, $\Delta(\rho_P)$ for the CH_3I 6*s* and 6*s*′ Rydberg transitions near the triple point density are identical to within experimental error for the perturbers argon and krypton, and differ only slightly (i.e., 30 meV) for CH_3I in xenon. The insensitivity of these CH_3I/perturber systems to the change in the CH_3I cationic core is probably caused by the large permanent dipole moment of CH_3I, which masks the effect of the quadrupole moment. Xenon, however, is extremely sensitive to electric fields because of its large polarizability. Therefore, the slight difference between the xenon induced shifts of the CH_3I 6*s* and 6*s*′ Rydberg transitions may well be caused by small changes in the permanent dipole moment of CH_3I influencing changes in the induced dipole or local quadrupoles in the xenon perturber.

A critical point effect on the 6*s* and 6*s*′ transition energies is also apparent in Fig. 14 for all three perturbers. The CH_3I 6*s* adiabatic transition in argon is blue-shifted by 20 meV near the critical temperature and critical density, while those in krypton and xenon are blue-shifted by 30 meV and 15 meV, respectively. Identical results are obtained for the CH_3I 6*s*′ adiabatic transitions in argon and krypton. However, a smaller critical effect of 5 meV is observed for the CH_3I 6*s*′ transition in xenon, which is related to the smaller overall blue shift of the CH_3I 6*s*′ transition in comparison to the 6*s* transition.

In the low to medium density range, the energy of the absorption maximum for the 6*s* and 6*s*′ CH_3I Rydberg states has a larger red shift in xenon, which is caused by the larger xenon polarizability. The CH_3I Rydberg states also broaden more quickly in xenon. This increased broadening is probably due to a combination of increased xenon polarizability and an increase in the probability of collisional de-excitation due to the size of xenon. However, $\Delta(\rho_P)$ is larger for argon than for krypton and xenon. This change is caused by an overall decrease in the total number of perturber atoms within the first solvent shell surrounding the CH_3I dopant as the perturber atoms become larger. The variation in the critical point effect, with krypton having a larger effect than argon and xenon, is caused by the strength of the perturber/CH_3I interactions in comparison to the perturber/perturber interactions, coupled with the differences in the ground-state and excited-state dopant/perturber interaction potentials. The CH_3I/Kr ground state potential well depth is close (i.e., 24 K) to the Kr/Kr potential well depth. This implies that the CH_3I/Kr interactions near the krypton critical point will be comparable to the Kr/Kr interactions, thereby leading to a large increase in the local perturber density near the critical point of the perturber, and a larger critical point effect. Similarly, the critical point effect decreases as one goes from krypton to argon to xenon because the difference in well depth for all intermolecular potentials increases.

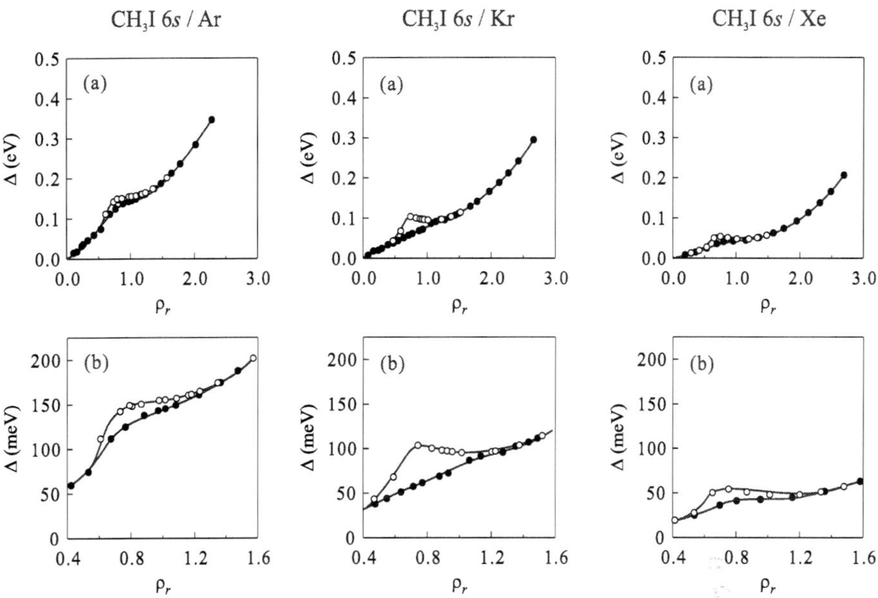

Figure 14. (a) The perturber induced shift $\Delta(\rho_P)$, as approximated by a moment analysis [i.e., eq. (21)], of the simulated primary transition for the CH_3I 6s Rydberg state as a function of the reduced perturber number density ρ_r for argon, krypton and xenon. (•), simulations obtained at noncritical temperatures; (○), simulations near the critical isotherm. (b) An expanded view of $\Delta(\rho_P)$ near the perturber critical point. $\rho_c = 8.0 \times 10^{21}$ cm^{-3} for argon, $\rho_c = 6.6 \times 10^{21}$ cm^{-3} for krypton and $\rho_c = 5.0 \times 10^{21}$ cm^{-3} for xenon [12]. The solid lines provide a visual aid. See text for discussion.

5. Conclusion

In this work, the structure of low-n Rydberg states doped into supercritical fluids was investigated in several atomic perturbers. Both the experimental absorption spectra and full line shape simulations over the entire perturber density range at non-critical temperatures and along isotherms near perturber critical isotherms were presented for all dopant/perturber systems. These accurate line shape simulations allowed us to extract the perturber-induced energy shift $\Delta(\rho_P)$ from the simulated primary low-n Rydberg transitions. These shifts showed a striking critical point effect in all dopant/perturber systems. Our group also performed similar absorption measurements of atomic and molecular low-n Rydberg states in molecular perturbers [39, 40] with similar results. Because of the brevity of this Chapter, the details of these measurements cannot be presented here.

In all of the systems investigated [37–40], the dopant low-n Rydberg states are extremely sensitive to the nature of the perturbing fluid. When these states are doped into supercritical fluids, the surrounding perturbers interact with the central dopant causing shifts both in the dopant ground state energy and in the excited state energy. At low perturber number densities, the dopant/perturber interaction stabilizes the dopant ground state and the low-n Rydberg state. As the perturber density increases, perturber/dopant interactions lead to the formation of a perturber solvent shell around the dopant core, thereby inducing local perturber density inhomogeneities. This solvent shell begins to shield the optical electron from the cationic core. Therefore, the dense perturber fluid increases the dopant excitation energy, resulting in a blue shift of the abosrption band, which is observed experimentally. The local density of the first perturber solvent shell is almost proportional to the perturber bulk density at non-critical temperatures. However, near the critical isotherm and critical density of the perturber, the dopant/perturber interactions strengthen due to the increased perturber/perturber correlation length. This increased order yields a corresponding increase of the local density in the solvent shell that, in turn, leads to a stronger shielding of the optical electron from the cationic core. Thus, increased blue shifts of the low-n absorption bands are observed in all dopant/perturber systems near the critical point of the perturber. The area of this critical effect is demarcated by the turning points that bound the saddle point in the thermodynamic phase diagram of the critical isotherm.

For fluids with similar compressibilities, the structures of low-n dopant Rydberg states in the perturbing fluid show systematic behaviors. At non-critical temperatures, $\Delta(\rho_P)$ is determined by the polarizability and size of the perturbing fluid. The larger the polarizability and, therefore, the larger the size, the smaller the perturber-induced energy shift of the dopant absorption bands. This is caused by the number of atoms that can exist between the optical electron and the dopant cationic core, coupled with the strength of the shielding. The large overall energy shift observed in the dopant low-n Rydberg states perturbed by CF_4 [39, 40], on the other hand, was caused by the larger compressibility of CF_4 in comparison to the other gases in this study [37, 38, 40]. This larger compressibility implies that CF_4 is closer together on average at high perturber number densities than are the other perturbers studied, which increases the local density of CF_4 and, therefore, increases the blue shift in this perturber.

The critical point effect, on the other hand, is dominated by the similarity of the perturber/perturber interaction with the dopant/perturber ground state and dopant/perturber excited state interactions, coupled with the overall local density of the system. In krypton, the well depth of the ground state perturber/perturber intermolecular potential and the dopant/perturber intermolecular potential shows greater similarity in comparison to that in Ar and Xe. Moreover, the excited state CH_3I/Kr interaction is slightly stronger than the ground state Kr/Kr interaction. These facts dictate that the largest critical point effect for CH_3I in atomic perturbers is in Kr. Similarly, the largest overall critical effect was observed in CH_3I/CH_4 [39, 40]. This large critical effect is caused by both the ground state and excited state CH_3I/CH_4 interactions having strengths comparable to the CH_4/CH_4 interaction. Although the excited state CH_3I/CF_4 interactions are comparable in strength to the CF_4/CF_4 interactions, the ground state CH_3I/CF_4 interactions are not close to those of CF_4/CF_4. Similarly, the Xe/CF_4 ground state interactions are comparable to the ground state CF_4/CF_4 interactions, but the excited state Xe/ground state CF_4 interactions are weaker. Moreover, the bulk critical density in CF_4 is small in comparison to the rest of the perturbers investigated here. This results in the CF_4 critical effect on $\Delta(\rho_P)$ being the smallest one observed [39, 40].

These data sets also allowed us to generate a consistent set of intermolecular potential parameters for various dopant/perturber systems, which are summarized in Appendix A. Several general trends in these parameters can be observed. For atomic perturbers, the steepness of the exponential-6 intermolecular potential (i.e., γ) used to model the dopant excited state/perturber intermolecular interaction decreases with increasing perturber size and polarizability. This trend is reversed in molecular perturbers, were the larger, more compressible CF_4 has a steeper repulsive component in comparison to CH_4. The excited vibrational states of CH_3I always have exponential-6 potentials with a smaller γ in comparison to the CH_3I adiabatic transition in the same perturbing gas. Moreover, the vibrational states always have an equilibrium collision radius that is identical or larger than the collision radius of the adiabatic transition. The excited state collision radii are always larger than the ground state collision radii, as one would expect. However, the interaction strength of the excited state (as gauged by the well depth) can be stronger or weaker than that for the ground state of the same system. These changing interactions are what dominate the variations observed in the critical effects for each of the dopant/perturber systems investigated here.

An understanding of the structure of low-n Rydberg states in supercritical fluids is an important tool in the investigation of solvation effects, since these studies can yield accurate dopant/perturber ground state and excited state intermolecular potentials. We conclude from the present work that the absorption line shapes can be adequately simulated within a simple semi-classical line shape analysis. However, this work focused on highly symmetric perturbers. Future studies should concern more asymmetric perturbers and polar perturbers. Such an extension will require changing the calculation techniques involved in determining the radial distribution functions as well as the type of Fourier transform used to simulate the line shape. Since the excited state is sensitive to the structure of the perturbing fluid, we anticipate that multi-site intermolecular potentials and angular dependent intermolecular potentials will be needed as the perturber complexity increases, in order to model the full line shape accurately.

Acknowledgements

All experimental measurements were made at the University of Wisconsin Synchrotron Radiation Center (NSF DMR-0537588), with support from the Petroleum Research Fund (PRF#45728-B6), the Professional Staff Congress - City University of New York, the Louisiana Board of Regents Support Fund (LEQSF(2006-09)-RD-A-33), and the National Science Foundation (NSF CHE-0956719).

Author details

Luxi Li and Xianbo Shi
Brookhaven National Laboratory, Upton, NY, USA

Cherice M. Evans
Department of Chemistry, Queens College – CUNY and the Graduate Center – CUNY, New York, NY, USA

Gary L. Findley
Chemistry Department, University of Louisiana at Monroe, Monroe, LA, USA

Appendix A. Intermolecular potential parameters

Below is a tablulated list of the intermolecular potential parameters used to simulate the absorption line shapes in the various dopant/perturber systems presented or summarized in this work [40].

	ε/k_B (K)	r_e (Å)	γ	Ref.
Ar/Ar	119.5	3.826	–	[37, 38]
Kr/Kr	172.7	4.031	–	[38]
Xe/Xe	229.0	4.552	–	[38]
CH_4/CH_4 [a]	141.5	3.704	–	[39]
CF_4/CF_4	181.02	4.708	–	[39]
Xe/Ar	200.0	4.265	–	[37]
XeAr/Ar	195.0	4.310	–	[37]
Xe/CF_4	199.3	4.629	–	[39]
CH_3I/Ar	162.2	4.572	–	[38]
CH_3I/Kr	196.7	4.676	–	[38]
CH_3I/Xe	297.5	4.896	–	[38]
CH_3I/CH_4	195.8	4.243	–	[39]
CH_3I/CF_4	256.0	5.016	–	[39]
Xe $6s$/Ar	300.0	5.20	16.00	[37]
Xe $6s/CF_4$	135.0	6.55	12.25	[39]
CH_3I $6s$/Ar	110.0	6.30	12.75	[38]
CH_3I $6s$/Kr	245.0	6.20	11.30	[38]
CH_3I $6s$/Xe	400.0	6.39	10.25	[38]
CH_3I $6s/CH_4$	145.0	6.55	10.10	[39]
CH_3I $6s/CF_4$	185.0	6.84	12.10	[39]
Xe $6s'$/Ar	400.0	4.98	16.00	[37]
CH_3I $6s'$/Ar	110.0	6.30	12.75	[38]
CH_3I $6s'$/Kr	245.0	6.20	11.30	[38]
CH_3I $6s'$/Xe	400.0	6.29	10.25	[38]
CH_3I $6s'/CH_4$	145.0	6.55	10.10	[39]
CH_3I $6s'/CF_4$	185.0	6.84	12.10	[39]
Xe($6s$)Ar/Ar	250.0	5.25	16.00	[37]
CH_3I $6s$ ν_2/Ar	150.0	6.30	12.15	[38]
CH_3I $6s$ ν_2/Kr	225.0	6.30	10.75	[38]
CH_3I $6s$ ν_2/Xe	360.0	6.50	9.50	[38]
CH_3I $6s$ ν_2/CH_4	105.0	6.65	9.95	[39]
CH_3I $6s$ ν_2/CF_4	135.0	6.84	11.90	[39]
CH_3I $6s'$ ν_2/Ar	150.0	6.30	12.15	[38]
CH_3I $6s'$ ν_2/Kr	225.0	6.30	10.75	[38]
CH_3I $6s'$ ν_2/Xe	360.0	6.35	9.50	[38]
CH_3I $6s'$ ν_2/CH_4	105.0	6.65	9.95	[39]
CH_3I $6s'$ ν_2/CF_4	135.0	6.84	11.90	[39]

[a] Two-Yukawa potential with $\kappa_0 = 8.50$ Å, $z_1 = 0.90$ Å^{-1}, and $z_2 = 4.25$ Å^{-1}.

6. References

[1] M. B. Robin. *Higher Excited States of Polyatomic Molecules Vol. I – Vol. III*. Academic Press, New York, 1974, 1975, 1985. and references therein.

[2] R. Reininger, U. Asaf, I. T. Steinberger, and S. Basak. Relationship between the energy v_0 of the quasi-free-electron and its mobility in fluid argon, krypton, and xenon. *Physical Review B*, 28(8):4426–4432, 1983.

[3] R. Reininger, U. Asaf, and I. T. Steinberger. The density dependence of the quasi-free electron state in fluid xenon and krypton. *Chemical Physics Letters*, 90(4):287–290, 1982.

[4] A. O. Allen and W. F. Schmidt. Determination of the energy level v_0 of electrons in liquid argon over a range of densities. *Zeitschrift für Naturforschung A*, 37(4):316–318, 1982.

[5] W. von Zdrojewski, J. G. Rabe, and W. F. Schmidt. Photoelectric determination of v_0-values in solid rare gases. *Zeitschrift fur Naturforschung A*, 35(7):672–674, 1980.

[6] B. Halpern, J. Lekner, S. A. Rice, and R. Gomer. Drift velocity and energy of electrons in liquid argon. *Physical Review*, 156(2):351–352, 1967.

[7] W. Tauchert, H. Jungblut, and W. F. Schmidt. Photoelectric determination of v_0 values and electron ranges in some cryogenic liquids. *Canadian Journal of Chemistry*, 55(11):1860–1866, 1977.

[8] J. R. Broomall, W. D. Johnson, and D. G. Onn. Density dependence of the electron surface barrier for fluid helium-3 and helium-4. *Physical Review B*, 14(7):2819–2825, 1976.

[9] J. Jortner and A. Gaathon. Effects of phase density on ionization processes and electron localization in fluids. *Canadian Journal of Chemistry*, 55(11):1801–1819, 1977.

[10] N. Schwenter, E. E. Koch, and J. Jortner. *Electronic Excitations in Condensed Rare Gases*. Springer-Verlag, Berlin, 1985.

[11] U. Asaf, R. Reininger, and I. T. Steinberger. The energy v_0 of the quasi-free electron in gaseous, liquid, and solid methane. *Chemical Physics Letters*, 100:363–366, 1983.

[12] C. M. Evans and G. L. Findley. Energy of the quasifree electron in argon and krypton. *Physical Review A*, 72:022717, 2005.

[13] C. M. Evans and G. L. Findley. Energy of the quasi-free electron in supercritical argon near the critical point. *Chemical Physics Letters*, 410:242–246, 2005.

[14] C. M. Evans and G. L. Findley. Field ionization of c_2h_5i in supercritical argon near the critical point. *Journal of Physics B: Atomic, Molecular and Optical Physics*, 38:L269–L275, 2005.

[15] Luxi Li, C. M. Evans, and G. L. Findley. Energy of the quasi-free electron in supercritical krypton near the critical point. *Journal of Physical Chemistry A*, 109:10683–10688, 2005.

[16] Xianbo Shi, Luxi Li, C. M. Evans, and G. L. Findley. Energy of the quasi-free electron in xenon. *Chemical Physics Letters*, 432:62–67, 2006.

[17] Xianbo Shi, Luxi Li, C. M. Evans, and G. L. Findley. Energy of the quasi-free electron in argon, krypton and xenon. *Nuclear Instruments and Methods in Physics Research A*, 582:270–273, 2007.

[18] Xianbo Shi, Luxi Li, G. M. Moriarty, C. M. Evans, and G. L. Findley. Energy of the quasi-free electron in low density ar and kr: extension of the local wigner-seitz model. *Chemical Physics Letters*, 454:12–16, 2008.

[19] Xianbo Shi, Luxi Li, G. L. Findley, and C. M. Evans. Energy of the excess electron in methane and ethane near the critical point. *Chemical Physics Letters*, 481:183–189, 2009.

[20] G. D. Mahan. Satellite bands in alkali-atom spectra. *Physical Review A*, 6:1273–1279, 1972.

[21] M. Lax. The franck-condon principle and its application to crystals. *Journal of Chemical Physics*, 20:1752–1760, 1952.

[22] R. Granier, M. C. Castex, J. Granier, and J. Romand. Perturbation of the xenon 1469 a. resonance line by various rare gases and hydrogen. *Comptes Rendus de l'Académie des Sciences B*, 264:778, 1967.

[23] M. C. Castex, R. Granier, and J. Romand. Perturbation of the 1236-a. resonance line of krypton and the 1295-a. resonance line of xenon by various rare gases. *Comptes Rendus de l'Académie des Sciences B*, 268:552, 1969.

[24] M. C. Castex. Absorption spectra of xenon-rare gas mixtures in the far uv region (1150-1500 å): high resolution analysis and first quantitative absorption measurements. *Journal of Chemical Physics*, 66:3854–3865, 1977.

[25] I. Messing, B. Raz, and J. Jortner. Medium perturbations of atomic extravalence excitations. *Journal of Chemical Physics*, 66:2239–2251, 1977.

[26] I. Messing, B. Raz, and J. Jortner. Solvent perturbations of extravalence excitations of atomic xenon by rare gases at high pressures. *Journal of Chemical Physics*, 66:4577–4586, 1977.

[27] I. Messing, B. Raz, and J. Jortner. Perturbations of molecular extravalence excitations by rare-gas fluids. *Chemical Physics*, 25:55–74, 1977.

[28] I. Messing, B. Raz, and J. Jortner. Medium effects on the vibrational structure of some molecular rydberg excitations. *Chemical Physics*, 23:351–355, 1977.

[29] A. M. Halpern. Iterative fourier reconvolution spectroscopy: van der waals broadening of rydberg transitions; the $\sim$b $\leftarrow$ $\sim$x (5p, 6s) transition of methyl iodide. *Journal of Physical Chemistry*, 96:2448–2455, 1992.

[30] T. Kalbfleisch, R. Fan, J. Roebber, P. Moore, E. Jacobson, and L.D. Ziegler. A molecular dynamics study of electronic absorption line broadening in high-pressure nonpolar gases. *Journal of Chemical Physics*, 103:7673–7684, 1995.

[31] R. Fan, T. Kalbfleisch, and L. D. Ziegler. A molecular dynamics analysis of resonance emission: optical dephasing and inhomogeneous broadening of ch_3i in ch_4 and ar. *Journal of Chemical Physics*, 104:3886–3897, 1996.

[32] T. S. Kalbfleisch, L. D. Ziegler, and T. Keyes. An instantaneous normal mode analysis of solvation: methyl iodide in high pressure gases. *Journal of Chemical Physics*, 105:7034–7046, 1996.

[33] S. A. Egorov, M. D. Stephens, and J. L. Skinner. Absorption line shapes and solvation dynamics of ch_3i in supercritical ar. *Journal of Chemical Physics*, 107:10485–10491, 1997.

[34] E. Morikawa, A. M. Köhler, R. Reininger, V. Saile, and P. Laporte. Medium effects on valence and low-n rydberg states: No in argon and krypton. *Journal of Chemical Physics*, 89:2729–2737, 1988.

[35] P. Larrégaray, A. Cavina, and M. Chergui. Ultrafast solvent response upon a change of the solute size in non-polar supercritical fluids. *Chemical Physics*, 308:13–25, 2005.

[36] C. N. Tiftickjian and S. A. Egorov. Absorption and emission lineshapes and solvation dynamics of no in supercritical ar. *Journal of Chemical Physics*, 128:114501, 2008.

[37] Luxi Li, Xianbo Shi, C. M. Evans, and G. L. Findley. Xenon low-n rydberg states in supercritical argon near the critical point. *Chemical Physics Letters*, 461:207–210, 2008.

[38] Luxi Li, Xianbo Shi, C. M. Evans, and G. L. Findley. Ch_3i low-n rydberg states in supercritical atomic fluids near the critical point. *Chemical Physics*, 360:7–12, 2009.

[39] Luxi Li, Xianbo Shi, G. L. Findley, and C. M. Evans. Dopant low-n rydberg states in cf_4 and ch_4 near the critical point. *Chemical Physics Letters*, 482:50–55, 2009.

[40] Luxi Li. *Atomic and Molecular Low-n Rydberg States in Near Critical Point Fluids*. PhD thesis, The Graduate Center of the City University of New York, New York, NY, 2009.

[41] S. C. Tucker. Solvent density inhomogeneities in supercritical fluids. *Chemical Reviews*, 99:391–418, 1999.

[42] P. Attard. Spherically inhomogeneous fluids. i. percus-yevick hard spheres: osmotic coefficients and triplet correlations. *Journal of Chemical Physics*, 91:3072–3082, 1989.

[43] P. Attard. Spherically inhomogeneous fluids. ii. hard-sphere solute in a hard-sphere solvent. *Journal of Chemical Physics*, 91:3083–3089, 1989.

[44] J. Zhang, D. P. Roek, J. E. Chateauneuf, and J. F. Brennecke. A steady-state and time-resolved fluorescence study of quenching reactions of anthracene and 1,2-benzanthracene by carbon tetrabromide and bromoethane in supercritical carbon dioxide. *Journal of the American Chemical Society*, 119:9980–9991, 1997.

[45] Y. P. Sun, M. A. Fox, and K. P. Johnston. Spectroscopic studies of p-(n,n-dimethylamino)benzonitrile and ethyl p-(n,n-dimethylamino)benzoate in supercritical trifluoromethane, carbon dioxide, and ethane. *Journal of the American Chemical Society*, 114:1187–1194, 1992.

[46] R. S. Urdahl, D. J. Myers, K. D. Rector, P. H. Davis, B. J. Cherayil, and M. D. Fayer. Vibrational lifetimes and vibrational line positions in polyatomic supercritical fluids near the critical point. *Journal of Chemical Physics*, 107:3747–3757, 1997.

[47] R. S. Urdahl, K. D. Rector, D. J. Myers, P. H. Davis, and M. D. Fayer. Vibrational relaxation of a polyatomic solute in a polyatomic supercritical fluid near the critical point. *Journal of Chemical Physics*, 105:8973–8976, 1996.

[48] R. P. Futrelle. Unified theory of spectral line broadening in gases. *Physical Review A*, 5:2162–2182, 1972.

[49] J. L. Lebowitz and J. K. Percus. Statistical thermodynamics of nonuniform fluids. *Journal of Mathematical Physics*, 4:116–123, 1963.

[50] J. L. Lebowitz and J. K. Percus. Asymptotic behavior of radial distribution function. *Journal of Mathematical Physics*, 4:248–254, 1963.

[51] J. L. Lebowitz and J. K. Percus. Integral equations and inequalitites in theory of fluids. *Journal of Mathematical Physics*, 4:1495–1506, 1963.

[52] M. Born and H. S. Green. A general kinetic theory of liquids. i. the molecular distribution functions. *Proceedings of the Royal Society of London Series A*, 188:10–18, 1946.

[53] R. Kubo and Y. Toyozawa. Application of the method of generating function to radiative and non-radiative transitions of a trapped electron in a crystal. *Progress of Theoretical Physics*, 13:160–182, 1955.

[54] H. C. Jacobson. Moment analysis of atomic spectral lines. *Physical Review A*, 4:1363–1368, 1971.

[55] T. R. Strobridge. The thermodynamic properties of nitrogen from 64 to 300 $^{\circ}$k between 0.1 and 200 atmospheres. *NBS Technical Note*, 129, 1962.

[56] A. L. Gosman, R. D. McCarty, and J. G. Hust. Thermodynamics properties of argon from the triple point to 300 k at pressures to 1000 atmostpheres. *NBS Technical Note*, 27, 1969.

[57] W. B. Streett and L. A. K. Staveley. Experimental study of the equation of state of liquid krypton. *Journal of Chemical Physics*, 55:2495–2506, 1971.

[58] W. B. Streett, L. S. Sagan, and L. A. K. Staveley. Experimental study of the equation of state of liquid xenon. *Journal of Chemical Thermodynamics*, 5:633–650, 1973.

[59] E. W. Grundke, D. Henderson, and R. D. Murphy. Evaluation of the percus-yevick theory for mixtures of simple liquids. *Canadian Journal of Physics*, 51:1216–1226, 1973.

[60] W. H. Press, S. A. Teukolsky, W. T. Vetterling, and B. P. Flannery. *Numerical Recipes in FORTRAN: The Art of Scientific Computing*. Cambridge University Press, New York, 1992.

[61] Xianbo Shi. *Energy of the Quasi-free Electron in Atomic and Molecular Fluids*. PhD thesis, The Graduate Center of the City University of New York, New York, NY, 2010.

Chapter 17

Mathematical Methods to Analyze Spectroscopic Data – New Applications

E.S. Estracanholli, G. Nicolodelli, S. Pratavieira, C. Kurachi and V.S. Bagnato

Additional information is available at the end of the chapter

http://dx.doi.org/10.5772/48318

1. Introduction

Absorption and fluorescence spectroscopies in the visible and/or infrared spectrum are good options when a fast and objective analysis is required. Spectroscopy is based on light-matter interactions. This interaction occurs in different ways, and each molecule or an ensemble of molecules will show a distinctive response. The vibrational spectroscopy provides a fingerprint of the vibrational levels of a molecule usually at mid-infrared (MIR) radiation (400-4000cm^{-1}). The optical spectroscopy uses the ultraviolet-visible (UV-VIS) region (200-1000nm) of the electromagnetic spectrum and interrogates the electronic levels of a molecule. The instrumentation used to generate and detect this radiation is less complex and cheaper compared to other spectroscopy techniques, such as nuclear magnetic resonance, X-rays, etc. An absorption spectrum is obtained by irradiating a sample and measuring the light which is transformed into other forms of energy, e.g. molecular vibration (heat). A fluorescence spectrum is obtained only from fluorescent molecules, those that absorb and then emit radiation, acquiring the intensity of light emitted as a function of the wavelength. These spectra are characteristic for each molecule, because each one has different electronic levels and vibrational modes. These levels and modes are also influenced by the solvent of the molecule.

Vibrational spectroscopy applied to the mid portion of the infrared spectrum provides the basis to develop several of the most powerful methods of qualitative and quantitative chemical analysis. Some of the advantages to use this technique are: the information is collected on a molecular level, almost any chemical group has IR bands, it is very environment-sensitive.[1-3]

Optical fluorescence spectroscopy is highly sensitive and can provide different information about the molecules and the molecular processes such as the molecular interaction with the environment, the molecular bonding and concentration.[3, 4]

Spectroscopic techniques in this range of the electromagnetic spectrum have shown applications in different areas, from analytical chemistry to the diagnosis of some types of cancer, detection of citrus diseases and of dental caries, with a high sensitivity and good specificity rates. This is possible because the analyzed systems are composed of different types and concentrations of molecules. Thus, the spectrum of samples obtained under different conditions will also be different. It is therefore possible to identify and also quantify different compounds. However, the spectral variation can be characterized and correlated only with difficulty. This is mainly due to the fact that other phenomena, such as scattering and/or absorption, happen with the emitted light. In some cases, there may be other molecules in the sample presenting absorption bands that overlap in the same spectral region of the compound of interest, this mainly happens for absorption spectroscopy. In other cases, as in the case of fluorescence spectroscopy, the excitation and emitted light can be absorbed by other molecules making the signal too weak to be detected. A solution to this problem may be statistical procedures applied where the spectral information is correlated with any parameter of interest. [3-5]

A new application of a statistical method to process multi-layer spectroscopy information will be presented in this chapter. A brief review of the mathematical methods to analyze these spectroscopy data will be shown here, followed by two distinct examples. The first example is UV-VIS fluorescence spectroscopy, applied to detect the postmortem interval (PMI) in an animal model. The spectroscopy and statistical methods of analysis presented can be extended to other samples, like food and beverage. Here, a MIR absorption spectroscopy of liquid samples will be presented to detect and quantify certain compounds during the production of beer. Another system to measure liquid samples, which consists of a sample holder, will also be presented. This system offers a cheaper technique with a better signal compared to techniques used to analyze liquid samples in the MIR region.

2. Pattern recognition in the complex spectral database – Example of fluorescence spectroscopy used in forensic medicine

One of the limitations of conventional methods to determine the Postmortem Interval (PMI) of an individual is the fact that the measurements cannot be performed in real time and *in situ*. Several factors, environmental and body conditions influence the tissue decomposition and the time evolution, resulting in a poor resolution. Considering this limitation to determine the PMI, a possible solution is a new more objective method based on a tissue characterization of the degradation phases through optical information using fluorescence spectroscopy. If proven sensitive enough, this method shows a main advantage over conventional methods: less inter-variance and quantitative tissue information. These characteristics are relevant because they are less influenced by individual skills.[6]

During the decomposition process a wide variety of organic materials are consumed by natural micro-organisms and other unknown compounds produced by them. Using an objective method based on the tissue characterization of the stages of degradation by means of optical information using ultraviolet-visible fluorescence spectroscopy, followed by a statistical method based on PCA (Principal Component Analysis) made it possible to identify well features with time progression. The characteristic pattern of time evolution presented a high correlation coefficient, indicating that the chosen pattern presented a direct linear relationship with the time evolution. The results show the potential of fluorescence spectroscopy to determine the PMI, with at least a similar resolution compared to conventional methods.[6]

Another attractive feature of optical technologies is the fact that in situ information is achieved through a noninvasive and nondestructive interrogation with a fast response. Conventional laboratory techniques to determine PMI are time consuming and also require the cadaver removal from the location where it was found to a forensic lab facility. This operation already introduces additional changes to the analysis.

Fluorescence spectroscopy has been presented as a sensitive technique to biochemical and structural changes of tissues. The investigation of biological tissues is quite complex. Photons interact with biomolecules in several ways, and depending on the type of the interactions, they can be classified into three groups. The absorbers are the biomolecules that absorb photon energy. The fluorophores are biomolecules that absorb and emit fluorescent light. The scatterers are biomolecules that do not absorb the photons but change their direction. Several endogenous cromophores contribute and modify the final tissue spectrum. Distinct fluorophores emit light but the collected spectrum will be modified depending on the presence of absorbers and scatterers in the microenvironment on the path of the emitted photons and the probe interrogator. Taking into account all these light interactions that occur within the biological tissues, it is important to keep in mind that a tissue fluorescence spectrum is a result of the combination of all these processes occurring in the pathway between excitation and collection: excitation absorption and scattering, fluorescence emission, and fluorescence absorption and scattering.[5, 7]

Tissue changes begin to take place in the cadaver as soon as there is cessation of life. Optical characteristics change, and these changes may be detected using fluorescence spectroscopy. With the cessation of the metabolic reactions, tissue modifications are induced by several distinct factors, e.g. lack of oxygen and adenosine triphosphate, and intestinal microorganism proliferation. In this type of analyses, we first aimed to establish a proof of concept that fluorescence spectral variations for distinct PMIs are higher than the variance observed within each PMI. If the results are positive, a spectral time behavior can then be determined, i.e. fluorescence spectral changes identifying each PMI. This proposed method can be used to determine an unknown PMI based on a comparative analysis of a spectral database pattern. There is a potential correlation of the tissue fluorescence changes and the PMI, even though the same limitations concerning the time course variability of the cadaveric phenomena are also present, the optical spectra information can provide a more objective estimation.

Taking into account the resolution limitation to determine PMI in biological tissues, where the degradation process is non-homogenous and influenced by environment and cadaver intrinsic factors, the optical techniques may show a better PMI prediction when compared to current techniques.

The process using the principal component analysis is necessary to change the space analysis. For any type of spectroscopy, where space is determined by an analysis of light intensity in wavelengths of absorption or fluorescence, a change of base can be accomplished, where the variables become the variance of the dataset. We can explore this idea mathematically using a practical example. The set of spectra shown in figure 1 is a typical result of multiple samples.

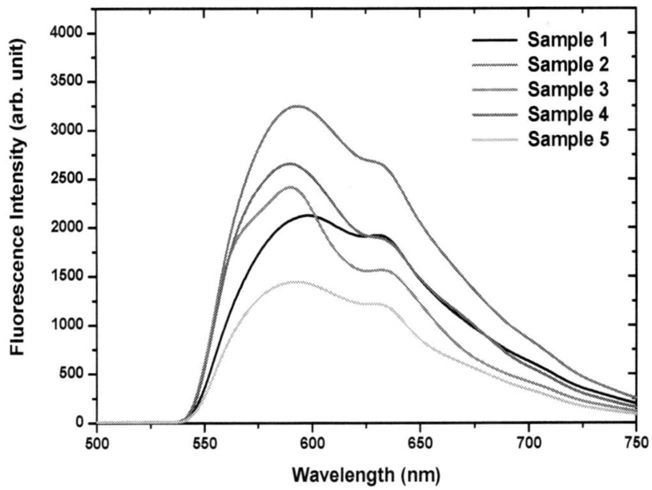

Figure 1. Fluorescence spectra of different samples.

In this case, we are dealing with intensities as a function of the wavelength, presented in a graphic form. These same curves can be represented in a matrix, as shown in table 1. In this form, each row corresponds to a single measurement, i.e., spectrum of a sample, and each column is the value of the wavelength considered, which makes up this spectrum. Thus, each array element is the intensity measured at a wavelength specific to a spectrum.

Wavelength (nm)	540	541	542	543	...	...	749	750
Intensity Sample1	27.5	30.6	33.9	37.7	...	...	199.9	198.4
Intensity Sample2	25.5	26.7	29.6	35.9	...	...	180.5	186.3
...	...	...	...	...	...	...	...	...
Intensity SampleN	24.1	25.3	27.9	30.2	...	...	176.8	180.6

Table 1. Fluorescence spectra in a matrix.

The next procedure to be performed after inserting the data set into this matrix representation is the centralization of the data around their average value. By fixing one column (wavelength) at a time we can calculate the average value for all lines (samples). Table 2 shows the mean values obtained.

Wavelength (nm)	540	541	542	543	...	...	749	750
Intensity Media	25.7	27.53	30.47	34.6	...	...	185.73	188.43

Table 2. Matrix of mean values for each wavelength.

In the next step, each of the intensity values of each sample should be subtracted from this average value in the respective wavelength. The results for our example are shown in table 3.

Wavelength (nm)	540	541	542	543	...	...	749	750
Intensity Sample1	1.8	3.07	3.43	3.1	...	...	14.17	9.97
Intensity Sample2	-0.2	-0.83	-0.87	1.3	...	...	-5.23	-2.13
...	...	...	...	...	...	...	...	...
Intensity SampleN	-1.6	-2.23	-2.57	-4.4	...	...	-8.93	-7.83

Table 3. Normalized fluorescence spectra.

It is important to note that this procedure resulted in a better match between the variables. The first values had higher intensities (approximately 8 times) than the wavelengths around 750nm in relation to values around 540 nm. If this normalization of the data had not been performed, the outcome would have had a greater influence for the longer wavelengths, as if they possessed some kind of "privilege", which would not be correct from the standpoint that all the measured variables are also important.

Since each element of the initial data array is represented by an element q_{ij}, we can consider this procedure performed using the equation 1:

$$X_{ij} = q_{ij} - \overline{q}_j \tag{1}$$

Where: x_{ij} is the element of our new data matrix; q_{ij} is the array element data corresponding to the i-th measurement variable j; $\overline{q}_j$ is the mean value of the variable j;

As the data were previously normalized, i.e. centered on their mean values, we proceed with the construction of the correlation matrix, where we obtain information about a dataset that indicates how the variables are correlated. This is possible by calculating the product of the transposed data matrix by itself. In mathematical terms, if x is our new array of standardized data x_{ij} composed of elements, then the correlation matrix R formed by these correlation coefficients is given by:

$$\mathbf{R}=\mathbf{X}^{T} \cdot \mathbf{X} \tag{2}$$

A matrix whose elements are given by:

$$r_{jj'} = \sum_{i=1}^{n} x_{ij} x_{ij'} = \sum_{i=1}^{n} \frac{(q_{ij} - \overline{q}_j).(q_{ij'} - \overline{q_{j'}})}{\sigma_j . \sigma_{j'}} \tag{3}$$

The r_{jj} value is a standardized covariance between -1 and 1. It should be noted that the matrix is Hermitian (symmetric in the case of real variables, which is our case). We can also confirm that the elements along the main diagonal of the correlation matrix (elements where $j = j'$) correspond to the variance of the variable q_j. As noted earlier, the correlation matrix R is Hermitian and therefore its eigenvalues are real and positive and its eigenvectors are orthogonal. For this procedure, it is important to note that the values of the wavelengths themselves are not taken into account in the mathematical calculation. The data selected for the next step are the rows and columns highlighted in table 3.

After calculating all the elements of the correlation matrix, the diagonalization is necessary. The diagonalization process provides two sets of data. The first are the eigenvectors: vectors which constitute a new base having the direction and sense in which the initial data set has more tendencies to vary, i.e. the maximization of the variance. The second sets of data are the eigenvalues, which provide the weight information, i.e. the relevance of each of the directions of the eigenvectors.

The eigenvalues are represented by the matrix K and the eigenvectors by the matrix V:

$$K = \begin{bmatrix} \lambda_1 & 0 & \dots & 0 \\ 0 & \lambda_2 & \dots & 0 \\ \dots & \dots & \dots & 0 \\ 0 & 0 & \dots & \lambda_n \end{bmatrix} \tag{4}$$

As the diagonalized matrix was a correlation matrix, we assume that this new base formed by the set of eigenvectors of each R represents a percentage of the total variance; and the information contained in each eigenvector are unique and exclusive, since they are mutually orthogonal. Each λ_i in the K matrix represent the weight of the specific eigenvector. Through these eigenvalues we can determine which of the principal components explains the greatest amount of data. For the simple relationship between the value of each eigenvalue divided by the sum of all eigenvalues, i.e. $\lambda_i / \sum_i \lambda_i$, we can determine the weight (or representation) of each eigenvector.

Once we determined the basis that maximizes the variance of the data set, which should be done by "projecting" the initial data matrix in this new basis through the product between the eigenvectors and the matrix of normalized data, we obtain:

$$\mathbf{S}=\mathbf{V}\cdot\mathbf{Q} \tag{5}$$

This data set designed in the new base (matrix S) is known as *Score*. The transformation of the basic matrix data S into the matrix Q by the base which maximizes the variances is known as Karhunen-Loève transformation.

The matrix of scores representing the data in our new base is expressed in such a way that each column represents the projection of the initial data into one of the eigenvectors, or in other terms, in either direction variance. Each line of the S matrix still represents a measure, or spectrum, as shown in table 4.

	PC1	PC2	PC3	PC4	...	...	PC N-1	PCN
Sample1	0.6	0.4	0.7	1.1	...	...	2.0	1.7
Sample2	0.8	1.2	1.4	2.2	...	...	1.7	0.6
...	...	...	...	...	...	...	...	...
SampleN	0.2	2.1	0.1	0.3	...	...	0.9	0.5

Table 4. Presentation of the available data with the new basis

Now, instead of analyzing the data obtained on the basis of the variables that were defined as the value of the intensity at each wavelength, these are considered in the space of the variances of these values. This change of base allows a significant reduction of the information in which the data are analyzed.

Spectroscopy experiments measure intensity values in hundreds or thousands of wavelengths, providing up to hundreds or thousands of variables to be analyzed. Depending on the experiment, when calculating the principal components of this system we represent around 90% of the information system, i.e. the spectra obtained in only two components of this new base. The graph of two major principal components (PC1 versus PC2) provides a much better view to then analyze these data rather than the hundreds of dimensions we had obtained before. In other words, we now work with a significantly reduced number of variables, wavelengths, with no loss of information. On this new basis, each sample, which was previously represented graphically by a curve with hundreds of points, shall be represented by a single point only. This significant reduction and simplification makes it much easier to detect spectral patterns.

We will now go back to the example of determining the postmortem interval - the set of measures shown in figure 1. Each sample is a fluorescence spectrum from different postmortem intervals. We apply the above procedure and obtain results on this new basis. For this case, the first two principal components show a representation greater than 91%. These results are shown in figure 2, where each point represents a spectrum.

Comparing data of figure 2 to those of figure 1 a temporal evolution of standard measurements becomes obvious. Based on this analysis, each region of space PC1 x PC2 is characterized by a postmortem interval. So in terms of a practical application, if we have a spectrum obtained from an unknown postmortem interval, we just design it based on this new basis and thus allow matching of the region of space to this spectrum, represented by a new point. Thus, this procedure can be used to determine a postmortem interval using fluorescence spectroscopy for situations where this value is unknown.

The present methodology and results shown by Estracanholli et al.[6, 8] demonstrated the use of fluorescence tissue spectroscopy to determine PMI as a valuable tool in forensic medicine. Two approaches were employed to associate the spectral changes with the time evolution of tissue modification. First, direct spectral changes were computed using inter-spectra analysis, allowing establishing a pattern of the sample distribution with a time evolution. Second, the use of a statistical method based on PCA helped to identify features over time. In both cases, the characteristic time evolution pattern presented a high correlation coefficient, indicating that the chosen pattern presented a direct linear relationship with time.

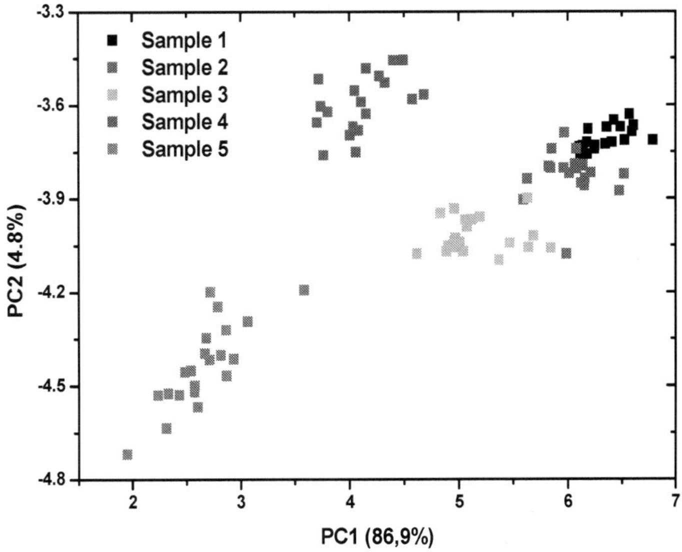

Figure 2. PC1 versus PC2 showing a more evident distinction between the samples.

However, other cases of application of spectroscopic techniques require a more robust processing. One such case is the other example cited above, where the goal is to quantify various compounds in a complex sample containing various interferences, and this most often occurs in the regions of overlapping absorption (or transmission and emission) of compounds of interest. For these cases, the application of artificial neural networks is a powerful solution.

3. Quantification of a composite with overlapping bands - Using vibrational spectroscopy in a beer sample

Beer brewing is a relatively long and complex biotechnological process, which can generate a range of products with distinct quality and organoleptic characteristics, all of which may be relevant to determine the type of product that should be made. Failures during important steps such as saccharification and fermentation can lead to major financial losses, i.e. to a loss of a whole batch of beer. Currently, analyses of the physic-chemical processes are carried out offline using traditional tests which do not provide any immediate response, e.g. HPLC (High Performance Liquid Chromatography). In the case of micro- breweries, which currently increase, some of these tests cannot be performed at all, due to the prohibitive cost of these tests. Therefore, many breweries do not have a possibility to identify errors during the production and to take corrective actions early-on. Today, problems are detected only at a later stage, towards the end of the brewing process. Currently, most systems used in the breweries consume time and can potentially compromise the quality of a whole batch. A solution to this problem is a new method consisting of a system to monitor in real time (online) the saccharification and fermentation steps of the wort. The amounts of alpha-amylase and beta-amylase in the grain are correlated with the time required to convert all grain starch into sugars. [3, 9-11]

The brewing-process is based on traditional recipes, a defined period of time and temperature. The amount of different types of enzymes in the grain when the wort is produced is, however, not known, since this amount depends on many factors, e.g. storage conditions, temperature, humidity, transport. Due to these factors, the saccharification step could be stopped, which would mean that a significant amount of starch would remain in the wort, and therefore the procedure would result in poor wort. Or, it is also possible that all starch may have been converted to sugar and that the process continues longer than necessary. It is therefore critical to obtain data concerning the amount of sugar and alcohol in the wort fast. It is possible to get these data, using absorbance data in the mid infrared region (MIR) and analyzing these statistically using PCA and Artificial Neural Network (ANN) to determine the amount of sugars and alcohol in the wort during the saccharification and the fermentation procedure. These optical techniques provide huge advantages because they can be easily adapted to the industrial equipment, providing real-time responses with a high specificity and sensitivity. By applying these techniques, the procedure of saccharification and fermentation can be modified in each brewing step to increase the quality of the wort and eventually of the beer. This routine analysis during processing can also be used for other liquid samples.

A main feature of ANN is its ability to learn from examples, without having been specifically programmed in a certain way. In the case of spectroscopy, satisfying results can be achieved when ANN is used with supervised training algorithms. The external supervisor (researcher) provides information about the desired response for the input patterns, i.e. where there is an "a priori knowledge" of the problem. A neural network can be defined as applying non-linear vector spaces between input and output. This is done through layers of neurons and activation functions, where the input values are added according to weight and "bias" specific, producing a single output value [12-14]. A network "feedforward" is progressive or shows no recursion, if the input vector and a layer formed by the values precede the output layer, as shown in figure 3.

Formally, the activation function of the i-th neuron in the j-th layer is denoted by $F_{i,j}(\times)$; its output itself, j, can be calculated from the output of the previous layer itself, j-1,the weights $W_{i,k,j-1}$ (the index k indicates the neuron connected to the preceding layer) and bias $b_{i,\ j}$ according to the following formula

$$s_{i,j} = F_{i,j}\left(b_{i,j} + \sum_k w_{i,k,j-1}\ s_{k,j-1} \right) \tag{6}$$

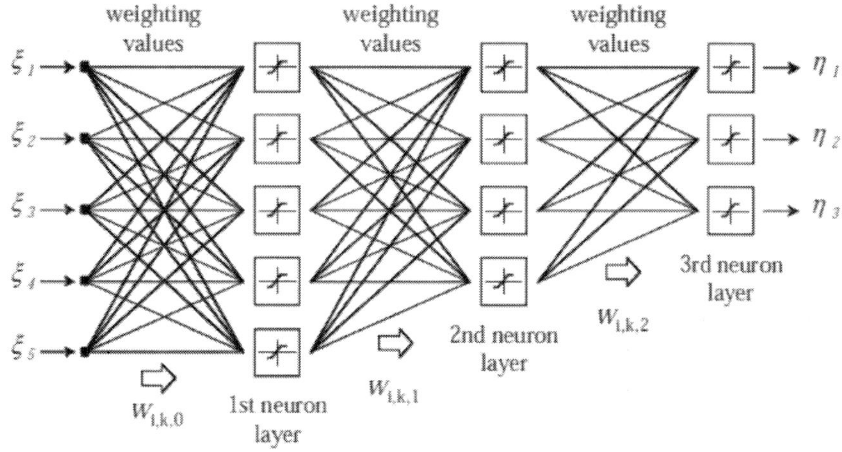

Figure 3. Schematic architecture of a neural network (perceptron multilayer).

The input and output values of the network being denoted by ξ_i and η_i respectively, the mapping can be determined due to a successive application of equation 6, which results for example in the following equation in the previous case:

$$\xi_i = F_{i,3}\left(b_{i,3} + \sum_{k=1}^{4} w_{i,k,2} F_{k,2}\left(b_{k,2} + \sum_{m=1}^{5} w_{k,m,1} F_{m,1}\left(b_{m,1} + \sum_{n=1}^{5} w_{m,n,0}\, \eta_n \right)\right)\right) \tag{7}$$

Since the choice of the activation function usually falls on the logistic sigmoid due to some of its mathematical properties (be class C ∞, for example), the above expression shows the relationship between ξ_i and η_i wich is defined by the weighing values and the bias. A very important characteristic of NN is its ability to learn, or the ability to reproduce the input-output pairs predetermined by properly adjusting the weights and the bias from training data and according to an adjustment rule. The method of a *"backpropagation"* rule is probably the best known training, and it is especially suited for progressive architectures. This rule is based on the successive application of the maximum slope algorithm determined from the first derivatives of the error between the desired outputs obtained by the parameters of the internal network. The backpropagation can be summarized in the following steps: (1) initialize the network parameters, $b_{i,j}$ and $w_{i,k,j}$ (2) select an entry ξ_i^p training data and form the pair (η_i^p, δ_i^p) , (3) calculate the error with a standard convenient Euclidean, e.g.

$$e = \sqrt{\Sigma_i(\delta_i^p - \eta_i^p)^2} \tag{8}$$

(4) Calculate the error derived from the above equation in relation to $b_{i,j}$ and $w_{i,k,j}$ (5) modify the parameters of the network according to the following rule and learning rate:

$$b_{i,j} \leftarrow b_{i,j} - \alpha \frac{\partial e}{\partial b_{i,j}} \quad and \quad w_{i,k,j} \leftarrow w_{i,k,j} - \alpha \frac{\partial e}{\partial w_{i,k,j}} \tag{9}$$

(6) Iterate steps (2) through (5) until a number of training cycles or stopping criteria has been achieved.[12, 13, 15, 16]

We can show in our case of beer analysis to which extent this processing technique is powerful. It has been applied widely in the interpretation of spectral data. In this case, an infrared absorption spectrum was obtained by Fourier Transform Infrared (FTIR) spectrometer [1, 2], the spectra is show in figure 4 and figure 5. In this case, the research objective was to provide a new method to determine the concentration of sugars and ethanol in beer wort during beer saccharification and fermentation in a short processing time. In our example, compounds of interest to be quantified can be separated into four main types of sugars present in the sample: glucose, maltose, maltotriose, dextrin (sugar chain length) and ethanol. It is important to note that the maltose binding is composed of two molecules of glucose, maltotriose three molecules of glucose sugar and that dextrins are composed of a large number of glucoses. Thus, the fundamental basis of these sugars is the same, the glucose, being differentiated only by the number of basic elements connected.

The absorption bands of these elements are expected to be so close that there is an overlap in the spectra, making the detection and quantification very complex. Figure 4 shows an example of absorption spectrum of a sample of ethanol, maltose 10%, and beer wort which contains some types of sugars. It is quite difficult to distinguish between the absorption spectra of the beer wort and the maltose, which contains certain types of sugars.

If we consider also the presence of ethanol (which has an absorption band in the same spectral region as the sugar) in the fermentation step, the procedure becomes even more complex. In figure 5 the extent of absorption during the fermentation step is shown, where the sample had initially all sugars without ethanol and ends up having only a part of dextrin (no fermentable sugar) and ethanol.

In this case, we first use the technique of principal component analysis in order to achieve a reduction of the number of variables to be analyzed. These spectra, which originally had about 1000 variables (wavelength where the absorbance is measured), can by these means represented by a few (two, in this case) variables, or principal components, with a high representation of information: 97.9%. The relationship between the two higher principal components is presented in figure 6 below.

Each spectrum of figure 5 is represented in figure 6 by a single point. In this new base of analysis, the wavelengths do not have any more significance, but the variance is important now. Each pair (PC1, PC2) represents a specific concentration of sugar and ethanol, which changes during the fermentation process. It is computationally feasible at this time, to apply an artificial neural network based on the values of the pairs for each of these points. For the first time, this experiment should be performed as previously described: as a case of a supervised NN e.g. a multilayer perception network). Therefore a method is required as the gold standard to calibrate or to train our neural network. One of the most widely accepted methods is the technique of HPLC (High Performance Liquid Chromatography). Using this technique, we can accurately quantify all the types of sugars of interest and the ethanol. The compounds of interest were measured using the standard method and assembling the ANN. In the neural network input (η_i) using ordered pairs of principal components and the output (ξ_i), the results obtained using the HPLC techniques show the amounts of compounds of interest, in this case, the sugars and the ethanol.

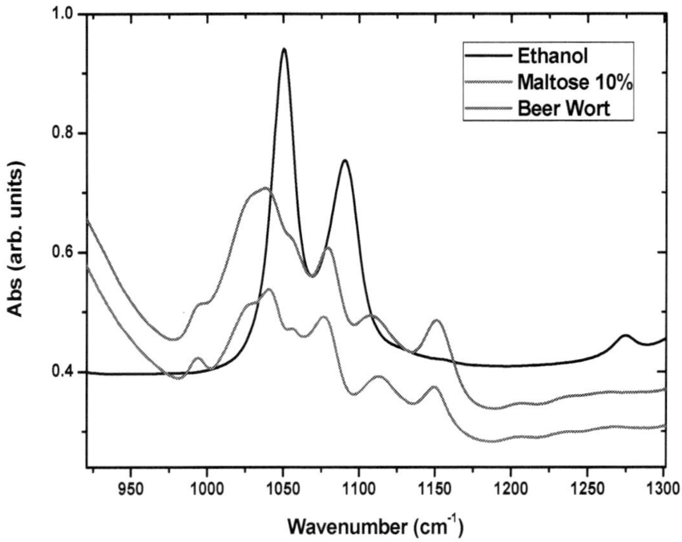

Figure 4. Absorption spectrum of a sample of ethanol, maltose 10%, and beer wort which contains certain types of sugars.

A certain part of the data (approximately 1/3), must be separated first in order to perform a further validation step. With 2/3 of the remaining data, the neural network is performed in the training stage following the equations and the structures described before, where the weight of each neural layer is adjusted in order to converge the network. The adjustment can be done as often as necessary, until the output (ξi) is as close to the true (real) value as required.

The training is complete, oncet the weight of the neurons has been adjusted, and the network has been converged with the desired error. The weight values should then be saved and stored before proceeding to the next step, which is the validation step: using the neural network to provide results of new spectra. With the data that were originally separated (1/3 data) and the values of the weights defined by the training stage, the neural network is run again. At this stage, note that the backpropagation system should not be executed. Simply use the matrix of the weights saved, and insert the data that have been separated for this validation step as inputs for the new network. Thus, the network will be performed only in the forward direction, supplying in a very short time of processing the output values ξi.

These output values are compared with the expected values using the HPLC technique, using a correlation curve between the two techniques. If these results are satisfactory, the process of mounting the system to quantify the compounds of interest is complete, and can be passed on for practical use.

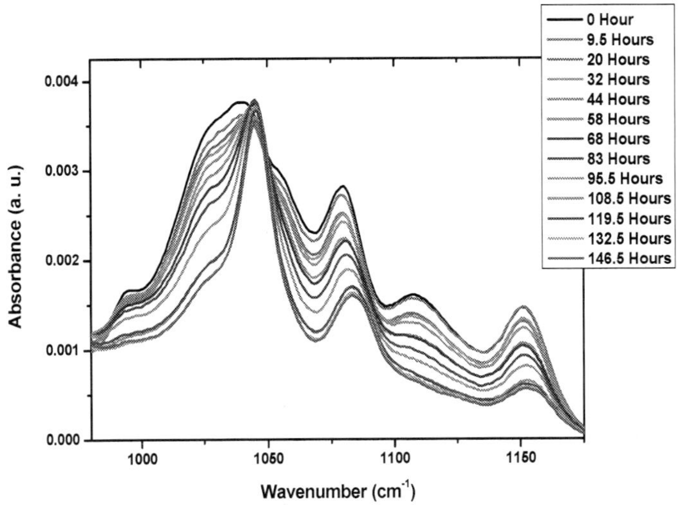

Figure 5. Absorption during the fermentation process.

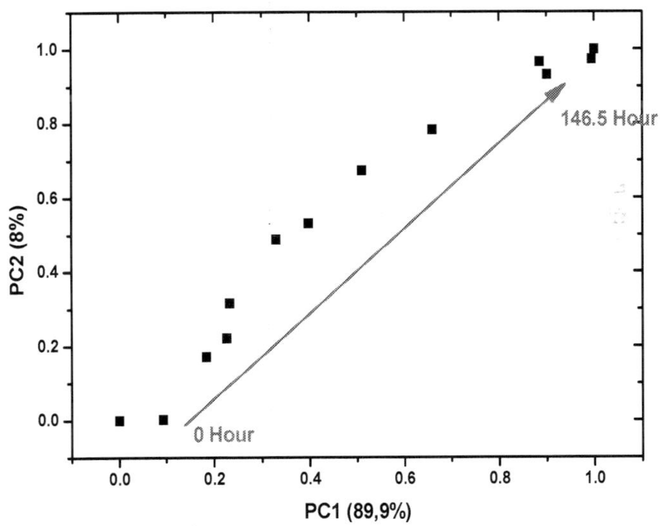

Figure 6. PC1 versus PC2 showing the time evolution of the fermentation process.

To be able to use spectroscopy with the neural processing requires using a standard method. We can simply use the PCA to reduce processing variables, entering the values of ordered pairs into the network, together with the weight values and collecting predetermined output results, in this case the amounts of sugars and ethanol. In the case of fermentation of the wort, using a number of principal components around three , a neural network comprising an input layer with 23 neurons and an output layer of 5 neurons , it is possible to quantify each type of sugar and ethanol with a quoted error of ± 0.2%. Here we exemplify our results showing the correlation between the value determined by the concentration of maltose using spectroscopy and HPLC technique (figure 7), where R^2 and the coefficient slope is 0.991 and 0.999 respectively. The results of a linear fit show a good agreement between the proposed new method and the standard procedure. This result allows the use of our technique in brewery, as it enables monitoring quality and making process control less time consuming.

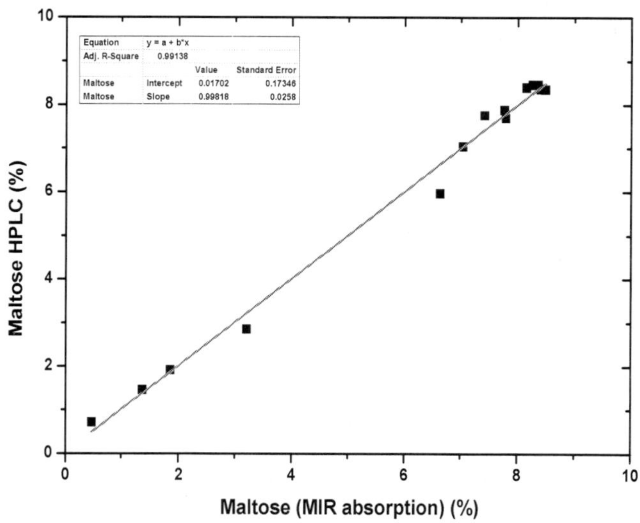

Figure 7. Correlation between the standard method (HPLC) and proposed procedure (MIR absorption).

4. Conclusion

In the analysis of spectroscopic data, not only the technique to obtain the values of different properties is important, but the correct mathematical processing of the data is actually the main issue to obtain the correct information. Especially the distinctions of multiple values which are correlated to a specific class of phenomena are the hide information that can be conveniently extracted. During our exposition in this chapter, we have concentrated in demonstrating how powerful the correct spectroscopy analysis can be when the first obtained data have been correctly arranged, allowing a mathematical procedure that treats the information as a whole instead of concentrations in individual values. Many techniques are today available for such procedures, but especially the Principal Component Analysis is quite powerful to be applied when spectral information is not restricted to a single wavelength, but rather to a large portion of the spectra.

We have concentrated on a relevant case where the UV-VIS portion of fluorescence spectrum is obtained and applied to determine its correlation with the postmortem interval in an animal model. The fluorescence in this case is subject to many effects due to the biological tissue modification as a natural evolution once the living metabolic action has been interrupted. This is clearly the case where biochemical modification causes alteration of spectrum as a whole and the attempt to concentrate the observation on individual features may fail. With the application of PCA to collect data, rich information patterns made a high correlation between extracted information and the real postmortem time interval possible. The classification of patterns and congregations of collections of information create a distinction into groups of distinct PMI. Even though we have used the method for PMI determination, the method has been shown to be as well powerful in applications in the field of cancer diagnostic, fermentation processing in beverage production, quality control in industry, identification of plagues and other features of interest in agriculture. The level of application of the PCA technique can go beyond the identification of pattern and correlation with values and can also provide specific quantification of individual chemical components of the system which is investigated.

To demonstrate this feature, we consider as an example the sugar quantification during beer production. These cases represent a bigger challenge to innumerous systems in several areas. Using the PCA procedure associated with a Neural Network (NN) we can quantify the composites in a sample, obtaining results comparably quickly. Here, we used the example of beer analysis. Using the MIR absorption spectroscopy of liquid samples, without any type of pre-procedure, we detected and quantified specific compounds (glucose, maltose, maltotriose, dextrin and ethanol) during the production of beer. The NN were used to determine the amount of these types of sugar and alcohol in the wort during the saccharification and fermentation. In the correlation between the values determined by the concentration of maltose spectroscopy with the HPLC technique we find the R2 and coefficient slope to be 0.991 and 0.999 respectively. Finally, the presentation of this chapter is to show the real power of the conjugation of spectroscopy techniques with data analyses. The field is clearly growing in diversity and importance.

Author details

E.S.Estracanholli, G.Nicolodelli, S.Pratavieira, C.Kurachi and V.S. Bagnato
Institute of Physics of São Carlos, University of São Paulo, SP, Brazil

Acknowledgement

The authors acknowledge the support provided by CNPq (INOF – INCT Program and scholarship) and FAPESP (CEPOF – CEPID Program and scholarship).

5. References

[1] Socrates G. Infrared and raman characteristic: group frequencies: J. Wiley 2001.

[2] Stuart BH. Infrared spectroscopy: Fundamentals and applications: J. Wiley 2004.

[3] Bauman RP. Absorption spectroscopy: John Wiley 1963.

[4] Lakowicz J. Principles of Fluorescence Spectroscopy. Springer 2006.

[5] Ramanujam N. Fluorescence spectroscopy of neoplastic and non-neoplastic tissues. Neoplasia. 2000 Jan-Apr;2(1-2):89-117.

[6] Estracanholli ES, Kurachi C, Bagnato VS. Tissue Fluorescence Spectroscopy in Death Time Estimation. *Forensic Pathology Reviews*: Springer Science 2011.

[7] Pratavieira S, Andrade CT, Salvio AG, Bagnato VS, Kurachi C. Optical Imaging as Auxiliary Tool in Skin Cancer Diagnosis. *Skin Cancers - Risk Factors, Prevention and Therapy* 2011.

[8] Estracanholli ES, Kurachi C, Vicente JR, de Menezes PFC, Silva OCE, Bagnato VS. Determination of post-mortem interval using in situ tissue optical fluorescence. Optics Express. 2009 May 11;17(10):8185-92.

[9] Schieberle HDBWGP. Food chemistry: Springer 2009.

[10] Woodcock T, Downey G, O'Donnell CP. Better quality food and beverages: the role of near infrared spectroscopy. Journal of near Infrared Spectroscopy. 2008;16(1):1-29.

[11] Bamforth CW. Food, fermentation and Micro-organisms. : Blackwell Science 2005.

[12] Kohonen T. Analysis of a Simple Self-Organizing Process. Biological Cybernetics. 1982;44(2):135-40.

[13] Kohonen T. Self-Organized Formation of Topologically Correct Feature Maps. Biological Cybernetics. 1982;43(1):59-69.

[14] Kandel ER, Schwartz JH, Jessel TM. Principles of neural science: McGraw-Hill 2000.

[15] Lefebvre JCPNREWC. Neural and Adaptive Systems: Fundamentals through Simulations: John Wiley & Sons 1999.

[16] Marks RDRRJ. Neural smithing: supervised learning in feedforward artificial neural networks. : MIT Press 1999.

Chapter 18

Injection and Optical Spectroscopy of Localized States in II-VI Semiconductor Films

Denys Kurbatov, Anatoliy Opanasyuk and Halyna Khlyap

Additional information is available at the end of the chapter

http://dx.doi.org/10.5772/48290

1. Introduction

Novel achievements of nano- and microelectronics are closely connected with working-out of new semiconductor materials. Among them the compounds II-VI (where A = Cd, Zn, Hg and B = O, S, Se, Te) are of special interest. Due to unique physical properties these materials are applicable for design of optical, acoustical, electronic, optoelectronic and nuclear and other devices [1-3]. First of all the chalcogenide compounds are direct gap semiconductors where the gap value belongs to interval from 0.01 eV (mercury chalcogenides) up to 3.72 eV (ZnS with zinc blende crystalline structure) As potential active elements of optoelectronics they allow overlapping the spectral range from 0.3 μm to tens μm if using them as photodetectors and sources of coherent and incoherent light. The crystalline structure of II-VI compounds is cubic and hexagonal without the center of symmetry is a good condition for appearing strong piezoeffect. Crystals with the hexagonal structure have also pyroelectric properties. This feature may be used for designing acoustoelectronic devices, amplifiers, active delay lines, detectors, tensile sensors, etc. [1-2]. Large density of some semiconductors (CdTe, ZnTe, CdSe) makes them suitable for detectors of hard radiation and α–particles flow [4-5]. The mutual solubility is also important property of these materials. Their solid solutions give possibility to design new structures with in-advance defined gap value and parameters of the crystalline lattice, transmission region, etc. [6].

Poly- and monocrystalline films of II-VI semiconductors are belonging to leaders in field of scientific interest during the last decades because of possibility of constructing numerous devices of opto-, photo-, acoustoelectronics and solar cells and modules [2-5]. However, there are also challenges the scientists are faced due to structural peculiarities of thin chalcogenide layers which are determining their electro-physical and optical characteristics. The basic requirements for structure of thin films suitable for manufacturing various microelectronic devices are as follows: preparing stoichiometric single phase monocrystalline layers or columnar strongly textured polycrystalline layers with low concentration of stacking faults (SF), dislocations, twins with governed ensemble of point defects (PD) [7-8]. However, an enormous number of publications points out the following features of these films: tend to departure of stoichiometric composition, co-existing two polymorph modifications (sphalerite and wurtzite), lamination morphology of crystalline grains (alternation of cubic and hexagonal phases), high concentration of twins and SF, high level of micro- and macrostresses, tend to formation of anomalous axial structures, etc. [2-3, 9]. Presence of different defects which are recombination centers and deep traps does not improve electro-physical and optical characteristics of chalcogenide layers. It restricts the application of the binary films as detector material, basic layers of solar energy photoconvertors, etc.

Thus, the problem of manufacturing chalcogenide films with controllable properties for device construction is basically closed to the governing of their defect structure investigated in detail. We will limit our work to the description of results from the examination of parameters of localized states (LS) in polycrystalline films CdTe, ZnS, ZnTe by the methods of injection and optical spectroscopy.

1.1. Defect classification in layers of II-VI compounds

Defects' presence (in the most cases the defects of the structure are charged) is an important factor affecting structure-depended properties of II-VI compounds [3, 5, 10]. Defects of the crystalline structure are commonly PD, 1-, 2-, and 3-dimensional ones [11-12]. Vacancies (V_A, V_B), interstitial atoms (A_i, B_i), antistructural defects (A_B, B_A), impurity atoms located in the lattice sites (C_A, C_B) and in the intersites (C_i) of the lattice are defects of the first type. However, the antistructural defects are not typical for wide gap materials (except CdTe) and they appear mostly after ionizing irradiation [13-14]. The PD in chalcogenides can be one- or two-charged. Each charged native defect forms LS in the gap of the semiconductor, the energy of the LS is ΔE_i either near the conduction band (the defect is a donor) or near the valence band (then the defect is an acceptor) as well as LS formed in energy depth are appearing as traps for charge carriers or recombination centers [15-16]. Corresponding levels in the gap are called shallow or deep LS. If the extensive defects are minimized the structure-depending properties of chalcogenides are principally defined by their PD. The effect of traps and recombination centers on electrical characteristics of the semiconductor materials is considered in [16]. We have to note that despite a numerous amount of publications about PD in Zn-Cd chalcogenides there is no unified theory concerning the nature of electrically active defects for the range of chalcogenide vapor high pressures as well as for the interval of high vapor pressure of chalcogen [13-14, 17-18].

Screw and edge dislocations are defects of second type they can be localized in the bulk of the crystalites or they form low-angled boundaries of regions of coherent scattering (RCS). Grain boundaries, twins and surfaces of crystals and films are defects of the third type. Pores and precipitates are of the 4th type of defects. All defects listed above are sufficiently influencing on physical characteristics of the real crystals and films of II-VI compounds due to formation of LS (along with the PD) in the gap of different energy levels [17-20].

2. Using injection spectroscopy for determining parameters of localized states in II-VI compounds

2.1. Theoretical background of the injection spectroscopy method

The LS in the gap of the semiconductor make important contribution to the function of the device manufactured from the material solar cells, phodetectors, γ-ray detectors and others), for example, carriers' lifetime, length of the free path, etc., thus making their examination one of them most important problems of the semiconductor material science [3-5, 8, 13, 14, 18].

There are various methods for investigation of the energy position (E_t), concentration (N_t) and the energy distribution of the LS [21-23]. However, their applicability is restricted by the resistance of the semiconductors, and almost all techniques are suitable for low-resistant semiconductors. At the investigation of the wide gap materials II-VI the analysis of current-voltage characteristics (CVC) at the mode of the space-charge limited current (SCLC) had appeared as a reliable tool [24-25]. The comparison of experimental and theoretical CVCs is carried out for different trap distribution: discrete, uniform, exponential, double-exponential, Gaussian and others [26-36]. This method is a so-called direct task of the experiment and gives undesirable errors due to in-advance defined type of the LS distribution model used in further working-out of the experimental data. The information obtained is sometimes unreliable and incorrect.

Authors [37-40] have proposed novel method allowing reconstructing the LS energy distribution immediately from the SCLC CVC without the pre-defined model (the reverse task), for example, for organic materials with energetically wide LS distributions [41-42]. However, the expressions presented in [37-40], as shown by our studies [43-45], are not suitable for analysis of experimental data for mono- and polycrystalline samples with energetically narrow trap distributions. So, we use the principle [37-40] and obtain reliable and practically applicable expressions for working-out of the real experiments performed for traditional II-VI compounds.

Solving the Poisson equation and the continuity equation produces SCLC CVC for rectangular semiconductor samples with traps and deposited metallic contacts, where the source contact (cathode) provides charge carriers' injection in the material [24-25]:

$$j = e\mu E(x) n_f(x), \tag{1}$$

$$\frac{dE(x)}{dx} = \frac{e\left[\left(n_f(x) - n_{f0}\right) + \sum_j \left(n_{t_j}(x) - n_{t_{j0}}\right)\right]}{\varepsilon\varepsilon_0} = \frac{en_s(x)}{\varepsilon\varepsilon_0}, \tag{2}$$

where j current density passes through the sample;
e electron charge;
μ drift carrier mobility;
ε_0 dielectric constant;
ε *permittivity of the material*

$E(x)$ is an external electric field changing by the depth of the sample; this field injects free carriers from the source contact (cathode) (x=0) to the anode collecting the carriers (x=d);

$n_f(x)$ is the free carriers' concentration at the injection;

n_{f0} is the equilibrium free carriers concentration;

$n_{t_j}(x)$ is the concentration of carriers confined by the traps of the j-group with the energy level E_{t_j};

n_{tj0} is the equilibrium carriers concentration trapped by the centers of the j-group;

$n_s(x)$ is a total concentration of the injected carriers.

The set of equations (1), (2) is commonly being solved with a boundary condition $E(0)=0$. The set is soluble if the function from n_f and n_t is known. We assume that all LS in the material are at thermodynamic equilibrium with corresponding free bands, then their filling-in by the free carriers is defined by the position of the Fermi quasi-level E_F. Using the Boltzmann statistics for free carriers and the Fermi – Dirac statistics for the localized carriers we can write [39-40]:

$$n_f(x) = N_{C(V)} \exp\left(\frac{E_{C(V)}(x) - E_F(x)}{kT}\right), \tag{3}$$

$$n_t(E,x) = \frac{h(E,x)}{1 + g\exp\frac{E_t(x) - E_F(x)}{kT}}, \tag{4}$$

where $N_{c(v)}$ are states density in conduction band (valence band);
$E_{c(v)}$ is energy of conduction band bottom (valence band top);
k is Boltzmann constant;
T is the temperature of measurements;
$E_F(x)$ is the Fermi quasi-level at injection;

g is a factor of the spin degeneration of the LS which depends on its charge state having the following values: –1/2, 1 or 2 (typically g = 1) [15, 39-40].

The zero reference of the trap energy level in the gap of the material will be defined relatively to the conduction band or valence band depending on the type (n or p) of the examined material: $E_{C(V)}$=0.

The set of equations (1)–(2) can also be reduced to integral relations. Detailed determination of these ratios presented in [37].

$$\frac{1}{j}=\frac{1}{e\mu d}\frac{\varepsilon\varepsilon_0}{e}\int_{n_{fc}}^{n_{fa}}\frac{dn_f}{n_f^2[(n_f-n_{f_0})+\sum_j(n_{t_j}-n_{t_{j0}})]}\equiv y, \tag{5}$$

$$\frac{U}{j^2}=\frac{\varepsilon\varepsilon_0}{e(e\mu)^2}\int_{n_{fc}}^{n_{fa}}\frac{dn_f}{n_f^3[(n_f-n_{f_0})+\sum_j(n_{t_j}-n_{t_{j0}})]}\equiv z, \tag{6}$$

where j, U are current density and voltage applied to the sample;
d is the sample thickness;
n_{fc}, n_{fa} are free carriers' concentration in cathode and anode, respectively.

Equations (5) and (6) determine SCLC CVCs in parametric form for an arbitrary distribution of LS in the gap of the material.

At thermodynamic equilibrium the total concentration (n_{s_0}), the carriers concentration for those localized on the traps (n_{t_0}), and the free carriers' concentration in the semiconductor (n_{f_0}) are in the function written as follows: $n_{s_0}=n_{t_0}+n_{f_0}$, де $n_{f_0}=N_{C(V)}\exp\left(-\frac{E_{C(V)}-E_{F_0}}{kT}\right)$ in case when $E_c - E_{F0} \geq 3kT$ ($3kT$= 0,078 eV at the room temperature; E_{F0} is the equilibrium Fermi level. It must be emphasized that this charge limits the current flow through the sample and determines the form of the SCLC CVC.

The carriers' injection from the source contact leads to appearance of the space charge ρ in the sample, formed by the free carriers and charge carriers localized in the traps, $\rho=en_i=e(n_s-n_{s_0})=e[(n_t+n_f)-(n_{t_0}+n_{f_0})]$, where n_i is the concentration of injected carriers.

Under SCLC mode the concentration of injected carriers is considerably larger than their equilibrium concentration in the material and, at the same time, it is sufficiently lower than the total concentration of the trap centers ($n_{f0} \ll n_i \ll N_t$) [24-25]. Thus, in further description we will neglect the second term in the expression written above (except some special cases). Then we have $\rho=en_s(x)\sim e\sum_j n_{t_j}(x)$.

Using (5) and (6) we find the first and second derivatives of z from y:

$$z'=\frac{dz}{dy}=\frac{d(U/j^2)}{d(1/j)}=\frac{d}{e\mu n_{fa}}, \tag{7}$$

$$z'' = \frac{d^2 z}{dy^2} = \frac{d}{d(1/j)} \frac{d(U/j^2)}{d(1/j)} = \left| \frac{\rho_a d^2}{\varepsilon\varepsilon_0} \right|. \qquad (8)$$

As the SCLC CVC are commonly represented in double-log scale [24-25], equations (7), (8) are rewritten with using derivatives: $\eta = \frac{d(\ln j)}{d(\ln U)}, \eta' = \frac{d^2(\ln j)}{d(\ln U)^2}, \eta'' = \frac{d^3(\ln j)}{d(\ln U)^3}$.

Then we have

$$n_{fa} = \frac{\eta}{2\eta - 1} \frac{jd}{e\mu U} = \frac{1}{\alpha} \frac{jd}{e\mu U}, \qquad (9)$$

$$\frac{\rho_a}{e} = \frac{2\eta - 1}{\eta} \frac{\eta - 1}{\eta} \left[1 - \frac{\eta'}{\eta(2\eta - 1)(\eta - 1)} \right] \frac{\varepsilon\varepsilon_0 U}{ed^2} = \alpha\beta \frac{\varepsilon\varepsilon_0 U}{ed^2}, \qquad (10)$$

where $\alpha = \frac{2\eta - 1}{\eta}$, $\beta = \frac{\eta - 1}{\eta} \left[1 - \frac{\eta'}{\eta(2\eta - 1)(\eta - 1)} \right] = \frac{\eta - 1}{\eta}(1 + B)$.

Further we will neglect the index *a*.

As a result, the Poisson equation and the continuity equation give fundamental expressions for a dependence of the free carrier concentration in the sample n_f (the Fermi quasi-level energy) and space charge density at the anode ρ on the voltage U and the density of the current j flowing through the structure metal-semiconductor-metal (MSM).

Now let us consider the practical application of expressions (7) and (8) or (9) and (10) for reconstructing the trap distribution in the gap of the investigated material. We would restrict with the electron injection into n-semiconductor.

If the external voltage changes the carries are injected from the contact into semiconductor; at the same time, the Fermi quasi-level begins to move between the LS distributed in the gap from the start energy E_{F0} up to conduction band. This displacement ΔE_F leads to filling-in of the traps with the charge carriers and, consequently, to the change of the conductivity of the structure. Correspondingly, under intercepting the Fermi quasi-level and the monoenergetical LS the CVC demonstrates a peculiarity of the current [24-25]. As the voltage and current density are in the function of the LS concentration with in-advanced energy position and the Fermi quasi-level value we obtain a possibility to scan the energy distributions. This relationship is a physical base of the injection spectroscopy method (IS).

Increase of the charge carriers dn_s in the material at a low change of the Fermi level position is to be found from the expression:

$$\frac{1}{e} \frac{d\rho}{dE_F} = \frac{dn_i}{dE_F} \approx \frac{dn_s}{dE_F} \approx \frac{dn_t}{dE_F}. \qquad (11)$$

The carrier concentration on deep states can be found from the Fermi-Dirac statistics

$$n_s = \int_{E_1}^{E_2} n_s(E)dE = \int_{E_1}^{E_2} h(E)f(E-E_F)dE + n_f, \tag{12}$$

where $dn_s(E)/dE$ is a function describing the energy distribution of trapped carriers;
$h(E)=dN_t/dE$ is a function standing for the energy trap distribution;
E_1, E_2 are energies of start and end points for the LS distribution in the gap of the material.

It is assumed that the space trap distribution in the semiconductor is homogeneous by the sample thickness then $h(E,x) = h(E)$.

After substitution of (12) in (11) we obtain a working expression for the functions $d\rho/dE_F$ and $h(E)$

$$\frac{1}{e}\frac{d\rho}{dE_F} \approx \frac{dn_s}{dE_F} = \frac{d}{dE}\int_E n_s(E)dE = \int_E h(E)\frac{d(f(E-E_F))}{d(E-E_F)} + \frac{n_f}{kT} \tag{13}$$

Thus, at arbitrary temperatures of the experiment the task of reconstructing LS distributions reduces to finding function $h(E)$ from the convolution (12) or (13) using known functions $n_s(E_F)$ or $d\rho/dE_F$. The expression (12) is the most preferable [39-40]. In general case the solution is complex and it means determining the function $h(E)$ from the convolution (12) or (13) if one of the functions n_s or dn_s/dE_F is known [43-45]. We have solved this task according the Tikhonov regularization method [46]. If the experiment is carried at low temperatures (liquid nitrogen) the problem is simplified while the Fermi-Dirac function in (13) may be replaced with the Heavyside function and, neglecting n_f, we obtain

$$\frac{1}{e}\frac{d\rho}{dE_F} \approx \frac{dN_t}{dE_F} \approx h(E). \tag{14}$$

This equation shows that the function $1/e\ d\rho/dE_F$ - E_F at low-temperature approximation immediately produces the trap distribution in the gap of the semiconductor. Using (7) and (8), we transform the expression (14) for practical working-out of the experimental SCLC CVC. As the free carrier concentration and the space charge density are to be written as follows:

$$n_f = \frac{d}{e\mu}\frac{j}{2U - U'j}, \tag{15}$$

$$\frac{\rho}{e} = (U''j^2 - 2U'j + 2U)\frac{\varepsilon\varepsilon_0}{ed^2}, \tag{16}$$

the expression (14) will be

$$h(E) \approx \frac{1}{e}\frac{d\rho}{dE_F} = \frac{1}{kT}\frac{\varepsilon\varepsilon_0}{ed^2}\frac{U'''j^3(2U - U'j)}{(U''j^2 - 2U'j + 2U)}. \tag{17}$$

Using derivatives η, η', η'' this expression is easily rewritten:

$$h(E) \approx \frac{1}{e}\frac{d\rho}{dE_F} = \frac{1}{kT}\frac{\varepsilon\varepsilon_0 U}{ed^2}\frac{2\eta-1}{\eta^2}\left[1+\frac{(3\eta-3)\eta\eta'-\eta\eta''+3\eta'^2}{\eta^2\left((2\eta-1)(\eta-1)-\eta'/2\right)}\right]. \tag{18}$$

The expression (18) is also can be written with the first derivative (η) only. Denote

$$C=\frac{(3\eta-3)\eta\eta'-\eta\eta''+3\eta'^2}{\eta^2\left[(2\eta-1)(\eta-1)-\eta'/2\right]}=(2-3\eta)B+\frac{d\ln(1+B)}{d\ln U}=\frac{(2\eta-1)B+(3\eta-2)B^2+\frac{d\left[\ln(1+B)\right]}{d\ln U}}{1+(\eta-1)B},$$

where $B=-\frac{1}{\eta(2\eta-1)(\eta-1)}\frac{d\eta}{d\ln U}$.

We obtained an expression used by authors [39-40] for analysis of energetically wide LS distributions in organic semiconductors.

$$h(E) \approx \frac{1}{e}\frac{d\rho}{dE_F} = \frac{1}{kT}\frac{\varepsilon\varepsilon_0 U}{ed^2}\frac{2\eta-1}{\eta^2}(1+C)=\frac{\alpha\beta}{kT}\frac{\varepsilon\varepsilon_0 U}{ed^2}\frac{1}{\eta-1}\left(\frac{1+C}{1+B}\right). \tag{19}$$

To make these expressions suitable for the working-out of SCLC CVC for the semiconductors with energetically narrow trap distributions we write them with reverse derivatives $\gamma=\frac{d(\ln U)}{d(\ln j)}$, $\gamma'=\frac{d^2(\ln U)}{d(\ln j)^2}$, $\gamma''=\frac{d^3(\ln U)}{d(\ln j)^3}$.

As a result:

$$h(E) \approx \frac{1}{e}\frac{d\rho}{dE_F} = \frac{1}{kT}\frac{\varepsilon\varepsilon_0 U}{ed^2}\left[\frac{(2\gamma-3)\gamma'+\gamma''}{(2-\gamma)(1-\gamma)+\gamma'}+\gamma\right](2-\gamma). \tag{20}$$

Solving the set of equations (3) and (7) gives energetical scale under re-building deep trap distributions. Using various derivatives we obtain

$$E_F = kT\ln\frac{e\mu N_{C(V)}}{d}+kT\ln\frac{2U-U'j}{j} = kT\ln\frac{e\mu N_{C(V)}}{d}+kT\ln\frac{j}{U}+kT\ln\frac{\eta}{2\eta-1}=$$

$$=kT\ln\frac{e\mu N_{C(V)}}{d}+kT\ln\frac{j}{U}+kT\ln\frac{1}{2-\gamma}. \tag{21}$$

Using sets of equations (17) - (18) or (20) - (21) allows to find a function describing the LS distribution in the gap immediately from the SCLC CVC. To re-build the narrow or monoenergetical trap distributions (typical for common semiconductors) the most suitable expressions are written with derivatives. The first derivative γ defines the slope of the CVC section in double-log scale relative to the current axis, the η defines the slope of the CVC section in double-log scale relative to the voltage axis. For narrow energy distributions this angle η is too large, and under complete filling-in of the traps it closes to [24-25]. However, it means the slope to the current axis is very small allowing finding the first and higher order derivatives with proper accuracy [44, 45, 48]. It is important that the narrowest trap distributions give the higher accuracy under determination of the derivatives γ, γ', γ'' !

If the distributions in the semiconductor are energetically broadened all expressions (17), (18), and (20) can be used as analytically identical formulas.

As is seen from the expressions written above, in order to receive information about LS distribution three derivatives are to be found at each point of the current-voltage function in various coordinates. Due to experimental peculiarities we had to build the optimization curve as an approximation of the experimental data with it's further differentiation at the sites. The task was solved by constructing smoothing cubic spline [47]. However, the numerical differentiation has low mathematical validity (the error increases under calculation of higher order derivatives). To achieve maximum accuracy we have used the numerical modeling with solving of direct and reverse tasks.

Under solving the direct task we have calculated the functions ρ - E_F and $1/e\ d\rho/dE_F$ - E_F on base of known trap distribution in the gap of the material (the input distribution) using the expressions (12) and (13). Then we have built the theoretical SCLS CVCs ((5), (6)). The mathematical operations are mathematically valid. To solve the reverse problem of the experiment CVCs were worked out using the differential technique based on expressions (17), (21), (18), (20). As a result we have again obtained the deep centers' distribution in the gap of the material (output distribution). Coincidence of the input and output trap distributions was a criterion of the solution validity under solving the reverse task. Further the program set was used for numerical working-out of the experimental CVCs [43-45, 48].

2.2. Determination of deep trap parameters from the functions $1/e{\cdot}d\rho/dE_F$ - E_F under various energy distributions

Now we determine how the energy position and the trap concentration under presence of the LS in the gap may be found for limit cases by the known dependence $1/e{\cdot}d\rho/dE_F$ - E_F. In the case of mono-level the LS distribution can be written as $h(E) = N_t\delta(E - E_t)$, where δ is a delta-function.

After substituting this relationship in (12), (13) we obtain

$$n_s \approx n_t = \frac{N_t}{1 + g\exp\left(\frac{E_F - E_t}{kT}\right)}, \tag{22}$$

$$\frac{1}{e}\frac{d\rho}{dE_F} \approx \frac{dn_t}{dE_F} = \frac{gN_t\exp(\frac{E_F - E_t}{kT})}{kT(1 + g\exp(\frac{E_F - E_t}{kT}))^2}, \tag{23}$$

The value of the last function at the maximum (E_F=E_t) is $\left(\frac{1}{e}\frac{d\rho}{dE_F}\right)_{E_{Fm}} = \frac{gN_t}{kT(1+g)^2}$,

or at g=1 - $\left(\frac{1}{e}\frac{d\rho}{dE_F}\right)_{E_{Fm}} = \frac{N_t}{4kT}$..

Than

$$N_t = 4kT\left(\frac{1}{e}\frac{d\rho}{dE_F}\right)_{E_{Fm}} . \quad (24)$$

Thus, building the function $1/e \cdot d\rho/dE_F$ - E_F and finding the maximums by using (24) gives the concentration of discrete monoenergetical levels. The energy position of the maximum immediately produces energy positions of these levels.

If the LS monotonically distributed by energy $h(E)$=AN_t= const are in the gap of the material it is easy to obtain

$$N_t = h(E) = \frac{1}{e}\frac{d\rho}{dE_F}. \quad (25)$$

In other words the trap concentration in the sample under such distributions is immediately found from the function $1/e \cdot d\rho/dE_F$ - E_F.

In general case when LS distribution in the gap of the material is described by the arbitrary function their concentration is defined by the area under the curve $1/e \cdot d\rho/dE_F$ - E_F and at low temperatures can be found from the relationship $N_t = \int_{E_1}^{E_2} h(E)dE$. Under reconstruction of such distributions from the SCLC CVC these distributions are energetically broadening depending on the temperature of experiment [43-45]. LS energy positions are again determined by the maximums of the curve.

The correct determination of the trap concentration from the dependence $1/e \cdot d\rho/dE_F$ - E_F may be checked out by using the function ρ - E_F. In case of the mono-level where the Fermi quasi-level coincides with the LS energy position, it is easy to obtain from (22) - $[n_t]_{E_{Fm}} = \frac{N_t}{1+g}$, then $N_t = (1+g)[n_t]_{E_{Fm}}$. If g=1 then $[n_t]_{E_{Fm}} = N_t/2$, $N_t = [2n_t]_{E_{Fm}}$.

If the LS distribution is a Gaussian function ($h(E) = \frac{N_t}{\sigma_t(2\pi)^{1/2}}\exp\left(-\frac{E-E_t}{\sqrt{2}\sigma_t}\right)^2$) the relationship for determination of N_t is analogous to that described above.

Earlier [43-45] we have described the effect of experimental factors on accuracy of determining parameters of the deep centers by IC method. In Ref. [44, 45, 48] it was shown that the neglecting third order derivative or even the second order derivative does not lead to considerable decrease of the accuracy in determination of the LS parameters. It was demonstrated that under neglecting the 3rd order derivative γ'' in (20) the error in definition of the function $h(E)$ at the point $E_F=E_t$ is no more than 0.4%. At the same time this error is somewhat larger in the interval E_F-E_t~kT but is not larger than (4–7)%. Such a low error of the calculation of the LS parameters is caused by the interception of zero point and the derivative γ'' near the point $E_F=E_t$ (commonly in the range of $\left|E_F - E_t\right| < 0,2kT$). As a result (regarding the absence of accurate experimental measurement of the 3rd derivative) it does not affect the differential working-out of CVCs in the most important section where the Fermi quasi-level coincides with the LS energy position.

If the second order derivative in the working expressions is neglected the error of the defining the function $h(E)$ in the most principal ($E_F \approx E_t$) is about (30-40)%. In both cases the simplification of the expression (21) does not contribute errors to the definition of energy position of the traps' level. Remember that the traditional method of SCLC CVC gives 60-100% error of the traps' concentration [24-25].

3. Methods of preparation and investigation of II-VI films

Thin films CdTe, ZnS, ZnTe were prepared on glass substrates in vacuum by close-spaced vacuum sublimation (CSVS) [49-50]. For further electrical investigations we have deposited hard-melted metal conductive layers on the main substrate by electron beam evaporation (Mo – for CdTe, ZnS; Cr, Ti – for ZnTe). The up-source contact (In(Ag) or Cr in dependence on the conductivity type of the semiconductor) was deposited by the vacuum thermal evaporation. Under condensation of the films of binary compounds the chalcogenide stoichiometric powders were used.

The common temperature of the evaporator was T_e = 973 K for zinc telluride, T_e = (1200÷1450) K for zinc sulfide and T_e = (933÷1023) K for cadmium telluride. The substrate temperature was changed in a wide range T_s = (323÷973) K. Time of deposition was varied: t = (10÷30) min.

Morphology of the samples' surfaces was investigated by optical and electron microscopy. Jeffries' method was used to determine the arbitrary grain size (D) in the condensates. The films' thickness (d) was measured by fractography and interferential methods. The element composition of the layers was studied by X-ray spectroscopy (XRS) analysis using the energy-dispersed X-ray analysis (EDAX) unit or by Rutherford back scattering (RBS) technique (if it was possible). Structural examinations of the films were carried out by the XRD-unit in Ni-filtered K_α radiation of Cu-anode. The XRD patterns were registered in the range of Bragg angles from 20^0 to 80^0. Phase analysis was provided by comparison of inter-plane distances and arbitrary intensities from the samples and the etalon according to the ASTM data [51]. Structural properties of II-VI films are investigated in [20, 49-50, 54-56].

Dark CVC at different temperatures and $\sigma - T$ dependencies of the sandwich-structures (MSM were examined in vacuum by standard techniques (Fig. 1) [21-22].

The power of electronic scheme was estimated by source of stable voltage AIP 120/0.75 that provided a possibility of precise voltage regulation in electric circle in the ranfe of U =0.1 ÷ 120 V.

A current that passed throught samples in the range of $I = (10^{-9} \div 10^{-5})$ A measured by digital nanoampermeter. Voltage drop on sample was fixed by digital multimeters APPA-108N and UT70B. Sample temperature at inveatigation of electro-physical properties controlled by authomatic feedback temperature controller "OVEN TRM10", input signal from it fed from chromel-alumel thermocouple.

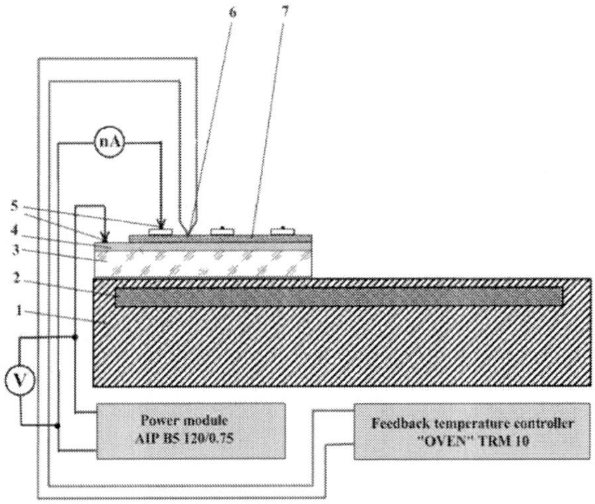

Figure 1. Typical electrical-type scheme for CVC and σ-T characteristic investigations of II-VI semiconductors films: 1 – heater holder; 2 – heater; 3 – glass substrate; 4 – lower conductive layer (Mo, Cr, Ti); 5 – collectors; 6 – thermocouple; 7 – II-VI film

The current mechanisms were identified by the differential method developed in [52-53]. This technique completely analyses j–U, γ–U and d(logγ)/d(log$U$)-U functions, where γ=d(log$j$)/d(log$U$) and differentiates satellite and concurrent current mechanisms in the structures and defines the high-field mechanisms among all of them. When the CVCs of multilayered structures were determined by unipolar injection from the source contact the experimental curves were numerically studied by using low-temperature and high-temperature approximations of the IS method [43-45, 48].

PL spectra of CdTe, CdSe and ZnTe films were studied using the spectrometer SDL-1 under excitation of the samples by Ar-laser (λ=514 nm for CdTe and λ=488,8 nm for ZnTe). PL spectra from ZnS films are registered by MPF-4 Hitachi and xenon bulb (λ=325 nm). The temperature in all experiments was stable in the range 4.7÷77 K by using the system "UTREX" [49]. The films CdTe, ZnTe were investigated in the range of edge luminescence, the films ZnS were studied in the impurity energy range.

At interpretation of the PL data it was suggested that the radiation had appeared as a result of electrons' transfer from the conduction (valence) band or shallow donor (acceptor) levels to the deep LS in the gap of the material. Then the activation energy of the processes are defined from the expression:

$$\Delta E = h\nu = E_g - E_i = E_g - (E_a + E_d), \tag{26}$$

where E_a, E_d are energy levels of the donors and acceptors in the gap of the material.

The set of methods for defining parameters of LS in the gap allowed to enhance the accuracy of data obtained and to examine traps and recombination centers with wide energy range.

4. Determination of LS parameters of polycrystalline chalcogenide films by injection spectroscopy method and analysis of $\sigma - T$ functions

4.1. General description of CVC and $\sigma - T$ functions

Dark CVC of sandwich structures current-conductive substrate-film-upper drain contact were measured at different temperatures for examining electrical properties of Zn and Cd chalcogenide films and determination of parameters for LS in the gap of material. Besides that, the function conductivity-temperature was studied in ohmic sections of the CVC and in some cases in the square section of the CVC. Energy positions of donor (acceptor) centers in the films were found from dependencies $\log\sigma = f(10^3/T)$ taking into account their Arrhenius-like character [21-22].

As was shown by the study, the CVC of multilayered structures MSM is defined by the condensation conditions of chalcogenide films, their crystal structure, and material of bottom and upper metallic contacts. CVC of multilayered structures based on low-temperature condensates of II-VI compounds were linear or sublinear. For ZnTe-based MSM structures the CVC were defined by the Pool-Frenkel mechanism, and the data were linearized in the coordinates $\log(I/U) - U^{1/2}$ [52].

Fig. 2 plots typical double-log CVC measured at different temperatures. This figure also shows the function $\sigma - T$ measured at the ohmic section of the CVC.

It is found out that the $\sigma - T$ function of low-temperature condensates are linear with the slope to the T axis decreasing at lowering the measurement temperature. These features are typical for the material with various types of donor (acceptor) impurities with different activation energy. The CVC of high-temperature condensates were somewhat others (Fig. 2). The linear sections are reveled, their slope to the T axis increases as the measurement temperature decreases. It is typical for compensated materials [21-22]. The compensation effect appears more visible under sufficiently low experimental temperatures when the electron concentration becomes close to that of acceptor centers. The slope of the straight lines to the T-axis increases from the value $E_a/2k$ up to the value E_a/k, making it possible to define activation energy for donor and acceptor centers [21-22].

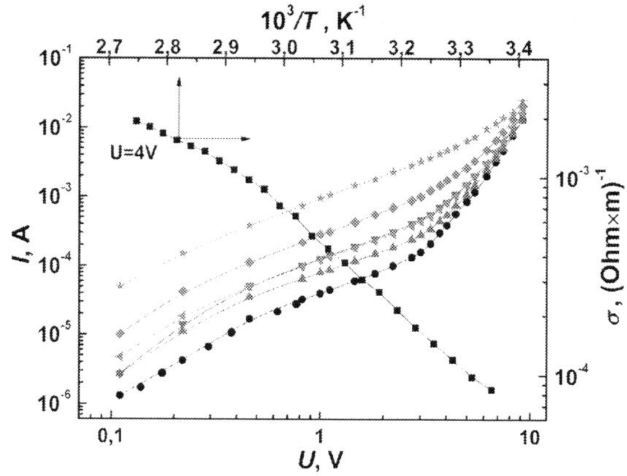

Figure 2. CVC of the structure Cr/ZnTe/Ag at various temperatures: • – T = 298 K; ▲ – T = 303 K; ▼ – T = 308 K; ▶ – T = 313 K; ♦ – T = 318 K; * – T = 323 K, and the dependence logσ – 1/T obtained from the ohmic section of the CVC. The film is prepared at T_e = 973 K and T_s = 823 K

CVC of multilayered structures where chalcogenide films are prepared at T_s > (500÷600) K were superlinear. As is analytically shown, they are determined by the unipolar injection from the drain contact. Typical SCLC CVCs of the examined films are plotted in Figs. 2-3. CVCs of high-temperature condensates in the range of high field strength a set of linear sections with various slopes to the U-axis was observed. As a rule, the sections with functions: $I - U$, $I - U^{2}$, $I - U^{3\text{-}5}$, $I - U^{8\text{-}10}$ were the most pronounced. In some cases after superlinear sections we have observed a square dependence I on U, which had further changed again to the supelinear one with a very large slope η (η ~ 13–25). The current jump was revealed and the samples were turn on the low-ohmic state as an irreversible process.

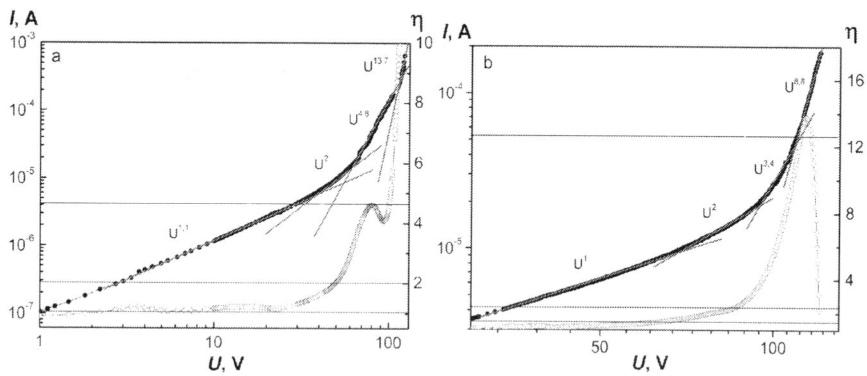

Figure 3. Double-log SCLC CVC of multilayered structures Mo/CdTe/Ag and results of their differentiation. CdTe films are prepared at T_e = 893 K and various T_s: 723 K (a); 823 K (b)

The features of the CVCs are clearly shown in functions $\eta - \log U$ giving a possibility to reveal a fine structure of the CVCs (Fig.3). Each point of this graph defines the slope of the CVC in double-log scale to the voltage axis. Dependencies $\eta - \log U$ were obtained by differentiating the CVC in every experimental point. As it was mentioned above, the problem mathematically reduces to the building smoothing cubic spline which approximates experimental data and its differentiation at the sites.

The curves $\eta - \log U$ resulting from the working-out of the SCLC CVC showed 1-4 maximums in correspondence to the sections of sharp current increase in the $I - U$ dependencies. The most often values of η_{ext} were 8–10. Sometimes the functions $\eta - \log U$ were practically revealed.

Horizontal sections with the almost constant slope $\eta > 2$ were also observed. It may be explained by the presence in the samples of sets of monoenergetical or quasi-monoenergetical levels traps of various energy position and concentration or by availability of the exponential (or other form) LS energy distribution. The specific points of CVCs were used for calculating trap parameters in the material, the ohmic sections helped to find specific conductivity of the layers σ = $(10^4 \div 10^5)$ Ω·m. As a result we obtained the concentrational distributions of the traps in the gap of material $h(E)$-E, their energy position (E_t) and concentration (N_t).

At the high voltage the CVCs are typical for the unipolar injection, but, according to [52-53] there some other current mechanisms leading to qualitatively similar current-voltage functions. Thus, we had to identify them additionally according to the procedure described in [52] by analyzing functions $\log I - \log U$, $\eta - \log U$ and $\log \eta - \log U$. It allowed identifying high-voltage current mechanisms in the samples and defining (in some cases) their type.

For further definition of the dominant current mechanism in the base chalcogenide layer we calculated the discrimination coefficient Q_{ext} in the extremum points of the function $\eta - \log U$ and compared it with coefficients typical for other mecahisms [52]. We have found $Q_{ext} > 10^6 \div 10^7$ almost in all cases, what is significantly larger than the values of Q_{ext} typical for the field trap ionization and the barrier –involved current mechanism in the material. This, in turn, points out [52-53] that the extremums in functions $\eta - \log U$ are caused by filling-in of the traps in the material with charge carriers injected from the metallic contact. Using various analytical methods allows to conclude with a good reliability that the CVC's features for multilayered structures with high-temperature chalcogenide layers ($T_s > 500$ K), were caused namely by the SCLC mechanism. Further we have worked out the CVCs due to injection currents only.

Fig. 4. illustrates a typical example of the CVC working-out. It is easy to see that the LS distributions are obtained under analysis of two different CVCs and they are in a good correlation.

To make the distribution more precise we have plotted in the same picture the Gaussian curve. It is seen that for examined polycrystalline CdTe films there are trap distributions in the gap with a form closed to that of the Gaussian one with a small half width σ_t.

Broadening energy levels in CdTe layers prepared by the vacuum condensation may be due to statististical dispersion of polar charge carriers' energy caused by fluctuative irregularities of the film crystalline lattice. This effect is enhanced near the substrate where the most defective layer of the film is grown. This region was an object for determining LS parameters by the method of SCLC CVC.

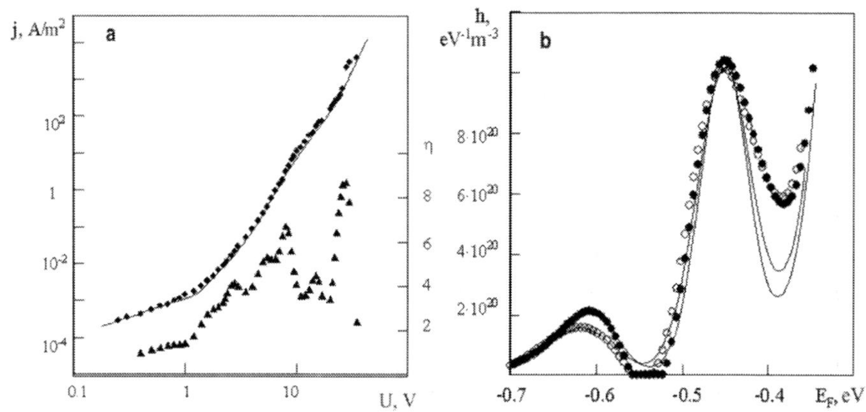

Figure 4. SCLC CVC and its derivative $\eta(U)$ for CdTe-based sandwich structures (a), and trap distribution in the gap of cadmium telluride (b): • – $j(U)$; ▲ – $\eta(U)$ (a); the energy trap distribution is resulted from the high-temperature IS method (b) (• – first measurement; o - repeating measurement at somewhat other temperature); the Gaussian distributions (solid line) are presented for comparison.

4.2. LS parameters from CVC and $\sigma - T$ functions

SCLC CVC was used for determination of trap parameters in the films. The low level of scanning LS spectrum was defined by the position of the equilibrium Fermi level E_{F0}, i.e. its position without charge carrier injection in the sample (ohmic section of the CVC), the upper limit was defined by the position of the Fermi quasi-level at the turn-on of the multilayered structures into low-ohmic state. The start position of the Fermi level was pre-defined by the equilibrium carrier concentration in the material, respectively, by the conductivity of the films. The calculations showed the position of the equilibrium Fermi level E_{F0} was coincide or was close to the energy of the deepest LS in the corresponding samples. The Fermi level is fixed by the traps because the concentration of free carriers in the films is close to the full concentration of LS located at grain boundaries and in bulk crystallites of condensates. As a result, the deepest trap levels located lower than the energy of the equilibrium Fermi level were not revealed in chalcogenide films by the SCLC CVC method.

The possibility of revealing shallow traps in the samples ($E_t \leq 0.21$ eV, for ZnTe films) is restricted by their turn-on into the low-ohmic state stimulated namely by the LS. Thus, the SCLC CVC method had revealed the traps with energy higher positions. However, the traps with different energies also may exist in the samples as shown by the data from the slope of conductivity-temperature functions in ohmic and square sections of the CVCs and luminescence spectra.

4.2.1. *CdTe films*

Table 1 presents some results of IS calculations for deep centers in polycrystalline and monocrystalline CdTe films. In the gap of the polycrystalline material are LS with E_1 = (0.68÷0.70) eV; E_2 = (0.60÷0.63) eV; E_3 = (0.56÷0.57) eV; E_4 = (0.51÷0.53) eV; E_5 = (0.45÷0.46) eV; E_6 = (0.39÷0.41) eV and concentration $N = (10^{18}\div10^{20})$ m^{-3}. The concentration of these LS is in the range $N_t = (10^{18} \div10^{21})$ m^{-3} and mostly increases with closing their energy positions to the bottom of the conduction band. The traps by the profile $h(E) = N/\sqrt{2\pi\sigma_t}\exp(\Delta E^2/2\sigma_t^2)$ are similar to the mono-energetical ones with a half width $\sigma_t = (0.011\div0.015)$ eV. The dominant LS affecting SCLC CVC are the LS with energies E_t = (0.60÷0.63) eV; E_t = (0.56÷0.57) eV; E_t = (0.45÷0.46) eV. Only the traps (if revealed) with E_t = 0.40 eV had the larger concentration.

The LS were registered not only in polycrystalline films but also in monocrystalline layers. We have resolved the traps with E_t = (0.56÷0.57) eV; E_t = (0.52÷0.53) eV; E_t = (0.45÷0.46) eV and E_t = (0.40÷0.41) eV in the gap of the material. The monocrystalline condensates had lower resistance that the polycrystalline layers (10÷100 times), the equilibrium Fermi level in these films was placed more closely to the conduction (valence) band than that in polycrystalline films. Thus, the deepest traps were not revealed by SCLC CVC method in monocrystalline layers. So, the traps E_t ~ 0.70 eV and E_t ~ 0.62 eV found in polycrystalline films may be presented in lower-resistive monocrystalline films.

Ionization energies of the defects in the gap of CdTe were determined from the slope of functions conductivity-temperature in coordinate's logσ-1/T [21-22]. Table 2 lists the results for polycrystalline and monocrystalline CdTe films. In high-temperature polycrystalline condensates the following activation energies were observed for conductivity: E_t=0.15; 0.33; 0.40÷0.41; 0.46; 0.60÷0.61, 0.80 eV. In the monocrystalline films the LS had smaller activation energy: E_t=0.06÷0.07; 0.13÷0.14; 0.22÷0.23; 0.29; 0.40; 0.46 eV. Activation energy E_t = (1.50÷1.52) eV is typical for high temperatures of the experiment and corresponds to the gap of the material. The comparison of the LS energy levels from the SCLC CVC and $\sigma - T$ functions is carried out in Table 2. The values E_t from the $\sigma - T$ functions correlate with those observed in CdTe films by SCLC CVC method.

The wide range of the traps revealed in CdTe condensates is obviously caused by investigation of disordered transition layer of the films formed under the film condensation near the substrate. In this layer may be presented foreign impurities adsorbed from the substrate and residual atmosphere under film condensation. Besides that for CdTe the concentration of uncontrolled residual impurities in the charge mixture can be $N_t = (10^{20}\text{-}10^{21})$ m^{-3} which is behind the sensitivity of the IS method. These impurities can form a chain of complexes impurity-native defect producing deep levels in the gap of the semiconductor.

Sample number	d, μm	T_s, K	T_e, K	E_t, eV	N_t, m^{-3}	σ_t, eV
1	8	743	1023	0.63	$4.4 \cdot 10^{19}$	0.030
2	19	748	948	0.61	$1.7 \cdot 10^{19}$	0.031
(1st measurement)				0.45	$7.3 \cdot 10^{19}$	0.028
2	19	748	948	0.62	$1.5 \cdot 10^{19}$	0.035
(2nd measurement)				0.45	$8.1 \cdot 10^{19}$	0.032
3	12	748	968	0.68	$7.8 \cdot 10^{18}$	0.023
				0.62	$1.5 \cdot 10^{19}$	0.023
				0.53	$6.1 \cdot 10^{19}$	0.027
4	9	723	893	0.62	$6.6 \cdot 10^{18}$	0.021
				0.56	$4.4 \cdot 10^{19}$	0.016
5	12	823	893	0.62	$2.0 \cdot 10^{18}$	0.023
				0.57	$1.7 \cdot 10^{19}$	0.015
6	11	753	933	0.62	$4.6 \cdot 10^{18}$	0.019
(monocrystalline)				0.52	$1.3 \cdot 10^{19}$	0.009
				0.41	$1.1 \cdot 10^{20}$	0.016
7	15	753	953	0.60	$2.3 \cdot 10^{18}$	0.019
				0.52	$3.6 \cdot 10^{18}$	0.020
				0.46	$8.6 \cdot 10^{18}$	0.020
				0.41	$1.4 \cdot 10^{19}$	0.015
8	26	758	978	0.61	$3.6 \cdot 10^{18}$	0.023
				0.56	$3.0 \cdot 10^{19}$	0.015
				0.52	$7.4 \cdot 10^{19}$	0.015

Table 1. Parameters of LS revealed in CdTe films by high-temperature IS

E_t, eV			Interpretation
From SCLC CVC	From σ-T dependencies		
Polycrystalline films	Polycrystalline films	Monocrystalline films	
0.68-0.70	0.80	-	V_{Te}^{2+} (0.71 eV [57-61]
0.60-0.63	0.60	-	Te_{Cd}^{2+} (0.59 eV) [57-61]
0.56-0.57	0.57	-	Cd_i^{c2+} (0.56 eV) [57-61]
0.51-0.53	-	-	V_{Te}^{2-} (0.50 eV) [57-61]
0.45-0.46	0.46	0.46	Cd_i^{c-} (0.46 eV) [57-61]
0.39-0.40	0.40÷0.41	0.40	$Te_{Cd}^{2+} . V_{Te}^{+}$ (0.40 eV) [57-61]
-	-	0.29	
-	-	0.22÷0.23	Cd_i^{2+} (0.20 eV) [57-61]
-	0.15	0.13÷0.14	
-	-	0.06÷0.07	

Table 2. Energetical positions of LS levels for defects in the gap of CdTe

As the chalcogenide films were not doped in-advance all LS found here are corresponding to native defects and their complexes with uncontrolled impurities. The interpretation is a challenge while the energy spectrum of PD in the gap of tellurium is studied not enough and identification in the most cases is not satisfactory (Table 3). For example, in [62] the levels E_t of LS are studied by photoinduced currents (PICTS) and authors give more than 150 values of deep levels, where the sufficient part of them is caused by the native defects. More reliables are some theoretical works where energies E_t are calculated («*ab anitio*») [57-61]. We have used namely the data Wei [57-58] obtained from the first principles. Table 3 summarizes our results.

According to calculations the deep centers with energy position 0.71 eV are belonging to V_{Te}^{2-}. We have experimentally observed the level E_t = (0.68÷0.70) eV which may be caused by this defect. Analogically, the LS with energies (0.60÷0.63) eV may be ascribed to the antistructural defect Te_{Cd}^{2-} (0.59 eV), and (0.56÷0.57) eV and (0.45÷0.46) eV to the interstitial cadmium in different charge states: Cd_i^{c2-} (0.56 eV), Cd_i^{c-} (0.46 eV). The level 0.29 eV is also formed by the native defect bound with cadmium Cd_i^{a2-} (0.33 eV). Different ionization energies of interstitial cadmium are due to its place in octo- or tetrahedral position in the crystal lattice of the material.

4.2.2. *ZnTe films*

Table 4.3 summarizes the results of calculations for ZnTe condensates in dependence on physical technical growth conditions. SCLC CVC method reveals set of trap groups with the most probable energy position E_{t1} = 0.21 eV; E_{t2} = (0.32÷0.34) eV, E_{t3} = 0.57 eV; E_{t4} = (0.41÷0.42) eV; E_{t5} = 0.89 eV. The concentration of the revealed LS is in interval $N_t = (10^{20} \div 10^{21})$ m^{-3}. The LS with energy E_t = (0.32÷0.33) eV are dominant in the most samples and they determine the CVCs of the films.

The trap spectrum in ZnTe films can be partially checked by investigation of temperature – conductivity functions for the condensates. As shown by analysis of $\sigma - T$ functions in the Ohmic section of the CVC for high-temperature ZnTe condensates the following conductivity activation energies are typical: 0.05 eV; (0.14÷0.15) eV; (0.20÷0.21) eV; (0.33÷0.34) eV; (0.42÷0.43) eV; (0.51÷0.52) eV; (0.57÷0.58) eV; (0.69÷0.70) eV and 0.89 eV (Table 4). Set of E_t values from the $\sigma - T$ functions is in a good correlation with those in ZnTe films defined by SCLC CVC method (Table. 3) and low-temperature luminescence (Table. 2).

As the films ZnTe as CdTe layers were not doped in-advance all the calculated LS are due to native PD, their complexes, uncontrolled impurities and their complexes with native defects.

The LS in monocrystals and films ZnTe were studied by SCLC CVC in [24, 63, 64]. Authors [24] have found the trap parameters in monocrystalline samples by the voltage of complete trap filling-in: E_t = 0.17 eV and $N_t = 10^{22}$ m^{-3}. On the other hand, measurements of $\sigma - T$ dependencies in the square section of the CVC gave E_t = 0.14 eV and $N_t = 10^{23}$ m^{-3} taking in mind the presence of traps in the material authors [64] calculated the LS density in ZnTe films prepared by laser evaporation technique: N_t = (4.2÷8.4)·10^{22} m^{-3}. The trap energy is not defined in this work. In ZnTe films obtained by the electro-deposition the trap concentration is N_t = 3.6·10^{21} m^{-3} [63].

Parameters of film condensation	From SCLC CVC		From σ-T dependencies	
	E_t, eV	N_t, m^{-3}	E_t, eV	N_t, m^{-3}
T_s = 623 K, T_e = 973 K	-	-	0.21	2.1·10^{20}
	0.34	8.6·10^{20}	0.34	7.3·10^{20}
T_s = 673 K, T_e = 973 K	0.33	5.3·10^{20}	-	-
T_s = 723 K, T_e = 973 K	0.34	2.9·10^{20}	0.33	4.1·10^{20}
	-	-	0.57	5.5·10^{20}
	-	-	0.89	8.4·10^{20}
T_s = 773 K, T_e = 973 K	0.32	5.3·10^{20}	-	-
T_s = 823 K, T_e = 973 K	0.42	2.1·10^{20}	-	-
T_s = 873 K T_e = 973 K	0.32	1.5·10^{21}	-	-
T_s = 623 K T_e = 973 K	0.35 0.37	8.8·10^{20}	-	-

Table 3. LS parameters in ZnTe films from SCLC CVC and σ-T functions

As is seen from the Table 3, the trap concentration in ZnTe films is significantly lesser than that in condensates prepared by laser evaporation, electro-deposition methods and even in the monocrystalline material [23, 63- 64]. It shows a high structural perfectness and stoichiometry of the layers.

Nevertheless, the most levels found in ZnTe films may be identified with some probability. The level E_1 = 0.05 eV is commonly bound with single-charged dislocation V_{Zn}^{-}, and the level E_2 = 0.15 eV is bound with a double-charged V_{Zn}^{2-} Zn vacancy [65, 66]. In later works the second level is ascribed to Cu as to a traditional residual impurity in ZnTe, and the double-charged Zn vacancy is supposed to have a more deeper energy level 0.21 eV [66]. It is thought that the energy activation (0.36÷0.40) eV [67, 68] is for the common substitution impurity in ZnTe, namely O_{Te}. The most deepest level 0.58 eV authors [67] ascribe to the Te vacancy V_{Te}^{2-} (interstitial zinc Zn_i^{2-}). The possible interpretation of LS in ZnTe films is listed in table 5.7. Other energy levels on our opinion are belonging to the uncontrolled impurities and complexes native defect-impurity.

We have revealed trap levels with energy position E_t = (0.22 – 0.25) eV and concentration N_t = (5.0·10^{20} ÷ 1.5·10^{21}) m^{-3} by the analysis of SCLC CVC in ZnS films. These LS may be localized in the gap due to presence of the interstitial Zn atom Zn_i^{2-}. LS with energy position E_t = (0.24÷0.25) eV were observed in [70] the thermo stimulated current technique.

T_s, K	623		673	723		773	823	873	Interpretation
Section / E_t, eV	Ohm.	Sq.	Ohm.	Ohm.	Sq.	Ohm.	Ohm	Ohm.	
E_9					0.89				–
E_8	0.70							0.69	–
E_7			0.58	0.58	0.57		0.58		V_{Te}^{2+} (Zn_i^{2+}) [67]
E_6	0.51			0.52		0.51	0.52		V_{Zn}^{2-} (0.50) [69]
E_5	0.43		0.42				0.42		O_{Te} (0.41) [67]
E_4	0.33	0.34	0.33		0.33		0.34	0.33	O_{Te} (0.36) [68]
E_3		0.21		0.20		0.20			V_{Zn}^{2-} (0.21) [67]
E_2						0.14		0.15	V_{Zn}^{2-}, Cu_{Zn}^{+} (0.15) [69]
E_1		0.05							V_{Zn}^{-} (0.05) [65]

Table 4. Energy positions of LS for defects in ZnTe gap determined from the slope of σ–$1/T$ functions

According to the Arrhenius equation σ- T functions allowed calculating conductivity activation energies in linear sections: E_1=0.03 eV; E_2 = (0.07÷0.08) eV, E_3 = 0.15 eV; E_4 = (0.23÷0.24) eV; E_5 = 0.33 eV; E_6 = 0.46 eV; E_7 = 0.87 eV.

4.2.3. ZnS films

Table 5 summarizes LS parameters calculated by SCLC CVC method and from the σ- T functions in ZnS condensates prepared under various physical technical conditions. Reference data are presented for comparison. The table shows a correlation between our results and data obtained by other authors [71-74]. Besides that, there is a coincidence of defect energy positions defined from the SCLC CVC and σ- T functions.

T_s, K (T_e = 1173 K)	From CVC SCLC		From σ-T dependencies	
	E_t, eV	N_t, m^{-3}	E_t, eV	Reference data
523 K	0.22	5,1·10^{20}	0.03	0.029 [71]
			0.07	0.06 [72]
590 K	0.25	1.5·10^{21}	0.07	0.06 [72]
			0.24	0.24; 0.25 [70, 72]
			0.33	0.31-0.33 [70]
			0.87	0.81-1.29 [74]
673 K	0.23	8,2·10^{20}	0.03	0.029 [71]
			0.23	0.24; 0.25 [70, 72]
			0.15	0.14 [72, 73]
			0.46	-

Table 5. LS parameters defined by analysis of SCLC CVC and σ- T functions at ohmic section of the CVC for ZnS films prepared under various physical technical condensation modes

All the LS found here were not identified because of absence of corresponding reference data. Only the levels with activation energy $E_1 = 0.15$ eV and $E_2 = (0.22\div0.25)$ eV may be bound with single- Zn_i^- and double charged Zn_i^{2-} interstitial Zn atom.

5. Determination of LS parameters of polycrystalline chalcogenide films by optical spectroscopy (low-temperature photoluminescence)

Low-temperature photoluminescence (PL) is one of the most reliable tools applied for investigation of longitudinal, native, impurity and point defect ensembles in semiconductors. High resolution of the method makes it possible to examine not only bulk materials (bulk chalcogenide semiconductor are now good studied [75-87, 90-93, 96, 99-100, 102-104]), but also thin films, in particular, chalcogenide semiconductor thin layers. In this part we present data obtained by studying low-temperature PL spectra of ZnTe, CdTe and ZnS films. These results allowed monitoring and adding new results to those given by the IS method.

5.1. CdTe films

Fig. 5 (a, b) illustrates the typical spectra of these films. As shown, the spectra for both types of the films are significantly similar. A modest energetical displacement of lines in spectra from epitaxial films comparing to those from the polycrystalline layers films deposited on glass may be caused by presence of sufficient macrodeformations in the layers CdTe/BaF$_2$. PL spectra from CdTe layers have lines originated from optical transfers with participation of free and bound excitons, transfers valence band – acceptor (e-A), donor-acceptor transfers (DAP), the radiation caused by presence of dislocations or DP (donor pairs, DP) (Y - stripes); the spectra also have a set of lines corresponding to optical transitions where phonons take place (LO - phonon replica) [87-99].

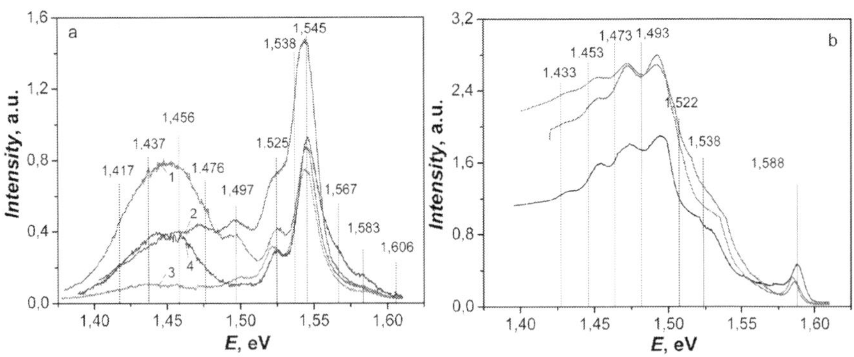

Figure 5. PL spectra registered at T=4,5 K for polycrystalline films CdTe/glass (a), prepared at T_e = 893 K and various T_s, K: 473 K (1); 523 K (2); 623 K (3); 823 K (4) and for the epiraxial layer CdTe/BaF$_2$ (b): T_e = 893 K, T_s = 798 K

Activation energies relative to the valence band (while the most samples were of p-type conductivity) were calculated using expression (26) (in analogy with description above).

The gap of CdTe at T=4.5 K was supposed to be E_g =1.606 eV. The data are presented in Table 6.

E_i, eV	Reference data, eV	ΔE_i, eV	Recombination type	Interpretation
1.583	1.589-1.588	0.023	Exciton	A^0X, A -Li, Na [90, 92, 95, 99, 100]
1.567	1.568	0.039	Exciton	(A^0X)-LO [90, 100]
1.545	1.546	0.061 (0.050) [13] (0.058-Li. Na)	e-A - - DAP	$A\text{-}V_{Cd}^{-}$ [13, 94, 95] $A - Li, Na$ [90, 92, 99] $(V_{Cd} - O_{Te})^-$[97, 102] $D - V_{Cd}^{-}$ [13, 103]
1.538	1.538	0.068 (0.067-V_{Cd})	DAP -	D - A [93] D - A (Na) [13, 92]
1.525		0.081	(e-A) LO	[97]
1.497	1.496 1.499 1.495	0.109 (0.107 [13]) (0.111 [13])	e-A - -	O [99] Ag_{Cd} [13] V_{Cd}^{2-} [13]
1.476	1.474 1.477	0.130	e-A	Y [96-99, 101] Y (α-dislocations [93]
1.457	1.459	0.149 (0.146 [13, 92])	e-A - - DAP	$(A\text{-}X)\text{-}LO$ [101] Y [91] Cu_{Cd} [13] $(In - V_{Cd}^{2-})^-$ [13, 92, 100]
1.438	1.436	0.168	e-A DAP	$(A\text{-}X)\text{-}2LO$ [101] $(In - V_{Cd}^{2-})^- - LO$ [13, 92, 100]
1.419	1.415	0.187	e-A DAP	$(A\text{-}X)\text{-}3LO$ [101] $(In - V_{Cd}^{2-})^- - 2LO$ [13, 92, 100]

Table 6. Principal lines of PL spectra of CdTe films and their interpretation

The lines due to exciton recombination in CdTe single crystals are well-known. Authors [13] show energy level diagram for the exciton localized on neutral donors or acceptors and possible transfers between these levels. Commonly the elements of 3rd Group (Ga, In, Al) and 7th Group (Cl, Br, I) are shallow donors in CdTe and ZnTe, and acceptors are the elements of 1st Group and 5th Group (Li, Na, Cu, Ag, Au, N, P, As). These elements are typical excessive impurities in compounds II-VI. Authors [13] also give the ionization energies of principal dopant impurities in CdTe: for the donors (13.67÷14.48) meV, for acceptors 56 (N) - 263 (Au) meV. We have used these values for further interpretation of the experimental results.

Unlike ZnTe condensates the peak bound with a free exciton recombination at energies E_i = 1.596 eV [12] had not been observed for CdTe films. However, the spectra showed a line caused by the recombination of exciton localized on neutral acceptor A^0X - E_i = (1.583÷1.588) eV (1.589 eV [90, 92, 95, 99, 100]). This line indirectly demonstrated that the investigated films were of *p*-type conductivity and correspondingly low concentration of dopant impurities. Maybe there is a reason for absence of the peak bound with the exciton localized on the neutral donor D^0X – 1.593 eV [13, 90, 100] in the registered spectra. The excessive impurity (Li, Na) commonly is an acceptor in II-VI compounds which produces shallow LS near the valence band.

In some PL spectra of CdTe films we have observed the peak due to the phonon repetition of the line from the bound exciton (A^0X)-LO at E_i = 1.567 eV. The similar peak with E_i=1.568 eV and E_i=1.570 eV was also observed in [90, 100]. It should be noted that the excitation energy of the longitudinal phonon in CdTe is LO(Γ) – 21.2 meV [13, 90, 92]. This value is almost coinciding with that observed experimentally (21 meV) showing our correct interpretation of the experimental data.

The most intensive peak of 1.545 eV was observed in PL spectra from polycrystalline films. The similar peak with energies E_i=1.55 eV and E_i=1.545 eV was registered by authors [92, 94, 97, 99, 103]. The common interpretation says that this peak is caused by the electron transition between the conduction band and acceptor (e-A) (a single-charged vacancy V_{Cd}^- [94] or other shallow acceptor [92, 99]). Nevertheless, authors [13, 95, 103] point out this radiation as a consequence of p resenting donor-acceptor pairs (DAP) where the acceptor is a native defect (V_{Cd}^-) [13, 103] or another uncontrolled shallow impurity [95]. Authors [97] have found the activation energy of corresponding donors and acceptors: 8 meV and 47meV.

Results of investigated polycrystalline CdTe films in hetero structures CdTe/ZnS under air and vacuum annealing have given [97] another interpretation. It is supposed that the luminescence with 1.55 eV is due to oxygen presence in the material. In which form it exists in the material (substitutional impurity or oxide phase) is not established. However, authors [102] have studied the LS in CdTe single crystals by thermo electronic spectroscopy and demonstrated the energy level 0.06 eV bound with a complex $(V_{Cd} - O_{Te})^-$.

Analyzing our results allows us to conclude that the peak E_i=1.545 eV is rather due to the electron transitions between the conduction band and acceptor (a single-charged vacancy or DAP). Really if this peak was caused by oxygen we could it observed in PL spectra from both polycrystalline and epitaxial films but there are no such a peak in PL spectra from the films CdTe/BaF_2. Besides that no structural method had revealed the oxygen in these compounds The films under investigation have shown no registered donor impurities of considerable concentration, so the interpretation of this peak as a consequence of the DAP presence is lesser probable than that a consequence of e-A transition.

In some cases the PL spectra from the polycrystalline films showed an asymmetric peak 1.545 eV indicating that in reality it may be a superposition of two nearest lines. Mathematical analysis showed that the most probable position of the additional peak is E_i = 1.538 eV. The similar peak was observed in spectra from the epitaxial films $CdTe/BaF_2$. The line with the same energy was revealed by authors [93] in PL spectra from deformed CdTe single crystals and is supposed to be caused by defects generated in the material due to slip of principal Cd(g)-dislocations. Authors [13, 92] explain the peak E_i= 1.538 eV as one of unknown nature. Similar interpretation is also in [89] where the line E_i= 1.539 eV is caused by DAP (here the acceptor is sodium, Na_{Cd}). The next peak E_i = 1.525 eV is likely is the phonon repetition of the previous one (e-A)-LO [96].

The PL line E_i = 1.497 eV was observed in [99] on monocrystalline CdTe samples under doping by ion implantation. As this line has appeared in the samples doped with oxygen only the authors suggest it is caused by the presence of this impurity. Other authors suppose this line is due to electron transitions between the conduction band and the level of the substitutional impurity acceptor Ag_{Cd} (E_V+0.107 eV) [13] or by native defect V_{Cd}^{-2} (E_v+0.111 eV) [13].

The wide radiation stripe in polycrystalline films at the energy 1,45 eV is separated in single peaks based on results from the PL of epitaxial films. They are shown in Fig. 5.

The peak 1.476 eV in [96-98] is due to longitudinal defects (dislocations and DP, a so-called Y-stripe). Authors [90, 98] assume the Y-stripe at (1.46-1.48) eV is caused by longitudinal defects (dislocations). Authors [99] make it more precicely: this peak is caused by the recombination of exciton localized on slipped Cd-dislocations. Authors [93] have investigated The photoluminescence of deformed CdTe single crystals and showed that the peak E_i=1.476 eV is not caused by the Cd-dislocations but is due to the electron states of 60^0 Te(g)-dislocations (α-dislocations). So that, the number of authors have the same opinion that this line in PL spectra is caused by the longitudinal defects. We also agree with this interpretation.

The lines 1.453 eV, 1.433 eV and 1.413 eV which are good resolved in spectra from the epitaxial films $CdTe/BaF_2$ are very similar to 1LO, 2LO, 3LO repetitions of the peak E_i = (1.473÷1.476) eV. However, the energy difference of these lines (ΔE=0.0200 eV) does not coincide with the energy of longitudinal optical phonons in CdTe 0.0212 eV making it difficult to interpret the corresponding peaks unambiguously. At the same time, the analogous set of lines with the LO structure and the energy difference 0.0200 eV in the range ΔE = (1.39÷1.45) eV was observed by authors [101]. They have studied polycrystalline CdTe films deposited by vacuum evaporation at T_s = (723÷823) K on glass and aluminum substrates.

Authors [100] have examined undoped and doped with donor impurities (Al, In) CdTe single crystals and also have observed the PL stripe in the energy range ΔE = (1.380÷1.455) eV containing four lines with LO structure. The authors interpreted them as electron transition between DAP and their phonon repetitions. Authors [81, 91] suppose the wide peak 1.46 eV is due to the excitons localized in longitudinal defects, probably dislocations

(*Y*-stripe). The lines 1.455 eV, 1.435 eV and 1.415 eV were observed in [94] from the polycrystalline CdTe films prepared by the gas-transport method.

As we see the most authors have an unique opinion: the set of lines in the range ΔE=(1.413÷1.476) eV is due to longitudinal defects (rather dislocations), and their intensity [93] can be a measurement unit of these defects in the material.

For polycrystalline films (Fig. 5, a) the LO structure of the stripe caused by the longitudinal defects at energies ~1.45 eV has practically not been observed, maybe because of superposition with additional lines of another origin.

The defect complexes in the material (A-centers) are also resolved by the PL in the same energy range (it can be considered as a partial case of DAP). According to [13, 97] A- centers $\left(V_{Cd}^{2-} - D^{+}\right)^{-}$ where Cl is a donor produce the line and its LO-phonon repetitions with energies 1.454, 1.433, 1.412, 1.391, 1.370, 1.349 and 1.328 eV. However, as is seen in Fig. 5, this stripe is displaced relatively to that observed experimentally, so the experimental PL spectra of CdTe films can be completely explained by these complexes only. The narrower stripe with peaks 1.458, 1.437, 1.417 and 1.401 eV produces the A-center where indium is a donor. This stripe has the better coincidence with experimental one but is also displaced. Besides that, it is difficult to explain why the A-complex is observed in the polycrystalline films and is not observed in the epitaxial layers while the charge mixture for both types of the films is the same. Thus we suppose the interpretation of the wide stripe in the energy range ΔE = (1.413÷1.4760) eV due to longitudinal defects is more reliable.

Under change of condensation conditions of polycrystalline samples we have observed the change of intensity for a stripe due to prolonged defects (~1.45 eV). As shows Fig. 5, as the substrate temperature increases from 473 K to 623 K the intensity of this stripe is decreasing and then it increases as the T_s increases. These results have a good correlation with data of investigation of CdTe film substructure [49], this fact points out an enhance of the structural quality (lowering vacancy concentration) of the bulk crystallites in condensates under elevating substrate temperature up to T_s=623 K, but this quality becomes lower as the substrate temperature increases over 623 K.

As the substrate temperature elevated (T_s>723 K) the optical properties of CdTe films were strongly degraded forming a number of additional peaks in the PL spectra which finally become a bell-like curve without possibility to identify the separate lines. Morphological studies demonstrated further increase of the crystallite sizes in this temperature range. However, the volume of these crystallites becomes a high-defective one.

Table 6 summarizes results of PL spectra interpretation for CdTe films showing their high optical quality.

5.2. ZnTe films

Fig. 6 illustrates typical PL spectra of ZnTe films registered at 4.5 K. A number of lines is observed, their energies are indicated in the Fig. 6 and are listed in Table 7. Analysis and interpretation of the PL peaks are carried out according to reference data [75-89].

The low-temperature PL spectra of ZnTe films show a set of peaks originated from: i) optical transitions under participation of free (X) and bound on neutral donor (D^0X) and acceptor (A^0X) excitons; ii) transitions valence band – acceptor impurity (e-A), iii) radiation due to presence of longitudinal defects (dislocations, *Y*-stripe); iv) optical transitions where phonons of different type are participating (LO (0.0253 eV), TO, LA (0.0145 eV), TA (0.007 eV) -repetition).

We calculated activation energies of corresponding processes using the expression (26). The gap of ZnTe crystal at 4.5 K was supposed to be $E_g = 2.394$ eV. As the examined material was of p-type conductivity the activation energies were counted down relative to the valence band. Table 7 summarizes these data.

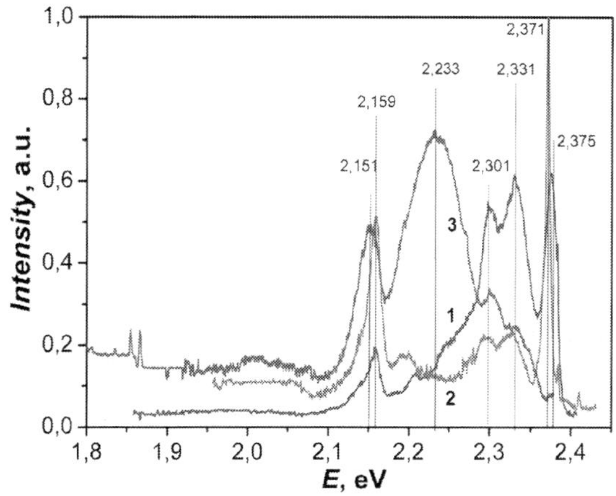

Figure 6. Photoluminescence spectra registered at T=4.5 K for ZnTe films prepared at T_e = 973 K and various T_s, K: 573 K (1); 673 K (2); 773 K (3)

Optical transitions with energy (2.381÷2.383) eV were observed in [68, 75-82, 84-86] where authors had studied monocrystalline or bulk polycrystalline ZnTe of high structural and optical quality. These transitions are commonly relating to a free exciton (X). Earlier [82] the PL line of E_i = (2.374÷2.375) eV was suggested to be caused by the exciton bound on neutral acceptor (zinc vacancy V_{Zn}). Further [76-78] it was shown that other acceptor centers take part in forming such an excitonic complex, in particular, acceptor centers due to uncontrollable impurities (Li, Cu) in ZnTe are of interest. However, in the most recent works [68, 81] this line is ascribed to the exciton localized on shallow neutral donor (atoms of uncontrollable impurities from 3rd and 7th Groups of the Periodical System (In, Ga, Al, Cl, Br. I)). These impurities form in the gap of the material more narrower levels than the acceptor ones. The line with E_i=2.371 eV which is energetically closed to that considered above is due to radiation of bound excitons [76-78, 81]; nevertheless the impurity (acceptor) in this complex has (obviously) somewhat larger energy level causing other energy of the stripe. These acceptors are native defects and uncontrollable excessive impurities (Li, Na, Ag, Cu).

According to [81] Li_{Zn} is the most probable candidate while it forms in the gap of the material energy level 60.6 meV.

Radiation line E_i, eV	Reference data, E_i, eV	Activation energy, ΔE, eV	Recombination type	Interpretation
2.383	2.381÷2.383	0.011	Exciton	X, n=1 [68, 76, 82]
2.375	2.375 2.379	0.019	Exciton	A^0X [76-77] A- V_{Zn} [82] D^0X, D- In [68, 86]
2.371	2.374; 2.375	0.023	Exciton	A^0X, A-Li,Cu [68]
2.331	2.334; 2.332	0.060 (0.061 – Li [75, 78]) (0.063 – Na [75])	e-A	A - Li_{Zn}, Na_{Zn} [75, 76]
2.301	2.307	0.093	(e-A)-LO	A - Li_{Zn} [76-77]
2.270	2.270	0.124	e-A	A - Ag_{Zn} [76]
2.233	2.230	0.161	e-A	A - Cu_{Zn} [75, 79]
2.208		0.186		
2.194	2.195; 2.19	0.200	e-A	Y_1 [75]
2.159		0.235		
2.151	2.155	0.243	e-A	Y_2 [75, 76, 78, 84]

Table 7. Principal lines in PL spectra of ZnTe films and their interpretation

It should be noted that the presence of excitonic lines in PL spectra from high-temperature ZnTe condensates points out their high optical and crystal quality. These lines are of sufficient intensity in the spectra from the films deposited at the substrate temperature T_s=573 K and the larger intensity for condensates prepared at T_s=673 K. Excitonic lines in the spectra from low-temperature condensates and layers manufactured at T_s>773 K were not registered. Thus, the results of PL studies indicate that the films deposited at the substrate temperatures T_s= (623÷673) K are the most optically perfect layers. These data are coinciding with the results of investigations of substructural characteristics of ZnTe films reported earlier [54]. According to these data the dependence of the CSD sizes on the substrate temperature is a curve with the maximum at T_s = (600÷650) K. The minimal dislocation concentration is also observed in the films under these temperatures.

The line E_i = 2.34 eV belonging [66] to V_{Zn} is not observed in spectra of the radiation recombination in ZnTe films. This fact is also confirming high stoichiometry of the films under study.

The set of nearest lines in the energy range ΔE = (2.30÷2.33) eV and ΔE = (2.17÷2.25) eV authors [76-79] ascribe to the electron transitions from the conductance band to the shallow acceptor levels formed by Li or Cu atoms and their phonon repetitions (LO – 25.5 meV). There are stripes 2SLi, $3S_b$Li, (e-A) Li, 2P Li, $4S_b$Li, $4S_b$Li-LO, 2SCu, $3S_b$Cu, $4S_b$Cu, $2S_b$ Cu - LO, $2S_b$ Cu - 2LO and others. Experimental and theoretical values of the activation energy for ground and excited states for the main excessive impurities in ZnTe (lithium and copper) are reported in [76]. They are in the energy range ΔE = (0.0009÷0.0606) eV for Li and ΔE = (0.001÷0.148) eV for Cu. However, in [65] the line E_i = 2.332 eV is supposed to be due to other excessive impurity Na_{Zn}, and in [82] this line is due to the native defect V_{Zn}. Another optical transition E_i = 2.27 eV authors [77] ascribe to the Ag impurity 2S Ag.

What about the peaks in the energy range E = (2.10÷2.21) eV. These transitions were for the first time observed in [75-79] and authors had called them Y_i-lines. They are ascribed to the distortions of the crystalline lattice of the material near incoherent twin boundaries, dislocations and other longitudinal defects where the dangling bonds are formed in the semiconductor material. So that, the lines E_i = 2.159 eV and E_i=2.194 eV can be interpreted as Y_2 (2.155 eV) and Y_1 (2.195 eV) [75]. They are due to longitudinal defects and the change of their intensity may point out the change of these defects concentration in the material. Somewhat other energy position of the line due to oxygen (2.06 eV) is reported in [66]. Thus, analysis of the reference data has forced us to conclude that PL lines in the energy interval ΔE = (1.835÷2.055) eV are rather caused by oxygen, its complexes and phonon repetitions. If it is true, the analysis of PL spectra from ZnTe films indicates the increase of the oxygen content in the samples under increasing the condensation temperature. Actually, if there is no oxygen in the samples prepared at 573 K, its concentration in high-temperature films (T_s=773 K) is sufficiently larger. Oxygen concentration in the material strongly depends on the vacuum conditions under the film preparation and the charge mixture quality.

5.3. ZnS films

Low-temperature photoluminescence is the most reliable tool for examining wide gap materials providing minimization of overlapping peaks due to various recombination processes. The typical PL spectra from ZnS films at 4.7 K are shown in Fig. 7. The detailed analysis of the PL spectra (identification of complex broadened lines) was carried out by ORIGIN program. Maximums of the peaks revealed by this analysis (Fig.7) are noted by vertical lines.

It should be noted that the PL spectra registered at various temperatures of experiment have no sufficient distinctions except those with somewhat larger line intensities in spectra obtained at 77 K. Analysis of the spectra shows that for ZnS films deposited at T_s = (393-613) K the peaks with λ_i = 396 nm (E_i = 3.13 eV) and λ_i = 478 nm (E_i=2.59 eV) are dominating. Further working-out of the spectra demonstrated that the peak λ_i= 396 nm is asymmetric (Fig. 7) what is may be explained by the superposition of two closely placed lines. The spectra also have low intensity peaks with λ_i=603 nm (E_i=2.06 eV) and λ_i=640 nm (E_i=1.94 eV).

PL spectra from the films prepared at higher T_s is sufficiently changed. There is a number of overlapping peaks where the most intensive ones are in the wavelength range $\Delta\lambda_i$ = (560÷620) nm.

Under interpretation of PL spectra from ZnS films we have calculated the activation energies of processes causing the corresponding lines. We also have suggested the PL radiation took place under transfers of electrons from the conduction band (or shallow donors) to the deep LS in the gap of the material. Then the optic depth of the energy level of the defect (ΔE) relative to the valence band causing the spectral peak may be found from (26) supposing the optical gap of the material at 4.5 K is $E_g = 3.68$ eV.

Taking into account that the chalcogenide films were not doped in-advance one can suggest that the lines in spectra are due to transfers of carriers between conduction band and LS caused by the native point defects, their complexes and uncontrolled impurities. We made an attemption to identify these LS according to reference data [104-108] (Table 8). As is shown there is a good correlation of our results and those obtained by other authors for ZnS single crystals.

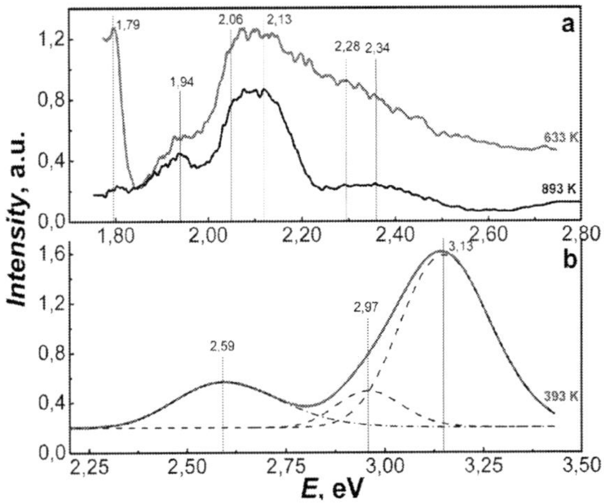

Figure 7. Typical PL spectra for ZnS films (a) and the example of the peak differentiation (b)

The investigations have shown that the Schottky defect V_{Zn} is a dominant defect type in ZnS films prepared at low substrate temperatures T_s = (393-613) K. As T_s increases the number of single-charged Zn vacancies in the condensates decreases, and concentration of double-charged Zn vacancies increases. In the films deposited at higher substrate temperatures T_s=(653-893) K single-charged S vacancies V_S^{-} and double-charged S vacancies V_S^{2+} and interstitial Zn atoms Zn_i are dominating.

Such features of the PD ensemble in the samples are obviously caused by processes of condensation and re-evaporation of Zn and S atoms from the substrate. Actually, at low T_s the defect formation in the films is determined by higher S pressure comparing to Zn pressure in the mixture vapor providing Zn vacancy formation in ZnS condensates. As T_s increase the PD ensemble in the material is determined by the more rapid re-evaporation of the same S atoms from the substrate resulting in production of Zn-beneficiated films. Sulfur vacancies and interstitial Zn atoms are being dominant defects in such condensates.

T_s, K	Measurement range λ, nm	λ_1, nm E_1, eV	λ_2, nm E_2, eV	λ_3, nm E_3, eV	λ_4, nm E_4, eV	λ_5, nm E_5, eV	λ_6, nm E_6, eV	λ_7, nm E_7, eV	λ_8, nm E_8, eV
393	360÷640	**396**	417	478	-	-	603	-	-
		3.13	2.97	2.59	-	-	2.06	-	-
613	450÷720	-	-	**478**	-	-	603	640	-
		-	-	**2.59**	-	-	2.06	1.94	-
653	450÷720	-	-	-	530	**582**	**603**	640	**690**
		-	-	-	2.34	**2.13**	**2.06**	1.94	**1.79**
893	450÷720	-	-	-	530	**582**	**603**	640	690
		-	-	-	2.34	**2.13**	**2.06**	1.94	1.79
ΔE_i, eV Exper.	-	**0.55**	**0.71**	**1.09**	**1.34**	1.55 **1.40**	1.62 **1.40**	**1.74**	1.89 **1.74**
Defect	-	V_{Zn}^-	$(O_s,V_{Zn}^{2-})S_i^-$	V_{Zn}^{2-}	Cu	V_S^{2-}	V_S^{2-}	V_S^-	V_S^-
ΔE_i, eV ref.	-	0.60	0.70	1.10	-	1.40	1.40	1.90	1.90
Reference data	-	[107]	[107, 108]	[104]	[104]	[104]	[104]	[104]	[104]

Table 8. Results of PL spectra working-out and their comparison with reference data (solid values are for peaks of maximum intensity)

The PL spectra of ZnS films have also revealed low intensity lines from the activator impurity (Cu) and, possible, S_i^- [106] or a complex defect (O_s,V_{Zn}^{2-}) [107]. Results of studying low-temperature PL allowed constructing energy position model of native point defects in zinc sulfide films prepared by quasi-closed space technique (Fig. 8).

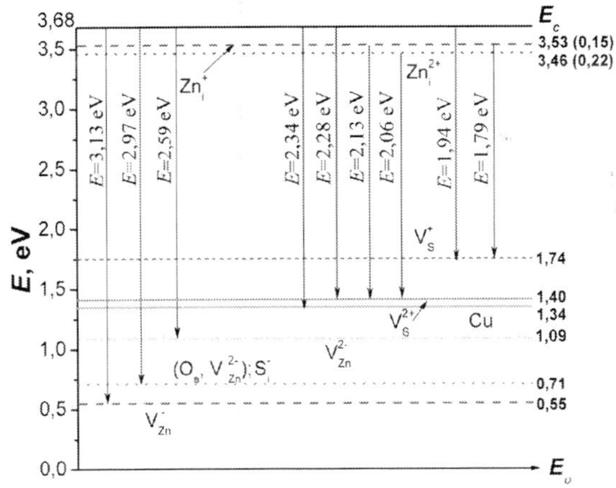

Figure 8. The model of the level positions for native point defects in band gap of ZnS films

To explain the experimental results we have used quasi-chemical formalism for modeling the point defects ensemble in the examined chalcogenide films in dependence on physical technical conditions of layer condensation. This method concerns all defects, electrons and holes as components of thermodynamic equilibrium in the bulk crystal (complete equilibrium of the point defects). Then the modeling procedure reduces to solving set of equations which describe penetration of point defects into solid from the gas state along with the equation of electroneutrality and intrinsic conductivity equation [109-110]. The most complete spectrum of the native defects was taken into account under modeling the point defects ensemble. Calculations were carried out for the complete defects' equilibrium as well as for their quenching. Under modeling we have used energies of native defects formation obtained «*ab initio*» in [56-60]. Reference data of ionization energies of acceptor and donor centers of point defects in CdTe, ZnS, and ZnTe were used along with results of our experiments. Main data of modeling are presented in [111-116].

6. Conclusions

1. Express-method of IS providing maximum information on deep centers in high-resistive films based on analysis of SCLC CVC is developed and allows without additional studies
- identifying current mechanism in the structure as SCLC;
- receiving correct information on LS parameters in the gap of material: energy position, concentration and energy distribution immediately from the experimental CVCs without model framework.
2. Using IS method the LS spectrum in the gap of polycrystalline (and monocrystalline) films of II-VI compounds is examined. These results are checked and made more accurately by analysis of σ - T – functions and low-temperature luminescence.
3. Using the methods mentioned above in the gap of polycrystalline material are revealed the LS with following energy positions: $E_{t1} = 0.05$, $E_{t2} = (0.14 \div 0.15)$, $E_{t3} = (0.20 \div 0.21)$, $E_{t4} = (0.32 \div 0.34)$, $E_{t5} = (0.42 \div 0.43)$, $E_{t6} = (0.51 \div 0.52)$, $E_{t7} = (0.57 \div 0.58)$, $E_{t8} = (0.69 \div 0.70)$ eV (ZnTe); $E_{t1} = (0.13 \div 0.15)$, $E_{t2} = (0.39 \div 0.40)$, $E_{t3} = (0.45 \div 0.46)$, $E_{t4} = (0.51 \div 0.53)$, $E_{t5} = (0.56 \div 0.57)$, $E_{t6} = (0.60 \div 0.63)$, $E_{t7} = (0.68 \div 0.70)$ eV (CdTe); $E_{t1} = 0.03$, $E_{t2} = (0.07 \div 0.08)$, $E_{t3} = 0.15$, $E_{t4} = (0.23 \div 0.24)$, $E_{t5} = 0.33$, $E_{t6} = 0.46$, $E_{t7} = 0.87$, $E_{t8} = 1.94$, $E_{t9} = 2.34$, $E_{t10} = 2.59$, $E_{t11} = 2.97$, $E_{t12} = 3.13$ eV (ZnS) and concentration $N_t=(10^{19}–10^{21})$ m^{-3}. Comparing reference data produced an identification of these levels as ones belonging to native point defects, uncontrolled impurities and their complexes. The wide range of LS revealed is due to high-sensitive methods used under investigations as well as because of examining traps in the intermediate layer of the films forming under condensation near the substrate.

Author details

Denys Kurbatov and Anatoliy Opanasyuk
Sumy State University, Sumy, Ukraine

Halyna Khlyap
TU Kaiserslautern, Kaiserslautern, Germany

Acknowledgement

This work is supported by the Ukraine State Agency for the Science, Innovation and Informatization and by the NRF grant funded by the MEST of Korea within the project «Advanced materials for low-cost high-efficiency polycrystalline hetero junction thin films solar cells» and by the Ministry of Education and Science, Youth and Sport of Ukraine (Grant №. 01110U001151). The authors wish to thank Prof. Yu.P. Gnatenko and P.M. Bukivskij from the Institute of physics NAS of Ukraine for the PL measuring of some II-VI film samples.

7. References

[1] Georgobiani, A. (1974). Wide Band gap II-VI Semiconductors and Perspectives of Their Usage (in Russian), *Uspekhi Fizicheskikh Nauk*, Vol.113, No.1, pp. 129-155, ISSN 0042-1294.

[2] Pautrat, J. (1994). II–VI Semiconductor Microstructures: From Physics to Optoelectronics, *Journal de Physique III*, Vol.4, – pp. 2413-2425, ISSN 1155-4320.

[3] Takahashi, K., Yoshikawa, A. & Sandhu, A. (2007). *Wide Bandgap Semiconductors. Fundamental Properties and Modern Photonic and Electronic Devices*, Springer, ISBN, Berlin, Heidelberg, NewYork, Germany, USA.

[4] Owens, A. (2004). Compound Semiconductor Radiation Detectors, *Nuclear Instrumental Methods*, Vol. 531, pp. 18-37, ISSN 0168-9002.

[5] Grinev, B., Ryzhikov, V. & Seminozhenko, V. (2007). *Scintillation Detectors and Radiation Control Systems on Their Base (in Russian)*, Naukova Dumka, ISBN, Kyiv, Ukraine.

[6] Berchenko, N., Krevs, V. & Seredin, V. (1982). *Reference Tables (in Russian)*, Voenizdat, ISBN, Moskov, USSR.

[7] Ohring, M. (1992). *The Materials Science of Thin Films*, Academic Press, ISBN, NewYork, USA.

[8] Poortmans, J. & Arkhipov, V. (2006). *Thin Film Solar Cells: Fabrication, Characterization and Application*, John Wiley & Sons, Ltd. IMEC, ISBN, Leuven, Belgium.

[9] Panchekha, P. (2000). Structure and Technology Problems of II-VI Semiconductor Films, *Functional materials*, Vol.7, No.2, pp. 1-5, ISSN 1616-3028.

[10] Holt, D. & Yacobi, B. (2007). *Extended defects in semiconductors. Electronic properties, device effects and structures*, Cambridge University press, ISBN, New York, Melbourne, Madrid, Cape Town, USA, Australia, Spain, South Africa.

[11] Milns, A. (1977). *Impurities with Deep Levels in Semiconductors (in Russian)*, Mir, ISBN, Moskow, USSR.

[12] Shaskolskaya, M. (1984). *Crystallography: Studying Reference for Technical Universities (in Russian)*, Vysshaya Shkola, ISBN, Moskow, USSR.

[13] Korbutyak, D. & Melnychuk, S. (2000). *Cadmium Telluride: Impurity-Defect States and Detector Properties*, Ivan Fedorov, ISBN, Kyiv, Ukraine.

[14] Fochuk, P. (Experimental identification of the Point Defects/ P. Fochuk, O. Panchuk // CdTe and related Compounds: Physics, Defects, Hetero- and Nano-Structure, Crystal

growth, Surfaces and Applications. [R. Triboulet, P.Siffert]. Netherlands: Elsevier, 2010. – P. 292-362.

[15] Stokman, F. (1973) On the Classification of Traps and Recombination Centres, *Physica Status Solidi A*, Vol.20. – pp. 217-220, ISSN 1862-6319.

[16] Serdyuk, V., Chemeresyuk, G. & Terek, M. (1982). *Photoelectric Processes in Semiconductors*, Vyshcha Shkola, Main Edition, ISBN, Odessa, USSR.

[17] Meyer, B. & Stadler, W. (1996). Native Defect Identification in II-VI Materials, *Journal of Crystal Growth*, Vol.161, pp. 119-127, ISSN 0022-0248.

[18] Neumark, G. (1997). Defects in Wide Band Gap II-VI Crystals, *Material Science and Engineering A*, Vol.R21, No.1, - pp. 1-46, ISSN 0921-5093.

[19] Grundmann, M. (2010). *The Physics of Semiconductors. An Introduction Including Nanophysics and Applications*, Springer-Verlag, ISBN, Berlin, Heidelberg, Germany.

[20] Kosyak, V., Opanasyuk, A. & Panchal, J. (2011). Structural and Substructural Properties of the Zinc and Cadmium Chalcogenides (review), *Journal of Nano and Electronic Physics*, Vol.3, No.1, – pp. 274-301, ISSN 2077-6772.

[21] Pavlov, A. (1987). *Methods of Measuring the Semiconductor Materials Parameters (in Russian)*, Vysshaya Shkola, ISBN, Moskow, USSR.

[22] Vorobjev, Ju., Dobropolskiy, V. & Strikha, V. (1988). *Methods of Semiconductors Investigation*, Vyshcha Shkola, ISBN, Kyiv, USSR.

[23] Gorokhovatskij, Ju. & Bordovskij, G. (1991). *Thermoactivation Current Spectroscopy of High-ohmic Semiconductors and Dielectrics*, Nauka, ISBN, Moskow, USSR.

[24] Lampert, M. & Mark, P. (1973). *Injection Currents in Solid States*, Mir, ISBN, Moskow, USSR.

[25] Kao, K. (1984). *Electrons Transport in Solid States*. Vol.1, Mir, ISBN, Moskow, USSR.

[26] Lalitha, S., Sathyamoorthy, R. & Senthilarasu, S. (2004). Characterization of CdTe Thin Film-dependence of Structural and Optical Properties on Temperature and Thickness, *Solar Energy Materials & Solar Cells*, Vol.82, – pp. 187-199, ISSN 0927-0248.

[27] Ibrahim, A. (2006). DC Electrical Conduction of Zinc Telluride Thin Films, *Vacuum*, Vol.81, pp. 527–530, ISSN.

[28] Rose, A. (1955). Recombination Processes in Insulators and Semiconductors, *Physical Review*, Vol.97, pp. 322—323, ISSN 1943-2879.

[29] Rose A. (1955). Space–charge–limited Currents in Solids, *Physical Review*, Vol.97, pp. 1538–1544, ISSN 1943-2879

[30] Nespurek, S. & Sworakowsky, J. (1980). Evolution of Validity of Analytical Equations Describing Steady–state Space–charge–limited Current–voltage–characteristics, *Czechoslovak Journal of Physics*, Vol.B30, No.10, pp. 1148–1156, ISSN 0011-4626.

[31] Mark, P. & Helfrich, W. (1962). Space–charge–limited Currents in Organic Crystals, *Journal of Applied Physics*, Vol.33, pp. 205–215, ISSN 0021-8979.

[32] Thomas, J., Williams, J. & Turton L. (1968). Lattice Imperfections in Organic Solids. Part 3, 4, *Transactions of the Faraday Society*, Vol.64, pp. 2496—2504, ISSN 0014-7672.

[33] Hwang, W. & Kao, K. (1976). Studies of the Theory of Single and Double Injections in Solids with a Gaussian Trap Distribution, *Solid State Electronics*, Vol.19, pp. 1045–1047, ISSN 0038-1101.

[34] Nespurec, S. & Semejtec, P. (1972). Space Charge Limited Currents in Insulators With Gaussian Distribution of Traps, *Czechoslovak Journal of Physics*, Vol.B22, pp. 160–175, ISSN 0011-4626.

[35] Boncham, J. (1973). SCLC Theory for a Gaussian Trap Distribution, *Australian Journal of Chemistry*, Vol.26, pp. 927–939, ISSN 0004-9425.

[36] Simmons, J. & Tarn, M. (1973). Theory of Isothermal Currents and the Direct Determination of Trap Parameters in Semiconductors and Insulators Containing Arbitrary Trap Distributions, *Physical Review*, Vol.B7, pp. 3706-3713, ISSN 1943-2879.

[37] Pfister, J. (1974). Note of Interpretation of Space–charge–limited Currents With Traps, *Physica Status Solidi A*, Vol.24, No.1, pp. K15-K17, ISSN 1862-6319.

[38] Manfredotti, C., de Blasi, C. & Galassini, S. (1976). Analysis of SCLC Curves by a New Direct Method , *Physica Status Solidi A*, Vol.36, No.2, pp. 569-577, ISSN 1862-6319.

[39] Nespurek, S. & Sworakowski, J. (1978). A Differential Method of Analysis of Steady-state Space-charge-limited Currents: an Extension, *Physica Status Solidi A.*, Vol.49, pp. 149-152, ISSN 1862-6319.

[40] Nespurek, S. & Sworakowski, J. (1980). Use of Space–charge–limited Current Measurement to Determine of Properties of Energetic Distributions of Bulk Traps, *Journal of Applied Physics*, Vol.51, No.4, pp. 2098-2102, ISSN 0021-8979.

[41] Nespurek, S. & Sworakowski, J. (1990). Spectroscopy of Local States in Molecular Materials Using Space-charge-limited Currents, *Radiation Physics and Chemistry*, Vol.36, No.1, pp. 3-12, ISSN 0969-806X.

[42] Nespurec, S., Obrda, J. & Sworakowsky, J. (1978). Study of Traps for Current Carriers in Organic Solids N,N' – Diphenyl–p–Phenylenediamine, *Physica Status Solidi A*, Vol.46, No.1, pp. 273–280, ISSN 1862-6319.

[43] Lyubchak, V., Opanasyuk, A. & Tyrkusova, N. (1999). Injection Spectroscopy Method for investigation of Deep Centers in Cadmium Telluride Films (in Ukrainian), *Ukrayinskyi Fizychnyi Zhurnal*, Vol.44, No.6, pp. 741-747, ISSN 0503-1265.

[44] Opanasyuk, A., Tyrkusova, N. & Protsenko, I. (2000). Some Features of the Distributions Deep States Reconstruction by Injection Spectroscopy Method (in Ukrainian), *Zhurnal Fizychnykh Doslidzhen*, Vol.4, No.2, pp. 208-215, ISSN 1027-4642.

[45] Opanasyuk, A., Opanasyuk, N. & Tyrkusova, N. (2003), High-temperature Injection Spectroscopy of Deep Traps in CdTe Polycrystalline Films, *Semiconductor Physics, Quantum Electronics & Optoelectronics*, Vol.6, No.4, pp. 444-449, ISSN 1560-8034.

[46] Tikhonov, A. & Jagola, A. (1990). *Numerical Nethods of Non-correct Problems Solution*, Nauka, ISBN, Moskow, USSR..

[47] Zavjalov, Ju., Kvasov, B. & Miroshnichenko, V. (1980). *Spline-functions Methods*, Nauka, ISBN, Moskow, USSR..

[48] Turkusova, N. (2002). *Injection Spectroscopy of Deep Trap Levels in Cadmium Telluride Films, PhD Thesis (in Ukrainian)*, Sumy State University, Sumy, Ukraine.

[49] Kosyak, V., Opanasyuk, A. & Bukivskij, P. (2010). Study of the Structural and Photoluminescence Properties of CdTe Polycrystalline Films Deposited by Closed Space Vacuum Sublimation, *Journal of Crystal Growth*, Vol.312, pp. 1726-1730, ISSN 0022-0248.

[50] Kurbatov, D., Khlyap, H. & Opanasyuk, A. (2009). Substrate–temperature Effect on the Microstructural and Optical Properties of ZnS Films Obtained by Close–spaced Vacuum Sublimation, *Physica Status Solidi A*, Vol.206, No.7, pp. 1549-1557, ISSN 1862-6319.

[51] (1988). *Selected Powder Diffraction Data for Education Straining (Search Manual and Data Cards)*, International Centre for Diffraction Data, USA.

[52] Zjuganov, A. & Svechnikov, S. (1981). *Injection-contact Effects in Semiconductors (in Russian)*, Naukova Dumka, ISBN, Kiev, USSR.

[53] Zjuganov, A., Smertenko, P. & Shulga, E. (1979). Generalized Method of Determination the Volume and Contact Semiconductor Parameters by Current-voltage characteristic (in Russian), *Poluprovodnikovaja Tekhnika i Elektronika*, Vol.29, pp.48-54, ISSN.

[54] Kurbatov, D., Kolesnyk, M. & Opanasyuk, A. (2009). The Substructural and Optical Characteristics of ZnTe Thin Films, *Semiconductor Physics, Quantum Electronics & Optoelectronics*, Vol.12, No.1, pp. 35-40, ISSN 1560-8034.

[55] Kurbatov, D., Denisenko, V. & Opanasyuk, A. (2008). Investigations of Surface Morphology and Chemical Composition Ag/ZnS/Glassceramic Thin Films Structure, *Semiconductor Physics, Quantum Electronics & Optoelectronics*, Vol.11, No.4, pp. 252-256, ISSN 1560-8034.

[56] Balogh, A., Duvanov, D. & Kurbatov, D. (2008). Rutherford Backscattering and X–ray Diffraction Analysis of Ag/ZnS/Glass Multilayer System, *Photoelectronics*, Vol.B.17, pp.140-143, ISSN 0235-2435.

[57] Wei, S. & Znang, S. (2002). Chemical Trends of Defect Formation and Doping Limit in II–VI Semiconductor: The Case of CdTe, *Physical Review B*, Vol.66, pp. 1-10, ISSN 0163-1829.

[58] Wei, S. & Zhang, S. (2002). First-Principles Study of Doping Limits of CdTe, *Physica Status Solidi B*, Vol.229, No. 1, pp. 305–310, ISSN 0370-1972.

[59] Soundararajan, R., lynn, K. & Awadallah, S. (2006). Study of Defect Levels in CdTe Using Thermoelectric Effect Spectroscopy, *Journal of Electronic Materials*, Vol.35, No.6, pp., ISSN 0361-5235.

[60] Berding, M. (1999). Native Defects in CdTe, *Physical Review B*, Vol.60, No.12, pp. 8943-8950, ISSN 0163-1829.

[61] Berding M. (1999). Annealing Conditions for Intrinsic CdTe, *Applied Physics Letters*, Vol.74, No.4, pp. 552-554, ISSN 0003-6951.

[62] X.Mathew. Photo-induced current transient spectroscopic studu of the traps in *CdTe// Solar Energy Materials & Solar Cells*. 76, pp. 225-242 (2003).

[63] Mahalingam, T., John, V. & Ravi, G. (2002). Microstructural Characterization of Electrosynthesized ZnTe Thin films, *Crystal Research and Technology*, Vol.37, No.4, pp. 329–339, ISSN 1521-4079.

[64] Ibrahim, A, El-Sayed, N. & Kaid, M. (2004). Structural and Electrical Properties of Evaporated ZnTe Thin Films, *Vacuum*, Vol.75, pp. 189–194, ISSN 0042-207X.

[65] Dean, P., Venghaus, H. & Pfister, J. (1978). The Nature of the Predominant Acceptors in High Quality Zinc Telluride, *Journal of Luminescence*, Vol.16, pp. 363-394, ISSN 0022-2313.

[66] Bhunia, S., Pal, D. & Bose, N. (1998). Photoluminescence and Photoconductivity in Hydrogen–passivated ZnTe, *Semiconductor Science and Technology*, Vol.13, pp. 1434-1438, ISSN 0268-1242.

[67] Sadofjev, Ju. & Gorshkov, M. (2002). Deep Level Spectra in ZnTe:Cr^{2+} Layers Obtained by Epitaxy from Molecular Beams (in Russian), *Fizika I Tekhnika Poluprovodnikov*, Vol.36, No.5, pp. 525-527, ISSN 0015-3222.

[68] Korbutyak, D., Vakhnyak, N. & Tsutsura, D. (2007). Investigation of Photoluminescence and Electroconductivity of ZnTe Grown in Hydrogen Atmosphere, *Ukrainian Journal of Physics*, Vol.52, No.4, pp. 378-381, ISSN 2071-0186.

[69] Feng, L., Mao, D. & Tang, J. (1996). The Structural, Optical, and Electrical Properties of Vacuum Evaporated Cu-doped ZnTe Polycrystalline Thin Films, *Journal of Electronic Materials*, Vol. 25, pp. 1422–1427, ISSN 0361-5235.

[70] Atakova, M., Ramazanov, P. &Salman, E. (1973). Local Levels in a Zinc Sulfide Film, *Izvestiya Vysshikh Uchebnykh Zavedenii, Fizika*, Vol.10, pp. 95 – 98, ISSN 0021-3411.

[71] Venkata Subbaiah Y., Prathap, P. & Ramakrishna Reddy, K. (2006). Structural, Electrical and Optical Properties of ZnS Films Deposited by Close-spaced Evaporation, *Applied Surface Science*, Vol. 253, pp. 2409 – 2415, ISSN 0169-4332.

[72] Venkata Subbaiah, Y., Prathap, P. & Ramakrishna Reddy, K. (2008). Thickness Effect on the Microstructure, Morphology and Optoelectronic Properties of ZnS Films, *Journal of Physics: Condensed Materials*, Vol.20, pp. 035205 – 035215, ISSN 0953-8984.

[73] Turan, E., Zor, M. & Aybek, A. (2007). Thermally Stimulated Currents in ZnS Sandwich Structure Deposited by Spray Pyrolysis, *Physica B: Condensed Materials*, Vol.395, No.1, pp. 57 – 64, ISSN 0921-4526.

[74] Abbas, J., Mehta, C. & Saini, G. (2007). Preparation and Characterization of n-ZnS and its Self-assembled Thin Film, *Digest Journal of Nanomaterials and Biostructures*, Vol.2, No.3, pp. 271 – 276, ISSN 1842-3582.

[75] Kvit, A., Medvedev, S. & Klevkov, Ju. (1998). Optical Spectroscopy of Deep States in ZnTe (in Russian), *Fizika I Tekhnika Poluprovodnikov*, Vol.40, No.6, pp. 1010-1017, ISSN 0015-3222.

[76] Bagaev, V., Zajtsev, V. & Klevkov, Ju. (2003). Influence of Annealing in Vapors and in Liquid Zn on ZnTe High-frequency Polycrustalline Photoluminescence (in Russian), *Fizika I Tekhnika Poluprovodnikov*, Vol.37, No.3, pp. 299-303, ISSN 0015-3222.

[77] Klevkov, Ju., Martovitskij, V., Bagaev, V. (2006). Morphology, Twins Formation and Photoluminescence of ZnTe crystals Grown by Chemical Synthesis of Components From Vapor Phase (in Russian), *Fizika I Tekhnika Poluprovodnikov*, Vol.40, No.2, pp. 153-159, ISSN 0015-3222.

[78] Klevkov, Ju., Kolosov, S. & Krivobokov, V. (2008). Electrical Properties, Photoconductivity and Photoluminescence of large-crystalline p-CdTe (in Russian), *Fizika I Tekhnika Poluprovodnikov*, Vol.42, No.11, pp. 1291-1296, ISSN 0015-3222.

[79] Bagaev, V., Klevkov, Ju. & Krivobokov, V. (2008). Photoluminescence Spectrum Change Near the Twins Boundaries in ZnTe Crystals Obtained by High-speed Crystallization (in Russian), *Fizika Tverdogo Tela*, V.50, No.5, pp. 774-780, ISSN 0042-1294.

[80] Makhniy V. & Gryvun, V. (2006). Diffusion ZnTe:Sn Layers with Electron Conductivity (in Russian), *Fizika I Tekhnika Poluprovodnikov*, Vol.40, No.7, pp. 794-795, ISSN 0015-3222.

[81] Tsutsura, D., Korbutyak, O. & Pihur, O. (2007). On Interaction of Hydrogen Atoms With Complex Defects in CdTe and ZnTe, *Ukrainian Journal of Physics*, Vol.52, No.12, pp. 1165-1169, ISSN 2071-0186.

[82] Taguchi, T, Fujita, S. & Inushi, Y. (1978). Growth of High-purity ZnTe Single Crystals by the Sublimation Travelling Heater Method, *Journal of Crystal Growth*, Vol.45, pp. 204-213, ISSN 0022-0248.

[83] Dean, P. (1979). Copper, the Dominant Acceptor in Refined, Undoped Zinc Telluride, *Luminescence*, Vol.21, pp. 75-83, ISSN 1522-7243.

[84] Garcia, J., Remon, A. & Munoz, V. (2000). Annealing-induced Changes in the Electronic and Structural Properties of ZnTe Substrates, *Journal of Materials Research*, Vol.15, No.7, pp. 1612-1616, ISSN 0884-2914.

[85] Uen, W., Chou, S. & Shin, H. (2004), Characterizations of ZnTe Bulks Grown by Temperature Gradient Solution Growth, *Materials Science & Engineering A*, Vol.B106, pp. 27-32, ISSN 0921-5093.

[86] Yoshino, K, Yoneta, V. & Onmori, K. (2004). Annealing Effects of High–quality ZnTe Substrate, *Journal of Electronic Materials*, Vol.33, No.6, pp. 579-582, ISSN 0361-5235.

[87] Yoshino, K., Kakeno, T. & Yoneta, M. (2005). Annealing Effects of High–quality ZnTe Substrate, *Journal of Material Science: Materials in Electronics*, Vol.16, pp. 445-448, ISSN.0957-4522.

[88] Ichiba, A., Ueno, J. & Ogura, K. (2006). Growth and Optical Property Characterizations of ZnTe:(Al, N) layers Using Two Co–doping Techniques, *Physica Status Solidi C*, Vol.3, No.4, pp. 789-792, ISSN 1610-1642.

[89] Bose, D. & Bhunia, S. (2005). High Resistivity In–doped ZnTe: Electrical and Optical Properties, *Bulletin of Material Science*, Vol.28, No.7, pp. 647-650, ISSN 0250-4707.

[90] Grill, R., Franc, J, & Turkevych, I. (2002). Defect-induced Optical Transitions in CdTe and $Cd_{0,96}Zn_{0,04}Te$ *Semiconductor Science and Technology*, Vol.17, pp. 1282-1287, ISSN 0268-1242.

[91] Ushakov, V. & Klevkov, Ju. (2007). Microphotoluminescence of Undoped Cadmium Telluride Obtained by Non-equilibrium Method of Direct Synthesis in Components Fluctuation (in Russian), *Fizika I Tekhnika Poluprovodnikov*, Vol.41, No.2, pp. 140-143, ISSN 0015-3222.

[92] Babentsov, V., Corregidor, V. & Castano, J. (2001). Compensation of CdTe by Doping With Gallium, *Crystal Research and Technology*, Vol.36, No.6, pp. 535-542, ISSN 1521-4079.

[93] Tarbayev, G. & Shepelskij. (2006). Two Series of "Dislocation" Photoluminescence Lines in Cadmium Telluride Crystals (in Russian), *Fizika I Tekhnika Poluprovodnikov*, Vol.40, No.10, pp. 1175-1180, ISSN 0015-3222.

[94] Aguilar-Hernandez, J., Contreras-Puente, J. & Vidal-Larramendi J. (2003). Influence of the Growth Conditions on the Photoluminescence Spectrum of CdTe Polycrystalline Films Deposited by the Close Space Vapor Transport Technique, *Thin Solid Films*, Vol.426, pp. 132-134, ISSN 0040-6090.

[95] Aguilar-Hernandez, J., Cardenas-Garcia, M. & Contreras-Puente, G. (2003). Analysis of the 1,55 eV PL Band of CdTe Polycrystalline Films, *Material Science & Engineering A*, Vol.B102, pp. 203-206, ISSN 0921-5093.

[96] Armani, N., Ferrari, C. & Salviati, G. (2002). Defect-induced Luminescence in High–resistivity High–purity Undoped CdTe Crystals, *Journal of Physics: Condensed Matterials*, Vol.14, pp. 13203-13209, ISSN 0953-8984.

[97] Gheluwe, J., Versluys, J. & Poelman, D. (2005). Photoluminescence Study of Polycrystalline CdS/CdTe Thin Film Solar Cells, *Thin Solid Films*,Vol.480–481, pp. 264-268, ISSN 0040-6090.

[98] Okamoto, T., Yamada, A. & Konagai, M. (2001). Optical and Electrical Characterizations of Highly Efficient CdTe Thin Film Solar Cells, *Thin Solid Films*, Vol.387, pp. 6-10, ISSN 0040-6090.

[99] Holliday, D., Potter, M. & Boyle, D. (2001). Photoluminescence Characterisation of Ion Implanted CdTe, *Material Research Society, Symposium Proceedings*, Vol.668, pp. H1.8.1-H1.8.6, ISSN 0272-9172.

[100] Palosz, W., Grasza, K. & Boyd, P. (2003). Photoluminescence of CdTe Crystals Grown Physical Vapor Transport, *Journal of Electronic Materials*, Vol.32, No.7, pp.747-751, ISSN 0361-5235.

[101] Corregidor, V., Saucedo, E. & Fornaro, L. (2002). Defects in CdTe Polycrystalline Films Grown by Physical Vapour Deposition, *Materials Science and Engineering A*, Vol.B 91–92, pp. 525-528, ISSN 0921-5093.

[102] Soundararajan, R., Lynn, K. & Awadallah, S. (2006). Study of Defect Levels in CdTe Using Thermoelectric Effect Spectroscopy, *Journal of Electronic Materials*, Vol.35, No.6, pp. 1333-1340, ISSN 0361-5235.

[103] Babentsov, V. & Tarbaev. (1998). Photoluminescence of Re-crystallized by Nano-second Laser Irradiation Cadmium Telluride (in Russian), *Fizika I Tekhnika Poluprovodnikov*, Vol.32, No.1, pp. 32-35, ISSN 0015-3222.

[104] Morozova, N. & Kuznetsov, V. (1987). *Zinc Sulfide. Obtaining and Optical Properties (in Russian)*, Nauka, ISBN, Moskow, USSR.

[105] Sahraei, R., Aval, J. & Goudarzi, A. (2008). Compositional, Structural and Optical Study of Nanocrystalline ZnS Thin Films Prepared by a New Chemical Bath Deposition Route, *Journal of Alloys and Compounds*, Vol.466, pp. 488-492, ISSN 0925-8388.

[106] Prathap, P. Revathi, N. & Venkata Subbaiah, Y. (2008). Thickness Effect on the Microstructure, Morphology and Optoelectronic Properties of ZnS Films, *Journal of Physics: Condensed Materials*, Vol.20, pp. 035205-035215, ISSN 0953-8984.

[107] Prathap, P. & Venkata Subbaiah, Y. (2007). Influence of Growth Rate on Microstructure and Optoelectronic Behaviour of ZnS Films, *Journal of Physics D: Applied Physics*, Vol. 40, pp. 5275-5282, ISSN 0022-3727.

[108] Lee, H. & Lee, S. (2007). Deposition and Optical Properties of Nanocrystalline ZnS Thin Films by a Chemical Method , *Current Applied Physics*, Vol.7, pp. 193-197, ISSN 1567-1739.

[109] Kroger, F. (1964). *The chemistry of Imperfect Crystals*, North-holland publishing company, ISBN, Amsterdam, Netherlands.

[110] Grill, R. & Zappetini, A. (2004). Point Defects and Diffusion in Cadmium Telluride, *Progress in Crystal Growth and Characterization of Materials*, Vol.48, pp .209-244, ISSN 0146-3535.

[111] Kosyak, V. & Opanasyuk, A. (2005). Point Defects Encemble in CdTe Single Crystals in the Case of full Equilibrium and Quenching (in Ukrainian), *Fizyka I Himiya Tverdych Til*, Vol.6, No.3, pp. 461-471, ISSN 1729-4428.

[112] Kosyak, V., Opanasyuk, A. & Protsenko, I. (2005). Ensemble of Point Defects in Single Crystals and Films in the Case of Full Equilibrium and Quenching, *Functional Materials*, Vol.12, No.4, pp. 797-806, ISSN 1616-3028.

[113] Kosyak, V. & Opanasyuk, A. (2007). Calculation of Fermi Level Location and Point Defect Ensemble in CdTe Single Crystal and Thin Films, *Semiconductor Physics, Quantum Electronics & Optoelectronics*, Vol.10, No.3, pp. 95-102, ISSN 1560-8034.

[114] Kolesnik, M., Kosyak, V. & Opanasyuk, A. (2007). Calculation of Point Defects Ensemble in CdTe Films Considering Transport Phenomenon in Gas Phase, *Radiation Measurements*, Vol.42, No.4–5, pp. 855-858, ISSN 1350-4487.

[115] Kosyak, V., Kolesnik, M. & Opanasyuk, A. (2008). Point Defect Structure in CdTe and ZnTe Thin Films, *Journal of Material Science: Materials in Electronics*, Vol.19, No.1, pp. S375-S381, ISSN 0957-4522.

[116] Kurbatov, D, Kosyak, V., Opanasyuk, A. (2009). Native Point Defects in ZnS Films, *Physica. B. Condensed Materials*, Vol.404, No.23–24, pp. 5002-5005, ISSN 0921-4526.